LONG-TERM ENVIRONMENTAL EFFECTS OF OFFSHORE OIL AND GAS DEVELOPMENT

LONG-TERM ENVIRONMENTAL EFFECTS OF OFFSHORE OIL AND GAS DEVELOPMENT

Edited by

DONALD F. BOESCH and NANCY N. RABALAIS

Louisiana Universities Marine Consortium
Chauvin, Louisiana, USA

ELSEVIER APPLIED SCIENCE
LONDON and NEW YORK

ELSEVIER APPLIED SCIENCE PUBLISHERS LTD
Crown House, Linton Road, Barking, Essex IG11 8JU, England

Sole Distributor in the USA and Canada
ELSEVIER SCIENCE PUBLISHING CO., INC.
52 Vanderbilt Avenue, New York, NY 10017, USA

WITH 66 TABLES AND 58 ILLUSTRATIONS

British Library Cataloguing in Publication Data

Long-term environmental effects of offshore oil and gas development.
1. Offshore oil industry—Environmental aspects
2. Offshore gas industry—Environmental aspects
I. Boesch, Donald F. II. Rabalais, Nancy N.
333.8′23 TD195.03

ISBN 1 85166 094 1

Library of Congress CIP data applied for

Phototypeset and printed in Great Britain by Page Bros (Norwich) Ltd, Norwich

PREFACE

With the expansion of exploration for oil and gas in offshore regions during the 1970s, there was much concern regarding the environmental effects of future development. In the United States legal and legislative actions have been taken to stop or slow development, in large part based on concerns that deleterious effects on the marine environment would result. Ambitious Federal programs of studies of the potentially affected environment were implemented to address these concerns and ensure environmental protection. Despite these efforts, controversies regarding the seriousness of potential effects still exist, particularly with regard to subtle, but long-term effects.

Despite several evaluations of existing knowledge of the effects of offshore oil and gas development, the concern over long-term effects was unfocused. What exactly are the effects which might occur and what is the relative seriousness of each? In response to the need to answer these questions and to develop a considered and carefully planned strategy to address the remaining concerns, a detailed assessment was undertaken by a group of experts, culminating in this book. These efforts were supported by the National Oceanic and Atmospheric Administration and the National Science Foundation.

The ultimate purpose of our efforts is to develop recommendations for the design of an environmental research and monitoring program to quantify and evaluate the significance of subtle and long-term effects of offshore oil and gas development activities. To accomplish this the participants decided that extensive background must be developed to support the conclusions and recommendations. Hence, detailed technical papers are included in addition to the overall assessment and research plan in Chapter 1.

A large number of individuals contributed diligently and significantly to the completion of the volume. In addition to the authors of the individual chapters, a Steering Committee consisting of Donald F. Boesch, James N. Butler, David A. Cacchione, Joseph R. Geraci, Jerry M. Neff, James P. Ray and John M. Teal defined the scope, selected the technical authors, reviewed their contributions and developed the consensus assessment and recommended research needs. Throughout their deliberations, William G. Conner and Douglas A. Wolfe of the National Oceanic and Atmospheric Administration and James Cimato of the Minerals Management Service participated as liaisons with their agencies. Glynis A. Duplantis, Veronica A. Lyons, Lisa M. Brunette, and Diane Zelasko performed the word-processing through the many revisions.

D. F. Boesch
N. N. Rabalais

CONTENTS

LIST OF CONTRIBUTORS

RONALD M. ATLAS

Department of Biology, University of Louisville, Louisville, Kentucky 40292, USA (Chapter 7)

RICHARD BARTHA

Department of Biochemistry and Microbiology, Rutgers University, New Brunswick, New Jersey 08903, USA (Chapter 7)

PAUL D. BOEHM

Battelle, New England Marine Research Laboratory, 197 Washington Street, Duxbury, Massachusetts 02332, USA (Chapter 6)

DONALD F. BOESCH

Louisiana Universities Marine Consortium, Chauvin, Louisiana 70344, USA (Chapters 1, 3, 4, 13)

JAMES N. BUTLER

Division of Applied Sciences, Harvard University, 29 Oxford Street, Cambridge, Massachusetts 02138, USA (Chapter 1)

DAVID A. CACCHIONE

U.S. Geological Survey, 345 Middlefield Road, Menlo Park, California 94025, USA (Chapter 1)

JUDITH M. CAPUZZO

Woods Hole Oceanographic Institution, Woods Hole, Massachusetts 02543, USA (Chapter 8)

ROBERT S. CARNEY

Coastal Ecology Institute, Center for Wetland Resources, Louisiana State University, Baton Rouge, Louisiana 70803, USA (Chapter 14)

JOSEPH R. GERACI

Wildlife Section, Department of Pathology, Ontario Veterinary College, Guelph, Ontario N1G 2W1, Canada (Chapters 1, 12)

WILSON HOM

Science Applications International Corporation, 476 Prospect Street, La Jolla, California 92037, USA (Chapter 5)

GEORGE L. HUNT, JR.

Department of Ecology and Environmental Biology, University of California, Irvine, California 92717, USA (Chapter 11)

JERRY M. NEFF

Battelle, New England Marine Research Laboratory, 397 Washington Street, Duxbury, Massachusetts 02332, USA (Chapters 1, 4, 10)

JAMES R. PAYNE

Science Applications International Corporation, 476 Prospect Street, La Jolla, California 92037, USA (Chapter 5)

CHARLES R. PHILLIPS

Science Applications International Corporation, 476 Prospect Street, La Jolla, California 92037, USA (Chapter 5)

NANCY N. RABALAIS

Louisiana Universities Marine Consortium, Chauvin, Louisiana 70344, USA (Chapters 3, 4)

JAMES P. RAY

Environmental Affairs Division, Shell Oil Company, P.O. Box 2463, Houston, Texas 77001, USA (Chapters 1, 2)

GORDON A. ROBILLIARD

ENTRIX, Inc., 1470 Maria Lane, Walnut Creek, California 94596, USA (Chapter 13)

DAVID J. ST. AUBIN

Wildlife Section, Department of Pathology, Ontario Veterinary College, Guelph, Ontario N1G 2W1, Canada (Chapter 12)

ROBERT B. SPIES

Environmental Sciences Division, Lawrence Livermore National Laboratory, P.O. Box 5507, L453, Livermore, California 94550, USA (Chapter 9)

JOHN M. TEAL

Department of Biology, Woods Hole Oceanographic Institution, Woods Hole, Massachusetts 02543, USA (Chapter 1)

CHAPTER 1

AN ASSESSMENT OF THE LONG-TERM ENVIRONMENTAL EFFECTS OF U.S. OFFSHORE OIL AND GAS DEVELOPMENT ACTIVITIES: FUTURE RESEARCH NEEDS

Donald F. Boesch, James N. Butler, David A. Cacchione, Joseph R. Geraci, Jerry M. Neff, James P. Ray and John M. Teal

CONTENTS

SUMMARY

Of the many issues raised regarding the potential effects of expanded development of offshore oil and gas resources, the potential for long-term and insidious effects on the marine environment has frustrated resolution. It is suspected that chronic effects are of greatest concern but, paradoxically, they are hard to detect and quantify.

This chapter presents a critical evaluation on the large body of information assembled and reviewed in succeeding chapters related to the long-term effects of offshore oil and gas development activities. We have attempted to focus on those marine environmental effects which are long-lasting (>two years) and significantly deleterious to human resources (such as fisheries) and ecosystem integrity. This evaluation is based on interpretation of relative risks based on the probability and severity of effects and on the potential that new scientific information or interpretation of existing information could contribute to resolution of an issue. We then provide recommendations for the studies required, their feasibility and the use of resulting information.

Because ecosystems are complex, open and dynamic, there are fundamental problems in identifying the nature and extent of environmental effects and in determining causality. Uncovering subtle effects in the coastal ocean requires long-term observation and difficult and imaginative experimentation to overcome the obstacles provided by natural variability, statistical limits of detection, the effects of other human activities, recognition of recovery, and unknown relationships within ecosystems and their role in supporting human resources.

The potential for long-term effects depends on the environment in which the development takes place or through which the oil and gas is transported and how the development is accomplished. In the United States, offshore oil and gas production has to date been limited to the northwestern Gulf of Mexico (the vast majority), southern California and Cook Inlet, Alaska. Although an ambitious program of exploration and development of previously undeveloped "frontier" areas was begun in the 1970s, no economically viable discoveries have yet been made outside of these historically producing regions. Based on indicators including proven reserves, current drilling activities, estimates of undiscovered resources and industry interest, it is now clear that, although some exploratory activity and potential production may take place off the Atlantic coast, Florida, the northwestern states and in the Gulf of Alaska, U.S. offshore oil and gas development will be concentrated in the northwestern Gulf of Mexico, off southern California and in the Beaufort and Bering Seas of Alaska for the remainder of the century. Drilling in the deeper waters of the continental slope and under heavy sea ice conditions will present new challenges to the industry in terms of environmental engineering and safety.

Modes of transportation of oil and gas from offshore will vary depending on the product and amount of production, the distance to shore, the nature of the intervening environment, and the capabilities of onshore facilities. The extent and duration of effects of oil spills resulting from pipeline ruptures or loss from transshipment will vary depending on the nature of the coastal ecosystems affected and the presence of colonies of birds and mammals. Similarly, dredging for pipelines and required navigational access will pose different threats to disparate coastal environments. Knowledge of the comparative sensitivity of marine ecosystems often limits extrapolation of results from one area to another.

Based on detailed consideration of the probability and severity of effects and the potential for resolution of uncertainties, we have identified ten categories of potential long-term environmental effects of offshore oil and gas development activities for future investigation. Of high priority are 1) chronic biological effects resulting from the persistence of medium and high molecular weight aromatic hydrocarbons and heterocyclics and their degradation products in sediments and cold environments; 2) the residual damage from oil spills to biogenically structured communities, such as coastal wetlands, reefs and vegetation beds; and 3) effects of channelization for pipeline routing and navigation in coastal wetlands. Of intermediate priority are 1) effects of physical fouling by oil of aggregations of birds, mammals and turtles; 2) effects on benthos of drilling discharges accumulated through field development rather than from exploratory drilling; and 3) effects of produced water discharges into nearshore rather than open shelf environments. Of lower priority are 1) effects of noise and other physical disturbances on populations of birds, mammals and turtles; 2) the reduction of fishery stocks due to mortality of eggs and larvae as a result of oil spills; and 3) effects of artificial islands and causeways in the Arctic on benthos and anadromous fish species.

For each of these major categories of effects, sequential approaches are developed for quantification of long-term effects. Recommended research includes generic experimental approaches, for example, on the persistence of medium and high molecular weight hydrocarbons in sediments and their metabolic fate in organisms; observational studies, for example, following the recovery of oiled communities and monitoring of potentially affected colonies of birds and mammals; carefully designed measurements of environmental processes, for example, transport of contaminated sediments and hydrologic flow in altered wetlands; and regionally focused field assessments.

For each stage of the recommended study sequence, an appraisal of the feasibility of the study is given based on whether it can be satisfactorily accomplished within a 10-year time frame using available methods or requires development of new methods or innovative approaches. The preferred regional focus, where appropriate, is also indicated.

Given the great diversity of potential effects and regional differences in potential effects, we recommend implementation of and commitment to a U.S. interagency program plan which guides regional research and monitoring efforts together with generic research programs. Of critical importance to the success of such a program are centralized management within agencies and sufficient interagency overview to assure compliance, iterative review of objectives and progress, emphasis on innovation and application of state-of-the-art methods, and multiyear research funding.

INTRODUCTION

Concerns regarding the effects of offshore oil and gas development activities on the marine environment have focused most sharply on oil spills and the operational discharge of materials, such as drilling fluids, during exploratory drilling. Such effects are generally perceived as acute and ephemeral, although potentially catastrophic in the case of oil spills. The acute effects of oil spills and drilling discharges have become increasingly well understood (National Research Council, 1983, 1985), due in part to heavy investment of public and private support of research.

In recent years, insidious effects have been uncovered for agents and activities once presumed harmless, for example polychlorinated biphenyls (PCBs) and carbon dioxide released into the atmosphere. As a consequence, environmental scientists and the general public turn their attention to the potential for less obvious and longer-lasting effects of human activities and byproducts. This leads, as Lewis (1982) pointed out, to "the apparent paradox that it is the unknown, the suspected but hard-to-detect chronic effects, that are the real cause for concern." It is against this background that the National Marine Pollution Program Plan (Interagency Committee on Ocean Pollution Research, Development, and Monitoring, 1981) concluded that the most significant unanswered questions for offshore oil and gas development are those regarding the effects on ecosystems of long-term, chronic, low-level exposures resulting from discharges, spills, leaks and disruptions caused by development activities.

True to this paradox, concerns about long-term and chronic effects are difficult to resolve, the issues contentious, and the angst high. In the summary of a British symposium on the long-term effects of oil pollution, Clark (1982) highlighted the considerably divergent views. Debates rage over appropriateness of methodologies, interpretation of results and the potential for undiscovered effects (e.g., Sanders and Jones, 1981).

An overall assessment of the potential environmental effects of existing and future offshore oil and gas development requires critical evaluation beyond that provided by the authors of the individual review chapters on which this synthesis is based. Specifically, we must determine whether the potential for an effect is realistic, of long duration, and significantly deleterious to human resources or ecosystem integrity. All of these characteristics are relative: how probable must the effect be, how long must it persist, and how pervasive must be its repercussions? All these evaluations call on our judgments.

The term "long-term effect" is almost always used without definition. Is it an effect that persists or one which results from a persistent activity? The latter includes the first, but an acute event may result in a persistent effect. As used here, long-term effects are those which either result from activities which extend over long time periods or persist as a result of brief activities. Because the recovery of marine communities from oil spills has been documented for periods ranging from two to ten years (Clark, 1982), long-term will be operationally considered to include time periods greater than two years.

Using this terminology, an oil spill resulting from a blowout or pipeline rupture may have long-term effects if the effects persist for more than two years. It is these residual effects which are the subject of our attention and the more immediate effects are of interest only insofar as they relate to an understanding of these residual effects. More pertinent to the offshore oil and gas development issue, however, are the cases of habitat disruptions or chronic petroleum contamination, either as a result of continuous or intermittent discharges (produced waters, drilling fluids containing oil, deck washings, etc.) or from repetitive, accidental spills (numerous small spills and a small number of major spills during the life of a field).

Setting some required level of significance of the effect (either to humans or the ecosystem) is more difficult, because it involves consideration of spatial extent, persistence and recoverability, as well as the value of the ecosystem components affected. In general, field assessments around point source discharges from oil and gas development structures have been able to document biological effects only well within 1 km of the source. Our present concern is focused primarily on effects which occur on much larger scales. It is unwise, however, to set an exact spatial threshold for concern because of the interaction of space, recovery time and resource value. For example, an effect which is elicited over 1 km^2 of a rare or exceptionally valuable habitat and persists for decades is certainly of greater concern than one which occurs over 2 km^2 of a more widespread habitat and lasts no more than two years.

Environmental resources of value to humans are the focus of our assessments of risks and severity of effects. These resources include those of direct economic value, such as fisheries, but also include those which may be of little or no economic value, but are of intrinsic value to human society. Examples of the latter include marine mammals, endangered species, and rare or aesthetically pleasing environments. In addition to direct effects on those resources, we have also to consider effects on the marine and coastal ecosystems which support these resources insofar as these effects place the resources of ultimate concern in jeopardy.

We include in this evaluation the environmental effects of oil and gas development activities in offshore environments, including the area which is legally defined in the United States as the Outer Continental Shelf (beyond state territorial waters and under Federal jurisdiction) as well as nearshore environments where ownership is vested in the states. Our considerations are limited to marine environments of the continental shelf and slope and to aquatic environments of the coastal fringe which are affected by offshore activities. Effects on terrestrial environments and social and economic impacts are not reviewed.

We begin our assessment with a consideration of the problems inherent in detecting and evaluating long-term environmental effects. Secondly, we identify the coastal and offshore ecosystems most likely to be affected and their relative susceptibility. We then deduce, based on the detailed evaluations of the supporting technical reviews and the above criteria of duration and significance, the potential long-term effects of offshore oil and gas development on resources of intrinsic and economic value, and the ecosystem functions which support these resources. In this assessment, we provide an evaluation of the relative risks of such effects based on consideration of their probability and severity, although the limits of our understanding and the diversity of environments under consideration do not allow these evaluations to be absolute. We also discuss, for each issue identified, the potential that new scientific information or the interpretation of existing information could contribute to the satisfactory resolution of that issue. Finally, we provide some more detailed recommendations regarding the studies required, their feasibility and the use of resulting information in decision-making.

IDENTIFYING LONG-TERM ENVIRONMENTAL EFFECTS

Most ecosystems are complex, open and dynamic. This results in fundamental problems in identifying the nature and extent of environmental effects of contaminants or human activities and in determining causality. These problems plague all environmental sciences, but become particularly difficult in the case of long-term effects in the coastal ocean. There effects may be subtle, the requirements for observation long-term, and the difficulties in relevant experimentation great. It is helpful here to consider in a general sense these fundamental problems in order to properly evaluate the limitations to current understanding and the requirements for improved study design.

Variability

Variations inherent to biological systems result from both the natural variability of the physical environment and of the biological processes themselves. Natural variation in space and time has been one of the greatest problems encountered in assessments of effects in the field (Chapter 14). Natural variability often overshadows impact effects or confounds the resolution of such effects. Variations in space exist on a variety of scales and have to be understood, at least at scales above that of the sample size, in order to determine if differences observed in contaminant levels or biota are attributable to a human activity. Understanding temporal variability is also important in "before-and-after" comparisons of environmental variables or biological response to an impact-producing activity. Particularly when the frame of reference is "long-term," one must compare the magnitude of an environmental effect to the concurrent range of natural variability.

The appropriate length of study may be difficult to predict *a priori*, but the generation time of important species would usually be a reasonable starting point. For some responses it is necessary for interactions to occur that may be a product of the generation times of the interactants, e.g., predator control of ecosystem structure. Note that for long-lived animals, such as some sea birds, the appropriate time frame for studies may well be decades. Identification of the nature and causes of variation should be an objective of ecosystem studies. It is not appropriate simply to consider variations as part of measurement error. They must also be recognized as an integral part of biological systems.

Benthic communities, at least in temperate waters, are less variable spatially than planktonic ones. The benthos is also generally less variable temporally than the plankton because benthic organisms are more fixed in place and generally longer-lived. For these reasons, as well as the relatively greater susceptibility of organisms exposed to contaminants accumulated in sediments, the identification of long-term pollution effects in the benthos has been more successful than in other ecosystem components.

The problems caused by natural variations in time and space for the identification of effects induced by human activities have sometimes discouraged the use of baseline and monitoring approaches (Burroughs, 1981). As discussed below, this problem is most constructively viewed in terms of setting limits of change, within which effects are either acceptable or simply undetectable within the constraints of practical design. Furthermore, even effects which can be definitely ascribed to a certain activity must be evaluated in the context of natural temporal variability to determine if they are significant.

Limits of Detection

The success and efficiency with which effects can be identified depend on assumptions about the degree of change in variables one wishes to detect. This may seem simple and obvious, but it is surprising how frequently these assumptions are not made explicity (Chapter 14). Insensitive methods using sampling designs with poor power are able to detect only the grossest effects and thus have little to contribute to determination of long-term, potentially subtle effects. It is important that the sampling design be capable of detecting the degree of change which is considered unacceptable or which nature forces us to accept as feasible. Furthermore, the sensitivity of methods to detect such a change should be clearly stated.

It is also advisable that studies be designed to measure biological and environmental variables of ecological or economic importance or special usefulness as indicators. Effects not considered in the design of a study can rarely be found through an unfocused, general survey. This is especially true in the oceans because marine ecosystems are too poorly understood and too inaccessible to be able to detect unanticipated effects. By contrast, terrestrial environments are more accessible to direct observation, allowing more timely modification of the course of study when confronted by the unexpected.

Effects of Other Human Activities

Man uses the oceans for many purposes and it is necessary to consider the effects of a specific activity such as offshore oil and gas development in the context of the effects of other uses. This is instructive both in terms of evaluating the environmental "costs" of various uses in the same currency and the capacity of the ecosystem or resource to absorb the impacts. Practical difficulties with such comparisons include limited knowledge about the effects of the various uses and assignment of cause of observed alterations among the uses. Although it may be appealing simply to compare the relative contribution of contaminants from different sources, this may be misleading because of variations in exposure mode and concentration. Furthermore, biological response may be non-linear. That is to say, the additional 5% contribution of a contaminant, for example, may overwhelm the capacity of an ecosystem to accommodate it.

Interrelationships in Ecosystems

The more that is learned, the more ecologists are surprised by how thoroughly and complexly the components of ecosystems are connected. Variations or alterations in one biotic component may have subtle repercussions in other seemingly unrelated components. This feature contributes to a lingering uncertainty about whether the effects of a contaminant or activity are understood well enough to be predictive. In addition, ecosystems may be highly connected to other ecosystems, particularly in the coastal ocean. Continental shelf ecosystems interact with coastal systems by environmental forces (e.g., runoff, storms, etc.) and movement of biota between them. Similarly, the continental shelf is influenced by the dynamics of the adjacent oceanic regime through boundary current variations, upwelling and similar phenomena (Chapter 3).

Recovery

There is remarkable ignorance about the processes and rates of recovery of living resources and ecosystems in coastal environments after perturbations caused either by natural events or human activities. Even defining recovery is difficult and covers a range of possibilities. If an economic resource is the prime consideration, then return of that resource to its previous productivity might be a suitable definition. In the extreme, complete recovery may require the restoration of the ecosystem to its pre-impact state, including the relative age distributions of its populations, occurrence of all species previously present, etc. In any practical sense, however, the definition of recovery must include some consideration of the normal variations in ecosystems; a system can seldom be expected to return to the identical state from which it started. It would be more appropriate to consider recovery complete when the system is again varying within the bounds exhibited by similar but undisturbed (control) systems. The time required for a system to recover, however defined, can be used as a measure of the significance of an effect. If this time is short, the action is less significant than another action resulting in a much longer recovery period.

Relationship of Ecosystems to Human Resources

Concern about environmental effects is ultimately based on resources of value to society, whether economic, inherent or aesthetic. The case of an activity deleteriously affecting a commercial or recreational fishery, for example, is relatively clear-cut. The value of the resources affected can be determined and weighed against the societal benefits of the activity. Although the simplicity of this process is greatly overstated, social valuation of effects on the ecosystems which support these resources is much more difficult than for effects on the resources themselves. The relationships of ecosystem components to the resources is poorly understood, and, consequently, evaluations of the resulting effects on the resources are usually conjectural. One approach is to focus on those factors which appear to be critical to the success of resource populations and on those other living components known to be important in supporting the resources, for example, prey populations. Even then, there are considerable uncertainties regarding overlooked or obscure population controls, on one hand, and the capacity of the resource species to accommodate ecosystem change (for example, by switching to alternate prey) on the other.

SUSCEPTIBILITY OF COASTAL AND OFFSHORE ECOSYSTEMS

Many concerns about environmental risks of offshore oil and gas development are raised from a regional perspective. Public officials, managers, and the general public, when confronted with the potential for oil and gas exploration and potential development off their coast, perceive a set of environmental issues of local relevance. A broader perspective must consider the large differences in development potential, proximity to shore, and transportation modes in various regions of the United States as well as in other parts of the world. Furthermore, the great diversity in the marine and coastal ecosystems which may be affected by such development must also be considered in this assessment of potential long-term effects.

Several questions must be addressed. Where is development most likely? How will it be accomplished? What is the relative susceptibility of the coastal and offshore resources and ecosystems involved? To what degree can experience or understanding about effects in one region be applied in assessing the potential for long-term effects in another?

Location of Development

Offshore oil and gas production in the United States is presently limited to the northwestern Gulf of Mexico, southern California, and Cook Inlet, Alaska. The vast majority of the past and present production is from the Gulf of Mexico.

Through 1984, over 10 billion barrels of oil and 81 trillion cubic feet of gas have been produced from the offshore United States (LaLiberté and Harris, 1986). Of this, approximately 4 billion barrels of oil and 15 trillion cubic feet of gas were produced from state waters. Of the offshore oil produced, 22% was from off California (only 14% of that from Federal waters) and 76% from off Louisiana (82% of that from Federal waters). The northwestern Gulf of Mexico off Louisiana and Texas has produced 97% of the offshore gas (84% of that from Federal waters).

An ambitious program of leasing the rights to oil and gas resources of the U.S. Outer Continental Shelf (OCS) began as a result of the 1973 Arab oil embargo and continues today at an accelerated pace. This has opened the prospect for oil and gas exploration and production off nearly all of the continental United States, including Alaska (Figure 1.1).

The location of offshore oil and gas development in the future will depend on resource estimates, successes in exploratory drilling, technological feasibility of drilling and transportation, market factors and governmental policy. Although predictions are speculative, several indicators may be examined to assess development scenarios during the next 10 to 20 years. The distribution of mobile exploratory drilling rigs under contract in spring 1984 (Figure 1.2, A) gives some indication of resulting production 5 to 10 years hence. The U.S. remains the world leader in offshore drilling because of the continuing activities in the Gulf of Mexico, but it is significant to note that offshore drilling is taking place virtually throughout the world. Some drilling will be required to exploit proven reserves, most of which are in the Gulf of Mexico OCS (Figure 1.2, B). Conditional estimates of undiscovered, recoverable resources (Figure 1.2, C) show that they may be located principally in the northwestern Gulf of Mexico, the Alaskan Arctic, off California, and in deep waters off the Mid-Atlantic states. Finally, estimates of the number of new wells (exploration and development) to be drilled during the next 10 years predict that 85% will be in the Gulf of Mexico (Figure 1.2, D). Although some exploratory activity and potential production will take place in other regions, it is clear that U.S. offshore oil and gas development will be concentrated overwhelmingly in the northwestern Gulf of Mexico, off southern and central California, and in the Beaufort and Bering Seas of Alaska for the remainder of the century.

In addition, there will be a general trend toward development in deeper water environments, although new development can be expected in shallow waters in California and Alaska. Any production off the Atlantic coast will probably be in the deep waters at the edge of or off the continental shelf. Exploration and development on the continental slope there and in the northwestern Gulf of Mexico is already active. Oil and gas development in deep water environments poses different technical and environmental considerations than in shallow waters. Production sites are generally farther from shore, reducing the potential of oil spills from blowouts reaching shore. Because of the large water volumes off the shelf, dispersion of contaminants released from the rig or platform is great. However, geohazards related to seabed slumping may be more likely and accidental spills may be more difficult to control on the slope. In addition, any

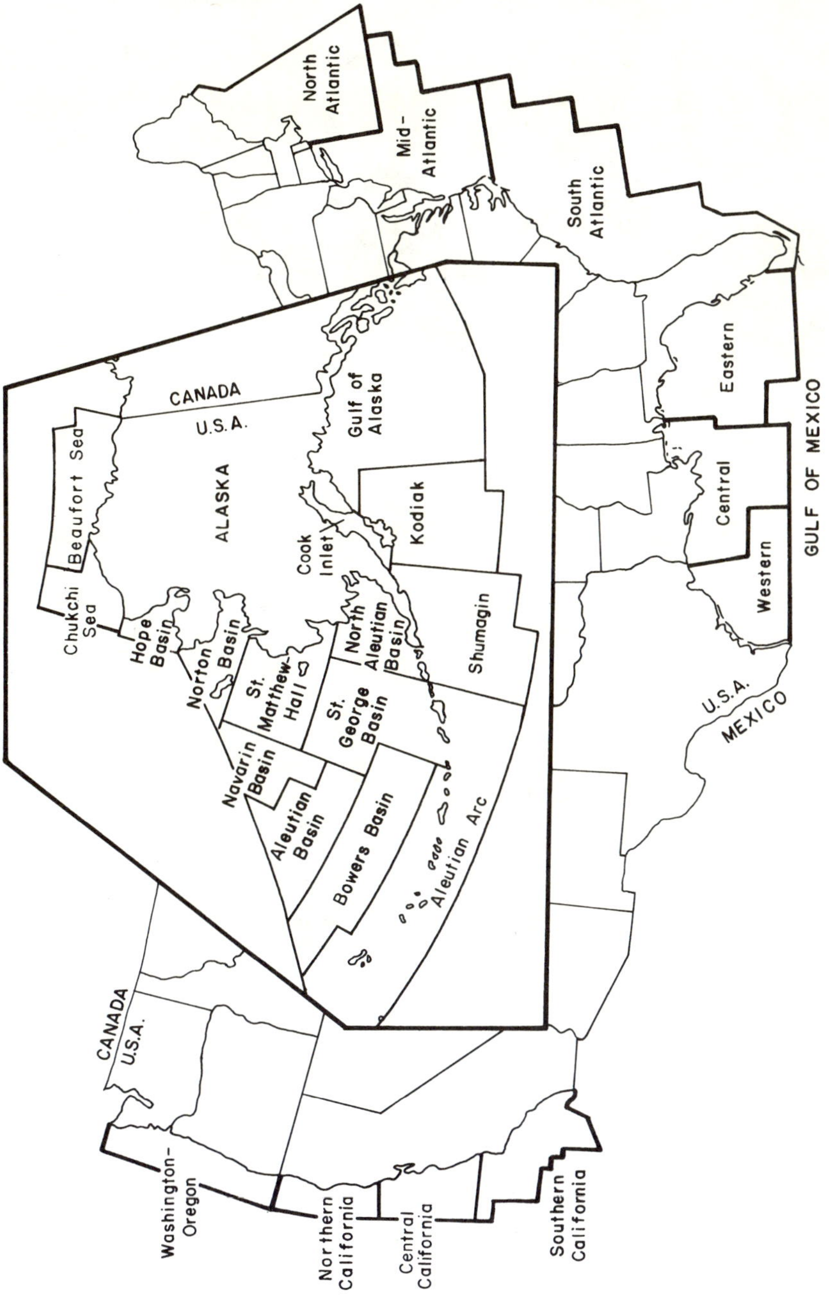

Figure 1.1. Federal offshore planning areas of the United States for which oil and gas development is underway or planned (Alaska not to same scale as contiguous 48 states).

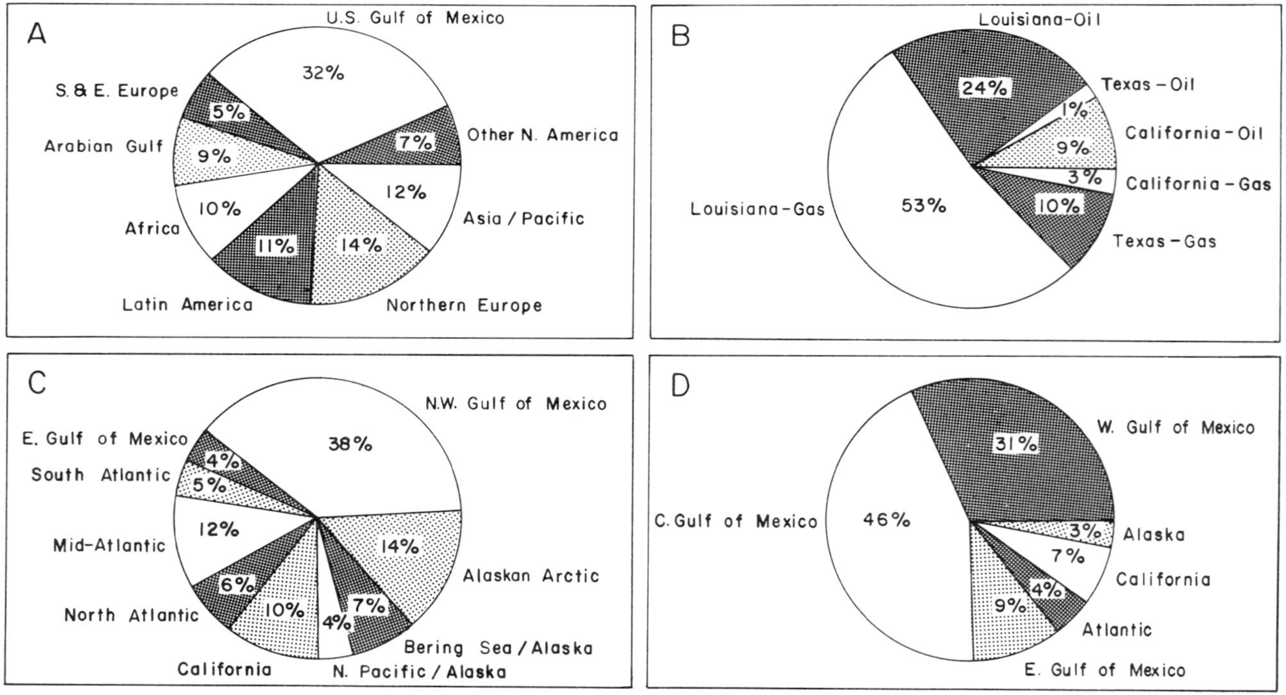

Figure 1.2. Indicators of future oil and gas development activities in U.S. offshore regions. A. Worldwide distribution of offshore mobile drilling rigs (exploratory or delineation drilling) under contract in April 1984 (Moore, 1984). B. Estimated recoverable reserves in U.S. Federal offshore waters (Essertier, 1983; Havran *et al.*, 1982). Gas reserves expressed in energy equivalents of oil. C. Conditional estimates of undiscovered economically recoverable resources of oil and gas (energy equivalents) (Essertier, 1983). D. Number of wells (exploration and production) predicted to be drilled, 1984–1993 (source, Minerals Management Service).

effects on the benthic environment may be longer lasting because of slower recovery rates of the deep-water biota (Chapter 3).

Sea ice introduces new technical and environmental issues in the arctic and subarctic regions of Alaska and requires the development of new technologies to ensure environmental safety. Sea ice also makes the containment and clean up of spilled oil very difficult. Spilled oil may remain toxic under ice because of the slow rate of physical/chemical degradation and biodegradation at low temperatures (Chapters 5 and 7).

Transport and Service Facilities

Oil, gas and condensates produced offshore must eventually be transported to shore for refining, processing and consumption. Transportation of product, which may be the aspect of offshore oil and gas development with the greatest environmental risks, will vary widely in means and geographic extent for different offshore areas. The means of conveyance will vary for different fields depending on the product and amount of production, the distance to shore, the nature of the intervening environment, and the capacities of onshore facilities. Virtually all of the oil and gas produced in the Gulf of Mexico flows through pipelines because the extensive development and existence of onshore facilities makes this feasible. Hydrocarbons from some frontier offshore fields may be transported by vessel, at least until production makes pipelines economically feasible. Oil produced in the Beaufort Sea will likely be transported ashore by pipeline and thence through the Trans-Alaska pipeline, loaded on tankers at Port Valdez, and shipped to ports in the U.S. and other countries.

The effects of oil spills which might occur as a result of pipeline ruptures and transshipment accidents and the effects of physical alterations due to pipeline installation are highly dependent on environmental characteristics which vary widely among regions. This variability is particularly true for coastal environments. In some areas pipelines would traverse a steep sandy or rocky intertidal zone. This has occurred in southern California, and the physical effects are very restricted and ephemeral. However, pipelines laid through the intertidal wetlands of coastal Louisiana have resulted in essentially permanent effects over large areas (Chapter 13). Similar coastal conditions exist in the South Atlantic Bight and the Yukon delta of Alaska.

The effects of servicing offshore production from onshore bases will also vary widely because of regional differences in coastal environments and industrial infrastructures. In some areas, ports and industrial bases are adequate; in others they are lacking or insufficient, and new development may produce significant effects on the coastal environment.

The Marine Environment

The long-term environmental effects which may result from offshore oil and gas development undoubtedly depend greatly on the characteristics of the environment and ecosystems in which they occur. Rabalais and Boesch (Chapter 3) reviewed the dominant environmental processes, particularly as they affect the benthic component of the ecosystem, for areas of the U.S. coastal ocean in which oil and gas development is proposed. These regions vary greatly in their physical regime (influence of oceanic currents, storms, tidal currents, wave climate, temperature and its variability, and ice formation), environmental geology (sediment type, sedimentation rates, resuspension and bed movements, presence of hard substrates, and ice scour), chemical contamination and biology.

There are, however, some common threads of biotic adaptation and response which allow a reasonable, if at this time primitive, ordering of ecosystems by their relative susceptibility and sensitivity to specific impacts. Using such general models of the relationship of biotic organization to environmental conditions in combination with models of the behavior of the physical environment (involving, for example, dispersion and sediment transport), one can, in at least a qualitative way, extend results from one environment to others. The suggestion that each offshore environment is different and, consequently, understanding of the long-term effects of offshore development in one region is irrelevant to another region is unduly pessimistic. To be sure, there are limits and sometimes formidable obstacles to such extrapolation, but placing observations of different environmental conditions and different regions in a coherent context, rather than treating these observations individually, is not only more efficient, but will yield a base for more confident predictions.

IDENTIFICATION OF POTENTIAL LONG-TERM EFFECTS

The objective of this discussion is to identify and evaluate the long-term effects which may be expected to occur, given the present level of scientific understanding, as a result of offshore oil and gas development. The analysis and identification process is based heavily on the detailed technical reviews presented in the later chapters of this volume. The potential long-term effects identified are summarized in Table 1.1. Three factors are considered: the probability that the effect may occur, the seriousness of the effect on valued resources and the duration of the effects. These cannot always be quantitatively expressed, but it is clear that they vary greatly among the issues. The integration of probability and severity is difficult. The loss of a year class of a fish stock as a result of an oil spill killing eggs and larvae is in our estimation highly improbable, but if this did occur, the effects would be severe. In contrast, exposure of marine mammals to noise from industrial activities is highly likely, but probably has little effect on the populations.

We have also evaluated the degree to which the issues might be resolved by additional research or information synthesis (Table 1.1). Some issues can be resolved satisfactorily: they would be found not to be significant or the steps that should be taken to mitigate undesirable effects will become clear. Other issues will remain not fully resolved, but substantially better understanding can be gained which contributes to decision making and regulation. Other issues will, in our estimation, remain very difficult or impossible to resolve. Our objective is to assess which of the issues concern most probable, severe and long-lasting effects and which are most subject to resolution.

TABLE 1.1

Assessment of the potential long-term effects of offshore oil and gas development activities by probability, severity and potential for resolution. The probability and severity of effects vary significantly among regions (see text for elaboration)

Potential effect	*Probability*	*Severity*	*Potential for resolution**
Effects on resources of intrinsic value			
Effects of physical fouling by oil of birds, fur-bearing mammals, and turtle eggs	medium	high	2
Effects resulting from ingestion and inhalation of noxious oil components and fouling of feeding apparatus	medium	medium	1
Disturbance by noise and other physical factors of birds, mammals and turtles	high	low	2
Effects on resources of economic value			
Reduction of fishery stocks by an oil spill damaging eggs and larvae	low	high	1
Effects on adults and juveniles of nearshore sediment contamination by medium molecular weight aromatic hydrocarbons and products	medium	medium	2
Effects on ecosystem support of resources			
Oil spills:			
Effects resulting from persistent contamination by medium and high molecular weight aromatic hydrocarbons, heterocyclics and their degradation products in sediments of low energy or cold environments	high	medium	2
Residual damage to biogenically structured communities by acute oiling	medium	high	2
Operational discharges:			
Effects of drilling discharges on benthos accumulated through field development	high	low	3

(*continued*)

TABLE 1.1—*contd.*

Potential effect	*Probability*	*Severity*	*Potential for resolution**
Effects of organic components of water-based drilling fluids	low	low	3
Effects of produced water discharges in nearshore environments	high	low	3
Physical alterations:			
Effects of pipeline routing through coastal wetlands	high	high	2
Effects of shore-based support facilities	high	medium	2
Effects of gravel islands and causeways in arctic environments	high	low	3

*Potential for resolution:
3 – can be resolved satisfactorily either by dismissing it as a significant issue or determining appropriate corrective action
2 – substantially better understanding can be developed which will contribute to decision-making
1 – advances possible, but issue will remain very difficult to resolve

EFFECTS ON RESOURCES OF INTRINSIC VALUE

Certain species or environments are deemed worthy of protection by our society not primarily because they furnish economic benefits but because of their aesthetic, cultural or social values. Included are species which may be rare or near extinction, as well as air-breathing, higher vertebrates such as birds, mammals and turtles. Other such intrinsically valued resources are environments which are subjects of human fascination (for example, coral reefs), are unique or nearly so, or are protected natural ecosystems which serve as wildlife sanctuaries and refuges. Risks of long-term effects to such intrinsically valued habitats are considered together with other ecosystems in a subsequent section; we here specifically consider intrinsically valued species.

Exposure of marine birds, mammals, and turtles to offshore oil development can threaten the survival of individuals and possibly large elements within a population (Chapters 11 and 12). The ultimate impact depends on the nature and extent of the contaminated area, the species and the dependence of the animal on the impacted area. Potential threats include physical fouling, ingestion and inhalation, noise, physical disturbance, and reduced abundance of food (Table 1.2).

TABLE 1.2
Potential effects of various threats of oil and gas development activities to marine turtles, birds, and mammals

Threat	*Effect*		
	Turtles	*Birds*	*Mammals*
Physical Fouling	Egg mortality	Mortality of eggs & all age classes	Otters-hypothermia/death Polar bears-hypothermia/death Seals (young)-hypothermia/death & rare immobilization
Ingestion	Fouling of feeding apparatus Toxicity Bioaccumulation	Bioaccumulation	Baleen whales-fouling of feeding apparatus Bioaccumulation in those which feed on surface or on benthos
Inhalation	Possible residual damage to respiratory system, affecting fitness and longevity		
Noise	Unknown	Variable, ranges from no effect to disturbance and death of eggs and young	Variable, ranges from interference with movement patterns in cetaceans to dispersion of pinnipeds from rookeries
Physical Disturbance	Lower reproductive success	Lower reproductive success	Interference with movement patterns
Ecosystem Disturbance	Reduced food supply in critical feeding areas		

Physical Fouling

Any animal exposed to spilled oil might suffer deleterious effects as a result of physical fouling. Those groups most threatened are birds and those marine mammals which rely on hair or fur for thermal insulation (for example, sea otters, polar bears, and newborn seals and sea lions). Such an impact if confined to a few animals within a population would have few long-term consequences, but could have a significant effect on a discrete, concentrated stock of animals such as a resting raft of sea birds in a feeding area, otters in an embayment, polar bears within an ice lead, or nursing seals on offshore rookeries. Present evidence suggests that cetaceans and adult pinnipeds are not threatened by physical fouling with oil (Chapter 12). The long-term population-level effects of physical fouling are estimated to be of low probability in the case of marine mammals but are of medium probability for birds. However, if heavy oiling of a spatially restricted population should occur, the effects might be severe. Better understanding of the circumstances and effects of physical fouling could contribute to decision-making, but the issues are unlikely to be completely resolved.

Inhalation and Ingestion

Birds and marine mammals that surface in an oil spill inhale petroleum vapors, possibly enough to cause residual damage to the respiratory system and to serve as a route of entry for systemic accumulation of petroleum compounds. The effects on birds may compound the more significant problem of increased thermal conductance associated with surface fouling. The actual effect on marine mammals of such exposure could be increased by pre-existing stress, parasites, and disease. Animals away from the immediate area of the spill, or exposed to weathered oils, would not be expected to suffer any consequences from inhalation. Mammals in the vicinity of a spill would have a high probability of inhaling noxious vapors, but birds would have a low probability. Turtles can respond to strong odors by breath-holding and thus reduce their exposure to petroleum vapors at the site of a spill. In general, effects resulting from inhalation are expected to be minor and not long-term.

Young turtles face a peculiar threat from oil spill residues. Tar becomes lodged in their mouths in such a way as to impair feeding. Baleen whales face a comparable threat in that oil adhering to the baleen plates may obstruct water flow, thereby impeding food-gathering efficiency. Whales most vulnerable to this threat would include the surface-feeders (e.g., right whales), especially those occupying contaminated calving grounds where their movements may be relatively confined.

Ingested oil can be harmful, either acutely when consumed in large quantities (which is unlikely) or by the action of metabolized products accumulated over time. Effects of ingestion on marine turtles are unknown; those on birds in an experimental setting are associated with a variety of physiological effects including those related to reduced hatchability of eggs and viability of offspring. There are few comparable data for free-ranging seabirds.

Polar bears, most seals, and odontocetes are predatory. They would not likely consume oil accidentally, nor scavenge food coated with it. A few of the bottom feeders, such as otters, walruses and bearded seals, would be expected to consume more contaminated food than pelagic and surface feeders. The same would be true of the bottom-scouring gray whale, whereas other baleen whales may ingest both fresh oil and floating residues. The quantities ingested may not cause acute toxicity, but could lead to deposition of hydrocarbons in tissues. The fate and consequences of these accumulated compounds is not known. Again, although ingestion is probable, predicted effects are not severe.

The potential for resolution of the issues concerning the effects of ingestion and inhalation is quite limited, although it appears more promising for birds than for mammals and turtles.

Noise and Other Disturbances

Noise is associated with all phases of offshore petroleum exploration and production. It accompanies seismic surveying, drilling, air and ship support, construction, and the operation of onshore and offshore facilities. The effects of noise on birds are variable—ranging from no obvious effect to dramatic and fatal startle reflexes in some cliff rookeries. Marine mammals, like other vertebrates, respond to sharp sound pulses with a startle reflex which can alter their behavior but otherwise become habituated to low-level background noise. The effects of noise on marine turtles has not been studied.

It is the startle reflex which may ultimately threaten the survival of an individual or a small colony of birds or mammals. Eggs, hatchlings, fledglings and young marine mammals are particularly vulnerable to a flurry of injurious activity, which may result in loss of eggs or young, dispersion from the nesting site or rookery, and disruption of vital parent-offspring bonds.

Any form of physical disturbance can have similar consequences. For example, human intrusions into colonies of birds can result in reduced reproductive success, and ill-timed illumination of turtle nesting areas may redirect the hatching period out of phase or cause misnavigation by adults and hatchlings.

Studies have concentrated on those effects which are most obvious and can be tested using conventional approaches. We recognize that there may be indirect effects which are more subtle and less easily recognized. For example, noise can stress non-auditory physiology by driving the stress response toward lowering resistance to disease and promoting hypertension and endocrine imbalance (Chapter 12). Although the probability of disturbance effects is high, the long-term severity of such effects is judged to be low for most birds, mammals and turtles. The bowhead whale, may be an exception. It has an extremely small remaining population and must migrate through the Alaskan Arctic where oil and gas development is expanding.

EFFECTS ON RESOURCES OF ECONOMIC VALUE

Fisheries (for both finfish and shellfish) involve the interaction of natural and human social systems, effectively including both the species harvested and the people harvesting them. This is important to understand when considering the conflicts between the utilization of renewable fishery resources and nonrenewable energy resources in such places as Georges Bank, the Bering Sea, California and the Gulf of Mexico. The fisherman is generally not concerned about total fishery yield but about his success in harvesting the resource of his choice in the environment of his choice. The sustained high yields of commercial and recreational fisheries in the northwestern Gulf of Mexico, the very area where offshore oil and gas development has been most intense, is frequently cited as evidence that development activities do not affect fishery resources. Great changes in fishing effort and in environmental conditions make this assertion very difficult to evaluate. The New England or Alaska fisherman is concerned about the effects of an additional stress on fishery resources already stressed by heavy fishing pressure.

Offshore oil and gas development may have long-term effects on fisheries in three ways: the effect of an oil spill coincident with a critical period of concentration of eggs and larvae near the water's surface; the effect of toxic compounds of petroleum origin chronically released from contaminated sediment on juveniles and adults of demersal or benthic species; and the effect of physical destruction or alteration of critical habitats. The last category is an indirect effect mediated through a change in the ecosystem and will be considered in the next section.

Effects of Oil Spills on Fishery Stocks

There exists no direct evidence that an oil spill has affected a stock as a result of mortality of eggs and larvae (Chapter 9). There is concern, however, that a large spill occurring during a critical recruitment period could seriously diminish recruitment to the stock for the year, particularly for those species in which eggs and larvae concentrate in the near-surface waters. For a species such as haddock in which only one year class out of every five to ten contributes substantially to the fishery, the loss of a good year class could be disastrous; for others such as cod, in which the contributions of the different year classes are much more uniform, the effect on the stock would be much less important.

Predictive models (e.g., Spaulding *et al.*, 1983) have demonstrated that if worst-case assumptions about mortality to eggs and larvae due to oil spills and the importance of larval recruitment to stock size are valid, significant effects on a stock could occur as a result of a large spill. Both assumptions, however, are tenuous. Exposure and toxicity likely to be experienced by epipelagic eggs and larvae resulting from surface slicks is poorly known. Based on existing data (Chapters 5 and 8), it seems that toxic concentrations would not be widespread. Factors controlling recruitment to the stock are also poorly known, which limits predictions of the effect of larval mortality on the adult stock. Effects on the stock, other than catastrophic effects, would be difficult to detect and attribute to an oil spill because of the great and largely unexplained year-to-year variability in recruitment. Thus, we conclude that the probability of such an effect occurring is low (on the basis of the improbable coincidence of a critical recruitment period and a large spill resulting in toxic effects), although if it did occur the effect on resources could be severe. Furthermore, given the variability question, the potential for resolution is low, although the issue might be more tractable if there were a better understanding of the controls of stock recruitment.

Sediment Contamination and Nearshore Fisheries

In nearshore environments, petroleum hydrocarbons resulting from an oil spill or operational discharges are more likely to reach the seabed and be incorporated into bottom sediments. Toxic hydrocarbons, particularly medium and high molecular weight aromatics and heterocyclics, may persist for long periods in anaerobic sediments (Chapters 6 and 7) to be released chronically or episodically into the environment and exert potentially toxic effects. Because juvenile forms of many economically important species live in inshore environments, events there may affect those species even though offshore adult populations are not directly or immediately affected. Anadromous species such as salmon are a specially sensitive case because they concentrate in estuaries and rivers during both spawning and seaward migrations. Of course, uncertainties similar to those discussed above regarding the development of toxic concentrations and the effects of larval mortality on adult stocks pertain to nearshore areas as well.

Exposure to toxic hydrocarbons, although more localized, may occur over more extended periods.

Consequently, the probability of this effect occurring is judged high and the severity low. A better understanding of effects of sediment contamination would contribute to sounder or more-confident decision-making (for example, regarding oil spill cleanup strategies), but is unlikely to fully resolve the question of impacts on nearshore fisheries.

EFFECTS ON ECOSYSTEM SUPPORT OF RESOURCES

In addition to the direct effects of offshore oil and gas development on resources of intrinsic and economic value, indirect effects on marine ecosystems may be significant. Because the relationship of ecosystem characteristics and functions to the economically and intrinsically valuable resources is complex and uncertain, consideration of indirect ecosystem effects is difficult. As a result, research has focused on the effects of oil and gas development activities on specific ecosystem components and much less on the significance of alterations in populations and communities to the total ecosystem function and renewable resources.

Oil Spills

Oil spills are by definition acute and episodic events. However, spills may exert long-term effects either as a result of residual contamination, slow recovery of damaged biota, or by their repeated occurrence. The following hypotheses regarding the long-term effects of oil spills on marine ecosystems emerge from the reviews:

1. After a spill, the relatively undegraded petroleum hydrocarbons are gradually or intermittently released from anaerobic sediments or from sediments or under ice in cold environments resulting in long-term effects on benthic and demersal species.
2. Aromatic hydrocarbons are often incompletely degraded, producing among other compounds oxygenated aromatics, which are highly toxic and persistent (Chapters 5, 7, and 8).
3. Sublethal effects, which have subtle consequences to populations of exposed species, result from sediment contamination by persistent aromatic hydrocarbons, heterocyclics and their degradation products (Chapter 9).

Petroleum hydrocarbons are evaporated, oxidized or biodegraded relatively rapidly in high-energy, oxygen-rich environments. However, if they are trapped in anaerobic sediments, sediments not subject to frequent resuspension, or under ice in cold environments, biodegradation proceeds slowly, and toxic hydrocarbons may persist in the environment for several years. Stable, fine-grained sediments which are anaerobic below the sediment-water interface are found in sheltered nearshore habitats, on the outer shelf, and on the continental slope. Unless oil were released at the seabed as a result of a subsurface blowout, however, large concentrations of petroleum hydrocarbons would not be expected to accumulate in deep water sediments. Thus concern is principally focused on nearshore environments directly impacted by oil spills.

Medium molecular weight aromatic hydrocarbons, such as naphthalenes, phenanthrenes, fluorenes and dibenzothiophenes, seem to be primarily responsible for toxic effects on benthos. A residuum of naphthalenes exceeding 0.01 ppm in interstitial waters appears to be a threshold for such an effect (Chapter 9). In addition, high molecular weight aromatic hydrocarbons and heterocyclics are known to be very persistent (Chapters 6 and 7). Understanding the biogeochemical processes which allow such persistent contamination, the availability of sequestered compounds to the biota, and effects on exposed populations, is critical to resolving this issue.

The formation of toxic oxygenated products is complex because of the extreme diversity of possible product compounds and the limited knowledge of their occurrence in the environment and their toxicity. Although it is highly likely that such compounds will be produced, it is not at all clear whether such production is significant, in terms of quantity or toxicity. Long-term effects are most likely to result from persistent contamination by aromatic hydrocarbons, heterocyclics and their degradation products in extensive shallow water habitats with fine sediments (northern Gulf of Mexico and Alaska) and where cold temperatures may delay weathering (Alaska). The severity, however, should be moderate because the impact would generally be localized, communities are resistant to modest contamination, and, consequently, it is unlikely that valuable resources would be severely damaged. The issue, like others concerning oil spills, is approachable but will remain difficult to resolve because knowledge is lacking about the dependence of valuable resources on benthic ecosystem components.

4. Long-term effects result from acute damage due to an oil spill on biogenically structured habitats such as coral reefs, mangrove swamps, salt marshes, oyster reefs, seagrass beds and kelp forests.

Here the concern is that even though oil may not persist following an oil spill, the time required for recovery of damaged populations of organisms which provide the physical structure of the habitat will be many years. In some cases where the structure-forming species actually stabilizes the habitat, it is conceivable that permanent modification of that habitat could result from an acute incident. Questions to be addressed are a) what are the exposure conditions under which toxic effects may be exerted on the primary structure-forming species and b) what are the population recovery rates? There is an overall moderate probability that such communities would be deleteriously affected by an oil spill. Some (e.g., mangroves) may be more susceptible than others, but should damage occur the impact may be severe because of the long time required for recovery.

Operational Discharges

The major operational discharges associated with offshore oil and gas exploration, development, and production are drilling fluids, drill cuttings, and produced water. Water-based drilling fluid, which is the only type permitted for discharge to U.S. coastal and offshore waters, is a freshwater or seawater slurry of clay (or natural organic polymer), barium sulfate, lignosulfonate, lignite, and sodium hydroxide, plus several minor additives (National Research Council, 1983). Cuttings are crushed formation solids produced by the grinding action of the drill bit as it penetrates into the earth. Produced water usually has been in contact with the fossil fuel-bearing formations and may contain elevated concentrations of hydrocarbons, dissolved inorganic ions, metals, and other uncharacterized soluble organic materials.

During drilling of an exploratory well, from 5,000 to 30,000 barrels of drilling fluid (containing 200 - 2,000 metric tons of solids) may be used and discharged. From 1,000 to 2,000 metric tons of drill cuttings may be generated and discharged. Development wells often are shallower, smaller in diameter, and are drilled more rapidly than exploration wells, and so smaller amounts of drilling fluid and cuttings are generated per well. However, as many as 100 wells may be drilled from a single development platform. During production of oil or gas from a platform, up to 10,000 barrels (1.6 million liters) of produced water may be discharged per day. If the produced water contained a mean of 48 ppm oil, as much as 77 liters of oil would be discharged per day with produced water from a well.

Because the acute toxicity of water-based drilling fluids is low, and concentrations of drilling muds and cuttings in the water column decline rapidly following discharge due to dilution and sedimentation to the bottom, adverse impacts on water column organisms are expected to be very slight and of short duration. Longer-term impacts of such discharges are restricted to the benthos near the discharge point, where significant amounts of mud and cuttings solids settle and persist on the bottom in low-energy environments (National Research Council, 1983; Chapter 10).

Most investigations of the effects of drilling fluids and cuttings have been in the context of exploratory drilling; effects on the benthos have been limited to within a few hundreds of meters of the discharge. In the few cases where recovery has been monitored, residual effects seem mainly caused by the accumulation of cuttings which attract a different fauna than the native seabed. Laboratory experiments have further suggested that the principal source of toxicity in drilling fluids is diesel fuel added to some fluids to lubricate the drill bit. There is little evidence, however, that hydrocarbons from diesel fuel accumulate in sediments as a result of exploratory drilling discharges. For these reasons, we conclude that discharges from exploratory drilling in offshore environments would not result in significant long-term effects except where they may directly impact a rare community which takes a particularly long time to recover.

The quantities of drilling fluids and cuttings discharged from multi-well development platforms are substantially larger than the quantities discharged from exploratory wells. Few studies have assessed the effects of discharges from multiple well platforms nor the effects of discharges from concentrations of platforms. Most of those studies were unable to separate effects of drilling discharges from those due to other emissions and physical alterations (Chapter 14).

Based on inventories of barium (an excellent tracer of drilling muds, but in itself essentially non-toxic), sediment contamination from multiple wells drilled from a single platform appears less than that resulting from a single well discharge multiplied by the number of wells, because contaminated sediments disperse with time. On the wave-influenced shallow continental shelf, sedimented contaminants disperse but where sediments are relatively stagnant, contamination (and the effects thereof) may persist. In deep continental slope environments, materials discharged near the surface tend to disperse greatly before settling to the seabed, resulting in only very slight contamination of bottom sediments. Consequently, the greatest uncertainty remaining concerns the long-term effects of discharges as a result of intense development level drilling in outer shelf, depositional environments.

Following discharge to the ocean, produced water is diluted rapidly with seawater, so that no significant biological impacts due to altered salinity, ion ratios, or oxygen concentrations are anticipated. In shallow areas with high suspended sediment loads, medium molecular weight hydrocarbons and metals can adsorb to suspended particles and be deposited in sediments. If this results in persistent elevation in sediment hydrocarbon and metal concentrations, modification of benthic communities can result within a few hundred meters of the discharge (Chapter 10).

In some cases, produced water is not separated from the oil and gas at the well head or on a production platform but is transported to some remote collection point where separators serve a number of wells. This may be at another offshore platform or at an onshore facility. In the Gulf of Mexico, a large portion of the produced water from offshore wells is discharged from onshore facilities into coastal waters which have lower rates of dispersion and more rapid sedimentation of contaminants.

Based on the above considerations, we conclude that the potential long-term effects of operational discharges on marine ecosystems are as follows (see also Table 1.1):

1. Benthic communities in the vicinity of multiple well platforms may be modified as a result of large and extended discharges of drilling fluids and cuttings, particularly in depositional, outer shelf environments. Although the probability of such effects occurring is high, such effects are expected to be only subtle alterations of communities outside of an area of a few hundred meters around the platform. Although these effects may not be adverse or severe, the potential for resolution of this controversial issue by well-designed studies appears high.
2. Organic components (petroleum, mineral-oil based lubricants and lignosulfonates) used as additives in water-based drilling fluids in both exploratory and development drilling are present at low levels, but may exert effects on the benthos. The probability and severity of such effects is judged low, but the potential for resolution of this issue is high.
3. Chronic produced water discharges may affect benthic organisms, particularly in shallow coastal waters. The probability appears high but because

of the restricted areal extent, the severity would be low. Satisfactory resolution of this issue can be achieved through well-designed quantitative research.

Habitat Alterations

While oil spills and operational discharges introduce contaminants that may be chemically toxic or physically disruptive to marine biota, oil and gas operations may also result in physical habitat alterations. The resulting biological effects to those ecosystems that support resources of intrinsic or economic importance may be longer lasting than those resulting from spills or operational discharges and, in some cases, permanent.

Habitat alterations offshore result from the placement of hard structures in the sea and the disruption of bottom substrates during pipeline emplacement or by anchors or other devices dragged across the seabed. Structures such as platforms, well jackets, subsea connectors, exposed pipelines and large discarded objects increase the spatial heterogeneity of a soft bottom continental shelf and slope and provide substrates for encrusting epibiota. The dense epibiota may result in heavy deposition of skeletal material (shells, tests, etc.) and fecal material under a rig or platform which may alter the natural benthic community in the immediate vicinity of the structure (Wolfson *et al.*, 1979). Structures may further attract fish and other motile animals which feed on the epibiota or other attracted animals or which find refuge there. The attraction of fishes to structures may increase the population carrying capacity of those species within the shelf environment or may merely concentrate the existing populations. If such concentration makes the species more susceptible to overfishing, it might be argued that the effect is deleterious. The populations of many fishes attracted to oil and gas structures in the Gulf of Mexico seem to be enhanced (Gallaway and Lewbel, 1982). It appears that the long-term physical effects of such offshore structures are usually beneficial and would only rarely be significantly deleterious to living resources. Consequently, the effects of oil and gas structures merit further assessment only in the development of designs to enhance their beneficial effects in their disposition after abandonment and in reduction of their interference with fishing activities.

Pipelines have been the conventional means of transporting offshore oil and gas to shore. Extensive pipeline construction in the next 20 years is expected in the U.S. only off southern California and in the northwestern Gulf of Mexico.

Pipelines are required to be buried below the sediment surface in water depths less than 60 m. Burial is usually accomplished by jetting a trench in which the pipe lies. This results in a large disturbance of surface sediments and their communities, and, in some cases, a long-lasting alteration of the bottom topography. The effects of sediments suspended during pipeline emplacement are of short duration and are not of concern in the context of long-term effects. Recovery of macrobenthos from small scale disturbances ranges from a period of weeks for temperate, shallow water communities to a year or more in continental shelf environments and to many years for continental slope communities (Boesch and Rosenberg, 1981). Recovery of hard substrate communities which are biogenically structured, such as coral reefs, may require longer periods than sediment-dwelling communities in comparable depths. The long-term effects of offshore pipeline emplacement are therefore greatest for hard substrate communities, and near the shelf edge or deeper; but pipeline burial is seldom attempted in either of these cases.

Oil and gas development activities which may affect the coastal zone are many and varied. Most pipelines terminate onshore and consequently must cross coastal environments. In addition, offshore activities generally require coastal support bases which may damage or displace wetlands and shallow water habitats. Since materials for offshore fields are usually supplied by vessels, there may be increased need for navigation channels and their associated alterations of coastal habitats.

The potential for significant long-term effects on coastal ecosystems varies widely in the U.S. depending on the nature of the coastal ecosystems and the techniques used during oil and gas development. In many regions of the U.S. coast, coastal wetlands and shallows are lacking or can be easily avoided. In others, onshore space and navigation demands can be easily met with existing facilities. However, in the northern Gulf of Mexico, which is the most heavily developed region and which will continue to experience the greatest drilling activity in the next decade, coastal effects are perceived as the major environmental concern related to offshore oil and gas activity (Chapter 13). This is a result of the extensive coastal wetlands, scarcity of fastlands and extensive shallow water bodies which characterize parts of the northern Gulf coast. Such conditions also exist to varying degrees in other U.S. coastal areas adjacent to potential offshore oil and gas development (South Atlantic coast and some regions of Alaska).

It has been estimated that at least half of the rapid loss of coastal wetlands in Louisiana (over 100 km^2/yr) is the direct or indirect result of channelization, mainly for oil and gas extraction and transportation (Scaife *et al.*, 1983). Most of the channelization is in support of oil and gas development in the wetlands and estuaries rather than offshore, but at least 128 pipelines from offshore production sites have landfalls in Louisiana, Mississippi or Alabama (Minerals Management Service, 1983a). All but a few of these cross the coast of Louisiana, and the majority cross wetlands at some point.

The conventional routing of pipelines through wetlands involves dredging a canal in which the pipeline lies, although newer technologies allow pulling the pipeline through a narrower trench. The actual width of the corridor of direct impacts usually exceeds 60 m. Pipelines emanating from offshore production sites traverse wetlands over lengths ranging from tens of meters, where there are only fringing marshes, to tens of kilometers in the case of the broad Mississippi Deltaic Plain marshes. The full extent of these direct impacts has not been quantified. Furthermore, the total effects of canal excavation on marsh loss may be as much as four times the direct removal due to the channel and banks of excavated material. Indirect effects are due to saltwater intrusion, enhanced subsidence caused by the placement of dredged material on the marsh, and interference with natural hydrologic flow patterns resulting in a deficiency in sediment accretion to counterbalance natural subsidence (Chapter 13).

Numerous navigation channels which support transportation for offshore oil and gas development also cross the Gulf coast. Channels parallel to the coast tie together the upper ends of natural drainage basins and produce altered freshwater flow patterns. Channels perpendicular to the coast generally result in saltwater intrusion in the otherwise shallow and convoluted estuaries. Furthermore, vessel wakes may cause rapid widening of navigation canals through erosion, increasing saltwater intrusion.

Despite the uncertain magnitude of the effects of coastal alterations and the degree to which future offshore development will result in such impacts, such physical alterations of the Gulf coastal zone constitute documentable and essentially permanent effects. Also, because of the relationship between high fisheries production and the extent of coastal wetlands (Boesch and Turner, 1984), these effects may be deleterious to valuable resources.

The effects of physical perturbations of coastal environments have not generally been regarded as highly significant issues in other OCS regions. Most of the coasts bordering potential shelf petroleum provinces have more limited intertidal areas and do not have extensive wetlands. Pipeline landfalls pass without much lasting disturbance across a beach or rocky shore to uplands; in California, most of Alaska, and the northeastern U.S., wetlands can be avoided in routing pipelines. The Sea Islands region of Georgia and South Carolina, the Yukon delta region of Norton Sound and Bristol Bay in Alaska may be other areas where wetlands are difficult to avoid. The northern Chukchi Sea and Beaufort Sea coasts may be susceptible to physical alterations, but the barrier island-lagoon-tundra shoreline system is highly dynamic because of seasonal ice activity, and effects would therefore not be expected to persist.

Offshore oil and gas development is also taking place in other regions of the world which are characterized by extensive coastal wetlands (West Africa, Latin America and Asia). Enhanced understanding of the northern Gulf of Mexico experience would be valuable in reducing the long-term impacts of physical alterations of the coastal zone overseas as well as in the developing sensitive regions of the U.S.

In the arctic environments of Alaska and Canada, islands and causeways are constructed as a base for drilling because of the ice hazards confronting conventional rigs and steel platforms. Gravel islands and causeways, of course, permanently displace the benthic community in their location. Furthermore, if the seabottom is dredged for the construction materials, at least a temporary perturbation of the dredging site will result, and the effect of the excavation may be long-lasting if the nature of the habitat is changed. The presence of islands and especially causeways may have a larger scale effect if it alters the flow regime, affecting sediment deposition or erosion as well as other environmental conditions. Resulting habitat alterations may, in turn, affect the distribution of benthos, fish and birds, especially in barrier-lagoon systems. Causeways perpendicular to the shore may also interfere with alongshore migratory patterns of anadromous fish.

All of these long-term effects of islands and causeways in arctic environments are judged to be of low severity, although the existence of at least some effect of dredging is deemed probable. Based on knowledge to date, the potential for disruption of migratory patterns is yet very speculative.

Although the long-term effects due to physical alterations on coastal ecosystems will be highly variable as a result of differences in coastal environments and development approaches, such effects are highly probable and may be severe in some areas. Better understanding of the nature and ramifications of such impacts will contribute to better decision-making and more effective mitigation of impacts.

FUTURE STUDY NEEDS

To this point, we have sought to determine the environmental effects resulting from offshore oil and gas development which have the potential for persistence over at least several years and a significant deleterious influence on marine resources. We have also provided our assessments of the relative probability and severity of these effects and the potential that such effects could be dismissed as improbable or insignificant or effectively mitigated based on additional research or information synthesis.

We have not concluded that any of these possible effects are or would be catastrophic, but our present state of knowledge is insufficient to dismiss the possibility of some serious but insidious effects. We have, on the other hand, implicitly dismissed many effects which have been suggested (for example, those resulting from exploratory drilling discharges) as being highly unlikely to result in long-term effects.

It is apparent to us that the limitations of contemporary science in providing confident predictions about the marine environment result in the existence of legitimate, unanswered questions concerning the long-term effects of offshore oil and gas development. How much emphasis should be placed on research on long-term effects compared to descriptive environmental studies which are often performed prior to offshore development? What are the priorities for useful and feasible research on long-term effects?

Over $344 million were spent between 1973 and 1983 by the U. S. Department of the Interior alone on environmental studies aimed at contributing to decision-making or resolving conflicts regarding offshore oil and gas development (Minerals Management Service, 1983b). Considerable additional sums have been expended by other Federal agencies, industry, the states, and other involved parties. The majority of these studies have had as their purpose the description of the environment to predict possible effects (for example, measuring currents to predict the trajectory of oil spills) or to measure subsequent change (baseline studies).

Direct measurement of effects in the environment during exploration and development or the experimental simulation of effects has been only a relatively small part of this effort. As a consequence, there has existed a "prediction gap," wherein the accumulation of descriptive environmental data has exceeded the ability to use it in predicting environmental effects. In addition, the underdeveloped status of predictive capabilities has generally not even allowed testing of the utility of this accumulated descriptive information to understanding environmental effects. For environmental assessment to mature, therefore, resources must be reallocated to studies which increase confidence in predicting effects. Such studies must be relevant to non-trivial effects in the natural environment and must emphasize environmental and biotic processes, not simply patterns.

Based on our considerations in the first part of this Chapter of probability and severity of effects and, secondarily, of the potential for resolution of the issues, nine general long-term effects issues have been identified as being of high priority for future investigation (Table 1.3). Determining priorities among such diverse

TABLE 1.3

Summary of the potential long-term environmental effects of offshore oil and gas development activities which are of high priority for future investigation

High priority:

- Chronic biological effects resulting from the persistence of medium and high molecular weight aromatic hydrocarbons and heterocyclics and their degradation products in sediments and cold environments
- Residual damage from oil spills to biogenically structured communities such as coastal wetlands, reefs and vegetation beds
- Effects of channelization for pipeline routing and navigation on coastal wetlands

Intermediate priority:

- Effects of physical fouling by oil of birds, mammals and turtles, especially in species in which a large percentage of the population aggregates at certain times
- Effects on benthos of drilling discharges accumulated through field development
- Effects of produced water discharges generated offshore but discharged into near-shore environments

Low priority:

- Effects of noise and other physical disturbances on populations of birds, mammals and turtles
- Reduction of fishery stocks due to mortality of eggs and larvae as a result of oil spills
- Effects of man-made, usually gravel, islands and causeways in the Arctic on benthos and anadromous fish species.

subjects is always difficult, but, based on the consensus of the authors, we have grouped these issues into three priority levels in an attempt to show their relative importance. Several related issues individually listed in Table 1.1 are combined in this list. In addition, we have indicated in Table 1.4 the regional relevance or recommended geographic focus by major region of the U.S. in which offshore oil and gas exploration or development is being planned or pursued.

For each of the potential long-term effects identified, we recommend study approaches in the following discussion. Furthermore, we evaluate the feasibility of these approaches with regard to the effort required and the present or

TABLE 1.4
Proposed regional focus of priority studies of the potential long-term effects of offshore oil and gas development activities

<table>
<thead>
<tr><th>Effect</th><th>North and Mid-Atlantic</th><th>South Atlantic</th><th>Gulf of Mexico</th><th>California</th><th>Alaska</th></tr>
</thead>
<tbody>
<tr><td>Chronic effects of persistent hydrocarbons</td><td colspan="5">← Generic →</td></tr>
<tr><td rowspan="6">Residual damage to biogenically structured communities</td><td colspan="3">← Salt Marshes →</td><td colspan="2">← Kelp →</td></tr>
<tr><td></td><td colspan="4">← Rock Reefs →</td></tr>
<tr><td></td><td colspan="2">← Oyster Reefs →</td><td></td><td></td></tr>
<tr><td></td><td colspan="2">← Coral Reefs →</td><td></td><td></td></tr>
<tr><td></td><td></td><td>← Mangroves →</td><td></td><td>← Sedge Marshes →</td></tr>
<tr><td></td><td></td><td>← Sea Grasses →</td><td></td><td>← Sea Grasses →</td></tr>
<tr><td>Effects of wetlands channelization</td><td></td><td>← Potentially Impacted Areas →</td><td>← Impacted Areas →</td><td></td><td>← Potentially Impacted Areas →</td></tr>
<tr><td rowspan="3">Physical fouling by oil of birds, mammals and turtles</td><td>← Birds →</td><td colspan="2">← Turtles →</td><td colspan="2">← Birds →</td></tr>
<tr><td></td><td></td><td></td><td colspan="2">← Sea Otters →</td></tr>
<tr><td></td><td></td><td></td><td colspan="2">← Pinniped Pups →</td></tr>
<tr><td>Effects of accumulated drilling discharges</td><td></td><td></td><td>← Texas-Louisiana Outer Shelf →</td><td>← Santa Maria Basin →</td><td></td></tr>
<tr><td>Effects of discharge of produced waters in nearshore environments</td><td></td><td></td><td>← Louisiana, Texas →</td><td>← Southern California →</td><td></td></tr>
<tr><td rowspan="3">Effects of physical disturbance on birds, mammals and turtles</td><td></td><td></td><td></td><td></td><td>← Bowheads →</td></tr>
<tr><td></td><td colspan="2">← Turtles →</td><td colspan="2">← Selected Birds →</td></tr>
<tr><td></td><td></td><td></td><td colspan="2">← Selected Mammals →</td></tr>
<tr><td>Reduction in fishery stocks by oil spills</td><td colspan="5">← Generic →</td></tr>
<tr><td>Effects of gravel islands and causeways</td><td></td><td></td><td></td><td></td><td>← Beaufort Sea →</td></tr>
</tbody>
</table>

foreseeable availability of methodology needed. The approaches are classed as having 1) limited feasibility within the next 10 years; 2) potential feasibility but requiring the development of methods or application of innovative approaches; or 3) high feasibility using available methods. Approaches directed to proximate environmental responses (e.g., fate and uptake of pollutants) are in general more feasible than the extended responses which are of ultimate concern (e.g., effects on populations and resources). This is merely symptomatic of the complexity of biotic interactions in ecosystems.

The study approaches are further evaluated by relevance to or recommended focus in regions of the United States and by study duration and timing. It is clear from this appraisal that, although a number of approaches could quickly yield very useful results with existing methods, many issues concerning long-term effects will require innovative study approaches, methods development, and extended effort to achieve the level of resolution deemed possible.

More emphases in future studies should be placed on a) processes coupling physical transport, chemical transformation, and environmentally realistic exposure of the biota; b) biological effects which influence population success; c) more rigorous experimental approaches; and d) the consequences of ecological change to resources valued by humans.

The recommended trends to increased experimentation and environmental realism suggest, in particular, experimental field and microcosm or mesocosm approaches which will require considerable innovation, and for which success is not assured. Experience with ecological experiments in the marine environment has grown rapidly during the last 10 years (see, for example, summaries of mesocosm experiments by Grice and Reeve, 1982; Oviatt *et al.*, 1982, 1984; Brockmann *et al.*, 1983). Experiments are increasingly feasible in continental shelf and simulated continental shelf environments; oil and gas platforms or large drilling rigs offer platforms of opportunity for relevant experiments directed at some of the questions we have discussed. Petroleum seeps on the continental shelf also provide opportunities for experiments for assessing the effects of hydrocarbons.

RECOMMENDED STUDY APPROACHES

For each major category of study need listed in Table 1.3, research approaches are recommended in Tables 1.5 through 1.13. The approaches are generally listed in sequential order—in sequence of logic if not chronology—for each of the nine effects categories. For each study approach, an appraisal of feasibility, duration and timing is given, together with the recommended regional focus, if any.

It should be understood that the adequate resolution is unlikely if individual approaches are pursued in isolation. In some cases, however, results of a particular study approach may render unnecessary other approaches later in the sequence. For example, a preliminary study may adequately dismiss a potential effect as unlikely to persist or to be extremely isolated.

TABLE 1.5

Recommended study approaches for the resolution of potential long-term effects resulting from the persistence of medium and high molecular weight aromatic hydrocarbons, heterocyclics, and their partial degradation products

Study approaches	*Feasibility*[1]	*Regional focus*	*Duration (years)*	*Sequencing*
1. Sedimentologic and geochemical dynamics of hydrocarbons and products				
a. Determine the persistence of parent compounds and their degradation products under various conditions (temperature, light, oxygen, microbial populations).	High	Generic experimental research	5	Immediate
b. Conduct mesocosm and field experiments and make long-term field observations on sedimentation, retention, degradation and erosional transport of petroleum hydrocarbons in sediments.	High-Potential	Generic experimental research	5	Laboratory studies concurrent with and dependent on results of 1, a
		Field research should emphasize areas with extensive nearshore, fine-graded sedimentary or cold environments, particularly Gulf of Mexico and Alaska.	10	Field studies dependent on results from 1, a and experimental studies; long-term for life of development or activity being assessed; low-level effort
c. Develop process models which relate the fate of contaminants to sediment type, erosion and deposition and likelihood of contamination	Potential	Generic research with emphasis on environments likely to be developed and contaminated	3	Initial descriptive models to direct the above research and quantitative models which employ results of 1, a and 1, b

2. Bioavailability of petroleum hydrocarbons and their degradation products to various trophic groups of benthos and valuable species				
a. Conduct experiments and make field observations to determine bioavailability of compounds from contaminated sediments.	High-Potential	Generic laboratory research Field research should emphasize areas with extensive nearshore, fine-grained sedimentary or cold environments, particularly Gulf of Mexico and Alaska; contingent on results of 1, b.	5 10	Laboratory studies concurrent with and dependent on 1, a Field studies follow at a low level from progress in 1, b
b. Determine the bioaccumulation potential via ingestion by birds, mammals and turtles.	High-Potential	Research should emphasize: Alaska and California (mammals, shearwaters, storm petrels, alcids); Gulf of Mexico and Atlantic (turtles, mammals).	5	Immediate

(continued)

TABLE 1.5—*contd.*

Study approaches	*Feasibility*[1]	*Regional focus*	*Duration (years)*	*Sequencing*
c. Relate bioavailability of major compounds to their persistence in sediments.	Potential	Generic research	1	Initial models following results of 1, a
			2	Subsequent refinement following results of 1, b and 2, a-b
3. Sublethal effects of petroleum hydrocarbons and their degradation products				
a. Determine biochemical and physiological responses to exposure in potentially susceptible biota.	Potential	Generic laboratory research	5	Following progress of 1, a-b and 2, a-b
		Field research as above in 1, b and 2, a-b and in chronically contaminated environments		
b. Determine the quantitative importance of detoxification reactions that produce more toxic intermediate metabolites.	Potential	Generic laboratory research	5	Following progress of 2, a-b
c. Assess the relationship between tissue concentrations and pathological condition of selected species.	High	Generic laboratory research	5	Concurrent with 2, a-b
		Field research (as in 2, a-b)	10	Long-term observations dependent on results of 2, a-b

d. Monitor the population dynamics of selected susceptible colonies of marine birds, turtles and mammals.	Potential	As in 2, b	10+	Long-term low-level effort coupled with 2, b and 3, a.
e. Determine relationship between biochemical and physiological indicators of stress and population-level responses.	Limited	Generic research	5	Following progress of 3, a, c and d
f. Evaluate the consequences of long-term effects on populations to living resources and ecosystem support of these resources.	Limited	Generic research	1	Initial models based on results from 1, a-b; 2, a-b; 3, a and c
			2	Refinement of models based on all results

[1] Feasibility of approach judged as "High" (can be satisfactorily accomplished within a 10-year time frame using available methods); "Potential" (requires development of methods or innovative approaches); and "Limited" (probably infeasible within a 10-year time frame).

Persistent Hydrocarbon Contamination

The long-term effects of oil spills are potentially the most serious of the effects of offshore oil and gas development activities, but also the hardest to define and control. There is a long history of research on the fate and effects of oil in the marine environment, yet viewpoints in the scientific community on some issues diverge widely (Royal Commission on Environmental Pollution, 1981; Clark, 1982; National Research Council, 1984). The research approaches outlined in Table 1.5 address the most important unresolved issues concerning effects of persistent contamination by petroleum hydrocarbons and degradation products in environments conducive to such retention—fine-grained sediments and cold environments. The study approaches recommended are far more complex and multifaceted than those addressing subsequent issues. They also involve more generic experimental research; the other issues are more heavily dependent on field observations.

The recommended approaches addressing the long-term effects of hydrocarbon contaminants fall into three groups: a) the sedimentologic and geochemical dynamics of hydrocarbon contaminants and their degradation products (what compounds persist, for how long, and where are they transported?) b) bioavailability (are the persistent compounds taken up by the biota, are they bioaccumulated by air-breathing animals, what insight can be provided by measuring body burdens in stranded animals?); and c) chronic and sublethal effects.

Research on the long-term fate of medium and high molecular weight aromatics, heterocyclics and their degradation products depends on understanding the conditions which allow them to persist. Most of the existing information is based on field observations or small-scale laboratory experiments. Controlled experiments are required to separate multiple factors, and experimental approaches should be generally scaled up for increased realism. Thus, mesocosm and field experimental approaches are particularly recommended. These will allow better extrapolation to natural conditions through application of site-specific sediment transport models.

Research on bioavailability should concentrate on the uptake and retention of contaminants from bottom sediments and on the potential for long-term build up of contaminants in the tissues of birds, mammals and turtles, in which the primary uptake route is probably direct ingestion.

Determination of chronic and sublethal effects on the biota is always difficult. During the past decade, there has been a proliferation of stress indicator techniques for evaluating responses of organisms to pollutants (McIntyre and Pearce, 1980), yet the relationship of the response to survival of the individual, much less the population, is frequently unknown. Particularly sensitive are biochemical responses that relate to energy metabolism and membrane function (such as lysosomal stability), biochemical responses that relate to detoxification (such as induction of mixed-function oxidases), and physiological responses (such as scope for growth) that measure the energy available for growth and reproduction.

Induction of mixed-function oxidase activity in marine organisms is a response to petroleum hydrocarbons which detoxifies and removes hydrocarbons but also produces more toxic primary and secondary metabolites. Recent evidence has

TABLE 1.6
Recommended study approaches for the resolution of potential long-term effects which result in residual damage due to acute oiling of biogenically structured communities

Study approaches	*Feasibility*[1]	*Regional focus*	*Duration (years)*	*Sequencing*
1. Appraisal of known recovery rates in different communities	High	North Atlantic: salt marshes South Atlantic: salt marshes, coral reefs, rock reefs, oyster reefs Gulf of Mexico: salt marshes, mangroves, coral reefs, rock reefs, seagrass beds, oyster reefs California: kelp forests, rock reefs Alaska: kelp forests, rock reefs, sedge marshes, seagrass beds	2	Immediate
2. Determine rates of recovery from various levels of disturbance.	Limited	As above	10+	
a. Site specific experimental disturbance of habitats (enclosed or natural)				Based on study needs identified in 1
b. Site specific long-term observations after accidental spills				As opportunities arise
3. Relate recovery rates to persistence of contaminants.	Potential	As above	10+	Concurrent with 2

[1] Feasibility of approach judged as "High" (can be satisfactorily accomplished within a 10-year time frame using available methods); "Potential" (requires development of methods or innovative approaches); and "Limited" (probably infeasible within a 10-year time frame).

suggested a direct relationship between detoxification processes and a) loss of reproductive effort (Spies *et al.*, 1983), b) developmental and energetic abnormalities (Capuzzo *et al.*, 1984), and c) histopathologic changes (Malins *et al.*, 1983). Other parameters that may prove to be useful in predicting significant effects on populations include changes in blood glucose, energy metabolism or hormonal levels, particularly when these measurements can be made in conjunction with estimates of mixed-function oxidase activity, metabolite formation and organismal effects, such as fecundity and development rate. No single index can provide the predictive capability to evaluate population change; hence future efforts should emphasize the relationship of multiple response indices.

Biogenically Structured Communities

Recovery from effects of an oil spill of a community in which organisms provide the physical structure of the habitat depends both on the persistence of contamination (addressed above) and on the inherent ability of the community to recover. Numerous observations have been made on the recovery of some communities, such as rocky intertidal communities and salt marshes, following an oil spill or some other disturbance. Less information is available concerning recovery of other community types. We felt that the literature on known recovery rates of biogenically structured communities and on the factors influencing those rates needs to be critically reviewed and synthesized (Table 1.6). Following this synthesis, new research should be performed in which community recovery is studied following experimental disturbance and following the acute effects of accidental spills. This research will be dictated by the recommendations of the critical synthesis and, of course, opportunities provided by the accidental spills.

Wetland Channelization

The construction of pipelines and navigation channels through extensive intertidal zones, particularly coastal wetlands, results in destruction of habitat and may cause a reduction in support of economically important living resources. A quantitative evaluation of direct effects (area dredged and spoil banks) which have resulted from offshore development in the Gulf of Mexico is required (Table 1.7). The area indirectly altered as a result of saltwater intrusion and by disruption of the hydrologic regime in wetlands should also be estimated. This can be accomplished in a one-year study through analysis of habitat maps, aerial imagery and construction records. Field studies in these altered habitats are required, particularly to determine effects on hydrology and sediment supplies. The effects on dependent fisheries can be estimated to a first approximation based on extrapolation from areal estimates of habitats modified, but thorough quantification of impact requires sampling of variously altered habitats and determination of the effects of alteration on the populations and productivity of the fishery resource. This will require longer-term field studies extending over approximately four years. The effectiveness of various mitigative measures (push-pull installation, backfilling, plugging, interruption of spoil banks) on preserving wetland area, maintaining hydrologic patterns and supporting living resources should be assessed in the field. The results will have value in planning pipeline

TABLE 1.7
Recommended study approaches for the resolution of potential long-term effects resulting from channelization of wetlands for pipeline routing and navigation

Study approaches	*Feasibility*[1]	*Regional focus*	*Duration (years)*	*Sequencing*
1. Evaluate area presently affected by pipeline routing and navigation canals. a. Map and imagery comparisons b. Assessment of areas of altered hydrology (flow and salinity)	High	Gulf of Mexico	1	Immediate
2. Determine how wetlands are physically changed and processes and mechanisms by which change has occurred (direct effects, altered hydrology, saltwater intrusion).	High	Gulf of Mexico	4	Following 1
3. Assess effects of habitat alterations and hydrologic changes on dependent fisheries.	Potential	Gulf of Mexico	4	Following 1 in conjunction with 2
4. Evaluate mitigation strategies, both existing and proposed.	Potential	Gulf of Mexico Alaska South Atlantic	3	Following 1 and concurrent with 2 and 3

(continued)

TABLE 1.7—*contd.*

Study approaches	*Feasibility*[1]	*Regional focus*	*Duration (years)*	*Sequencing*
5. Identify coastal habitats susceptible to likely future developments and develop coastal vulnerability models to guide future coastal modifications to minimize undesirable effects.	Potential	Gulf of Mexico Alaska South Atlantic	3	Immediate with refinement based on 4

[1] Feasibility of approach judged as "High" (can be satisfactorily accomplished within a 10-year time frame using available methods); "Potential" (requires development of methods or innovate approaches); and "Limited" (probably infeasible within a 10-year time frame).

TABLE 1.8

Recommended study approaches for the resolution of potential long-term effects resulting from the physical fouling by oil of birds, mammals, and turtles

Study approaches	*Feasibility*[1]	*Regional focus*	*Duration (years)*	*Sequencing*
1. Determine locations in which concentrations of animals occur with high frequency and where habitats are susceptible to development.				
a. Literature review	High	National	1	Immediate
b. Field observations of aggregations of selected species	High	Focus dependent on 1, a	10	Continuous, follows 1, a

(continued)

TABLE 1.8—*contd.*

Study approaches	*Feasibility*[1]	*Regional focus*	*Duration (years)*	*Sequencing*
2. Assess population dynamics of selected species and colonies.	Potential	Alaska: alcids, storm petrels, kittiwakes, sea otters, polar bears, pinniped pups California: alcids, storm petrels, gulls, sea otters, pinniped pups Hawaii: shearwaters Gulf of Mexico: turtles South Atlantic: turtles	10	Continuous, follows 1, a
3. Monitor physiology and behavior of animals associated with spills. a. Comparison of impacted and non-impacted areas and populations b. Site specific and species specific experimental oiling of habitats and native species	Potential	As in 2	5	As opportunities arise
4. Isolate effects of natural fluctuations, fisheries interactions, and oil and gas development.	Limited	Focus dependent on 3, a	10	Continuous but dependent on results from 3, a

[1] Feasibility of approach judged as "High" (can be satisfactorily accomplished within a 10-year time frame using available methods); "Potential" (requires development of methods or innovative approaches); and "Limited" (probably infeasible within a 10-year time frame).

routing to serve new development both in areas known to be sensitive (e.g., Louisiana) and frontier areas (South Carolina, Georgia, and certain parts of Alaska). They could also assist in mitigating existing damage. In addition, coastal regions susceptible to potential pipeline routes, such as the Yukon delta and portions of the southeastern U.S., should be evaluated using existing data for their vulnerability to physical alterations.

Fouling of Birds, Mammals and Turtles

Fouling is an inevitable consequence of oil coming in contact with birds and some fur-bearing mammals such as otters and polar bears. Locations of large concentrations of vulnerable species should be identified and the ability of feeding or resting birds and turtles to detect and avoid floating oil determined (Table 1.8).

There should be emphasis on identifying specific regions where vulnerable animals congregate, such as feeding grounds for walruses, rafting sites for large concentrations of migrating birds, nesting and breeding areas. This information should identify the time and place of greatest vulnerability to each species or group of turtles, birds, and marine mammals. When integrated with knowledge of environmental conditions (e.g., spill trajectory models) in regions where offshore oil production occurs or is likely, predictions of vulnerability can be made. Some of the information for such a model exists, and it should be brought

TABLE 1.9

Recommended study approaches for the resolution of potential long-term effects of operational discharges from offshore oil and gas development drilling on the benthos

Study approaches	*Feasibility*[1]	*Regional focus*	*Duration (years)*	*Sequencing*
1. Measure chemical tracers of drilling discharges in sediments (barium, chromium, lignosulfonate, medium weight aromatic hydrocarbons).	High	Developed area: Gulf of Mexico, Texas-Louisiana outer shelf Developing area: Central California, Santa Maria Basin	2 5 10	Initial intensive reconnaissance, Gulf and California Assess effects, annual, Gulf Continuous assessment of effects for duration of development, California
2. Assess sedimentologic and geochemical dynamics (sediment texture, erosion and deposition, exchanges with water column).	Potential	As in 1	As in 1	As in 1

(*continued*)

TABLE 1.9—*contd.*

Study approaches	*Feasibility*[1]	*Regional focus*	*Duration (years)*	*Sequencing*
3. Determine bioavailability over time from sediments of potentially toxic components of drilling fluids to various trophic groups of benthos.	High	Generic research	3	Immediate
4. Measure biological effects at three levels: a. Individual-induction of enzyme systems, biochemical and physiological stress indices b. Population-age and size structure, reproduction and recruitment, microbes c. Community-biological interactions resulting from population alterations	High	As in 1	As in 1	As in 1

[1] Feasibility of approach judged as "High" (can be satisfactorily accomplished within a 10-year time frame using available methods); "Potential" (requires development of methods or innovative approaches); and "Limited" (probably infeasible within a 10-year time frame).

together through a penetrating review of this literature, which will at the same time expose gaps in our understanding of temporal variability and other factors associated with these events. Additional data and monitoring of selected areas of critical importance will be necessary.

Central to the question of vulnerability to fouling is whether an animal will detect and avoid floating oil. This question has been answered for a representative odontocete, but the same approach is not applicable to mysticetes. An experimental study of avoidance in pinnipeds might help to clarify their apparent equivocal response to oil. Additional controlled field studies on select species of seabirds will help clarify their response. There is a need for controlled studies to determine the reaction of turtles and selected species of seabirds to oil slicks and tar balls.

Drilling Discharges

The cumulative effects of operational discharges during all phases of field development present a problem central to the question of long-term impact, yet such effects have been difficult to assess thus far. Study approaches 1, 2, and 4 in Table 1.9 call for well-designed field studies of the effects of drilling discharges during development in continental shelf environments where long-term accumulation of contaminants in bottom sediments is most likely. These environments should be depositional and be characterized by relatively little transport of bottom sediments. One recommended study area is the deep continental shelf off southwestern Louisiana and southeastern Texas, which is a region already heavily developed but removed from the potentially confounding influence of the Mississippi River. Another is the deep shelf in the Santa Maria Basin area off central California, which is not yet developed, but in which significant discoveries have been made. The specific study sites should be carefully selected to avoid some of the problems inherent in previous studies (see Chapters 9, 10, and 14). The locations should have relatively uniform conditions (topography and sediment composition), little contemporary terrestrial input, low seasonal variation at the seabed and relatively little turbulence above the seabed.

To link biological responses with chemical effects, the chemical tracers, sedimentological and geochemical dynamics, and biological aspects must be integrated. An initial high resolution survey should establish the general pattern of faunal-environmental covariance. The approach should not be a complete characterization, which is fruitless, but should emphasize "target species" (Chapter 14). With information on the dominant benthic fauna available, selected subsets can be used to study population and community level effects. These observations should seek to answer a three-part question: a) Has there been an exposure (e.g., chemical tracer in the sediments)? b) Has there been an individual response (e.g., induced enzyme systems, physiological stress indices)? c) Has there been a population and community response (e.g., age and size structure, recruitment potential, changes in the hydrocarbon-degrading bacteria in the sediments)?

The exact sampling design should be developed on the basis of the initial reconnaissance to determine general patterns of variance. General linear models (analyses of variance and covariance or multiple regression) should be employed and analyzed. Field studies should be supplemented by generic experimental research to determine the bioavailability of potentially toxic components of drilling fluids from sediments as indicated under approach 3.

We do not recommend additional long-term studies on the effects of discharges from exploratory drilling in continental shelf environments. The direct exposure of susceptible, rare shelf communities may still merit study, and many questions remain regarding the effects of drilling discharges in enclosed, shallow water environments, but current information seems sufficient for sound environmental management concerning discharges during exploratory drilling under most continental shelf conditions.

Nearshore Discharges of Produced Waters

Although relatively little is known of the nature and effects of produced water discharges, general considerations suggest that toxic concentrations of

TABLE 1.10

Recommended study approaches for the resolution of potential long-term effects of produced water discharges into nearshore environments

Study approaches	*Feasibility*[1]	*Regional focus*	*Duration (years)*	*Sequencing*
1. Conduct additional inventories and detailed chemical analysis of a variety of produced waters to determine potential toxicants and environmental tracers and predict fate.	High	National	2–3	Immediate
2. Assess sedimentologic and geochemical dynamics (as in Table 1.9, 2).	High	Gulf of Mexico Southern California	2 5	Initial, intensive reconnaissance Assess effects, annual
3. Measure biological effects at three levels—individual, population, and community (as in Table 1.9, 4).	High	As in 2	As in 2	As in 2

[1] Feasibility of approach judged as "High" (can be satisfactorily accomplished within a 10-year time frame using available methods); "Potential" (requires development of methods or innovative approaches); and "Limited" (probably infeasible within a 10-year time frame).

components would not persist following discharge into open shelf waters. Long-term discharges might contaminate bottom sediments through sorption onto natural suspended sediment particles, but this potential is much greater where produced waters are discharged into shallow coastal waters, where smaller volumes are available for dilution and sediment loads are higher. Produced waters from offshore are frequently transported to coastal bases before separation from the oil or gas product. Since this is the situation with the greatest potential impact, it is recommended that research on the effects of produced water discharges first address such coastal discharges.

Recommended study approaches (Table 1.10) are conceptually similar to those which address offshore drilling discharges (Table 1.9). One difference is the chemical analyses required to assess the composition of produced waters, especially aromatic hydrocarbons, non-volatile soluble organics, and the chemical speciation of trace metals upon discharge into the environment. Adsorption/desorption processes of metals and organics from produced waters on suspended particles in receiving waters should be investigated. Biological effects are also an important part of the study, as described in Table 1.9, item 4.

Noise and Other Disturbances

The effects of noise and physical disturbances on most species of birds, mammals and turtles are normally brief. In those relatively few cases where a significant probability of long-term effects remains, a concerted effort should first be made to draw together what is known, to recommend "safe distance" guidelines for regulatory use and to identify the most important remaining questions (Table 1.11). Additional field observations may be required following

TABLE 1.11

Recommended study approaches for the resolution of potential long-term effects resulting from disturbance by noise and other physical factors on birds, mammals and turtles

Study approaches	*Feasibility*[1]	*Regional focus*	*Duration (years)*	*Sequencing*
1. Determine locations in which concentrations of animals occur with high frequency and where habitats are susceptible to disturbance.				
a. Literature review	High	National	1	Immediate
b. Field surveys of selected species in selected areas	High	Focus dependent on 1, a	2	To fill gaps identified in 1, a

(*continued*)

TABLE 1.11—*contd.*

Study approaches	*Feasibility*[1]	*Regional focus*	*Duration (years)*	*Sequencing*
2. Assess population of potentially affected bird species and colonies, turtle nesting sites, pinniped rookeries, inshore cetaceans, including bowhead whales.	Potential	Alaska and California: alcids, shearwaters, sea otters, pinniped rookeries, inshore cetaceans, bowhead whales Gulf of Mexico and South Atlantic: turtle nesting areas	10	Following 1, a
3. Monitor physiology and behavior of animals associated with disturbance. a. Compare impacted and non-impacted areas and populations. b. Monitor specific sites and species where there is controlled experimental disturbance.		As in 2	5	As opportunities arise

[1] Feasibility of approach judged as "High" (can be satisfactorily accomplished within a 10-year time frame using available methods); "Potential" (requires development of methods or innovative approaches); and "Limited" (probably infeasible within a 10-year time frame).

this synthesis, to be followed by longer-term assessments of population dynamics and physiological and behavioral studies. Controlled studies can be undertaken for some birds, pinnipeds and turtles: comparing fitness, behavior, and reproductive success between disturbed and undisturbed areas. Dolphins and whales are generally more elusive; yet some, such as the gray whale, the right whale, and the bowhead whale have migratory and residence patterns which could allow observational studies.

Effect of Oil Spills on Fishery Stocks

Laboratory and field studies attempting to relate the effects of oil spills to mortality of eggs and larvae and hence to fishery stocks have a low feasibility and are not recommended. Rather, our recommendations are limited to synthetic approaches in which existing and forthcoming information concerning the distributions and abundance of eggs and larvae of commercially significant fishes and invertebrates are mapped and analyzed to identify the locations and timing important to stock recruitment (Table 1.12). Coupling of distributional data with regional oil spill trajectory models will allow development of vulnerability models.

TABLE 1.12
Recommended study approaches for the resolution of potential long-term effects resulting from oil spills which reduce fishery stocks by killing eggs and larvae.

Study approaches	*Feasibility*[1]	*Regional focus*	*Duration (years)*	*Sequencing*
1. Develop a catalog of fish and commercial invertebrate eggs and larval distributions based on existing knowledge and data.	High	National	2	Immediate
2. Identify areas in time and space where various life stages, particularly eggs and larvae, are likely to occur in areas susceptible to oil spills from offshore oil and gas development.	Potential	National	2	Following 1
3. Develop vulnerability models based on distributions and oil spill models	Potential	Focus dependent on 2	2	Following 2

[1] Feasibility of approach judged as "High" (can be satisfactorily accomplished within a 10-year time frame using available methods); "Potential" (requires development of methods or innovative approaches); and "Limited" (probably infeasible within a 10-year time frame).

Gravel Islands and Causeways

The construction of gravel islands, causeways and other structures may deleteriously affect adjacent shallow water ecosystems in the Beaufort Sea and other nearshore regions of Alaska.

Islands and especially causeways will alter the local current flow, resulting in changes in the habitat conditions in surrounding areas. Resolution of this issue requires measurements of the flow field around structures and assessment of alterations in distribution patterns of suspended and bottom sediments, temperature, salinity, nutrients, and other physico-chemical parameters of the water column and benthic habitats (Table 1.13). Effects should be interpreted in terms of impacts on the benthos and anadromous fish. Causeways and islands may affect the long-coast movement or migration patterns of fish and some mammals. Field sampling and intensive tagging of fishes combined with long-term areally extensive monitoring of tag returns is an effective technique.

Dredging of the seabed for construction materials may alter benthic habitats and diminish the food resources of important species such as walrus, gray whale and bearded seals, especially in the Chukchi and northern Bering Seas. Bottom

TABLE 1.13

Recommended study approaches for the resolution of potential long-term effects of man-made islands and causeways in arctic environments

Study approaches	*Feasibility*[1]	*Regional focus*	*Duration (years)*	*Sequencing*
1. Evaluate alteration of longshore currents and its effect on sediment and hydrological patterns	High	Beaufort Sea	3–4	Dependent on timing of activity to be monitored
2. Determine extent of disruption of migration patterns.	High	Beaufort Sea	3–4	As above
3. Evaluate the long-term effects of dredging on bottom topography and benthos.	High	Beaufort Sea	3–4	As above

[1] Feasibility of approach judged as "High" (can be satisfactorily accomplished within a 10-year time frame using available methods); "Potential" (requires development of methods or innovative approaches); and "Limited" (probably infeasible within a 10-year time frame).

surveys will describe the alterations of bottom topography and substrate type, as well as sediment infilling of excavations. In assessing effects on benthic communities, emphasis should be placed on the prey of bottom feeding fishes and mammals.

A LONG-TERM EFFECTS STUDY PROGRAM

This assessment of potential long-term environmental effects of offshore oil and gas development, and studies appropriate to their evaluation, was conducted in response to the recommendation of the National Marine Pollution Program Plan that a "10-year interagency research program should be planned and implemented to investigate the long-term, low-level adverse effects of OCS and other ocean use activities" (Interagency Committee on Ocean Pollution Research, Development, and Monitoring, 1981). Furthermore, it follows an attempt to develop such a plan during a 1981 workshop (National Marine Pollution Program Office, 1982).

In light of the reviews and evaluations of issues included in the supporting technical chapters and this synthesis, we must now address several questions. Given the general lack of conclusively demonstrated, serious long-term effects and the disparate nature of the issues regarding potential long-term effects, is a coherent research program required? Are there realistic and useful goals which can be met by such a program? If required, how should a long-term effects program be organized to insure efficient progress toward resolution of the issues and use of results?

Should There Be a Long-Term Effects Program?

Despite considerable research on the environmental issues related to offshore oil and gas development, there remain a number of unresolved concerns about the long-term effects of offshore development which merit continued investment in scientific understanding. The reasons for this current status are many: the heretofore poor definition of the long-term effects which may occur, the paucity of well-designed studies in the marine environment, and limited understanding of environmental processes and biological dynamics are a few important reasons. To be sure, the large bulk of this book shows we have learned quite a lot which is relevant. We have resolved some concerns and, perhaps for the first time, stated and rated specific unresolved issues.

How serious are the risks of long-term, deleterious effects of offshore oil and gas development? This is impossible to answer in absolute terms. Even in relative terms, an answer is difficult because we have not attempted to similarly evaluate other serious ocean pollution issues, such as problems of persistent contamination by highly toxic and persistent xenobiotic substances, eutrophication of coastal waters, or estuarine habitat modification. Offshore oil and gas development has the potential to be a particularly widespread activity in the coastal ocean, is planned in environments where there are considerable environmental hazards (polar regions and continental margins), and inspires considerable public and professional concern about long-term environmental effects.

Research generally related to assessing or predicting the effects of offshore oil and gas development, including research on the fate and effects of oil in the marine environment, expanded greatly in the mid 1970s but has been reduced in effort in the early 1980s (Koons and Gould, 1984). Despite this trend, it is clear that considerable research on the long-term effects of offshore oil and gas development will continue in response to information needs of decision-makers and society in general.

A substantial number of the long-term effects issues listed in Table 1.1 are amenable to some degree of resolution: either dismissal of the issue as unlikely or insignificant or development of strategies for mitigation. Using the approach outlined, we have directed attention to those issues amenable to resolution. Other issues listed are likely to remain poorly resolved. Based on the evaluations in Tables 1.5 through 1.13 concerning the feasibility of study approaches, we conclude that there are useful and attainable goals of a coherent, but multi-faceted, research and monitoring program on the long-term effects of offshore oil and gas development.

A coherent study plan, such as that outlined here and appropriately revised with time, would be useful in guiding not only Federally-sponsored research and monitoring, but industry-sponsored efforts as well. The National Marine Pollution Program Plan suggested that the 10-year research program recommended "should be jointly implemented by the Federal Government and private industry as OCS development takes place" (Interagency Committee on Ocean Pollution Research, Development, and Monitoring, 1981). We also believe that a close cooperation between government and industry is required in terms of study planning, sponsorship and logistical support for the program outlined to be successful. A unified joint study plan could also serve as a central focus for debates concerning the potential of long-term environmental effects of offshore oil and gas development.

Program Organization

In the National Marine Pollution Program Plan and the subsequent planning workshop (National Marine Pollution Program Office, 1982), it was proposed or assumed that a discrete interagency research program on long-term effects of offshore oil and gas development would be implemented. The workshop proposed that coordinated, multifaceted programs be carried out in at least one historically developed offshore region and one frontier region. Our present views diverge somewhat in that, given the diversity of issues, regional differences in potential effects, and the better focus on real issues we now have, we do not recommend discrete programs focusing on only two offshore regions. Rather, we recommend implementation of and commitment to an interagency program plan which guides regional research and monitoring efforts together with generic research programs.

Of critical importance to the success of such a program are the following attributes:

1. Centralized management within agencies and sufficient interagency overview to assure compliance with the program plan. Interagency and inter-

regional communication and application of results. Allocation by the participating agencies of a sufficient level of staff support and program authority to insure implementation of the plan.

2. Iterative review of objectives and progress in meeting the objectives in order to insure efficiency of effort, usefulness of results, and evolution of understanding in those areas where critical conceptual development is incomplete.
3. Emphasis on a high degree of innovation and application of state-of-the-art scientific methods.
4. Within the framework of program evolution, multiyear research funding commitments for those program elements based on long-term experimental programs and field observations and successive experiments.

LITERATURE CITED

Boesch, D.F. and R. Rosenberg. 1981. Response to stress in marine benthic communities. Pages 179-200 *in* G.W. Barrett and R. Rosenberg (eds.), Stress Effects on Natural Ecosystems. John Wiley & Sons, Ltd., New York.

Boesch, D.F. and R.E. Turner. 1984. Dependence of fisheries on salt marshes: The role of food and refuge. Estuaries 7: 460-468.

Brockmann, U.H., E. Dahl, J. Kuiper and G. Kattner. 1983. The concept of POSER (Plankton Observation with Simultaneous Enclosures in Rosfjorden). Mar. Ecol. Prog. Ser. 14: 1-8.

Burroughs, R.H. 1981. OCS oil and gas: relationships between resource management and environmental research. Coastal Zone Manage. J. 9: 77-88.

Capuzzo, J.M., B.A. Lancaster and G. Sasaki. 1984. The effects of petroleum hydrocarbons on lipid metabolism and energetics of larval development and metamorphosis in the American lobster (*Homarus americanus*). Mar. Environ. Res. 14: 201-228.

Clark, R.B. 1982. The impact of oil pollution on marine populations, communities and ecosystems: a summing up. Phil. Trans. R. Soc. Lond. B 297: 433-443. Reprinted in R.B. Clark (ed.). 1982. The Long-Term Effects of Oil Pollution in Marine Populations, Communities and Ecosystems. The Royal Society, London.

Essertier, E.P. 1983. Federal Offshore Statistics. Leasing, Exploration, Production Revenue. Minerals Management Service, U.S. Department of the Interior, Washington, D.C., 103 p.

Gallaway, B.J. and G.S. Lewbel. 1982. The Ecology of Petroleum Platforms in the Northwestern Gulf of Mexico: A Community Profile. U.S. Fish and Wildlife Service, Office of Biological Services, Publ. No. FWS/OBS-82/27. Washington, D.C., 91 p.

Grice, G.D. and M.R. Reeve (eds.). 1982. Marine Mesocosms. Biological and Chemical Research in Experimental Ecosystems. Springer-Verlag, New York, 430 p.

Havran, K.J., J.D. Wiese, K.M. Collins and F.N. Kurz. 1982. Gulf of Mexico Summary Report 3. U. S. Geological Survey Open-File Rep. 82-242.

Interagency Committee on Ocean Pollution, Research, Development, and Monitoring. 1981. Second Federal Plan for Ocean Pollution Research, Development, and Monitoring. National Oceanic and Atmospheric Administration, National Marine Pollution Program Office, Rockville, Maryland, 185 p.

Koons, C.B. and H.R. Gould. 1984. Worldwide Status of Research on Fate and Effects of Oil in the Marine Environment—1982. Special Report, Exxon Production Research Company.

LaLiberté, P. and W.M. Harris. 1986. Federal Offshore Statistics: 1984. Leasing, Exploration, Production, & Revenues. Minerals Management Service, U.S. Department of the Interior, Washington, D.C., 100 p.

Lewis, J.R. 1982. The composition and functioning of benthic ecosystems in relation to the assessment of long-term effects of oil pollution. Phil. Trans. R. Soc. Lond. B 297: 257-267. Reprinted in R.B. Clark (ed.). 1982. The Long-Term Effects of Oil Pollution on Marine Populations, Communities and Ecosystems. The Royal Society, London.

Malins, E.C., M.S. Myers and W.T. Roubal. 1983. Organic free radicals associated with idiopathic liver lesions of English sole (*Parophrys vetulus*) from polluted marine environments. Environmental Sci. Technol. 17: 679-685.

McIntyre, A.D. and J.B. Pearce (eds.). 1980. Biological Effects of Marine Pollution and the Problem of Monitoring. Rapp. P.-V. Réun. Cons. Int. Explor. Mer 179: 1-346.

Minerals Management Service. 1983a. Regional Environmental Assessment of Pipeline Activities. Metairie, Louisiana, 195 p.

Minerals Management Service. 1983b. Outer Continental Shelf Studies Program. Contract Projects—Fiscal Year 1973 through 1983. Fourth Edition. Washington, D.C., 236 p.

Moore, W.D., III. 1984. Offshore drilling increases, attrition, cut mobile rig surplus. Oil & Gas J. 82: 103-105.

National Marine Pollution Program Office. 1982. OCS Long-Term Effects Program: Implementation Status Report. National Oceanic and Atmospheric Administration, Rockville, Maryland, 7 p., 4 Appendices.

National Research Council. 1983. Drilling Discharges in the Marine Environment. National Academy Press, Washington, D.C., 180 p.

National Research Council. 1985. Oil in the Sea. Inputs, Fates, and Effects. National Academy Press, Washington, D.C., 601 p.

Oviatt, C.A., J. Frithsen, J. Gearing and P. Gearing. 1982. Low chronic additions of No. 2 fuel oil: Chemical behavior, biological impact and recovery in a simulated estuarine environment. Mar. Ecol. Prog. Ser. 9: 121-136.

Oviatt, C.A., M.E.Q. Pilson, S.W. Nixon, J.B. Frithsen, D.T. Rudnick, J.R. Kelly, J. F. Grassle and J.P. Grassle. 1984. Recovery of a polluted estuarine system: A mesocosm experiment. Mar. Ecol. Prog. Ser. 16: 203-217.

Royal Commission on Environmental Pollution. 1981. Eighth Report. Oil Pollution of the Sea. Cmnd. 8358. H. M. S. O., London.

Sanders, H.L. and C.C. Jones. 1981. Oil, science, and public policy. *In* T.C. Jackson and D. Reische (eds.), Coast alert: Scientists speak out. Friends of the Earth Publishers, San Francisco.

Scaife, W.W., R.E. Turner and R. Costanza. 1983. Coastal Louisiana recent land loss and canal impacts. Environmental Management 7: 433-442.

Spaulding, M.L., S.B. Saila, E. Lorda, H. Walker, E. Anderson and J.C. Swanson. 1983. Oil-spill fishery impact assessment model: Application to selective Georges Bank fish species. Estuar. Coastal Shelf Sci. 16: 511-541.

Spies, R.B., D.W. Rice and R.R. Ireland. 1983. Preliminary studies of growth, reproduction and activity of hepatic mixed function oxidase in *Platichthys stellatus*. Second International Symposium on Responses of Marine Organisms to Pollutants, Woods Hole, Massachusetts (Abstract).

Wolfson, A., G. Van Blaricom, N. Davis and G.S. Lewbel. 1979. The marine life of an offshore oil platform. Mar. Ecol. Prog. Ser. 1: 81-89.

CHAPTER 2

PETROLEUM INDUSTRY OPERATIONS: PRESENT AND FUTURE

James P. Ray

CONTENTS

INTRODUCTION

One of the most difficult, and at the same time important, challenges in predicting and planning for the studies of long-term effects is to determine what industry activities will take place, where they will be located, when they will occur, and what will be their magnitudes. Information from a variety of industry sources was utilized in assembling the following predictive scenarios for future industry activity. As is evidenced by the in-depth discussions throughout this volume, determination of perturbations to the marine environment are difficult to detect in the environment, especially when little is known of the natural variability which often masks the more subtle long-term effects which may be associated with offshore oil and gas development.

The many experts who have combined their knowledge and experience in studying the marine environment have predicted the types of impacts most likely to occur and have given judgments on those that may be measured using currently available technologies. As noted, most of the changes can only be measured in close proximity to the perturbation over relatively short periods of time. In properly planned long-range research programs, it is important to have a full understanding of the industrial activity of interest. This is the key to selecting proper study areas and is important in determining the perturbations and pollutants to be monitored.

In this chapter I have focused on the major areas of activity in the United States for the next decade. In part, I find a paradox in recommending study areas. In terms of numbers of wells drilled and amount of production expected, the Gulf of Mexico by far exceeds the activity levels of other areas. This in itself suggests a need for further study in the Gulf of Mexico to determine if, in fact, long-term effects can be measured. The other side of the argument is that because of the extensive development in most of the Gulf of Mexico over the past 30 years, it would be difficult to isolate an area and determine new impacts due to a specific location. On the other hand, new frontier areas, although they will be less heavily developed, may provide the opportunity for conducting long-term studies on the possible low-level effects of exploration and development activities. However, it may be many years before the effects of production-level development would be manifest. Merit probably lies in studies in both historically developed and frontier areas.

An important factor in selecting appropriate areas in which to evaluate the long-term effects of offshore development is the likelihood of extensive exploration and production activities. This is sometimes difficult to predict because opinions of the oil and gas potential of offshore basins vary greatly and may change rapidly with the results of exploratory drilling (for example, the greatly reduced industry interest off the Atlantic coast). In addition, knowledge of the technology to be employed (for example, the types of platforms or artificial islands) and its potential impacts is important. Finally, there is a need to understand the operational procedures to be followed, including the types and quantities of pollutants which may be released.

In the past, many of the expensive and time-consuming studies have not achieved their desired goals because of poor planning and a lack of understanding of industry operations. Determination and quantification of long-term, low-level effects from oil and gas operations will be extremely difficult, and careful preplanning will be essential if the future programs are to be successful. The following description of industry locations, activities, and impacts is cursory at best. Planning for future research should include industry experts so that much more detailed input can be included with the study designs.

TECHNOLOGICAL DEVELOPMENTS

Arctic

The Arctic is currently experiencing a very rapid evolution in technology related to the structures needed in the severe ice conditions of the Beaufort, Chukchi and Bering Seas. Each of the areas presents unique problems, from large ice movements as seen in parts of the Bering and Chukchi Seas, to thick ice movements of the nearshore Beaufort, and the even more imposing problems of pack ice movement in the offshore Beaufort.

During the next decade, the use of artificial islands will continue to evolve, with the eventual movement toward offshore dredging for gravel when nearby shore sources are not available. New artificial structures, of both concrete and steel, are being designed or built for the Arctic. These are reusable drilling structures which can be moved from location to location. New to the Alaskan Beaufort in the next decade will be the development of structures that will operate in deeper waters (>20 m). This will extend the current capabilities beyond the shear zone and present more difficult engineering problems because of the movement of the multiyear pack ice. Also moving into this zone during the open water part of the year will be drill ships. In 1985, drill ship operations will begin in the eastern sector of the Alaskan Beaufort. This new type of operation for Alaska brings associated problems relating to ice breaking, oil spill control, and acoustical interference of the fall bowhead whale migration.

New drill ship technologies will be employed for deep-water drilling in the Navarin Basin, especially in the northern parts of the basin where there will be considerable ice in the late fall. Concurrent with the potential development of the Navarin Basin will be many logistical problems of supplying the offshore operations, and of retrieving and transporting the oil to suitable shore bases for further transportation and processing. They may also include a range of engineering considerations for shore-based stations in the Alaska Peninsula and Aleutian Islands where considerations of seismic activity are important.

Arctic operations also provide unique challenges for offshore production. Pipelines coming ashore either have to be buried beneath the ocean floor to protect them from ice scour, or above the bottom on causeways. In addition to these concerns, allowances must be made for the effects of permafrost on the pipeline, and for the effects of the pipeline on the permafrost (e.g., problems related to melting).

Deep Water

With the recent exploration discoveries in deep-water regions, such as the

Santa Maria Basin (California), Cognac and Green Canyon (Gulf of Mexico), the industry will continue to move into the deeper waters of the continental slope over the next decade. Recent engineering successes off the Atlantic coast have pushed achievable drilling depths beyond 2000 meters. Although these depths are achievable with current technology, one of the major challenges still remaining is how to produce oil and gas from these depths. Recent achievements in water depths of slightly over 300 meters have included the successful installation of new tension leg platforms. These are floating platforms which are held firmly to the sea floor by a series of tensioned cables. This precludes the engineering problems and high materials costs associated with designing solid structures such as Shell's Cognac platform.

Also, still to be completely developed are subsea completion systems for retrieving deep-water petroleum. Once these techniques are perfected, a deep-water well could be completed and the product piped into shallower waters to a conventional platform for processing and transshipment.

Pollution Control

The characteristics of pollution from the oil and gas industry will change significantly over the next decade, especially in comparison to what has been common over the first 30 years of offshore activity. This is the result of two factors: first, the continual improvements resulting from environmental regulation; second, the improvement in the technologies relating to chemicals used and the equipment associated with their use and discharge.

Due in part to the increases in our environmental understanding of the fate and effects of pollutants in the marine environment, we have continually been evolving towards improved management of the types of materials used and discharged, and the types and locations of these discharges. For example, recent research has shown that the most toxic component commonly used in water-based drilling fluids is diesel fuel. Under new Environmental Protection Agency (EPA) regulations, future use of diesel will be closely regulated. Correspondingly, close attention is now being paid to the proximity of sensitive biological environments to industry operations.

The Environmental Protection Agency, in cooperation with the industry, is working on a system whereby detailed information on all chemical additives used for both drilling fluids and produced water treatment will be available. This will allow for a more reasoned management of the materials being released into the environment. These types of controls were for the most part nonexistent in the early years of offshore development.

All facets of offshore engineering include considerations for improvements in the quality and quantity of materials being discharged into the environment. Throughout the system, equipment is continually being improved to prevent the accidental release of oil to the environment. This includes improved blow out preventers and improved fuel transfer equipment and procedures. New equipment is being designed for the cleaning of oil from cuttings and the removal of residual oil from produced waters.

In summary, the potential impacts of offshore oil and gas over the next decade will be continually decreasing. This, of course, precludes the low probability of catastrophic oil spills, either from blow outs or transportation accidents.

REGIONAL RESOURCE POTENTIALS

Estimation of the oil and gas resource potential of offshore regions is a contentious and dynamic issue. Although geophysical data for the offshore regions of the United States are voluminous, they are subject to interpretation by petroleum geologists which can vary widely among companies and government agencies. Even then, the resource estimates are speculative until exploratory drilling and subsequent delineation are completed. Regions which have appeared very promising based on interpretation of geophysical data have yielded no discoveries or discoveries of less than commercially viable quantities of oil and gas. On the other hand, fortunes have been made by exploring areas which were deemed of low potential or were written off after disappointing initial exploration (e.g., Prudhoe Bay, Alaska).

With this caveat in mind, it is useful to consider the most recent U.S. government estimates of the undiscovered oil and gas resource potential of U.S. offshore regions in general terms, without paying too much heed to the absolute quantities. Table 2.1 presents the most recent U.S. Department of the Interior estimates of the undiscovered, economically developable resources by Outer Continental Shelf planning area. These estimates include those yet undiscovered resources underlying leased tracts as well as those in areas yet to be leased. These estimates are based on analyses by Department of the Interior geologists of the presence and size of hydrocarbon bearing structures within each region. Conditional mean estimates represent average estimates of the volume of resources which may be present (in billions of barrels of oil and trillion cubic feet of gas). The marginal probability of success represents the chance that one or more geological conditions exist such that the planning area is considered to contain a commercial accumulation of hydrocarbons. This probability is high in regions already known to contain such accumulations (e.g. the Central and Western Gulf, Southern California, and Beaufort Sea) and low where no discoveries have yet been made. To yield a composite picture, an estimate of the risked oil equivalent (in billion barrels) is provided by multiplying the conditional estimates by the marginal probability of success and converting the resulting "risked" gas estimate to Btu equivalent of oil (5,620 cubic feet of gas = one barrel of oil).

In addition, the Department of the Interior has solicited the evaluations of oil companies in terms of their estimation of the resource potential and their exploration interests. Industries assigned ranks to the 24 planning areas (Central and Northern California were treated as one area). The overall industry rankings are also presented in Table 2.1.

Based on the Department's conditional estimates, the greatest oil resources may lie in the Navarin Basin, Central Gulf of Mexico, Chukchi Sea, Western Gulf of Mexico and Beaufort Sea, and the greatest gas resources may be in the Central and Western Gulf, South Atlantic and Chukchi Sea. However, when

TABLE 2.1

Estimates of undiscovered economically developable resources in U.S. outer continental shelf planning areas as of July 1984 (Minerals Management Service, 1985a) and overall rank of industry interest in terms of resource potential and exploration interest (Minerals Management Service, 1985b).

Planning area	*U.S. Department of the Interior estimates*						*Industry interest rank*
	Conditional mean oil (10^9 bbl)	*Conditional mean gas (10^{12} cu ft)*	*Marginal probability of success*	*Risked oil equivalent (10^9 bbl)*	*Percentage risked oil equivalent*	*Rank*	
Atlantic							
North Atlantic	0.35	7.14	0.30	0.49	1.72	11	11
Mid-Atlantic	0.35	6.02	1.00	1.42	5.03	6	14
South Atlantic	0.86	16.18	0.25	0.93	3.31	9	18
Total Atlantic	1.56	29.34		2.84	10.06		
Gulf of Mexico							
Eastern Gulf	0.41	2.19	1.00	0.79	2.80	10	6
Central Gulf	3.72	30.69	1.00	9.18	32.49	1	1
Western Gulf	1.90	26.76	1.00	6.66	23.57	2	2
Total Gulf	6.03	59.64		16.63	58.86		
Pacific							
Southern California	1.54	2.42	1.00	1.97	6.97	3	4
Central California	0.56	0.79	0.65	0.46	1.61	12	4
Northern California	0.42	1.86	0.60	0.45	1.59	13	*
Washington-Oregon	0.18	3.26	0.20	0.15	0.54	16	13
Total Pacific	2.70	8.33		3.03	10.72		

Alaska							
Gulf of Alaska	0.54	8.34	0.08	0.16	0.57	15	19
Cook Inlet	0.21	0.35	0.03	negligible		21	16
Kodiak	0.15	2.92	0.05	0.03	0.12	18	21
Shumagin	0.05	1.42	0.03	0.01	0.03	20	17
Aleutian Arc	negligible					25	23
North Aleutian Basin	0.36	2.62	0.20	0.17	0.58	14	8
St. George Basin	1.69	15.76	0.22	0.99	3.50	8	9
Bowers Basin	negligible					23	22
Aleutian Basin	negligible					22	23
St. Matthew-Hall	negligible					24	20
Navarin Basin	4.80	5.84	0.27	1.58	5.58	5	7
Norton Basin	0.64	2.94	0.12	0.14	0.49	17	12
Hope Basin	0.17	1.81	0.02	0.01	0.03	19	15
Chukchi Sea	2.68	15.10	0.20	1.07	3.80	7	10
Beaufort Sea	1.28	5.62	0.70	1.60	5.65	4	3
Total Alaska	12.57	62.72		5.75	20.36		

*Central and Northern California were ranked as one unit in this survey.

viewed as the probability-weighted, risked assessments of oil and gas resources, the Central and Western Gulf overwhelmingly dominate, followed by a secondary group including Southern California, Beaufort Sea, Navarin Basin and Mid-Atlantic.

Keeping in mind that even within the industry evaluations vary, it is interesting to compare the industry and government rankings. Both agree in the importance of the Central and Western Gulf, Southern California and the Beaufort Sea and in the low potential of many of the Gulf of Alaska and Bering Sea basins. Industry interest in the Atlantic, after poor success of exploratory drilling, is lower than government estimates would predict. The reverse is the case for Central and Northern California and the North Aleutian Basin. Viewed in total, these data suggest that offshore oil and gas exploration and development activities will continue to focus predominantly in the historically developed regions of the Gulf of Mexico and California, or in extensions of those regions, and that exploration will expand in the Alaskan Arctic and western (deepwater) portions of the Bering Sea.

ATLANTIC

The future for offshore drilling on the Atlantic coast is currently uncertain (Anonymous, 1985). With the departure of the Discoverer Seven Seas drill ship that was drilling the deep-water wells for Shell Oil Company, there is currently no drilling activity off the U.S. Atlantic coast. There is a possibility that Chevron will drill a well in deep water off the coast of North Carolina in 1986. In September 1984, the Georges Bank lease sale (#82) was canceled by the Minerals Management Service because of a lack of bids from the industry. There has been a similar lack of interest in bidding on the South Atlantic area. With a limited number of future sales planned for the Atlantic coast, and the higher priorities in other outer continental shelf (OCS) areas, there is little likelihood that there will be much activity on the Atlantic continental shelf during the next decade.

GULF OF MEXICO

General

The Gulf of Mexico will continue to be the busiest of offshore areas over the next decade. It is expected that over 90% of all offshore drilling and production will occur in this area. The general trends will be for a continued increase in activity to the east of the Mississippi River, off the Mobile Bay area (Alabama), Panama City, and farther south along the lower Florida coast. Off the Louisiana coast, there will be an increase in the exploration and development of deep water-tracts (e.g. Green Canyon). In addition, there will be a continued development of historically exploited areas in the shallower, nearshore waters.

Eastern Gulf

The eastern Gulf of Mexico has become an area of increasing activity since 1983. Significant gas finds in the deep pay zones (>6100 m) in the Mobile Bay area have caused a flurry of activity. Mobil Oil Company is developing the four block Mary Ann field at the mouth of Mobile Bay. The discovery is estimated at 600 billion cubic feet of reserves. Three platforms will be installed on Block 76, and one platform each on Blocks 75 and 95.

Exxon is also very active in the area and recently announced significant gas shows in the vicinity of the Mary Ann field. It appears that there will be significant development in the Mobile Bay area over the next decade. Operationally, this area is slightly different from typical Gulf of Mexico production fields. Due to the deep drilling depths, high temperatures and sour gas (hydrogen sulfide), the quantity of drilling discharges will be slightly higher per well, and the composition of drilling fluids will be slightly different. There will be an increased need for oil-based fluids, or fluids with high lubricity agent content. However, this may not warrant additional research on the environmental effects of these fluids, because the new EPA regulations will not allow the discharge of these materials.

Off Florida, drilling activity is on the increase following a several year lapse since the early disappointments in the Destin Dome area. In 1985, Shell is expected to drill four wells in the Destin Dome Block 160 area, assuming drilling rigs are available. Sohio has exploration plans for up to three wells 20 miles off Panama City. Chevron also plans to drill exploratory wells on Blocks 422 and 617. All of these exploratory plans for the northwest Florida coast are contingent upon working out arrangements with the Department of Defense over activities in this disputed zone. The military currently uses much of this area for defense testing, missile range and aircraft carrier operations.

Central Gulf

The major trend for Gulf of Mexico development during the next decade will be the exploration of deeper waters on the continental slope. During 1984, several discoveries were announced in the Green Canyon area which lies in water depths predominantly >300 m. Shell has announced strikes on Blocks 65, 63, 10 and 19. It has been estimated that the potential reserves are in the 200 to 300 million bbl range. A multiple well program is planned for further delineation of this field. In the same general area, discoveries were also announced by Mobil, Conoco, Placid, Odeco, Marathon, Amerada Hess and Sohio. It is difficult to predict the exact number of wells expected in this one area of the Gulf, but at the present, it appears that this will be a major area of activity over the next several years. When major finds are made, the platforms designed for these water depths will generally have a large number of wells per platform because of the high construction and operation costs per platform. Shell's Cognac platform has over 60 wells, and as seen off California, large subsurface structures may be drilled with an excess of 100 wells per platform. In the deeper waters, the exploratory drilling will be done from semisubmersibles and drill ships. Although there is the possibility that fixed leg development/production platforms will be built, the most likely scenario is the future use of new designs such as the tension leg platforms and subsea completions.

Although much of the drilling will occur in deeper water, a large number of wells will continue to be drilled on the inner and mid shelf. Chevron will be concentrating on nearshore shallow drilling with interest being shown in the Ship Shoal Block 69 area and the South Timbalier Block 51. Additional activities are expected in the following areas: Champlin, a gas field on Blocks A-185, A-193, and A-194 on High Island East Addition; Tenneco, gas production from Brazos Block A-16 and Matagorda Island Block 712; Santa Fe Minerals has planned eight wells on East Breaks Block 173; CNG will be developing a gas find on East Cameron Block 299; Transco discovery on West Cameron Block 556. These are but a few of the development activities planned for the next 10 years. The central Gulf of Mexico will continue to be by far the most active offshore drilling area.

Western Gulf

Activity continues to increase in the western Gulf lease area, with new exploration occurring both inshore and offshore. The area is predominantly a gas province, with production coming ashore via pipeline. During the past several years, development has increased at the edge of the continental slope. One area of such development is the sometimes controversial Flower Garden Banks area which contains the northernmost coral reefs in the U.S. Although an area of less activity than the central Gulf, there will still be considerable activity over the next decade.

PACIFIC

General

California will be the second busiest offshore area over the next decade after the Gulf of Mexico. Two primary areas will be the focus of the exploration drilling. The Santa Barbara Channel area will continue to have considerable exploration and development. The new area that will be extremely active during the next ten years will be the Santa Maria area off Point Conception. The activity will be characterized by fairly large platforms, most with over 50 wells. Water depths will tend to be deeper, most ranging in the 150 to 500 m range. Most transportation will be to shore via pipeline with numbers limited by unitization programs. This will require common use of pipelines for both economic and environmental reasons.

There are currently 25 platforms in Federal waters off California. Nineteen of these are in the Santa Barbara Channel area (primarily from Ventura north to Point Conception). Six platforms are currently in the southern California area. Also included in the current count are five artificial islands. Plans for the near future project 15 new platforms, six of these are planned for the Santa Maria basin, eight for the Santa Barbara Channel, and one for the Los Angeles basin. The projected additional production that these new platforms may represent is approximately 400,000 bbl/day. That is one third of what the state currently produces from all sources and is twice the current offshore production.

Central California

The Point Arguello field is located approximately 16 km off Point Conception, and is 72 km west of Santa Barbara. This field is believed to be the largest ever discovered in U.S. outer continental shelf waters. Chevron discovered this field in 1981 and estimates the reserves to range from 300 to 500 million bbl. A three platform program is designed for this field, which is expected to produce up to 180,000 bbl/day. Current plans are for a 48 slot (well) platform (Hermosa) to be installed in 183 m of water on OCS-P-0316 (first production expected in 1986); a 56 slot platform (Hidalgo), located 8.7 km to the northwest of Hermosa, to be installed in 122 m of water on OCS-P-0450, with first production expected in 1987; and Texaco's platform Harvest (50 slot) to be installed in 204 m of water on OCS-P-0315, with first production in 1986. Within this 25 tract area between Point Arguello and Point Conception, up to five more platforms could be installed within the next ten years.

The Point Pedernales field is located 14.1 km north of Pt. Arguello in the Santa Maria basin and has estimated reserves of 350 million barrels of oil. The field is estimated at 8 miles by 2.4 km. Two production platforms are planned for this field. Union has planned platform Irene, which will have 72 slots and be placed in 73 m of water. Exxon is planning a 60 slot platform for the same area. The San Miguel field will be developed by Cities Service which plans to install a 70 slot platform (Julius) which will be located 12.9 km off the coast (28 km southwest of Pismo Beach).

Southern California

Eight new platforms are planned for the Santa Barbara Channel. Exxon is planning three to four platforms to develop their Santa Ynez Unit which is estimated to have a producing potential of 80,000 to 90,000 bbl/day. The field has already been defined by the Hondo, Pescado and Sacate fields which where discovered in 1968 and 1969. The estimates for this area are 400 million barrels and 700 billion cubic feet of gas. Currently the only platform in the area is Hondo which was installed in 1981. Exxon is currently planning Hondo B which will be located 4.8 km west of Hondo. The Sacate Platform is planned for development of the Sacate field on OCS-P-0193; Pescado A or Pescado B-2 platform for OCS-P-0182, and the Pescado B-1 for OCS-P-0183. Both the Pescado and Sacate Fields lie approximately 9.7 km west of the Hondo field. The Hondo B platform may be installed in water depths greater than 350 m.

Arco is planning to develop the Coal Oil Point field which is located in state waters directly off the beach at Goleta. The reserves estimated are 100 million barrels. Two platforms are planned for State Leases 308 and 309. The possible locations are approximately 2.4 km offshore, directly off the University of California at Santa Barbara.

Chevron is planning platform Gail which will have 36 slots. This will be installed in the Sockeye Field which is located 8 km south of the Santa Clara Field. It will be installed in 222 m of water and is estimated to possibly produce 10,000 bbl/day. The oil produced would be shipped to platform Grace and then to shore through the existing pipelines. Chevron also plans to replace their island Esther in the Belmont field off Seal Beach which was severely damaged in the winter storm of March, 1983.

ALASKA

General

Alaska will represent a major focal point for exploration in the offshore during the next decade. Due to the extreme environmental conditions and high costs involved, the activity levels (i.e., number of offshore wells drilled) will be less than in southern California and the Gulf of Mexico, but much effort will be expended because several areas in Alaska are perceived as important on the industry list (Table 2.1).

Due to the high costs of operating in Alaska environments, it will take the discovery of exceedingly large oil fields (by comparison to the lower 48 states) to economically justify development in Alaskan offshore regions. The National Petroleum Council estimates that the minimum economic reserve for a 10% rate of return varies from 500 million barrels in Norton Sound to 900 million barrels in the Navarin Basin. Favorable reservoir characteristics—regardless of field size—will be required to produce the required rates of return. Shallow reservoirs, low initial well productivities and thin pay zones may make field development marginal. Because most gas development scenarios involve a liquified natural gas system to transport the product to market, gas will not be economical to develop in the foreseeable future, barring a dramatic reduction in liquefaction costs and a significant increase in the price of gas.

Development Scenarios

Projections on the nature of future petroleum development in offshore Alaska are speculative and are very sensitive to a number of geologic, technical, economic and environmental regulations and stipulations.

Due to the high costs of operating in these areas, unitization of facilities (i.e., sharing facilities whenever feasible, such as pipelines, gravel islands, etc.) will be common practice for both exploratory and development/production activities.

The ability to share infrastructure (pipelines, tankers, shore bases) with other fields and even with other basins will be an important factor affecting the economic variability of oil discoveries. The economics of operating in Alaska will dictate that fields will be developed with relatively few multiwell production platforms, and operators can be expected to share trunk pipelines and other facilities with adjacent fields. In addition, the number of wells drilled to discover and delineate fields or basins will be relatively few compared with offshore California and the Gulf of Mexico.

There will not be any appreciable change in the quantity and the nature of drilling muds and cuttings. Water-based muds from offshore operations are generally disposed into the sea or under the sea ice. Oil-based muds are not discharged as a result of environmental regulations. The quantities of produced water cannot be predicted in advance, and a determination has not been made yet as to whether they will be reinjected or discharged on site. If discharged, the produced waters will have to meet regulatory requirements for oil and grease content, and the discharge regulations over the next decade will closely regulate the treatment chemicals allowed in the produced water. Domestic waste water will be treated on site and discharged as currently practiced in the lower 48 states. Waste hydrocarbon fluids may be combined with production, reinjected, burned as fuel, incinerated or brought to an onshore disposal area. Sewage sludge and combustible solid wastes would be incinerated and the ashes brought to an onshore disposal area.

Production/Transportation Scenarios

Each of the potential petroleum basins presents a unique combination of oceanographic and geological characteristics that will determine the development strategies should commercial oil and gas discoveries be made. Presently, Cook Inlet is the only offshore area in the world where sea ice is a major design consideration.

Petroleum development within the leased outer continental shelf regions is well within the present state-of-the art technologies with regard to the major oceanographic design parameters of bathymetry, wind and wave regimes, the presence of pack ice, and the possibility of severe structural icing. In each lease sale area, sea ice is a major design consideration for offshore facilities, necessitating special platform designs. As environments differ from basin to basin, the type of drilling structures and transportation methods will vary.

Three types of drilling/production platforms are feasible in Alaskan OCS waters depending upon site specific conditions and economics:

1. Modified upper Cook Inlet structure such as a monopod with no external bracing.
2. Artificial islands such as the gravel structures constructed in the Beaufort Sea.
3. Large concrete or steel gravity structures such as the Exxon CIDS now in place in the Beaufort Sea.

Gulf of Alaska

During the next decade, little activity is expected in the Gulf of Alaska. Due to the failure of previous lease sales and the high levels of activity expected in the more promising Beaufort and Bering Sea basins, little exploration and production activity is predicted for the Gulf of Alaska over the next decade.

Bering Sea

General

The St. George and North Aleutian Shelf basins and the southern part of the Navarin basin lie near the southern limit of the Bering Sea seasonal ice. In Norton Sound where ice is present up to eight months of the year, ice loading is the overriding platform design and transportation consideration. Only the St. George basin and the North Aleutian Shelf basin have significant earthquake exposure that will be a major consideration in the design of production platforms, pipelines and shore facilities. Seismic intensity increases toward the Aleutian Islands and Alaska Peninsula.

Water depths in the St. George and Navarin basins for the most part are comparable to the central and northern North Sea ranging from 90 to 150 m and

70 to 200 m, respectively. Norton Sound and the North Aleutian shelf water depths are generally comparable to upper Cook Inlet. The wave regime in the Bering Sea is generally less severe than the North Sea and Gulf of Alaska, with only the Navarin basin approaching North Sea conditions.

All of the Bering Sea basins are remote from existing petroleum transportation and processing facilities. The most remote is the Navarin basin which is over 600 km from the western Alaska mainland and 1000 km northwest of Dutch Harbor in the Aleutians. Because of these difficult logistics, it is probable that an at-sea loading to tankers is probable.

Norton Sound

Most of the Norton Sound is within the operational range of conventional jackup rigs for open water season exploration (June-October). Drill ships will be limited to water depths greater than 30 m. In shallower waters, gravel islands would permit year-round drilling, but would be limited by the availability of gravel sources. Gravel islands, cone structures or modified upper Cook Inlet structures such as the monopods, are feasible in Norton Sound depending upon the specific site conditions.

Due to shallow shelfing nature of the Norton Sound nearshore, it would be difficult to build a deep water port for a crude oil terminal or LNG plant. Most of the lease area is within 65 km of land and would be serviced by pipeline. Crude oil terminal designs for the Norton Sound area include the possibility of a long causeway to a conventional dock (up to 5 km long). Other options are a sea island pier connected to the shore by a pipeline or a monobuoy single point mooring for loading tankers.

St. George

Water depths in St. George basin range from 90 to 150 m and are comparable to the North Sea. The wave regime is more severe than upper Cook Inlet and Norton Sound but less than the Gulf of Alaska. The area is in a zone of high seismicity (Zone 3).

Although exploratory drilling in St. George Basin is within the operational capabilities of semisubmersibles, drill ships, and, in some areas, jackups, there will be a need to develop year-round capabilities in the northern part of the lease sale area where ice incursions can be anticipated between January and April during some years. In order to withstand ice conditions, future production platforms will probably be of the North Sea monopod design with the drilling conductors passing through the center of the structure. Current limitations on the number of wells is in the range of 50. The two primary transportation scenarios currently under consideration are pipelines to the Aleutians and Alaska Peninsula, or the onsite offloading to tankers in the lease area. Sea and fog conditions make this a questionable option at this time (e.g., >40% fog during summer). Support operations would probably be staged out of the Dutch Harbor/ Unalaska harbors.

North Aleutian Shelf

The conditions in this region are similar to those described for the St. George basin. The exploration systems would also be similar. The pipeline distances to the Alaska Peninsula terminal sites would be <130 km. Onshore pipeline distances across the peninsula to the possible Pacific terminal sites would range from as little as 10 km (Cold Bay) to 105 km (Makushin Bay).

Navarin Basin

The Navarin basin will be the most difficult of the Bering Sea regions to explore and develop. This is due both to its very remote location and very severe weather conditions. Water depths in the lease area range from 70 to 200 m, with the most severe wave regime of the Bering Sea basins. Conditions occasionally approach those of the North Sea with a maximum 100-year wave of 27 to 30 m. An ice coverage of 6 to 8 oktas (okta = 12.5%) is common in the northern part of the lease area. Exploration by conventional rigs will be limited to the summer season. To drill in the winter season will require ice-strengthened dynamically positioned semisubmersibles or drill ships with possible ice breaker support. Resupply of bulk materials will probably be via sea from Dutch Harbor in combination with storage barges moored at St. Matthew Island (if allowed).

Steel jacket, concrete or steel gravity structures are some of the designs being considered for the Navarin basin. These would contain the drilling conductor pipes within their structure to protect against ice. Proper soil conditions are thought to exist in the Navarin area to support the gravity type structures.

Due to the remote location and harsh conditions, a field of giant proportions would have to be found before the Navarin could be economically developed. Two main transportation options appear available: 1) a long pipeline to St. Paul Island, or 2) development of an offshore loading system, with transshipment to a Very Large Crude Carrier (VLCC) location in the Aleutians.

Beaufort/Chukchi Sea

The Beaufort and Chukchi Seas together rank first in oil and gas potential in the offshore regions of Alaska. Three lease sales are currently scheduled for this area in the next four years in water depths to 213 m. The harsh conditions and remote locations will require long lead times in any development. As in the Bering Sea, sea ice is a major design consideration. Along the coast, sea ice retreats during July through September, although heavy pack ice can be blown in against the coast at any time.

Artificial gravel islands in water depths up to 20 m have year-round drilling capability. The economic use of artificial islands would be dependent upon the availability and cost of gravel sources. Jackup rigs in water depths up to 70 m provide seasonal drilling capability. Open sea conditions with 50% ice coverage may last four to five months. Gravity platforms have year-round drilling capability in depths from 20 to 70 m. During the 1985 season, drill ships will be used for the first time in the Alaskan Beaufort Sea (near Barter islands). This technology has been used extensively by the Canadians for several years. During the next decade, production structures will be a combination of artificial gravel islands and gravity based cone structures in depths to 70 m. The prime limitation is the resistance to multiyear ice floes and pressure ridges.

Pipeline distances in the Beaufort Sea could extend up to 200 km, with distances in the Chukchi ranging up to 550 km. Most of the distance would be over land. Marine pipelines would be entrenched in the bottom to protect them from ice scour and buried near shore to protect them from the winter land fast ice. The nearshore pipe may encounter permafrost and would have to be insulated.

Two offshore discoveries during 1984 will result in development and production operations over the next decade. The Endicott project off the Sag River Delta will include two islands and a gravel causeway connecting the nearest island to shore. Preplanning has already identified the need to build breaches in the causeway so that nearshore migration of anadromous fish is not inhibited (see Chapter 13). Shell Oil Company and partners announced a discovery on Seal Island to the west of Prudhoe Bay. This will also be the location of much development activity over the next several years. This will probably include a buried pipeline to shore.

Gravel islands will still be a major construction technique for the nearshore shallow waters. Although most gravel to date has come from onshore gravel pits, future island building activities may use offshore dredging techniques for obtaining the necessary gravel.

LITERATURE CITED

Anonymous. 1985. Special reports - U.S. offshore. Offshore, January 1985, pp 51-66.

Minerals Management Service. 1985a. Estimates of Undiscovered Economically Recoverable Oil and Gas Resources for the Outer Continental Shelf as of July 1984. Minerals Management Service, Washington, D.C., Report No. MMS 85-0012.

Minerals Management Service. 1985b. Draft Proposed Program. 5-Year Outer Continental Shelf Oil and Gas Leasing Program for Mid-1986 through Mid-1991. Minerals Management Service, Washington, D.C.

CHAPTER 3

DOMINANT FEATURES AND PROCESSES OF CONTINENTAL SHELF ENVIRONMENTS OF THE UNITED STATES

Nancy N. Rabalais and Donald F. Boesch

CONTENTS

INTRODUCTION

Continental shelves constitute relatively small areas of the ocean (7.6% of ocean area) where water depths are generally less than 200 m. Continental shelves are those relatively gently grading (average slope of 0.2%; Emery and Uchupi, 1972) platforms which extend from the shore to the point at which the bottom exhibits a more rapid inclination towards the continental slope. Despite their relatively small area, continental shelves provide most of the ocean's fisheries harvest and oil and gas production. Ironically, though, continental shelf environments have been somewhat of a neutral zone between the "blue water" investigations of the open seas and those of more accessible littoral and estuarine environments. As a consequence, many facets of the oceanography and ecology of continental shelves are less well known than for environments farther inshore or offshore. With the potential of expanded offshore oil and gas development and concerns about dumping of wastes, continental shelves have been the subject of greatly intensified study during the last 10 to 15 years, particularly in the United States. As yet, however, there are few emerging syntheses which provide a scientific framework for a comparative understanding of continental shelf environments.

This chapter summarizes and compares the ecologically important features and environmental processes of the continental shelves of the North American United States. These shelf environments are exceptionally diverse. They range over 48 degrees latitude, extend from polar seas nearly to the tropics, and vary greatly in their hydrodynamic processes. For example, the Bering Sea shelf is greater than 500 km wide; the Atlantic and Gulf of Mexico shelves are also quite wide (150 to >200 km), but the California shelf is exceptionally narrow (generally, <50 km). Open oceanic processes are less important than local processes (the effects of land runoff and weather) in bounded, shallow epicontinental shelf areas compared to narrow marginal continental shelves (Nio and Nelson, 1982). Physical processes off the Pacific coast tend to be less energetic than those off the Atlantic coast where the shelves are wide and shallow and influenced by a western boundary current (Pietrafesa, 1983). The shelf off the northeastern United States has been heavily influenced by erosional processes due to glaciation, while portions of the Gulf of Mexico have a long history of active deposition. Sea ice is a physically, geologically and biologically important feature of the shelves of the Bering Sea and Arctic Ocean, while hurricanes have major environmental consequences in the Gulf of Mexico and South Atlantic Bight.

In this review we will particularly focus on benthic organisms and the features and processes which influence them. This focus has several reasons. First, the benthos of U.S. shelves has been extensively surveyed during this 10 to 15 year period of increased study. Second, the benthos has been the principal emphasis of studies of the effects of oil spills and other discharges because of their susceptibility, longevity and relative immobility. Thus, future studies of long-term effects of oil and gas development activities will undoubtedly concentrate on the benthos. Despite the large body of data on benthos of U.S. shelves now available, comparative ecological synthesis is handicapped by variable methodologies and the lack of both thorough analysis of these data and interpretive publications.

This comparative synthesis is organized by regions as defined in Table 3.1 and

TABLE 3.1
Continental shelf regions of the United States as used in this Chapter compared with planning areas designated by the U.S. Minerals Management Service

Regions (this Chapter)	*Coastal boundary*	*Minerals Management Service offshore planning areas*
Atlantic Coast		
New England		North Atlantic
	Montauk Point	
Middle Atlantic Bight		Middle Atlantic
	Cape Hatteras	
South Atlantic Bight		South Atlantic
Gulf of Mexico		
West Florida Shelf		Eastern Gulf
	Cape San Blas	
North Central Gulf of Mexico		Eastern Gulf (in part) Central Gulf (in part)
	Louisiana-Mississippi border	
Northwestern Gulf of Mexico		Central Gulf Western Gulf (in part)
	Matagorda Bay	
South Texas		Western Gulf
Pacific Coast		
Southern California		Southern California
	Point Conception	
Central and Northern California		Southern Cal. (in part) Central-Northern California
	California-Oregon border	
Washington-Oregon		Washington-Oregon
Alaska		
Gulf of Alaska		Gulf of Alaska Kodiak Cook Inlet Shumagin
	Aleutian Islands	
Bering Sea		North Aleutian Basin St. George Basin St. Matthew Hall Navarin Basin Norton Basin
	Bering Strait	
Alaskan Arctic		Hope Basin Chukchi Sea Beaufort Sea

Figure 3.1. Federal offshore planning areas of the United States (solid lines). Regions considered in this Chapter delineated by dashed lines or in parentheses if different from Department of Interior (1985) designations. (Alaska not to same scale as contiguous 48 states.)

TABLE 3.2
Matrix comparing the dominant features and processes of U.S. Continental shelf areas

Region	*Shelf morphology*	*Stage of sedimentary evolution*	*Sedimentary regime*	*Depocenters*
New England	Dominated by shallow plateau (Georges Bank) bordered by deeper channels; Gulf of Maine with irregular bedrock and basins; large sand shoals and waves of Nantucket Shoals; shelf edge incised by submarine canyons	Autochthonous	Low depositional, dominantly sands with gravels in areas, sand shoals and waves in shallower areas, grade to medium to fine sands mid shelf	Major area SW of Georges Bank, "Mud Patch," heads of submarine canyons, some eventually to Long Island Sound
Middle Atlantic Bight	Broad, gently sloping platform with complex mesoscale topography of ridges and smaller sand waves, sediment-filled channels	Autochthonous	Low depositional, dominantly sands >75%, mostly >90%, fine-grained sediments generally absent except for some accumulation in depressions	Deposited in salt marshes and estuaries or carried to shelf edge, slope, and heads of submarine canyons
South Atlantic Bight	Broad, gently sloping platform with cross-shelf shoals, sand waves, rocky outcrops; calcareous reefs at shelf break; no submarine canyons	Autochthonous	Low depositional, coarse to medium sands with finer sands at depth, little fine fraction except in cape-associated "shadows"	Deposited in marshes and estuaries or carried to shelf edge

(continued)

TABLE 3.2—*contd.*

Region	*Shelf morphology*	*Stage of sedimentary evolution*	*Sedimentary regime*	*Depocenters*
West Florida Shelf	Broad, gently sloping platform of carbonate sand sheet with sub-surface or low relief exposed hard strata at mid and outer shelf and algal nodule areas at shelf break	Autochthonous	Predominantly sands with little fine fraction, some areas of finer sediments nearshore, mid shelf and near Florida Keys	Off shelf and into Florida current, may be removed from shelf area
North Central Gulf of Mexico	Broad, gently sloping shelf except at Mississippi River prodelta and De Soto Canyon; ridges and pinnacles on outer shelf and at shelf break	Autochthonous/ Allochthonous	Quartz sand sheet with increasing fine fraction towards Mississippi River prodelta and off major estuaries	Mississippi River prodelta, sounds, off Mobile Bay, also to outer shelf of West Florida shelf
Northwestern Gulf of Mexico	Gently sloping, wide shelf with relict shorelines, distributary ridges, relict reefs and numerous diapirs on shelf, shelf break and slope; very wide, hummocky slope	Allochthonous/Climax Grading	Sand decreases and silt increases seaward; patchy occurrence of sand areas, increase in clay W of delta, poorly to very poorly sorted sediments inshore	Mississippi and Atchafalaya deltas and prodeltas, outer shelf and slope
South Texas	Gently sloping, wide shelf with deltaic bulges at Rio Grande and off Matagorda Island with broad, ramp-like indentation between; relict shorelines and reefs, sediment-filled channels, diapirs on shelf	Climax Grading	Mostly silty sands with decrease in sand and increase in silt offshore, patches of sands on ancient deltas, higher clay content in area mid shelf off Port Aransas	Bays, outer shelf and slope

Southern California	Narrow shelf, complex borderland with southeast-northwest trending ridges and basins, shallow rocky banks at shelf edge, indentation by submarine canyons	Climax Grading	Progrades from sands to silty sands to sandy silts and then silts at shelf edge, more complex sediments around rock outcrops	Deposition mostly in nearshore basins and on slopes but also in outer basins
Central and Northern California	Narrow, gradually sloping shelf incised by several submarine canyons along entire length, bands of bedrock habitats on inner shelf, rock outcrops on outer edge of basins at shelf edge	Climax Grading/ Allochthonous	Coarser, sandy sediments nearshore to finer silts and clays offshore, fine sediments between Eel and Klamath Rivers and off Russian River	Submarine canyons and slopes
Washington-Oregon	Narrow, gradually sloping and generally featureless shelf, some major submarine canyons incise shelf	Climax Grading/ Allochthonous	Inner shelf sands to 60 m, mid shelf silts to 120 m, some sandy areas near canyons	Mid-shelf sediments of silt similar to Columbia River suspended load
Gulf of Alaska	Broad, gently sloping marginal shelf, dynamic bedforms of sand waves (8 to 15 m high); major submarine valleys and basins in NE, series of flat banks (50 to 100 m) and troughs down to 200 m off Kodiak	Autochthonous/ Climax Grading	Mostly reworked glacial off Kodiak, shelly sands nearshore, coarse sediments on banks, fine in troughs; silts/clays and high sedimentation rates inshore in NE, sand and gravel mixed with silts and clays at shelf break	Troughs off Kodiak (except Stevenson), volcanic ash as indicator of present day dispersal patterns; removed from shelf and slope in NE

(continued)

TABLE 3.2—*contd.*

Region	*Shelf morphology*	*Stage of sedimentary evolution*	*Sedimentary regime*	*Depocenters*
Bering Sea	Broad (>500 km), gently sloping marginal shelf in SE to broad, shallow epicontinental shelf in NE; relatively flat and featureless in SE, undulating hummocky in W, modern and relict sand ridges and shoals in NE, prograding Yukon delta in Norton Sound	Climax Grading, SE; Autochthonous, N; Allochthonous, Norton Sound	Fine-grained transgressive sands with decrease in sands offshore in SE; glacial gravel and transgressive sands in W; thin transgressive sands facies in N, nondepositional; thick, muddy deposits in Norton Sound	Major center in southern Chukchi Sea for Yukon-derived sediments, also eastern sector of Norton Sound, localized areas on southeastern shelf
Alaskan Arctic	Broad, gently sloping marginal shelf in Chukchi Sea to very narrow shelf in Beaufort Sea; little bathymetric relief except for Barrow Sea Valley and Hanna Shoal in Chukchi Sea	Climax Grading	Majority of Chukchi Shelf with 5ϕ sediments; Beaufort shelf with primarily sandy muds and silt and clay at shelf edge, gravels at shelf break, some input from river run-off, tundra slumping, and river ice	Low sediment transport nearshore and predominantly westward, possible depositional site in SW Harrison Bay for Colville River delta muds

Region	Tides	Wave climate	Dominant processes and events	Benthic boundary layer dynamics
New England	Semidiurnal of magnitude, 4 m at mouth of Bay of Fundy; tidal current 30–40 cm/s in channels, >55 cm/s on Georges Bank proper	30–40% of waves >1.5 m, 50% of waves > 1 m	Constant reworking by fast tidal currents and waves down to 60 m, winter storms	Surficial sediments under almost constant scour, dynamic sand waves, low suspended particles over Georges Bank, 750–800 μg/l over mid shelf, 250 μ/l over outer shelf
Middle Atlantic Bight	Semidiurnal of moderate magnitude; tidal current <25 cm/s	~30% of waves >1.5 m, ~5–10% of waves >3.5 m	Wind-influenced currents on inner shelf, storm wave generated currents affect outer shelf, winter storms	Ripple marks down to 200 m; winter storms main cause of disturbance >60 m, more frequent disturbance <60 m by winds and tidal currents; sand waves not active
South Atlantic Bight	Semidiurnal of moderate magnitude; tidal current <25 cm/s	Nearshore waves <0.5 m, ~20% of waves >1.5 m, ~5% of waves >3.5 m	Tidal currents and wind regimes affect depths to 20 m, freshwater inflow off Georgia and S. Carolina, hurricanes, Gulf Stream intrusions	Evidence of ripple marks in <20 m, no active sand waves
West Florida Shelf	Predominantly diurnal, but mixed and semidiurnal, low magnitude 0.3–1.2 m; tidal currents generally <8 cm/s	Generally low, most 0.3–1.3 m (summer-winter)	Seasonal wind-generated currents and waves, hurricanes, winter frontal passages	Localized turbidity fronts, seasonal sand ripples, near-bottom nepheloid layers influenced by storms, frontal passages, bottom currents; strong nepheloid layer at mid shelf

(continued)

TABLE 3.2—*contd.*

Region	*Tides*	*Wave climate*	*Dominant processes and events*	*Benthic boundary layer dynamics*
North Central Gulf of Mexico	Predominantly diurnal, but mixed and semidiurnal low magnitude 0.3–1.2 m; tidal currents generally <8 cm/s	Generally low wave regime, most 0.5–1.0 m (summer-winter)	Seasonal wind-generated currents and waves, particularly in winter, high frequency of hurricanes, winter frontal passages	Spring turbid bottom waters (surface also) near Mississippi River, water column turbid whole depth in winter
Northwestern Gulf of Mexico	Diurnal and mixed, low magnitude 0.5 to 1.5 m; tidal currents low, ≤14 cm/s	Low wave regime 0.5 m in summer to 1.0 m in winter, historical maximum of 7.3 m in summer	River discharge, inshore effects as far west as Galveston, wind-generated currents and waves, high frequency of hurricanes, winter frontal passages, widespread seasonal hypoxia	Nepheloid layers and turbidity currents, near-bottom nepheloid layer present all seasons except winter, in <20 m associated with wind-generated currents
South Texas	Mixed diurnal and semidiurnal, low magnitude 0.4–0.5 m; tidal current velocity ≤14 cm/s	Low wave regime, most 0.9–1.8 m; summer, most 0–0.6 m; winter, most 1.2–1.8 m; waves up to 2–3 m in hurricanes	Persistent SE winds predominate, along with winter northerly winds cause wind-generated currents and waves, high frequency of hurricanes, winter frontal passages	Near-bottom turbidity all seasons, variable in thickness and distribution, related to thermal mixing, internal waves, nearshore suspension by waves and tidal mixing
Southern California	Mixed semidiurnal, of moderate magnitude, 0.3 to 2.5 m	Average significant wave heights 1–2 m, tsunamis with waves of 6.3 m	Wind-generated nearshore water movements, prevailing winds from NW (16–32 km/hr), more onshore in summer, spring and summer upwelling, occasional small tsunamis, periodic El Niño events, seismicity	Sediment erosion from exposed shallow shelf with deposition in basins of southern California borderland

Central and Northern California	Mixed semidiurnal of moderate magnitude 0.3 to 2.3 m	Average significant wave heights of 1–3 m, storm waves of 5–11 m, tsunamis with waves of 6.3 m	Severe extratropical winter storms, moderate storm surges of 0.5 to 1 m more common, spring and summer upwelling, occasional tsunamis, seismicity	Winter storm dominated sedimentary regime with sediment deposition off major northern California rivers
Washington-Oregon	Mixed semidiurnal of moderate magnitude, up to 3 m in spring off Oregon	Average wave heights greater than to south, tsunamis with waves of 6.3 m	Severe extratropical winter storms, spring and summer upwelling, occasional tsunamis	Mid-shelf silt deposits transient and undergo repeated resuspension and redeposition by storm-induced bottom currents
Gulf of Alaska	Mixed semidiurnal, mean range at Kodiak 0.9 to 3.3 m, in NE 2 to 4 m; swift tidal current in Cook Inlet averages 300 cm/s	High waves in storms (9 m and higher), 3 m waves prevalent in winter, maximal tsunami waves up to 33 m	Extreme storm conditions with winds up to 160 km/hr, floe ice and pack ice in Cook Inlet in winter, tsunamis, seismicity and volcanism	In general, little suspended matter due to strong currents and winter storm waves which prevent sediment accumulation, water exiting Cook Inlet with high suspended sediments (5–200 mg/l) which flows into Shelikof Strait
Bering Sea	Mixed semidiurnal tides with marked diurnal inequalities; in southern area near peninsula and land restrictions up to 5 m, in NE tides <0.5 m	Larger waves in SE up to 10–20 m, lower in NE up to 7 m, maximum wave heights 32 m, waves most influential in shallow Norton Sound and Bristol Bay	Severe storms may double strength of wind-generated currents and waves, N-S differences in storms; ice movements, dense coverage, ice presence, N-S differences; tsunamis in North Aleutian shelf; potential seismicity and volcanism	In NE, major storm surges every several years resuspend and transport sediments; dynamic sediment transport by ice gouging, storm waves and ice loading

(continued)

TABLE 3.2—*contd.*

Region	*Tides*	*Wave climate*	*Dominant processes and events*	*Benthic boundary layer dynamics*
Alaskan Arctic	Mixed semidiurnal tides with mean range of 10 to 30 cm, small tidal currents of 0.3 to 0.5 cm/s	Extreme storm waves not as likely as in Bering Sea or Gulf of Alaska because generation hampered by pack ice, 2.5 m waves during storms, only 22% > 0.5 m	Severe storms with high winds but waves tempered by sea ice, gouging by offshore pack ice, offshore subsea permafrost, active ice flow lead along Chukchi Sea coast, coastal erosion on Beaufort Sea coast	Storm affected sediments in Chukchi Sea, ice gouging on seabed out to 60 m in whole area; low coastal sediment transport inshore on Beaufort Shelf, some reworking of coarse sediments by hydrodynamic forces on inner and central shelf

Region	*Bottom temperature regime (winter °C to summer °C)*			*Benthic studies*	*Biogeographic affinities*
	Inner shelf	*Outer shelf*	*Shelf break*		
New England	0.5–4° to 10–12°	4–5° to ~8°	10°	Wigley 1965, Wigley & McIntyre 1964, Wigley & Theroux 1976, 1981, Maurer & Leathem 1981a,b, Michael 1977, Michael *et al.* 1983, Battelle/W.H.O.I. 1983, 1984	Boreal or cold water assemblage
Middle Atlantic Bight	1–4° to 14–17° with latitudinal differences	5–8° to 13–14°	10–12°	Gross 1976, Wenner & Boesch 1979, Schaffner & Boesch 1982, Boesch & Bowen in press	Transitional between cold water assemblage to north and warm, temperate south of Cape Hatteras, shallow water Carolinian affinities, offshore boreal affinities

South Atlantic Bight	6–10° to 28° with latitudinal differences	12–17° to 26–27° with latitudinal differences	15–17° [14 to 21°]	Williams *et al.* 1968, Day *et al.* 1971, Tenore 1979, Herbst *et al.* 1979	Warm, temperate Carolinian affinities inshore with subtropical and tropical affinities offshore
West Florida Shelf	20° to 26° with latitudinal differences	20° to 24° with latitudinal differences	16° to 22° with latitudinal differences	State Univ. Syst. Flor. 1977, Dames & Moore 1979, Woodward-Clyde & Cont. Shelf Assoc. 1983, Lyons 1980, Collard & D'Asaro 1973, Lyons & Collard 1974	Shallow shelf with Carolinian affinities, deep shelf with West Indian/tropical affinities
North Central Gulf of Mexico	~12–15° to 29°	~18–19° to 21–25°	~19° to 22°	State Univ. Syst. Flor. 1977, Dames & Moore 1979, Lyons & Collard 1974	Warm, temperate Carolinian affinities, separated by some from northern Gulf W of Mississippi River, outer shelf tropical affinities
North-western Gulf of Mexico	15–16° to 27–28°	18–19° to 21–25°	19–20° to 22°	Ward *et al.* 1979, Bedinger 1981, Middleditch 1981, Jackson 1977, Hann & Randall 1980, DeRouen *et al.* 1982	Warm, temperate Carolinian affinities, separated by some from northern Gulf E of Mississippi River, outer shelf tropical affinities
South Texas	14–15° to 28° with latitudinal differences	15–17° to 25° with latitudinal differences	19–22° with latitudinal differences	Holland 1977, 1979, Flint 1981, Flint & Holland 1980, Flint & Rabalais 1980a,b, 1981	Mostly warm, temperate Carolinian affinities with more subtropical influence, considered by some as Texas transitional between temperate and tropical outer shelf tropical affinities
Southern California	12–14° to 17–19.5° with latitudinal differences	10–13° to 15–17°	10–13° to 15–17° @ 200 m 8–9°C throughout year	Jones 1969, Barnard & Hartman 1959, Barnard 1963, Fauchald & Jones 1977, 1978, Balcom 1981	Transitional between southern subtropical Panamanian province and northern temperate Oregonian province
Central and Northern California	10° to 14° with latitudinal differences	9–10° to 12–13°	9–10° to 12–13°	Jones & Stokes, Assoc., Inc. 1981	Temperate, Oregonian province

(continued)

TABLE 3.2—*contd.*

Region	*Bottom temperature regime (winter °C to summer °C)*			*Benthic studies*	*Biogeographic affinities*
	Inner shelf	*Outer shelf*	*Shelf break*		
Washington-Oregon	9–10° to 12–13°	9–10° to 12–13°	9–10° to 12–13°	Carey 1972, Lie & Kelley 1970, Lie & Kisker 1970, Richardson *et al.* 1977	Temperate, Oregonian province
Gulf of Alaska	<2° in May, 1–2° to 6–9° in <100 m over shoals on Kodiak where water nearly isothermal	2–3° in May	4–5°C throughout year at Kodiak shelf break	Feder & Matheke 1979, Feder & Jewett 1980	
Bering Sea	<−1.5 to 8–10° with latitudinal and depth differences	−1–0° to −1–2° with latitudinal differences	1–3° to 2–4° with latitudinal differences	Hood & Calder 1981, Stoker 1981, Haflinger 1981, Feder & Jewett 1981, Nelson *et al.* 1981	Primarily boreal Pacific
Alaskan Arctic	−1–4° in summer to −1° in fall with most values at −1–10°	2° in summer to −1.5° in fall	0° in summer to −1° in fall	Carey *et al.* 1974, 1984, Carey & Ruff 1977, Bilyard & Carey 1979, 1980, Broad *et al.* 1981, Stoker 1981	Primarily boreal Pacific in Chukchi with more Arctic species than Bering Sea; dominance of amphiboreal species in Beaufort, many also occur in temperate or tropical latitudes, or both

Region	Inner shelf	Dominant macrofauna Outer shelf	Shelf break
New England	60 m—top of bank: *Tellina agilis* (b), *Pseudunciola, Protohaustorius* (a), *Polygordius, Exogene, Hebes, Spiophanes* (p), *Echinarachnius* (ecd)	80 m: syllid polychaetes, *Phallodrilus* (o), *Unciola, Erichthonius, Byblis* (a)	Paraonid, capitellid, cirratulid polychaetes, *Ampelsica* (a); in another study: *Pseudohaustorius, Protohaustorius, Trichophoxus* (a)
Middle Atlantic Bight	~20–30 m, dynamic sands: interstitial feeders, *Tanaissus* (t), *Polygordius, Goniadella, Lumbrinerides* (p)	~40–100 m: fewer interstitial feeders and more amphipods, *Unciola, Ampelisca, Erichthonius, Photis, Rhepoxinius* (a), *Spiophanes, Goniadella, Lumbrinerides, Euchone, Lumbrineris* (p)	~110–200 m: increase in burrowers and subsurface deposit feeders, *Onuphis, Lumbrineris, Spiophanes, Aricidea* (p) *Harbansus* (os), *Amphioplus* (op) *Ampelisca, Unciola* (a)
South Atlantic Bight	<40 m "turbulent zone": *Paleanotus, Polygordius, Lumbrineris* (p), *Protohaustorius, Maera* (a)	40–124 m, more quiescent zone: *Onuphis, Spiofillicornis, Chaetozone, Pomatoceros* (p), *Siphonocoetes* (a)	160–205 m: *Odontosyllis, Onuphis, Lumbrineris, Prionospio, Chaetozone* (p), *Paraphoxus, Unciola* (a), *Nucula* (b)
West Florida Shelf	<40 m: *Fabricia, Vermiliopsis, Prionospio, Hydorides, Lumbrineris, Goniadides, Ehlersia* (p), *Photis, Maera* (ga), *Cyclaspis* (cu), *Phtisica* (cp), *Lucina* (b)	40–100 m: *Magelona, Ceratocephale, Synelmis, Schistomeringos, Tharyx, Prionospio, Ampharete* (p), *Lucina* (b), *Selenaria* (br)	100 m: *Glycera, Prionospio, Synelmis, Terebellides* (p)
North Central Gulf of Mexico	Not delineated	Dominants for entire shelf: *Syllis, Sphaerosyllis, Websterinereis, Glycera, Lumbrineris, Prionospio, Paraprionospio, Mediomastus* (p)	Not delineated

(*continued*)

TABLE 3.2—*contd.*

Region	*Inner shelf*	*Dominant macrofauna* *Outer shelf*	*Shelf break*
Northwestern Gulf of Mexico	*Lumbrineris, Paraprionospio, Magelona, Sigambra, Diopatra* (p), *Golfingia* (s), *Ampelisca* (a)	*Cossura, Ninoe, Nephtys, Notomastus, Lumbrineris, Paraprionospio* (p), *Corbula, Nuculana* (b), *Volvulella* (g)	No shelf break stations
South Texas	~15–30 m: *Magelona, Nereis, Mediomastus, Aricidea, Prionospio, Paraprionospio* (p), *Tellina* (b), *Ampelisca* (a)	~40–90 m: *Paraprionospio, Cossura, Nephtys, Paraonis, Magelona, Asychis, Notomastus, Mediomastus* (p), *Ampelisca* (a), *Apseudes* (t), *Eudorella* (cu)	~100–135 m: *Paralacydonia, Tharyx, Sternaspis, Paraonis, Sigambra* (p), *Xenanthura* (i), *Amygdalum, Nuculana, Pitar* (b), *Alternochelata* (os)
Southern California	<25 m, dynamic sands: *Sipunculus* (s), *Loimia, Ophelia, Nephtys, Nothria, Prionospio, Diopatra* (p), *Tellina* (b) *Amphipholus* (op), *Diastylopsis* (cu), *Paraphoxus* (a)	28–109 m: *Pectinaria* (p), *Heterophoxus, Paraphoxus, Westwoodilla* (a), *Parvilucina, Tellina* (b), *Euphilomedes* (os) *Amphiodia* (op), *Diastyllis* (cu) *Listriolobus* (ech)	161–620 m: *Pectinaria, Prionospio, Maldane* (p), *Ampelisca* (a), *Limifossor* (g), *Allocentrotus, Brissopsis* (ecd), *Amphiodia* (op), *Cyclocardia* (b)
Central and Northern California	Sands of shallow shelf: *Diopatra, Nothria* (p), *Dendraster* (as), amphipods	Not delineated	Not delineated
Washington-Oregon	~35 m, inner shelf sands: *Diastylopsis* (cu), *Ampelisca, Paraphoxus* (a), *Tellina, Macoma* (b), *Owenia, Nephtys, Chaetozone* (p)	~95 m, intermediate shelf sands: *Magelona, Sternaspis, Nephtys, Haploscoloplos* (p), *Yoldia, Macoma, Axinopsida* (b), *Paraphoxus* (a)	~150 m deep-water muds: *Prionospio, Sternaspis, Ninoe* (p), *Axinopsida, Adontorhina, Macoma* (b), *Heterophoxus* (a), *Brisaster* (ecd), *Ophiura, Amphioplus* (op)

Gulf of Alaska	NE: inshore, deposit feeders (worms) predominate (61–65% of sample), *Nucula, Nuculana, Yoldia, Thyasira, Axinopsida* (b), *Terebellides, Nereis, Lumbrineris, Sternaspis* (p), *Heterophoxus* (a), *Chaetoderma* (ap), *Diamphiodia* (op)	Same as inner and mid shelf	NE: shelf break and slope, more suspension feeders (clams), comprise 32% of sample, deposit feeders 26%, *Notoproctus, Glycera,* sabellids, syllids (p), sipunculids, brachiopods, *Astarte* (b), *Anonyx* (a)
Bering Sea	*Byblis, Ampelisca* (a), *Capitella, Ampharete, Haploscoloplos, Myriochele, Sternapsis* (p), *Cylinchna* (g), *Serripes* (b), *Diamphiodia, Gorgonocephalus* (op)	*Axinopsida, Tellina, Yoldia, Clinocardium* (b), *Byblis, Bathymedon, Protomedeia* (a), *Capitella, Haploscoloplos, Brada, Artacama* (p) *Echinarachnius* (ecd), *Chaetoderma* (ap)	*Aricidea, Glycera, Asabellides, Harmothoe, Maldane* (p), *Ampelisca* (a), *Ophiura* (op), *Golfingia* (s), *Astarte, Hiatella,* (b)
Alaskan Arctic	In Chukchi similar to Chukchi outer shelf; in Beaufort, inshore zone <20 m, polychaetes, gammarid amphipods, isopod, bivalves, priapulid	Chukchi: *Maldane* (p), *Astarte, Macoma, Nucula, Yoldia* (b), *Ophiura* (op), *Golfingia* (s), *Pontoporeia* (a); Beaufort: offshore zone (20–200 m), fauna not delineated	In Chukchi, similar to outer shelf; in Beaufort, not studied

Abbreviations for taxa are: a—amphipod, ap—aplocophoran, as—asteroid, b—bivalve, br—bryozoan, cu—cumacean, ca—caprellid, ech—echiuran, ecd—echinoid, g—gastropod, ga—gammarid amphipod, i—isopod, o—oligochaete, op—ophiuroid, os—ostracod, p—polychaete, s—sipunculan, t—tanaidacean.

(continued)

TABLE 3.2—*contd.*

Region	*Macrofaunal variability*	*Epifaunal predators*	*Hard substrates and rare habitats*
New England	Regional stations distinct by depth and sediment type across seasons during two years, little seasonal variability with differences not repeatable over two years	Gadid fishes, flatfishes, cancroid crabs, starfish	Epibenthic and soft coral communities at heads of submarine canyons
Middle Atlantic Bight	Variability related to mesoscale topography, bathymetric gradients related to temperature, temp. variability, frequency and magnitude of sediment disturbance, variability of population and community structure decreases with depth	Gadid fishes, flatfishes, cancroid crabs, starfish	Epibenthic and soft coral communities at heads of submarine canyons
South Atlantic Bight	High seasonal variability and year-to-year differences nearshore, variability with depth related to frequency and magnitude of sediment disturbance as well as sediment distribution		Shallow coral patches near shore, calcareous reefs at shelf break, scleratinian coral reefs at shelf break off Florida
West Florida Shelf	Seasonal differences in community structure, particularly nearshore, faunal differences related to bathymetry and sediment distributions	Penaeid and sicyonid shrimp, portunid crabs, sea basses, flatfishes, holothurians (bioturbation)	Numerous hard substrates with epibenthic fauna and hermatypic corals, algal nodule substrates, nearshore seagrass beds, Florida Middle Grounds
North Central Gulf of Mexico	Seasonal differences with decreases in polychaete taxa and density and crustacean density in winter, variability in polychaete density with depth related to sediment differences	Penaeid and sicyonid shrimp, portunid crabs, sciaenid fishes, flatfishes	Exposed hard substrate areas on outer shelf and at shelf break
North-western Gulf of Mexico	Seasonal changes in abundance and dominance, variability in community structure related to sediment differences both with depth and physiography and to seasonal hypoxia on inner shelf	Penaeid shrimp, portunid crabs, sciaenid fishes, flatfishes	Exposed salt dome banks of carbonate sediments on outer shelf and shelf break, many with hermatypic corals, most with top environmental priority ranking, East and West Flower Gardens

South Texas	Some seasonal differences but not consistent over depth or latitude gradients, large-scale community differences related more to sediment distributions and hydrographic variability	Penaeid shrimp, portunid crabs, sciaenid fishes, flatfishes	Exposed salt dome banks of carbonate sediments on outer shelf most with low diversity epibenthic communities, most with low environmental priority ranking
Southern California	Seasonal recruitment shown by aggregations of *Amphiodia* in summer on mainland shelf, no apparent seasonality in basins; large-scale differences related to complex topography	Sea basses, sciaenid fishes, sea urchins and starfish, cancroid crabs, pandalid shrimp	Scattered submerged ridges and topographic highs in complex borderland, rocky banks at shelf edges, some in shallower water with purple coral (*Allopora*), others with rich epibiota; kelp forests, Tanner and Cortez Banks
Central and Northern California		Sea basses, rock fish, cancroid crabs, flatfishes	Scattered hard substrates on shelf and at shelf edge, some *Allopora* in Central California, kelp forests in Central California
Washington-Oregon	Increases in density, diversity and standing stock with depth	Cancroid crabs, flatfishes, rock fish	
Gulf of Alaska	In NE, east/west differences, increase in diversity at shelf break, differences in community structure related to sediment differences and sedimentation rate	Lithodid and cancroid crabs, pandalid shrimp, flatfishes, sea urchins and starfish	Area of bedrock extending from Kayak Island to 200 m, gravel areas at shelf break with communities similar to those of hard substrate areas
Bering Sea	No seasonal or annual standing stock differences over 5 years, "relatively stable"; most nearshore/offshore differences or large-scale differences related to sediment distributions, latitudinal differences related to sea ice dynamics	Lithodid crabs, pandalid shrimp, flatfishes, codfish, starfish, marine mammals	Yukon River delta area
Alaskan Arctic	In Chukchi, no seasonal or annual standing stock differences over 5 years, "relatively stable"; in Beaufort, longitudinal variability related to warmer Bering-Chukchi water in W and E-W sediment differences, seaward differences related to sea ice dynamics	Whitefish, cisco, grayling, arctic char, arctic cod, flatfishes, sculpin, marine mammals	Boulder areas on inner shelf with epibiotic communities, Boulder Patch in Stefansson Sound, kelp forests, Colville River delta area

Figure 3.1. The dominant features and processes and attributes of the benthic biota are summarized and compared in Table 3.2.

ATLANTIC COAST

General Oceanography

The shelf waters of the eastern United States are influenced by the major currents of the western North Atlantic clockwise gyre, particularly the Florida Current and the Gulf Stream (Emery and Uchupi, 1972). Inshore of the Gulf Stream, the cold Labrador Current flows toward the equator, outside of and counter to the gyre. Seasonally variable gyres develop in the Gulf of Maine and on Georges Bank. North of Cape Hatteras, cold water generally flows southwestward on the shelf and is eventually entrained in the slope water and Gulf Stream. This cooler and freshened shelf water seldom penetrates around Cape Hatteras; consequently, there is a sharp temperature discontinuity at Cape Hatteras. Shelf water in the South Atlantic Bight flows northeasterly north of Cape Fear, but southerly south of Cape Fear except during the spring (Ingham, 1982). The South Atlantic Bight is subject to periodic intrusions of Gulf Stream water onto the shelf (Allen *et al.*, 1983).

The tides along the eastern U.S. are semidiurnal (Emery and Uchupi, 1972) with a range less than 0.5 m in the open ocean increasing across the continental shelf to a range of 2-3+ m at the coast. Resonance permitted by geographic features along the northeastern continental margin increases the tides to 14.5 m (e.g., the head of the Bay of Fundy). As the high tide crest moves shoreward more or less simultaneously along the entire Atlantic coast, it is slowed markedly as it crosses wide, shallow areas such as Georges Bank and the continental shelves off the southeastern U.S. and the Gulf of Mexico. The highest tidal current velocities (more than 55 cm/s) occur on Georges Bank and Nantucket Shoals (Emery and Uchupi, 1972). A smaller area of high velocity is found at the mouth of Delaware Bay and a few other bay mouths. Otherwise, tidal current velocities are less than 28 cm/s.

Shelf physical conditions are strongly influenced by weather on short time scales. In summer, when winds are mainly from the south and southwest, winds exceed 20 km/h less than 30% of the time over virtually the entire Atlantic coast. In contrast, this velocity is exceeded over 50% of the time during the winter, and off New England winds exceed 63 km/h more than 20% of the time. Wind speed influences wave height; consequently, in the winter off New England waves exceed 1.5 m more than 50% of the time and 3.5 m 10% of the time (Emery and Uchupi, 1972). Wave height at the shoreface is greatest in the Middle Atlantic Bight because Georges Bank and islands provide some protection north of Cape Cod. Higher waves (12 m and possible 17 m) are generated by the high winds and low pressures of hurricanes (Emery and Uchupi, 1972). The shape of the coast, shelf topography and direction of hurricane advance affect the height of storm surges. These waves have effects not only on coastal environments, but may also transport sediments on the shelf.

New England

Physical Processes

Only a small portion of the cold and slightly dilute water from the Labrador Current enters the Gulf of Maine, mostly in spring and early summer (Emery and Uchupi, 1972). River runoff from the adjacent land and net precipitation is minor. The chief source of water entering the Gulf of Maine is underflow from the slope water, itself a mixture of shelf and Gulf Stream water. The counterclockwise flow in the Gulf of Maine and Bay of Fundy is strongest in the spring and early summer when flow into the Gulf from the Scotian shelf and stream discharges in Maine, Nova Scotia and New Brunswick are greatest (Emery and Uchupi, 1972; Ingham, 1982). During fall and winter the gyre weakens, allowing water to drift southward onto Georges Bank and into Great South Channel (Ingham, 1982).

Water moves onto Georges Bank from the Gulf of Maine in the late fall and early winter, as well as from the slope to the south, from the Scotian shelf to the northeast and from Nantucket Shoals to the west (Hopkins and Garfield, 1981; Butman *et al.*, 1982). A persistent clockwise gyre on Georges Bank is defined by a southwestward flow along the southern flank, a northward flow on the eastern side of the Great South Channel, a northeastward flow on the northern flank and an eastward, southeastward and southward flow on the northeast peak (Butman *et al.*, 1982). Current speeds are typically 5-10 cm/s, thus a water parcel at the 60-m isobath, could circuit the Bank in two months. Flow along the southern flank diverges to the southwest, with most of the flow continuing to the west into the Middle Atlantic Bight, particularly during winter, and some of the flow returns northward toward the Gulf of Maine, particularly during summer.

Current flow on Georges Bank is highly energetic and strongest near the surface. Mean current flow on the southern flank is typically 15 cm/s at 10-15 m with speeds declining with depth to generally less than 5 cm/s above the sea bed (Butman *et al.*, 1982; Allen *et al.*, 1983). Mean flow is highest at the end of the summer and lowest at the end of winter. Superimposed on these currents are strong, semidiurnal, clockwise rotary tidal currents with maximum speeds ranging from 35 cm/s on the flank to 100 cm/s on the shallow crest (Emery and Uchupi, 1972; Knebel, 1981; Allen *et al.*, 1983). The strong tidal currents cause effective vertical mixing at shallower depths (<60 m), resulting in little or no stratification in the spring and summer compared to stratified conditions over the deeper portions of the Bank.

The open New England shelf region is subjected to frequent gales and north-easterly storms, particularly in the winter, creating high storm waves (Emery and Uchupi, 1972). Wave heights greater than 1 m occur more than 50% of the time on Georges Bank, and 1-yr and 100-yr maximum wave heights are projected at 11-12 m and 19 m, respectively. Waves are reduced in the Gulf of Maine (Emery and Uchupi, 1972).

Upwelling is active along the northern and northeastern edge of Georges Bank (Ingham, 1972). Warm core rings propagate southeast of Georges Bank, where Gulf Stream meanders often reach high amplitudes. Injections of slope water and modified Gulf Stream water onto southwestern Georges Bank and southern

New England have been well-documented. Rings and meanders which approach Georges Bank do so when younger, larger and stronger in rotary flow than when they later enter the Middle Atlantic shelf, and these intrusions are about 35% more frequent.

Geology

The Gulf of Maine is characterized by 21 basins with depths as great as 311 m and as much as 135 m deeper than their sills (Emery and Uchupi, 1972). These basins occupy 30% of the area of the Gulf. Between the basins, the floor of the Gulf of Maine is irregular owing, in part, to outcrops of bedrock and to concentrations of boulders. Two features branch from the Gulf of Maine—Northeast Channel, a U-shaped, broadly curving channel with a deep (230-270 m) hummocky floor, and Great South Channel, a broad, triangular southward rising reentrant into Georges Bank that nearly separates the bank from the continental shelf to the southwest.

The top of Georges Bank has two aspects—the southern half, which slopes smoothly toward the 110-m shelf break, and the northern half, which is irregular with elongate sand shoals (10 km apart and as long as 75 km) that trend northwesterly and are separated by flat-floored troughs (Emery and Uchupi, 1972; Shepard, 1973). In water depths <60 m, where tidal currents are particularly strong, large sand waves are ubiquitous (Knebel, 1981). Storm waves break across the shoal crests and create highly turbulent conditions. Secondary short-crested sand waves (from 10 to 20 m in height and 100 to 200 m apart) on the tops and sides of the shoals trend approximately east-west. Tertiary sand waves nearly parallel the secondary waves and are topped with still smaller sand ripples. The shallowest part of the top of Georges Bank undergoes constant reworking by swift tidal currents causing a shifting of all but the largest sand waves. Local variations in the thickness and composition of the sand sheet are common (Knebel, 1981).

Another large area of sand waves or ridges, closely akin to those on Georges Bank, is Nantucket Shoals (Emery and Uchupi, 1972; Shepard, 1973). The features are caused by the tidal ebb and flood of large volumes of water to and from the Gulf of Maine. The sand shoals off New England represent, in part, recessional moraines and in part sand ridges formed by littoral drift during the past rise in sea level. Since then, tidal currents have reshaped both types of ridges into their present form.

Overall, sandy sediments predominate on the New England continental shelf (Wigley, 1961; Maurer and Leathem, 1981a). Sediments below the 100-m isobath in the Gulf of Maine are reworked and poorly-sorted glacial material and range from gravels to clays. On Georges Bank, 75% of the sediments are sands. Gravelly sediments are found on the top of the bank, on the northeastern corner and in the Great South Channel. Dominant sediment types along the southern flank are medium and fine sands with some coarser sediments near the edge of the shelf. Finer silt and clay sediments occur in patches in deep water off the northwestern corner of the Bank. Sediments in the heads of canyons, which indent the southern edge of the Bank, vary from silty fine sand to fine sand and increase in silt content down the canyons. On Nantucket Shoals, sands make up greater than 95% of the sediments with very little gravel. To the south of this area in deeper water (60-200 m), is a large muddy area ("the Mud Patch") which is a site of active sediment deposition. Finer silts and clays winnowed from Georges Bank and Nantucket Shoals are thought to be deposited in this area (Twichell *et al.*, 1981; Bothner *et al.*, 1981a).

Suspended particles are low in concentration in the waters overlying Georges Bank, except after major storms (Bothner, 1981b). Most of the finer sediments have already been winnowed away. Over shallower portions of central Georges Bank, suspended particles range from 750 to 800 μg/1 and decrease to 250 μg/1 in deeper slope waters to the south where there is less influence of tides, currents, waves, storm-driven currents and internal waves.

Benthos

Several surveys of benthic populations on the New England continental shelf have been made (Table 3.2). Most concentrated on Georges Bank because of the large commercial fisheries supported by the area. Methodologies differed considerably among these studies, and different faunal groups were emphasized; therefore, little comparable or comprehensive information is available.

Maurer and Leathem's study (1981a) covered only two seasons and polychaetes but was the most geographically comprehensive. Numerical classification of polychaete collections showed five major habitats: 1) Nantucket Shoals and the greater part of Georges Bank, 2) the southern flank at depths of 40 to 100 m, including the heads of some submarine canyons, 3) the muddy sediments to the southeast and in deeper water, including the Mud Patch, 4) the northern flank and 5) the Gulf of Maine.

The Battelle/Woods Hole Oceanographic Institution study (Battelle/W.H.O.I., 1983, 1984), while more restricted geographically, was a more complete survey of the benthic infaunal community. Use of 0.3-mm sieve size improved sampling efficiency for several species of small polychaetes, especially syllids (*Exogone* and *Sphaerosyllis*), the paranoid (*Paradoneis*), the recently hatched young of the most common arthropods and thin arthropods such as *Tanaissus* and *Erichthonius*. Battelle/W.H.O.I. found that replicate samples from any particular regional station were distinct from those found at any other station, and stations grouped consistently by depth and sediment types across seasons during two years. The community at 60 m was dominated by the archiannelid *Polygordius*, a bivalve *Tellina* and the amphipods *Pseudunciola* and *Protohaustaurius*. Those at 80 m were dominated by syllid polychaetes and an oligochaete *Peosidrilus*. Amphipods were also abundant—*Unciola* and *Erichthonius* at one station and *Byblis* at another. At 100 m the community was somewhat distinct with capitellid polychaetes dominating but was similar to communities at 80 m in the abundance of *Ampelisca*, *Polygordius* and the polychaete *Protodorvillea*. Finer sediments in the Mud Patch southwest of Georges Bank were dominated by several species of polychaetes, including *Cossura*, paranoids and sabellids and an oligochaete *Tubificoides*. Communities at the shelf break and at the head of submarine canyons were dominated by paranoid, capitellid and cirratulid polychaetes and the amphipod *Ampelisca*.

Community parameters vary between studies, in part because of methodological differences. Wigley (1965, 1968) reported a faunal composition (1-mm sieve size) for Georges Bank of 66% crustaceans, 20% annelids, 3% echinoderms and 3% molluscs. Maurer and Leathem (1981a) estimated that polychaetes accounted for 54% of all infaunal species and 53% of total individuals (0.5-mm sieve size). Battelle/W.H.O.I. (1983) reported that polychaetes accounted for 39%, arthropods 20% and molluscs 17% of all taxa identified (0.3-mm sieve size). Reported densities range from 432 to 20,553 polychaetes/m^2 (Michael, 1977; Maurer and Leathem, 1981a) to 2500 to 55,000 individuals/m^2 on Georges Bank (Battelle/W.H.O.I., 1983). There is a general trend of increasing density with depth on the shelf (Maurer and Leathem, 1981a; Battelle/W.H.O.I., 1983).

Wet weight biomass of polychaetes for Nantucket Shoals and Georges Bank averaged 19.5 g/m^2 for winter-spring samples (Maurer and Leathem, 1981a). Polychaete biomass on Georges Bank ranged from 3-24.5 g/m^2 but mostly was less than 10 g/m^2 (Battelle/W.H.O.I., 1983). Maurer and Leathem reported a decrease in biomass with depth from Georges Bank proper to the southern flank. Just the opposite was found by Battelle/W.H.O.I. (1983) where highest biomass of polychaetes was at the shelf break and in the Mud Patch.

Middle Atlantic Bight

Physical Processes

Water leaving the Gulf of Maine and escaping the Georges Bank gyre rounds Cape Cod in a southwestward flow along the shelf of the Middle Atlantic Bight, which lies between Cape Cod and Nantucket Shoals to the northeast and Cape Hatteras to the south (Beardsley *et al.*, 1976; Beardsley and Boicourt, 1981; Allen *et al.*, 1983). This down-shelf mean flow is persistent on the outer shelf, intensifying in the winter, but current reversals on the inner shelf may be forced by winds, particularly in the winter. Nearshore, low salinity estuarine outflows form persistent southward flowing coastal jets 20 km wide or less. The southward-flowing shelf waters, diluted by river runoff, eventually become entrained in the Gulf Stream off Cape Hatteras except when strong northeast winds force these waters around Cape Hatteras and into Raleigh Bay (Ingham, 1982). Mean currents flow approximately parallel to the isobaths with speeds of 5 to 10 cm/s at the surface and 2 cm/s or less above the seabed (Ingham, 1982). Low frequency events, particularly winter storms, can cause much more energetic variations in the flow field (of the order of 50 cm/s). Tidal currents in the Middle Atlantic Bight are relatively weak, except over Nantucket Shoals and off some embayments, with a maximum amplitude of 10 to 15 cm/s oriented primarily in a cross-shelf direction. Thus, where transport processes on Georges Bank are tidally-dominated, the Middle Atlantic tends to be storm-dominated.

The Middle Atlantic Bight receives considerable freshwater runoff averaging 157 km^3/yr, nearly half of which occurs during the spring. This reduces the surface salinities on the inner shelf to less than 32‰. The Bight undergoes a marked seasonal change in stratification as a result of freshwater runoff, vernal warming and wind mixing (Ingham, 1982; Allen *et al.*, 1983). Stratification occurs in spring and summer, with a warmer surface mixed layer sharply separated from cooler bottom water, called the "cold pool," on the mid and outer shelf. This feature can be traced from the southern flank of Georges Bank to Cape Hatteras and serves to extend the distribution of boreal benthic fauna along the Middle Atlantic shelf to Cape Hatteras (Bowen *et al.*, 1979). Deepening of the surface mixed layer begins in late summer with atmospheric cooling and strong wind mixing, and by November thermal stratification is completely broken down.

The shelf break is bathed by slope water seaward of the shelf-slope front. The front is highly dynamic, changing location and form in response to meteorological forcing, gravitational flow and mesoscale ocean dynamics (Mooers *et al.*, 1979). The intersection of the front and the seabed tends not to move from the 100-m isobath; consequently, the bottom boundary along the shelf break and slope is characterized by constant temperatures which gradually decline with depth.

Upwelling and interactions of shelf water with warm core rings resulting from Gulf Stream meanders are features of shelf waters in the Middle Atlantic Bight (Ingham, 1982). During the periods of prevailing southwesterly winds, mostly in the summer months, upwelling can occur along the New Jersey-Virginia coast. South of Long Island, Walsh *et al.* (1978) have shown indirect evidence for outer shelf upwelling and episodic mixing with the passage of storms. Warm core rings of the Middle Atlantic shelf are less frequent than in the vicinity of Georges Bank; older, smaller and weaker in rotary flow; and also remain close to the continental slope.

Geology

The continental shelf of the Middle Atlantic Bight is a broad, gently sloping platform varying in width from 160 km south of Cape Cod to 140 km off New Jersey and narrows to 25 km off Cape Hatteras (Emery and Uchupi, 1972). The depth of the shelf break decreases from about 150 m south of Georges Bank to about 50 m off Cape Hatteras (Allen *et al.*, 1983).

The major features of the shelf are partially submerged end moraines of Wisconsin Age, terraces which mark sea level stands during or after the Pleistocene, submerged former stream channels, and the shelf edge incisions of numerous submarine canyons, the largest of which is Hudson Canyon (Emery and Uchupi, 1972). Associated with the submerged shores are several submerged former stream channels. The longest is the Hudson Shelf Valley which extends from off New York Bay to the outer part of the shelf as a linear feature that has been partly filled with marine sediments so that the axis is a series of elongate depressions with maximum depth of 88 m below sea level and 56 m below the adjacent shelf surface. The channel does not connect directly with the head of Hudson Canyon that indents the shelf break, but is separated by a delta or apron built of stream-contributed sediments at a period of lower sea level stand. Other more completely filled channels extend from Block Island, Delaware Bay, Chesapeake Bay, Great Egg River (south of Atlantic City) and Vineyard Sound.

The continental shelf is covered by a surficial sand sheet 0 to 30 m thick which may be locally eroded, exposing finer, semi-consolidated sediments or a lag of clay lumps and oyster shells (Swift *et al.*, 1972; Knebel and Spiker, 1977). Over most of the shelf, surface sediments are sands (>75% and mostly >90%) or gravelly sands to water depths of at least 200 m. Because of the rapid Holocene transgression and limited input of modern detrital sediment escaping coastal plain estuaries, shelf sediments contain few silts and clays. On the upper continental slope, shelf sands grade abruptly into clayey silts.

Superimposed on the relict, large scale physiographic features are smaller scale features of a variety of sizes. Linear sand ridges and intervening swales cover most of the Middle Atlantic shelf (Duane *et al.*, 1972; Swift *et al.*, 1972). Ridges trend roughly northeast to southwest and have a mean crest-to-crest spacing of 1.4 km and a mean relief of 4.7 m on the inner shelf off New Jersey. On the central shelf these respective dimensions are 2.5 km and 6 m, and on the outer shelf 6.1 km and 6 m. These large fields of ridges and swales are thought to have originated at the shoreface and have been stranded by transgression. Sand waves (Knebel and Folger, 1976), current lineations (McKinney *et al.*, 1974) and wave and current ripple patterns may be superimposed on the mesoscale ridge and swale topography. Contemporary hydrodynamic processes interact with the seabed features redistributing surficial sediments. Typically, moderately-sorted medium sands are found on ridge crests, better-sorted medium-fine sands are found on the ridge flanks and fine sands in the swales, except where the swale has eroded down to the earlier Holocene or Pleistocene strata, leaving an erosional coarse lag (Stubblefield *et al.*, 1975).

Most of the fine-grained river sediments that reach the Atlantic continental shelf are transported back into the estuaries and coastal wetlands (Meade, 1972) or bypass the shelf to be deposited on the slope. Swift *et al.* (1976), Butman *et al.* (1979) and Vincent *et al.* (1981) documented the importance of storm events in causing large sediment transports with sustained southwestward near-bottom currents of greater than 50 cm/s for about 12 h. These short, efficient, storm-related, large scale transports are separated by longer periods of quiescent, minimal transport. Transport is westward off Long Island and southward off New Jersey (Ingham, 1982). Internal waves may resuspend or transport bottom sediments on the Middle Atlantic outer shelf during summer (Knebel, 1981). Internal waves are generated at the shelf break by the diurnal or semidiurnal tide and move across the area in packets. The main effect of these processes is to prevent the accumulation of fine-grained sediments at the shelf break.

Studies in the Hudson Shelf Valley and Canyon system (Keller *et al.*, 1973; Nelsen *et al.*, 1978) indicated a long-term down canyon (seaward) transport of fines to the continental rise. A large depositional area is located within the apex of the New York Bight, the Christiansen Basin (Swift *et al.*, 1976; Freeland *et al.*, 1976). This topographic low area contains higher levels of silt and clay than surrounding sand areas and may act as a temporary deposition center trapping fine sediments which are only periodically resuspended by major storms. Smaller topographic depressions on the shelf also accumulate muddy fine sands winnowed from the surrounding sea bed (Stubblefield *et al.*, 1975).

Benthos

On the Middle Atlantic shelf, polychaetes dominate the macrobenthos numerically, accounting for 40 to 60% of the individuals and occasionally up to 90% (Boesch, 1979). Pericaridean crustaceans, particularly amphipods, are also important constituting 10 to 30% of individuals, and in some topographic depressions more than 70%. Densities of macrobenthos are highest in topographic depressions (6800 to 14,000 individuals/m^2 on the outer shelf and 5000 to 8200 individuals/m^2 on the inner shelf) (Boesch, 1979). Remaining outer shelf and shelf break habitats support greater densities (3800 and 3600 individuals/m^2, respectively) than inner and mid shelf environments (2900 and 2500 individuals/m^2, respectively).

Distribution patterns of macrobenthos across the continental shelf and upper slope were described by Boesch and Bowen (in press). Although distribution patterns were generally continuous with depth, sharper biotic changes allowed division of the community gradient into inner shelf (~20 to 30 m), mid shelf (30 to 50 m), outer shelf (50 to 100 m), shelfbreak (50 to 100 m) and slope (>200 m) habitats. The sharpest changes were evident near the shelf break. The dynamic sandy bottoms of the inner shelf were dominated by small interstitial feeders, such as the tanaidacean *Tanaissus*, the polychaetes *Polygordius*, *Goniadella* and *Lumbrinerides* and burrowing deposit feeders. On the central shelf there were fewer interstitial feeders and fossorial amphipods were abundant because of the somewhat finer sands. Dominant taxa here were the amphipods *Pseudunciola*, *Byblis*, *Rhepoxinius* and *Protohaustorius*, the tanaidacean *Tanaissus* and the polychaetes *Spiophanes*, *Goniadella* and *Lumbrinerides*. The outer shelf habitats were dominated by tubicolous amphipods—six of the top ten species were amphipods and five of these were surface deposit-feeding tube dwellers, *Unciola*, *Ampelisca*, *Byblis* and *Erichthonius*. Assemblages in outer shelf topographic depressions were even more heavily dominated by pericaridean crustaceans. The shelf break fauna was highly diagnostic—the polychaetes *Lumbrineris latreilli*, *Kinbergonuphis*, *Mooreonuphis*, *Aricidea neosuecica* and *Spiophanes wigleyi*, the bivalve *Thyasira*, the ostracod *Harbansus*, the amphipods *Ampelisca* and *Unciola* and the ophiuroid *Amphioplus*. Complex factors related to depth are important in controlling faunal assemblages across the Middle Atlantic shelf. In particular, temperature and temperature variability, the frequency and magnitude of bottom sediment disturbance, and the deposition of fine sediments are important.

Complex mesoscale topography (100 to 1000 m horizontally and 10 m vertically) create differences in sediment characteristics even though sediments were all predominantly sand with little gravel, silt or clay (Boesch, unpubl.). Consequently, variations in benthic community structure are strongly correlated to variations in the percentage of coarse and fine sand. Shallow terraces and linear sand ridges with medium to coarse sands are inhabited mostly by small interstitial feeding polychaetes. Many tubicolous species and subsurface deposit feeders are excluded from these coarse sediments which are frequently transported by storm-generated currents. In the topographic depressions the prevalent fine sands excluded interstitial feeders in favor of tubicolous, surface deposit-feeding amphipods. Sediments in the depressions are more stable and support richer, denser and more trophically diverse communities.

The fauna of the Middle Atlantic shelf is transitional between cold water assemblages to the north (boreal) and warm, temperate assemblages to the south (Carolinian) with the faunal affinities varying with depth. The shallow water and estuarine benthic fauna has more southern affinities, and the central and outer shelf are part of a boreal continuum without a faunal barrier at Cape Cod (Bowen *et al.*, 1979; Boesch and Bowen, in press). Cape Hatteras, the southern boundary of the Middle Atlantic Bight, and Cape Lookout, farther to the south, provide sharp discontinuities to tropical and subtropical fauna (Cérame-Vivas and Gray, 1966; Herbst *et al.*, 1979).

South Atlantic Bight

Physical Processes

The circulation and hydrographic conditions of the South Atlantic Bight are influenced by the Gulf Stream which follows the shelf edge from the Straits of Florida to Cape Hatteras. Meanders and spin-off eddies, often induced by topographic irregularities at the shelf edge, result in considerable exchange of water with the outer shelf (Allen *et al.*, 1983). These result in the intrusion of deeper, cooler and more nutrient-rich waters onto the shelf, particularly off northern Florida and in the Carolina embayments (Blanton *et al.*, 1981). On the middle shelf (20-40 m), the predominant forcing is from the wind rather than the Gulf Stream. Mean flows in the northern half of the South Atlantic Bight are northward but are complicated by counterclockwise eddies within coastal embayments (Emery and Uchupi, 1972). In the southern half, flow is northward during the spring but southward or variable during the rest of the year. Occasionally, cooler and fresher Middle Atlantic Bight water is forced around Cape Hatteras by winter gales, but usually the exchange of shelf water around the cape is small. The Florida shelf is narrow, and currents flow northward over much of the shelf as a fringe of the Florida Current (Gulf Stream).

A more or less distinct water mass of reduced salinity resides on the inner shelf of the South Atlantic Bight, essentially throughout the year off Georgia and South Carolina. This coastal water is freshened by the many rivers of the region which discharge 3 to 8 km^3/month (Blanton, 1980; Atkinson *et al.*, 1983). The zone of reduced salinity extends 10 to 15 km offshore and is bounded by a sharp front, especially in seasons when the winds blow southward. During spring, northward wind stress spreads the coastal water offshore (Allen *et al.*, 1983). In the northern and southern parts of the Bight, runoff is slight and nearshore salinities are high.

Seasonal temperature changes are buffered by the shelf water mass, consequently bottom water temperatures on the inner shelf range more widely (12 to 28°C) than on the outer shelf (16 to 26°C) or at the shelf break (15 to 18°C) (Emery and Uchupi, 1972; Atkinson *et al.*, 1979; Wenner *et al.*, 1983). Nutrients are supplied to the shelf primarily by intrusions of the Gulf Stream, which stimulate subsurface phytoplankton blooms over a two- to four-day pulse period (Atkinson *et al.*, 1978; Pomeroy *et al.*, 1983). The discrete coastal water mass has the effect of retaining nutrients contributed from runoff and coastal tidal exchange on the inner shelf most of the time (Pomeroy *et al.*, 1983).

Geology

At Cape Hatteras, the continental shelf narrows to 30 km, but to the south it is broad, exceeding 100 km in width near the center off Georgia, with a narrowing to 80 km off Cape Fear (Day *et al.*, 1971; Shepard, 1973; Allen *et al.*, 1983). Most of the shelf is relatively shallow, 50 to 55 m out to the shelf break, until the sudden increase in depth to the continental slope which starts between 80 and 160 m. The shelf narrows south of Jacksonville, Florida until it virtually disappears at Palm Beach (Shepard, 1973).

The most prominent topographic features of the South Atlantic shelf are long, sinuous shoals that reach almost across the continental shelf from the major promotories (Emery and Uchupi, 1972). Large shoal areas extend from Cape Hatteras, Cape Lookout and Cape Fear; smaller ones, from Cape Romain and Cape Canaveral. Semicircular embayments are present between the four northern coastal projections. Sand waves also are prominent features along the entire length of the shelf (Emery and Uchupi, 1972; Shepard, 1973). Those between Cape Fear and Cape Romain have a fan-shaped pattern that radiates from a point on the outer part of the shelf. Between Cape Romain and Cape Canaveral, the sand waves radiate from a landward point.

The South Atlantic shelf is neither traversed by shelf channels nor incised by the heads of submarine canyons (Emery and Uchupi, 1972). The broad, flat region between 32 and 36 m depth off Cape Romain may be a submerged delta formed by the Santee River during lower sea level. As many as two to seven terraces, or submerged shores, occur across the continental shelf. Some are as deep as 120 m and cut into the slope well below the shelf break. South of Miami, the shelf break is abrupt and is bordered by calcareous ridges of ancient lithothamnion algal aggregations which support tropical epifauna (Menzies *et al.*, 1966). The reef extends more or less unbroken from Cape Hatteras to Miami and supports scleratinian coral colonies and associated fauna off the central eastern Florida shelf (Avent *et al.*, 1977).

Additional reef areas are associated with the terraces or submerged shores that cross the continental shelf (Wenner *et al.*, 1983; Peckol and Searles, 1983, 1984). In the nearshore environment at depths of 5 to 15 m there is a patchy occurrence of ledges, rock outcrops and submerged reefs overgrown with calcareous organisms (Pearse and Williams, 1951). Reef-forming corals occur in the outcrops in the mid-shelf areas of Onslow Bay (Macintyre, 1970) together with a rich community of sessile invertebrates.

Several small topographic depressions are present in the form of spring holes, for example one 5 km off Crescent Beach, Florida, which is 30 m in diameter and has a maximum depth of 42 m below the general shelf level of 17 m (Emery and Uchupi, 1972). Water at a temperature of 22°C rises to the ocean surface carrying with it much sediment and forming a surface boil.

The Florida Keys serve as a transition between the physiographic provinces in the Atlantic Ocean and those in the Gulf of Mexico (Emery and Uchupi, 1972).

The seaward side of the keys are considered part of the South Atlantic shelf; the landward side, part of the West Florida shelf. The Keys are outcrops of the Pleistocene Key Largo limestone and the Miami oolite. The shelf on their seaward side averages about 7 km to the shelf break and is only about 10 m deep. The surface of the shelf is quite irregular with abundant coral reefs that are most concentrated at the shelf break.

Fine shelly sand occupies more than 90% of the South Atlantic shelf's surface area (Day *et al.*, 1971). The silt and clay content generally ranges from 0.1 to 10% across most of the shelf with higher values very near shore (Windom and Betzer, 1979). In an area off Cape Lookout, Day *et al.* (1971) observed ripple marks on the sand bottom down to 20 m indicative of frequent wave-induced sediment movement. They referred to this area as the "turbulent zone" and noted differences in the faunal community between 20 and 40 m as a function of the influence of waves. Their 40 m station had substantially finer sands, however, and this may have exaggerated the effect of diminished turbulence (Boesch and Bowen, in press). The sediment of the outer shelf off North Carolina is fine to medium sand with 1.2 to 2.1% silt, (Day *et al.*, 1971). The substrate on the upper slope grades to a fine muddy sand (1.5 to 4.2% silt) mixed with pteropod shells.

The low depositional environments are similar to those of the Middle Atlantic Bight. The continental shelf is largely characterized by extensive, smooth expanses of sand. Along the capes are localized areas where fines appear to be migrating seaward. Turbid plumes extend across the shelf in these areas (Buss and Rodolfo, 1972). Several lines of evidence similar to those discussed for the Middle Atlantic Bight indicate beach and estuarine sands from the southeastern U.S. coast are derived in part from the adjacent continental shelf (Pilkey and Field, 1972). There are gradations to slightly finer sands with depth and localized sediment variations. Silts and clays comprise less than 5% of the sediment and are present only near the coast and over the upper continental slope (Buss and Rodolfo, 1972). Cape Hatteras marks a sedimentary boundary separating northern carbonate-poor sediments with higher silt and clay contents from well-sorted, silt-poor carbonate sands to the south (Buss and Rodolfo, 1972).

Benthos

Few studies have focused on the soft substratum benthic infaunal communities of the South Atlantic Bight. Day *et al.* (1971) studied infauna (≥1 mm) along a cross-shelf transect off Cape Lookout (Beaufort, North Carolina); Frankenberg and Leiper (1977), infauna (≥1 mm) on the inner shelf off Georgia; and Tenore (1979), infauna (≥0.5 mm) in a large area of the shelf from Cape Fear to north of Daytona Beach.

Polychaetes dominate the infauna—about 40% in Day *et al.* (1971) and greater than 50% of the density and biomass in most samples in Tenore (1979). Tenore (1979) found that biomass was variable on the inner shelf, relatively high in the "large middle region" and low on the upper slope. Densities in the 20-200 m depth range were 1500 to 23,600/m^2, but most were between 3500 and 8500/m^2. Low mean density and biomass were characteristic of the macrofaunal community throughout the area.

Day *et al.* (1971) found a faunal continuum across the shelf, but with major divisions obvious between 20 and 40 m and between 125 and 160 m. These divisions marked the "turbulent zone," the outer shelf and the upper continental slope. The turbulent zone traversed stations of 3 m to 40 m with characteristic species included in the polychaete genera *Paleanotus*, *Lumbrineris*, *Magelona* and *Macroclymene*, the archiannelid *Polygordius*, the amphipods *Platyischnopus* and *Maera* and the echinoderm *Mellita*. The inner shelf benthic community studied by Frankenberg and Leiper (1977) was characterized by both temporal and spatial variability with variations in density of more than four orders of magnitude temporally and of three orders of magnitude only 5.5 km apart. Peaks in densities of dominant species varied through the year with some being most numerous in January through April and others dominating through the summer. Dominant species in these inner shelf communities (Frankenberg and Leiper, 1977) were the polychaetes *Spiophanes*, *Glycera* and *Magelona*, the cumacean *Oxyurostylis*, the bivalves *Tellina*, *Ensis* and *Solen*, the ophiuruid *Hemipholis*, the cephalochordate *Branchiostoma* and the amphipods *Paraphoxus* and *Acanthohaustorius*.

The division between the turbulent zone (≤20 m) and the more quiescent outer shelf (>40 m) was attributed by Day *et al.* (1971) to the reduced wave energy felt at the greater depths. As noted earlier, however, finer sediments were found at the 40 m station. On the outer shelf (40-124 m), there was generally reduced abundance of the characteristic fauna, including the polychaetes *Notomastus*, *Ampharete*, *Amphicteis* and *Chone*, the amphipod *Siphonoecetes*, the brachiopod *Glottidia* and the sipunculid *Aspidosiphon*. In the depths of 160-205 m ("upper continental slope" according to Day *et al.*, 1971), characteristic species were the polychaetes *Scolaricia*, *Notomastus*, *Lumbrineris* and *Chaetozone*, the amphipods *Paraphoxus*, *Siphonoecetes* and *Unciola*, scaphopods and the bivalves *Ledella* and *Thyasira*.

Tenore (1979) found no clearcut dominance of one or several species, either throughout or in geographic portions of the shelf. *Spiophanes* and *Unciola* were the only species composing more than 5% relative abundance of the fauna. Sporadically at a few stations, there were high densities of particular species (e.g., in spring, *Spiophanes* constituted up to 22% of the fauna at 5 stations). Most of the species could be considered rare; only 12 species (all polychaetes) constituted more than 0.2% of the mean total density in all seasonal samples: *Spiophanes*, *Spio*, *Prionospio*, *Parapionosyllis*, *Exogone*, *Typosyllis*, *Sphaerosyllis*, *Synelmis*, *Protodorvillea*, *Paleanotus* and *Goniadides*. Latitudinal differences were not seen in the benthic communities (Tenore, 1979). Variation in benthic community composition was related to factors associated with the depth gradient, i.e., temperature and temperature variability, freshwater plumes, changes in sediment particle sizes and the decreasing effects of wind-forced hydraulic factors, including hurricanes, with depth (Day *et al.*, 1971; Tenore, 1979).

The fauna of the South Atlantic Bight is considered part of the warm temperate Carolinian province. In deeper waters, however, there are numerous southern species of the Caribbean zone, which indicates the influence of the Gulf Stream near the northern extreme of the Carolinian zone (Williams *et al.*, 1968). Cape Lookout also marks a zone of zoogeographic change, being characterized by thermal barriers, as is Cape Hatteras, but also by the presence of substrates which support a more diverse warm water fauna (Herbst *et al.*, 1979).

GULF OF MEXICO

General Oceanography

Large scale water circulation in the Gulf of Mexico is influenced by the Loop Current and associated eddies, the semipermanent gyre in the western Gulf, winds, freshwater input and the density structure of the water column (Huh *et al.*, 1981; Sturges and Horton, 1981; Sturges and Evans, 1983). Water enters the Gulf of Mexico through Yucatan Strait and forms the Loop Current. Part of the current bends to the right, flows through the Straits of Florida and joins the Florida Current. Some of the water flows farther north into the Gulf and then veers to the right to form a clockwise gyre which is bounded by two or more smaller counterclockwise gyres off West Florida. The remaining water turns left after traversing most of the width of the Gulf and contributes to a complex series of anticyclonic warm eddies which travel west across the Gulf in a process of decay that typically lasts 4 to 10 months. The Loop Current has an annual cycle of growth and decay, but the variability in patterns from year to year is significant.

Gulf of Mexico tides are of reduced amplitude compared to those of the eastern U.S. (Murray, 1972; Emery and Uchupi, 1972) and range from 0.3 to 1.2 m. The tide is delayed many hours in the Gulf of Mexico compared to the Atlantic coast as it is slowed across wide, shallow areas. The Gulf tides are predominantly diurnal but major variations create mixed or semidiurnal tides along certain shores. Tidal currents are typically much slower in the Gulf of Mexico (<28 cm/s) than the open continental shelf of the Atlantic Ocean (Emery and Uchupi, 1972), especially the more northerly areas. Around inlets, keys, or barrier islands, however, they may frequently reach a velocity of 150 cm/s.

The Azores-Bermuda atmospheric high pressure cell dominates wind circulation over the Gulf, particularly during the spring and summer months (Brower *et al.*, 1972). During the relatively constant summer conditions, winds are predominantly southeasterly but are more southerly in the northern Gulf. In October there is a generally easterly flow throughout the Gulf. Winter winds usually blow from easterly directions with fewer southerlies but more northerlies. Winds in the summer season fall mostly between 2 to 5 km/h, but the winter winds dominate over a wider range of 2 to 12 km/h.

Concurrently, wave heights during both summer and winter are predominantly in the 0 to 3 m range but there is a shift in dominance towards larger wave heights during the winter season. Winter storm systems frequently cause moderately high winds (28 to 37 km/h) and waves that mask local tides. These conditions are occasionally harsh (>89 km/h), yet the most extreme conditions are associated with tropical storms. The largest and most destructive storms affecting the Gulf of Mexico and adjacent coastal zones are tropical cyclones which have their origin (during mid-season of June through October) over the warm, tropical waters of the central Atlantic Ocean, Caribbean Sea or southeastern Gulf of Mexico. There is a high probability that tropical cyclones will travel through the Gulf each year.

West Florida Shelf

Physical Processes

The major hydrographic influence on the West Florida shelf is the Loop Current (Huh *et al.*, 1981; Sturges and Horton, 1981; Sturges and Evans, 1983). There is little effect on the shelf in fall, and eddies rarely appear in the winter. In March a warm eddy forms on the shelf south and east of Cape San Blas. In mid to late spring, the Loop Current impinges onto the shelf break and outer shelf along nearly the entire length of the West Florida shelf. According to Chew (1953, 1955) and Hela (1956), a permanent cyclonic eddy exists on the Southwest Florida shelf which is driven by the Loop Current. The Loop Current is known to reach speeds of greater than 200 cm/s. For currents on the West Florida shelf, Mooers and Price (1975) and Niiler (1976) have found extreme velocities associated with storms, 100 cm/s, but flow was typically 20 cm/s in approximately 30 m on the Florida Middle Grounds (Hopkins and Schroeder, 1981).

River discharge along the west Florida shelf is low compared to the North Central Gulf area (State Univ. System of Florida, 1977). In the area of Cape San Blas to Tampa, the inflow is 1258 m^3/s with most of the input from Apalachicola Bay and Suwanee Sound. From Tampa Bay to the Florida Keys, the river discharge is negligible (149 m^3/s).

Winter storms and atmospheric disturbances (e.g., hurricanes) may force considerable resuspension of bottom sediments (State Univ. System of Florida, 1977; Fanning *et al.*, 1982). Atmospheric cold fronts beginning in October and November induce mixing by winds and water surface cooling. In January and February (1976), the shelf waters were turbid over long periods reflecting the repeated suspension of the fine fractions of bottom sediments as a result of winter storms, and the waters were vertically well mixed. Beginning in April-May, a restratification of the water column begins with a decrease in turbulent forces and a gradual warming of surface waters. This is usually coupled with impingement of warmer oceanic waters at the offshore stations. In September-October 1975, on transects along the West Florida shelf from Charlotte Harbor to south of Cape San Blas, there was strong vertical stratification and the presence of bottom nepheloid layers, especially evident at some stations after a hurricane (September, 1975) which caused strongly developed turbid layers. In another series of samples on the Southwest Florida shelf in October (Woodward-Clyde Consultants and Continental Shelf Assoc., Inc., 1983), localized turbidity fronts from 150 to 2700 m long were present in water depths of 53 to 75 m and in July were present in 95 to 120 m. Along the West Florida shelf overall transmissometry is reduced markedly towards the Cape San Blas area.

Geology

The West Florida shelf is a broad platform with slopes generally 0.02 to

0.04° on the inner and mid shelf and 0.2° on the outer shelf (Woodward-Clyde Consultants and Continental Shelf Assoc., Inc., 1983). The Florida Bay area of the shelf is a broad, shallow (<3 m deep) area of mud bottom that is being encroached by a mangrove shore (Emery and Uchupi, 1972). Wave energy is low, and the floor of the bay contains many mud banks. North of Florida Bay, depths are greater and wave energy is higher closer to shore; thus, fine-grained sands form small waves that prograde into mud-filled bays. Submerged shores indicated by a series of ridges oriented diagonal to the isobaths are found at 20, 60 and 160 m (Emery and Uchupi, 1972). These features extend along the entire West Florida shelf. The depth of the shelf break is variable because the shelf does not end abruptly, but steepens gradually between 50 and 100 m (Emery and Uchupi, 1972; Shepard, 1973). The continental slope is steep and smooth to the north but has many valley-like indentations to the south (Shepard, 1973).

Several areas of potential mass movements of sediments have been identified on the West Florida shelf (State Univ. System of Florida, 1977; Minerals Management Service, 1983). Surface expression of noncontinuous but widespread porous limestone features is evident in some areas between the Florida Middle Grounds and the Florida Keys. Unstable slopes with apparent sediment slumping are present on the continental slope off of Tampa Bay. In addition, there are minor near-surface shallow faults at mid shelf off of Fort Meyers.

The West Florida shelf has a thin surface veneer of unconsolidated sediments with little active sedimentation since the Pleistocene. The majority of the West Florida shelf is characterized by a carbonate sand sheet with carbonate values generally ≥80% (Doyle and Sparks, 1980; Woodward-Clyde Consultants and Continental Shelf Assoc., Inc., 1983). This sheet extends from Cape San Blas on the middle and inner shelf to Florida Bay and to the Dry Tortugas on the outer shelf. Kaolinite dominates the clay mineralogy of the sand sheet. The innermost portion of the West Florida shelf is a relatively pure quartz sand. Very fine sands are found at the more nearshore stations while the offshore stations are more variable with coarse to medium sands. Sand ripples are present in the sand sheet during spring, but not summer on the Southwest Florida shelf (Woodward-Clyde Consultants and Continental Shelf Assoc., Inc., 1983). Most of these sand ripples occur at stations less than 32 m in depth, but some occur at 48 and 52 m. The ripple axes are oriented north-south with estimated heights of 5-10 and 10-20 cm and wave lengths of 20-64 cm. The location, frequency and seasonality of the sand ripples points to regional storms as a probable cause of formation.

The sand sheet off of Florida Bay and north of the Florida Keys contains a greater silt-clay fraction (70-79%) and sediments are sandy silt with a very fine sand fraction (Woodward-Clyde Consultants and Continental Shelf Assoc., Inc., 1983). The sediments of this area are probably derived from the modern carbonate sediments accumulating in Florida Bay although they appear to be similar to the west Florida lime mud facies present on the continental slope (Ludwick, 1964). Another area of fine sediments is centered mid shelf off of Tampa Bay (Doyle and Sparks, 1980). The offshore side of the carbonate sand sheet is characterized by limey muds with a high carbonate (>75%) content dominated by planktonic foraminifera and fine carbonate nannoplankton (Doyle and Sparks, 1980).

Part of the sand sheet is thin (<1 m) and underlain by hard substrates (Tertiary rocks) characterized by large attached epibiota such as sponges and soft corals. This bottom type occurs on approximately 40% of the sea floor shallower than 63 m between Charlotte Harbor and the Florida Keys (Woodward-Clyde Consultants and Continental Shelf Assoc., Inc., 1983) but is less common elsewhere. In other areas, the hard substrates are exposed through the sand sheet in the form of ledges or exposed, low-relief rocks. The prevalence of these habitats is low on the Southwest Florida shelf (Woodward-Clyde Consultants and Continental Shelf Assoc., Inc., 1983). The Florida Middle Grounds northwest of Tampa Bay is a larger area of exposed hard substrate. The Middle Grounds support reef-building corals and both a Caribbean eurythermal complex and a Caribbean restricted species complex of algae, invertebrates and fishes (Rezak and Bright, 1981). Patches of coralline algal nodule layers over sand and algal nodule pavement with dead *Agaricia* accumulations over a sand bottom are characteristic of a few areas at the 100-m contour of the West Florida shelf from Fort Meyers to the Dry Tortugas (Woodward-Clyde Consultants and Continental Shelf Assoc., Inc., 1983). These substrates are of low relief (Woodward-Clyde Consultants and Continental Shelf Assoc., Inc., 1983) and have the potential for burial by sands and for movement by bottom currents.

Benthos

The soft bottom communities of the West Florida shelf were studied as part of the Southwest Florida Ecosystems Study (Woodward-Clyde Consultants and Continental Shelf Assoc., Inc., 1983) and as part of the MAFLA program which extended to the shelf off Mississippi and Alabama (State Univ. System of Florida, 1977; Dames & Moore, 1979). There was some overlap in these two studies in the Charlotte Harbor/Fort Meyers area. Polychaetes dominated the taxa representing 47-51% of the total nominal species (Woodward-Clyde Consultants and Continental Shelf Assoc., Inc., 1983). Crustaceans were 28-29% of the total species and molluscs, 10-17%. Polychaetes also dominated by abundance of individuals. Faunal density ranged from 2000-8000 individuals/m^2 in the fall and up to 11,000 individuals/m^2 in the spring (0.5-mm mesh sieve). Generally, faunal density decreased with depth although the trend was less obvious in the stations nearest the Florida Keys. Patterns of community composition varied by season (two studied), and these differences were more obvious in the more northerly transect off Naples and less obvious with the stations near the Florida Keys (Woodward-Clyde Consultants and Continental Shelf Assoc., Inc., 1983). Seasonal differences in number of individuals and number of species were also noted in the MAFLA study (Dames & Moore, 1979).

On the southwestern shelf (Woodward-Clyde Consultants and Continental Shelf Assoc., Inc., 1983), the inner shelf stations (mostly <40 m) were dominated by the polychaetes *Vermiliopsis*, *Fabricia*, *Hydroides*, *Lumbrineris*, *Goniadides*, *Pisione*, *Ehlersia* and *Prinospio*, the gammarid amphipods *Maera* and *Photis*, the cumacean *Cyclaspis* and the caprellid *Phtisica*. The inner mid-shelf stations (40-100 m) were dominated by the polychaetes *Synelmis* (dominant), *Magelona*, *Ceratocephale*, *Schistomeringos*, *Tharyx*, *Ampharete*, *Prionospio*, *Paraprionospio*, *Cossura* and *Sigambra*, the bivalve *Lucina* and the bryozoans *Selenaria*. The offshore mid-shelf stations (>100 m) were characterized by polychaetes *Glycera*, *Prionospio*, *Synelmis* and *Terebellides*. There was a distinct fauna group in the mid-shelf stations near the Florida Keys which included oligochaetes and the polychaetes *Ampharete*, paraonids, *Minuspio*, *Prionospio*, *Sigambra* and *Magelona*.

On the West Florida shelf, there are numerous exposed rock outcrops, rocky ledges, and low relief (<1 m) rock areas but these account for little areal coverage. Other more prevalent hard substrate areas are frequently covered by a thin, mobile veneer of sand. Both hard bottom types are characterized by epibenthic flora and fauna including macroalgae, such as *Halimeda* and coralline algae, sponges, soft corals and often patches of *Agaricia* (Woodward-Clyde Consultants and Continental Shelf Assoc., Inc., 1983). Other areas of soft bottom sands are covered with coralline algal nodule layers or an algal nodule pavement with *Agaricia* accumulations. By far, though, the greatest areal coverage of sea floor is by sandy sediment habitats.

The faunal affinities of the West Florida shelf (exclusive of the Middle Grounds as discussed above) are Carolinian for the shallow shelf assemblage. The deeper shelf assemblages show West Indian (or tropical) affinities. In addition to these two depth-related faunal provinces, Collard and D'Asaro (1973) outlined an additional slope assemblage.

North Central Gulf of Mexico

Physical Processes

For our considerations, the North Central Gulf of Mexico includes the area from the Mississippi River delta east to Cape San Blas. Many of the processes important in influencing shelf circulation on the West Florida shelf are similar on the North Central shelf. Differences lie in the influence of the Loop Current and in runoff from the Mississippi River. This portion of the shelf is seldom under the direct influence of the Loop Current but warm eddies often break off from it and move northward particulary in May or June (Woodward-Clyde Consultants and Continental Shelf Assoc., Inc., 1983).

The 80-km protrusion of the Mississippi delta into the Gulf alters and affects the currents, tides and wave fields in the local coastal waters (Murray, 1976). The shallow sound and shelf east of the delta, the severe bathymetric curvature of the delta itself, and the long, regular coast and shelf extending to the west create large scale topographic controls in the flow field. The Mississippi River is also the major source of river discharge. This discharge flows mostly to the shelf edge of the prodelta and westward averaging 13,528 m^3/s. The area between the Mississippi River and Cape San Blas also receives substantial river runoff (3837 m^3/s) with most contributed from the Mobile Bay (2068 m^3/s) and Mississippi Sound (943 m^3/s) systems.

Geology

The sediment discharge of the Mississippi River averages 200 to 500 million tons per year (reviewed by Milliman and Meade, 1983). During the period of stable sea level over the last 7000 years, the delta has rapidly prograded and switched several times. Most of the former delta lobes are to the west, but the Chandeleur Islands mark the end of the abandoned St. Bernard delta active 1700 to 4700 years ago. When deposition is active, as during the formation of natural levees along distributaries or in river mouth bars, the delta rises above sea level. Where deposition is less active, as within the embayments between distributaries, subsidence as a result of compaction causes the delta surface to fall below sea level. When the river flow reaches the ocean and expands laterally, most of its load of sediment is quickly deposited. Part of it progrades the distributaries about 50 m/yr, but most of it is deposited beyond the distributaries.

The fine-grained material of the large sediment mass composed of sands and clays with high water content and organic matter is deposited so rapidly that gases from the decayed organic matter becomes trapped forming a highly under-consolidated gas-saturated sediment. This results in a variety of sediment instabilities at the shelf edge (Coleman and Prior, 1983). These sediments may be moved initially by high river flood stages, storm wave action, degassing and faulting. Subsequent material movement will flow under the influence of gravity and form natural extensions of the Mississippi delta and cover much of the adjacent continental shelf areas.

Bathymetric relief features from the Mississippi delta to Cape San Blas are relict spur-like ridges and pinnacles that are common on the outer shelf and at the shelf break (State Univ. System of Florida, 1977). The most well-developed pinnacles are around the margins of the DeSoto Canyon. Unidentified structures, appearing to result from salt dome intrusion, are located in the area immediately south of Mobile Bay. Farther offshore from the Florida panhandle in 21 to 27 m are large (10 m high) sand waves (Emery and Uchupi, 1972). Bathymetric contours (Emery and Uchupi, 1972) also suggest the presence of several incompletely filled channels and associated deltas mostly at a depth of 60 m between Cape San Blas and Mobile Bay. Faults are numerous in the area between Horn Island and Pensacola from nearshore to the shelf break (Minerals Management Service, 1983). A few small faults exist on the shelf off Panama City, extending from mid shelf to nearshore. Unstable sediments exist around the upper slope in the vicinity of De Soto Canyon, particularly on the steeper western side.

The sediment regime of the North Central Gulf is more complex than that of the West Florida shelf with variations more pronounced in an east-west direction than with depth. The Mississippi River delta system forms a continental margin province which dominates the north central portion of the Gulf of Mexico. Most of the sediment of the Mississippi River is delivered directly to the shelf edge or is transported to the west due to the distribution of the major distributaries and the Coriolis force acting on the plume. Sediments on the eastern margin of the delta change from mud to a sand sheet of predominantly quartz off Alabama and northwest Florida (Doyle and Sparks, 1980). The sediments to the west are finer with a low carbonate content (<25%) and have Mississippi-type heavy mineral and clay mineral suites, the latter dominated by smectite. The relative kaolinite content increases towards Cape San Blas. Sediments with higher carbonate content (>75%) and finer particles are found at the shelf edge along the margins of De Soto Canyon.

Benthos

The North Central Gulf shelf areas were studied as part of the MAFLA program (State Univ. System of Florida, 1977; Dames & Moore, 1979). These studies treated major faunal groups separately, and no inclusive community patterns were presented. Thus, characterization of the benthos is difficult for this area. The area west of Cape San Blas was characterized by a low species diversity, abundance and biomass compared to the Florida carbonate sand sheet (State Univ. System of Florida, 1977). This was attributed to the finer-grained sediments and higher sedimentation rates. Density of polychaetes ranged from 200 to 2000 individuals/m^2 and decreased with depth off Pensacola, where sediments graded from coarse to very coarse sands nearshore to silts at the shelf break. On a transect off Mobile Bay, there was some decrease in density with depth, but density was more directly related to sediment distribution. Lower abundance of polychaetes (150 to 650 individuals/m^2) was found in silty sediments in shallow shelf areas near the Mississippi River prodelta (20 to 25 m). Seasonal changes in density were not consistent across stations. Depth was found to be the major factor influencing species affinities and dominant species assemblages of both infaunal and epifaunal taxa. Variations among the differing depth zones along the same transect were usually greater than variations between the same depth zones of different transects. Dominant genera were *Syllis*, *Sphaerosyllis*, *Websterinereis*, *Glycera*, *Lumbrineris*, *Paraprionospio*, *Prionospio* and *Mediomastus*.

Dames & Moore (1979) noted the influence of the Mississippi River on mollusc populations on transects off Mississippi and Alabama. High sedimentation, turbidity and resuspension of fine sediments resulted in small populations of molluscs, predominantly deposit-feeding bivalves. Large seasonal fluctuations as well as yearly fluctuations were apparent. Polychaete density was low at deeper stations and stations with very fine sediments (Dames & Moore, 1979). Dames & Moore (1979) reported an east-west pattern in polychaete distributions on the outer shelf. Both unstable mud bottoms off Mobile Bay and in deep water off Cape San Blas and medium to coarse foraminiferal sands along the eastern slope of De Soto Canyon supported fewer species. More stable bottoms in deep water along the western slope of the De Soto Canyon supported a highly diverse polychaete fauna. Disjunct distributions of several polychaete species and congeneric replacements occurred at the De Soto Canyon (Dames & Moore, 1979). Possible causes for these patterns are change from quartz to calcareous sediments, the impact of the canyon on circulation and river influences to the west.

The faunal affinities of the North Central Gulf shelf are mostly temperate at all depths with a diminished tropical fauna on rock outcrops in shallower depths (Lyons and Collard, 1974). Organisms with more tropical affinities occur on rocky areas in deeper parts of the shelf and near the shelf edge (Lyons and

Collard, 1974). Species richness and faunal density were lower in this region than on the West Florida shelf (State Univ. System of Florida, 1977).

Northwestern Gulf of Mexico and South Texas

Physical Processes

Knowledge of circulation on the continental shelf of the northwestern Gulf of Mexico is less well developed than most other regions of the U.S. coastal ocean. The region is not directly influenced by major ocean currents, except for the passage of anticyclonic gyres which spin off the Loop Current and travel westward along the outer shelf. Circulation on the shelf proper is more affected by wind forcing, tides and river discharges (Murray, 1972, 1976). A net westward (Louisiana) and southwesterly (Texas) flow along the shelf characterizes the predominant conditions from fall to early spring (Smith, 1980). In summer, the flow is to the west and southwest from Louisiana to about 95°W where it converges with an opposing flow to the north and northeast. A clockwise eddy is frequently found just west of the Mississippi River delta. This eddy advects part of the river's plume back toward shore where it may be entrained in a coastal boundary layer. An easterly flowing countercurrent and energetic cross-shelf currents were also observed near the shelf break by McGrail and Carnes (1983). A counterclockwise gyre has been observed off South Texas during the winter which migrates along the shelf edge to the north during spring and summer (Smith, 1980; Gallaway, 1981). The gyre may cause transient summer upwelling of cooler water onto the shelf during the summer.

The large freshwater discharges of the Mississippi and Atchafalaya Rivers influence the hydrography of the northwestern Gulf shelf. The influence is especially prominent in the reduced salinity of inner shelf waters as far west as Galveston (Nowlin, 1971). Occasionally, during the late spring this influence may extend farther offshore and down the Texas coast (Smith, 1978). Related to the density stratification influenced by late spring river discharges and the inorganic and organic nutrient inputs is the development of depressed levels of dissolved oxygen in bottom waters of the inner shelf during summer (Turner and Allen, 1982). Hypoxia in bottom waters seems to be a recurrent summer phenomenon off Louisiana (Gaston, 1985) and is known to occasionally extend at least to Freeport, Texas (Harper *et al.*, 1981a).

As a result of the broad, shallow shelf and abundance of fine sediments, nepheloid layers of resuspended sediments are common in the water column of the northwestern Gulf shelf (Brooks *et al.*, 1981; Kamykowski and Bird, 1981; McGrail and Carnes, 1983). Typically these are located in a bottom mixed layer above the bottom, but mid-depth nepheloid layers may also exist in association with density discontinuities. These probably represent turbid, near-bottom water masses which have been transported offshore and have overridden clearer water (McGrail and Carnes, 1983).

Although surface waters undergo seasonal temperature fluctuations (typically 20 to 28°C) and may have reduced salinity as a result of river discharges, bottom waters over the outer half of the shelf exceed 34‰ and are very homeothermal—

20 to 24°C from 50 to 100 m and 15 to 17°C at the shelf break (Etter and Cochrane, 1975).

Geology

The continental shelf from the Mississippi River delta to the Rio Grande is gently sloping and wide, over 200 km off of the Texas-Louisiana border (Emery and Uchupi, 1972; Shepard, 1973). There are many more physiographic irregularities in the central part of the shelf than to the east and southwest. Topographic features include many channels, most of which are associated with longitudinal ridges, and largely filled extensions of large rivers across the shelf. A detailed map of the nearshore (<20 m) South Texas shelf (Rusnak, 1960) outlines many sand ridges. Some are nearly parallel with the shore and are probably remnants of submerged barrier islands. Others, oriented at a steep angle to the shore, were interpreted by Rusnak as remnants of former distributaries of the Rio Grande. Off northern Padre Island, the nearshore seabed is essentially smooth with minor irregularities (Rusnak, 1960). The surface of the shelf farther offshore is relatively featureless (Emery and Uchupi, 1972). Smoothly projecting areas may represent submerged deltas of the Brazos, Colorado and Rio Grande Rivers. The deltas may be associated with former sea level stands at about 60 and 160 m depth, corresponding to those on the West Florida shelf.

The topography of the northwestern Gulf north of Matagorda Bay is marked by numerous protuberances which have been shown in most cases to be caused by salt or shale diapirs (Emery and Uchupi, 1972; Rezak *et al.*, 1983). The depth trends of these protuberances, 17, 60, and 85 m, may correspond to still stands of postglacial sea level. The mid-shelf banks arise from 80 m or less and have a relief of 15 to 50 m. These are all associated with salt diapirs and may outcrop as relatively bare, bedded Tertiary limestones, sandstones, claystones and siltstones. The shelf-edge carbonate banks and reefs of the northwestern Gulf are located on complex diapiric structures and have well-developed carbonate caps. Where suitable hard substrates exist in the absence of chronically turbid water, salinities are high, and water temperatures range from 18 to 30°C, conditions are favorable for the growth of tropical reef communities dominated by corals or coralline algae. The two largest of more than 130 banks that form topographic elevations are East and West Flower Garden Banks. Diapirs are also found on the continental slope creating an unusually hummocky physiography (Shepard, 1973).

The stage of sedimentary evolution (Curray, 1965) grades from allochthonous at the Mississippi River delta area to a climax grade on the South Texas continental shelf. The result is a complex of sediment regimes with a decrease in the silt/clay content in the nearshore regions to the west and south where the percentage sand increases (Minerals Management Service, 1983).

In general, the sediment sand content decreases across the shelf. There are exceptions to this, associated mostly with topographic features and the allochthonous sedimentary regime near the Mississippi River delta. A high percentage of silt and clay is found in the area nearshore south-southwest of Timbalier and Barataria Bays (Huang, 1981). The predominant clay mineral is smectite (Huang, 1981) as it is on the east side of the Mississippi River delta. This pattern suggests contemporary transport of Mississippi River clays by longshore currents along the coastal boundary. There is an area of increased sand content off Terrebonne Bay on Ship Shoal, which emerges 4 to 6 m above the surrounding silty floor and nearshore off Vermilion and Atchafalaya Bays. These sandy shoals represent the distal portions of abandoned Mississippi River delta lobes (Penland and Boyd, 1982). The sediments of the southeastern Louisiana shelf are poorly- to very-poorly sorted (Huang, 1981).

Off southwestern Louisiana, the sediment regime of the inner shelf (10 m) is soft mud, with the sand content never exceeding 48% and frequently less than 20% (Weston and Gaston, 1982). With few exceptions, the substrate is silty clay or sandy mud. To the west, the drowned Pleistocene deltaic plain of the Brazos-Colorado River and the Pleistocene Beaumont formation contribute compacted silts and clays to an area otherwise characterized by sands and muddy sands. Within the 11-m contour, the sediments are primarily firmly packed silts and clays that may be covered by a thin veneer of very fine silt, depending on preceding weather conditions (Harper and McKinney, 1980). At 17 to 20 m, the sediments are mostly very fine sand with occasional patches of clay or silt (Harper and McKinney, 1980; Anderson *et al.*, 1981). A coarser shell hash forms a subordinate fraction. The nearshore shelf off Texas is a highly dynamic sedimentary environment in which relict deposits are actively being exposed, eroded and redistributed by bottom currents (Anderson *et al.*, 1981; Harper *et al.*, 1981b).

The sediments of the South Texas shelf include a wide range of textures from muddy sands to silty clays with a decrease in abundance of sand-sized sediments seaward. Sand is transported seaward from the high energy zone of the inner shelf, and the thin, discrete sand layers in the subsurface sediments extending to at least 18 km offshore suggest this transport of sand is influenced by short-lived events (Berryhill, 1977). Silt is the predominant mud constituent of sediments in this area with clay restricted mostly to areas off Port Aransas and Matagorda Island. Off South Texas the fine sediments are thought to have been derived primarily from the Rio Grande.

On the northwestern Gulf shelf, extensive slumping of sediments occurs at the shelf break and on the upper slope. Although there is generally a low cohesiveness of sediments over the South Texas shelf, the only extensive slumping of sediments occurs on the slope (Berryhill, 1977). Small scale faulting occurs along the outer edge of the South Texas shelf (45 to 180 m) (Berryhill, 1977).

Benthos

Baker *et al.* (1981) studied the benthic macrofauna around oil and gas platforms on the Louisiana shelf in an area extending 320 km west from the Mississippi River delta in 6 to 98 m depth. Polychaetes dominated the macrobenthos (29% of nominal species and 69% of density), followed by crustaceans (15% of nominal species) and bivalves (7% of individuals). Among the top ten macroinfaunal taxa common to each collection period were the polychaetes *Paraprionospio*, *Sigambra*, *Cossura*, *Magelona*, *Nephtys*, *Lumbrineris*, *Tharyx* and *Nereis*. Community composition closely corresponded to sediment patterns and depth zonation.

Information on the benthic fauna of the inner shelf of the northwestern Gulf (10 to 20 m depth) is available from brine disposal monitoring studies (Harper and McKinney, 1980; Weston and Gaston, 1982) and the Buccaneer Oil and Gas Field study (Harper *et al.*, 1981b). Off southwestern Louisiana, the numerically dominant species were the polychaetes *Sabellides*, *Magelona*, *Paraprionospio* and *Mediomastus*, the bivalve *Mulinia* and the phoronid *Phoronis* (Weston and Gaston, 1982). Variability in the community structure was attributed to spatial variability due to the patchiness of macrobenthic organisms and temporal variability due to a period of larval recruitment. Seasonal hypoxic events, which were documented in a later study (Gaston, 1985), accounted for significant changes in the benthic community structure. Off Galveston, Texas the polychaete *Paraprionospio pinnata* was numerically dominant and its population fluctuations largely determined the total population density with the exception of large sets of the bivalves *Mulinia lateralis* and *Abra aequalis*, both in winter (Harper and McKinney, 1980). Amphipods, like the bivalves, displayed a pronounced seasonality, increasing principally in the spring, with lesser increases in the fall. The polychaetes did not exhibit well-defined seasonal fluctuations. Off Freeport, Texas, polychaetes again dominated but the community was not dominated by one or a few species (Harper *et al.*, 1981b). Average faunal abundance was 5000 to 7000 individuals/m^2 with decreases in July through January and increases through April. The benthos on the shallow Texas shelf has been occasionally affected by seasonal hypoxia, and changes in the benthic community structure reflected varying responses by taxonomic groups to recovery following differential effects of the hypoxic conditions (Harper *et al.*, 1981a).

On the South Texas shelf, polychaetes were also the dominant taxa comprising about 60% of the species with crustaceans accounting for 15% and molluscs, 12% (Flint and Rabalais, 1981). The inner shelf (15 to 30 m) is characterized by variability in hydrography and poorly-sorted sandy sediments which provide an unstable habitat in which a few species exhibit dominant abundance (low evenness). Characteristic organisms were the polychaetes *Magelona*, *Nereis*, *Mediomastus*, *Aricidea*, *Paraprionospio* and *Prionospio*, the bivalve *Tellina* and the amphipod *Ampelisca*. The macrobenthos averaged 18,000 individuals/m^2. Another area with coarse sediments, associated with the Rio Grande deltaic bulge in deeper, less variable bottom waters, supported the most diverse fauna on the shelf. Deeper habitats exhibited less bottom water variability and an increase in the silt content of the sediments. Density on the mid shelf (40 to 90 m) averaged 3300 individuals/m^2 and was characterized by the polychaetes *Paraprionospio*, *Cossura*, *Nephtys*, *Paraonis*, *Magelona*, *Asychis*, *Notomastus* and *Mediomastus* and the pericaridean crustaceans *Ampelisca*, *Apseudes* and *Eudorella*. On the outer shelf (100 to 135 m), the clay content of the sediment increased and density of macrobenthos averaged 2300 individuals/m^2, characterized by the polychaetes *Paralacydonia*, *Tharyx*, *Sternaspis*, *Paraonis* and *Sigambra*, the isopod *Xenanthura*, the bivalves *Amygdalum*, *Nuculana* and *Pitar* and the ostracod *Alternochelata*. Across the shelf there was a proportional decrease in surface feeding polychaetes and suspension-feeding amphipods and increase in subsurface deposit-feeding polychaetes. Over two years, seasonal (three periods per year) and yearly variability were not consistent over depth or latitude.

The northwestern Gulf and South Texas benthic fauna is an extension of the warm temperate Carolinian province with divisions at the Rio Grande and just east of the Mississippi River delta (Hedgpeth, 1953). The southern portion of the South Texas continental shelf is inhabited by a more tropical, Caribbean fauna. The fauna of the outer shelf of the northwestern Gulf has more tropical affinities than the warm, temperate inner shelf.

PACIFIC COAST

General Oceanography

Oceanographic conditions over the relatively narrow continental shelf of the western U.S. are heavily influenced by circulation and properties of the North Pacific Ocean (reviewed by Hickey, 1979). The dominant ocean current is the California Current, the eastern boundary current of the clockwise North Pacific gyre. Because eastern boundary currents are weaker than western boundary currents such as the Gulf Stream in the Atlantic, separation between coastal and oceanic circulation tends to be diffuse off the Pacific coast. Four water masses contribute to the California Current system: 1) sub-Arctic waters from offshore and to the north, 2) central Pacific water, 3) equatorial Pacific water entering from the south as a subsurface, counter current and 4) mid-depth oceanic waters upwelled at the shelf edge. As the California Current flows southward from Vancouver Island, its surface waters are warmed by the sun and mixing with central Pacific waters. Eventually it veers to the west to join the North Equatorial Current. The warm, saline California Undercurrent flows northwestward usually below 200 m from Baja California to north of Cape Mendocino (Reid *et al.*, 1958). This warm current occasionally flows at the surface in fall and winter to north of Point Conception, wherein it is called the Davidson Current. The Southern California Countercurrent refers to the northward flow which is found south of Point Conception inshore of the Channel Islands in the Southern California Bight. Another important feature in the large-scale regional oceanography is the Columbia river effluent which contributes 77% of the total drainage into the Pacific Ocean between San Francisco and the Strait of Juan de Fuca (Hickey, 1979). The runoff flows northward off Washington during winter and southward off Oregon during the summer.

The speed of the California Current is relatively constant, averaging 10 cm/s in summer and 30 cm/s in winter (Allen *et al.*, 1983). The current is largely wind-driven, and the position and intensity of atmospheric pressure cells determine the current speed and direction. Speeds are greatest under northerly and northwesterly winds and greatest off northern California in June and July and off central and southern California in May and June (Jones & Stokes Assoc., Inc., 1981).

Alongshore currents on the shelf north of 43°N have a strong and repeatable seasonal cycle (Allen *et al.*, 1983). The mean current is northward in winter, southward in spring and southward at the surface but northward at the bottom in summer (Huyer *et al.*, 1978). Off southern California when the water column is well-stratified (spring and summer), mean currents at the surface and bottom are opposed. In other seasons the vertical structure is quasi-barotropic, sheared in the vertical, but without reversals (Allen *et al.*, 1983).

Wind-induced Ekman circulation causes coastal upwelling of cool, nutrient-laden waters along the Oregon and California shelf, especially during spring and summer (Reid *et al.*, 1958; Wickett, 1967; Komar *et al.*, 1972). Major upwelling events, defined as lasting more than six days with surface temperature reduction of more than 3°C from the long-term mean, usually occur twice a summer off southern California (Dorman and Palmer, 1981). Weaker upwelling usually occurs in the spring. Upwelling is most intense to the south of capes and peninsulas and in association with submarine canyons. Unusually warm waters associated with periodic El Niño events may disrupt these annual upwelling patterns, reducing nutrient inputs into the euphotic zone (Dayton and Tegner, 1984).

Water temperature along the Pacific coast responds to seasonal currents, winds, insolation and upwelling. The northward flowing Davidson Current results in winter temperatures which are higher than might be expected and spring and summer upwelling may cause pools of low temperature surface water surrounded by waters warmed by insolation (Jones & Stokes Assoc., Inc., 1981). North of 34°N strong upwelling reduces the seasonal range and the cool period is lengthened; the temperature range is 2-3°C (Godshall and Williams, 1981). At 40°N the range is only 1°C with a March average of 10°C and an August average of 11°C. Between 28 and 34°N, the seasonal range is greater with a March average of 12°C and an August average of 19°C. Few comprehensive descriptions of near-bottom temperatures on the shelf exist. Godshall and Williams (1981) described the following bottom temperature regimes: the inner shelf of central and northern California with a range of 10 to 14°C, the inner shelf of southern California with a range of 13 to 18°C, and the outer shelf of the two areas with ranges of 9 to 13°C and 12 to 16°C, respectively.

The typical maritime climate which exists over much of California buffers it from severe storms (Godshall and Williams, 1981). Occasionally, tropical storms will enter the southern California area, but their effect is drenching rain rather than damaging wind. Winter storms of varying magnitude characterize the shelf north of 40°N. Waves 5 to 11 m high may result from severe extratropical storms off the central and northern California coast. The severity of winter storms and wave height decrease to the south (Emery, 1960). Still some winter storms off southern California may be severe (Dayton and Tegner, 1984). Winter storms during 1982 and 1983 coincided with an El Niño event which brought unusually warm surface waters, diminution of the California Current and depression of the thermocline (Chelton *et al.*, 1982). Additional periods of extreme wave heights occur during tsunamis resulting from submarine seismic activity. The tsunami generated by the 1964 Alaskan earthquake arrived at high tide and reached 6 m above mean high water in northern California.

Intense winds and waves may result in acceleration of bottom currents and sediment transport. Mean current velocities tend to be very low (2-3 cm/s) (Komar *et al.*, 1972) but current speeds of 40 cm/s during the winter and 80 cm/s during severe storms were recorded by Sternberg and McManus (1972) on the Washington continental shelf. In addition to acceleration of currents due to wind stress, waves themselves may result in considerable bottom sediment transport. Komar *et al.* (1972) presented evidence for bottom sediment reworking due to surface waves to depths as great as 204 m off Oregon.

Diurnal tide curves for the California coast are mixed with one cycle of greater range and one of lesser range, unlike the more symmetrical tide curves of the Atlantic coast (Emery, 1960). The mean tide range for southern California is 1 m, but the extreme is 2.5 m during the spring tides of the solstices (Emery, 1960). Spring tides off Oregon are 3 m (Komar *et al.*, 1972).

Southern California

Geology

The continental margin off southern California from the U.S.-Mexican border to Point Conception consists of a narrow continental shelf and a complex continental borderland characterized by highly irregular topography of channels, ridges, basins and islands (Emery, 1960; Shepard, 1973). The roughly parallel rows of basins and ridges are oriented northwest to southeast. The coastline is markedly curved, running east-west at Santa Barbara and north-south at San Diego. Along the eastern boundary, the mainland shelf extends to depths of 80 to 150 m and is between 0.8 and 22 km wide (Emery, 1960). Insular shelves range from 0.2 to 35 km wide. Toward the slope (Patton Escarpment), topographic highs and lows are numerous. Eighteen basins and several open troughs, 600 to 2000 m deep, are enclosed by topographic highs, including deeply submerged sills, shallow flat-topped banks, ridges or islands (Emery, 1960). The basins are progressively shallower and less irregular from offshore to nearshore, indicating progressively thicker filling of basins closer to shore (Emery, 1960). Numerous submarine canyons border the mainland (20), islands (20) and submarine banks (2) (Emery, 1960). Others exist near the mainland but are buried beneath thick sediments. The continental slope lies 80 to 250 km off the mainland shore (Emery, 1960).

Due to the variable submarine topography, a complex series of substrate characteristics exists offshore (Shepard, 1973). In general, the sediments prograde from sands to silty sands and then to silts at the shelf edge. The mainland shelf areas are characterized by coarse, terrigenous sediments (Balcom, 1981). In deeper slope habitats, finer sediments are present. More complex sediments occur around rock outcrops and on the offshore banks and ridges where erosion of biogenic calcium carbonate deposits results in coarse-grained substrates. Nearshore basins typically have high sedimentation rates dominated by land-derived detritus (Emery, 1960). The outer basins, basin slopes and canyon walls are also sites of sediment deposition (Greene *et al.*, 1975; Dept. of Interior, 1983a). In all, flat basins and trough floors comprise about 17% of the area of the continental borderland (versus 70% in slopes, bank tops and shelves), but are the chief areas of sediment accumulation (Emery, 1960).

Downslope mass movements of sediments (slumps and slides) are common throughout the borderland. Many conditions giving rise to seafloor instability are characteristic of the region—localized thick accumulations of unconsolidated, water-saturated sediments, steep slopes, and seismic and storm activity (Field and Edwards, 1980). High seismicity characterizes all of the California coastal region. More than 20 earthquakes of magnitude 6.0 or greater have been recorded in southern California since 1912 (Dept. of Interior, 1983a). Offshore southern California the shelf as well as the slope, is cut by numerous faults, many of which are active (Dept. of Interior, 1983a) and several of which are considered capable of generating large magnitude earthquakes (U.S. Geological Survey, 1976; Yerkes *et al.*, 1980).

Benthos

The southern California area is a zone of biotic transition between two larger biogeographic regions, the Oregonian province north of Point Conception and the subtropical Panamanian province south of Magdalena Bay, Baja California (Valentine, 1963, 1966). Mixtures of cooler California Current waters and warmer waters from the south create conditions in which species of both provinces are found with California forms occupying cooler open coastal sites and Panamanian forms chiefly occupying the warmer embayments.

The complex topography and sediments of the southern California continental margin provide a complex array of benthic habitats. Fauchald and Jones (1977) indicated that the single most important environmental variable governing the distribution of species was depth, which was significantly more important than sediment and areal location, at least on the shelves and slopes. Two macrofaunal zones were identified in shelf depths less than 100 m. In shallow nearshore areas (<25 m) with coarse-grained sediments, the brittle star *Amphipholis* dominated the fauna (Balcom, 1981). This is also the area of the *Nothria-Tellina* association described by Jones (1969), in which *Diopatra ornata* and *Prionospio malmgreni* were also conspicuous elements. Other common taxa were the gastropod *Olivella*, the cumacean *Diastylopsis* and the amphipod *Paraphoxus*. On the finer shelf sediments (28-109 m), another brittle star *Amphiodia urtica* dominated the fauna (Balcom, 1981). The echiuran *Listriolobus*, the brachiopod *Glottidia*, the pelecypods *Axinopsida*, *Mysella* and *Parvilucina*, the ostracod *Euphilomedes* and gammarid amphipods were also common (Barnard, 1963; Jones, 1969; Balcom, 1981). Throughout the mainland shelf most of the species were polychaetes and amphipods, 42% and 36%, respectively; most individuals were polychaetes and small crustaceans, 42% and 38%, respectively (Emery, 1960). Because of their small sizes, these crustaceans and polychaetes form only a small percentage of the total biomass, in contrast to the larger but less numerous echinoderms. At the shelf break and beyond in deeper water (between 100 and 200 m), the sea urchins *Allocentrotus* and *Brissopsis* were prevalent, and a distinct zone between 100 and 150 m was dominated by large numbers of the pelecypod *Cyclocardia ventricosa* (Jones, 1969; Balcom, 1981).

Other delineations of the borderland macrobenthos have been made for the insular shelves (0-100 m), slopes and irregular areas (100 m - basin floor), basins (depths variable) and ridge and bank tops (100-300 m) (Balcom, 1981). Dominant invertebrates of the insular shelves were the gastropods *Amphissa* and *Alvinia* and the ostracod *Euphilomedes*, which was also characteristic of the finer mainland shelf sediments. No consistent insular shelf habitats were observed as was the case for the mainland shelf; rather, communities varied by locale, i.e., island. The slope areas had fewer species, lower abundance and lower diversity than adjacent shelf environments. The fauna was characterized by the urchins *Allocentrotus* and *Brissopsis*, the polychaetes *Pectinaria* and *Maldane*, the crustaceans *Ampelisca* and *Euphilomedes* and the aplacophoran *Limifossor* (Fanchald and Jones, 1978; Balcom, 1981). The fauna for each basin was distinct including species not common to other basins (Balcom, 1981). The shallowest basins exhibited extremely low abundance and low species diversity (Emery, 1960; Balcom, 1981). This may be due in part to low oxygen concentrations (Balcom, 1981) which were normally less than 0.3 ml/l below the basin sill in the shallower basins and about 2 ml/l in the deeper basins (Emery, 1960). The outer borderland basins, on the other hand, had an abundance of polychaetes, ophiuroids, holothurians and comatulid crinoids (Emery, 1960). Also frequent were brachiopods, siliceous sponges, urchins and sea whips.

The banks and ridges of the borderland are covered by sand and shell debris and rock cobble and are exposed to strong currents and continuous wave action. Wave-induced ripple marks in coarse sediments were evident to depths of 90 m. The primarily sessile, epibenthic communities varied by bank and with depth on the bank—crustose and erect coralline algae, brown and red algae, anemones, encrusting and massive sponges, fan corals, stony coral and the hydrocoral *Allopora*. Soft bottom fauna of the bank tops was characterized by *Amphiodia* and *Parvilucina* (Fauchald and Jones, 1978).

The giant kelp *Macrocystis* forms extensive forests along the mainland shelves, particularly along rocky shores and along insular shelves where there is a general downward displacement of floral zones due to reduced sedimentation and increased light penetration. Both *Macrocystis* and the elkhorn kelp *Pelagophycus porra* live in depths as great as 30 m (Emery, 1960). Nearer shore the smaller *Pelagophycus* occurs in depths of 3 to 10 m where the surf is strong.

Central and Northern California

Geology

Unlike southern California, the continental shelf of central and northern California is more gradually sloping. Although it is periodically cut by submarine canyons or interrupted by shallow banks or sea mounds, it lacks the complexity of the southern California borderland. The shelf is narrow, generally less than 50 km wide with an average seaward inclination of 3° (Jones & Stokes Assoc., Inc., 1981).

North of the Gorda Escarpment (~40°N), the shelf is narrow (19 to 40 km wide) and is cut by the Eel Submarine Canyon located 10 km offshore. Between Punta Gorda (~40°N) and Point Reyes (~38°N) the shelf is narrower (11 km wide) in the northern part to 40 km near Point Reyes. Several large submarine canyons cut the shelf, including the Mendocino and Mattole Canyons at the Gorda Escarpment, and the Delgado Canyon just north of Point Delgado. The head of Delgado Canyon is 2 km offshore, and the canyon transports sediment to the Delgado deep sea fan. The submarine canyons in the vicinity of Point Reyes are farther offshore on the continental slope. The widest area of the shelf (about 50 km) whereon are located the Farallon Islands lies between ~38°N to 37°N and is a broad bank of sandy and silty sediments and shell fragments. The shelf narrows to about 6 km south of 37°N and is cut by a number of submarine canyons, the largest of which is Monterey Submarine Canyon which begins about 4 km offshore. The shelf between 36°N and Point Conception is not cut by submarine canyons but grades gently to the Arguello Plateau on the upper continental slope.

Sediments of the central and northern California shelf generally grade from sands in shallow water nearshore to silt and clay substrates in the deeper waters along the outer shelf (Dept. of Interior, 1983b). Sand generally occurs to depths of 55 to 76 m. Between the Eel and Klamath Rivers, the mid-shelf sediment is mostly silt with a thickness of about 14 m; the surficial sediment grades to silt-clay on the outer shelf. In the vicinity of the Russian River the substrate is muddy (Jones & Stokes Assoc., Inc., 1981).

Northern California has more rainfall than other parts of California and in this area major rivers flow throughout the year and carry sediments to the shoreline. South of San Francisco (Cape Mendocino, according to some accounts), many streams and rivers are blocked by stream-mouth spits which are breached only occasionally by short periods of heavy discharge. Sediment deposition from central California, therefore, is minor throughout much of the year.

Major seismically active faults are capable of producing earthquakes larger than magnitude 7. Considerable mobilization of offshore unconsolidated sediment may be expected to accompany seismic activity on both the continental shelf as well as on the steeper continental slope. Shallow gas and gas-charged sediments may also contribute to shallow slope failures (tens of meters of unconsolidated sediments); these conditions exist in or adjacent to Santa Maria Basin (Dept. of Interior, 1983b). Massive slumping of sediments occurs at the head of Monterey Canyon (Shepard, 1973).

Rock outcrops are located in deeper water between 38° and 39°N along the outer edges of Bodega and Santa Cruz Basins near San Francisco and along the periphery of Santa Maria Basin (Dept. of Interior, 1983b). The largest of these banks is Santa Lucia Bank off Santa Maria. Cordell Bank off San Francisco is the shallowest and has a diverse and abundant community, including the hydrocoral *Allopora californica*. Bedrock habitats occur in a 0.5 to 1.5 km band along the coast (Jones & Stokes Assoc., Inc., 1981). The inner limit occurs from 12.5 to 30 m deep; the outer limit occurs at 18 to 55 m where the rock bottom gives way to unconsolidated bottom.

Benthos

Two kelps, *Macrocystis* and *Nereocystis*, have overlapping forest-forming ranges in central California. *Macrocystis* is distributed from Sitka, Alaska to Baja California but does not form extensive forests north of Point Año Nuevo (37°N) (Dept. of Interior, 1983b). Small forest patches have been reported as far north as Mendocino County. *Nereocystis* is distributed from Alaska to Santa Barbara but forms forests only north of Point Conception (Smith, 1969).

The subtidal mud habitats on the continental shelf and slope from just north of Point Conception to the California-Oregon border is virtually unstudied. Most studies have been in the sand habitat on the shallow shelf, where communities of tubeworms *Diopatra ornata* can be found in coarse sandy habitats associated with high organic content and another tubeworm *Nothria elegans* is characteristic of finer sands (Jones & Stokes Assoc., Inc., 1981). *Dendraster excentricus*, a sand dollar, is found in coarse to medium sand. Amphipod communities are also characteristic of a variety of sand habitats. The fauna of central and northern California is part of the Oregonian province which extends from Alaska to about Point Conception (Valentine, 1963).

Washington-Oregon

Geology

The continental shelf topography of Washington and Oregon is generally featureless and quite uniform compared to other U.S. shelf regions (Allen *et al.*, 1983). There are only a few major estuaries and bays, but some major submarine canyons, including Rogue Canyon, Astoria Canyon and Juan de Fuca Canyon. The width of the shelf is typically 50 km and the shelf break is at approximately 180 m, ranging from 16 km off southern Oregon at 185 m to 75 km at the Washington-Oregon border at 150 m (Shepard, 1973; Komar *et al.*, 1972).

Modern sediment forms two major deposits on the continental shelf: 1) inner shelf sand, extending from shore to 40-60 m water depth and 2) mid-shelf silt extending to about 120 m (McManus, 1972). Some areas of the outer shelf are covered by sand, especially north and south of Astoria Canyon and between Guide and Grays Canyons (Gross *et al.*, 1967). Mid-shelf sediments are composed primarily of coarse silts and closely resemble the sediment texture of the Columbia River suspended load. The Columbia River discharges approximately 10^7 metric tons of sediment (mostly fine-grained) annually and at least 50% of this sediment is estimated to accumulate on the Washington shelf (Nittrouer and Sternberg, 1981). The silt deposit trends from the Columbia River toward the head of Quinault Canyon (north-northwesterly in distinct bands) and is found on the outer shelf (>90 m) north of the canyon. The deposit progressively thins and becomes finer-grained away from the Columbia River as a result of decreasing accumulation rate (Nittrouer and Sternberg, 1981). The mid-shelf silt deposits are transient to some extent and often undergo repeated resuspension and redeposition by storm-induced bottom currents prior to final burial (Nittrouer and Sternberg, 1981; Smethie *et al.*, 1981).

Benthos

Lie and Kisker (1970) described benthic infaunal communities on the Washington continental shelf north of the Columbia River: 1) A shallow water

(<36 m) sand-bottom community was found where sand averaged 96%. The most abundant species were the cumacean *Diastylopsis*, the amphipods *Ampelisca* and *Paraphoxus*, the lamellibranch bivalves *Tellina* and *Macoma* and the polychaete *Owenia*. 2) An intermediate depth (96 m) sand-bottom community was found where sand averaged 68%. The most abundant species were the polychaetes *Sternaspis*, *Magelona*, *Nephtys* and *Haploscoloplos*, the lamellibranch bivalves *Yoldia* and *Axinopsida* and the amphipod *Paraphoxus*. 3) A deep water mud-bottom community was found at mean depths of 154.5 m in sediments with 50% muds. The most abundant species were the polychaetes *Prinospio*, *Sternaspis* and *Ninoe*, the lamellibranch bivalves *Axinopsida*, *Adontorhina* and *Macoma* and the amphipod *Heterophoxus*.

The mean standing crop increased with depth on the Washington shelf (Lie and Kisker, 1970) as did the abundance and diversity of organisms on the Oregon-Washington shelf (Carey, 1972; Richardson *et al.*, 1977; Nittrouer and Sternberg, 1981). Near the Columbia River, the deeper assemblage differed in species composition and community structure from that farther north (Richardson *et al.*, 1977). Polychaetes were generally the most abundant macrofauna on the Washington shelf, representing >70% of the individuals (Smethie *et al.*, 1981). Nittrouer and Sternberg (1981) classified the polychaetes of inner shelf sands, mid-shelf silt and outer shelf sands and found motile burrowers (primarily capitellids) to dominate the mid-shelf silts and outer shelf sands. The polychaete community within the inner shelf sands was less motile and contained a larger fraction of filter feeders and surface deposit feeders than found offshore. The fauna of Oregon and Washington is part of the Oregonian province which extends from Alaska to Point Conception (Valentine, 1963).

ALASKA

General Oceanography

Vast continental shelves surround Alaska on three sides. To the south, along the northern rim of the Pacific Basin, the Gulf of Alaska forms a broad embayment. To the west, an exceptionally wide shelf underlies most of the Bering Sea, between the Aleutian Island chain and the Bering Strait. Extensive arctic environments are found in the Chukchi and Beaufort Seas to the north.

Waters influencing the Bering Sea and the Gulf of Alaska originate from subarctic waters of the North Pacific (Sverdrup *et al.*, 1942). The Subarctic Current is formed within the large water mass lying north and east of the North Pacific Current. One branch of the Subarctic Current flows to the north and enters the Bering Sea, where it flows along the Aleutian chain, circling counter-clockwise. The Subarctic Current further divides before it reaches North America, sending a branch to the south to form the California Current and a branch to the north as the Alaska Current. The Alaska Current recurves westward along the shelf break and dominates circulation in the Gulf of Alaska. Its flow is intensified off Kodiak Island in the western Gulf, with speeds of 100 cm/s. Farther to the southwest, part of the Alaska Current enters the Bering Sea and the rest flows along the southern border of the Aleutian Island chain. A mixture of Bering Sea water and Alaskan Coastal Water moves through the Bering Strait, into the Arctic Ocean and flows eastward along the shelf break.

Gulf of Alaska

Physical Processes

Allen *et al.* (1983) summarized general circulation patterns on the continental shelf of the Gulf of Alaska. A narrow intense coastal current, the Alaska Coastal Current, flows from southeastern Alaska, beyond Kodiak Island and into the Bering Sea. Currents on the northeastern Gulf inner shelf generally flow along isobaths, and the eddy kinetic energy increases toward the Alaska Current at the shelf break. In the northwestern Gulf, although the mean current energy increases, the influence of eddies decreases. Shelf-break eddies are generally transient, but a permanent eddy has been located west of Kayak Island which directs a portion of the relatively fresh coastal current southward into the Alaska Current. Circulation in lower Cook Inlet is a continuous channel flow connected with the coastal region north of Kodiak Island. The Kenai Current, a seasonal, density-driven coastal flow, passes through Shelikoff Strait between Kodiak Island and the mainland (Cannon and Lagerloef, 1983). The bathymetry of the Kodiak shelf creates two hydrologic environments. Over shallow banks (<100 m) tidal mixing plus wind mixing and thermal convection effectively mix the water column. Over troughs the water column is stratified because wind mixing, winter overturn and tidal mixing are insufficient to mix the deeper water column (>150 m) (Sobey, 1980).

The dominant physical phenomenon in the Gulf of Alaska is the seasonal change in the position of the Aleutian Low. In early autumn it migrates out of the northern Bering Sea and crosses the Alaska Peninsula. In winter the low is usually centered in the Gulf of Alaska (~55°N, 155°W). The Aleutian Low is continuously reinforced from October to March by lows moving into the area from the Pacific. Frequent storms with highly variable, gusty winds move rapidly through the region. A weak high occurs in summer. The weather over the continental shelf provides large seasonal signals in temperature, wind, pressure and precipitation (Allen *et al.*, 1983). The consequence is strong meteorological forcing over the shelf. Shelf circulation here is further influenced by wind stress and runoff. Winds over the northern Gulf of Alaska, with the possible exception in summer, are alongshore from east to west and create a coastal convergence and downwelling (Royer, 1983; Allen *et al.*, 1983). Maximum downwelling occurs in January and minimum in February (Royer, 1983). The circulation of deeper water (>100 m) responds to the seasonal wind stress with renewal of bottom water in fjords in late summer (Allen *et al.*, 1983). Absence of strong winds in summer allows a relaxation of downwelling and the onshore intrusion of relatively warm, salty water from the central Gulf of Alaska (Allen *et al.*, 1983).

The wave climate in the Gulf of Alaska is severe, with North Pacific storms generating waves to 9 m or higher. Seas with 3-m waves are prevalent throughout the winter. Waves ⩾6 m occur frequently (⩾9% of the time) in all months except June-August. Tides are mixed semidiurnal (with marked diurnal inequalities).

Diurnal tidal ranges are 2 to 4 m with a maximum of 4.4 m on the northeastern coast and 3 m off Kodiak Island.

The low pressure system over the Gulf interacts with coastal topography and marine and continental air masses and causes high precipitation rates in the coastal drainage region (Royer, 1983). Maximum annual discharge occurs in October, although precipitation is maximum later in the fall. Precipitation decreases both to the west of 150°W and offshore. Discharge is minimized by freezing air temperatures and the storage of precipitation as snow. A smaller discharge peak represents spring melting. The mean discharge is 730 km^3/yr, 20% of which is in glacial fields. The coastal flow current is apparently isolated from the effects of the Alaska Current and, even hundreds of kilometers from its sources, the coastal flow remains quite narrow, less than 25 km wide. Thus the Alaska Coastal Current is a major avenue for influx of fresh water into the North Pacific and also an important source of low salinity water for the Bering Sea (Royer, 1983).

Geology

The continental shelf in the northern and northeastern Gulf of Alaska is broad, up to 200 km, and contains numerous deep troughs in excess of 200 m deep beginning within several kilometers of shore (Allen *et al.*, 1983). Major submarine valleys are located off the Alsek River, Yakutat Bay and Seal River; a submarine basin occurs west of Kayak Island off the Copper River delta and Controller Bay. Twenty percent of the mountainous coastal region in the northern Gulf is covered with glaciers, and the only major river is the Copper River (Allen *et al.*, 1983).

The sediments of the northern Gulf are primarily sands on the inner shelf. Silts and clays dominate the mid and outer shelves. The principal sediment sources to the northern Gulf of Alaska are the Copper River and the coastal streams draining the Bering, Guyst and Malaspina Glaciers. As this material enters the Gulf of Alaska, westward currents transport it to the west except near Kayak Island where it is deflected to the south, then trapped in a counterclockwise gyre west of Kayak Island. Across most of the shelf area west of Yakutat Bay, sediments are fine and the sedimentation rate is high. On topographic highs such as Tarr Bank, strong bottom currents and frequent winter storm waves prevent sediment accumulation, and bedrock, gravel and sand dominate the sediments. The westward transport of suspended sediments by the Alaska Current also prevents the accumulation of sediment along the shelf break and on the continental slope. At the shelf break, gravel (3 to 19%) is mixed with sand (8-50%), silt and clay (Feder and Matheke, 1979).

In the western Gulf of Alaska along the Kodiak shelf, the seafloor consists of a series of flat banks (50 to 100 m deep) that are cut by transverse troughs (>200 m); low hills and shallow depressions exist on the banks and depressions in the troughs. There are three major sand-wave fields on the Kodiak shelf: in Stevenson Trough, on northern Albatross Bank, and on southern Albatross Bank between Chirikof and Trinity Islands (Hampton, 1983). The occurrence of large sand waves up to 8-15 m high and 300 m long indicates the action of strong bottom currents. It is uncertain whether these currents are active at present or existed during the Holocene sea level rise.

Most of the surface sediment on the Kodiak shelf consists of Pleistocene glacial material that is being reworked under the present sedimentary regime (Peterson, 1980). The influx of additional sediment to the shelf is low, compared to the northern and northeastern Gulf. Small amounts are added and incorporated into shelf sediments from volcanic ash. Other modern sources of detrital influx are the Copper River and, to a lesser extent, the erosion of seafloor outcrops and runoff from Kodiak Island. The direction of sediment transport on the Kodiak shelf is generally from the banks into the troughs. Accumulation of fine sediments in the troughs is enhanced by the presence of sills which are about 30 m shallower than adjacent landward sections of the troughs and which restrict transport of sediments offshore. An exception to this general pattern is Stevenson Trough which contains abundant sand-sized material, sandwave bedforms and low percentage of volcanic ash. Sediments on the slope off this trough are similar, suggesting a route for high-energy bedload transport of sediment across the Kodiak shelf.

The Gulf of Alaska-Aleutian Area is one of the most seismically active on earth. During the last 75 years, five major earthquakes have occurred in the Gulf of Alaska Tertiary province. Identified areas of sediment slumps or slides on the northern Gulf shelf are associated with some troughs and basins, a mid-shelf region between Icy and Yakutat Bays, and some areas of the continental slope. Since 1902, at least 95 potentially destructive events have occurred in the vicinity of the Kodiak shelf (Hampton, 1983). Indications of sediment slides are rare on the Kodiak shelf but abundant on the adjacent upper continental slope. This is in contrast to the northeastern Gulf where large slumps of fine-grained underconsolidated sediments occur on low slopes.

Benthos

Faunal distributions in the northern Gulf of Alaska are related primarily to sediment distribution which is controlled by the deposition of predominantly glacially derived fine sediments (Feder and Matheke, 1979). Polychaetes represented 29% of the species collected followed by molluscs (15%), arthropod crustaceans (14%) and echinoderms (5%). Motile deposit feeders mainly polychaetes, predominated (61-65%) in the fine sediment of the mid to outer shelf where sediments were fine and sedimentation rates high (Feder and Matheke, 1979). Those species which successfully occupied the muddy environments were usually widely distributed. As the sediment changed from silt to clay and to sand and gravel mixed with silt and clay at the shelf break, the numbers of sessile and suspension feeding organisms increased. Suspension feeders, such as bivalves, comprised 32% of the macrofaunal organisms; deposit feeders were 26%. The diversity and species richness of the fauna in the Tarr Bank and shelf break stations were among the highest found. Diversity was greater in areas where the sedimentation rate was reduced and the presence of sand and gravel substrates increased environmental heterogeneity. There was a slight change in benthic fauna from east to west.

Strauch *et al.* (1980) reported on the marine benthos of the Kodiak shelf but noted little information on the infauna. Most studies conducted in the area have been limited to shallow inner shelf areas and bays, emphasized commercially important species, especially the king crab, and are limited mostly to epibenthic sampling (e.g., Feder and Jewett, 1980). Additional information has been obtained from stomach analyses of epifauna taken around Kodiak Island. In addition to extensive epifaunal trawl collections in the nearshore Kodiak shelf, Feder and Jewett (1980) summarized collections taken by pipe dredge and sediment sweep. Bivalves *Axinopsida*, *Psephidia*, *Nucula*, *Nuculana* and *Macoma* dominated the dredge samples. Polychaetes collected by dredge were deposit feeders *Haploscoloplos*, *Heteromastus*, *Nephtys*, *Glycinde* and *Myriochele*. The sediment sweeps of fine sandy sediments in 11 m water depth contained snowcrab (*Chionoecetes*) megalopae and juveniles, gammarid amphipods, gastropods (*Lacuna*) and bivalves (*Chinocardium*, *Hiatella* and *Mya*).

Bering Sea

Physical Processes

The eastern Bering Sea connects with the Gulf of Alaska by Unimak Pass and with the Arctic Ocean through Bering Strait. Water enters from the Subarctic Current along western passes of the Aleutian Islands and from the Alaska Current and Alaska Coastal Current through Unimak Pass. A greater volume of water flows through Bering Strait ($\sim 1 \times 10^6$ m^3/s) than through Unimak Pass ($\sim 0.15 \times 10^6$ m^3/s) (Allen *et al.*, 1983). The Bering Sea is isolated from the direct effects of circulation in the Pacific Ocean by the Alaska Peninsula and sheltered from the Arctic Ocean by the narrow, shallow Bering Strait. There is a net excess of precipitation over the shelf and river discharges, principally from the Kvichak, Kuskokwim, and Yukon Rivers, add about 1.5×10^4 m^3/s (Allen *et al.*, 1983). Much of the remaining transport required to make up the Bering Strait outflow apparently comes across the shelf south of Cape Navarin (Allen *et al.*, 1983).

Over the southeastern shelf there are three identifiable water masses separated by fronts at the 50, 100 and 170 m isobaths (Allen *et al.*, 1983). In the coastal domain, tidal mixing exceeds buoyancy input and, away from the direct influence of river discharge, the water is vertically mixed. In the middle shelf domain when seasonal input of buoyancy (either from melting ice or insolation) exceeds tidal mixing, there is a two-layered structure. Surface cooling in winter and increased frequency and strength of storms destroys the structure over the middle shelf, but a stronger density gradient is maintained across the 50-m isobath. The water column undergoes a broad transition (50 km) between the middle shelf domain and the outer shelf domain. Within this area middle shelf waters extend seaward near the surface and outer shelf waters intrude landward near the bottom. The outer shelf domain is characterized by well-mixed upper and lower layers separated by an intermediate layer with much fine structure. The outer shelf is bathed by slope waters which are warmer and more saline than the waters of the middle shelf. On the northern shelf there are also three identifiable water masses, with 1) a more saline water mass to the west of St. Lawrence Island and to the west of Bering Strait (Anadyr water), 2) a less saline, coastal water mass with pronounced seasonal salinity changes (Alaska Coastal Water) and 3) an intervening water mass of intermediate salinity (Bering shelf water) (Nelson *et al.*, 1981). In Norton Sound the water column in summer is strongly two-layered in both temperature and salinity, and the eastern and western portions of the Sound are isolated.

The tidal wave for the Bering Sea enters from the North Pacific Ocean through the central and eastern passages of the Aleutian Islands and then propagates eastward onto the shelf. Tides are semidiurnal and dominate the kinetic energy on the southeastern shelf but become less energetic farther north. Along the Alaskan coast at Cape Romanzof (Central Bering Sea) the tidal amplitude is 2.1 m (Brower *et al.*, 1977). In the northern Bering Sea, tidal ranges are small (<0.5 m) (Larsen *et al.*, 1981).

Three current regimes have been identified over the southeastern shelf and are nearly coincident with the hydrographic domains. Coastal waters from the Gulf of Alaska enter the Bering Sea through Unimak Pass and then continue northeastward along the Alaska Peninsula. Within Bristol Bay the flow becomes counterclockwise and then follows the 50-m isobath past Nunivak Island and continues northward. Currents are strongest near the front between the coastal and mid-shelf water masses with maximum speeds of 5.5 cm/s occurring in winter. Although tides dominate the kinetic energy, significant pulses of flow are wind driven. The middle shelf current regime has wind-driven pulses but the mean current is insignificant except near the front boundaries. On the outer shelf flows are significant with speeds up to 11 cm/s to the northwest and up to 5.5 cm/s to the northeast. Flow along the slope averages between 5.5 to 14 cm/s toward the northwest. Circulation on the northern shelf is dominated by a generally northward flow toward the Arctic Ocean, but part of the Alaskan Coastal Water affecting the Yukon River plume moves in a counterclockwise gyre around the margin of Norton Sound (Nelson and Creager, 1977). This pattern can be reversed due to large-scale meteorological forcing, particularly in early winter. East and west of St. Lawrence Island and through Bering Strait, flow reaches 14 cm/s or more; south of the island the flow is weaker. In Norton Sound the northward mean flow appears only in the western portion; currents in the remainder of the sound are weak. Wind-driven currents in Norton Sound with instantaneous speeds up to 100 cm/s have been observed.

The Aleutian Low, normally located in the vicinity of the Aleutian Islands, dominates the climatology. During winter there are two storm tracks, one parallel to the Aleutian Islands and one curving northward along the Siberian coast. Mean winter winds are from the northeast, and outbreaks of cold polar air which continue for 1 to 2 weeks are common. The mean winter winds are stronger than those of summer and result in stronger subtidal flows over much of the southeastern shelf. They also have a dramatic impact on water temperature and ice production. Ice cover is a seasonal feature of the eastern Bering Sea shelf varying from none in summer to greater than 80% coverage of 0.5 to 2.0 m thick ice during its maximum extent in mid-March. During October the ice edge advances rapidly to the south, moving into the Bering Sea during November. The mean ice edge reaches almost to 57°N (Weeks and Weller, 1984). In the northern Bering Sea, ice is preferentially produced along south-facing coasts (Allen *et al.*, 1983). During the open water season the sea is subject to occasional strong northerly winds (Larsen *et al.*, 1981). Summer storms tend to migrate northward into the Bering Sea and mean winds are from the south. In the fall, strong south-southwesterly winds cause high waves and storm surges along the entire west Alaska coast (Larsen *et al.*, 1981). Storm generated waves range up to 30 m in the southern Bering Sea (Weeks and Weller, 1984).

Geology

The continental shelf bordering the west coast of Alaska in the eastern Bering Sea is the second largest shelf of the world's oceans, surpassed only by that of the Arctic. It exceeds 150 km at its narrowest point (Allen *et al.*, 1983) and over 500 km at its widest (Ingraham, 1981). The shelf is bounded on the south by the Alaska Peninsula and on the north by the Seward Peninsula and Siberia. The shelf slopes gradually to about 170 m where it terminates abruptly as the continental slope drops somewhat precipitously into the Aleutian Basin. The shelf is indented by several submarine canyons (probably the largest in the world; Shepard, 1973) and characterized by several large embayments (e.g., Bristol Bay and Norton Sound) and islands (e.g., St. Lawrence and Nunivak Islands). The Bering Sea receives freshwater input from two major Alaskan rivers (the Yukon and Kuskokwim). St. Matthew and Nunivak Islands at ~60°N provide an artificial division into northern and southern areas of the eastern Bering Sea.

Sand-sized sediments dominate the southeastern Bering Sea shelf, constituting 20 to 100%, with an average of 68% (Burrell *et al.*, 1981). At 30 m depth sediments average in excess of 90% sand and at 45 m, 80% sand. Sands are coarser than 125 μm shallower than 35 m but become finer with increasing depth. Sediments coarser than 250 μm are restricted to depths of <50 m, suggesting that resuspension and transport of fine-grained sediments occur at least for that depth. Gravel is found mostly in nearshore areas especially Bristol Bay, Kuskokwim Bay and Unimak Pass. The central portion of the St. George Basin contains finer sediments which are very poorly sorted, indicating the lack of significant winnowing and a sink for the fine-grained materials (Gardner *et al.*, 1979). Moderately sorted sediments are found on the northwestern border of the Bering Canyon, the head of Pribilof Canyon and the topographic high of the Pribilof ridge (Gardner *et al.*, 1979). Across the remainder of the shelf (>50 m depth) the relative amount of the sand component decreases. Well-sorted, very fine sands are found near the shelf break at 150 m.

Areas of potentially unstable sediments are found on the continental slope and rise and the walls of the major submarine canyons, Pribilof and Bering Canyons (Gardner *et al.*, 1979). One of the most active seismic and volcanic zones in the world borders the southern Bering Sea along the Alaska Peninsula and eastern Aleutian Islands arc. There is high potential for earthquakes with strong ground motion and local tsunamis of 30 m height.

The sediments and sedimentary processes of the northeastern Bering Sea shelf, including Norton Sound, have been more extensively studied. The northeastern shelf is a complex system of sand ridges, sand wave fields and shoals with fine- to medium-grained sands inherited from a transgressive Pleistocene-Holocene transgressive nearshore environment (Field *et al.*, 1981; Nelson *et al.*, 1982). The ridge and swale morphology is similar in many respects to that of the U.S. Atlantic shelf. The eastern part, a broad, flat marine re-entrant (Norton Sound), is covered by silt and very fine sand (Larsen *et al.*, 1981; Olsen *et al.*, 1982). Sand-sized material dominates in the outer reaches of the sound, as on the southeastern shelf, and is also the major component within Norton Bay (Burrell *et al.*, 1981). Within the most eastern part of the embayment and over a considerable portion of the central region, mud constitutes the dominant sediment size (Burrell *et al.*, 1981). These substantially finer-grained, weak and highly compressible sediments of Holocene age are derived from the Yukon River and frcm local rivers and streams (Olsen *et al.*, 1982). The Yukon River contributes nearly 90% (96.8 x 10^6 tons/yr) of the river sediment entering the Bering Sea. The muddy sediments deposit low energy environments with negligible ice loading, low waves and weak bottom currents. Areas of central and western Norton Sound with silty fine sand and sandy silts are high energy environments with extensive ice loading, high waves and strong bottom currents (Olsen *et al.*, 1982). Most of the immense quantity of sediment derived from the Yukon River is transported northward through the Bering Strait (Nelson and Creager, 1977; Burrell *et al.*, 1981). As much as 15 to 90 x 10^6 tons/yr of suspended clay- to sand-sized sediment may be carried into the Chukchi Sea (Nelson and Creager, 1977).

The sediments of the northern Bering Sea shelf are affected by a number of dynamic conditions—winter sea ice, sea level setup, storm waves and strong currents (geostrophic, tidal and storm). Larsen *et al.* (1981) summarized the active sedimentary processes in this region which include thermogenic gas seeps, seafloor gas cratering, sediment liquefaction, ice gouging, scour depression formation, coastal and offshore storm surge and associated deposition of sand, and movement of large-scale bedforms. Erosional and depositional processes are most intense in the shallower parts of the shelf and along the coastline during storm surge flooding. In the Yukon prodelta area and in central Norton Sound, where currents are constricted by shoal areas and made turbulent by local topographic irregularities, storm-induced currents have scoured large, shallow depressions. Storm surge and waves generate bottom-transport currents that deposit layers of sand as thick as 20 cm in Yukon prodelta mud as far as 100 km from land. Ice gouges to the depth of 1 m are numerous and ubiquitous in the area of the Yukon prodelta. Although less common than in the prodelta, ice gouges are present on the rest of the northern Bering Sea shelf where water depths are less than 20 to 30 m. The homogeneous fine sands in Chirikov Basin support abundant populations of the tubicolous amphipod *Ampelisca macrocephala*, which are an important prey of the gray whale. While feeding on these amphipods, gray whales leave extensive feeding pits on the sea floor (Johnson and Nelson, 1984). Modifications of these feeding pits occur by sediment infilling, by further feeding or by current scour enlargement.

Benthos

Benthic macroinfauna has been reported for the northern Bering Sea by Stoker (1981) and for the southeastern part by Haflinger (1981). In addition, biological sedimentary structures of the Bering shelf (Nelson *et al.*, 1981) and epifaunal invertebrates (Jewett and Feder, 1981; Feder and Jewett, 1981) have been described. Stoker's (1981) study concentrated primarily on standing stock and those organisms retained on a coarse 3-mm sieve, but was much more extensive in aereal coverage than Haflinger's (1981) study. Quantitative results from the combined 3-mm and 1-mm sieve fractions showed that the most ubiquitous major taxonomic groups in terms of frequency of occurrence and comprising the most species were polychaetes, followed by bivalves, gastropods and amphipods (Stoker, 1981). Problems in sampling, however, precluded taking populations of the large, deep-burrowing bivalves *Mya* and *Spisula*. Assemblage distribution patterns based on the 3-mm fraction were related to depth on the southeastern shelf but were more complex on the northeastern shelf between St. Lawrence Island and Bering Strait. Amphipods *Ampelisca macrocephala*, which comprise the main prey of gray whales, *A. birulai* and *Byblis gaimardi* were dominant fauna in the northwestern area. Echinoderms *Ophiura* and *Strongylocentrotus* were also dominant in this area. In the southeastern Bering shelf, polychaetes, bivalves and echinoderms were the dominant taxa. Haflinger (1981) also found that major boundaries for infaunal communities on the southeastern shelf follow 50 and 100 m isobaths, coinciding with hydrographic frontal zones. A fourth faunal community is found at the head of Bristol Bay in gravel and sand substrates (Haflinger, 1981).

Stoker (1981) found little seasonal or annual fluctuation in density or biomass across the Bering and Chukchi shelves and pointed to a reliable and fairly uniform benthic food supply and life histories, with direct larval development or brooding behavior, as possible reasons for this population stability. The reduced standing stock south of St. Lawrence Island (compared to the Bering Strait and southern Chukchi Sea) were attributed to heavy predation by bottom-feeding fishes and marine mammals, particularly walruses, on the central and southern Bering shelf, as well as trawling activities of commercial fisheries.

Alaskan Arctic

Physical Processes

Bering Sea water enters the Chukchi and Beaufort Seas through Bering Strait and travels eastward along the shelf break as far as 150°W (Matthews, 1983). Alaskan Coastal Water also moves through Bering Strait, mixes with ambient surface water as it moves eastward, and has been identified as far east as 148W. The Bering Sea water is more saline than the Alaska Coastal Water. The cross-shelf circulation on the Beaufort Shelf in Alaska is characterized by the advection of these more saline waters onto the shelf or the sinking of brine produced by the freezing process on the inner shelf and its movement seaward in the lower layer (Matthews, 1983). There is also a suggestion of upwelling along the Beaufort Sea shelf break. Beyond the barrier islands and out to 60 m depths on the Beaufort shelf, a 10-m thick, bottom layer of saline water is delineated. On the inner shelf (<40 m) during the ice-free season, the prevailing winds are easterly and the surface currents are generally eastward. There are, however, periods of strong westerly wind associated with storm systems which produce major positive surges and account for the greatest coastal erosion rates. In the eastern Chukchi Sea, nearshore currents predominantly move water upcoast, generally following local wind patterns (Dept. of Interior, 1983c). There are only a few small rivers draining into the Alaskan Beaufort Sea, the largest of which is the Colville; however, to the east is the Mackenzie River which dominates coastal oceanography (Craig, 1984). There are no major rivers along the northeastern Chukchi coastline.

Except of course for sea ice, the oceanographic conditions in the Chukchi and Beaufort Sea are less severe than in the Bering Sea and the Gulf of Alaska. Tidal amplitude is small and tidal currents are weak (Matthews, 1983), mostly dampened by sea ice. The tides are generally mixed semidiurnal with mean ranges from 10 to 30 cm. The mean spring tidal range at Barrow is 13.6 cm and in Stefansson Sound is 15.2 cm (Matthews, 1983). In winter, dense pack ice prevents waves and wave-induced turbulence. In summer open pack ice in the marginal sea ice zone and the limited fetch of open water dampen the generation of significant waves (Carey *et al.*, 1984). Extreme wave heights are 15 m in the eastern Beaufort, but more important are surges which can occur when a major storm approaches or crosses the coast (Weeks and Weller, 1984). These can generate a 3-m surge plus 3-m waves which cause flooding over 1 km inland and occasionally produce significant bottom currents. Normally, storm surges are an order of magnitude larger than astronomic tides (Matthews, 1983). These storm surges contribute to coastal erosion in summer and ice override in winter. The normal wave field has virtually no swell because of the ice cover, and the relatively low waves approaching the shoreline produce only moderate longshore currents and sediment transport.

The presence of ice is a dominant environmental feature of the northern Chukchi and Beaufort Seas where heavy ice is always a possibility, even during the peak of the summer melt (Weeks and Weller, 1984). The position of the ice edge in the Beaufort and Chukchi Seas, although highly variable during minimum extent, runs roughly east-west. During some years the ice edge is as far as 250 km north of Barrow but in others is pressed tightly against the coast by onshore winds. Most of the ice is pack ice which drifts as a result of wind and current forcing. Fast ice in the Chukchi is limited to a few protected bays, the most notable of which is Kotzebue Sound. Along the Beaufort coast there is a more extensive belt of fast ice whose stability is enhanced by the presence of small barrier islands and grounded pileups of sea ice.

Geology

The Beaufort Sea shelf extends east for 600 km from Point Barrow and is relatively narrow, typically 86 km wide with the western part being wider than the eastern. The shelf grades steadily from the coastline to the shelf break. East of 152°W the shelf break occurs at about 60 m and is relatively close to shore; west of this longitude the shelf break is less well-defined and occurs over a greater range of water depths. The inner shelf is shallow: 20 km offshore the depth is still typically 10 m (Matthews, 1983). The coastline is characterized by several long shallow lagoons and large shallow open embayments.

West of Point Barrow the Chukchi Sea shelf is extensive and extremely flat, interrupted in a few places by topographic features such as the Hanna and Barrow Sea Valleys and Hanna Shoal. The Barrow Sea Valley on the northern Chukchi Sea shelf is a flat-bottomed channel, 200 km long and 2 to 8 km wide, which separates the Bering and Chukchi shelves. About 20 km south of the shelf break, the channel becomes the Barrow Canyon. Hanna Shoal overlies a structural high in the northeastern part of the Chukchi Sea and rises to within 25 m of the surface. The shelf of the southern Chukchi Sea is a continuation of the epicontinental shelf of the Bering Sea and is less than 60 m deep.

The inshore zone of the Beaufort Sea to about 10 m has wave- and current-worked muddy sands (Barnes *et al.*, 1981). The short-range variability of sediment types is high in water depths less than 15 m due to the interplay of ice gouging and hydraulic reworking. Intensive sediment reworking in this area occurs on an interval of 5-10 years, during open water seasons with intense fall storms when the prevalent ice-gouged terrain is infilled and reworked and replaced by sand waves a meter or more in height. The resultant bedform is a combination of smooth-surface sand waves or linear, current-shaped sand bars resting on highly jagged relief forms carved into overconsolidated silty clay which outcrops in the troughs between sand bodies. On the crests of the shoals, wave-formed ripples are found in clean gravel of 2 cm diameter indicating current orbital velocities of 100 cm/s which were necessary to shape gravel into these bedforms. The seafloor of Boulder Patch in the western part of Stefansson Sound is characterized by a veneer of pebbles, cobbles and boulders up to 2 m in diameter. The sediments in the vicinity consist of muds and sands with some gravelly sands and patches of overconsolidated clays. The area also has a low deposition rate. Fine-grained sediments dominate the shallow delta platform of the Colville River, and there is a patchily distributed sand-silt substrate in 3 to 4 m of water off Harrison Bay (Broad *et al.*, 1981). With distance offshore there is a general decline in percent sand and increase in percent mud.

Sediments from coastal erosion and flooding of rivers are initially deposited on delta front platforms and along coastal shallows. In mid to late summer much of the turbid coastal water transport is initiated in inshore regions by wave resuspension. The turbid water plume is carried westward, but wave refraction along the coastline and offshore islands may cause local reversals. The south-western corner of Harrison Bay may be a deposition site for muds resuspended off the Colville River mouth (Naidu *et al.*, 1981). The amount of sediment transport is probably small because of the low wave energy and short open water season. Suspended sediments often become incorporated into sea ice out to and including the stamuhki zone, where transport is minimal until melting in the spring.

Information about the sediments of the Chukchi shelf is limited. Sediments of the Chukchi shelf are predominantly mud, apparently from the Mackenzie and other rivers, along with a scattering of sand and gravel that is rafted into the sea by drift ice (Shepard, 1973). Deposits of coarse-grained sediments (gravel and sand) are located near shore and on Hanna Shoal (Creager and McManus, 1967).

Several geologic and physical processes (Jackson and Kurz, 1983) influence the sedimentary regime and the benthos. A major factor is the sea ice regime which can be divided into the landfast ice zone, the shear or stamuhki zone, and the pack ice zone. The boundaries between these zones vary geographically, seasonally, and yearly and are strongly influenced by bathymetry and the position of offshore islands and shoals. The interaction between the seafloor and keels of drifting ice masses, particularly ridges, produces gouges in the sediments. The greatest density of gouges is found in the sediments that underlie the stamuhki zone where the common incision depth is greater than 1 m. Gouging in the Beaufort generally occurs in a zone from the coastline to the 60 m isobath; the highest density of gouges occurs in the 15- to 30-m depth range. In the northeastern Chukchi Sea, ice gouging appears to be more intense shoreward of the Barrow Sea Valley and in the vicinity of Hanna Shoal (Dept. of Interior, 1983c).

Seismic activity on the Beaufort continental shelf is confined to an area of young faulting and Holocene uplifts off Camden Bay (Jackson and Kurz, 1983). Areas adjacent to this, as well as the Beaufort shelf, are underlain by unconsolidated or poorly consolidated Holocene and Pleistocene sediments which have low shear strength and are susceptible to tectonic instability. Active slumping and sliding is common on the outer shelf and upper slope seaward of the 50 to 65-m isobaths. Limited data suggests that the outer part of the Chukchi shelf and the upper part of the slope have some features which indicate downslope movement of sediment masses. Shoreward of the Beaufort outer shelf where slumping occurs, Holocene sediments thin and slopes become gentler. Coastal bluffs, however, are subject to mass wasting and slumping. The Beaufort Sea coastline retreats 0.9 to 3.0 m/yr as a consequence of the erosion of ground ice and frozen soil by surface water.

Benthos

Stoker (1981) concluded that the benthic macrofauna of the Bering and Chukchi shelves were clearly similar and interdependent and that a distinction between the two shelves was artificial. Major noncontiguous elements were present between the fauna of the Chukchi shelf and that of the Bering shelf. The benthic infaunal communities of the Chukchi shelf did not show the bathymetric groupings of the central and southern Bering shelf, but rather resembled the outer shelf and shelf break fauna of the Bering shelf. Dominant species (3-mm sieve fraction) were the polychaete *Maldane*, the ophiuroid *Ophiura*, the sipunculan *Golfingia*, the bivalves *Astarte*, *Macoma*, *Nucula* and *Yoldia* and the amphipod *Pontoporeia*. Standing stock in the southern and central Chukchi shelf is higher than the adjacent Bering shelf and the diversity is at about the same level as the southern Bering (adjacent to a decline in the Chivikov Basin and Bering Strait areas). These trends are somewhat at odds with theories of high latitude fauna and are possibly attributed to the large influx of food from high primary productivity rates in the Bering Strait region (Sambrotto *et al.*, 1984), the attenuation of bottom currents after passage through the Bering Strait which allows for settling of fine organic detritus, and reduced predation pressures by bottom-feeding fishes, mammals and commercial fisheries compared to areas of the Bering shelf. In the northern Chukchi, reduced food availability and increased environmental stress may account for low standing stocks in this area. Although more arctic species are represented in the southeastern Chukchi shelf, the benthic fauna is primarily boreal Pacific in origin. Low bottom water temperatures may preclude reproduction of many species, which depend on recruitment of larvae swept north from the Bering Sea.

Based on the distribution of macrobenthic assemblages, habitats of the Beaufort Sea shelf may be divided into an inshore zone (2 to 20 m depth), the offshore, central or shelf zone (20 to 100 m depth) and the slope zone (>100 m) (Broad *et al.*, 1981). Carey and coworkers have studied the benthos of the western Beaufort Sea with primary emphasis on the polychaetes and bivalves of the central shelf zone and the slope zone (Carey *et al.*, 1974; Carey and Ruff, 1977; Bilyard and Carey, 1979, 1980). In the inshore zone, principal infaunal organisms were polychaetes, amphipods, an isopod, bivalves and the priapulid *Halicryptus spinulosus*. The dominant motile epifaunal invertebrates were *Mysis littoralis* and *M. relicta*, amphipods *Pontoporeia*, *Apherusa*, *Gammarus* and *Onisimus* and isopods *Saduria*. Biomass and diversity generally increased with depth in the inshore zone except in the shear zone at 15 to 25 m where the moving ice pack disturbs the sediments, thereby minimizing the abundance of infaunal organisms (Dept. of Interior, 1983c). Within the bivalve fauna in the inshore zone, there were no spatial or depth patterns in overall abundance, species richness or species composition (Carey *et al.*, 1984). Diversity and biomass of infauna increased beyond the minimum abundance zone with distance offshore, at least as deep as the 200-m isobath (Dept. of Interior, 1983c).

The principal infaunal organisms of the offshore zone are polychaetes, bivalves, ophiuroids, holothuroids and crustaceans (Broad *et al.*, 1981). The polychaetes were most abundant and comprised 32 to 87% of the total macrobenthos. Molluscs and arthropods were next in abundance (~5 to 50%). No dominant compositional trends were apparent with increasing depth or distance from shore, but the biomass and density of the macrofauna (≥1.0 mm) increased, reaching a maximum on the upper continental slope (Carey *et al.*, 1974; Carey and Ruff, 1977). Within the polychaete fauna, species richness and abundance were maximal along the outer continental shelf and upper continental slope but the composition was similar across the 20 to 200 m collections (Bilyard and Carey, 1979).

Because mud and sand sediments predominate in the Beaufort Sea, the rocky Boulder Patch in Stefansson Sound is identified as an important habitat because of its diverse assemblage of plants and invertebrates (Craig, 1984). Fairly large stands of both red and brown kelp exist in this area (Dept. of Interior, 1983c). Other macrophytic communities are sparse and found in small scattered patches because of the paucity of hard substrates, the shorefast ice and ice gouging, and the high rate of sediment deposition in some areas (Dept. of Interior, 1983c).

A biologically important feature of the Beaufort Sea is the occurrence of a band of relatively warm and brackish water (5-10°C, 10-25‰) that lies adjacent to the shoreline in summer (Craig, 1984). It extends the length of the coast (750 km), is narrow (usually 2-10 km), and is often distinctly different from adjacent marine waters (−1 to 3°C, 27-32‰). The band provides important feeding habitat for anadromous fishes (cisco and char) which enter the waters each summer and disperse parallel to shore, feed extensively on an abundant supply of epibenthic mysids and amphipods. Dominant marine species (cod and sculpin) enter the nearshore waters later in summer as salinities increase. In winter the estuarine band is absent and most anadromous species return to North Slope rivers; marine species remain under nearshore ice but vacate shallow waters which freeze to 2 m depth.

COMPARISON OF VULNERABILITY OF SHELF ENVIRONMENTS

The bewilderingly rich variation in oceanographic and biologic conditions on U.S. continental shelves defies clear and comprehensive comparisons among regions. Even the highly simplified summary of dominant environmental features and characteristics of the benthos presented in Table 3.2 is not easy to comprehend. Some patterns related to latitude, depth and sediment type do emerge in comparison of benthic faunas among regions. These are interesting in terms of comparative ecology and evolution but must await fuller interpretation elsewhere. More pertinent here are observations and inferences regarding the relative sensitivity of continental shelf ecosystems to long-term effects of oil and gas development.

The theory of comparative response of different ecosystems to pollutants and other perturbations is not well developed. There are several schemes which describe the relative sensitivity of coastal habitats and communities to oil spills, based on the persistence of stranded oil and recoverability of the affected biota. For example, Gundlach and Hayes (1978) and Owens and Robilliard (1981) developed habitat vulnerability indices which can be used in guiding protection and cleanup responses to oil spills on a local scale. Minerals Management Service (1985) developed a methodology for comparing the environmental sensitivity of entire planning regions to be used in guiding oil and gas leasing policies. The composite sensitivity index developed included components representing sensitivity of coastal habitats, marine (continental shelf) habitats and biota (e.g., birds, mammals and fisheries) to oil spills. The coastal and marine habitat sensitivity components are sums of products of the proportional extent of habitat types (e.g., wetlands, sandy and rocky shores for coastal habitats and submerged vegetation, submarine canyons, coral reefs, and sediment covered bottom for shelf habitats) and a sensitivity coefficient.

Although there is a reasonable basis of experience and theory for coefficients of susceptibility to oil spills for coastal habitats of differing geologic nature and latitude, there is not a similar basis for shelf environments. Mud and sand covered shelf habitats, which comprise the vast majority of the continental shelves (>90% in all regions except the Eastern Gulf of Mexico), are all scored with equally low sensitivity in the Department of Interior (1985) scheme. Consequently, the small differences evident in sensitivities among the regions are due to the presence of relatively small areas of submerged vegetation which is deemed highly sensitive. Furthermore, the Department of Interior (1985) environmental sensitivity analysis does not consider sensitivity to operational discharges and physical impacts. Is there, then, any basis from our synthesis for judging the relative sensitivities of continental shelf environments to the composite, long-term effects of oil and gas development activities?

From the viewpoint of effects on benthic communities, four factors emerge which can underpin such a comparison of sensitivities: 1) sedimentary regime, 2) temperature, 3) depth and 4) prevalence of biogenically structured communities.

Sedimentary Regime

The sedimentary regime affects the nature of bottom deposits (e.g., sediment grain size and organic content), the frequency of disturbance of the seabed, and the ultimate depositional fate of fine particles on which particle-reactive pollutants, including many trace metals and the more persistent petroleum hydrocarbons (Chapter 6), are concentrated. Although all continental shelves were influenced by the relatively recent transgression of sea level during the Holocene interglacial period, the contemporary sedimentary regimes vary tremendously among the continental shelves of the U.S. It is helpful to categorize the shelf regions using Curray's (1965; see also Swift, 1970) three stages in sedimentary evolution of shelves following eustatic sea level rise (Table 3.2). Under conditions of autochthonous sedimentation, the shelf is veneered with sediments left by the erosional process of shoreface retreat during transgression and modern, fluvial (fine) sediments are trapped in estuaries. Contemporary erosional processes on the shelf are sufficient to prevent the accumulation of fine, modern sediments on the shelf except in isolated areas, e.g., the New England "Mud Patch" (Chapter 6). An autochthonous sedimentation regime exists on the continental shelf of the U.S. Atlantic coast and in the eastern Gulf of Mexico. Although there might be an acute effect of an oil spill or drilling discharge, contaminants would not be expected to reside in the sediments for long periods.

Under allochthonous sedimentation, modern, river-supplied silts and clays overlie the autochthonous veneer except on the outer shelf. Finally, in the third stage, climax grading exists wherein sediments are progressively finer with depth. Under intense allochthonous sedimentation, such as off the Mississippi River delta, oil and gas related contaminants might be deposited, but may be overshadowed as a result of heavy deposition and river-borne contaminants. Outer shelf environments under climax grading are probably those most susceptible to long-term contamination by particle-borne pollutants.

Temperature

Bottom water temperatures influence biogeographic distributions, population dynamics and the biodegradation of pollutants. All of these are related to the sensitivity to and recovery from the impacts of oil and gas development activities. Warmer waters are conducive to more rapid biodegradation of contaminants and, generally speaking, are populated by benthic species with shorter generation times and broader dispersal abilities. These life history characteristics facilitate more rapid recovery from local population mortality.

Depth

Water depth across the continental shelf affects the temperature variability of bottom waters and bottom stresses resulting from waves and tides which resuspend or transport sediments. The benthos of the outer shelf and upper slope confronts a less variable temperature regime and less frequent and substantial disturbance of its sediment habitat. Consequently, indigenous populations may be older, more slowly growing and reproducing, and more biologically structured than those of the inner shelf. This would result in slower recovery time from community disruptions (Boesch and Rosenberg, 1981). Continental shelves vary in the energy of their benthic boundaries as a result of exposure to open oceanic conditions and storms and tides. Thus, the degree of "biological accommodation" as opposed to "physical control" (Sanders, 1969) in the benthic communities at similar depths will vary among regions.

Biogenically Structured Communities

Benthic communities which are integrally dependent on physical structures, which are themselves formed by living organisms, may be inherently slow to recover from severe impacts. This is because the organisms creating the structure, such as corals, sea grasses or kelp, are often slow growing and themselves must recover to accommodate complete recovery of the associated plant and animal communities.

The susceptibility of such biogenically structured communities as coral reefs (Gulf of Mexico and South Atlantic Bight), soft coral communities in submarine canyons (New England and Middle Atlantic Bight), seagrass beds (Eastern Gulf of Mexico and Bering Sea), and macroalgal beds (California and Alaska) has been a particularly great concern related to offshore oil and gas development. Although recovery of these communities if severely damaged would be protracted, most of these communities exist in nondepositional environments. Consequently, particle-borne contaminants would not be expected to accumulate in their habitats except if the inputs were particularly concentrated (e.g., a bulk discharge of drilling fluids on top of the habitat in question). Biogenically structured communities are, on the other hand, highly susceptible to direct physical impacts such as pipeline emplacement and anchor dragging. The distribution of biogenically structured communities varies widely among continental shelf environments of the U.S. (Table 3.2).

LITERATURE CITED

Allen, J.S., R.C. Beardsley, J.O. Blanton, W.C. Boicourt, B. Butman, L.K. Coachman, A. Huyer, T.H. Kinder, T.C. Royer, J.D. Schumacher, R.L. Smith, W. Sturges and C.D. Winant. 1983. Physical oceanography of continental shelves. Reviews of Geophysics and Space Physics 21: 1149-1181.

Anderson, J.B., R.B. Wheeler and R.R. Schwarzer. 1981. Surficial sediments and suspended particulate matter. Pages 59-67 *in* B.S. Middleditch (ed.), Environmental Effects of Offshore Oil Production. The Buccaneer Gas and Oil Field Study. Plenum Press, New York.

Atkinson, L.P., R.T. Barber and M.J. Wade. 1979. Hydrography, nutrients, chlorophyll, and organic carbon of South Atlantic/Georgia Bight. Pages 15-81 *in* South Atlantic Benchmark Program, Outer Continental Shelf (OCS) Environmental Studies, Volume 3, Results of Georgia Bight of South Atlantic Ocean. Final Rept. Contract No. AA550-CT7-2 to Bur. of Land Management, Texas Instruments, Inc., Dallas, Texas.

Atkinson, L.P., T.N. Lee, J.O. Blanton and W.S. Chandler. 1983. Climatology of the southeastern United States continental shelf waters. J. Geophys. Res. 88(C8): 4705-4718.

Atkinson, L.P., G.A. Paffenhöfer and W.M. Dunstan. 1978. The chemical and biological effect of a Gulf stream intrusion off St. Augustine, Florida. Bull. Mar. Sci. 28: 667-679.

Avent, R.M., M.E. King and R.H. Gore. 1977. Topographic and faunal studies on shelf edge prominences off the central eastern Florida coast. Int. Rev. ges. Hydrobiol. 62: 185-208.

Baker, J.H., K.T. Kimball, W.D. Jobe, J. Janousek, C.L. Howard and P.R. Chase. 1981. Part 6, Benthic Biology. Pages 1-137 *in* C.A. Bedinger, Jr. (ed.), Ecological Investigations of Petroleum Production Platforms in the Central Gulf of Mexico. Pollutant Fate and Effects Studies. Volume I. Rept. to Bur. Land Management, Contract No. AA551-CT8-17, Southwest Research Institute, San Antonio, Texas.

Balcom, B. 1981. Results of the 1975-78 BLM Baseline Studies and Analysis Program: Subtidal Benthic Environments of the Southern California Bight. Bureau of Land Management, Pacific OCS Office, Los Angeles, California, 73 p.

Barnard, J.L. 1963. Relationship of benthic Amphipoda to invertebrate communities of inshore sublittoral sands of southern California. Pac. Natur. 3: 439-466.

Barnard, J.L. and O. Hartman. 1959. The sea bottom off Santa Barbara, California: Biomass and community structure. Pac. Natur. 1: 1-16.

Barnes, P.W., E. Reimnitz and A.S. Naidu. 1981. Sediments. Pages 99-100 *in* D.W. Norton and W.M. Sackinger (eds.), Proc. of a Synthesis Meeting: Beaufort Sea - Sale 71 - Synthesis Report. National Oceanic and Atmospheric Administration, Outer Continental Shelf Environmental Assessment Program, Juneau, Alaska.

Battelle/Woods Hole Oceanographic Institution. 1983. Georges Bank Benthic Infauna Monitoring Program. Final Rept., Year 1, Contract No. 14-12-001-29192, U.S. Dept. of Interior, Minerals Management Service, New York OCS Office, New York, 153 p.

Battelle/Woods Hole Oceanographic Institution. 1984. Georges Bank Benthic Infauna Monitoring Program. Final Rept., Year 2, Contract No. 14-12-001-29192, U.S. Dept of Interior, Minerals Management Service, Atlantic OCS Office, Vienna, Virginia, 173 p.

Beardsley, R.C. and W.C. Boicourt. 1981. On estuarine and continental shelf circulation in the Middle Atlantic Bight. Pages 198-233 *in* B.A. Warren and C. Wunsch (eds.), Evolution of Physical Oceanography. MIT Press, Cambridge, Massachusetts.

Beardsley, R.C., W.C. Boicourt and D.V. Hansen. 1976. Physical oceanography of the Middle Atlantic Bight. Pages 20-43 *in* M.G. Gross (ed.), Middle Atlantic Continental Shelf and the New York Bight. Amer. Soc. Limnol. Oceanogr. Spec. Symp. 2.

Bedinger, C.A., Jr. (ed.). 1981. Ecological Investigations of Petroleum Production Platforms in the Central Gulf of Mexico. Rept. to Bur. Land Management, Contract No. AA551-CT8-17, Southwest Research Institute, San Antonio, Texas.

Berryhill, H.L., Jr. (ed.). 1977. Environmental Studies, South Texas Outer Continental Shelf, 1975: An Atlas and Integrated Summary. Final Rept. to Bur. Land Management, U.S. Geological Survey, Corpus Christi, Texas, 303 p.

Bilyard, G.R. and A.G. Carey, Jr. 1979. Distribution of western Beaufort Sea polychaetous annelids. Mar. Biol. 54: 329-339.

Bilyard, G.R. and A.G. Carey, Jr. 1980. Zoogeography of western Beaufort Sea Polychaeta (Annelida). Sarsia 65: 19-26.

Blanton, J.O. 1980. The transport of freshwater off a multi-inlet coast. Pages 49-64 *in* P. Hamilton and K.B. MacDonald (eds.), Estuarine and Wetland Processes with Emphasis on Modeling. Plenum Press, New York.

Blanton, J.O., L.P. Atkinson, L.J. Pietrafesa and T.N. Lee. 1981. The intrusion of Gulf Stream water across the continental shelf due to topographically induced upwelling. Deep-Sea Res. 28: 393-405.

Boesch, D.F. 1979. Benthic ecological studies: Macrobenthos. *In* Middle Atlantic Outer Continental Shelf Environmental Studies. Volume IIB. Chemical and Biological Benchmark Studies. Rept. to Bur. Land Management, Contract No. AA550-CT6-62, Virginia Institute of Marine Science, Gloucester Point, Virginia.

Boesch, D.F. and M.A. Bowen (in press). Bathymetric distribution of assemblages of macrobenthos in the Middle Atlantic Bight, U.S.A. Mar. Ecol. Prog. Ser.

Boesch, D.F. and R. Rosenberg. 1981. Response to stress in marine benthic communities. Pages 179-200 *in* G.W. Barrett and R. Rosenberg (eds.), Stress Effects on Natural Ecosystems. John Wiley & Sons, Ltd., New York.

Bothner, M.H., E.C. Spiker, P.P. Johnson, R.R. Rendigs and P.J. Aruscavage. 1981a. Geochemical evidence for modern sediment accumulation on the continental shelf off southern New England. J. Sediment. Petrol. 51: 281-292.

Bothner, M.H., C.M. Parmenter, R.R. Rendigs, B. Butman, L.J. Poppe and J.D. Milliman. 1981b. Studies of Suspended Matter along the North and Middle Atlantic Outer Continental Shelf. U.S. Geological Survey Open-File Report 82-938, 34 p.

Bowen, M.A., P.O. Smyth, D.F. Boesch and J. van Montfrans. 1979. Comparative biogeography of benthic macrocrustaceans of the Middle Atlantic (U.S.) continental shelf. Bull. Biol. Soc. Wash. 3: 214-255.

Broad, A.C., W. Griffiths and A.G. Carey, Jr. 1981. Invertebrates. Pages 27-31 *in* D.W. Norton and W.M. Sackinger (eds.), Proc. of a Synthesis Meeting: Beaufort Sea - Sale 71 - Synthesis Report. National Oceanic and Atmospheric Administration, Outer Continental Shelf Environmental Assessment Program, Juneau, Alaska.

Brooks, J.M., D.A. Weisenburg, C.R. Schwab, E.L. Estes and R.F. Shokes. 1981. Surficial sediments and suspended particulate matter. Pages 69-115 *in* B.S. Middleditch (ed.), Environmental Effects of Offshore Oil Production. The Buccaneer Gas and Oil Field Study. Plenum Press, New York.

Brower, W.A., J.M. Meserve and R.G. Quayle. 1972. Environmental Guide for the U.S. Gulf Coast. U.S. National Climate Center, Asheville, North Carolina, 177 p.

Brower, W.A., H. Searby and J. Wise. 1977. Climatic Atlas of the Outer Continental Shelf Water and Coastal Regions of Alaska, Bering Sea. Vol. 2. U.S. Dept. of Commerce, National Oceanic and Atmospheric Administration, Boulder, Colorado, 443 p.

Burrell, D.C., K. Tommos, A.S. Naidu and C.M. Hoskin. 1981. Some geochemical characteristics of Bering Sea sediments. Pages 305-319 *in* D.W. Hood and J.A. Calder (eds.), The Eastern Bering Sea Shelf: Oceanography and Resources. Volumes 1&2. National Oceanic and Atmospheric Administration, Office of Marine Pollution Assessment, Washington, D.C.

Buss, B.A. and K.S. Rodolfo. 1972. Suspended sediments in continental shelf waters off Cape Hatteras, North Carolina. Pages 263-280 *in* D.J.P. Swift, D.B. Duane and O.H. Pilkey (eds.), Shelf Sediment Transport: Process and Pattern. Dowden, Hutchinson & Ross, Inc., Stroudsburg, Pennsylvania.

Butman, B., R.C. Beardsley, B. Magnell, D. Frye, J.A. Vermerch, R. Schlitz, R. Limeburner, W.R. Wright and M.A. Noble. 1982. Recent observations of the mean circulation on Georges Bank. J. Phys. Oceanogr. 12: 569-691.

Butman, B., M. Noble and D.W. Folger. 1979. Long-term observation of bottom sediment movement on the Mid-Atlantic continental shelf. J. Geophys. Res. 84: 1187-1205.

Cannon, G.A. and G.S.E. Lagerloef. 1983. Topographic influences on coastal circulation: A review. Pages 235-252 *in* H.G. Gade, A. Edwards and H. Svendsen (eds.), Coastal Oceanography. Plenum Press, New York.
Carey, A.G. 1972. Ecological observations on the benthic invertebrates from the central Oregon continental shelf. Pages 422-443 *in* A.T. Pruter and D.L. Alverson (eds.), The Columbia River Estuary and Adjacent Ocean Waters. Univ. of Washington Press, Seattle.
Carey, A.G., Jr. and R.E. Ruff. 1977. Ecological studies of the benthos in the western Beaufort Sea with special reference to bivalve molluscs. Pages 505-530 *in* M.J. Dunbar (ed.), Polar Oceans. Proc. Polar Oceans Conf., McGill Univ., Montreal, May 1974. Arctic Inst. N. Amer., Calgary, Canada.
Carey, A.G., Jr., R.E. Ruff, J.G. Castillo and J.J. Dickinson. 1974. Benthic ecology of the western Beaufort Sea continental margin: Preliminary results. Pages 665-680 *in* J.C. Reed and J.E. Sater (eds.), The Coast and Shelf of the Beaufort Sea. Proc. Sympos. Beaufort Sea Coast and Shelf Res., Arctic Inst. N. Amer., Arlington, Virginia.
Carey, A.G., Jr., P.H. Scott and K.R. Walters. 1984. Distributional ecology of shallow southwestern Beaufort Sea (Arctic Ocean) bivalve Mollusca. Mar. Ecol. Prog. Ser. 17: 125-134.
Cérame-Vivas, M.J. and I.E. Gray. 1966. The distributional patterns of benthic invertebrates of the continental shelf off North Carolina. Ecology 47: 260-270.
Chelton, D.B., P.A. Bernal and J.A. McGowan. 1982. Large-scale interannual physical and biological interactions in the California Current. J. Mar. Res. 40: 1095-1125.
Chew, F. 1953. Results of hydrographic and chemical investigations in the region of the "red tide" bloom on the west coast of Florida in November, 1952. Bull. Mar. Sci. 2: 610-625.
Chew, F. 1955. On the Offshore Circulation and a Convergence Mechanism in the Red Tide Region off the West Coast of Florida. Rept. to Flor. State Bd. Conservation, Univ. of Miami Mar. Fisheries Res. Tech. Rept. No. 55-5.
Coleman, J.M. and D.B. Prior. 1983. Deltaic influences on shelfedge instability processes. Pages 121-137 *in* D.J. Stanley and G.T. Moore (eds.), The Shelfbreak: Critical Interface on Continental Margins. Soc. Economic Paleontologists and Mineralogists, Spec. Publ. No. 33, Tulsa, Oklahoma.
Collard, S.B. and C.N. D'Asaro. 1973. Benthic invertebrates of the eastern Gulf of Mexico. Pages IIIG-1 to IIIG-28 *in* J.I. Jones, R.E. Ring, M.O. Rinkel and R.E. Smith (eds.), A Summary of the Knowledge of the Eastern Gulf of Mexico. State Univ. System of Florida, Institute of Oceanography, St. Petersburg, Florida.
Craig, P.C. 1984. Fish use of coastal waters of the Alaskan Beaufort Sea: A review. Trans. Amer. Fish. Soc. 113: 265-282.
Creager, J.S. and D.A. McManus. 1967. Geology of the floor of the Bering and Chukchi Seas-American Studies. *In* D.M. Hopkins (ed.), The Bering Land Bridge. Stanford Univ. Press, Stanford, California.
Curray, J.R. 1965. Quaternary history, continental shelves of the United States. Pages 723-735 *in* H.E. Wright and G. Frey (eds.), The Quaternary of the United States. Princeton University Press, Princeton, New Jersey.
Dames & Moore. 1979. Mississippi, Alabama, Florida Outer Continental Shelf Baseline Monitoring Survey; MAFLA, 1977/78. Volume II, Compendium of Work Element Reports. Final Rept., Contract No. AA550-CT7-34 to Bur. of Land Management, New Orleans, Louisiana.
Day, J.H., J.G. Field and M.P. Montgomery. 1971. The use of numerical methods to determine the distribution of the benthic infauna across the continental shelf of North Carolina. J. Anim. Ecol. 40: 93-125.
Dayton, P.K. and M.J. Tegner. 1984. Catastrophic storms, El Niño, and patch stability in a southern California kelp community. Science 224: 283-285.
Dept. of Interior. 1983a. Final Environmental Impact Statement. Proposed Southern California Lease Offering, April 1984. Department of the Interior, Minerals Management Service, Pacific OCS Region, Los Angeles, California.

Dept. of Interior. 1983b. Draft Environmental Impact Statement. Proposed 1983 Outer Continental Shelf Oil and Gas Lease Sale Offshore Central California. OCS Sale No. 73. Department of the Interior, Minerals Management Service, Pacific OCS Region, Los Angeles, California.

Dept. of Interior. 1983c. Draft Environmental Impact Statement. Diapir Field Lease Offering (June 1984). Dept. of the Interior, Minerals Management Service, Alaska OCS Region, Anchorage, Alaska.

Dept. of Interior. 1985. 5-Year Outer Continental Shelf Oil and Gas Leasing Program for Mid-1986 through Mid-1991. Draft Proposed Program. U.S. Dept. of Interior, Minerals Management Service, Reston, Virginia.

DeRouen, L.R., R.W. Hann, D.M. Casserly and C. Giammona (eds.). 1982. West Hackberry Brine Disposal Project Pre-Discharge Characterization. Rept. to Dept. Energy, Contract No. DE-AC96-80P010228, McNeese State Univ., Lake Charles, Louisiana, and Texas A&M Univ. Research Foundation, College Station.

Dorman, C.E. and D.P. Palmer. 1981. Southern California summer coastal upwelling. Pages 44-56 *in* F.A. Richards (ed.), Coastal upwelling. American Geophysical Union, Washington, D.C.

Doyle, L.J. and T.N. Sparks. 1980. Sediments of Mississippi, Alabama, and Florida (MAFLA) continental shelf. J. Sediment. Petrology 50: 905-916.

Duane, D.B., M.E. Field, E.P. Meisburger, D.J.P. Swift and S.J. Williams. 1972. Linear shoals on the Atlantic inner continental shelf, Florida to Long Island. Pages 447-498 *in* D.J.P. Swift, D.B. Duane and O.H. Pilkey (eds.), Shelf Sediment Transport: Process and Pattern. Dowden, Hutchinson & Ross, Inc., Stroudsburg, Pennsylvania.

Emery, K.O. 1960. The Sea Off Southern California. John Wiley and Sons, Inc., New York, 366 p.

Emery, K.O. and E. Uchupi. 1972. Western North Atlantic: Topography, Rocks, Structure, Water, Life, and Sediments. American Assoc. Petroleum Geologists, Tulsa, Oklahoma, 532 p.

Etter, P.C. and J.D. Cochrane. 1975. Water temperature on the Texas-Louisiana shelf. Marine Advisory Bulletin, Commerce. Texas A&M Sea Grant Publ. No. TAMU-SG-75-604, College Station, Texas, 22 p.

Fanning, K.A., K.L. Carder and P.R. Betzer. 1982. Sediment resuspension by coastal waters: a potential mechanism for nutrient re-cycling on the ocean's margins. Deep-Sea Res. 29: 953-965.

Fauchald, K. and G.F. Jones. 1977. Benthic Macrofauna. Chapter 24 *In* Year I Southern California Baseline Study, Rep. to Bur. Land Management, Pacific OCS Office, Los Angeles, California, Contract No. AA551-CT5-52, Science Applications, Inc., La Jolla, California.

Fauchald, K. and G.F. Jones. 1978. Variation in Community Structure of Shelf, Slope and Basin Macrofaunal Communities of the Southern California Bight. Chapter 19 *in* Year II Benthic Study, Rep. to Bur. Land Management, Pacific OCS Office, Los Angeles, California, Contract No. AA551-CT6-40, Science Applications, Inc., La Jolla, California.

Feder, H.M. and S.C. Jewett. 1980. Distribution, Abundance, Community Structure and Trophic Relationships of the Nearshore Benthos of the Kodiak Continental Shelf. Final Report to National Oceanic and Atmospheric Administration, Outer Continental Shelf Environmental Assessment Program. Inst. of Mar. Sci., Univ. of Alaska, Fairbanks, 243 p.

Feder, H.M. and S.C. Jewett. 1981. Feeding interactions in the eastern Bering Sea with emphasis on the benthos. Pages 1229-1261 *in* D.W. Hood and J.A. Calder (eds.), The Eastern Bering Sea Shelf: Oceanography and Resources. Volumes 1&2. National Oceanic and Atmospheric Administration, Office of Marine Pollution Assessment, Washington, D.C.

Feder, H.M. and G.E.M. Matheke. 1979. Distribution, Abundance, Community Structure, and Trophic Relationships of the Benthic Infauna of the Northeast Gulf of Alaska. Final Report, RU #281, to National Oceanic and Atmospheric Administration, Outer

Continental Shelf Environmental Assessment Program. Inst. of Mar. Sci., Univ. of Alaska, Fairbanks, 247 p.

Field, M.E. and B.D. Edwards. 1980. Slopes of the Southern California Borderland: A regime of mass transport. Pages 169-184 *in* M.E. Field, A.H. Bouma, I.P. Colburn, R.G. Douglas and J.C. Ingle (eds.), Quarternary Depositional Environments of the Pacific Coast. Society of Economic Paleontologists and Mineralogists, Pacific Section, Pacific Coast Paleogeography Symposium 4.

Field, M.E., C.H. Nelson, D.A. Cacchione and D.E. Drake. 1981. Sand waves on an epicontinental shelf: northern Bering Sea. Mar. Geol. 42: 233-258.

Flint, R.W. 1981. Gulf of Mexico outer continental shelf benthos: Macroinfaunal environment relationships. Biol. Oceanogr. 1: 135-155.

Flint, R.W. and J.S. Holland. 1980. Benthic infaunal variability on a transect in the Gulf of Mexico. Estuar. Coast Mar. Sci. 10: 1-14.

Flint, R.W. and N.N. Rabalais. 1980a. Polychaete ecology and niche patterns: Texas continental shelf. Mar. Ecol. Prog. Ser. 3: 193-202.

Flint, R.W. and N.N. Rabalais (eds.). 1980b. Environmental Studies, South Texas Outer Continental Shelf, 1975-1977. Volume III, Study Area Final Reports. Rept. to Bur. Land Management, Contract No. AA551-CT8-51, Univ. of Texas Marine Science Inst., Port Aransas, 648 p.

Flint, R.W. and N.N. Rabalais (eds.). 1981. Environmental Studies of a Marine Ecosystem South Texas Outer Continental Shelf. Univ. of Texas Press, Austin, 240 p.

Frankenberg, D. and A.S. Leiper. 1977. Seasonal cycles in benthic communities of the Georgia continental shelf. Pages 383-397 *In* B.C. Coull (ed.), Ecology of Marine Benthos. Univ. of South Carolina Press, Columbia.

Freeland, G.L., D.J.P. Swift, W.L. Stubbefield and A.E. Cook. 1976. Surficial sediments of the NOAA-MESA study areas in the New York Bight. Pages 90-101 *in* M.G. Gross (ed.), Middle Atlantic Continental Shelf and the New York Bight. Am. Soc. Limnol. Oceanogr. Spec. Symp. 2.

Gallaway, B.J. 1981. An Ecosystem Analysis of Oil and Gas Development on the Texas-Louisiana Continental Shelf. U.S. Fish and Wildlife Service, Office of Biological Services, Publ. No. FWS/OBS-81/27, 89 p.

Gardner, J.V., T.L. Vallier and W.E. Dean. 1979. Sedimentology and geochemistry of surface sediments, and the distribution of faults and potentially unstable sediments, St. George Basin region of outer continental shelf, southern Bering Sea. *In* Environmental Assessment of the Alaska Continental Shelf, Final Rept., RU #206, U.S. Dept. of Commerce, National Oceanic and Atmospheric Administration, Outer Continental Shelf Environmental Assessment Program.

Gaston, G.R. 1985. Effects of hypoxia on macrobenthos of the inner shelf off Cameron, Louisiana. Estuar. Coast. Shelf Sci. 20(5): 603-613

Godshall, F.A. and R.G. Williams (eds.). 1981. A Climatology and Oceanographic Analysis of the California Outer Continental Shelf Region. U.S. Dept. Commerce, Natl. Oceanic and Atmospheric Admin., Environmental Data and Information Service, Washington, D.C.

Greene, H.G., S.H. Clarke, Jr., M.E. Field, F.I. Linker and H.C. Wagner. 1975. Preliminary Report on the Environmental Geology of Selected Areas of the Southern California Continental Borderland. U.S. Geological Survey Open File Report No. 75-596, 70 p.

Gross, M.G. (ed.). 1976. Middle Atlantic Continental Shelf and the New York Bight. Amer. Soc. Limnol. Oceanogr. Spec. Symp. Vol. 2, 441 p.

Gross, M.G., D.A. McManus and H.-Y. Ling. 1967. Continental shelf sediment, northwestern United States. J. Sediment Petrol. 37: 790-795.

Gundlach, E.R. and M.O. Hayes. 1978. Vulnerability of coastal environments to oil spill impacts. Mar. Tech. Soc. J. 12: 18-27.

Haflinger, K. 1981. A survey of benthic infaunal communities of the southeastern Bering Sea Shelf. Pages 1091-1103 *in* D.W. Hood and J.A. Calder (eds.), The Eastern Bering

Sea: Oceanography and Resources. Volumes 1&2. National Oceanic and Atmospheric Administration, Office of Marine Pollution Assessment, Washington, D.C.
Hampton, M.A. 1983. Geology of the Kodiak Shelf, Alaska: Environmental considerations for resource development. Cont. Shelf Res. 1: 253-281.
Hann, R.W., Jr. and R.E. Randall (eds.). 1980. Evaluation of Brine Discharge from the Bryan Mound Site of the Strategic Petroleum Reserve Program. Rept. to Dept. Energy, Contract No. DE-FC96-79P010114, Texas A&M Univ. Research Foundation, College Station, Texas.
Harper, D.E., Jr. and L.D. McKinney. 1980. Benthos. Chapter 5 *in* R.W. Hann, Jr. and R.E. Randall (eds.), Evaluation of Brine Disposal from the Bryan Mound Site of the Strategic Petroleum Reserve Program. Rept. to Dept. Energy, Contract No. DE-FC96-79P010114, Texas A&M Univ. Research Foundation, College Station, Texas.
Harper, D.E., Jr., L.D. McKinney, R.R. Salazar and R.J. Case. 1981a. The occurrence of hypoxic bottom water off the upper Texas coast and its effects on the benthic biota. Contr. Mar. Sci. 24: 53-79.
Harper, D.E., Jr., D.L. Potts, R.R. Salzer, R.J. Case, R.L. Jaschek and C.M. Walker. 1981b. Distribution and abundance of macrobenthic and meiobenthic organisms. Pages 133-177 *in* B.S. Middleditch (ed.), Environmental Effects of Offshore Oil Production. The Buccaneer Gas and Oil Field Study. Plenum Press, New York.
Hedgpeth, J.W. 1953. An introduction to the zoogeography of the northwestern Gulf of Mexico with reference to the invertebrate fauna. Publ. Inst. Mar. Sci., Univ. of Texas 3: 107-224.
Hela, I. 1956. A pattern of coastal circulation inferred from synoptic salinity data. Bull. Mar. Sci. 6: 74-83.
Herbst, G.N., A.B. Williams and B.B. Boothe, Jr. 1979. Reassessment of northern geographic limits for decapod crustacean species in the Carolinian province, U.S.A.; some major range extensions itemized. Proc. Biol. Soc. Wash. 91: 989-998.
Hickey, B.M. 1979. The California Current system—hypotheses and facts. Prog. Oceanogr. 8: 191-279.
Holland, J.S. 1977. Invertebrate epifauna and macroinfauna. Chapter 9 *in* Environmental Studies, South Texas Outer Continental Shelf, Biology and Chemistry. Final Rept., Contract No. AA550-CT6-17 to Bur. of Land Management. Univ. of Texas Marine Science Institute, Port Aransas, Texas.
Holland, J.S. 1979. Benthic invertebrates: Macroinfauna. Chapter 17 *in* Environmental Studies, South Texas Outer Continental Shelf, Biology and Chemistry. Final Rept., Contract No. AA550-CT7-11 to Bur. of Land Management. Univ. of Texas Marine Science Institute, Port Aransas, Texas.
Hood, D.W. and J.A. Calder (eds.). 1981. The Eastern Bering Sea Shelf: Oceanography and Resources. Volumes 1&2. National Oceanic and Atmospheric Administration, Office of Marine Pollution Assessment, Washington, D.C., 1339 p.
Hopkins, T. and W. Schroeder. 1981. Physical and chemical oceanography. Pages 21-47, Section 2, Chapter 12, Florida Middle Ground *in* R. Rezak and T.J. Bright (eds.), Northern Gulf of Mexico Topographic Features Study. Rept. to Bur. Land Management, Contract No. AA551-CT8-35, Texas A&M Research Foundation, College Station, Texas.
Hopkins, T.S. and N. Garfield, III. 1981. Physical origins of Georges Bank water. J. Mar. Res. 39: 465-500.
Huang, W.H. 1981. Part 2, Sediment Physical Characterization. Pages 53a-132 *in* C.A. Bedinger, Jr. (ed.), Ecological Investigations of Petroleum Production Platforms in the Central Gulf of Mexico. Volume I, Pollutant Fate and Effects Studies. Rept. to Bur. Land Management, Contract No. AA551-CT8-17, Southwest Research Institute, San Antonio, Texas.
Huh, O.K., W.J. Wiseman, Jr. and L.R. Rouse, Jr. 1981. Intrusion of Loop Current waters onto the West Florida continental shelf. J. Geophys. Res. 86, C5, May 20: 4186-4192.

Huyer, A., R.L. Smith and E.J.C. Sobey. 1978. Seasonal differences in low-frequency current fluctuations over the Oregon continental shelf. J. Geophys. Res. 83: 5077-5089.
Ingham, M.C. (ed.). 1982. Summary of the Physical Oceanographic Processes and Features Pertinent to Pollution Distribution in the Coastal and Offshore Waters of the Northeastern United States, Virginia to Maine. National Oceanic and Atmospheric Administration, Northeast Monitoring Program, NOAA Technical Memorandum NMFS-F/NEC-17, Woods Hole, Massachusetts, 166 p.
Ingraham, W.J., Jr. 1981. Shelf environment. Pages 455-469 *in* D.W. Hood and John A. Calder (eds.), The Eastern Bering Sea Shelf: Oceanography and Resources. Volumes 1&2. National Oceanic and Atmospheric Administration, Office of Marine Pollution Assessment, Washington, D.C.
Jackson, J.B. and F.N. Kurz. 1983. Arctic Summary Report. Outer Continental Shelf Oil and Gas Activities in the Arctic and their Onshore Impacts. Rept. to U.S. Dept. Interior, Minerals Management Service, Contract No. 14-08-0001-19719, Rogers, Golder & Halpern, Inc., Reston, Virginia, 81 p.
Jackson, W.B. (ed.). 1977. Environmental Assessment of an Active Oil Field in the Northwestern Gulf of Mexico, 1976-1977. National Oceanic and Atmospheric Administration, National Technical Information Service, No. PB283890, Springfield, Virginia.
Jewett, S.C. and H.M. Feder. 1981. Epifaunal invertebrates of the continental shelf of the eastern Bering and Chukchi Seas. Pages 1131-1153 *in* D.W. Hood and J.A. Calder (eds.), The Eastern Bering Sea Shelf: Oceanography and Resources. Volumes 1&2. National Oceanic and Atmospheric Administration, Office of Marine Pollution Assessment, Washington, D.C.Johnson, K.R. and C.H. Nelson. 1984. Side-scan sonar assessment of gray whale feeding in the Bering Sea. Science 225: 1150-1152.
Jones & Stokes Associates, Inc., 1981. Ecological Characterization of the Central and Northern California Coastal Region. Volume II, Part 1, Regional Characterization. U.S. Fish and Wildlife Service Publ. No. FWS/OBS-80/46.1.
Jones, G.F. 1969. The benthic macrofauna of the mainland shelf of southern California. Allan Hancock Monogr. Mar Biol. No. 4: 1-219.
Kamykowski, D. and J.L. Bird. 1981. Phytoplankton associations with the variable nepheloid layer on the Texas continental shelf. Estuar. Coastal Shelf Sci. 13: 317-326.
Keller, G.H., D. Lambert, G. Rowe and N. Staresinic. 1973. Bottom currents in the Hudson Canyon. Science 180: 181-183.
Knebel, H.J. 1981. Processes controlling the characteristics of the surficial sand sheet, U.S. Atlantic outer continental shelf. Mar. Geol. 42: 349-368.
Knebel, H.J. and D.W. Folger. 1976. Large sand waves on the Atlantic outer continental shelf around Wilmington Canyon, off eastern United States. Mar. Geol. 22: M7-M15.
Knebel, H.J. and E. Spiker. 1977. Thickness and age of the surficial sand sheet, Baltimore Canyon trough area. Bull. Amer. Assoc. Petrol. Geol. 61: 861-871.
Komar, P.D., R.H. Neudeck and L.D. Kulm. 1972. Observations and significance of deep-water oscillatory ripple marks on the Oregon continental shelf. Pages 601-619 *in* D.J.P. Swift, D.P. Duane and O.H. Pilkey (eds.), Shelf Sediment Transport: Process and Pattern. Dowden, Hutchinson & Ross, Inc., Stroudsburg, Pennsylvania.
Larsen, M.C., C.H. Nelson and D.R. Thor. 1981. Sedimentary processes and potential geologic hazards on the sea floor of the northern Bering Sea. Pages 247-261 *in* D.W. Hood and J.A. Calder (eds.), The Eastern Bering Sea Shelf: Oceanography and Resources. Volumes 1&2. National Oceanic and Atmospheric Administration, Office of Marine Pollution Assessment, Washington, D.C.
Lie, U. and J.C. Kelley. 1970. Benthic infauna communities off the coast of Washington and in Puget Sound: Identification and distribution of communities. J. Fish. Res. Bd. Canada 27: 621-651.
Lie, U. and D.S. Kisker. 1970. Species composition and structure of benthic infauna communities off the coast of Washington. J. Fish. Res. Bd. Canada 27: 2273-2285.
Ludwick, J.C. 1964. Sediments in the northeastern Gulf of Mexico. Pages 204-238 *in* R.L. Miller (ed.), Papers in Marine Geology. MacMillan Co., New York.

Lyons, W.G. 1980. Molluscan communities of the west Florida shelf. Bull. Amer. Malacol. Union 1979: 37-40.
Lyons, W.G. and S.B. Collard. 1974. Benthic invertebrate communities of the eastern Gulf of Mexico. Pages 157-165 *in* R.E. Smith (ed.), Proc. of Marine Environmental Implications of Offshore Drilling in the Eastern Gulf of Mexico. State Univ. System of Florida, Inst. of Oceanography, St. Petersburg, Florida.
Macintyre, I.G. 1970. New data on the occurrence of tropical reef corals on the North Carolina continental shelf. J. Elisha Mitchell Sci. Soc. 86: 178.
Matthews, J.B. 1983. Some aspects of circulation along the Alaskan Beaufort Sea coast. Pages 475-497 *in* H.G. Gade, A. Edwards and H. Svendsen (eds.), Coastal Oceanography. Plenum Press, New York.
Maurer, D. and W. Leathem. 1981a. Ecological distribution of polychaetous annelids from the New England outer continental shelf, Georges Bank. Int. Revue ges. Hydrobiol. 66: 505-528.
Maurer, D. and W. Leathem. 1981b. Polychaete feeding guilds from Georges Bank, U.S.A. Mar. Biol. 62: 161-171.
McGrail, D.W. and M. Carnes. 1983. Shelfedge dynamics and the nepheloid layer in the northwestern Gulf of Mexico. Pages 251-264 *in* D.J. Stanley and G.T. Moore (eds.), The Shelfbreak: Critical Interface on Continental Margins. Society of Economic Paleontologists and Mineralogists Spec. Publ. No. 33, Tulsa, Oklahoma.
McKinney, T.F., W.L. Stubblefield and D.J.P. Swift. 1974. Large-scale current lineations in the central New Jersey shelf: Investigations by side scan sonar. Mar. Geol. 17: 79-102.
McManus, D.A. 1972. Bottom topography and sediment texture near the Columbia River mouth. Pages 241-253 *in* A.T. Pruter and D.L. Alverson (eds.), The Columbia River Estuary and Adjacent Ocean Waters. Univ. of Washington Press, Seattle.Meade, R.H. 1972. Sources and sinks of suspended matter on continental shelves. Pages 249-262 *in* D.J.P. Swift, D.B. Duane and O.H. Pilkey (eds.), Shelf Sediment Transport: Process and Pattern. Dowden, Hutchinson & Ross, Inc., Stroudsburg, Pennsylvania.
Menzies, R.J., O.H. Pilkey, B.W. Blackwelder, D. Dexter, P. Huling and L. McCloskey. 1966. A submerged reef off North Carolina. Inter. Rev. ges. Hydrobiol. 51: 393-431.
Michael, A. 1977. Benthic infauna, 4th quarterly progress report. Pages 288-327 *in* New England Outer Continental Shelf Environmental Benchmark. Fourth Quarterly Summary Report to Bur. of Land Management, Energy Resources Co., Inc., Cambridge, Massachusetts.
Michael, A.D., C.D. Long, D. Maurer and R.A. McGrath. 1983. Georges Bank Benthic Infauna Historical Study. Final Rept. to Minerals Management Service, Washington, D.C., 171 p.
Middleditch, B.S. (ed.). 1981. Environmental Effects of Offshore Oil Production. The Buccaneer Gas and Oil Field Study. Plenum Press, New York, 446 p.
Milliman, J.D. and R.H. Meade. 1983. World-wide delivery of river sediment to the oceans. J. of Geol. 91:1-21.
Minerals Management Service. 1983. Draft Regional Environmental Impact Statement Gulf of Mexico. Dept. of Interior, Washington, D.C., 735 p. and 14 visuals.
Mooers, C.N.K. and J.F. Price. 1975. General shelf circulation. Pages 41-50 *in* Compilation of and summation of historical and existing physical oceanography data from the eastern Gulf of Mexico. State Univ. System of Florida, Inst. of Oceanography, St. Petersburg, Florida.
Mooers, C.N.K., R.W. Garvine and W.W. Martin. 1979. Summertime synoptic variability of the Middle Atlantic shelf water/slope water front. J. Geophys. Res. 84: 4837-4854.
Murray, S.P. 1972. Observations on wind, tidal, and density-driven currents in the vicinity of the Mississippi River delta. Pages 127-142 *in* D.J.P. Swift, D.B. Duane and O.H. Pilkey (eds.), Shelf Sediment Transport: Process and Pattern. Dowden, Hutchinson & Ross, Inc., Stroudsburg, Pennsylvania.
Murray, S.P. 1976. Currents and Circulation in the Coastal Waters of Louisiana. Sea

Grant Publ. No. LSU-T-76-003, Center for Wetland Resources, Louisiana State Univ., Baton Rouge, 52 p.

Naidu, A.S., J. Ray and E. Reimnitz. 1981. Sediment resuspension, lateral transport, and depocenters. Pages 150-151 *in* D.W. Norton and W.M. Sackinger (eds.), Proc. of a Synthesis Meeting: Beaufort Sea - Sale 71 - Synthesis Report. National Oceanic and Atmospheric Administration, Outer Continental Shelf Environmental Assessment Program, Juneau, Alaska.

Nelsen, T.A., P.E. Gadd and T.L. Clarke. 1978. Wind-induced current flow in the upper Hudson shelf valley. J. Geophys. Res. 83(C12): 6073-6082.

Nelson, C.H., W. Dupré, M. Field and J.D. Howard. 1982. Variation in sand body types on the eastern Bering Sea epicontinental shelf. Geologie en Mijnbouw 61: 37-48.

Nelson, H. and J.S. Creager. 1977. Displacement of Yukon-derived sediment from Bering Sea to Chukchi Sea during Holocene time. Geology 5: 141-146.

Nelson, H., R.W. Rowland, S.W. Stoker and B.R. Larsen. 1981. Interplay of physical and biological sedimentary structures of the Bering continental shelf. Pages 1265-1296 *in* D.W. Hood and J.A. Calder (eds.), The Eastern Bering Sea Shelf: Oceanography and Resources. Volumes 1&2. National Oceanic and Atmospheric Administration, Office of Marine Pollution Assessment, Washington, D.C.

Niiler, P.P. 1976. Observations on low-frequency currents on the west Florida continental shelf. Memoires Societe Royale des Sciences de Liege 6: 331-358.

Nio, S.-D. and C.H. Nelson. 1982. The North Sea and northeastern Bering Sea: A comparative study of the occurrence and geometry of sand bodies of two shallow epicontinental shelves. Geologie en Mijnbouw 61: 105-114.

Nittrouer, C.A. and R.W. Sternberg. 1981. The formation of sedimentary strata in an allochthonous shelf environment: The Washington continental shelf. Mar. Geol. 42: 201-232.

Nowlin, W. 1971. Water masses and general circulation of the Gulf of Mexico. Oceanography International 12: 23-33.

Olsen, H.W., E.C. Clukey and C.H. Nelson. 1982. Geotechnical characteristics of bottom sediment in the northeastern Bering Sea. Geologie en Mijnbouw 6: 91-103.

Owens, E.H. and G.A. Robilliard. 1981. Shoreline sensitivity and oil spills—a reevaluation for the 1980's. Mar. Pollut. Bull. 12: 75-78.

Pearse, A.S. and L.G. Williams. 1951. The biota of the reefs off the Carolinas. J. Elisha Mitchell Sci. Soc. 67: 133-161.

Peckol, P. and R.B. Searles. 1983. Effects of seasonality and disturbance on population development in a Carolina continental shelf community. Bull. Mar. Sci. 33: 67-86.

Peckol, P. and R.B. Searles. 1984. Temporal and spatial patterns of growth and survival of invertebrate and algal populations of a North Carolina continental shelf community. Estuar. Coastal Shelf Sci. 18(2): 133-143.

Penland, S. and R. Boyd. 1982. Assessment of geological and human factors responsible for Louisiana coastal barrier erosion. Pages 14-38 *in* D.F. Boesch (ed.), Proceedings of the Conference on Coastal Erosion and Wetland Modification in Louisiana: Causes, Consequences, and Options. U.S. Fish and Wildlife Service, Office of Biological Services, Publ. No. FWS/OBS-82/59.

Peterson, R.E. 1980. Geology. Pages 11-43 *in* Environmental Assessment of the Alaskan Continental Shelf. Kodiak Interim Synthesis Report. National Oceanic and Atmospheric Administration and Bureau of Land Management, Outer Continental Shelf Environmental Assessment Program, Juneau, Alaska.

Pietrafesa, L.J. 1983. Shelfbreak circulation, fronts and physical oceanography: East and west coast perspectives. Pages 233-250 *in* D.J. Stanley and G.T. Moore (eds.), The Shelfbreak: Critical Interface on Continental Margins. Society of Economic Paleontologists and Mineralogists Spec. Publ. No. 33, Tulsa, Oklahoma.

Pilkey, O.H. and M.E. Field. 1972. Onshore transportation of continental shelf sediment: Atlantic southeastern United States. Pages 429-446 *in* D.J.P. Swift, D.B. Duane and O.H. Pilkey (eds.), Shelf Sediment Transport: Process and Pattern. Dowden, Hutchinson & Ross, Inc., Stroudsburg, Pennsylvania.

Pomeroy, L.R., L.P. Atkinson, J.O. Blanton, W.B. Campbell, T.R. Jacobsen, K.H. Kerrick and A.M. Woods. 1983. Microbial distribution and abundance in response to physical and biological processes on the continental shelf of southeastern U.S.A. Continental Shelf Res. 2: 1-20.

Reid, J.L., Jr., G.I. Roden and J.G. Wyllie. 1958. Studies of the California current system. Calif. Coop. Oceanic. Fish. Invest. Progr. Rep. 1 July-1 Jan. 1958, p. 27-56.

Rezak, R. and T.J. Bright (eds.). 1981. Northern Gulf of Mexico Topographic Features Study, Vols. 1-5. Rept. to Bur Land Management, Contract No. AA551-CT8-35, Texas A&M Research Foundation, College Station, Texas.

Rezak, R., T.J. Bright and D.W. McGrail. 1983. Reefs and Banks of the Northwestern Gulf of Mexico. Their Geological, Biological, and Physical Dynamics. Rept. to Minerals Management Service, Contract No. 14-12-001-29145, 501 p.

Richardson, M.D., A.G. Carey and W.A. Colgate. 1977. The Effects of Dredged Material Disposal on Benthic Assemblages. Appendix C, Tech. Rep. D-77-30, U.S. Army Corps of Eng. Waterways Exper. Stn., Vicksburg, Mississippi, 411 p.

Royer, T.C. 1983. Observations of the Alaska Coastal Current. Pages 9-30 *in* H.G. Gade, A. Edwards and H. Svendsen (eds.), Coastal Oceanography. Plenum Press, New York.

Rusnak, G.A. 1960. Sediments of Laguna Madre, Texas. Pages 153-196 *in* F.P. Shepard, F.B. Phleger and T.H. van Andel (eds.), Recent Sediments, Northwest Gulf of Mexico. American Assoc. Petroleum Geologists, Tulsa, Oklahoma.

Sambrotto, R.N., J.J. Goering and C.P. McRoy. 1984. Large yearly production of phytoplankton in the western Bering Strait. Science 225: 1147-1150.

Sanders, H.L. 1969. Benthic marine diversity and the stability-time hypothesis. Brookhaven Symp. Biol. 22: 71-81.

Schaffner, L.C. and D.F. Boesch. 1982. Spatial and temporal resource use by dominant benthic Amphipoda (Ampeliscidae and Corophiidae) in the Middle Atlantic Bight outer continental shelf. Mar. Ecol. Prog. Ser. 9: 231-243.

Shepard, F.P. 1973. Submarine Geology, 3rd edition. Harper & Row Publishers, New York, 517 p.

Smethie, W.M., Jr., C.A. Nittrouer and R.F.L. Self. 1981. The use of Radon-222 as a tracer of sediment irrigation and mixing on the Washington continental shelf. Mar. Geol. 42: 173-200.

Smith, G.M. 1969. Marine Algae of the Monterey Peninsula California, 2nd edition. Stanford Univ. Press, Stanford, California, 752 p.

Smith, N.P. 1978. Hydrography project. Pages 7-37 *in* R.W. Flint (ed.), Environmental Studies, South Texas Outer Continental Shelf, Biology and Chemistry. Rept. to Bur. Land Management, Contract No. AA550-CT7-11, Univ. of Texas Marine Science Institute, Port Aransas, Texas.

Smith, N.P. 1980. Hydrographic project. *In* R.W. Flint and N.N. Rabalais (eds.), Environmental Studies, South Texas Outer Continental Shelf, 1975-1977. Volume III, Study Area Final Reports. Rept. to Bur. Land Management, Contract No. AA551-CT-51, Univ. of Texas Marine Science Institute, Port Aransas, Texas.

Sobey, E.J.C. 1980. Physical oceanography. Pages 47-86 *in* Environmental Assessment of the Alaskan Continental Shelf. Kodiak Interim Synthesis Report. National Oceanic and Atmospheric Administration and Bureau of Land Management, Outer Continental Shelf Environmental Assessment Program, Juneau, Alaska.

State Univ. System of Florida. 1977. Baseline Monitoring Studies, Mississippi, Alabama, Florida Outer Continental Shelf, 1975-1976. Rept. to Bur. Land Management, Contract No. 08550-CT5-30, State Univ. Syst. Flor., Inst. of Oceanography, St. Petersburg, Florida.

Sternberg, R.W. and D.A. McManus. 1972. Implications of sediment dispersal from long-term, bottom-current measurements on the continental shelf of Washington. Pages

181-194 *in* D.J.P. Swift, D.B. Duane and O.H. Pilkey (eds.), Shelf Sediment Transport: Process and Pattern. Dowden, Hutchinson & Ross, Inc., Stroudsburg, Pennsylvania.
Stoker, S. 1981. Benthic invertebrate macrofauna of the eastern Bering/Chukchi continental shelf. Pages 1069-1090 *in* D.W. Hood and J.A. Calder (eds.), The Eastern Bering Sea Shelf: Oceanography and Resources. Volumes 1&2. National Oceanic and Atmospheric Administration, Office of Marine Pollution Assessment, Washington, D.C.
Strauch, J.G., Jr., G.R. Tamm, W.H. Lippincott, B.R. Mate and K.W. Fucik. 1980. Biology. Pages 103-232 *in* Environmental Assessment of the Alaskan Continental Shelf. Kodiak Interim Synthesis Report. National Oceanic and Atmospheric Administration and Bureau of Land Management, Outer Continental Shelf Environmental Assessment Program, Juneau, Alaska.
Stubblefield, W.L., J.W. Lavelle, T.F. McKinney and D.J.P Swift. 1975. Sediment response to the present hydraulic regime on the central New Jersey Shelf. J. Sediment. Petrol. 45: 337-358.
Sturges, W. and J.C. Evans. 1983. On the variability of the Loop Current in the Gulf of Mexico. J. Mar. Res. 41: 639-653.
Sturges, W. and C. Horton. 1981. Circulation in the Gulf of Mexico. Pages 41-88 *in* Proc. Symp. Environmental Research Needs in the Gulf of Mexico (GOMEX), Key Biscayne, Florida, 30 September—5 October 1979, Volume IIA. U.S. Dept. of Commerce, National Oceanic and Atmospheric Administration, Atlantic Oceanographic and Meteorological Laboratories, Miami, Florida.
Sverdrup, H.U., M.W. Johnson and R.H. Fleming. 1942. The Oceans. Their Physics, Chemistry, and General Biology. Prentice-Hall, Inc., Englewood Cliffs, New Jersey, 1087 p.
Swift, D.J.P. 1970. Quaternary shelves and the return to grade. Mar. Geol. 8: 5-30.
Swift, D.J.P., G.L. Freeland, P.E. Gadd, G. Han, J.W. Lavelle and W.L. Stubblefield. 1976. Morphologic evolution and coastal sand transport, New York-New Jersey shelf. Pages 69-89 *in* M.G. Gross (ed.), Middle Atlantic Continental Shelf and the New York Bight. Am. Soc. Limnol. Oceanogr. Spec. Symp. 2.
Swift, D.J.P., J.W. Kofoed, F.P. Saulsbury and P. Sears. 1972. Holocene evolution of the shelf surface, central and southern Atlantic shelf of North America. Pages 499-574 *in* D.J.P. Swift, D.B. Duane and O.H. Pilkey (eds.), Shelf Sediment Transport: Process and Pattern. Dowden, Hutchinson & Ross, Inc., Stroudsburg, Pennsylvania.
Tenore, K.R. 1979. Macroinfaunal benthos of South Atlantic/Georgia Bight. Pages 281-308 *in* South Atlantic Benchmark Program, Outer Continental Shelf (OCS) Environmental Studies, Volume 3, Results of Georgia Bight of South Atlantic Ocean. Final Rept. Contract No. AA550-CT7-2 to Bur. of Land Management, Texas Instruments, Inc., Dallas, Texas.
Turner, R.E. and R.L. Allen. 1982. Bottom water oxygen concentration in the Mississippi River Delta Bight. Contr. Mar. Sci. 25: 161-172.
Twichell, D.C., C.E. McClennen and B. Butman. 1981. Morphology and processes associated with the accumulation of the fine-grained sediment deposit on the southern New England Shelf. J. Sediment. Petrol. 51: 269-280.
U.S. Geological Survey. 1976. Oil and Gas Development in the Santa Barbara Channel, Outer Continental Shelf, California. U.S. Geological Survey, Final Environmental Statement, Vol. I, 226 p.
Valentine, J.W. 1963. Biogeographic units as biostratigraphic units. Bull. Amer. Assoc. Petrol. Geol. 47: 457-466.
Valentine, J.W. 1966. Numerical analysis of marine molluscan ranges on the extratropical northeastern Pacific shelf. Limnol. Oceanogr. 11: 198-211.
Vincent, C.E., D.J.P. Swift and B. Hillard. 1981. Sediment transport in the New York Bight, North American Atlantic shelf. Pages 369-398 *in* C.E. Nittrouer (ed.), Sedimentary Dynamics of Continental Shelves. Elsevier, New York.
Walsh, J.J., T.E. Whitledge, F.W. Barvenik, C.D. Wirwick, S.O. Howe, W.E. Esais

and J.T. Scott. 1978. Wind events and food chain dynamics within the New York Bight. Limnol. Oceanogr. 23: 659-683.
Ward, C.H., M.E. Bender and D.J. Reish (eds.). 1979. The Offshore Ecology Investigation. Effects of Oil Drilling and Production in a Coastal Environment. Rice Univ. Studies 65: 1-589.
Weeks, W.F. and G. Weller. 1984. Offshore oil in the Alaskan Arctic. Science 225: 371-378.
Wenner, E.L. and D.F. Boesch. 1979. Distribution patterns of epibenthic decapod Crustacea along the shelf-slope coenocline, Middle Atlantic Bight, U.S.A. Bull. Biol. Soc. Wash. 3: 106-113.
Wenner, E.L., D.M. Knott, R.F. Van Dolah and V.G. Burrell, Jr. 1983. Invertebrate communities associated with hard bottom areas in the South Atlantic Bight. Estuar. Coastal Shelf Sci. 17: 143-158.
Weston, D.P. and G.R. Gaston. 1982. Benthos. Chapter 5 *in* L.R. DeRouen, R.W. Hann, D.M. Casserly and C. Giammona (eds.), West Hackberry Brine Disposal Project Pre-Discharge Characterization. Rept. to Dept. Energy, Contract No. DE-AC96-80P010228, McNeese State Univ., Lake Charles, Louisiana and Texas A&M Univ. Research Foundation, College Station, Texas.
Wickett, W.P. 1967. Ekman transport and zooplankton concentrations in the North Pacific Ocean. J. Fish. Res. Bd. Canada 24: 581-594.
Wigley, R.L. 1961. Bottom sediments of Georges Bank. J. Sediment. Petrol. 31: 165-188.
Wigley, R.L. 1965. Density-dependent food relationships with references to New England ground fish. ICNAF Spec. Publ. 6: 51-514.
Wigley, R.L. 1968. Benthic invertebrates of the New England fishing banks. Bull. Amer. Littoral Soc. 5: 8-13.
Wigley, R.L. and A.D. McIntyre. 1964. Some quantitative comparisons of offshore meiobenthos and macrobenthos south of Martha's Vineyard. Limnol. Oceanogr. 9: 485-493.
Wigley, R.L. and R.B. Theroux. 1976. Macrobenthic Invertebrate Fauna of the Middle Atlantic Bight Region: Part II. Faunal Composition and Quantitative Distribution. National Marine Fisheries Center, Woods Hole, Massachusetts, 395 p.
Wigley, R.L. and R.B. Theroux. 1981. Macrobenthic Invertebrate Fauna of the Middle Atlantic Bight Region. Faunal Composition and Quantitative Distribution. Prof. Pap. U.S. Geological Survey 529-N.
Williams, A.B., L.R. McCloskey and I.E. Gray. 1968. New records of brachyuran decapod crustaceans from the continental shelf off North Carolina, U.S.A. Crustaceana 15: 41-66.
Windom, H.L. and P.R. Betzer. 1979. Trace metal chemistry of South Atlantic/Georgia Bight. Pages 153-195 *in* South Atlantic Benchmark Program, Outer Continental Shelf (OCS) Environmental Studies, Volume 3, Results of Georgia Bight of South Atlantic Ocean. Final Rept. Contract No. AA550-CT7-2 to Bur. of Land Management, Texas Instruments, Inc., Dallas, Texas.
Woodward-Clyde Consultants and Continental Shelf Asso., Inc. 1983. Southwest Florida Shelf Ecosystems Study - Year I. Final Rept., Contract No. 14-12-0001-29142 to Minerals Management Service, Gulf of Mexico OCS Region, Metairie, Louisiana.
Yerkes, R.F., H.G. Greene, J.C. Tinsley and K.R. Lajoie. 1980. Seismotectonic Setting of Santa Barbara Channel Area, Southern California. U.S. Geological Survey Open-File Report No. 80-299, 24 p.

CHAPTER 4

OFFSHORE OIL AND GAS DEVELOPMENT ACTIVITIES POTENTIALLY CAUSING LONG-TERM ENVIRONMENTAL EFFECTS

Jerry M. Neff, Nancy N. Rabalais and Donald F. Boesch

CONTENTS

INTRODUCTION

There often is much confusion about activities involved in offshore oil and gas development which might potentially result in adverse environmental effects. In the forefront of the minds of most people are, of course, large oil spills. Yet experience has shown that these occur rarely. Smaller accidental oil spills and discharges which are purposely made during normal operations are more pervasive. The regulation of operational discharges has come under increasing scrutiny in recent years. What exactly do they consist of, and how much is discharged? Finally, construction and transportation of equipment, materials and product are frequently not considered as causes of adverse environmental effects. What do they entail, and how might they affect the marine environment?

This chapter presents a description of activities involved in the exploration for and production of oil and gas in offshore environments. Special emphasis is placed on characterizing the nature and amount of discharges, both operational and accidental, based on recent experience. The chapter is intended as background and an information source for the chapters to follow, which will specifically consider the potential long-term effects of these activities. Excluded from this review, as they are from throughout the book, are those activities related to direct effects on human society and its economy and on air quality. Rather, this perspective is limited to effects on marine and coastal environments.

SEQUENCE OF ACTIVITIES

The sequence of activities necessary for evaluation of the resource potential, exploration, development, production and refining of oil and gas from offshore regions are summarized in Table 4.1, together with an abbreviated description of their potential effects.

The first industrial activity which takes place on the continental shelf is geophysical surveying to evaluate the oil and gas resource potential based on evidence of sources, reservoirs and traps in the geological strata. This involves the pulsing of high intensity acoustic signals through the ocean and sedimentary strata. Although early use of explosives resulted in some destruction of marine life, current technology is thought to be safe for fishes and invertebrates. In any case, long-term effects are unlikely. Concern continues to be raised about the effects of seismic surveying on marine mammals, particularly cetaceans which communicate with an elaborate repertoire of acoustic signals.

Once the geophysical data indicate a potential for recoverable oil and gas resources and leases for drilling rights are obtained, the exploratory drilling phase may be entered. The mobile drilling rigs, including barges, jackups, drill ships and semisubmersibles, which are used for exploratory drilling offshore, are fabricated at coastal or inland shipyards, not necessarily in the region of the exploratory drilling. In fact, under present economic conditions, most new large mobile drilling rigs are constructed in Asia and Europe. Rigs are towed or move under their own power to the site of exploratory drilling and anchored at multiple mooring points (may be dynamically positioned in deeper waters).

Initial drilling into the seabed (spudding in) in order to place risers to the surface results in the direct discharge of sediment, cuttings and drilling fluids at the seafloor. Thereafter, recirculated drilling fluids, necessary to cool, lubricate and transport solids from the drill bit, are separated from cuttings (the pulverized formation) on the rig. The cuttings are usually discharged overboard continuously, while the drilling fluids are reused and disposed of later, again generally overboard at the drilling location. The composition of drilling fluids is discussed in detail below. In addition to drilling discharges, water drainage from the deck of the rig may contain drilling fluids, oil and small quantities of industrial chemicals used aboard the rig. Sanitary wastes are usually discharged at sea after treatment.

The transport of materials and men to the rig may be accomplished by vessel or aircraft. As a result of increased vessel traffic, discharges of oily and sanitary wastes may increase. Coastal ports may be expanded, and navigation channels constructed or deepened. Helicopters and fixed wing aircraft may disturb nesting or aggregating birds or mammals.

Because the probability of discovery of an economically viable resource is low for any given exploratory well, there is typically a sparse distribution and brief duration of operational discharges during exploration. In many frontier regions, this may be all that occurs. If highly pressured gas or oil strata are drilled, the possibility of a blowout exists, but redundant blowout preventers make this a remote possibility.

TABLE 4.1
Major activities in the development of an offshore oil and gas field and their potential effects on marine and coastal environments

Activity	*Potential effects*
Evaluation	
Seismic surveying	Noise effects on fishes and mammals
Exploration	
Rig fabrication	Dredging and filling of coastal habitats (mostly overseas)
Rig emplacement	Seabed disturbance due to anchoring
Drilling	Discharge of drilling fluids and cuttings; risk of blowouts
Routine rig operations	Deck drainage and sanitary wastes
Rig servicing	Discharges from support vessels and coastal port development
Development and production	
Platform fabrication	Land use conflicts and increased channelization in heavily developed areas
Platform installation	Coastal navigation channels, seabed disturbance resulting from placement and subsequent presence of platform
Drilling	Larger and more heavily concentrated discharges of drilling fluids and cuttings; risk of blowouts
Completion	Increased risk of oil spills
Platform servicing	Dredges and coastal port development; discharges from vessels
Separation of oil and gas from water	Chronic discharges of petroleum and other pollutants
Fabrication of storage facilities and pipelines	Coastal use conflicts
Offshore emplacement of storage and pipelines	Seabed disturbances; effects of structures
Transfer to tankers and barges	Increased risk of oil spills; acute and chronic inputs of petroleum
Construction of onshore facilities for transportation and storage	Coastal use conflicts; alterations of wetlands in pipeline corridors
Pipeline operations	Oil spills; chronic leaks
Refining	
Construction and expansion	Coastal use conflicts
Operations	Increased pollutant loading; depends on regional demands, imports, etc.

In case a discovery of promising oil or gas shows, further drilling from mobile drilling rigs to delineate the bearing reservoirs is usually necessary before recovery of the resource begins. Should these results warrant, a fixed platform may be placed from which several development wells may be drilled. In deeper waters, development drilling may take place from platforms which are not founded on the seabed and subsea connections may be used to obviate a permanent surface platform installation. In some nearshore locations, particularly in the Arctic, an artificial island will be built of sand or gravel to support drilling and production. Drilling of multiple wells from a fixed location will, of course, result in larger and more heavily concentrated discharges of drilling fluids and cuttings. The risk of oil spills may increase as the producing well is completed.

Fabrication of platforms may take place onshore nearby the well location (Gulf of Mexico) or at some remote location and the platform assembled at the site. Platform fabrication and equipment storage yards in the Gulf of Mexico and North Sea regions are large and require a waterfront location, perhaps at the expense of marine intertidal or subtidal habitats. Transportation of the platforms from the yards may require navigation channel widening, deepening or straightening. Similar navigational requirements may result from the greatly enhanced service vessel traffic which accompanies development and production. Transportation of oil and gas across coastal wetlands often entails dredging channels for laying pipelines. Dredging activities affecting coastal habitats are discussed in Chapter 13 and are not further described here.

Fluids and gases recovered from the well may include crude oil, natural gas, petroleum condensates, nonhydrocarbon waste gases, and water produced from the bearing formations. This complex mixture must be separated, either on the producing platform, a collector platform or ashore. Crude oil and condensates must flow in one stream, natural gas in another, waste gases vented or burned, and produced waters discharged or reinjected. The composition of produced waters is discussed in detail below.

Transport of oil and gas usually involves pipelines buried in the seafloor, except where economic conditions make this infeasible. Then, storage of the product offshore and onloading onto tankers or barges may be employed. All transport methods involve some risk of accidental spills; however, pipelines generally have a relatively safer record than vessel transport, particularly if offshore transfer to vessels is involved. The environmental effects of refinery operations are beyond the scope of this review because, at that stage, oil from numerous sources other than offshore production is refined. Because offshore oil and gas production in the U.S. satisfies only a small portion of the nation's demand and this condition is not likely to change, offshore discoveries and production are not likely to influence the distribution or expansion of refineries.

OPERATIONAL DISCHARGES

During well drilling and during production of oil and gas offshore, a wide variety of liquid, solid and gaseous wastes are produced on the platform, some of which are discharged to the ocean (Table 4.2). Such discharges are regulated by the Environmental Protection Agency (EPA) through issuance of National Pollutant Discharge Elimination System (NPDES) permits. Liquid and solid wastes that may be permitted for discharge to the ocean include cooling water from machinery, deck drainage, domestic sewage, drill cuttings, drilling fluids and produced waters. In addition, submerged parts of the platform may be

TABLE 4.2
Major permitted discharges and potential impact-causing agents associated with offshore oil and gas exploration and production

- Drill cuttings—1100 MT/exploration well, less for development well
- Drilling fluids—900 MT/exploration well, 25% less for development well
- Cooling water, deck drainage, ballast water—may be treated in an oil/water separator
- Domestic sewage—primary activated sludge treatment
- Sacrificial anodes, corrosion, antifouling paints—may release small amounts of several metals (Al, Cu, Hg, In, Sn, Zn)
- Produced water (production only)—treated in oil/water separator to reduce total HC to mean of 48 ppm, daily maximum 72 ppm

protected against biofouling and corrosion with antifouling paints and sacrificial electrodes. These may release small amounts of toxic heavy metals to the water column (Al, Cu, Hg, In, Sn, Zn) (Dicks, 1982). Produced water and oily wastes from deck drainage are passed through an oil-water separator, and domestic sewage is treated in an activated sludge treatment system before discharge. Treated waste water containing up to 48 ppm oil and grease is permitted for discharge to the ocean.

Drilling Discharges

The major discharges associated with exploratory and development drilling are drill cuttings and drilling fluids. Drill cuttings are particles of crushed sedimentary rock produced by the action of the drill bit as it penetrates into the earth. Drill cuttings range in size from clay to coarse gravel and have an angular configuration (as compared to the rounded shape of most weathered natural sediments). Their chemistry and mineralogy reflect that of the sedimentary strata being penetrated by the drill. Cuttings are considered relatively inert; nevertheless, they represent a potential input of trace metals, hydrocarbons and suspended sediments to the receiving waters, and, in addition, may account for continuous losses of small amounts of drill muds which are removed by normal cuttings washing procedures.

Drilling fluids are specially formulated mixtures of natural clays and/or polymers, weighting agents and other materials suspended in water or a petroleum material. Discharge to the ocean of water-based, but not oil-based, drilling fluids may be allowed by NPDES permit. Water-based drilling fluids (in which the major liquid phase is fresh or sea water) are used almost exclusively for drilling in U.S. coastal and outer continental shelf waters. In other parts of the world, such as the North Sea, oil-based drilling fluids are used frequently in offshore drilling operations.

Drilling fluids perform several functions integral to the rotary drilling process. The most important of these include transport of cuttings to the surface, balance of subsurface and formation pressures thus preventing a blowout, and cool, lubricate and support part of the weight of the drill bit and drill pipe. Drilling fluids are formulated to perform these functions optimally (McGlothlin and Krause, 1980). During drilling, the mud engineer continually tests the drilling

TABLE 4.3
Specialty additives and their functions in water-based drilling fluids (from Moseley, 1981)

Additive	*Function*	*Concentration range in mud (g/l)*	*Frequency of use*
Sodium bicarbonate	Eliminate excess calcium ions due to cement contamination.	0.3–11.4	very common
Sodium chloride	Minimize borehole washout in salt zone.	28.5–357	rare
Ground nut shells, mica or cellophane	Minimize loss of drilling mud to formation.	14.3–142	common
Cellulose polymers or starch	Counter thick, sticky filter cake; decrease filtrate loss to formation.	0.7–14.3	very common
Aluminium stearate or alcohols	Minimize foaming.	0.1–0.3	common
Sodium chromate	Reduce viscosity increase in high temperature wells—aid deflocculation capability of lignosulfonate.	0.3–5.7	rare
Diesel or vegetable oil lubricant	Reduce torque and drag on drilling string.	5.7–142	common
Pill of oil-based mud spotting fluid	Counter differential pressure sticking of drill string. Pill is placed downhole opposite contact zone to free pipe.	570–1140	common
Paraformaldehyde bactericide	Retard bacterial degradation in polymer fluid systems. In development drilling, added to fluid left behind in casing to prevent casing string corrosion.	0.57–5.7	very common
Zinc compounds	Counter hydrogen sulfide contamination by precipitating sulfides.	1.4–14.3	common
KCL-polymer drilling fluid system	Improve wellbore stability in water-sensitive shale formation. Prevents shale swelling and sloughing.	—	rare
Asbestos*	Improve solids carrying capacity; lift formation drill solids out of hole.	2.9–28.5	very rare

*Use prohibited in most OCS regions under EPA's NPDES program.

fluid and adjusts its composition to counteract changes in down-hole conditions. Thus, the composition of a drilling fluid is changed continually as the well is drilled. No two drilling fluids are identical in composition.

The five major ingredients in water-based drilling fluids (barite, clay, lignosulfonate, lignite and caustic) account for over 90% of the total mass of additives used in water-based drilling fluids (National Research Council, 1983). The other major ingredient is fresh water or sea water. There are more than one thousand additional tradename or generic materials available for drilling fluid formulation (World Oil, 1980). Most of these materials, however, are designed for use in oil-based muds and rarely more than 10 to 20 specialty additives are used to formulate a typical offshore water-based drilling mud.

Barite (barium sulfate) is used as a weighting agent in drilling fluids. It has a density of 4.1-4.3 g/cc and a solubility in sea water of about 50-52 μg/l as Ba (Burton *et al.*, 1968; Chan *et al.*, 1977). The amount of barite added to a drilling mud may vary from 0 to about 700 lb/bbl (0-2 kg/l) and usually increases with depth of the well (National Research Council, 1983).

Bentonite clay (sodium montmorillonite), or sometimes attapulgite clay, is the major ingredient of most water-based drilling fluids. It is used to maintain the gel strength required to suspend and carry drill cuttings to the surface. It also helps coat the wall of the bore-hole to prevent loss of drilling fluids to permeable formations.

Lignosulfonates are organic polymers derived from the lignin of wood and are byproducts of the wood pulp and paper industry. When complexed with certain inorganic ions such as chromium, iron or calcium, they are effective in preventing flocculation of clays. They are used to control the viscosity of drilling fluids. Chrome or ferrochrome lignosulfonate is used most frequently in water-based muds for offshore drilling. Lignite (a soft coal) is used with lignosulfonate as a clay deflocculant and filtration control agent.

Caustic (sodium hydroxide) is used to maintain the pH of drilling fluid in the range of 10 to 12. A high pH is needed for optimum clay deflocculation by chrome lignosulfonate and to inhibit corrosion of drill pipe and growth of hydrogen sulfide-producing bacteria.

Specialty chemicals, formulations and processes are used to solve particular technical problems encountered down-hole during the drilling operation. The most frequently used specialty chemicals, their functions and frequency of use are summarized in Table 4.3.

Several metals are found in drilling fluids (Table 4.4). The metals of major environmental concern, because of their potential toxicity and/or abundance in drilling fluids, include arsenic, barium, chromium, cadmium, copper, iron, lead, mercury, nickel and zinc. Some of these metals are added intentionally to drilling muds as metal salts or organometallic compounds. Others are trace contaminants of major drilling mud ingredients. The metals most frequently present in drilling fluids at concentrations significantly higher than in natural marine sediments include barium, chromium, lead and zinc (Table 4.4). Barium in drilling fluids is almost exclusively in the form of barite. Bentonite clay also may contain some barium.

TABLE 4.4
Concentration ranges of several metals in drilling fluids from different sources and in typical marine sediments (concentrations in mg/kg dry wt, ppm)

Metal	*Concentration in drilling fluids*[1]	*Concentration in marine sediments*[2]
Barium	720–449,000	60–8100
Chromium	0.1–5960	10–200
Cadmium	0.16–54.4	0.3–1
Copper	0.05–307	8–700
Iron	0.002–27,000	20,000–60,000
Mercury	0.017–10.4	0.05–3.0
Lead	0.4–4226	6–200
Zinc	0.06–12,270	5–4000
Nickel	3.8–19.9	2–10 (10–1000)[3]
Arsenic	1.8–2.3	2–20
Vanadium	14–28	10–500
Aluminum	10,800	10,000–90,000
Manganese	290–400	100–10,000

[1] from Neff, 1982
[2] from Robertson and Carpenter, 1976
[3] deep-sea sediments

Chromium in drilling fluids is derived primarily from chrome and ferrochrome lignosulfonates. Different brands of chrome or ferrochrome lignosulfonate may contain from 1000 to 45,000 mg/kg chromium (Neff, 1982). Barite and lignite also may contain some chromium. In addition, inorganic chromate salts sometimes are added to drilling fluids for stabilization of chrome lignosulfonate at high temperatures, corrosion control, or H_2S scavenging. Frequently used offshore drilling fluids may contain 0.1 to about 1400 mg/kg dry weight, and exceptionally to 6000 mg/kg, total chromium. Chromium complexed to lignosulfonate is trivalent (Skelly and Dieball, 1969). Hexavalent chromium added to drilling muds is reduced quickly to the trivalent state by the lignosulfonate and other organic compounds in the mud, particularly at elevated temperatures. During use of a drilling fluid, the chrome lignosulfonate becomes adsorbed to the clay fraction (McAtee and Smith, 1969). Chrome-lignosulfonate-clay complexes are quite stable at normal operating temperatures. Above about 150°C, these complexes begin to break down due to thermal degradation of lignosulfonate.

Most of the other metals detected in some drilling fluids (mercury, lead, zinc, nickel, arsenic, cadmium and copper) are present primarily as trace impurities in barite, bentonite and sedimentary rocks in the formations penetrated by the drill. The average concentrations of these metals in marine sediments are as high as or higher, in most cases, than their concentrations in drilling muds (Table 4.4). The metallic impurities in barite are in the form of highly insoluble metal sulfides (Kramer *et al.*, 1980; MacDonald, 1982). Mercury is of particular concern because of its high toxicity. Although mercury from mercuric sulfide can be methylated to highly mobile and toxic methylmercury compounds by sediment bacteria, the speed and efficiency of this transformation is only 10^{-3} times that of methylation of ionic Hg^{+2} (Fagerstrom and Jernelov, 1971) and the rate-limiting step appears to be oxidation of sulfide to sulfate (Gavis and Ferguson, 1972). This reaction will be oxygen limited in most marine sediments. Pipe thread compound (pipe dope) and drill collar dope may contain several percent metallic lead, zinc and copper (Ayers *et al.*, 1980a). Some pipe dope gets into the drilling mud; however, metals from this source are in the form of fine metallic granules and are relatively inert biologically. Finally, inorganic zinc salts, such as zinc carbonate, zinc chromate or zinc sulfonate, may be added to drilling muds as H_2S scavengers. In such cases, zinc is precipitated as zinc sulfide.

The drilling fluid-handling system is an important part of any modern drilling rig and consists of several components (Figure 4.1). Drilling fluid is pumped under high pressure from the drilling fluid holding tanks on the platform down through the drill pipe and exits through nozzles on the drill bit. There it hydraulically removes cuttings generated by the grinding action of the drill bit. The drilling fluid, carrying cuttings with it, then passes up through the annulus (area between the drill pipe and the borehole wall or casing) to the drilling fluid return line. The drilling fluid passes through several screens and other devices which remove the cuttings from the fluid. The drilling fluid is returned to the holding tanks for recirculation down-hole, and the cuttings are discharged to the ocean.

During a normal exploratory drilling operation, several drilling fluid and cuttings-related effluents are discharged to the ocean. Typical discharges and discharge rates from an offshore platform are summarized in Table 4.5. The only, more or less, continuous discharge during normal drilling is of cuttings from the shale shakers. Although most of the drilling fluid is removed from the cuttings during passage through the shale shakers, discharged cuttings may contain 5 to 10% drilling fluid solids. Discharges of finer fractions of cuttings by the other solids-control equipment is more intermittent. The rate of cuttings discharge per day depends on the vertical distance drilled that day and the diameter of the drill hole. Hole diameter decreases in stages with depth from about 92 cm near the surface to about 16.5 cm at a depth of about 4600 m. Thus, the rate of cuttings discharge decreases as drill depth increases. In addition, drilling may actually occur only one-third to one-half the time during a two-three month drilling operation (National Research Council, 1983).

Whole used drilling fluids may be discharged intentionally in bulk quantities several times during a drilling operation. Small amounts (100-200 bbl; 15,900-31,800 l) of drilling fluid may be discharged to make space in the mud tanks for addition of water or drilling fluid ingredients added to change fluid properties. Changeover of mud programs, from one drilling fluid type to another, may require bulk discharge of most of the drilling fluid in the mud system. At the end of an exploratory drilling operation, most of the drilling fluid not left in the hole is discharged in bulk to the ocean. Bulk drilling fluid discharges may involve 1000-2500 bbl (159,000-397,500 l).

Unless restricted by NPDES permit, the rate of bulk drilling mud discharges ranges from 500 to 2000 bbl/h and may require 0.5 to 3 h (Ayers *et al.*, 1980b;

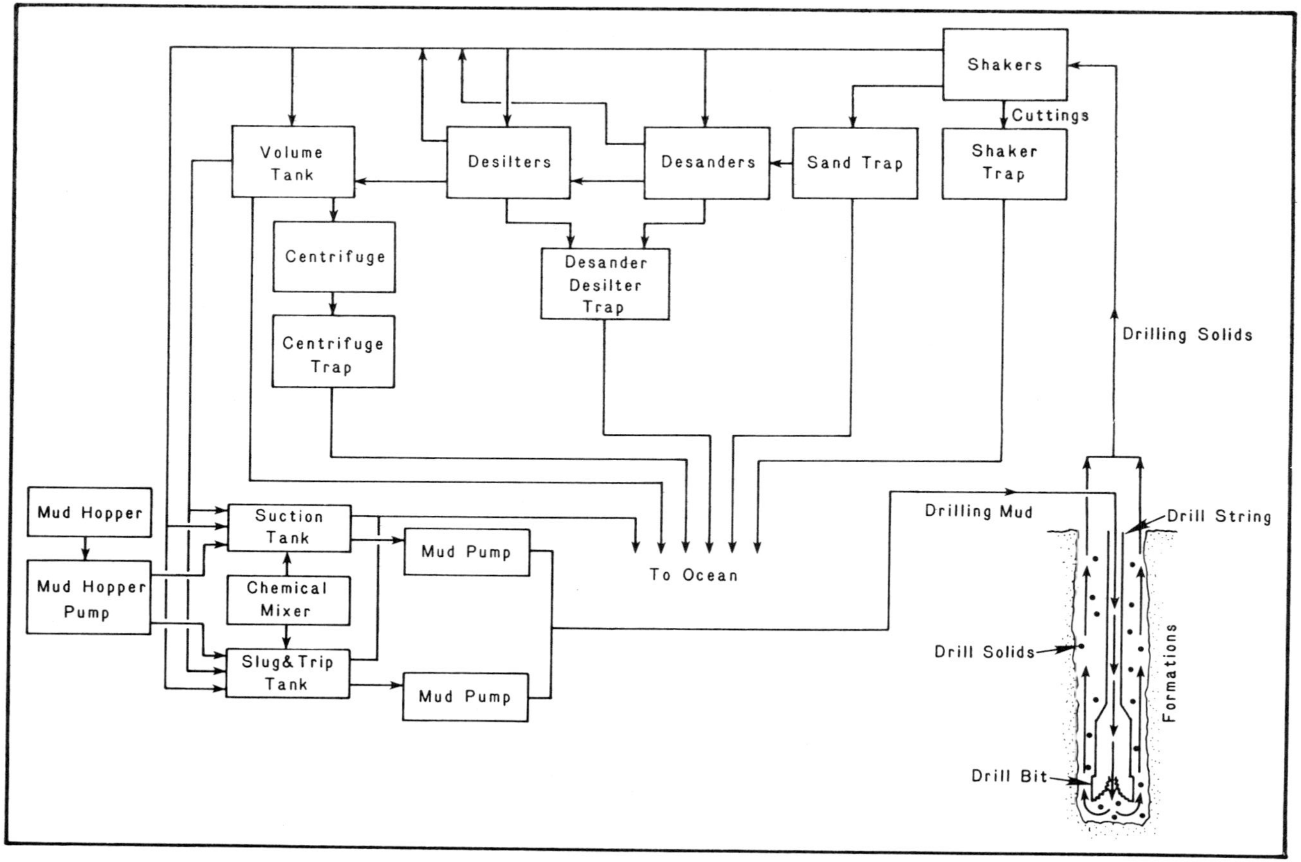

Figure 4.1. A generalized schematic diagram of the drilling fluid-handling system of an offshore oil rig (modified from Miller, 1983).

TABLE 4.5
Sources, discharge rates and discharge frequencies of continuous discharges of drilling fluids and drill cuttings (from Petrazzuolo, 1983)

Source	*Rate (l/h)*	*Frequency*
Shale shakers	160–320	Continuous during drilling
Desander	480	2–3 h/day during drilling
Desilter	2545–2700	2–3 h/day during drilling
Centrifuge	4770	1–3 h; every 2–3 days
Sand trap	87,500–420,000	2–10 min; every 2–3 days
Sample trap	240–480	5–10 min; every 2–3 days

Ray and Meek, 1980). Over the life of an exploratory well, from 5000 to 30,000 bbl of drilling fluids (200 to 2000 metric tons of solids) may be used. From 50 to 80% of this drilling fluid may be discharged to the ocean during or after drilling. Development wells, are usually shallower, smaller in diameter, and require less time to drill than exploration wells, and so the quantity of mud discharged per development well is usually somewhat smaller.

As many as 50 to 100 wells may be drilled from a single offshore development platform. After completion of each well, some of the drilling mud may be retained on board for use in drilling the next well. During drilling of a 10,000-ft (3048-m) production well, approximately 900 metric tons of drill cuttings will be generated and approximately 1000 tons of drilling fluid solids will be discharged. One or two wells at a time may be drilled from a development platform, each well requiring 2 to 6 months to complete. During the 4 to 20 years required to drill 50 wells from such a platform, approximately 95,000 metric tons of drilling fluid and cuttings solids would be discharged to the ocean.

Produced Water

Petroleum and natural gas may accumulate in commercial quantities where a layer of permeable sedimentary rock, such as sandstone or limestone, is sandwiched between layers of impermeable rock, such as shale, and lateral migration of the hydrocarbons is prevented by folding, faulting or salt dome intrusion of the sedimentary layers (Figure 4.2). Connate or fossil water (water that has been buried and out of contact with the atmosphere for at least a large part of a geologic period; White, 1957) may also accumulate in such reservoirs. Within the reservoir, natural gas accumulates at the shallowest depths, liquid petroleum is in the middle and water is at the greater depths. The relative proportions of the three materials may vary substantially in different reservoirs and one or more components may be absent.

During production of oil or gas, some of the connate water may be pumped up as well. This water is called formation water, produced water or oilfield brine effluent. Over the life of a well, the amount of water produced with the oil or gas often increases as the amount of oil produced decreases (Read, 1978). In older fields, production may be 95% water and 5% oil and gas.

Produced water may be reinjected into the reservoir to enhance recovery of

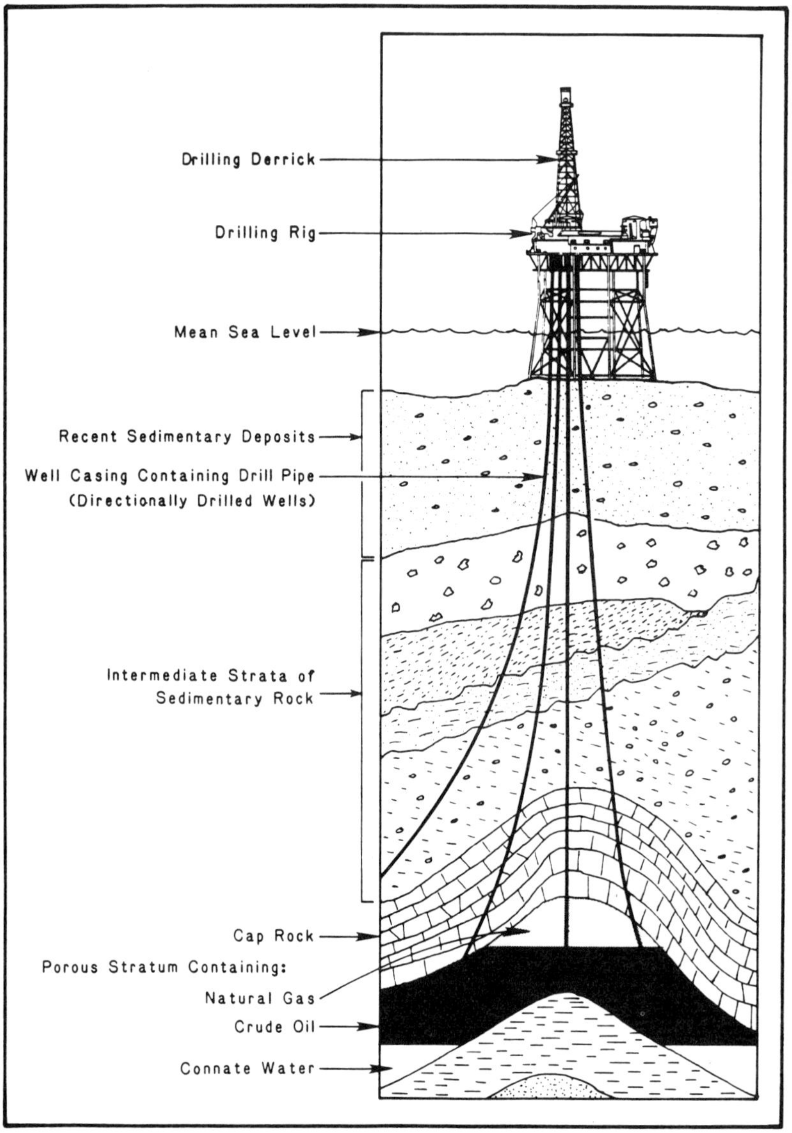

Figure 4.2. Schematic of an offshore development/production platform showing spatial relationship between the platform, drill string and the petroleum reservoir containing natural gas, crude oil and connate water.

the remaining hydrocarbons (secondary recovery), as a disposal mechanism for this potential pollutant or to control land subsidence. Surface fresh water or sea water also may be injected. In addition, water may leak into the well from shallower strata through a leaky casing or faulty completion. This water may find its way back to the surface as produced water. In 1970, daily production of produced water and oil in the United States were 3.78 and 1.51 trillion liters, respectively (Collins, 1975). In the northwestern Gulf of Mexico, an estimated 47.7 million l/day of produced water were discharged to outer continental shelf waters. An additional 47 million l/day were being treated onshore and discharged to coastal waters (Brooks *et al.*, 1977). Discharges of treated produced water to Cook Inlet, Alaska from two onshore treatment plants (at Kenai and Trading Bay) and from one offshore production platform (Dillon) were estimated in 1981 at 12.9 million l/day (Lysyj, 1982). The produced water discharge from a single platform usually is less than about 1.5 million l/day, whereas discharges from large facilities handling several platforms may be as high as 25 million l/day (Menzie, 1982).

The concentration of total dissolved solids (salinity) in produced water from different locations in the United States and Canada ranges from less than 3 to about 300 g/l (parts per thousand) (Rittenhouse *et al.*, 1969). Most produced waters are more concentrated than sea water (35 ppt) and are thought to be of marine origin (Collins, 1975). They have an ionic composition similar to an evaporate of sea water, although actual ion ratios vary substantially depending on the geologic period from which they come and the chemistry of the sediments with which they are associated (Table 4.6). As in sea water, sodium and chloride are the most abundant ions, with some exceptions. The ratio of calcium to

TABLE 4.6

Concentrations of several elements in sea water and oil field waters of several geologic ages (Tertiary-Cambrian); concentrations in mg/kg (ppm) (data from Collins, 1975)

Element	*Concentration in sea water*	*Concentration in oil field waters of different geologic ages*	
		Highest (Age[1])	*Range of means*
Sodium	10,560	120,000 (J)	23,000–57,300
Chloride	18,980	270,000 (P)	46,100–141,000
Calcium	400	205,000 (P)	2530–25,800
Magnesium	1270	26,000 (D)	530–4300
Potassium	380	11,600 (D)	130–3100
Sulfate	880	8400 (T)	210–1170
Bromide	65	6000 (J)	46–1200
Strontium	8	4500 (P)	7–1000
Ammonium	—	3300 (P)	23–300
Bicarbonate	—	3600 (T)	77–560
Iodide	0.05	1410 (P)	3–210
Barium	0.03	670 (C)	5–60
Boron	4.8	450 (T)	8–40
Carbonate	—	450 (M)	30–450
Lithium	0.2	400 (J)	3–50
Manganese	0.002	200 (D)	12–175
Rubidium	0.12	11 (D)	0.1–4
Cesium	0.0005	8.5 (P)	0.1–0.6
Copper	0.003	3 (M)	0.6–3

[1]C, Cretaceous; D, Devonian; J, Jurassic; M, Mississippian; P, Pennsylvanian; T, Tertiary

magnesium concentration often is reversed compared to sea water. The concentration of elemental sulfur may be quite high. In produced water from the Buccaneer gas and oil field in the northwestern Gulf of Mexico, the maximum concentration of sulfur was 1200 ppm and the mean was 460 ppm (Middleditch, 1981b). Several potentially toxic metals may be present at elevated concentrations in some produced waters (Table 4.7). The accuracy of analyses by several investigators of the concentrations of metals in produced water from the Buccaneer field has been questioned by Middleditch (1984). Several of these values are out of the range of values reported by Collins (1975) for produced waters in general. Analysis of metals in concentrated saline brines by atomic absorption spectrophotometry is technically difficult because of significant matrix interferences. Much of the data for metals concentrations in produced water probably is unreliable. Based on the available analyses, metals that may be present in produced water at substantially higher concentrations than in sea water include barium, beryllium, cadmium, chromium, copper, iron, lead, nickel, silver and zinc.

Produced water may contain small amounts of radionuclides, primarily in the form of radium (^{226}Ra and ^{228}Ra). The radium apparently is derived from the normal concentrations of uranium and thorium associated with the clay minerals

TABLE 4.7
Concentration ranges of metals in sea water and in produced waters discharged to the Gulf of Mexico; concentrations in μg/kg (ppb)

Metal	*Concentration in sea water*[1]	*Concentration in produced water general*[2]	*Concentration in produced water Buccaneer Field*[3]
Antimony	0.5	—	0.061
Arsenic	3.0	0–10,000	0.32
Barium	10–63	to 100,000	6.4–3500
Beryllium	0.0005	20–200	0.25–2.7
Cadmium	0.11	0–1	0.057–32
Chromium	0.13–0.25	—	0.83–260
Copper	0.5–3.5	500–3000	0.55–120
Iron	1.7–150	trace–1,000,000	260–2900
Lead[4]	0.6–1.5	trace–100,000	0.78–760
Manganese	0.1–8.0	1000–6000	0.84–4.3
Mercury	0.15–0.27	0–150	0.16–0.4
Nickel	2.0	<1–15	0.35–1200
Silver	0.145	—	0.028–110
Strontium	8000	to 3,500,000	230–71,000
Thallium	0.01	—	0.33
Zinc	1.5–10	trace–500,000	15–610

[1] from Goldberg, 1963; Hood, 1963
[2] from Collins, 1975
[3] from Middleditch, 1984
[4] Patterson *et al.* (1976) report 0.02–0.10 μg/l total Pb for southern California coastal waters.

and quartz sands that make up the matrix of the hydrocarbon/water reservoir (Reid, 1983). These radium isotopes are derived from radioactive decay of ^{230}Th and ^{232}Th. Radiodecay daughters of ^{226}Ra and ^{228}Ra of possible environmental interest include ^{210}Pb, ^{210}Po, and ^{228}Th and ^{224}Ra. In 32 samples of produced water from gas, oil and geothermal wells in coastal Louisana and Texas, Reid (1983) observed a direct correlation between salinity of the produced water (10-274 g/kg) and total radium concentration (30 to 2800 pCi/l) (Figure 4.3). Similar concentrations of radium have been reported in produced water from oil fields in the midwestern U.S. (Gott and Hill, 1953; Pierce *et al.*, 1955; Armbrust and Kuroda, 1956). However, there are no other data on radionuclide concentrations in produced waters from coastal and offshore waters of California and Alaska. The background concentration of total radium isotopes in coastal and marine waters is generally less than 1 pCi/l (Reid, 1983) or 2-17 x 10^{-14} g/l with concentration increasing with water depth (Szabo, 1967).

The U.S. EPA *Best Practicable Treatment Guidelines* restrict the concentration of oil and grease in produced water destined for ocean disposal to a monthly average of 48 ppm and a daily maximum of 72 ppm. New Source Performance Standards that have been proposed by the EPA include a daily maximum of 59 mg/l and a monthly average of 23 mg/l oil and grease (William Tilliard, EPA Washington, D.C., personal communication). The oil/water mixture produced from the well is either treated on the platform or transported to shore by pipeline to an onshore treatment plant. The oil and water phases are allowed to separate in a gravity separator and treated to remove additional dispersed oil before being discharged to the ocean or coastal waters. The produced water treatment system

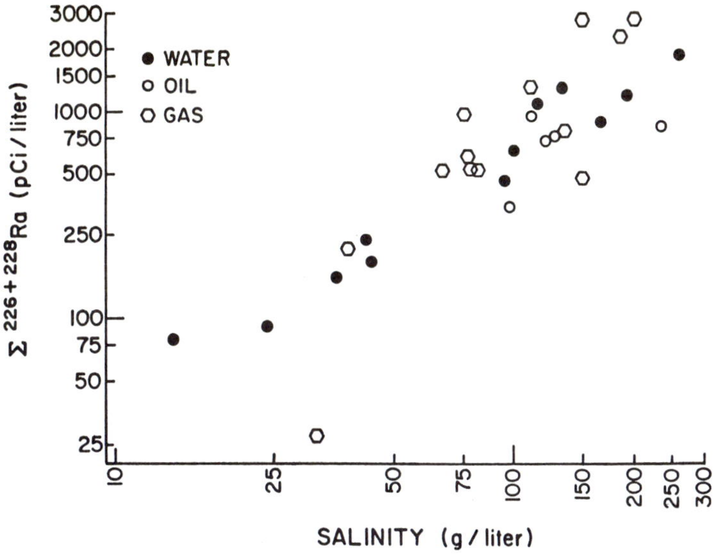

Figure 4.3. Relationship between total dissolved solids concentration (salinity) and concentration of total radium (Ra-226 plus Ra-228) in produced water from oil, gas and geothermal wells in Texas and Louisiana (from Reid, 1983).

TABLE 4.8
Some chemical characteristics of final produced water effluents from production systems on 10 offshore production platforms in the Gulf of Mexico (adapted from Jackson *et al.*, 1981)

Parameter	*Platform*									
	1	2	3	4	5	6	7	8	9	10
Water cut, percent	35	64	47	27	32	91	87	10	90	19
Produced water properties										
pH	6.9	7.0	6.3	6.6	6.3	6.7	6.6	7.1	6.3	6.2
Total dissolved solids (g/l)	105	80.5	203	114	138	108	112	114	162	163
Specific gravity	1.086	1.073	1.151	1.093	1.129	1.095	1.095	1.106	1.140	1.133
Surface tension (dynes/cm)	67	60	67	60	61	61	63	66	57	68
Oil concentration (mg/l ± S.D.)										
By Gravimetric	77 ± 73	63 ± 95	64 ± 74	22 ± 6.7	12 ± 13	26 ± 6.9	7.6 ± 5.2	18 ± 9.2	52 ± 24	48 ± 16
By Infrared	106 ± 99	81 ± 109	95 ± 103	39 ± 4.2	37 ± 19	36 ± 8.3	15 ± 3.7	36 ± 7.8	76 ± 38	48 ± 16
"Dispersed Oil"	38 ± 80	66 ± 106	92 ± 126	4.9 ± 5.1	5.9 ± 13	26 ± 8.6	1.6 ± 1.5	5.7 ± 7.7	63 ± 30	23 ± 13
"Soluble Oil"	61 ± 15	30 ± 32	21 ± 13	36 ± 4.1	28 ± 3.1	10 ± 2.3	13 ± 2.7	31 ± 2.7	13 ± 13	25 ± 4.7
"Soluble" Oil as Percent IR Oil	58	37	22	92	76	28	87	86	17	52

is designed primarily to remove particulate or dispersed oil and therefore has little effect on the concentration of dissolved petroleum hydrocarbons, other organics and metal ions in the produced water (Jackson *et al.*, 1981; Lysyj, 1982). Concentrations of soluble nonvolatile organic compounds in produced water may be as high as 500-600 mg/l and they are not removed by conventional treatment methods (Lysyj, 1982). The composition of this organic material is not known. Some chemical and physical characteristics of treated produced water from 10 platforms in the northwestern Gulf of Mexico are summarized in Table 4.8. Produced water represented from 27 to more than 90% of total liquids produced by these wells. The pH of these waters was near neutrality and salinity

TABLE 4.9

Concentrations of selected petroleum hydrocarbons in produced water effluents from the Buccaneer platform in the northwestern Gulf of Mexico; concentrations in μg/l (ppb)

Compound	*From Middleditch, 1981*	*From Sauer, 1981*
Aromatic Hydrocarbons:		
Benzene	6100	1150
Toluene	5460	7460
Ethylbenzene	1200	850
m- *p*- and *o*-oxylenes		3570
Total C_3-Benzenes	24.2	5590
C_4-Benzenes	22.2	830
C_5-Benzenes	4.5	NA*
C_6-Benzenes	3.2	NA
C_7-Benzenes	0.9	NA
C_8-Benzenes	1.2	NA
C_9-Benzenes	2.4	NA
C_{10}-Benzenes	1.0	NA
C_{11}-Benzenes	0.3	NA
Naphthalene	11.1	170
Methylnaphthalenes	7.2	20
C_2-Naphthalenes	10.4	NA
C_3-Naphthalenes	4.3	NA
C_4- and C_5-Naphthalenes	0.9	NA
Biphenyl or Acenaphthene	2.8	NA
C_1- and C_2-Biphenyls	2.9	NA
Alkanes:		
C_1–C_{13}	3120	3100
Cycloalkanes	2580	1060
Alkenes	580	NA
C_{14}–C_{29}	1476	NA
Total Aromatics Measured	12,860	16,070
Total Alkanes Measured	4596	3170
Total Cycloalkanes Measured	2580	1060
Total Alkenes Measured	580	NA
Total Hydrocarbons Measured	20,616	20,300

*Not Analyzed

ranged from 80 to 203 ppt. The concentration of total oil by infrared analysis ranged from 15 to 106 mg/l (ppm) and was quite variable from platform to platform and from a single platform over time. The concentration of oil in solution or colloidal suspension ranged from 10 to 61 ppm.

The solubility of petroleum hydrocarbons in sea water decreases logarithmically as hydrocarbon molecular weight increases (McAuliffe, 1966). Aromatic hydrocarbons are more water soluble than aliphatic hydrocarbons of similar molecular weight. Therefore, the soluble fraction of oil in produced water is greatly enriched in light aliphatic and especially aromatic hydrocarbons compared to the dispersed oil fraction (Neff and Anderson, 1981). Hydrocarbons in produced water from the Buccaneer gas and oil field in the Gulf of Mexico

TABLE 4.10

Hydrocarbon composition of oil from the C-2 separator platform, Trinity Bay, Texas, full strength effluent from the C-2 separator platform, water collected near the bottom at station 1 and bottom sediment at station 1 (from Armstrong *et al.*, 1979)

Compound	*Oil (%)*	*Produced water effluent (ppm)*	*Water (ppb)*	*Sediment (ppm)*
Alkanes:				
Total C_7–C_{31}	25.03	13.43	0.00	61.90
Aromatics:				
Benzene	1.27	3.30	1.5	—
Toluene	1.22	3.50	3.2	—
Xylene	1.60	2.40	*	—
C_2-Benzene	—	—	3.1	—
C_3-Benzene	1.00	0.65	0.8	0.6
C_4-Benzene	—	—	—	0.4
Naphthalene	0.21	0.30	0.4	0.2
1-Methylnaphthalene	0.29	0.37	0.2	0.5
2-Methylnaphthalene	0.39	0.50	0.6	0.5
Dimethylnaphthalenes	1.24	0.03	*	0.8
Trimethylnaphthalenes	1.52	0.26	*	10.0
C_4-Naphthalenes	—	—	—	0.4
Biphenyl	0.03	0.015	*	*
Methylbiphenyls	0.06	0.011	*	0.2
Dimethylbiphenyls	0.01	0.03	*	0.8
Fluorene	0.01	0.014	*	0.1
Methylfluorenes	0.07	0.063	*	0.7
Dimethylfluorenes	0.04	0.042	*	0.9
Trimethylfluorenes	—	—	—	0.6
Phenanthrene	0.02	0.036	*	0.1
Methylphenanthrenes	0.05	0.063	*	0.6
Dimethylphenanthrenes	0.03	0.020	*	1.0
Trimethylphenanthrenes	—	—	—	0.5
Total Aromatics	9.06	11.76	10.50	34.20
Total Hydrocarbons Measured	34.09	25.19	10.50	96.10

*Not detected by methods used

were dominated by monoaromatic hydrocarbons and light alkanes (Table 4.9). Naphthalenes were present in trace amoun s, and higher molecular weight aromatics were not measured. Middleditch (1981) estimated the concentration of benzo(a)pyrene in produced water at 1 to 5 μg/l, somewhat in excess of its estimated solubility in sea water: 0.1 to 1.0 μg/l (Neff, 1979).

Produced water from a separator platform in Trinity Bay, Texas contained traces of phenanthrenes (Table 4.10). The C_{14}-C_{29} alkanes and high molecular weight aromatic hydrocarbons probably were present primarily in the dispersed phase of the produced water, since they are virtually insoluble in sea water. Other organics sometimes encountered in produced waters include ketones (from solvents used to clean rig structures), phenols and organic acids from the produced water and biocides used to inhibit hydrogen sulfide, sulfuric acid and scale formation in the production system (Collins, 1975; Middleditch, 1981). The estimated amount of petroleum hydrocarbons in produced water discharged to the British and Norwegian sectors of the North Sea in 1978-1980 was 1435 metric tons/year (Read, 1978; Schreiner, 1978). The National Research Council (1985) estimated that the worldwide input of petroleum hydrocarbons from produced water discharges to be between 7500 and 11,500 metric tons per year, with about one-fourth of this in U.S. waters.

Deck drainage, which may contain a variety of chemicals such as detergents, solvents and metals, is processed through the oil/water separator before discharge to the ocean. In addition, a wide variety of chemicals may be added to the process stream of the oil/water separator and ultimately appear in the effluent water (Middleditch, 1984). These may include biocides, coagulants, corrosion inhibitors, cleaners, dispersants, emulsion breakers, paraffin control agents, reverse emulsion breakers, and scale inhibitors. The concentrations of these materials in produced water effluent are not well known.

OIL SPILLS

Offshore oil and gas development carries with it the risk of oil spills at the platform and in transporting the oil from the platform to shore. Spills at the platform result from leaks or blowouts during both exploratory and production drilling. Most oil and gas produced offshore is transported ashore through pipelines. Oil spills result from pipeline ruptures or chronic leaks. Where technologically difficult or economically infeasible, transport of oil by pipelines is replaced by storage of the product offshore, then transfer to tankers or barges. This method is commonly viewed as less safe than pipelines in that it creates an increased risk of oil spills, both acute spills and chronic inputs.

Lanfear and Amstutz (1983) presented data on accidental spills on the U.S. outer continental shelf (OCS). Although they provided analyses based only on spills of 1000 bbl or greater, the values allow for comparison of the rates of occurrence of the above mentioned types of oil spills associated with offshore development. The average spill rate for OCS platforms from 1964 to 1980 was 2.05 spills per billion barrels produced. The comparable value for spills for pipelines in the OCS was 1.6 spills per billion barrels. A value was not available

TABLE 4.11

Transportation modes and large spill information for U.S. OCS planning areas based on unleased resources as of July 31, 1986 (adapted from Department of Interior, 1985)

Planning area	*Undiscovered recoverable oil risked mean estimated (billion bbl)*	*Transportation mode (P, pipelines; T, tankers)*	*Probability of spill reaching land*	*Expected number of spills >1000 bbl*		*Estimated spills reaching shore (bbl)*
				Occurring	*Reaching shore*	
Central Gulf of Mexico	2.05	P/T	0.772	5.94	4.59	131,743
Western Gulf of Mexico	1.55	P	0.728	4.03	2.93	98,451
Southern California	1.23	P/T	0.500	2.30	1.15	69,766
Central California	0.36	P/T	0.444	1.84	0.82	21,032
Navarin Basin	0.45	P/T	0.150	2.87	0.43	76,851
Beaufort Sea	0.41	P/T	0.231	1.99	0.46	40,286
Chukchi Sea	0.40	P/T	0.225	1.75	0.39	39,195
Northern California	0.25	P/T	0.444	1.84	0.82	14,597
Eastern Gulf of Mexico	0.25	P/T	0.432	1.21	0.52	33,839
South Atlantic	0.13	P/T	0.039	0.63	0.02	17,661
St. George Basin	0.07	P/T	0.144	0.66	0.10	9572
Mid-Atlantic	0.04	T	0.106	0.37	0.04	3359
North Atlantic	0.02	P/T	0.100	0.16	0.02	1960
North Aleutian	0.01	P/T	0.342	0.16	0.05	893
Washington-Oregon	0.01	T	—	0.09	0.09	1306
Gulf of Alaska	0.01	P/T	0.851	0.10	0.09	496
Norton Basin	0.01	P/T	0.250	0.06	0.02	496
Kodiak	0.01	P/T	0.545	0.04	0.02	0
Hope Basin	negl.	P	NA	negl.	negl.	0
Shumagin	negl.	P	NA	negl.	negl.	0
Rest of Alaska	negl.	P	NA	negl.	negl.	0

for tanker accidents involved in transshipment to shore; however, one for general maritime transport of oil (3.87 spills per billion barrels transported) indicates that tanker transport creates a greater risk of oil spillage than pipelines. If those spills which occurred in harbors or piers are deleted from the analysis (i.e., less representative of those subjected to OCS winds and currents), then the expected spill rate for worldwide tanker accidents for 1974-1980 is 1.3 spills per billion barrels transported. The evidence now points to a sharp drop in oil spill occurrences from production platforms and tankers since 1974 (Lanfear and Amstutz, 1983). This better safety record could result from a number of factors—greater industry concern, increased public pressure, stricter government regulation, and better technology.

Large spills from OCS production are rare. No spills over 1000 bbl have occurred since 1981, and only three such spills since 1979 (Department of Interior, 1985). Nonetheless, large spills do occur, and the potential for damage from such spills is serious. The reduction in the oil spill rate since 1974 (Lanfear and Amstutz, 1983) and availability of data for spills >1000 bbl has made it difficult to predict spill rates needed in oil and gas resource management decisions. Nonetheless, based on the available data and models, predictions have been made for the various OCS regions (Table 4.11).

Although the number of small spills is larger, the total amount of oil from these is relatively small compared to the total amount attributable to large spills. For example, 934 small spills (<1000 bbl) constituted >99% of all production platform and pipeline incidents recorded in the Gulf of Mexico from 1974 to 1983 (Department of the Interior, 1985). Yet, these spills accounted for only about 28% of the volume of oil spilled during the period.

Based on new estimates by the National Research Council (1985), offshore oil and gas development contributes only a very small fraction of the petroleum entering the marine environment (Table 4.12). Other sources include river and terrestrial runoff from municipal, urban and industrial sources, natural seeps and atmospheric transport. A significant source is bilge cleaning of tankers.

Of the 0.04 to 0.07 x 10^6 metric tons per annum (mta) attributed to offshore production, major spills (>7 metric tons) contributed 0.03 to 0.05 x 10^6 mta, minor spills (<7 metric tons) 0.003 to 0.004 x 10^6 mta, and operational discharges 0.007 to 0.011 x 10^6 mta (National Research Council, 1985). Less than 0.01% of worldwide offshore petroleum production (658 x 10^6 mta in 1979) is accidentally spilled or operationally discharged into marine waters (Koons, 1984; National Research Council, 1985).

A compilation of minor crude oil spills by the U.S. Geological Survey for Gulf of Mexico OCS oil and gas operations indicates that 0.00024% of the total crude oil produced was spilled between 1971 and 1978. The spill rate for Lower Cook Inlet was less (0.0001%). Although similar data are not available for other U.S. areas, the spill rates are expected to be comparable. As offshore operations move into more severe environments, such as the Arctic, or into deeper waters, the incidence of minor spills may increase. On the other hand, technological advances, such as warning systems and improved blowout preventers, will help to reduce spills of all sizes.

TABLE 4.12
Input of petroleum into the marine environment. Units are 10^6 metric tons per annum (adapted from National Research Council, 1985)

Source	*Probable range*	*Best estimate*	*Percentage of total*
Natural Sources (Marine seeps, sediment erosion)	0.025–2.5	0.25	7.7%
Offshore Production (Spills and operational discharges)	0.04–0.06	0.05	1.5%
Transportation (Tanker operations, dry-docking, marine terminals, bilge and fuel oils, tanker accidents, non-tanker accidents)	0.95–2.62	1.47	45.2%
Atmosphere	0.05–0.5	0.30	9.2%
Municipal and Industrial Wastes and Runoff (Municipal wastes, refineries, non-refining industrial wastes, urban runoff, river runoff, ocean dumping)	0.585–3.12	1.18	36.5%
Totals	1.7–8.8	3.25	

The spill rate for major spills in the Gulf of Mexico OCS for the same period as above was 0.002% of the oil produced per year. This average is assumed to apply nationwide. Worldwide, however, the spill rate, with the exception of the United Kingdom, is probably higher. This assumption is based on less restrictive regulation of blowout prevention outside of the U.S. and United Kingdom.

Major spills which occur outside U.S. territorial waters were not included in these estimates, but as shown by the IXTOC-I spill, can be of consequence to U.S. coastal and offshore environments. A blowout occurred on an exploratory well, IXTOC-I, in the Bay of Campeche, Mexico in June 1979. Before capping of the well in March of the next year, an estimated 454 x 10^3 to 1.4 x 10^6 tons of oil were spilled (Atwood, 1981; Teal and Howarth, 1984; National Research Council, 1985). Oil reached both Mexican and Texas beaches. It was estimated that 10^5 tons of oil came ashore in Texas. Less than 10% of the oil from the blowout was recovered.

Even those areas free of oil exploration and production activity are subject to potential pollution resulting from petroleum transportation. A large portion (45.3%) of the petroleum entering marine waters is from this source which includes tanker operations, dry docking, marine terminals, bilge and fuel oils from all ships, and accidental spills from tankers and nontankers (Table 4.12). Discharges and accidents are, of course, more likely to occur in the normal tanker and shipping routes, with accidents more prevalent in congested areas, as well as in coastal areas where terminals are located.

The number of individual hydrocarbon components which enter the marine environment as a result of spills is quite large. The chemical composition of crude oils is complex and varies among different producing regions and even within a formation. A summary of the chemical composition of hydrocarbon sources is provided by the National Research Council (1985). Specific examples related to fates and effects of these hydrocarbons are provided in greater detail in Chapters 5, 6, 7, and 8.

LITERATURE CITED

Armbrust, B.F. and P.K. Kuroda. 1956. On the isotopic constitution of radium (Ra-224/Ra-226 and Ra228/Ra226) in petroleum brines. Trans. Amer. Geophys. Union. 37: 216-220.

Armstrong, H.W., K. Fucik, J.W. Anderson and J.M. Neff. 1979. Effects of oilfield brine effluent on sediments and benthic organisms in Trinity Bay, Texas. Mar. Environ. Res. 2: 55-69.

Atwood, D.K. (convener). 1980. Proceedings of a Symposium on Preliminary Results from the September 1979 *Researcher/Pierce* IXTOC-I Cruise. Key Biscayne, Florida, June 9-10, 1980. National Oceanic and Atmospheric Administration, Boulder, Colorado, 591 p.

Ayers, R.C., Jr., T.C. Sauer, Jr., R.P. Meek and G. Bowers. 1980a. An environmental study to assess the impact of drilling discharges in the Mid-Atlantic. I. Quantity and fate of discharges. Pages 382-418 *in* Symposium, Research on Environmental Fate and Effects of Drilling Fluids and Cuttings. Lake Buena Vista, Florida, January 21-24, 1980. American Petroleum Institute, Washington, D.C.

Ayers, R.C., Jr., T.C. Sauer, Jr., D.O. Stuebner and R.P. Meek. 1980b. An environmental study to assess the effect of drilling fluids on water quality parameters during high rate, high volume discharges to the ocean. Pages 351-381 *in* Symposium, Research on Environmental Fate and Effects of Drilling Fluids and Cuttings. Lake Buena Vista, Florida, January 21-24, 1980. American Petroleum Institute Washington, D.C.

Brooks, J.M., B.B. Bernard and W.M. Sackett. 1977. Input of low-molecular-weight hydrocarbons from petroleum operations into the Gulf of Mexico. Pages 373-384 *in* D.A. Wolfe (ed.), Fate and Effects of Petroleum Hydrocarbons in Marine Ecosystems and Organisms. Pergamon Press, New York.

Burton, J.D., N.J. Marshall and A.J. Phillips. 1968. Solubility of barium sulfate in sea water. Nature 217: 834-835.

Chan, L.H., D. Drummond, J.M. Edmond and B. Grant. 1977. On the barium data from the Atlantic GEOSECS expedition. Deep-Sea Res. 24: 613-649.

Collins, A.G. 1975. Geochemistry of Oilfield Waters. Elsevier Scientific Publishers, New York, 496 p.

Department of Interior. 1985. 5-Year Outer Continental Shelf Oil and Gas Leasing Program for Mid-1986 through Mid-1991. Draft Proposed Program. U.S. Dept. of Interior, Minerals Management Service, Reston, Virginia.

Dicks, B.M. 1982. Monitoring the biological effects of North Sea platforms. Mar. Poll. Bull. 13: 221-227.

Fagerstrom, T. and A. Jernelov. 1971. Formation of methyl mercury from pure mercuric sulfide in anaerobic organic sediment. Wat. Res. 5: 121-122.

Gavis, J. and J.F. Ferguson. 1972. The cycling of mercury through the environment. Water Res. 6: 989-1008.

Goldberg, E.D. 1963. Section I, Chemistry. Part I, The ocean as a chemical system. Pages 3-25 *in* W.N. Hill (ed.), The Seas, Volume 2. Interscience Publ., New York.

Gott, G. and J.W. Hill. 1953. Radioactivity in some oil fields of southeastern Kansas. U.S.G.S. Bull. 988E: 69-122.

Hood, D.W. 1963. Chemical oceanography. Oceanog. Mar. Biol. Ann. Rev. 1: 129-155.
Jackson, G.F., E. Hume, M.J. Wade and M. Kirsch. 1981. Oil content in produced brine on ten Louisiana production platforms. Report to U.S. Environmental Protection Agency, EPA-600/2-81-209. Municipal Environmental Research Lab., Cincinnati, Ohio.
Koons, C.B. 1984. Input of petroleum to the marine environment. Mar. Tech. Soc. J. 18: 4-10.
Kramer, J.R., H.D. Grundy and L.G. Hammer. 1980. Occurrence and solubility of trace metals in barite for ocean drilling operations. Pages 789-798 *in* Symposium, Research on Environmental Fate and Effects of Drilling Fluids and Cuttings. Lake Buena Vista, Florida, January 21-24, 1980. American Petroleum Institute, Washington, D.C.
Lanfear, K.J. and D.F. Amstutz. 1983. A reexamination of occurrence rates for accidental oil spills on the U.S. outer continental shelf. Pages 355-359 *in* Proceedings 1983 Oil Spill Conference. American Petroleum Institute, Washington, D.C.
Lysyj, I. 1982. Chemical composition of produced water at some offshore oil platforms. Report to U.S. Environmental Protection Agency, EPA-600/2-82-034. Municipal Environmental Research Laboratory, Cincinnati, Ohio.
MacDonald, R.W. 1982. An examination of metal inputs to the southern Beaufort Sea by disposal of waste barite in drilling fluid. Ocean Manage. 8: 29-49.
McAtee, J.L. and N.R. Smith. 1969. Ferrochrome lignosulfonates. I. X-ray absorption edge fine structure spectroscopy. II. Interaction with ion exchange resin and clays. J. Colloid Interface Sci. 29: 389-398.
McAuliffe, C.D. 1966. Solubility in water of paraffin, cycloparaffin, olefin, acetylene, cycloolefin and aromatic hydrocarbons. J. Phys. Chem. Wash. 70: 1267-1275.
McGlothlin, R.E. and H. Krause. 1980. Water base drilling fluids. Pages 30-37 *in* Symposium, Research on Environmental Fate and Effects of Drilling Fluid and Cuttings. Lake Buena Vista, Florida, January 21-24, 1980. American Petroleum Institute, Washington, D.C.
Menzie, C.A. 1982. The environmental implications of offshore oil and gas activities. Environ. Sci. Technol. 16: 454A-472A.
Middleditch, B.S. 1981. Hydrocarbons and sulfur. Pages 15-54 *in* B.S. Middleditch (ed.), Environmental Effects of Offshore Oil Production. The Buccaneer Gas and Oil Field Study. Plenum Press, New York.
Middleditch, B.S. 1984. Ecological Effects of Produced Water Discharges from Offshore Oil and Gas Production Platforms. Final Report on API Project No. 248. American Petroleum Institute, Washington, D.C., 160 p.
Miller, R.C. 1983. Fate and effect of drilling fluid and solids disposal in Beaufort Sea and Diapir Field. *In* Testimony of the Alaska Oil and Gas Association on Draft General NPDES Permits for the Beaufort Sea, submitted to Region X, U.S. Environmental Protection Agency, 74 p.
Moseley, H.R., Jr. 1981. Chemical components, functions, and uses of drilling fluids. Page 43 *in* Proceedings of UNEP Conference, Paris, France, June 2-4, 1981.
National Research Council. 1983. Drilling Discharges in the Marine Environment. National Academy Press, Washington, D.C., 180 p.
National Research Council. 1985. Oil in the Sea. Inputs, Fates, and Effects. National Academy Press, Washington, D.C., 601 p.
Neff, J.M. 1979. Polycyclic Aromatic Hydrocarbons in the Aquatic Environment. Sources, Fates, and Biological Effects. Applied Science Publ., Barking, Essex, England, 262 p.
Neff, J.M. 1982. Fate and Biological Effects of Oil Well Drilling Fluids in the Marine Environment: A Literature Review. U.S. Environmental Protection Agency, Environmental Research Laboratory, Gulf Breeze, Florida, EPA-600/53-82-064.
Neff, J.M. and J.W. Anderson. 1981. Response of Marine Animals to Petroleum and Specific Petroleum Hydrocarbons. Halsted Press, New York, 177 p.
Patterson, O., D. Settle and B. Glover. 1976. Analysis of lead in polluted coastal waters. Mar. Chem. 4: 305-319.

Petrazzuolo, G. 1983. Draft Final Technical Report Document: Environmental Assessment: Drilling Fluids and Cuttings Released onto the OCS. Submitted to Office of Water Enforcement and Permits, U.S. Environmental Protection Agency, Washington, D.C. by Technical Resources, Inc., Bethesda, Maryland.

Pierce, A.P., J.W. Mytton and G.B. Gott. 1955. Radioactive elements and their daughter products in the Texas panhandle and other gas fields in the U.S.A. Pages 494-498 *in* Proc. 1st U.N. Intern. Conf. Peaceful Uses of Atomic Energy, Geneva. Pergamon Press, New York.

Ray, J.P. and R.P. Meek. 1980. Water column characterization of drilling fluids dispersion from an offshore exploratory well on Tanner Bank. Pages 223-258 *in* Symposium, Research on Environmental Fate and Effects of Drilling Fluids and Cuttings. Lake Buena Vista, Florida, January 21-24, 1980. American Petroleum Institute, Washington, D.C.

Read, A.D. 1978. Treatment of oily water at North Sea oil installations - a progress report. Pages 127-136 *in* C.S. Johnston and R.J. Morris (eds.), Oily Water Discharges. Regulatory, Technical and Scientific Considerations. Applied Science Publishers, Barking, Essex, England.

Reid, D.F. 1983. Radium in formation waters: How much and is it of concern? Pages 187-191 *in* Proc. 4th Annual MMS Gulf of Mexico Regional Office Information Transfer Meeting. New Orleans, Louisiana, November, 1983.

Rittenhouse, G., R.B. Fulton, III, R.J. Grabowski and J.L. Bernard. 1969. Minor elements in oil field waters. Chem. Geol. 4: 189-209.

Robertson, D.E. and R. Carpenter. 1976. Activation analysis. Pages 93-160 *in* E.D. Goldberg (ed.), Strategies for Marine Pollution Monitoring. Wiley Interscience, New York.

Sauer, T.C., Jr. 1981. Volatile liquid hydrocarbon characterization of underwater hydrocarbon vents and formation waters from offshore production operations. Environ. Sci. Technol. 15: 917-923.

Schreiner, O. 1978. Discharge of oil-bearing waste water from the production of petroleum on the Norwegian continental shelf. Pages 137-153 *in* C.S. Johnston and R.J. Morris (eds.), Oily Water Discharges. Regulatory, Technical and Scientific Considerations. Applied Science Publishers, Barking, Essex, England.

Skelly, W.G. and D.E. Dieball. 1969. Behavior of chromate in drilling fluids containing chromate. Proc. 44th Ann. Meeting Society of Petroleum Engineers of AIME. Paper No. SPE 2539, 6 p.

Szabo, B.J. 1967. Radium content in plankton and sea water in the Bahamas. Geochim. Cosmochim. Acta 31: 1321-1331.

Teal, J.M. and R.W. Howarth. 1984. Oil spill studies: A review of ecological effects. Environ. Manage. 8: 27-44.

White, D.E. 1957. Magmatic, connate and metamorphic waters. Bull. Geol. Soc. Am. 68: 1669.

World Oil. 1980. World Oil's 1980-81 Guide to Drilling, Workover and Completion Fluids. Gulf Publ. Co., Houston, Texas.

CHAPTER 5

TRANSPORT AND TRANSFORMATIONS: WATER COLUMN PROCESSES

James R. Payne, Charles R. Phillips and Wilson Hom

CONTENTS

INTRODUCTION

This chapter reviews the transport and physical and chemical transformation of contaminants released from offshore oil and gas development activities while these contaminants are at the sea surface or in the water column. The contaminants considered include those released from both accidental discharges, such as oil spills, and operational discharges, including the disposal of drilling fluids and produced waters. An indepth review of the fate and effects of drilling fluids and produced waters is presented by Neff (Chapter 10), therefore the primary emphasis in this chapter is on oil spills and chronic releases of petroleum compounds. In addition, Boehm (Chapter 6) treats the transport and transformation of contaminants once they reach bottom sediments, although there is some overlap in our chapter with regard to the partitioning of contaminants between the dissolved and particulate phases.

When crude oil, refined petroleum products or produced waters and drilling fluids (which may contain petroleum components) are released at sea, the individual and combined materials are immediately subjected to a wide variety of chemical, physical and biological alterations. These alterations include evaporation, dissolution of specific components, dispersion, ingestion of oil droplets by pelagic organisms, oil adsorption onto suspended particulate material, photochemical oxidation of surface films, microbial degradation and removal by advective processes. Many of these processes have been studied in great detail and, in some cases, reasonable success has been achieved in modeling and predicting the fate of a petroleum mixture after it is released at sea. Table 5.1 presents a summary of these major oil-weathering phenomena and describes our current state of knowledge. The table specifically identifies areas which are adequately understood and areas where additional research may be warranted. The references cited in Table 5.1 include those from both the reviewed and "gray" literature; emphasis has been placed on articles which may not have been covered in the recent National Academy of Sciences review (National Research Council, 1985).

The major portion of this chapter covers specific areas which are less well understood or for which predictive capabilities do not currently exist. To the extent possible, the subjects covered in the text are outlined in Table 5.1. For brevity, reference is made to the tables whenever possible, and only areas which have not been reviewed or summarized elsewhere, such as time-dependent changes in physical properties of bulk oil after release at sea and potential impacts from chronic discharges (with a summary of exploratory drilling and production platform monitoring studies completed to date), are discussed in depth. The chapter then concludes with a presentation of gaps in our understanding of water column processes and other areas requiring further research.

TRANSFORMATION PROCESSES

Evaporation

At present, a number of models have been developed for predicting evaporative losses of specific compounds and pseudo-compounds (distillate cuts by

TABLE 5.1

Summary of water column oil weathering processes with regard to current knowledge and requirements for further research. Expanded discussions are presented in corresponding sections of the text. Processes important in influencing long-term effects are denoted by asterisks

Topic/processes	*Current state of understanding*	*Additional research warranted*	*Key references*
Major Oil Transformation Processes Occurring at the Air/Sea Interface and in the Water Column			
Evaporation—compound-specific and pseudocompound (distillate cut) rates *Surface spills* (case histories and wave tank simulations) Temperature effects Sea state	Very good, predictive capabilities currently available for numerous oils and temperatures in NOAA Oil Weathering Model (Payne *et al.*, 1984) (compilation of approach by Mackay *et al.*, 1980).	Distillation experiments on fresh and weathered crudes from real spill events for model verification. Controlled bubble flow experiments on density and viscosity alterations due to evaporative loss. Improved modeling and verification of surface sea state effects.	McAuliffe, 1977 Payne *et al.*, 1983a, 1984 NRC, 1985 Calder and Boehm, 1981 Grahl-Nielsen, 1978 Audunson, 1978 Raj, 1977
Subsurface releases (case histories)*** Dissolution and advective removal of lower molecular weight aromatics competes effectively with evaporation (e.g., IXTOC-I; Valdez oil terminal subsurface releases of treated tanker ballast; subsurface release of treated drilling fluids and produced waters).	Good in qualitative sense but trajectories not predictable without additional subsurface current data.	Modeling of the dissolution/evaporation behavior of specific hydrocarbons to predict biological effects.	Payne *et al.*, 1980a,b Boehm and Fiest, 1980a,b, 1982 Lysyj *et al.*, 1981 Walter and Proni, 1980 Payne and Phillips, 1985a

(continued)

TABLE 5.1—*contd.*

Topic/processes	*Current state of understanding*	*Additional research warranted*	*Key references*
Dissolution—compound-specific rates *Surface spills* and wave tank simulations	Good. Evaporation more important than dissolution in overall mass balance. Dissolution of LMW aromatics more pronounced.	Component-specific modeling with emphasis on surface versus subsurface release.	McAuliffe, 1977 Sutton and Calder, 1974, 1975 Grahl-Nielsen, 1978 Payne *et al.*, 1980b, 1983a, 1984 Brooks *et al.*, 1980
Subsurface releases			
Bulk oil-into-water dispersion***	Qualitatively, very good. Quantitatively, marginal except for data from relatively low turbulence (<5 knot) wind speed wave tank simulations.	Prediction of droplet recoalescence at sea surface. Effect of weathering on droplet size and dispersion rate. Importance of specific oil composition. Modeling thin or thick slick dispersion.	Mackay and Hossain, 1982 Payne *et al.*, 1983a Forrester, 1971 Raj, 1977
Effects of environmental conditions			
Temperature (oil viscosity)	Empirical observations for variety of oils available.		Twardus, 1980 Johansen, 1983 Gordon *et al.*, 1973
Sea state (turbulence)	Sea state dependence cannot be reliably predicted.		Johansen, 1983 Mackay *et al.*, 1979, 1980 Wheeler, 1978 Huang and Elliot, 1977 Gearing *et al.*, 1979
Degree of oil weathering (viscosity and oil/water interfacial surface tension; increase in oil density and influence of suspended particles)	Kinetic data available for Prudhoe Bay crude oil.		Payne *et al.*, 1983a, 1984

Dispersants	Numerous lab and limited field tests on a variety of oils and dispersant mixtures.	Effectiveness of dispersants on weathered oil. Diffusion control of dispersant mixing with various oils and washing off of emulsified oils.	Berridge *et al.*, 1968b Bridie *et al.*, 1980b Mackay *et al.*, 1980 Meeks, 1980 Sprague *et al.*, 1980 Bocard and Gatellier, 1981 Lee *et al.*, 1981 Nagy *et al.*, 1981
Subsurface releases			
Turbulence at the well head or pipeline rupture	Primarily observational.	Subsurface dispersion modeling for periods longer than several days.	Johansen, 1983 Conomos, 1975 Boehm and Fiest, 1982 Walter and Proni, 1980 Jordan and Payne, 1980
Subsurface turbulence and advection	Not studied.		
Oil density (oil type)	Empirical observations available from several spills.		Payne *et al.*, 1980a
Water-in-oil emulsification (mousse formation)	Empirically very good (see Table 5.2 for overall summary).	Improved mathematical modeling for variety of oils. Polar products identification and role in oil-specific emulsion stability. More work on photochemical products influence.	Payne and Phillips, 1985a Berridge *et al.*, 1968a,b Payne *et al.*, 1983a Davis and Gibbs, 1975 Bridie *et al.*, 1980a,b Klein and Pilpel, 1974a Mackay *et al.*, 1979, 1980 Bocard and Gatellier, 1981
Dependence on oil composition			
Degree of weathering—loss of light ends	Well-characterized, important for some oils.		
Presence of asphaltenes, waxes, metallo-porphyrins, and other surface active agents (NSO compounds), as well as products of photooxidation	Well understood and very oil dependent (see Table 5.2).		
Oil viscosity and percent water content	Well understood and oil specific.	Modelling efforts should include oil composition in predicting emulsification behavior.	

(continued)

TABLE 5.1—*contd.*

Topic/processes	*Current state of understanding*	*Additional research warranted*	*Key references*
Environmental effects on mousse stability	Viscosity versus temperature is predictable.		Twardus, 1980 Dickens *et al.*, 1981 Payne *et al.*, 1983a, 1984
Temperature Water content Bacteria			
Influence of mousse formation on oil-weathering processes			
Retarded diffusion of water soluble components	Suspected and measured with artificially generated mousse.		Payne *et al.*, 1981a Payne, in press
Retarded evaporation rates	Diffusion controlled.		Twardus, 1980 Payne *et al.*, 1984
Retardation of oil-slick dispersion and spreading	Observed with laboratory generated mousse.		Payne *et al.*, 1981a
Retardation of microbial degradation	O_2 and nutrient diffusion limited in more viscous mixtures.		
Flash point/burn point	Known to be affected by % water present (oil specific).		Twardus, 1980
Treatment of mousse with dispersants	No effective breakup of stabilized mousse achieved with viscosities >4000–7000 cP; ineffectiveness attributed to lack of dispersant penetration and rapid removal of dispersant by wave action.		Berridge *et al.*, 1968b Bridie *et al.*, 1980a,b

Oil adsorption onto suspended particulate material (SPM)			
Influence of SPM type	Well established.	Modeling oil/SPM interactions still in progress.	Mackay and Hossain, 1982
Mineralogy	Bentonite > Kaolinite > Montmorillonite.		Jordan and Payne, 1980 (review) Payne *et al.*, 1983b
Percent organic carbon	Most important factor; soil:water partition coefficients available on wide variety of pollutant groups (conflicting results in marine studies).		Karickhoff, 1981 Gearing *et al.*, 1979 Gearing and Gearing, 1983 Wade and Quinn, 1980 Zurcher and Thuer, 1978 Winters, 1978
Surface area, surface charge	Important for flocculation.		Meyers and Quinn, 1973 Bassin and Ichiye, 1977
Particle loading in water column	Believed to be important in Santa Barbara and IXTOC-I blowouts.		See Boehm, Chapter 6
Influence of oil weathering on SPM adsorption		SPM interaction with weathered oil. Effects of droplet size and degree of weathering.	
Density, viscosity, oil/water interfacial surface tension	Buoyancy considerations important in limiting sedimentation.		Mackay and Hossain, 1980
Change in oil/SPM agglomerate density	Empirical.		
Oil/water and water/particulate partition coefficients	Measured for four Alaskan SPM types and numerous PAHs and saturates. Sorption of truly dissolved components not considered important.	Observations of major spills to verify experimental results.	Payne *et al.*, 1981b Gearing and Gearing, 1983 Zurcher and Thuer, 1978

(*continued*)

TABLE 5.1—*contd.*

Topic/processes	*Current state of understanding*	*Additional research warranted*	*Key references*
Role of water column turbulence in oil/SPM interaction	Empirical.		Boehm and Fiest, 1980b Nelson, 1980 Payne *et al.*, 1980a Boehm *et al.*, 1982
Effects of dispersants on oil/SPM adsorption		Field studies.	Mackay and Hossain, 1982
Photooxidation of surface oil***			See Table 5.3 for detailed review.
Component-specific susceptibility and product identification	See Table 5.3.	Improved predictive capability for modeling photooxidation effects.	Klein and Pipel, 1974a,b Burwood and Spears, 1974 Zajic *et al.*, 1974
Photochemical mechanisms and reaction rates	See Table 5.3.		Payne and Phillips, 1985b
Observations of photooxidation products in the field	Several sulphoxides and phthalates isolated. Most products removed from slick by dissolution as evidenced by lack of photooxidized products in mousse. Limited data suggest enhanced toxicity. Photooxidation may be the same order of magnitude as microbial oxidation.	Analyses of larger samples. Microbial degradation of photooxidized substrates.	Patel *et al.*, 1979 Overton *et al.*, 1980 Calder *et al.*, 1978
Influence of photooxidation on mousse formation	Lab data available (see Tables 5.2 and 5.3).		Payne and Phillips, 1985a
Inhibition and sensitization of photooxidation	Partially characterized (see Table 5.3).	Humic acid-sensitized oxidation of water soluble aromatics.	Mabey *et al.*, 1983 Zepp *et al.*, 1977, 1981a,b

Ingestion of dispersed oil droplets and fecal material	Zooplankton believed to be important factor in short-term removal of petroleum residues from open ocean surface waters. Removal rate estimated to be same order of magnitude as input of petroleum to oceans.	Fate of encapsulated hydrocarbons.	Conover, 1971 Sleeter and Butler, 1982
Time-Dependent Changes in Physical Properties of Bulk Oil after Release at Sea			Payne *et al.*, 1983a, 1984
Oil/air and oil/water surface tensions	Time series decrease in oil/water surface tension measured. Lowered oil/water surface tension believed to be due to photooxidation and microbial degradation products.	Thermodynamic data on individual HC and NSO compounds to fully predict behavior of oil spilled at sea.	Payne and Phillips, 1985a
Percent water uptake (mousse formation)	Empirically well characterized. See Table 5.2 for similarities and differences among known mousse forming oils and Prudhoe Bay crude oil.		Berridge *et al.*, 1968b Twardus, 1980
Oil viscosity			Bridie *et al.*, 1980a,b
Oil density			Mackay *et al.*, 1979
Slick Drift, Spreading and Advection			
Wind and current effects	Spreading understood but not really predictable; too many variables.		NRC, 1985
Surface release			Jordan and Payne, 1980 Liu and Leendertse, 1978 1981a,b, in press

(continued)

TABLE 5.1—*contd.*

Topic/processes	*Current state of understanding*	*Additional research warranted*	*Key references*
Subsurface release	Subsurface trajectories characterized (IXTOC-I) but not predictable.	Predictions of subsurface trajectories.	Boehm and Fiest, 1980b Walter and Proni, 1980
Breakup of oil slick into patches	Observed but not well understood. Mass balance closure poor.	Additional modeling along lines of Johansen (1983).	
Simulation Models			
Spill trajectory models	Limited by lack of reliable wind speed and direction data and lack of detailed ocean current data. Subsurface releases not affected by wind, and predicted trajectories may be off by many kilometers.	Important topic area for additional work. Detailed wind and current data needed.	Liu and Leendertse, 1981a,b Anderson, 1983 Cornillion *et al.*, 1979 Audunson *et al.*, 1980 Torgrimson, 1981 Galt and Payton, 1983 Huang and Monastero, 1982 Huang, 1983
Oil weathering models	Models exist for individual aspects of oil weathering, but to date there is no single comprehensive model which includes all features.	Dispersion models are still least precise, particularly due to lack of verification at sea in higher turbulence conditions.	Mackay *et al.*, 1979 Johansen, 1983 Payne *et al.*, 1983a,b, 1984 Huang, 1983
Chronic Discharges			
Drilling fluids and cuttings	Major components and additives of fluids well known. The immediate fate of operational discharges has been described for a variety of offshore environments; results consistently demonstrate rapid dispersion and sinking of upper and lower plume components,	Partitioning behavior of constituent hydrocarbons.	Gray *et al.*, 1980 Perricone, 1980 Strosher, 1980 SAI, 1983 Crippen and Hood, 1980 Kramer *et al.*, 1980 NRC, 1983 Neff, Chapter 10 Trocine and Trefry, 1983

	respectively. Background levels are achieved within a period of several hundred meters of the discharge point.		Pierce *et al.*, 1985 Menzie, 1982 Ray and Meek, 1980 Ayers *et al.*, 1980a Liss *et al.*, 1980
Produced waters	General characteristics of produced water previously reported. Specific trace metal and hydrocarbons contents of selected discharges available, although the amount of variability between platforms and over time not well known. Limited data available for resultant metal and hydrocarbon concentrations in receiving waters and specific component partitioning behavior.	Routine analyses of produced water discharges during monitoring studies to relate composition of discharge to concentrations in receiving waters and sediments.	Middleditch, 1981 Sauer, 1981 Rose and Ward, 1981 Jackson *et al.*, 1981 Koons *et al.*, 1977
Dispersion models	Existing models based on empirical information, characteristically capable of predicting the short-term, near-field behavior of discharge materials; no long-term models have been validated.	Data necessary to develop models for long-term (>1 day) behavior, particularly aspects of sediment dispersion and resuspensive transport.	Runchal, 1983 Auble *et al.*, 1981 Brandsma *et al.*, 1980 Armstrong, 1981 Zemel, 1980 Brandsma and Sauer, 1983
Monitoring studies	Few monitoring studies have investigated long-term or chronic effects from development drilling operations. Data available indicate that no significant effects result in water column.	Monitoring studies should: — include analyses of drilling fluids, cuttings, produced waters — include predrilling (baseline) characteristics — employ appropriate sampling design to facilitate impact analysis — evaluate the importance of the point of discharge (surface versus subsurface).	Trocine and Trefry, 1983 Middleditch, 1981 Wheeler *et al.*, 1980 Menzie, 1982 Hrudey, 1980 Northern Technical Services, 1981, 1982

weight) from crude oil based on a Raoult's law approach and Henry's law constants. Unfortunately, it is impossible to completely characterize each distillate cut with respect to its mole fraction in the oil because all of the distillate cut components are never fully identified (Payne *et al.*, 1984). Therefore, assumptions concerning the composition of various distillate cut fractions have to be made. Nevertheless, reasonably good agreement has been obtained in predicted versus observed distillation curves for oil weathering in wave simulations (Payne *et al.*, 1983a, 1984). Temperature effects are fairly well predicted, whereas the effects of sea state on evaporation are only empirically approached at best (Payne *et al.*, 1984).

Dissolution

Dissolution of oil components is relatively unimportant to the mass balance of an oil slick, as less than 2 to 5% of the material present in the slick actually dissolves. Dissolution of aromatic hydrocarbons is critical, however, for evaluating possible exposures of affected organisms to specific compounds. Modeling compound-specific dissolution, with an emphasis on surface versus subsurface releases, is therefore warranted. Quantitative dissolution data have been obtained for lower and intermediate molecular weight aromatic and alkyl-substituted aromatic compounds from surface spills (McAuliffe, 1977; Grahl-Nielsen, 1978; Payne *et al.*, 1980b), wave tank simulations (Payne *et al.*, 1983a, 1984), and from subsurface releases (Payne *et al.*, 1980b; Brooks *et al.*, 1980). Rates of oil dissolution into tap water and sea water, and accompanying mathematical models of oil dissolution, have been presented by Osamor and Ahlert (1981).

With regard to subsurface releases, our experience is based mostly on case histories in which compound specific dissolution competes with evaporation, and lower molecular weight aromatics are effectively partitioned into the water column. Solubilization of aromatics was observed in the IXTOC-I blowout (Boehm and Fiest, 1980a,b; Payne *et al.*, 1980b; Walter and Proni, 1980). In addition, similar levels of dissolved aromatic hydrocarbons have also been reported in subsurface waters receiving tanker ballast at the Valdez oil terminal in Alaska (Lysyj *et al.*, 1981). In these situations, where the hydrocarbons are introduced at depth and mixing to the surface is inhibited due to density gradients or other factors, subsurface releases can yield concentrations upwards of 100 micrograms of total dissolved aromatic hydrocarbons per liter. The biological effects of water-accommodated or dissolved aromatics are discussed in Chapters 8 and 9.

Bulk Oil-in-Water Dispersion

Effects of Environmental Conditions on Dispersion Processes

Perhaps one of the most important processes for the ultimate determination of slick behavior is bulk oil-into-water dispersion. Although a very good qualitative understanding exists for the behavior of dispersed slick droplets in the water column, our quantitative understanding of this area is still marginal and modeling efforts are often empirical at best. Bulk oil dispersion is affected by temperature, which affects oil viscosity, and certainly by the sea state. Predictions of the effects of sea state are difficult, and verification of modeled oil dispersion processes have not been completed for open ocean spills. Excellent agreement has been obtained, however, between observed oil dispersion in wave tank simulations of low energy (<5 knots) sea states and model-predicted dispersion fluxes (Payne *et al.*, 1983a, 1984). Time dependent changes in viscosity and oil-water interfacial surface tension as a function of oil weathering have not been refined to permit adequate predictions for dispersion modeling of most oils. Also, prediction of oil-droplet recoalescence at the sea surface, knowledge of the effects of weathering on droplet size and dispersion rates, and our ability to model thick and thin patches of oil are still in their infancy (for reviews, see Mackay *et al.*, 1979, 1980; Johansen, 1983).

Considerable data describing the effectiveness of chemical dispersants for promoting oil-into-water dispersion are available from numerous laboratory and limited field tests. Additional work is needed, however, on the efficiency of dispersants for treating weathered or emulsified oils. The acute toxicity of specific dispersants to exposed organisms is an additional major concern. Mackay and Wells (1983) recently proposed a set of equations for describing specific processes associated with chemical dispersion of an oil slick, including methods for predicting the final toxicity of multiple toxicants in the water column following treatment of oil slicks with dispersants. The authors suggested that additional experimental data are needed to further refine the predictive equations.

Dispersion Following Subsurface Release

Oil dispersion within the water column following subsurface release is clearly affected by the following: 1) the turbulence at the well head or point of release (e.g., pipeline rupture), 2) natural subsurface turbulence from currents, breaking waves or storm surges, 3) advection and 4) the oil density (oil type). Our understanding of these processes is primarily observational, and at this time subsurface dispersion modeling is not very sophisticated.

Water-in-Oil Emulsification

Compositional Effects

Water-in-oil emulsification or mousse formation has been well studied (Payne and Phillips, 1985a). As the data in Table 5.2 illustrate, mousse formation is known to be extremely dependent on oil composition. The degree of weathering is also important for some oils. The ability to predict both the percent water uptake and product (mousse) viscosities has progressed significantly for specific oils (Mackay *et al.*, 1979, 1980); yet the overall ability to predict emulsification with a wide variety of oils is still somewhat lacking. Additional polar product identification and elucidation of the role of these materials in predicting emulsion stability is possibly warranted. More work on the photochemical products generated during slick weathering and their influence on emulsion stability may also be needed.

A number of laboratory studies have measured changes in physical properties which occur to oil after its release in sea water. Most of the experimentation

TABLE 5.2

Mousse formation experiments using a variety of fresh and artificially weathered (topped) crude oils in laboratory, outdoor test tank, and field experimental spills

Product tested	*Initial oil properties*								*Water-in-oil emulsion (mousse) properties*					
	Specific gravity	*Viscosity*	*Pour point °C*	*Wax % weight*	*% Asphaltenes*	*% Sulfur*	*V (ppm)*	*Ni (ppm)*	*Stable mousse formed/ appearance*	*Final % water*	*Viscosity*	*Pour point °C*	*Bacterial growth noted (6 weeks)*	*References*
Libyan (Brega)	0.829	4.13*	7.2	11.4	0.13	0.21	5		Borderline/dark brown-waxy	78.3			Fairly heavy	Berridge *et al.*, 1968a,b
Nigerian Light	0.867	5.16*	−15.0	8.5	0.05	0.19	5		Borderline/yellow brown-granular	77.3			Heavy infestation	Berridge *et al.*, 1968a,b
Iranian Light (Agah Jari)	0.854	5.6 *	−20.6	7.0	0.7	1.33	36		Rigid/foamy	79.1			Fairly heavy	Berridge *et al.*, 1968a,b
Iranian Heavy (Gach Saran)	0.869	8.83*	−12.2	6.7	1.9	1.58	107	37	Rigid/mid-brown	77.3			Very few	Berridge *et al.*, 1968a,b
Iraq (Kirkuk)	0.845	4.75*	−34.4	6.5	1.3	1.88	25		Rigid/dark brown	78.3			Very few	Berridge *et al.*, 1968a,b
Kuwait	0.869	9.6 *	−31.7	5.5	1.4	2.5	27	9	Rigid/mid-brown	79.1			Some present	Berridge *et al.*, 1968a,b
Venezuela (Tia Juana Medium)	0.896	25.0 *	−34.4	4.8	3.05	1.54	170	16	Rigid/dark oily brown	73.8			Very few	Berridge *et al.*, 1968a,b
Bunker C	0.990	2.8×10^7 cP @10°C	+7						Rigid and sticky	67.0	2.9×10^7 cP			Berridge *et al.*, 1968a,b Twardus, 1980
Light Arabian Crude	0.990	6.04 cS @38°C									500–800 cS @20°C			Solsburg, 1976 Twardus, 1980
Norman Wells	0.83 0.93	8.68 cP @10°C	−85 −8						Unstable	50.0	110 cP @10°C 240**	−48		Twardus, 1980 (additional data in original reference)
Sweet Blend	0.83 0.94	14.2 cP @10°C	−35 −12						Unstable	60.0	450 cP @10°C 520 cP**	−48		Twardus, 1980
Sour Blend	0.83 0.94		−50 +18								3000 cP**			Twardus, 1980
Bow River	0.90 0.99		−27 0								2150 cP**			Twardus, 1980
Lloydminster	0.90 0.98**	86.8 cP @10°C	−52 −9**						Stable	60 @10°C 2675**	2800 cP			Twardus, 1980

Weyburn Midale	0.89 0.99**	29 cP @10°C	−28 −3**						Stable	60 @10°C 1250**	4150 cP	−2.5	Twardus, 1980 Twardus, 1980
				Added % wax & % asphaltenes									
Topped 200+ Kuwaite (dewaxed & deasphaltenized)				0 5 0 5 5	0 0 1.4 1.4 0.14				Unstable*** emulsion Unstable**** emulsion Stable mousse Stable mousse				Bridie *et al.*, 1980a,b Bridie *et al.*, 1980a,b Bridie *et al.*, 1980a,b
Lube Oil									No emulsion	<1			Bridie *et al.*, 1980a,b
Lube Oil + 5% asphaltenes/wax mix									Stable emulsion	54	0.01–0.03 μm water droplet size		Bridie *et al.*, 1980a,b
Lube Oil + 10% asphaltenes/wax mix									Stable emulsion	67	0.01 μm water droplet size		
Venezuelan (Guanipa)	0.859[20]			—	5.2	1.66	105	18	Mousse and tarballs and flakes formed after few days in wave tank at 20°C (85% water-in-oil in first few days; dropped to 75% over 4 months). Specific gravity approached 1.0.				MacGregor and McLean, 1977
Libyan (Saria)	0.843[16]			20	0.15	0.14	0.5	5					MacGregor and McLean, 1977
Algerian (Zarzaitine)	0.818[16]			5.3	0.08	0.09	1	1	Unstable emulsion				MacGregor and McLean, 1977
Nigerian Medium	0.892[16]			4.6	0.1	0.25	0.8	7					MacGregor and McLean, 1977
Alberta Crude	0.829	8.25 cP @20°C							Stable mousse	70–80% in 30-minute wave tanks	100–200 cP		Mackay *et al.*, 1979
350°C Topped Kuwait	0.958	208 cS @50°C	+8		2.0	3.7	7% total 50	15	Stable mousse	50% after 8–100 weeks	2350 cS @50°C	13–15	Davis and Gibbs, 1975
Prudhoe Bay	0.893	19 cS	−10	—	23	—	28.3	13.5	Stable mousse/ light brown	55%	2800 cS after 12 days 7800 cS after 4 months in subarctic wave tank experiments		Payne *et al.*, 1981a; 1983a

(*continued*)

TABLE 5.2—*contd.*

Product tested	*Initial oil properties*								*Water-in-oil emulsion (mousse) properties*					
	Specific gravity	*Viscosity*	*Pour point °C*	*Wax % weight*	*% Asphaltenes*	*% Sulfur*	*V (ppm)*	*Ni (ppm)*	*Stable mousse formed/ appearance*	*Final % water*	*Viscosity*	*Pour point °C*	*Bacterial growth noted (6 weeks)*	*References*
Gasoline									No emulsion or mousse	Nil				Berridge *et al.*, 1968b
Kerosene									No emulsion or mousse	Nil				Berridge *et al.*, 1968b
Auto Diesel									No emulsion or mousse	Nil				Berridge *et al.*, 1968b
Marine Diesel	0.83 9.87**	10 cP @10°C	−15 −8**						No emulsion or mousse	Nil		17**		Twardus 1980; Berridge *et al.*, 1968b
Lube Oil 600									Unstable emulsion					Berridge *et al.*, 1968b
Lube Oil 2500									Fluid emulsion but no mousse					
Heavy naph Lube									Fluid emulsion but no mousse					

*Kinematic viscosity (cS) at 100°F
**Specific gravity and pour point after 4 weeks pan evaporation under atmospheric conditions (no water added except for occasional precipitation).
***93% of water shed after standing 15 minutes.
****86% of water shed after standing 15 minutes.

has been completed in mixing chambers and wave tanks, and in many cases evaporation and dissolution have been allowed to occur to further simulate ambient environmental conditions. In almost all instances, hydrocarbons with molecular weights less than nC_{11} to nC_{12} (distillation range 200-225°C) were lost during the initial states of weathering, similar to the behavior noted in the open ocean and near-coastal spills. The results of these studies, and the noted changes in physical properties and selected chemical characteristics of the crude oils (and resultant water-in-oil emulsions), are summarized in Table 5.2.

Results from laboratory studies evaluating changes in rheological properties of the spilled oil and water-in-oil emulsions are very dependent on the unique chemical compositions of the different crude oils and petroleum products tested. Heavier crudes with higher viscosities generally form the more stable emulsions (Bocard and Gatellier, 1981), and the presence of asphaltenes and higher molecular weight waxes are positively correlated with emulsion stability (Berridge *et al.*, 1968a,b; Davis and Gibbs, 1975; MacGregor and McLean, 1977; Mackay *et al.*, 1979, 1980; Twardus, 1980; Bridie *et al.*, 1980a,b; Bocard and Gatellier, 1981). Slightly different results have been obtained in various investigations, but it has generally been found that asphaltenes and waxes act together in the emulsification process although the asphaltenes appear to play a more significant role (Bridie *et al.*, 1980a,b; Berridge *et al.*, 1968b). The crystallizing properties of the component waxes (near the pour points of the oils tested) are believed to be important in affecting the internal oil/mousse structure and viscosity; the asphaltenes are believed to act as surfactants preventing water-water droplet coalescence in the more stable mixtures (Berridge *et al.*, 1968b; Canevari, 1969; Mackay *et al.*, 1973; Cairns *et al.*, 1974; Bridie *et al.*, 1980a,b). Other indigenous surface active agents, such as metallo-porphyrins and sulfur and oxygen compounds, may be equally important in emulsification. The products of photochemical and microbial oxidation have also been identified as having important roles as stabilizing agents. In instances where the above primary stabilizing components were not present, stable mousse could only be formed with photochemically or microbially weathered oils; for example, Brega, Nigerian, Zarzatine and light Arabian crudes exhibited this behavior (Berridge *et al.*, 1968b; Friede, 1973; Guire *et al.*, 1973; Klein and Pilpel, 1974a; Burwood and Spears, 1974; Zajic *et al.*, 1974; Bocard and Gatellier, 1981).

No stable water-in-oil emulsions could be formed in laboratory studies at any temperatures with light petroleum distillates such as gasoline, kerosenes, and several diesel fuels (Berridge *et al.*, 1968b; Twardus, 1980). Interestingly, stable mousse formation could only be obtained with several light lube oils when they were fortified with wax and asphaltene mixtures obtained from known mousse forming oils, such as Kuwait crude (Bridie *et al.*, 1980a,b). This asphaltene mixture could also contain other higher molecular weight surface active agents.

Environmental Effects on Mousse Emulsion Stability

Temperature is also a factor in mousse formation. In several instances at temperatures approaching the pour point of the heavier oils, stable emulsions have been generated regardless of wax or asphaltene content. Conversely, some destabilization and separation of water and oil has been noted in stable water-in-oil emulsions repeatedly exposed to freeze-thaw cycles (Twardus, 1980; Dickens *et al.*, 1981). Similar results have been obtained when laboratory-generated and real spill water-in-oil emulsions are subjected to prolonged heating on removal from the water column.

The absolute water content and sizes of water droplets incorporated into various mixtures of mousse significantly affect mousse stability and viscosity (Berridge *et al.*, 1968b; Mackay *et al.*, 1980; Twardus, 1980; Bocard and Gatellier, 1981). Positive correlations of percent water versus mousse stability and viscosity have been noted for several of the crude oils studied (Mackay *et al.*, 1979, 1980). In general, the most stable emulsions from laboratory and field observations contain water droplets in a size range from less than one to ten micrometers. Stable mousse can be formed with many oils in the range of 20 to 80% water; however, above an oil-specific critical point, significant destabilization of the emulsions occurs, presumably due to enhanced water-water contact and coalescence, which results in ultimate phase separation (Berridge *et al.*, 1968b; Twardus, 1980).

In most of the laboratory studies, the presence or absence of bacteria and suspended particulate material does not appear to affect emulsion behavior (Berridge *et al.*, 1968a,b; Davis and Gibbs, 1975). Bacterial growth is generally limited to the surface of the mousse products tested and is believed to be inhibited by limited oxygen and nutrient diffusion into the mousse. Toxic materials inherent to the oils themselves may also be responsible for these observations, although water content (and in particular the size of the water droplets encapsulated within the mixtures) has also been correlated with bacterial infestation on the less stable emulsions (Berridge *et al.*, 1968a,b). In several laboratory studies, significant bacterial utilization of the mousse only occurs after treatment with dispersants, which result in breakup of the material with concomitant increased surface-to-volume ratios (Bocard and Gatellier, 1981).

Influence of Mousse Formation on Oil Weathering

The physical properties of stable emulsions are appreciably different from those of the starting crudes. Increases in the specific gravity and viscosity of the emulsions affect spreading, dispersion and dissolution rates (Twardus, 1980; Payne *et al.*, 1981a). In addition, emulsification may affect evaporation rates of intermediate molecular weight (C_9 to C_{12}) hydrocarbons from the parent slick (Twardus, 1980; Payne *et al.*, 1981a). In general, these effects are more important in emulsions containing greater than 50% water. Emulsions with smaller percentages of incorporated water have physical properties which are proportionately similar to those of the starting crudes (Twardus, 1980; Mackay *et al.*, 1980). The influence of mousse formation on oil weathering processes has important implications for subsequent cleanup activities, such as skimming, mopping and pumping (Payne and Phillips, 1985a). In addition, the efficiencies of various sorbant materials reportedly decrease as the water contents of mousse mixtures increase (Twardus, 1980).

Treatment of Mousse with Dispersants

Pretreatment of oil or sea water with dispersants or demulsifiers generally inhibits laboratory mousse formation with most of the oils and petroleum products tested (Berridge *et al.*, 1968b; Bridie *et al.*, 1980a,b). In these studies, only 0.1 to 1% dispersant was required, and with several of the products tested, similar results were obtained when the dispersant was added to either the water or oil. Previously stabilized mousse is much more difficult to break up with commercially available dispersants, although some success has been obtained with various products when sufficient mixing energy is utilized in the laboratory to thoroughly mix the dispersant into the water-in-oil mixture (Bridie *et al.*, 1980b; Bocard and Gatellier, 1981; Lee *et al.*, 1981). In general, however, it has been noted that no effective breakup of stabilized mousse could be achieved for water-in-oil emulsions with viscosities greater than 4000-7000 cP (Mackay *et al.*, 1980; Lee *et al.*, 1981).

The ineffectiveness of several of the dispersants studied to break up stable mousse formations at sea has been attributed both to the lack of penetration of the dispersant into the mousse and to its rapid removal from the mousse surface into the water column by waves and sea surface turbulence (Lee *et al.*, 1981). In several planned sea tests, mousse forming crudes, such as La Rosa, were effectively dispersed before mousse formation occurred (JBF/API, 1976; McAuliffe *et al.*, 1981). Thicker lenses or patches of oil were observed to move along the leading (downwind) edge of these slicks, and dispersants were most effective when applied directly to the thicker lenses rather than the trailing sheen or thinner slick. Again, in the at-sea tests, mixed results have been obtained depending on the type of dispersant/demulsifier used and the oil/mousse mixture tested. It has been noted, however, that all dispersants work better when applied to the emulsions in an undiluted form, rather than when diluted with sea water.

Many of the mousse formations have not been effectively broken up by additions of demulsifiers in laboratory tests, although significant and near immediate decreases in viscosities are often noted. In several cleanup operations, injection of demulsifiers and dispersants into oil/mousse mixtures greatly enhanced pumping efficiency (Bridie *et al.*, 1980b; Bocard and Gatellier, 1981).

Oil/Suspended Particulate Material Interactions

Influence of SPM Type

Oil adsorption onto suspended particulate material (SPM) has been investigated since the early 1970s. Meyers and Quinn (1973) first examined adsorption of specific oil components on a variety of mineral types and evaluated the role of organic material on the suspended particulate material in oil adsorption. At this time the importance of mineralogy has been well characterized, with the oil adsorption capability decreasing in the order of bentonite > kaolinite > illite > montmorillonite. The presence of organic carbon on the mineral particles is believed to be the most important factor affecting oil adsorption, although some conflicting results have been reported. Meyers and Quinn (1973) reported an increase in oil adsorption potential after treatment of Narragansett Bay sediments with hydrogen peroxide to remove the organic coating, thereby freeing the surface area on the clay particles for oil adsorption. In contrast, other authors (Gearing *et al.*, 1979; Karickhoff, 1981; Gearing and Gearing, 1983) have suggested that surface organic carbon is required for significant oil adsorption. Modeling of oil/SPM interactions generally requires data on particle surface area, percent organic coating per unit surface area and the water partition coefficient for the component of interest. Surface charge is also important in oil adsorption and sediment flocculation, particularly with regard to clay material (Bassin and Ichiye, 1977). Finally, particle loading in the water column is critical because higher SPM concentrations provide a greater amount of material for oil adsorption. Characteristics of suspended particulate materials and their affinities for pollutants are discussed by Boehm (Chapter 6).

Influence of Oil Weathering on SPM Adsorption

At present the influence of oil-weathering on adsorption by suspended particulate material is only partly understood and additional work may be required. Oil droplet size has been shown to be important (Mackay and Hossain, 1982), but effects of further changes in density, viscosity, and oil/water interfacial surface tension on the oil/SPM interactions as the oil weathers require additional investigation. Buoyancy considerations are also important once oil/particle interaction has occurred. Some evidence suggests that the added buoyancy of the oil/particle mixture may actually limit the ultimate sedimentation process (Mackay and Hossain, 1982). The effects of dispersants on oil/suspended particulate material interactions have been investigated to a limited extent (Mackay and Hossain, 1982), but the need for additional studies is indicated.

Partition Coefficients

Results from previous laboratory and field studies have demonstrated that the differences in affinities of hydrocarbon fractions for adsorption onto suspended particulates may account for partitioning of lower and higher molecular weight compounds between dissolved and particulate pools. In particular, data from Payne *et al.* (1984), Gearing *et al.* (1979), and Boehm and Fiest (1980b) suggest that higher molecular weight saturated, acyclic and polynuclear aromatic hydrocarbons are preferentially associated with suspended materials, whereas lower molecular weight aromatics, including the relatively soluble naphthalenes, are preferentially partitioned into the dissolved phase. Compound-specific oil/water and oil/particulate partition coefficients have been determined for a variety of materials, including four Alaskan suspended particulate material types with a number of polynuclear aromatic compounds and high molecular weight saturates (Payne *et al.*, 1981b).

Role of Water Column Turbulence in Oil/SPM Interaction

The influence of water column turbulence and the point of oil release (surface versus subsurface) in enhancing oil droplet-SPM contact is only empirically understood. A time-dependent kinetic model to describe these observed phenomena (dispersed oil/SPM interactions) has not been fully developed. Since it is becoming apparent that this process is the first stage of one of the more important oil sedimentation mechanisms, it may be important to derive such a model. The resuspension and offshore transport of oil/sediment aggregates from oil stranded in the intertidal zone also represent an area where further research may be warranted.

The parameters and conditions that might influence the rate of "reaction" between dispersed oil and SPM are numerous. The concentrations of dispersed oil and SPM, size distribution of the droplets and SPM, composition of the oil and SPM, and the density of the oil and SPM will all have some effect on the rate of association. In contrast, field and laboratory studies suggest that sorption of truly dissolved components is not important. While most laboratory studies indicate that oil/SPM interactions are important, there is little evidence at this time that large amounts of oil are sedimented from major oil spill incidents (Boehm and Fiest, 1980b; Boehm *et al.*, 1982; Nelson, 1980).

Effects of Dispersants on Oil/SPM Adsorption

Chemical (dispersant) treatment of oil reduces the adhesion tendency of dispersed oil droplets both for other oil droplets and for suspended particulates (McAuliffe, 1977). Reductions in the amounts of dispersed oil adsorbing onto suspended sediments subsequently lowers the total fraction of the oil mass associated with the sinking of sorbed oil.

Photooxidation of Petroleum

Numerous reports describing the specific chemical changes in petroleum due to photochemical weathering processes have appeared in the open literature since the late 1960s (Payne and Phillips, 1985b). Berridge *et al.* (1968a) were among the first to speculate that the photooxidation of petroleum could lead to the formation of oxygenated products such as carboxylic acids, alcohols, peroxides, sulfoxides and related compounds. Kawahara (1969) used infrared spectroscopy to demonstrate that sunlight had indeed caused a chemical effect on petroleum. Further, Freegarde *et al.* (1971) used mercury lamps with various selected wavelengths less than 600 nm to demonstrate that a variety of organic acids and esters could be formed from the oxidation of petroleum. Since these early studies, the effects of photooxidation processes, using different crudes and individual components present in petroleum hydrocarbon mixtures, have been studied in laboratory and simulated field experiments.

A summary of previous photooxidation studies of crude petroleum and individual components is presented in Table 5.3. A variety of substrate types, identified products, light source types, and the presence or absence of sensitizers are identified. As noted in the table, some of the experiments are slightly flawed due either to the absence of an aqueous phase or to selection of light sources generating wave lengths (less than 295 nm) below those normally found in ambient sunlight. Nevertheless, a wide variety of substrates have been considered and numerous oxidation products identified.

In total, the results from previous research have demonstrated that photooxidation processes may have a considerable importance in the long-term weathering of spilled oil, both by enhancing dissolution of products and by

TABLE 5.3
Summary of more significant studies of photooxidation of petroleum and petroleum components

Substrate	*Products identified*	*Light source (wavelength)*	*Sensitizer used*	*Mechanisms proposed*	*Reaction rates identified*	*Comments*	*References*
Middle East Crude Oil	Thiocyclane-1-oxides (sulfoxides); C_1-C_4 alkyl substituted phenols (trace); C_{15}-C_{23} UCM of sulfoxides (build up with time/month); C_5-C_7 substituted thiocyclopentanes and thiocyclohexanes; and C_2-C_4 substituted dicyclothiocyclanes	Laboratory lighting Subdued daylight		Light catalyzed formation of benzyl hydroperoxides (isopropyl and aracyclane aromatics).	No	Air or other oxidant (peroxide, hydroperoxide or ozone) required. N_2 versus air atmosphere differences. Light formed hydroperoxides suspected as oxidant	Burwood and Spears, 1974
Phenanthrene	Fluorene; fluorenone; 9-phenanthrol; 9, 10,-phenanthrenequinone and 2, 2′-diformylbiphenyl via 9, 10-epoxy-9, 10-dihydrophenanthrene		Rose Bengal	$^1O_2$	No	9,10-epoxy could not be isolated directly	Dowty *et al.*, 1974
n-Hexadecane	Hexadecanol and alkyl-peroxides (1%)	400-W Hg lamp 313–335 nm	Xanthone	Xanthone triplet abstracts H˙ starting free radical process. O_2 required and alcohol formed via peroxide radical.	Initially 7.3×10^{-8} mole/min	$\phi = 0.16$–0.6; ratios of peroxide/alcohol less with xanthone versus 1-naphthol	Gesser *et al.*, 1977
Lybian Crude	Aromatic (20%) and aliphatic acids (28%), alcohols (23%) and aliphatic ketones (18%)	400-W Hg lamp >200 nm		Light absorption by aromatics → radicals 111°H abstraction → radical + O_2 → peroxi-radical˙ + ROOH → + ˙OH HO˙ + *n*-alkanes → R˙ + ROH		19 days constant irradiation. Thick and thin lenses noted with higher MW components in thicker lenses. nC_8 reduced to <1% of initial value. Photooxidation of oil with >300 nm $h\nu$ low. Photooxidation of oil with <300 nm $h\nu$ increased by 100×. Aromatic and aliphatic acids quite water soluble; therefore, further CO_2 loss unlikely. Ox products surface and may enhance water-in-oil emulsions and oil-in-water dispersions.	Hansen 1977, 1975

Hexadecane, low boiling Tijuana Crude, high boiling Tijuana Crude, Peruvian Crude		125-W Hg lamp 300–600 nm 1.5×10^{-7} Einsteins/min cm^2	1-Naphthol			Suggest products important for water-in-oil emulsions, viscosity and spreading properties. Vacuum distilled oil used. Oil lens diameter measured as f (time) $S = \sigma_w - (\sigma_0 + \gamma)$, σ_w = water:air, σ_0 = oil:air, γ = oil:water interfacial surface tension. γ changes due to adsorption of h [ox] products at interface. Decrease is due to diffusion control of ox-products in oil. Viscous oil versus nonviscous oil behavior different.	Klein and Pilpel, 1974a Berridge *et al.*, 1968a
Alkyl-benzenes cumuene	Benzyl-peroxide acetophenone + 2-phenyl-propanol	>290–375 nm hν Hg lamp with Ox-1 filter	1-Naphthol hν	1-Naphthol $\xrightarrow[O_2]{h\nu}$ naphthoxy radical + $HO^{\cdot}$ $ArOH \xrightarrow[O_2]{h\nu} ArO^{\cdot} + HO^{\cdot}$ $RH + ArO^{\cdot}$ $R^{\cdot} + ArOH$ $R^{\cdot} + O_2 \rightarrow ROO^{\cdot}$ $ROO^{\cdot} + RH$ $ROOH + R^{\cdot}$ $R''OOH + R'OOH +$ $ArO^{\cdot} \rightarrow R'OH +$ R′ | C═O + ArOR″ | R″ $+HO^{\cdot}$	Rates for cumene, s-butylbenzene and t-butyl-benzene, 1-naphthol	Naphthol also consumed.	Klein and Pilpel, 1974b
No. 2 Fuel Oil methylene (23% aromatic and low in sulfur) and individual HC components (most experiments done without water)	Hydroperoxides, acids, phenols	280 nm Hg vapor lamp	No	Yes	No, initiation period for ROOH production. $k = 2.1 \times 10^{-6}$ l/mole sec.	Benzylic methylene groups constrained in ring system yielded most hydroperoxides. Partition coeff. exps of hex:ben (2:1) and water and various peroxides completed. t-butyl-hydroperoxides 77% water extractable; tetralinhydroperoxides 8% water extractable; cumenehydroperoxides 11% water extractable; No. 2 fuel oil hydroperoxides 6% water extractable.	Larson *et al.*, 1976

(*continued*)

TABLE 5.3—*contd.*

Substrate	*Products identified*	*Light source (wavelength)*	*Sensitizer used*	*Mechanisms proposed*	*Reaction rates identified*	*Comments*	*References*
No. 2 Fuel Oil (no water present)	Hydroperoxides, phenols, acids, (ketones, trace of aldehydes); IR evidence of lactones, alcohols	Hg vapor lamp (pyrex filter used to eliminate <280 nm light)	Fluorenone generated *in situ* 1-naphthol also cited	Light catalyzed radical formation and O_2 incorporation to yield hydroperoxides and other product materials with active hydrogens.	Time series development of products noted.	Photooxidation product was highly toxic to yeast (acidic fraction most active). Addition of thiocyclohexane to 20-h irradiated oil reduced toxicity by factor of 5. Presumably due to decomposition of hydroperoxides required for other oxidants. Organic sulfur compounds also inhibit other component oxidation by radical processes.	Larson *et al.*, 1977
Refined petroleum	Products other than hydroperoxides inhibited by β-carotene. Significant lag time required for β-carotene destruction.	Hg vapor lamp	Inhibition by β-carotene and role of singlet $^1O_2$	Singlet oxygen and radical processes both important.		$^1O_2$ acceptors greatly inhibit photooxidation of hydrocarbons. $^1O_2$ important for initial hydroperoxide formation.	Larson and Hunt, 1978
No. 2 Fuel Oil layered on water (1:40 oil:water)	Highly alkyl substituted hydroperoxides, phenols and carboxylic acids (benzoic and naphthanoic). Extensive work with tetralin hydroperoxide. Products separated into "very weak acid fractions"—hydroperoxides (benzylic, tetrahydronaphthol structures). "Weak acids"—alkylated phenols, tetrahydronaphthol backbones. "Strong acids"—alkyl substituted benzylic acids.	>280 nm Pyrex filtered Hg vapor lamp or sunlight (for algal studies)	Anthracene, pyrene, etc. present in oil necessary for $^1O_2$ generation	$^1O_2$ + 2,3-dimethyl indene (or phenols), H-abstraction ↓ hydroxy radicals; other products ↓ H-abstraction aromatic sub.	ROOH conc. in water over time; reported linear increase over first 108 h. Initially ROOH in oil phase, then partitioned into water.	Hydroperoxides not analyzed directly, converted to alcohol by addition of thiocyclopentane. Peroxide content determined iodometrically. Toxicity to yeast and algae: hydroperoxide > carboxylic acids > phenols.	Larson *et al.*, 1979
No. 6 Fuel Oil in pentane (250 mg/l)	Not identified	Sylvania type G8T (254, 313 and 365 nm) light source 4 cm from oil source	No	Auto-catalytic proposed, but first and second order kinetics also possible.	260 nm abs. + $_{1/2}$ = 5–6 days 300 nm abs. + $_{1/2}$ = 5–8 days	Changes in UV spectra of oil noted as f (time).	Majewski *et al.*, 1974

Cumene Pyridine	Cumene hydroperoxide, cumyl alcohol, aceto-phenone, 2- and 4-isopropylphenols, Pyridine N-oxide, hydroxypyridines	Xenon lamp	No. Dark reactions catalyzed by azo-bis (2-carbomethyloxypropane) (source of $RO_2^{\cdot}$) and H_2O_2 + 290 nm $h\nu$ for $HO^{\cdot}$	Steady state conc. of radicals derived from kinetics	$t_{1/2}$ for $RO_2^{\cdot}$ rxn w/phenols, aromatic amines, hydroxylamines, hydroquinones est. at several days. $HO^{\cdot}$ too dilute to be important.	$RO_2^{\cdot}$ attacks cumene side chain; $HO^{\cdot}$ attacks benzene ring. $RO_2^{\cdot}$ yields pyridine N-oxides only; $HO^{\cdot}$ attacks ring (evidence of both radicals present in soln.). Steady state conc. est $[RO_2^{\cdot}]$ = 10^{-9} mole/l. Steady state conc. $[HO^{\cdot}] = 10^{17}$ mol/l	Mill *et al.*, 1980
*Benzo(a)anthracene Benzo(a)pyrene Quinoline Benzo(f)quinoline 9H-Carbazole 7H-Dibenzo (o,g) carbazole Benzo(b)thiophene Dibenzothiophene	30% 7,12-quinone 3,6- and 1,6-quinones (1:3) not identified two unidentified products (diff. in MeOH and CH_3-CN) seven unidentified carbazoles (+ added H, O) 2 unidentified products Sulfoxides suspected but not isolated	400-W medium pressure Hg lamp (313 or 316 nm bands or >300 nm broad spectrum light) and sunlight	Humic acids	None	Loss of PAH fit to first order regression and quantum yield estimated. $t_{1/2}$ at 40% N lat. estimated after Zepp and Cline (1977).	Strong absorption of PAHs affect 0 and yield $t_{1/2}$ values of minutes to hours. Seasonal effects minimal. N, S or O heteroatom substitution affects [OX] in some situations. Humic acids exhibit small effects suggesting importance of singlet states in PAHs.	Mill *et al.*, 1981
Light Arabian Crude (cargo of *Amoco Cadiz*) oil dissolved in hexane, layered on 4% saline solution	Alkyl substituted DBT sulfoxides	Natural sunlight on environmental samples from France and Sylvania EHC 500-W tungsten lamp with uranium glass filter for lab studies	No				Overton *et al.*, 1979
IXTOC-I Crude Oil dissolved in heptane then spread on sea water	nC_9-nC_{11} fatty acid methyl ester (FAME) + branched fatty acid methyl ester, C_1 and C_2 alkylated naphthols, substituted benzoic acids (C_1-C_6) substituted phenanthoic, benzothiophenoic and dibenzothiophenoic acids	Sunlight	No	Intermediate aromatic hydroperoxides involved in FAME formation as opposed to direct $^1O_2$ interaction. Rather propose $^3O_2$ [ox] or free radical reaction. No attempt made to assess rates or total amount of HC removed by this process.	No	Photooxidation products in sea water and oil samples (flakes) collected near well head and from rooftop aquarium systems with fresh IXTOC-I crude and artificial sea water (4-day exposure). Water samples from well head yielded FAMEs and phthalate esters (limited sample size precluded further identifications).	Overton *et al.*, 1980

(continued)

TABLE 5.3—*contd.*

Substrate	*Products identified*	*Light source (wavelength)*	*Sensitizer used*	*Mechanisms proposed*	*Reaction rates identified*	*Comments*	*References*
Phenanthrene	9, 10-Epoxy-9, 10-dihydrophenanthrene plus phenols and carboxylic acids, other products various: fluorene; fluorenone 2,3: 4,5-dibenzoxepin; 2,2′-diformyl-biphenyl-2′-formyl-biphenyl-2-carboxylic acid; 3,4-benzocoumarin; diphenic acid, 9-phenanthiol; 9,10-phenanthrenequinone; diphenic acid anhydride	500-W tungsten lamp with uranium filter	Methylene blue	No Singlet oxygen required. $h\nu$ + sensitizer $\rightarrow$ $^1S^*$ $^1S^* \rightarrow 1SC \rightarrow {}^3S^*$ $^3S^* + {}^3O_2 \rightarrow S + {}^1O_2$	No	Various reaction products prepared and MS fragmentation patterns of OX products from phenanthrene are presented to verify photooxidation products of phenanthrene.	Patel *et al.*, 1978
Dibenzothiophenes in *Amoco Cadiz* and medium Arabian Crude Oil	Sulfoxides	Natural (field samples); tungsten (lab)	No	No	No	Oil and aqueous phases combined before analysis	Patel *et al.*, 1979
Naphthalene 1-Methyl-naphthalene 2-Methyl-naphthalene Phenanthrene Anthracene 9-Methylanthracene 9,10-Dimethylanthracene Pyrene Fluoranthene Chrysene Naphthacene Benzo(a)pyrene Benz(a)anthracene	No Disappearance of PAH used to determine kinetics	450-W Hg lamp 313 nm 366 nm 436 nm	No	Solvent polarity (water versus isooctane) studies suggest radical cation intermediate. 9-Methylanthracene decomposition believed to proceed through excited singlet or short-lived triplet	First order at low aromatic concentrations $^1O_2$ photosensitized rates many orders of magnitude slower than direct photochemical transformation.	Half-lives of specific PAHs determined as function of depth. Depth dependence (reduction in rates) more pronounced for PAH absorbing shorter wavelength light. ϕ_r ranged from 0.001 to 0.01 and was roughly wavelength independent. Photoreactivity increases with increasing MW due to enhanced spectral overlap of absorption spectra and solar radiation. Several PAHs observed to degrade in absence of O_2. Results suggest $^1O_2$ accounts for negligible fraction of photoreaction of PAH dissolved in water (as opposed to surface slicks—see Larson and Hunt, 1978). Reduction in photoreactivity rates noted for sediment and SPM-bound PAH: most pronounced effects observed for higher MW PAH (9-methylanthracene through benzo(a)pyrene which have the greatest tendency to adsorb onto SPM.	Zepp and Schlotzhauer, 1979

increasing the general toxicity of the water soluble fraction. The majority of the products of photooxidation are removed from the parent oil by dissolution, which may represent losses similar in magnitude to those associated with microbial oxidation.

Photooxidation is further responsible for discernible changes in both the composition and physical properties of the exposed parent oil. Detectable increases in the nonvolatile residual fractions of the weathered oil accompany increases in the water soluble extractable components in the underlying waters. Changes in viscosity, spreading or contracting rates, and water-in-oil emulsification tendencies may also occur as a function of oil photooxidation.

Several mechanisms for the photooxidation of petroleum have been described, including free radical oxidation in the presence of oxygen, singlet oxygen initiation of hydroperoxide formation, and ground-state triplet oxygen combining with free radicals to form peroxides. Rates of photooxidation are considered wavelength dependent, but are also affected to some extent by turbidity levels and SPM concentrations (particularly for higher molecular weight aromatics). Photosensitized reactions are described by first-order kinetics.

The presence of inhibitors, such as sulfur compounds (e.g., thiocyclanes) or *beta*-carotenes, can restrict the formation of radical species or inhibit singlet oxygen-mediated peroxide formation. Humic substances may reduce the photolysis rates of UV-sensitive compounds, but they can also photosensitize transformations of organic compounds through an intermediate transfer of energy to molecular oxygen.

Field studies at spills of opportunity have detected the presence of several photo-oxidized products, including alkyl-substituted dibenzothiophene sulfoxides in oil samples, and benzoic acids and fatty acid methyl esters in seawater extracts. These photooxidized compounds had an enhanced water solubility and consequently were removed from surface slicks and diluted in ambient waters.

Additional research is needed to further characterize the products derived from photooxidation of weathered oil, as well as their eventual fate and chemical transformation. Similarly, additional data on the toxicity of the photochemical products are needed to characterize the environmental impacts associated with long-term weathering. Further study is also needed to define the possible affect of photooxidation processes on water-in-oil emulsification. Continued research in these areas will improve the predictive capabilities for future modeling of photochemical effects on oil weathering.

Ingestion of Dispersed Oil Droplets and Fecal Material

Ingestion of dispersed oil by zooplankton is believed to be an important factor in the short-term removal of petroleum residues from surface waters. Conover (1971) reported finding oil droplets in zooplankton feces after the tanker *Arrow* spill in Chadebucto Bay, Nova Scotia, in 1971. More recently, Sleeter and Butler (1982) observed significant levels of dispersed petroleum residues in fecal material collected in the Sargasso Sea, and they concluded that the removal rate of particulate/dispersed oil by zooplankton grazing may be of the same order of magnitude as the overall input of petroleum to the oceans. Encapsulated hydrocarbons in zooplankton feces are presumed to be sedimented, and the vertical flux of hydrocarbons can subsequently increase several-fold due to the significantly higher density and sinking rate of fecal pellets. The ultimate fate of hydrocarbons during and after this sedimentation process, however, still remains largely unknown.

TIME-DEPENDENT CHANGES IN THE PHYSICAL PROPERTIES OF BULK OIL AFTER RELEASE AT SEA

In a study where the weathering of Prudhoe Bay crude oil was examined under ambient subarctic weather regimes, Payne *et al.* (1983a) used 2800-l flow-through wave tank systems to evaluate changes in chemical and rheological properties of the oil with time. Component specific concentrations in the oil and water column in the wave tank systems were measured, and changes in density, viscosity, percent water incorporated and interfacial (oil/air and oil/water) surface tensions were reported. The changes in rheological properties of the oil/mousse observed during the first 12 days of the experiments are summarized in Figure 5.1. Water was not significantly entrained in the oil for the first 12 hours of the spill, and during this time significant dispersion of oil droplets into the water column was noted. After approximately 12 hours the water content in the oil increased in a smooth fashion, reaching a maximum of 55% water after 12 days. Correspondingly, the density increased from 0.88 g/ml to 0.99 g/ml over this time period. After an additional four months of weathering, 10 to 15 cm size balls of mousse were noted in the tanks along with a syrup-like water-in-oil mixture which had a higher water content (and density) and a slightly lower viscosity than that observed for the discrete mousse balls. The oil/water interfacial surface tension decreased from 27 dynes/cm in the fresh oil to 13 dynes/cm in the water-in-oil emulsion obtained after a 12 day period. After four months, the oil/water interfacial surface tension had decreased only slightly to a value of 12 dynes/cm. The oil/air interfacial surface tension did not change significantly over the four month period, although a very slight increase was indicated from 34 dynes/cm to 37 dynes/cm. Viscosity changed significantly, with an initial crude oil viscosity of 16 centistokes increasing to 2800 centistokes after 12 days. Four months later, the viscosity of the discrete balls of emulsified oil had reached 7200 centipoise. Simple pan evaporation experiments conducted in parallel to the wave tank studies showed an increase of viscosity from approximately 26 to 100 centistokes over the time frame of day 4 through day 12.

Prudhoe Bay crude oil has approximately 23% asphalts (Coleman *et al.*, 1978) and nickel and vanadium concentrations of 13.5 ppm and 28.3 ppm, respectively. These concentrations of surface active compounds should promote stable water-in-oil emulsification; however, data from the wave tank experiments demonstrated that this behavior did not occur (even at 0°C) without significant evap-

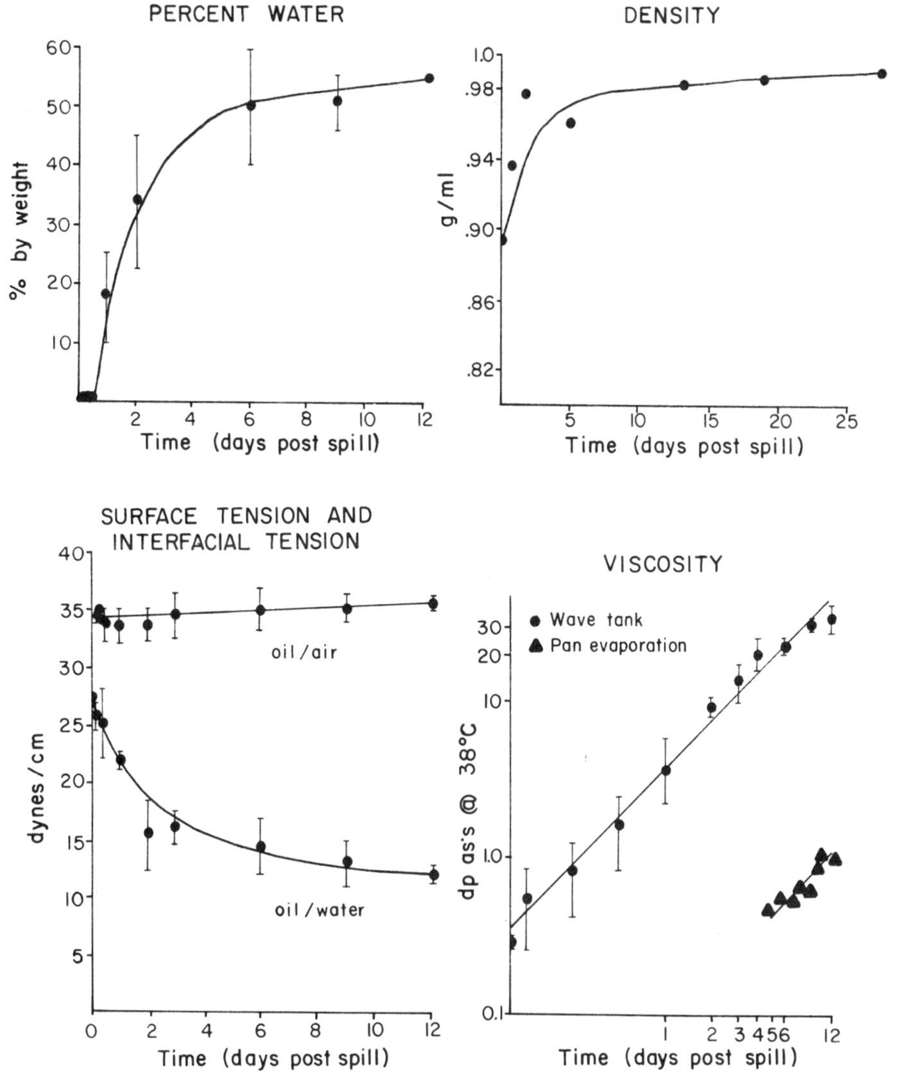

Figure 5.1. Rheological properties data on the Prudhoe Bay crude oil weathering in the wave tank systems. Values are means from the three ± S.D. (from Payne *et al.*, 1983a).

oration and dissolution weathering first removing the lower molecular weight components (Payne *et al.*, 1983a). Even after four months, the stability of the mousse was observed to be extremely temperature dependent, as a melting or thawing behavior was observed when the mousse temperature was increased from 0°C to 38°C. Significant quantities of air were also entrapped in the resultant mousse, but many of the air bubbles were lost during the warming process. Nevertheless, the resultant mixture had extremely high viscosities (at 38°C) and additional separation of water and oil was not observed.

SLICK DRIFT, SPREADING AND ADVECTION

Wind and Current Effects

Processes involved in the physical transport and dispersion of spilled oil in the marine environment are summarized in Jordan and Payne (1980) and also recently reviewed by the National Research Council (1985). In general, mechanisms affecting spill movement are fairly well understood; however, the magnitude and duration of simultaneous processes cannot be accurately predicted. Consequently, while several oil spill trajectory models currently exist for many specific geographical areas, the models have typically not been validated and are somewhat restricted by the lack of physical (wind and current velocities) data.

Drift is a large scale phenomena, measured by the movement of center of mass of an oil slick, and is primarily controlled by wind, waves and surface currents. When winds are the dominant force in drift movement, a slick can move at a rate of up to 3.6% of the wind speed (Nelson-Smith, 1973; Smith, 1977). However, prediction of slick drift by evaluating wind patterns alone is difficult because of the accompanying effects of current and wave perturbations.

Spreading of oil on the sea surface is governed by gravitational forces, surface tension, inertial forces and frictional forces (Wheeler, 1978), and is probably the dominant process affecting a slick during the first six to ten hours following a spill. The gravitational spreading force is proportional to the slick thickness, the thickness gradient and the density difference between the water and the oil. Simultaneous evaporation and dissolution processes alter the composition of the spilled oil, thus further affecting the oil density and spreading characteristics.

Subsurface movement of oil following a surface spill has been observed by Conomos (1975), after portions of a Bunker C crude oil spill sank and were eventually transported farther up an estuary in bottom density currents, while the remaining portions of the spill associated with surface waters were transported into the lower estuary and adjacent coastal waters via surface currents. During the IXTOC-I oil spill, Boehm and Fiest (1980a,b) characterized the subsurface oil plume for distances up to 20 kilometers from the spill site, and Walter and Proni (1980) used sonar techniques to track the movement of this subsurface plume. Payne *et al.* (1980b) measured high levels of dissolved aromatic hydrocarbons in subsurface waters resulting from the subsurface oil release. Nevertheless, the ability to completely model this dispersion behavior is still incomplete at this time.

Breakup of Slicks into Patches

Eventually, surface oil will spread into nonuniform patches which vary from thick patches to thin sheens. Wind effects will cause the thicker patches to drift faster than the sheen, resulting in slicks with higher densities of thick patches at the leading edge and a trailing sheen in the windward direction. Attempts to describe the slick area as a function of time after the spill event are typically expressed as a power function in time, and proportional to oil viscosity and interfacial tension. Further attempts to model spreading using a three-regime spreading theory (Fay, 1971), turbulent diffusion theory (Murray, 1972), Fickian diffusion (Hunter, 1980) and empirical approaches (Karpen and Galt, 1979) are reviewed in Huang (1983). The ultimate complexity of the spreading process, however, limits the accuracy of such spill area predictions.

SIMULATION MODELS

Spill Trajectory Models

A number of the existing oil spill simulation models are listed in Huang (1983). The majority of these numerical models were designed to predict spill trajectories or advection both for use in deploying cleanup equipment and for protecting important resource areas. Oil spill trajectory models and submodels have been developed by the Rand Corporation (e.g., Liu and Leendertse, 1979, 1981a,b, 1986) for possible spills in the Bering and Chukchi Seas during different oceanic seasons and from various locations, which represent hypothetical platforms, pipelines and transportation route sources. These are perhaps among the more comprehensive of the existing simulation models. These dispersion models are based on three techniques:

1) use of a three-dimensional model to compute local diffusion coefficients by determining tidal currents, residual circulation, subgrid scale turbulent diffusion, and the vorticity-gradient related dispersion coefficients; the model is formulated according to equations of motion for water and ice, continuity, state, the balances of heat and salt, and turbulent energy densities on a three dimensional grid;
2) the three-dimensional hydrographic model is then coupled with a two-dimensional stochastic weather (storm track) model to compute trajectories of hypothetical spills. Interrelationships of the two models are shown in Figure 5.2;
3) based on the solutions to one-dimensional (horizontal) diffusion equations, the concentrations of oil along the dispersion trajectories are predicted using local governing parameters identified in the three-dimensional hydrographic model.

Additional weathering processes, such as evaporation and dissolution, are also incorporated into the model to provide more realistic oil concentration versus distance predictions.

A common deficiency of trajectory models is the lack of local wind and current field data for describing the effects of real-time changes in shear stresses on slick advection. Many models must rely on wind data from onshore facilities, which may be significantly different than actual conditions at the spill site. Another problem with the use of most trajectory models for predicting the environmental fate of spilled oil is the failure to adequately account for the effects of simultaneous weathering processes.

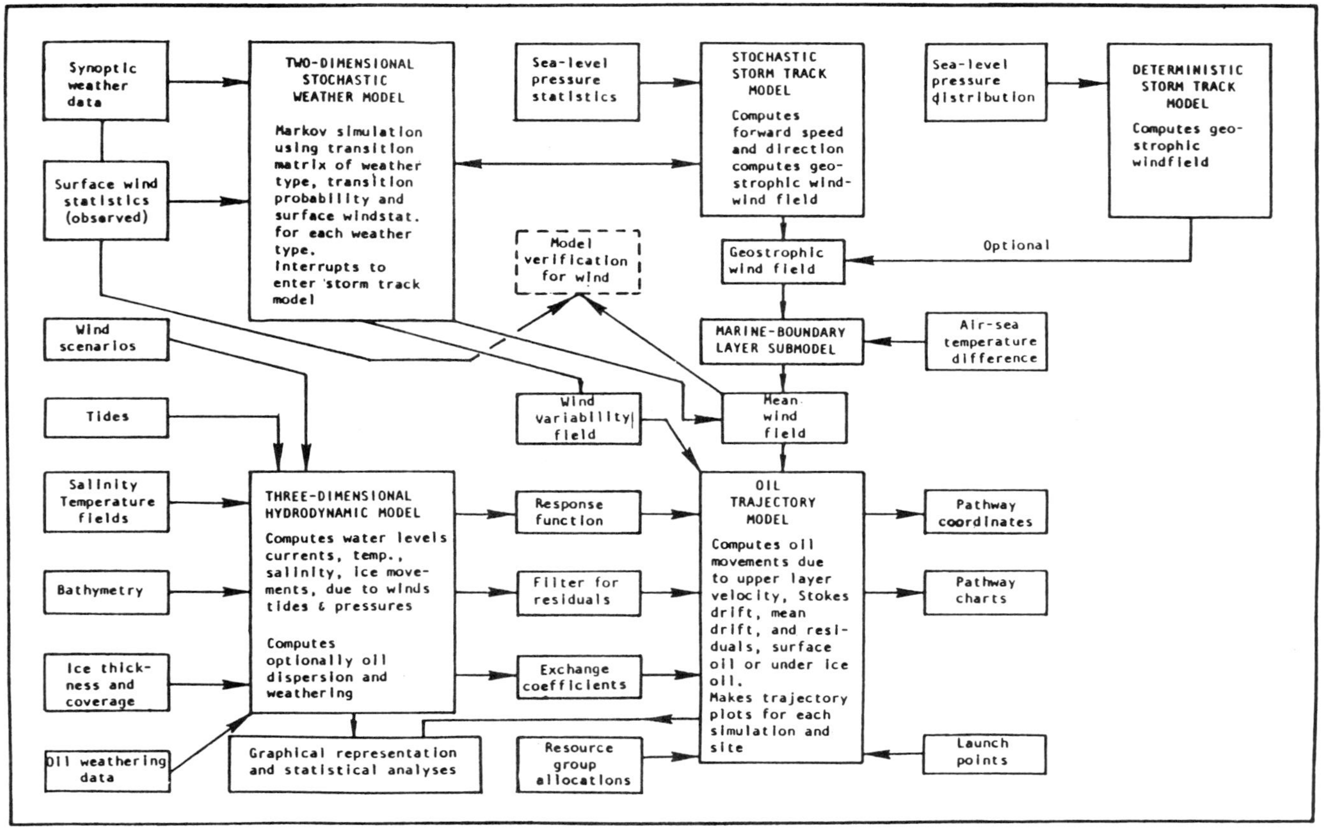

Figure 5.2. Essential components of the two-dimensional stochastic weather simulation model and interrelationships with the three dimensional hydrodynamic model and oil spill trajectory model (from Liu and Leendertse, 1986).

Oil Weathering Models

Composite models, reviewed by Huang (1983), have recently been developed to provide more realistic predictions on the environmental fate and behavior of oil spills. In addition to slick advection, algorithms describing weathering processes, such as spreading, dispersion and emulsification, have been derived for incorporation into simulation models. Corresponding models do not presently exist for describing sinking/sedimentation and autooxidation processes.

The primary objectives for a mathematical oil weathering model are to predict both the mass of oil remaining in a slick over time and the chemical composition and physical properties of the slick. Payne *et al.* (1983a, 1984) have developed predictive oil weathering models which generate material balances for both specific compounds and pseudocompounds (true boiling point distillation cuts) in a crude oil spill. These models are applicable to open ocean oil spills, spills in estuaries and lagoons and spills on land. The oil weathering processes included in the model are evaporation, dispersion into the water column, dissolution, water-in-oil emulsification (mousse formation) and slick spreading. The model is based on physical properties, such as oil/air interfacial surface tension, oil/water interfacial surface tension and oil viscosity, as well as mass transfer (rate) coefficients which were obtained from the open literature or from measurements made from simulated spills in outdoor wave tanks.

In general, reasonable correlations between predicted oil weathering behavior and observed chemical changes have been obtained (Payne *et al.*, 1984). Changes in predicted and observed chemical and physical properties of the oil slick also accompany changes in oil slick behavior, especially during the early weathering stages (from a freely flowing slick), through the water-in-oil emulsion or mousse formation stage, to the subsequent formation of tarballs stage. Models developed by Payne *et al.*, are presently capable of predicting oil weathering behavior in real spill situations. However, investigations of oil weathering at spills of opportunity have measured concentrations of specific components, whereas determinations of the overall mass balance using a pseudocomponent approach is needed for model verification.

At present, further work is needed to validate existing oil weathering models under higher turbulence regimes. The oil weathering models should also be expanded to predict oil/SPM interactions, the behavior of oil in various stages of ice growth and decay, and the transport and deposition of oiled sediments.

CHRONIC DISCHARGES

Routine discharges of drilling fluids, cuttings and produced waters from offshore oil and gas activities contribute to the mass input of petroleum hydrocarbons and trace metals to continental shelf waters. The environmental implications of these routine discharges have recently been reviewed by Menzie (1982) and by the National Research Council (1983). Discharges of bilge, ballast and cleaning waters from vessels, discharges of industrial and municipal effluents, river inputs from inland sources and natural oil seeps also add to the chronic input of hydrocarbons to coastal areas. However, comparatively fewer data are available to characterize the fate and environmental effects of these discharge sources.

Drilling Fluids and Cuttings

The use, composition and discharge of drilling fluids and cuttings are described in detail in Chapter 4 of this book. The discussion here will center on the environmental fate of the potential pollutants associated with these materials once discharged into the ocean. Neff (Chapter 10) also presents a review of studies on the environmental fate and effects of drilling discharges.

Discharged spent drilling fluids and cuttings represent a potential source of trace metals, hydrocarbons and suspended solids to the water column. Contributing to the total metals levels in the discharge source are Ba, which is from the fluid component barite, Cr, which is associated with the additive lignosulfonate, and As, Hg, Cd, Pb, Ni and Zn as potential contaminants present in the barite (e.g., Crippen and Hood, 1980; Kramer *et al.*, 1980).

Hydrocarbons may be derived from formation strata or present as contaminants from mud additives (such as diesel oil). Hydrocarbon levels in selected spent drilling fluids were reported by Science Applications, Inc. (1983). Total resolved saturates in whole mud extracts ranged from 10 to 2700 mg/l, whereas total resolved aromatics were present at 7 to 640 mg/l. Saturated hydrocarbons from nC_9 to nC_{31} were present in some whole fluids samples, although several formulations contained no *n*-alkanes larger than nC_{26}. Aromatic hydrocarbons included alkyl-benzenes, naphthalenes, phenanthrene, and alkyl-phenanthrene. Furthermore, analytical evidence suggested that petroleum hydrocarbons present in the drilling fluids were introduced in a chemically refined form (i.e., as an additive) rather than from crude oil contamination from the hole. Nevertheless, the specific chemical compositions of discharged spent drilling fluids will reflect the composition and concentrations of the various additives present and, to a certain extent, the formation conditions encountered during drilling.

Investigations by Pierce *et al.* (1985) suggest that saturated and aromatic hydrocarbons in spent drilling fluids are partitioned between dissolved and particulate phases following discharge; saturates are strongly associated with particulates (with a calculated distribution coefficient of 160 ± 27), whereas the aromatics were more evenly distributed between dissolved and particulate phases ($K_d = 38 \pm 24$). Boehm (Chapter 6) reported that unpublished data from laboratory partitioning experiments demonstrated that the majority (>80%) of the hydrocarbons associated with a diesel oil additive to a drilling fluid was partitioned into the dissolved or fine particulate phase, whereas <20% of the oil was associated with larger, sinking particulates. The majority of metals in the drilling mud discharge is associated with particulate material, although slight increases in the dissolved Cr and Fe levels may be apparent in receiving waters due to complexing with soluble organics, including lignosulfonate. At the pH of normal seawater, however, concentrations of dissolved Cr and Fe in the receiving waters will decrease with time as these metals adsorb onto suspended clays (Liss *et al.*, 1980).

Results from previous studies (Trocine and Trefry, 1983; Pierce *et al.*, 1985) have shown that petroleum hydrocarbons and the trace metals barium and chromium may be useful as "tracers" for monitoring dispersion of drilling fluids discharged into the water column. Ayers *et al.* (1980b) measured particulate Ba, Al and Cr concentrations in the Gulf of Mexico following high rate, high volume discharge of muds. They estimated that decreases in metal concentrations of three to four orders of magnitude occurred within 100 m, and that decreases of five to six orders of magnitude occurred within 500 to 1000 m of the discharge source. Dissolved metal levels decreased with distance (or time) from the source at a rate two orders of magnitude less than that of the particulate phases. Because of the rapid dilution observed during this study, the authors concluded that drilling fluids have a negligible effect on open ocean water quality despite the high discharge rates and volumes. Trocine and Trefry (1983) noted comparably high dilution rates of discharged drilling fluids in the Gulf of Mexico, but also detected a barium "haze" or particulate Ba enrichment in near surface waters due to the presence of suspended barite in a microparticulate (<4 μm in diameter) form. Similar attempts to trace a drilling fluid discharge by monitoring levels of dissolved lignosulfonate in receiving waters (Pierce *et al.*, 1985) were unsuccessful.

Predictions of postdischarge concentrations, and subsequent fate, of drilling fluid/cuttings constituents in the water column is particularly difficult because of the possible effects of variable current fields, density stratification, predischarge dilution, variable discharge rates, and variations in the composition and characteristics of the discharged material. Nevertheless, results from previous monitoring programs of oil and gas development activities, summarized by the National Research Council (1983), Menzie (1982) and Neff (Chapter 10), consistently demonstrate a rapid dispersion of drilling related discharges in continental shelf waters. Discharge plumes are typically diluted to background levels within a period of several hours and/or within several hundred meters of the discharge source. Therefore, accumulation of toxic trace metals and hydrocarbons in exposed shelf waters, due to periodic releases of water-based generic muds and cuttings, are unlikely, and cumulative impacts or long-term degradation of the water column from operational discharges are not major concerns (National Research Council, 1983).

Produced Waters

Few field studies have been conducted to characterize the behavior and fate of discharged produced waters (see Chapter 4 for information concerning the physical and chemical characteristics). Middleditch (1981) reported detectable levels of petroleum alkanes (as a produced water tracer) in waters directly below the discharge pipe, at the air/sea interface, and at nearby water column sampling stations; however, no obvious concentration gradients were apparent. In a related study, Rose and Ward (1981) noted that although discharged produced waters may be considered relatively nontoxic the potentials for aquatic hazards are case-specific and dependent upon the toxicity of the water, volumes discharged, and the fate subsequent to release.

Discharged produced waters are rapidly diluted within the immediate vicinity of an ocean outfall or diffuser. Minor, site-specific differences in dilution rates may reflect the relative density characteristics of the produced water discharge and ambient receiving waters, the local current regime, and wave effects. Nevertheless, significant increases in water concentrations of dissolved and particulate hydrocarbons and trace metals due to produced water discharges are not expected outside of the initial mixing zone or immediate vicinity of the discharge source. In particular, rapid removal of waste water-associated metals and hydrocarbons is promoted by particulate scavenging, advection, evaporation of lower molecular weight saturates, and additional weathering processes.

Although produced waters will be rapidly dispersed following discharge into an open coastal or shelf environment, variable discharge volumes will be released continuously throughout the duration of any particular oil and gas production operation. Thus, long-term effects to water column processes, consisting of localized increases in particulate metal and soluble lower molecular weight aromatic hydrocarbon (e.g., benzene, toluene and xylenes) concentrations, may be implicated within the mixing zone of the discharge. In addition, trace metals and hydrocarbons associated with the discharge may be scavenged from the water column and subsequently deposited within the sediments near the discharge point. The potential toxicity of produced waters to exposed organisms is reviewed by Neff (Chapter 10).

Dispersion Models

Over the past several years, several attempts have been made to model the dispersion of drilling fluids and cuttings discharges from drilling rigs in coastal situations. The strengths and limitations of available drilling effluent dispersion models were recently reviewed in Runchal (1983). At present, existing models based on empirical data from several field monitoring studies may adequately describe short-term dispersion processes. In contrast, models have not been successful in adequately predicting the long-term dispersion of discharged drilling materials because of insufficient data on transport rates, current patterns and the long-term behavior of the discharge components.

A simulation discharge/fate model for dispersion of drilling fluids and cuttings from an open ocean platform was described by Auble *et al.* (1981). Conceptually, the discharge separated into an upper and a lower plume; the lower plume contains the majority of the cuttings and drill fluid mass, whereas the upper plume comprises the liquid fraction and some fine-grained silts and clays which are separated from the lower plume by turbulent mixing. The lower plume sinks rapidly to the bottom with little horizontal displacement of dispersion by local currents. The upper plume spreads laterally and vertically, and is transported in the direction of the net current after reaching a depth of neutral density. Similar behavior has been observed during actual discharges from drilling platforms at Tanner Banks (Ray and Meek, 1980; Meek and Ray, 1980), in the Gulf of Mexico (Ayers *et al.*, 1980b), in the mid-Atlantic (Ayers *et al.*, 1980a) and in Cook Inlet, Alaska (Houghton *et al.*, 1980). Results of these field studies, and the output of the simulation model, both indicated relatively localized effects from routine discharges from drilling rigs in open coastal environments. Far-field effects or long-term accumulations were restricted by the high dilution and dispersion rates.

Conclusions from the field studies listed above are summarized in Table 5.4 (synopses of other field studies which focused on benthic effects are presented in Chapters 6, 10, and 14). Despite the differences in discharge depths, current regimes, and discharge rates, plume dilution rates were fairly consistent, and the measured levels of suspended solids and particulate trace metal constituents were typically reduced to background concentrations within a few hundred meters of the source. Subsequently, Auble *et al.* (1981) calculated a dispersion ratio from a multiple regression with transport time and discharge rates as independent variables. The regression equation:

$$\text{DISPR} = 10^{4.4495} * (1/\text{DISCHR})^{0.35674} * (\text{TT})^{1.1001}$$

where DISPR = dispersion ratio
DISCHR = discharge rate
TT = transport time (min)

is based on field measurements from studies listed in Table 5.4. The correlation coefficient (r^2) for the regression was 0.74.

The Auble *et al.* (1981) simulation model predicted that the bottom area affected by the deposited lower plume materials would be proportional to the bottom depth, current velocity and the inverse of the particle settling rate. Actual field studies have shown, however, that materials are not deposited evenly within a circular area, but are deposited in patterns aligned with the predominant current direction. Successive changes in varying current regimes may result in starburst depositional patterns, with greatest accumulations of materials near the platform.

A similar short-term dispersion model, developed by Brandsma *et al.* (1980), was based on earlier models of dredged material disposal operations. The dispersion of discharged materials was divided into three phases: convective descent of a jet of material, dynamic collapse and long-term passive diffusion. Several modifications were required to account for previously observed pre-discharge dilution and the formation of a surface plume composed of fine-grained materials. Plume behaviors predicted by the computer model were later compared to actual field results. The predicted high initial dilution rates (1000:1 after one minute) were verified, although agreement between predicted and observed plume behaviors further downstream was more erratic.

As mentioned previously, Runchal (1983) reported the results of a workshop held to evaluate the applicability of some of the available drilling fluid dispersion models. Workshop participants concluded that most of these models may provide an adequate prediction of the short-term (less than one day) fate of discharged materials, but no longer-term approaches are currently tenable. Furthermore, although the short-term models may provide a reasonable prediction of near-field behavior, the models have not been validated with field or laboratory data in the low densimetric Froude number range characteristic of most drilling mud

TABLE 5.4
Summary of continental shelf discharge monitoring studies

Reference	*General area*	*Chemical analyses*	*Sample type*	*General conclusions/ comments*	*Distance/radius of influence*	*Analyses of*	
						Produced water	*Drilling muds*
Ayers *et al.*, 1980a	Mid-Atlantic (off New Jersey coast)	Trace metals, suspended solids, light transmittance	Suspended particulates	Discharged fluids and cuttings transported from rig in direction of prevailing current; suspended solids concentrations reached background levels at distances of 0.3–0.6 km, transmittance reached background at 0.8–1.0 km.	1 km	—	No
Ayers *et al.*, 1980b	Gulf of Mexico	Trace metals, suspended solids, dissolved oxygen, temperature, salinity	Water column suspended sediments	High volume, high rate discharge of drilling fluids and cuttings. Suspended solids and particulate trace metals in upper plume reached background levels at distances of 0.5 km during 275 bbl discharge and at 1.0 km during 1,000 bbl discharge. Light transmittance still above background at a distance of 1.0 km.	0.5–1.0 km (>1.0 km for light transmittance)	—	No

Houghton *et al.*, 1980	Cook Inlet, Alaska	Suspended solids, trace metals	Sea water, sediments	Dilution rates of 10,000:1 within 100 m of the discharge point result in calculated increase in suspended solids concentrations of 100 mg/l.			
Hrudey, 1980	MacKenzie Delta, Canada (Arctic)	Na, K, Cl, Ba, Al, Cr, sulfate, organic and inorganic carbon, alkalinity, specific conductance	Water sump	Evidence of pollution of surface waters by sump fluids found; no evidence of ongoing leaching of sump fluids from holding areas was found.	100 m	—	—
Middleditch, 1981	Gulf of Mexico Buccaneer Oil Field	Hydrocarbons	Sea water, sediments	Alkanes detected in immediate mixing zone, and air/sea interface, of produced water discharge; no concentration gradients observed in water column. Gradients of alkanes in sediments were apparent, decreasing with increased distance from platform.		Yes	—
Northern Technical Services, 1981	Reindeer Island, Beaufort Sea, Alaska	Trace metals (Ba, Co, Cr, Cu, Fe, Mn, Ni, Pb, V, Zn)	Biota, benthic; water column	For below ice discharge, maximum concentrations occurred beneath discharge point with a decreasing radial gradient. Natural current variations dispersed the	100 m	—	—

(continued)

TABLE 5.4—*contd.*

Reference	*General area*	*Chemical analyses*	*Sample type*	*General conclusions/ comments*	*Distance/radius of influence*	*Analyses of*	
						Produced water	*Drilling muds*
Northern Technical Services, 1981 *contd.*	Reindeer Island, Beaufort Sea, Alaska	Trace metals (Ba, Co, Cr, Cu, Fe, Mn, Ni, Pb, V, Zn)	Biota, benthic; water column	accumulated materials over a season. Above-ice disposal delayed release until ice breakup facilitating maximum dispersion and dilution. Trace metal analyses of sediments and biota showed that effluents did not remain at drilling site.			
Northern Technical Services, 1982	Resolution, Endeavor and Ardlenge Islands, Beaufort Sea, Alaska	Trace metals (Ba, Cd, Cr, Cu, Hg, Pb, Zn), hydrocarbons	Biota, surface; sediments, water	Above-ice disposal regions melt earlier than adjacent ice areas with thick areas melting slower than thin areas. Salinity effects due to effluents were not observed. Changes in grain size distribution are likely natural variations. Trace metal and hydrocarbon fluctuations could not be attributed to effluent.			
Pierce *et al.*, 1983	Gulf of Mexico, Texas, Flower Gardens	Dissolved and particulate hydrocarbons, ferrochrome, lignosulfonate	Sea water, sediments	Petroleum hydrocarbons may be associated with the spent drilling fluid discharge;		—	Yes

				attempts to use dissolved lignosulfonate as a tracer were unsuccessful although petroleum hydrocarbons in sediments may be useful instead for monitoring dispersion. No evidence of petroleum hydrocarbons in water column.			
Ray and Meek, 1980	Tanner Banks, California	Trace metals, suspended solids, temperature, salinity dissolved oxygen, pH	Suspended particulates	Transmittance was the only hydrographic parameter significantly affected by drilling fluid discharge. Suspended solids and trace metal concentrations approached background levels at a distance of 0.2 km from the rig.	0.2 km	—	No
Trocine and Trefry, 1983	NW Gulf of Mexico	Trace metals	Sea water, sediments	Barite (Ba) was first fluid component to settle from discharge in large quantities, followed by bentonite (Fe); enriched Cr (likely associated with lignosulfonate) observed only in upper water column; Ba "haze" also observed in upper water column.	0.8 km (water) 0.8–1.9 km (sediments)	—	Yes

and produced water discharges. Many of the key dispersion processes, such as plume separations, flocculation, and dynamic collapse are still poorly understood and insufficient data are available to better define mechanisms and process rates. Recommendations from the workshop included the following: 1) identify a suitable methodology for developing a long-term model; 2) refine and incorporate descriptions of short-term phenomena such as initial and predischarge mixing and plume separation processes into existing models; 3) perform laboratory studies to investigate flocculation, deflocculation, sedimentation and resuspension processes under controlled conditions; 4) further investigate interactions of the discharge plume with the bottom, wake effects on rates of initial dilution and initial partitioning behaviors of certain discharge constituents; and 5) initiate a program to collect suitable field data for verifying short- and long-term model predictions. Further model refinements will also be needed to describe the effects of density stratification, variable current regimes and differences in discharge depths and rates on plume dispersion.

These recommendations for model refinements were defined for purposes of developing predictive tools which could be used to describe the physical dispersion of operational discharges, but not necessarily for evaluating long-term impacts to either water column or benthic processes. Environmental impacts within the water column are considered with respect to: 1) the upper plume materials, 2) lower plume materials which reach neutral buoyancy prior to deposition, and 3) materials resuspended from the bottom. As mentioned previously (Table 5.4), field studies have consistently demonstrated a rapid dilution of discharged drilling fluid within the upper plume. Long-term impacts associated with upper plume materials are therefore considered insignificant (National Research Council, 1983). In contrast, limited laboratory and field data are available to describe the fate of neutrally buoyant or resuspended lower plume materials. Brandsma and Sauer (1983) calculated that drilling fluid components in a buoyant lower plume may be present in concentrations an order of magnitude higher than those associated with the corresponding upper plume. Despite the relatively higher concentrations, long-term impacts to the water column are unlikely (National Research Council, 1983). Consequently, further model development and field validation for describing the fate of lower plume materials may be of interest for predicting concentrations of drilling fluid constituents encountered by exposed organisms and the areal extent of subsequent deposition and accumulation in the benthic environment. However, due to the localized and temporary nature of the impacts associated with operational discharges, there is no indication that discharges into high energy marine environments have real potentials for long-term effects to water column processes. Therefore, dispersion modeling may offer limited value for further evaluation of long-term impacts. In contrast, predictive models applied to the dispersion of sediment-associated contaminants may be more relevant for predicting possible long-term impacts from operational discharges (National Research Council, 1983).

Monitoring Studies

The majority of the previous offshore oil and gas operation field monitoring programs (summarized in Menzie, 1982) were intended only to sample the short-term impacts associated with platform discharges. Results from these studies, summarized in Table 5.4, were fairly consistent and indicated that the major portion of the dense, solid material sank rapidly, whereas the lighter materials in the upper plume were diluted to near background levels within a few hundred meters of the source. Although insufficient data were available to derive a mass balance for discharged muds and cuttings, empirical evidence indicated that the majority (>90%) of the discharge is initially deposited within 100 m of the drill rig. Subsequent dispersion was then a function of the frequency of particle resuspension and/or transport in near-bottom currents. Nevertheless, it is apparent that the relatively long-term and far-field impacts from routine drilling and production related discharges are associated with accumulations of contaminants in the benthic environment, whereas cumulative effects in the water column are unlikely (Menzie, 1982). Unfortunately, few long-term monitoring studies were designed to measure the actual rates and patterns of dispersion or accumulation of discharged contaminants in surficial sediments and epifaunal and infaunal tissues (further discussed in Chapters 9 and 10).

The results of a multiyear study to identify the types and extent of biological, chemical and physical alterations of the marine ecosystem associated with production operations in the Buccaneer Gas and Oil Field are presented in Middleditch (1981) and Wheeler *et al.* (1980). Measurable quantities of petroleum hydrocarbons were present in the water column and immediate air/sea interface within the initial mixing zone of the produced water discharge. The absence of detectable hydrocarbon concentration gradients in adjacent waters indicated rapid and efficient effluent dilution and dispersion. Sufficient information to suggest critical contaminant concentrations such that residual concentrations in the mixing zone exceed levels potentially harmful to marine organisms was unavailable.

In contrast to the absence of detectable concentration gradients in the water column, concentration gradients of petroleum hydrocarbons, which decreased with increasing distance from the platform, were detected in surficial sediments. Ratios of component hydrocarbons in sediments were similar to corresponding ratios in the produced water, suggesting rapid deposition of discharged hydrocarbons. Middleditch (1981) concluded that resultant concentrations in bottom sediments depend on the total quantities of contaminants released rather than the component concentrations in the discharge source. Results from these studies also suggest that near-field impacts to water quality are restricted to the zone of initial effluent mixing; far-field and cumulative effects are dependent upon the dispersion and subsequent deposition of sediments, which are the primary sink for released contaminants.

As an alternative approach to monitoring the effects of petroleum and other waste discharges on marine water quality, the concentrations of metals and hydrocarbons in mussels (*Mytilus californianus* and *M. edulis*) have been measured as part of the National (Goldberg *et al.*, 1983) and California (Stephenson *et al.*, 1979; Risebrough *et al.*, 1980) Mussel Watch programs. Organisms collected from areas near large population centers and from coastal bays and harbors typically had elevated contaminant body burdens when compared to levels in organisms from other remote, open coastal regions. In particular, mussels from collection stations exposed to sewage effluent had relatively higher levels of Pb and Ag (Stephenson *et al.*, 1979). Mussels from California bays and harbors accumulated petroleum hydrocarbons at levels only slightly below those collected near natural oil seeps at Coal Oil Point, near Santa Barbara. Nearly all of the California Mussel Watch samples collected along the entire coast contained levels of hydrocarbons suggestive of chronic oil pollution (Risebrough *et al.*, 1980).

A series of related studies have also recently been conducted to evaluate the effects of chronic oil input from natural oil seeps both to the adjacent water and sediment quality (Steurmer *et al.*, 1982; Reed *et al.*, 1977; Reed and Kaplan, 1977) and on local marine organisms (Steurmer *et al.*, 1981; Spies and Davis, 1979; Spies *et al.*, 1980). Results of these studies have also been summarized by Spies (Chapter 9). Although not considered monitoring studies, results from these investigations may be useful for predicting the behavior and consequences of other chronic petroleum hydrocarbon input sources to a nearshore environment.

INFORMATION GAPS AND AREAS REQUIRING FURTHER RESEARCH

Table 5.5 summarizes topics of specific areas identified in the preceding sections for which sufficient data are not currently available. Initial evaluations of the potential significance in terms of long-term effects are made for each of the listed subject areas.

In general, the processes of evaporation, dissolution and slick drift are relatively well understood, and the effects of these processes on spilled oil can be predicted using existing mathematical models. Similarly, the initial short-term dilution and dispersion of routine discharges from offshore drilling activities can also be approximated by existing models; however, the physical mixing processes governing dispersion are less well understood. Obviously, the accuracy of these predictive models is dependent upon the existence and quality of the input data and the ability to validate model predictions with actual field data.

In contrast, while limited information is currently available to adequately describe several oil transformation processes (e.g., mousse formation, photooxidation, and sedimentation), our modeling and predictive capabilities for the areas are not as well developed at this time. Further, subsurface advection processes, chemical and physical properties, changes associated with oil released in the presence of sea ice, and the long-term fates and environmental effects of chronic discharges are less well understood. Mathematical models for these processes and effects are currently limited by an incomplete understanding of the governing mechanisms, the absence of suitable rate constants or coefficients, and the lack of sufficient field data for model verification. Consequently, further refinement of composite models (i.e., Huang, 1983) for predicting simultaneous oil transport and weathering processes may await results from research on some of the proposed topics listed in Table 5.5.

As indicated in Table 5.5, several water column processes are currently: 1) incompletely understood, 2) relevant to considerations of long-term effects and 3) represent areas that warrant additional research. In order of their priority for predictive assessments of long-term effects on water column processes, these topics are as follows: 1) validation of extant oil weathering models under higher turbulence regimes, 2) additional weathering models under higher turbulence regimes, 3) additional investigations of weathering behavior of petroleum in intertidal and nearshore regimes, including processes of along-shore transport, oil/SPM interactions, sediment resuspension and offshore transport of oiled beach substrates and suspended particulates, 4) changes in the physical and chemical properties of spilled petroleum in the presence of sea ice at various stages of growth and decay, 5) development and validation of models describing long-term rates of accumulation and resuspension and transport of deposited contaminants associated with operational discharges of drilling fluids and produced waters, 6) additional data characterizing pseudocomponent and specific component mass transfer rates between a slick and the underlying water column, and 7) development and validation of algorithms for modeling photooxidation and emulsification processes.

Validation of oil weathering models under varying turbulence regimes can realistically be achieved only during investigations of spills of opportunity. Specific validation data are needed for comparing predicted versus observed rates of dispersion, component mass transfer, oil/SPM interactions and rates of emulsification. For the purposes of existing oil weathering models (e.g., Payne *et al.*, 1984), additional research is required for defining pseudocomponent and specific component mass transfer rates from a diffusion controlled oil phase. In most open ocean spills to date, component-specific measurements have been made, whereas descriptions of the overall mass balance using a pseudocomponent or distillate cut approach have not been attempted. A pseudocomponent approach would be useful for describing the thin and thick oil patches observed during actual spills and for development and validation of algorithms of oil/SPM interactions.

The environmental fate and long-term effects of spilled oil are expected to vary with respect to the energy regime or degree of exposure at the spill site. In particular, additional data are needed to characterize the weathering behavior of petroleum in low energy estuarine or protected bay environments where residues may persist in association with fine-grained sediments over a period of several years (i.e., Gundlach *et al.*, 1983). Furthermore, rates of the release of sedimented hydrocarbons to interstitial or boundary layer waters should be investigated in relationship to substrate type and rates of physical sediment dispersion (see Boehm, Chapter 6). Potentials for releases of sediment-associated lower molecular weight aromatics to the water column may have implications for the long-term effects associated with nearshore oil spills.

The presence of sea ice may significantly alter the magnitude of simultaneous weathering processes and specific influences on the weathering behavior and mass balance of spilled oil. In particular, processes associated with ice formation and ice flow may enhance rates of oil dispersion and emulsification while enhanc-

TABLE 5.5
Summary of current information gaps, areas requiring further research, and their importance for predicting long-term effects

Topics	*Importance relative to long-term effects*
A. Topics fairly well understood and predictable via modeling efforts	
1. Evaporation (compound specific and pseudocompound—distillation cuts)	No, in that more toxic lower molecular weight components are removed by this process
2. Dissolution (component specific)	Yes
3. Slick drift, if wind and current regimes are known *a priori*	Possible, if slick or chronic discharges drift into sensitive intertidal habitat
B. Topics less well understood where predictive capabilities are tenuous at best	
1. Dispersion (oil in water) and slick recoalescence	Yes, as it affects dissolution and direct exposure to pelagic and benthic organisms
a. Flocculation/deflocculation (drilling fluids) b. Settling/Sedimentation c. Physical mixing/breakup of oil	
2. Weathered oil slick viscosity—water-in-oil emulsion viscosity extremely oil type (source) dependent, presence of asphaltenes, waxes, NSO, etc.	No
3. Thick versus thin oil slick formation	
a. Investigate/characterized oil types which lead to thick/thin slicks	No, except as oil behavior effects clean up and attempts at chemical dispersion
b. Evaluate differences in dispersion of thin versus thick slicks—attempt to model differences	Yes

4. Trajectories of subsurface releases	
a. Problems of coupling with surface slick trajectory models	Possible
b. Investigate the possibility of density gradient modeling	Yes
c. Investigate atmospheric and thermal models for examples of density change considerations in physical models	Yes
5. Nearshore sedimentation, transport and long-term weathering	Yes
a. Few systematic data collected for nearshore processes that can be used for direction and guidance in oil spill cleanup efforts	
6. Chemical and physical properties changes in oil released in the presence of moving and growing sea ice	Yes, one season accumulation and transport of oil with little weathering before ice melt and release of oil.
a. Component characterization data for these conditions are limited	
b. Controlled laboratory experiments to characterize component-specific evaporation, dissolution, partitioning, etc. needed during ice growth and ice breakup	Yes
7. Additional photochemical products characterization and toxicity determinations needed along with detailed study of possible role of water-in-oil emulsification	Yes
C. General Comments	
1. Longer-term (weeks versus days) modeling approaches need to be explored-especially for spent drilling fluid discharges and produced waters	Probably not
2. Field data should be systematically collected for model verification	No
3. Longer-term monitoring efforts should focus on rates of dispersion and accumulation of discharge-related constituents in the benthic environment	Yes

ing emulsion stability and inhibiting rates of microbial degradation. However, few data are presently available to characterize changes in the weathering rates due to the presence of sea ice. The role of ice in influencing the long-term weathering behavior of petroleum could realistically be evaluated with the use of exposed flow-through seawater wave tanks with ice formation capabilities.

Realistic predictive models for characterizing the long-term effects associated with operational discharges of drilling fluids and produced waters are presently limited by the paucity of information describing the composition of some of these materials and sedimentation and sediment dispersion processes (see Chapters 6 and 10).

Water column dispersion processes can be approximated with existing models; however, the predicted dispersion rates and partitioning behavior of discharge components are not necessarily germane to evaluations of long-term impacts. Consequently, additional research on the water column effects associated with planned discharges from present offshore oil and gas operations is not critical.

LITERATURE CITED

Anderson, E.L. 1983. Study of wind and current data sets for IXTOC oil spill hindcasts. Pages 293-299 *in* Proceedings of the 1983 Oil Spill Conference. American Petroleum Institute, Washington, D.C.

Armstrong, R.S. 1981. Transport and dispersion of potential contaminants. Pages 403-420 *in* B.S. Middleditch (ed.), Environmental Effects of Offshore Oil Production. The Buccaneer Gas and Oil Field Study. Plenum Press, New York.

Atwood, D.K., J.A. Benjamin and J.W. Farrington. 1980. The mission of the September 1979 *Researcher/Pierce* IXTOC-I cruise and the physical situation encountered. Pages 1-16 *in* Proceedings of a Symposium Preliminary Results from the September 1979 *Researcher/Pierce* IXTOC-I Cruise (June 9-10, 1980, Key Biscayne, Florida). Office of Marine Pollution Assessment, National Oceanic and Atmospheric Administration, Washington, D.C.

Auble, G.T., A.K. Andrews, R.A. Ellison, D.B. Hamilton, R.A. Johnson, J.E. Roelle and D.R. Marmorek. 1981. Results of an adaptive environmental assessment modeling workshop concerning potential impacts of drilling muds and cuttings on the marine environment. *In* Proceedings from U.S. EPA Workshop September 14-18, 1981 at Gulf Breeze, Florida. Prepared by U.S. Fish and Wildlife Service, Fort Collins, Colorado, 64 p.

Audunson, T. 1978. The fate and weathering of surface oil from the Bravo blowout. Pages 445-475 *in* Proceedings of the Conference on Assessment of Ecological Impacts of Oil Spills. American Institute of Biological Sciences, Washington, D.C.

Audunson, T., V. Dalen, J. Mathisen, J. Haldorson and F. Krogh. 1980. SILKF-ORCAST - A simulation program for oil spill emergency tracking and long-term contingency planning. *In* Proceedings, Petromar '80, Monaco.

Ayers, R.C., Jr., T.C. Sauer, Jr., R.P. Meek and G. Bowers. 1980a. An environmental study to assess the impact of drilling discharges in the mid-Atlantic. I. Quantity and fate of discharges. Pages 382-418 *in* Symposium, Research on Environmental Fate and Effects of Drilling Fluids and Cuttings. Lake Buena Vista, Florida, January 21-24, 1980. American Petroleum Institute, Washington, D.C.

Ayers, R.C., Jr., T.C. Sauer, Jr., D.O. Stuebner and R.P. Meek. 1980b. An environmental study to assess the effects of drilling fluids on water quality parameters during high rate, high volume discharges to the ocean. Pages 351-361 *in* Symposium, Research on Environmental Fate and Effects of Drilling Fluids and Cuttings. Lake Buena Vista, Florida, January 21-24, 1980. American Petroleum Institute, Washington, D.C.

Bassin, J.J. and T. Ichiye. 1977. Flocculation behavior of suspended sediments and oil emulsions. J. Sed. Petrol. 47: 671-677.

Berridge, S.A., R.A. Dean, R.G. Fallows and A. Fish. 1968a. The properties of persistent oils at sea. J. Inst. Petrol. 54: 300-309.

Berridge, S.A., M.T. Thew and A.G. Loriston-Clarke. 1968b. The formation and stability of emulsions of water in crude petroleum and similar stocks. J. Inst. Petrol. 54: 333-357.

Bocard, C. and C. Gatellier. 1981. Breaking of fresh and weathered emulsions by chemicals. Pages 601-607 *in* Proceedings of the 1981 Oil Spill Conference. American Petroleum Institute, Washington, D.C.

Boehm, P.D. and D.L. Fiest. 1980a. Surface water column transport and weathering of petroleum hydrocarbons during the IXTOC-I blowout in the Bay of Campeche and their relation to surface oil and microlayer compositions. Pages 267-338 *in* Proceedings of a Symposium on Preliminary Results from the September 1979 *Researcher/Pierce* IXTOC-I Cruise (June 9-10, 1980, Key Biscayne, Florida). Office of Marine Pollution Assessment, National Oceanic and Atmospheric Administration, Washington, D.C.

Boehm, P.D., and D.L. Fiest. 1980b. Aspects of the transport of petroleum hydrocarbons to the offshore benthos during the IXTOC-I blowout in the Bay of Campeche. Pages 207-236 *in* Proceedings of a Symposium, Preliminary Results from the September 1979 *Researcher/Pierce* IXTOC-I Cruise (June 9-10, 1980, Key Biscayne, Florida). Office of Marine Pollution Assessment, National Oceanic and Atmospheric Administration, Washington, D.C.

Boehm, P.D. and D.L. Fiest. 1982. Subsurface distributions of petroleum from an offshore well blowout - the IXTOC-I blowout. Environ. Sci. Tech. 16: 67-74.

Boehm, P.D., D.L. Fiest and A. Elskus. 1982. Comparative weathering patterns of hydrocarbons from the *Amoco Cadiz* oil spill observed at a variety of coastal environments. Pages 159-173 *in* Proceedings of the International Symposium, *Amoco Cadiz* Fates and Effects of the Oil Spill. Centre Oceanologique de Bretagne, Brest, France.

Brandsma, M.G., L.R. Davis, R.C. Ayers, Jr. and T.C. Sauer, Jr. 1980. A computer model to predict the short-term fate of drilling discharges in the marine environment. Pages 588-610 *in* Symposium, Research on the Environmental Fate and Effects of Drilling Fluids and Cuttings. Lake Buena Vista, Florida, January 21-24, 1980. American Petroleum Institute, Washington, D.C.

Brandsma, M.G. and R.C. Sauer. 1983. The OOC model: prediction of short term fate of drilling fluids in the ocean. Part two: model results. *In* Proceedings of Minerals Management Service Workshop on Discharges Modeling, February 7-10, 1983, Santa Barbara, California. Prepared for Minerals Management Service by MBC Applied Environmental Sciences and Analytical and Computational Research, Inc.

Bridie, A.C., T.H. Wanders, W. Zegveld and H.B. Van der Heijde. 1980a. Formation, prevention and breaking of seawater-in-crude-oil emulsions "chocolate mousse". Marine Poll. Bull. 2: 343-348.

Bridie, A.C., T.H. Wanders, W. Zegveld and H.B. Van der Heijde. 1980b. Formation, prevention and breaking of seawater-in-crude-oil emulsions "chocolate mousse". Pages 33-39 *in* International Research Symposium on Chemical Dispersion of Oil Spills, Toronto, Canada. November 17-19, 1980. Institute of Environmental Studies, University of Toronto, Canada.

Brooks, J.M., D.A. Weisenburg, R.A. Burke, M.C. Kenicutt and B.B. Bernard. 1980. Gaseous and volatile hydrocarbons in the Gulf of Mexico following the IXTOC-I blowout. Pages 53-85 *in* Proceedings of a Symposium on Preliminary Results from the September 1979 *Researcher/Pierce* IXTOC-I Cruise (June 9-10, 1980, Key Biscayne, Florida). Office of Marine Pollution Assessment, National Oceanic and Atmospheric Administration, Washington, D.C.

Burwood, R. and G.C. Spears. 1974. Photo-oxidation as a factor in the environmental dispersal of crude oil. Estuar. Coast. Mar. Sci. 2: 117-135.

Cairns, R.J.R., D.M. Grist and E.L. Neustadter. 1974. The effect of crude oil-water interfacial properties on water-crude oil emulsion stability. Pages 135-151 *in* A.L. Smith (ed.), Theory and Practice of Emulsion Technology. Academic Press, New York.

Calder, J.A. and P.D. Boehm. 1981. The chemistry of *Amoco Cadiz* oil in Aber Wrac'h. Pages 149-158 *in* Proceedings of the International Symposium, *Amoco Cadiz* Fates and Effects of the Oil Spill. Centre de Bretagne, Brest (France).

Calder, J.A., J. Lake and J. Laseter. 1978. Chemical composition of selected environmental and petroleum samples from the AMOCO CADIZ oil spill. NOAA/EPA Special Report: The AMOCO CADIZ Oil Spill, a Prcliminary Scicntific Rcport. 283 p.

Canevari, G. P. 1969. General dispersant theory. Pages 171-177 *in* Proceedings of the 1969 Oil Spill Conference. American Petroleum Institute, Washington, D.C.

Coleman, H.J., E.M. Shelton, D.T. Nichols and C.J. Thompson. 1978. Analyses of 800 crude oils from the United States oil fields. BETC/RI-78/14, Bartlesville Energy Technology Center, Bartlesville, Oklahoma.

Conomos, T. J. 1975. Movement of spilled oil as predicted by estuarine nontidal drift. Limnol. Oceanogr. 20: 159-173.

Conover, R.J. 1971. Some relations between zooplankton and Bunker C oil in Chedabucto Bay following the wreck of the tanker Arrow. J. Fish. Res. Bd. Canada. 28: 1327-1330.

Cornillion, P., M.L. Spaulding and K. Hansen. 1979. Oil spill treatment strategy modeling for Georges Bank. Pages 685-692 *in* Proceedings of the 1979 Oil Spill Conference. American Petroleum Institute, Washington, D.C.

Crippen, R.W. and S.L. Hood. 1980. Metal levels in sediment and benthos resulting from a drilling fluid discharge into the Beaufort Sea. Pages 636-669 *in* Symposium, Research on Environmental Fate and Effects of Drilling Fluid and Cuttings. Lake Buena Vista, Florida, January 21-24, 1980. American Petroleum Institute, Washington, D.C.

Davis, S.J. and C.F. Gibbs. 1975. The effect of weathering on a crude oil exposed at sea. Water Research 9: 275-289.

Dickens, D.F., I.A. Buist and W.M. Pistruzak. 1981. Dome's petroleum study of oil and gas under sea ice. Pages 183-189 *in* Proceedings of the 1981 Oil Spill Conference. American Petroleum Institute, Washington, D.C.

Dowty, B.J., N.E. Brightwell, J.L. Laseter and G.W. Griffin. 1974. Dye-sensitized photooxidation of Phenanthrene. Biochem. Biophys. Res. Comms. 57: 452-455.

Fay, J. A. 1971. Physical processes in the spread of oil on a water surface. Pages 463-467 *in* Proceedings of the Joint Conference on the Prevention and Control of Oil Spills. American Petroleum Institute, Washington, D.C.

Forrester, W.D. 1971. Distribution of suspended oil particles following the grounding of the tanker ARROW. Jour. Mar. Res. 29: 151-170.

Freegarde, M., C.G. Hatchhard and C.A. Parker. 1971. Oil spilt at sea: Its identification, determination and ultimate fate. Laboratory Practice 20-4: 35-40.

Friede, J.D. 1973. The Isolation and Chemical and Biological Properties of Microbial and Emulsifying Agents for Hydrocarbons. Progress Report. AD 770-630. National Technical Information Service, U.S. Dept. of Commerce, Springfield, Virginia, 5 p.

Galt, J.A. and D.L. Payton. 1983. The use of receptor mode trajectory analysis techniques for contingency planning. Pages 307-312 *in* Proceedings of the 1983 Oil Spill Conference. American Petroleum Institute, Washington, D.C.

Gearing, J.N. and P.J. Gearing. 1983. Suspended load and solubility affect sedimentation of petroleum hydrocarbons in controlled estuarine ecosystems. Can. Jour. Fish. Aquatic Sci. 40 (suppl. 2): 54-62.

Gearing, J.N., P.J. Gearing, T. Wade, J.G. Quinn, H.B. McCarty, J. Farrington and R.F. Lee. 1979. The rates of transport and fates of petroleum hydrocarbons in a controlled marine ecosystem and a note on analytical variability. Pages 555-565 *in*

Proceedings of the 1979 Oil Spill Conference. American Petroleum Institute, Washington, D.C.
Gesser, H.P., T.A. Wildman and Y.B. Tewori. 1977. Photo-oxidation of n-hexadecane sensitized by xanthone. Environ. Sci. Tech. 11: 605-608.
Goldberg, E.D., M. Koide, V. Hodge, A.R. Flegal and J.H. Martin. 1983. U.S. Mussel Watch: 1977-1978, results of trace metals and radionuclides. Estuar. Coast. Shelf Sci. 16: 69-93.
Gordon, D.C., Jr., P.D. Keizer and N.J. Prouse. 1973. Laboratory studies on the accommodation of some crude and residual fuel oils in seawater. J. Fish. Res. Bd. Canada 30: 1611-1618.
Grahl-Nielsen, O. 1978. The Ekofisk Bravo Blowout. Petroleum hydrocarbons in the sea. Pages 477-499 *in* Proceedings of the Conference on Assessment of Ecological Impacts of Oil Spills. American Institute of Biological Sciences, Washington, D.C.
Gray, G.R., H.C.H. Darley and W.F. Rogers. 1980. Composition and Properties of Oil Well Drilling Fluids. Gulf Publishing Company, Houston, Texas, 618 p.
Guire, P.E., J.D. Friede and R.K. Gholson. 1973. Production and characterization of emulsifying factors from hydrocarbonoclastic yeast and bacteria. Pages 229-231 *in* D.G. Ahern and S.P. Meyers (eds.), The Microbial Degradation of Oil Pollutants. Publ. No. LSU-SG-73-01. Center for Wetland Resources, Louisiana State University, Baton Rouge, Louisiana.
Gundlach, E.R., P.D. Boehm, M. Marchand, R.M. Atlas, D.M. Ward and D.A. Wolfe. 1983. The fate of *Amoco Cadiz* oil. Science 221: 122-129.
Hansen, H.P. 1975. Photochemical degradation of petroleum hydrocarbon surface films on seawater. Mar. Chem. 3: 183-195.
Hansen, H.P. 1977. Photodegradation of hydrocarbon surface films. Rapp. P-V. Réun. Cons. Int. Explor. Mer 171: 101-106.
Houghton, J.P., R.P. Britch, R.C. Miller, A.K. Runchal and C.P. Falls. 1980. Drilling fluid dispersion studies at the Lower Cook Inlet, Alaska, continental offshore stratigraphic test well. Pages 285-308 *in* Symposium, Research on Environmental Fate and Effects of Drilling Fluid and Cuttings. Lake Buena Vista, Florida, January 21-24, 1980. American Petroleum Institute, Washington, D.C.
Hrudey, S. E. 1980. Sources and characteristics of liquid process wastes from arctic offshore hydrocarbon exploration. Arctic 32: 3-21.
Huang, J.C. 1983. A review of state-of-the-art oil spill fate/behavior models. Pages 313-322 *in* Proceedings of the 1983 Oil Spill Conference. American Petroleum Institute, Washington, D.C.
Huang, C.P. and H.A. Elliot. 1977. The stability of emulsified crude oils as affected by suspended particles. Pages 413-420 *in* D.A. Wolfe (ed.), Fate and Effects of Petroleum Hydrocarbons in Marine Ecosystems and Organisms. Pergamon Press, New York.
Huang, J.C. and F.C. Monastero. 1982. Review of the State-of-the-Art of Oil Spill Simulation Model: Final Report Submitted to the American Petroleum Institute, Washington, D.C.
Hunter, J. R. 1980. An interactive computer model of oil slick motion. Pages 42-50 *in* Proceedings, Oceanology International '80 (U.K.).
Jackson, G.F., M.J. Wade and M. Kirsch. 1981. Oil Content in Produced Brine in 10 Louisiana Production Platforms. Report for Municipal Environmental Research Laboratory, Office of Research and Development, U.S. Environmental Protection Agency, Cincinnati, Ohio.
JBF/API. 1976. Physical and Chemical Behavior of Crude Oil Slicks on the Ocean. API Publication 4290. American Petroleum Institute, Washington, D.C.
Johansen, O. 1983. Dispersion of Oil from Drifting Oil Slicks. Paper presented at the Arctic Marine Oil Spill Program Technical Seminar. June 14-16, 1983. Edmonton, Canada.
Jordan, R.E. and J.R. Payne. 1980. Fate and Weathering of Petroleum Spills in the Marine

Environment: A Literature Review and Synopsis. Ann Arbor Science Publishers, Ann Arbor, Michigan, 174 p.

Karickhoff, S.W. 1981. Semi-empirical estimation of sorption of hydrophobic pollutants on natural sediments and soils. Chemosphere 10: 833-846.

Karpen J. and J. Galt. 1979. Modeling of oil migration in Puget Sound. *In* Proceedings of the Oceans '79 Conference. Marine Technology Society, Washington, D.C.

Kawahara, F. K. 1969. Identification and differentiation of heavy residual oil and asphalt pollutants in surface waters by comparative ratios of infrared absorbances. Environ. Sci. Technol. 3: 150-153.

Klein, A.E. and N. Pilpel. 1974a. The effects of artificial sunlight upon floating oils. Water Res. 8: 79-83.

Klein, A.E. and N. Pilpel. 1974b. Photo-oxidation of alkylbenzenes initiated by 1 napthol. J. Chem. Soc: Faraday Trans., 1, 70: 1250-1256.

Koons, C.B., C.D. McAuliffe and F.T. Weiss. 1977. Environmental aspects of produced waters from oil and gas extraction operations in offshore and coastal waters. Pages 247-257 *in* Proceedings of the Offshore Technology Conference, Paper No. DTC 2447.

Kramer, J.R., H.D. Grundy and L.G. Hammer. 1980. Occurrence and solubility of trace metals in barite for ocean drilling operations. Pages 189-198 *in* Symposium, Research on Environmental Fate and Effects of Drilling Fluids and Cuttings. January 21-24, 1980, Lake Buena Vista, Florida, American Petroleum Institute, Washington, D.C.

Lacaze, J.C. and O. Villedon de Naide. 1976. Influence of illumination on phytotoxicity of crude oil. Mar. Pollut. Bull. 7: 73-76.

Larson, R.A. and L.L. Hunt. 1978. Photo-oxidation of a refined petroleum oil: Inhibition by b-carotene and role of a singlet oxygen. Photochem. Photobiol. 28: 553-555.

Larson, R.A., D.W. Blankenship and L.L. Hunt. 1976. Toxic hydroperoxides: Photochemical formation from petroleum constituents. *In* A.B.I.S. Symposium on Sources, Effects and Sinks of Hydrocarbons in the Aquatic Environment. American Institute of Biological Sciences, Washington, D.C.

Larson, R.A., T.L. Bott, L.L. Hunt and K. Rogenmuser. 1979. Photooxidation products of a fuel oil and their antimicrobial activity. Environ. Sci. Tech. 13: 965-969.

Larson, R.A., L.L. Hunt and D. W. Blankenship. 1977. Formation of toxic products from a No. 2 fuel oil by photo-oxidation. Environ. Sci. Technol. 11: 492-496.

Lee, M., F. Martinelli, B. Lynch and P.T. Morris. 1981. The use of dispersants on viscous fuel oils and water in crude oil emulsions. Pages 31-35 *in* Proceedings of the 1981 Oil Spill Conference. American Petroleum Institute, Washington, D.C.

Liss, R.G., F. Knox, D. Wayne and T.R. Gilbert. 1980. Availability of trace elements in drilling fluids to the marine environment. Pages 691-722 *in* Symposium, Research on Environmental Fate and Effects of Drilling Fluids and Cuttings. Lake Buena Vista, Florida, January 21-24, 1980. American Petroleum Institute, Washington, D.C.

Liu, S.K. and J.J. Leendertse. 1978. Multidimensional numerical modeling of estuaries and coastal seas. Pages 95-165 *in* Advances in Hydro-Science, Volume 11. Academic Press, New York.

Liu, S.K. and J.J. Leendertse. 1979. A Three-Dimensional Model for Estuaries and Coastal Seas: Volume VI, Bristol Bay Simulations. Report prepared for the National Oceanic and Atmospheric Administration, Washington, D.C., 121 p.

Liu, S.K. and J.J. Leendertse. 1981a. A Three Dimensional Model of the Eastern Bering Sea. Coastal Engineering, American Society of Civil Engineering, New York.

Liu, S.K. and J.J. Leendertse. 1981b. A 3-D oil spill model with and without ice cover. *In* La Mechanique des Napps d'Hydrocarbures. Assoc. Amicale de Ingenieur, Paris, France.

Liu, S.K. and J.J. Leendertse. 1986. A Three-Dimensional Model of the Gulf of Alaska. Coastal Engineering. XX. American Society Civil Engineering, New York.

Lysyj, I., G. Perkins, J.S. Farlow and R.W. Morris. 1981. Distribution of aromatic hydrocarbons in Port Valdez, Alaska. Pages 47-54 *in* Proceedings of the 1981 Oil Spill Conference. American Petroleum Institute, Washington, D.C.

Mabey, W.R., D. Tse, A. Baraze and T. Mill. 1983. Photolysis of nitroaromatics in aquatic systems. 1,2,4,6,-Trinitrotoluene. Chemosphere 12: 3-16.

MacGregor, C. and A.Y. McLean. 1977. Fate of crude oil spilled in a simulated Arctic environment. Pages 461-463 *in* Proceedings, 1977 Oil Spill Conference. American Petroleum Institute, Washington, D.C.

Mackay, D. and K. Hossain. 1980. Studies of oil sedimentation. Pages 120-172 *in* Proceedings of the Arctic Marine Oil Spill Program Technical Seminar. June 3-5, 1980, Edmonton, Canada.

Mackay, D. and K. Hossain. 1982. Interfacial tensions of oil, water chemical dispersant systems. Canadian J. Chemical Engineering 60: 546-550.

Mackay, D. and P.G. Wells. 1983. Effectiveness, behavior, and toxicity of dispersants. Pages 65-71 *in* Proceedings of the 1983 Oil Spill Conference. American Petroleum Institute, Washington, D.C.

Mackay, D., I. Buist, R. Mascarenhas and S. Patterson. 1979. Experimental studies of dispersion and emulsion formation from oil slicks. Pages 1.17-1.40 *in* Workshop on the Physical Behavior of Oil in the Marine Environment. Princeton Univ., prepared for the National Weather Service, Silver Spring, Maryland.

Mackay, D., I. Buist, R. Mascarenhas and S. Patterson. 1980. Oil Spill Processes and Models. A report submitted to Environmental Emergency Branch, Environmental Impact Control Directorate, Environment Protection Service, Environment Canada, (December) Ottowa, Ontario K1A 1C8.

Mackay, D., A.Y. McLean, O.J. Betancourt and B.C. Johnson. 1973. The formation of water-in-oil emulsions subsequent to an oil spill. J. Inst. Petroleum 59: 164-172.

Majewski, J., J. O'Brien and E. Barry. 1974. A kinetic study of a fuel oil undergoing photochemical weathering. Environ. Letts. 7: 145-161.

McAuliffe, C.D. 1977. Dispersal and alteration of oil discharged on a water surface. Pages 19-35 *in* D.A. Wolfe (ed.), Fate and Effects of Petroleum Hydrocarbons in Marine Ecosystems and Organisms. Pergamon Press, New York.

McAuliffe, C.D., D.E. Fitzgerald, B.L. Steelman, J.P. Ray, W.R. Leek and C.D. Barker. 1981. The 1979 Southern California dispersant treated research oil spills. Pages 268-282 *in* Proceedings of the 1981 Oil Spill Conference. American Petroleum Institute, Washington, D.C.

McLafferty, F.W. 1976. Interpretation of Mass Spectra, an Introduction. W. A. Benjamin, Inc., Massachusetts, New York, 229 p.

Meek, R.P. and J.P. Ray. 1980. Induced sedimentation, accumulation and transport resulting from exploratory drilling discharges of drilling fluids and cuttings on the Southern California Outer Continental Shelf. Pages 259-284 *in* Symposium, Research on Environmental Fate and Effects of Drilling Fluid and Cuttings. Lake Buena Vista, Florida, January 21-24, 1980. American Petroleum Institute, Washington, D.C.

Meeks, D.G. 1980. Performance of some oil dispersants on oil slicks of varying thickness. Mar. Poll. Bull. 11: 348-351.

Menzie, C.A. 1982. The environmental implications of offshore oil and gas activities. Environ. Sci. Tech. 16: 454A-472A.

Meyers, P.A. and J.G. Quinn. 1973. Association of hydrocarbons and mineral particles in saline solutions. Nature 244: 23-24.

Middleditch, B.S. (ed.). 1981. Environmental Effects of Offshore Oil Production. The Buccaneer Gas and Oil Field Study. Plenum Press, New York, 446 p.

Mill, T., D.G. Hendry and H. Richardson. 1980. Free radical oxidants in natural waters. Science 107: 886-887.

Mill, T., W.R. Mabey, B.Y. Lan and A. Baraze. 1981. Photolysis of polycyclic aromatic hydrocarbons in seawater. Chemosphere 10: 1281-1290.

Murray, S.P. 1972. Turbulent diffusion of oil in the ocean. Limnol. Oceanogr. 17: 651-660.

Nagy, E., B.F. Scott and J.H. Hart. 1981. The Fate and Dispersant Mixtures in Freshwater. Report EPS4-EC-81-3 prepared for the Environmental Emergency Branch, Environment Canada, 66 p.

National Research Council. 1983. Drilling Discharges in the Marine Environment. National Academy Press, Washington, D.C.

National Research Council. 1985. Oil in the Sea. Inputs, Fates, and Effects. National Academy Press, Washington, D.C., 601 p.

Nelson, T.A. 1980. Mineralogy of suspended and bottom sediments in the vicinity of the IXTOC-I blowout, September 1979. Pages 189-204 *in* Proceedings of a Symposium, Preliminary Results from the September 1979 *Researcher/Pierce* IXTOC-I Cruise (June 9-10, Kay Biscayne, Florida). Office of Marine Pollution Assessment, National Oceanic and Atmospheric Administration, Washington, D.C.

Nelson-Smith, A. 1973. Oil Pollution and Marine Ecology. Plenum Press, New York, 260 p.

Northern Technical Services. 1981. Beaufort Sea Drilling Effluent Disposal Study. Prepared for the Reindeer Island Stratigraphic Test Well Participants. 329 p.

Northern Technical Services. 1982. Above-Ice Effluent Disposal Tests SAG Delta No. 7, SAG Delta No. 8, and Challenge Island No. 1 Wells, Beaufort Sea, Alaska. Report Prepared for Sohio Petroleum Co. 183 p.

NRC, see National Research Council.Osamor, C.A. and R.C. Ahlert. 1981. Oil Slick Dispersal Mechanics. Report EPA-600/2-81-199 prepared for U.S. Environmental Protection Agency. 237 p.

Overton, E.G., J.L. Laseter, W. Mascarella, C. Rashke, I. Noiry and J.W. Farrington. 1980. Photochemical oxidation of IXTOC-I oil. Pages 341-386 *in* Proceedings of a Symposium on Preliminary Results from September, 1979 *Researcher/Pierce* IXTOC-I Cruise (June 9-10, 1980, Key Biscayne, Florida). Office of Marine Pollution Assessment, National Oceanic and Atmospheric Administration, Washington, D.C.

Overton, E.B., J.R. Patel and J.L. Laseter. 1979. Chemical characterization of mousse and selected environmental samples from the *Amoco Cadiz* oil spill. Pages 168-174 *in* Proceedings from 1979 Oil Spill Conference. American Petroleum Institute, Washington, D.C.

Patel, J.R., J.A. McFall, G.W. Griffin and J.L. Laseter. 1978. Toxic photo-oxygenated products generated under environmental conditions from phenanthrene. Pages 1-32 *in* E.P.A. Symposium on Carcinogenic Polynuclear Aromatic Hydrocarbons in the Marine Environment. August 14-18, 1978, Pensacola Beach, Florida.

Patel, J.R., E.B. Overton and J.L. Laseter. 1979. Environmental photo-oxidation of dibenzothiophenes following the AMOCO CADIZ oil spill. Chemosphere 8: 557-561.

Payne, J.R. and C.R. Phillips. 1985a. Petroleum Spills in the Marine Environment: Chemistry and Formation of Water-in-Oil Emulsions and Tar Balls. Lewis Publishers, Chelsea, Michigan. 148 p.

Payne, J.R. and C.R. Phillips. 1985b. Photochemistry of petroleum in water. Environ. Sci. Technol. 19: 569-579.

Payne, J.R., N.W. Flynn, P.J. Mankiewicz and G.S. Smith. 1980b. Surface evaporation/dissolution partitioning of lower molecular weight aromatic hydrocarbons in a down-plume transect from the IXTOC-I wellhead. Pages 239-265 *in* Proceedings of a Symposium on Preliminary Results from the September 1979 *Researcher/Pierce* IXTOC-I Cruise (June 9-10, 1980, Key Biscayne, Florida). Office of Marine Pollution Assessment, National Oceanic and Atmospheric Administration, Washington, D.C.

Payne, J.R., B.E. Kirstein, G.D. McNabb, Jr., J.C. Lambach, C. De Oliveira, R.E. Jordan and W. Hom. 1983a. Multivariate analysis of petroleum hydrocarbon weathering in the subarctic marine environment. Pages 423-434 *in* Proceedings 1983 Oil Spill Conference. American Petroleum Institute, Washington, D. C.

Payne, J.R., B.E. Kirstein, G.D. McNabb, Jr., J.L. Lambach, R.T. Redding, R.E. Jordan, W. Hom, C. De Oliveira, G.S. Smith, D.M. Baxter and R. Gaegel. 1984. Multivariate Analysis of Petroleum Weathering in the Marine Environment - Subarctic. Final Report on Contract No. NA8ORAC00018, submitted to National Oceanic and

Atmospheric Administration/Outer Continental Shelf Environmental Assessment Program, Juneau, Alaska.

Payne, J.R., B.E. Kirstein, R.F. Shokes, N.L. Guinasso, L. Carver, K.R. Fite, R.E. Jordan, P.J. Mankiewicz, G.S. Smith, W.J. Paplawsky, T.G. Fanora and J.L. Lambach. 1981a. Multivariate Analysis of Petroleum Weathering in the Marine Environment - Subarctic. Annual report submitted to National Oceanic and Atmospheric Administration, Office of Marine Pollution Assessment, Juneau, Alaska.

Payne, J.R., J.L. Lambach, R.E. Jordan, C.R. Phillips, G.D. McNabb, Jr., M.K. Beckel, G.H. Farmer, R.R. Sims, Jr., J.G. Sutton and A. Abasumara. 1983b. Georges Bank Monitoring Program: Analysis of Hydrocarbons in Bottom Sediments and Analysis of Hydrocarbons and Trace Metals in Benthic Fauna During the Second Year of Monitoring. Prepared for U.S. Minerals Management Service, Washington, D.C., 151 p.

Payne, J.R., G.S. Smith, J.L. Lambach and P.J. Mankiewicz. 1981b. Chemical weathering of petroleum hydrocarbons in sub-arctic sediments: Results of chemical analyses of naturally weathered sediment plots spiked with fresh and artificially weathered Cook Inlet crude oils. Cited in: Griffiths, R.P. and R.Y. Morita. 1981. Study of Microbial Activity and Crude Oil-Microbial Interaction in the Water and Sediments of Cook Inlet and the Beaufort Sea. Final Report RU 190, Submitted to National Oceanic and Atmospheric Administration/Outer Continental Shelf Environmental Assessment Program, Juneau, Alaska.

Payne, J.R., G.S. Smith, P.J. Mankiewicz, R.F. Shokes, N.W. Flynn, V. Moreno and J. Altamirano. 1980a. Horizontal and vertical transport of dissolved and particulate-bound higher molecular weight hydrocarbons from the IXTOC-I blowout. Pages 119-167 *in* Proceedings of a Symposium on Preliminary Results from the September 1978. *Researcher/Pierce* IXTOC-I Cruise (June 9-10, 1980, Key Biscayne, Florida). Office of Marine Pollution Assessment, National Oceanic and Atmospheric Administration, Washington, D.C.

Perricone, C. 1980. Waterbase drilling fluids. Pages 30-37 *in* Symposium, Research on Environmental Fate and Effects of Drilling Fluids and Cuttings. Lake Buena Vista, Florida, January 21-24, 1980. American Petroleum Institute, Washington, D.C.

Pierce, R.H., D.C. Anne, I. Saksa and B.A. Weicheet. 1985. The fate of select organics from spent drilling fluid discharged to the marine environment. Chapter 7 *in* I. Duedall (ed.), Energy Wastes in the Ocean. John Wiley Interscience and Sons, New York (in press).

Raj, P.P.K. 1977. Theoretical Study to Determine the Sea State Limit for the Survival of Oil Slicks on the Ocean. U.S. Department of Transportation, USCG Report No. CG-D-90-77.

Ray, J.P. and R.P. Meek. 1980. Water column characterization of drilling fluids dispersion from an offshore exploratory well on Tanner Bank. Pages 223-252 *in* Symposium, Research on Environmental Fate and Effects of Drilling Fluid and Cuttings. Lake Buena Vista, Florida, January 21-24, 1980. American Petroleum Institute, Washington, D.C.

Reed, W.E., I.R. Kaplan, M. Sandstrom and P. Mankiewicz. 1977. Petroleum and anthropogenic influence on the composition of sediments from the Southern California Bight. Pages 183-188 *in* Proceedings of the 1977 Oil Spill Conference. American Petroleum Institute, Washington, D.C.

Reed, W.E. and I.R. Kaplan. 1977. The chemistry of marine petroleum seeps. J. Geochem. Explor. 7: 255-293.

Risebrough, R.W., B.W. de Lappe, E.F. Letterman, J.L. Lane, M. Firestone-Gillis, A.M. Springer and W. Walker II. 1980. California Mussel Watch: 1977-1978. Volume II. Organic Pollutants in mussels *Mytilus californianus* and *M. edulis* along the California coast. Water Quality Monitoring Report No. 79-22. Prepared for State Water Resources Control Board, Sacramento, California.

Rose, C.D. and T.J. Ward. 1981. Acute toxicity and aquatic hazard associated with discharged formation water. Pages 302-328 *in* B.S. Middleditch (ed.), Environmental

Effects of Offshore Oil Production. The Buccaneer Gas and Oil Field Study. Plenum Press, New York.
Runchal, A.K. 1983. An evaluation of effluent dispersion and fate models for OCS platforms. Volume 1: Summary and recommendations. Proceedings of the Minerals Management Service Workshop on Discharges Modeling 7-10 February 1983. Santa Barbara, California. Prepared for Minerals Management Service by MBC Applied Environmental Sciences and Analytical and Computational Research, Inc., 69 p.
SAI, 1983. See Science Applications, Inc.
Sauer, T.C., Jr. 1981. Volatile liquid hydrocarbon characterization of underwater hydrocarbon vents and formation waters from offshore production operations. Environ. Sci. Tech. 15: 917-923.
Science Applications, Inc. 1983. Drill Mud Assessment: Chemical Analysis Reference Volume. Prepared for Environmental Protection Agency, Environmental Research Laboratory, Gulf Breeze, Florida.
Sleeter, T.D. and J.N. Butler. 1982. Petroleum hydrocarbons in zooplankton fecal pellets in the Sargasso Sea. Mar. Poll. Bull. 13: 54-56.
Smith, G.L. 1977. Determination of leeway of oil slicks. Pages 351-362 *in* D. A. Wolfe (ed.), Fate and Effects of Petroleum Hydrocarbons in Marine Ecosystems and Organisms. Pergamon Press, New York.
Solsburg, L.B. 1977. A field evaluation of oil spill recovery devices. Pages 303-307 *in* Proceedings 1977 Oil Spill Conference. American Petroleum Institute, Washington, D.C.
Spies, R.B. and P.H. Davis. 1979. The infaunal benthos of a natural oil seep in the Santa Barbara Channel. Mar. Biol. 50: 227-238.
Spies, R.B., P.H. Davis and D.H. Steurmer. 1980. Ecology of a submarine petroleum seep off the California coast. Pages 208-263 *in* R. Geyer (ed.), Environmental Pollution. I. Hydrocarbons. Elsevier Press, Amsterdam.
Sprague, J.B., J.H. Vandermeulen and P.G. Wells. 1980. Oil and Dispersants in Canadian Seas - Research Appraisal and Recommendations. Report prepared for the Environmental Emergency Branch, Environment Canada. 182 p.
Stephenson, M.D., M. Martin, S.E. Lange, A.R. Flegal and J.H. Martin. 1979. California Mussel Watch: 1977-1978. Volume II. Trace Metal Concentrations in the California Mussel *Mytilus californianus*. State Water Resource Control Board Water Quality Monitoring Report. No. 79-22. Sacramento, California, 110 p.
Strosher, M.T. 1980. Characterization of organic constituents in waste drilling fluids. Pages 70-97 *in* Symposium, Research on Environmental Fate and Effects of Drilling Fluids and Cuttings. Lake Buena Vista, Florida, January 21-24, 1980. American Petroleum Institute, Washington, D.C.
Stuermer, D.H., R.B. Spies and P.H. Davis. 1981. Toxicity of Santa Barbara seep oil to starfish embryos. I. Hydrocarbon composition of test solutions and field samples. Mar. Environ. Res. 5: 275-286.
Stuermer, D.H., R.B. Spies, P.H. Davis, D.J. Ng, C.J. Davis and S. Neal. 1982. The hydrocarbon chemistry of the Isla Vista seep environment. Mar. Chem. 11: 413-426.
Sutton, C. and J.A. Calder. 1974. Solubility of higher molecular weight n-paraffins in distilled water and seawater. Environ. Sci. Tech. 8: 654-657.
Sutton, C. and J.A. Calder. 1975. Solubility of alkylbenzenes in distilled water and seawater at 25°C. J. Chem. Eng. Data 20: 320-322.
Torgrimson, G. M. 1981. A comprehensive model for oil spill simulation. Pages 423-428 *in* Proceedings of the 1981 Oil Spill Conference. American Petroleum Institute, Washington, D.C.
Trocine, R.P. and J.H. Trefry. 1983. Particulate metal tracers of petroleum drilling mud dispersion in the marine environment. Environ. Sci. Tech. 17: 507-512.
Twardus, E.M. 1980. A Study to Evaluate the Combustibility and Other Physical and Chemical Properties of Aged Oils and Emulsions. R and D Division, Environmental

Emergency Branch, Environmental Impact Control Directorate, Environmental Protection Service, Environment Canada, Ottawa, Ontario.

Wade, T.C. and J.G. Quinn. 1980. Incorporation, distribution and fate of saturated petroleum hydrocarbons in sediments from a controlled marine ecosystem. Mar. Environ. Res. 3: 15-33.

Walter, D.J. and J.R. Proni. 1980. Acoustic observations of subsurface scattering during a cruise at the IXTOC-I blowout in the Bay of Campeche, Gulf of Mexico. Pages 525-541 *in* Proceedings of a Symposium on Preliminary Results from the September 1979 *Researcher/Pierce* IXTOC-I Cruise (June 9-10, 1980, Key Biscayne, Florida). Office of Marine Pollution Assessment, National Oceanic and Atmospheric Administration, Washington, D.C.

Wheeler, R.B. 1978. The Fate of Petroleum in the Marine Environment. Exxon Production Research Company Special Report. Exxon Production Research Company. Houston, Texas, 31 p.

Wheeler, R.B., J.B. Anderson, R.R. Schwarzer and C.L. Hokanson. 1980. Sedimentary processes and trace metals contaminants in the Buccaneer oil/gas field, Northwestern Gulf of Mexico. Environ. Geol. 3: 163-175.

Winters, J.K. 1978. Fate of Petroleum Derived Aromatic Compounds in Seawater Held in Outdoor Tanks. South Texas Outer/Continental Shelf Study, Chapter 12. Draft Final Report for Bureau of Land Management, New Orleans, Louisiana.

Zajic, J.E., B. Supplisson and B. Volesky. 1974. Bacterial degradation and emulsification of No. 6 fuel oil. Environ. Sci. Technol. 8: 664-668.

Zemel, B. 1980. The use of radioactive tracers to measure the dispersion and movement of a drilling mud plume of Tanner Bank, California. Pages 812-827 *in* Symposium on Research on the Environmental Fate and Effects of Drilling Fluids and Cuttings. Lake Buena Vista, Florida, January 21-24, 1980. American Petroleum Institute, Washington, D.C.

Zepp, R.G. and P.R. Schlotzhauer. 1979. Photoreactivity of selected aromatic hydrocarbons in water. Pages 141-158 *in* P.W. Jones and P. Labor (eds.), Polynuclear Aromatic Hydrocarbons, Third International Symposium on Chemistry and Biology-Carcinogenesis and Mutagenesis. Ann Arbor Science Publ., Ann Arbor, Michigan.

Zepp, R.G., G.L. Baughman and P.F. Schlotzhauer. 1981a. Comparison of photochemical behavior of various humic substances in water: Sunlight induced reactions of aquatic pollutants photosensitized by humic substances. Chemosphere 10: 109-117.

Zepp, R.G., G.L. Baughman and P.F. Schlotzhauer. 1981b. Comparison of photochemical behavior of various humic substances in water: II. Photosensitized oxygenations. Chemosphere 10: 119-126.

Zepp, R.G., N.L. Wolfe, G.L. Baughman and R.C. Hollis. 1977. Singlet oxygen in natural waters. Nature 3: 421-423.

Zurcher, F. and M. Thuer. 1978. Rapid weathering processes of fuel oil in natural waters. Analysis and interpretations. Environ. Sci. Tech. 12: 838-843.

CHAPTER 6

TRANSPORT AND TRANSFORMATION PROCESSES REGARDING HYDROCARBON AND METAL POLLUTANTS IN OFFSHORE SEDIMENTARY ENVIRONMENTS

Paul D. Boehm

CONTENTS

INTRODUCTION

If offshore oil and gas activities have any long-term impacts in marine environments, those impacts would probably be manifested in the benthos and, particularly, in the sedentary benthic populations. Such impacts would be related to the transport of contaminated particulate material from the water column to the benthos. Chemical contaminants may be on solid particles as part of source material, or they may become sorbed to particles. Resultant biological impacts would then be related directly to: 1) the change in sediment texture or habitat, 2) the bioavailability of these chemicals and 3) the residence time of the pollutant substrate or chemical in the system. The residence time is determined by those processes which physically remove the pollutants from the system (e.g., resuspension or burial) as well as those processes which degrade the sediment-associated chemicals and render them less harmful or less available.

The coupling of the water column particulates and contaminants to the benthos involves both deposition and resuspension. Additionally, several other factors have a great bearing on whether these chemicals, once deposited have the potential to effect "changes" in biota. These factors which determine the bioavailability of chemicals include the extent of oxygenation of the sediment, the release of chemicals to the aqueous phases and the ability of the organisms to exchange the substance from a sorbed state to an area of lipid storage within the animal.

Several important issues relate to the transport and transformations to, from, and within sediments:

1. To what degree are chemicals from offshore oil and gas development activities deposited in sediments? Are contaminants once deposited remobilized and transported (i.e., to areas of accumulation) and vertically mixed below the bioavailable zone of the sediments? What are the mechanisms of transport and remobilization? What are the rates of transport?
2. Where do particle-bound contaminants accumulate in offshore and coastal regions?
3. How are contaminants, transformed (i.e., metabolized and degraded, released to overlying waters, assimilated and metabolized by benthic biota)? Do these transformations enhance or inhibit the possible effects of these chemicals?
4. Are chemicals in sediments bioavailable? Do they become bioavailable with time (burial or transformation)? Are they taken up directly from the sediment particles or from the water within the sediments?
5. How long do they reside in a bioavailable zone?

This chapter will address the state of knowledge of these and other issues and how to apply this knowledge in meaningful scientific assessments of the impacts of offshore oil and gas development activities.

THE NATURE OF PARTICULATE MATERIAL IN THE OCEAN

Types of Particles and Their Pollutant Affinities

Natural particulate matter in the water column of coastal and offshore systems has a great influence on the behavior of pollutants. Particulate material can act to remove pollutants from the water column by adsorption, flocculation, and complexation; can inhibit or facilitate pollutant bioavailability; can transport pollutants from one area to another; and can, under the right conditions, release contaminants into the water column. The affinity of trace organic and metal pollutants for particle surfaces tends to concentrate insoluble pollutant compounds and elements at the seafloor in areas of sediment deposition. Particulate material reduces the residence time of relatively insoluble chemicals in the water column. The more water soluble organic compounds and metals tend to partition into the water column. However, these more soluble materials may also be transported to the sediment in relatively lesser quantities than those remaining in the aqueous phase, nevertheless significantly affecting the chemical composition of the surface sediment.

Suspended sediment, suspended particulate matter, total suspended matter, and suspended solids are all terms that have been used to describe the total amount of solid particles suspended within sea water. Suspended sediment itself, irrespective of any associated chemical contaminant materials, can be classified as a pollutant if introduced levels cause a deleterious effect on planktonic or benthic biota. Effects of elevated fine particle concentrations on metabolic processes of filter-feeding bivalves, on egg and larval development, and on fish behavior and feeding have been well documented (see Schubel, 1982, for a review). These effects can be direct (e.g., gill clogging, smothering) or indirect (e.g., oxygen depletion caused by high oxygen demand of organic rich particles, reduction in available light, etc.).

The transport of contaminants to the benthos is a function of: 1) the existing pollutant loadings on discharged particles, 2) the concentration of particles which offer a surface for pollutant adsorption and 3) the affinity of pollutants for various particles.

Two broad factors determine a pollutant's affinity for a particular particle: 1) the nature of the particle's surface, vis-a-vis electrostatic charge, surface area (particle size) and organic content and 2) the pollutant's distribution coefficient $(K_d) = C_p/C_w$ where C_p is the concentration of a pollutant associated with a given mass of particles and C_w is the concentration of pollutant in an equal mass of water.

Particle types with which pollutants may associate include: 1) inorganic particles (e.g., clays) introduced via fluvial transport or resuspension of existing bottom sediments; 2) inorganic particles introduced locally through ocean disposal activities (e.g., dredged materials and drilling fluids and cuttings); 3) detrital particles, i.e., the remains of planktonic organisms; 4) living planktonic particles consisting of organic, silicate or carbonate materials; and 5) flocculated organic substances.

Particle size is a very important characteristic of naturally occurring suspended particulate material with regard to sorption of pollutants and residence time in the water column (i.e., settling rate). Fine particles occur in the coastal ocean as individual or composite particles (aggregates and agglomerates) (Schubel, 1982). The large fraction of the total number of suspended particles (>90%) is probably accounted for by individual particles. Most of the volume and mass is probably accounted for by composite particles (Schubel, 1982). With increasing distance from nearshore inputs of particles (e.g., rivers), a greater percentage of suspended particles are biogenic. Resuspended sediment is periodically introduced into continental shelf waters through a number of mechanisms to be discussed later.

Agglomeration of particles by organisms can profoundly affect the residence time of particles in the water column and, hence, the sedimentation of associated pollutants. The sinking of biogenic particles, primarily in the form of fecal pellets, has been suggested as a major depositional mechanism for pollutants sorbed on suspended sediments. Polychlorinated biphenyls (PCBs) and petroleum hydrocarbons are two classes of hydrophobic organic pollutants which have been shown to be transported vertically by zooplankton fecal pellets (Elder and Fowler, 1977; Johanssen *et al.*, 1980). Suspension feeders in the photic zone or at the sediment-water interface can actively or passively trap suspended pollutants in mucous nets.

Other major mechanisms which increase the affinities of pollutants for particles include:

1. Organic flocculation and ionic sorption-desorption. Metal-hydroxide and organic coatings facilitate metal sorption on particles or coprecipitation of metals.
2. Organic complexation and flocculation. This includes metal chelation and transport within organic colloids, as well as micellar entrapment of organic compounds.
3. Hydrophobic associations with surfaces. These affect the extent of sorption of non-polar organics (e.g., hydrocarbons) as influenced by particle surface area and organic content of the particles. Hydrocarbons in water associate with suspended particulate material in inverse proportion to their aqueous solubility. Increased organic content of particles tends to increase sorption and the particle-pollutant association (Meyers and Quinn, 1973; Rubenstein *et al.*, 1983)
4. Electrostatic charge of the surface of clay particles.

Sources and Concentrations of Particulate Sediments

The total assemblage of suspended particles in offshore regions which become permanently or temporarily included in surface sediments are, at any given time, a combination of primary and composite particles from riverine discharges, planktonic remains or detritus, living biomass of phyto- and zooplankters, resuspended bottom sediment (induced by currents and organisms), solid pollutant materials (tarballs, marine litter, industrial and municipal wastes), dry aerosol deposition (eolian input), marine fecal material, shoreline erosion, drilling inputs and other human activities.

Inputs of suspended particulates from river discharges affect large areas of the coastal ocean and can result in concentrations of suspended inorganic particles of several hundred milligrams per liter or more (Schubel, 1982; Milliman, 1980). Storm induced resuspension of fine-grained sediments can also cause elevated levels (tens to hundreds of milligrams per liter) for short periods. The particle loadings due to biotic production in the photic zone of the coastal ocean can

TABLE 6.1
Suspended particulate concentrations in continental shelf waters (adapted from Schubel, 1982)

Location	*Concentration (mg/l)*	*Reference*
East China Sea (off Changjing River)	0.77–38.9	Qinchun *et al.*, 1983
Argentine Shelf	0.1–40.6	Pierre and Siegel, 1979
United States		
Georges Bank	0.1–4	Bothner *et al.*, 1981
	0.5–15	Milliman *et al.*, 1980
Mid-Atlantic Bight	0.1–1	Meade *et al.*, 1975
New York Bight	0.05–3.2	Biscaye and Olsen, 1976
		Drake, 1977
Delaware Bay	10–100	Biggs *et al.*, 1983
Blake Plateau	0.15–2.2	Manheim *et al.*, 1970
Cape Canaveral (hurricane)	7.0	Manheim *et al.*, 1970
Mississippi River Delta	250–670	Scruton and Moore, 1953
Gulf of Mexico	3–40	Hausknecht, 1980
	0.3–1.8	Brooks *et al.*, 1981
California Coast	1–4	Karl, 1980
Alaska		
Northon Sound	1–20	Sharma, 1979
Cook Inlet	1–2; 2000 (extremes)	Burbank, 1974; Britch, 1976
Beaufort Sea	>35 (storms)	Naidu, 1979
Beaufort Sea	2–36 (mean 7)	Northern Technical Services, 1981
Yukon River	980	Strakov, 1967
Beaufort Sea	0.1–1.2 (under ice)	Barnes *et al.*, 1977

contribute significantly to the total suspended particle concentrations. This is especially true during planktonic blooms when living plankton and its detrital remains can represent most of the particulate matter in the upper water column.

Turbidity maxima exist in areas of intense riverine inputs and also as semipermanent features of the near-bottom environments in some continental shelf areas as nepheloid layers (Ewing and Thorndike, 1965). The continental shelf of the Gulf of Mexico west of the Mississippi River delta is an area known for these semipermanent layers of resuspended sediments in which particulate concentrations of 10 to 100 ppm are found. McGrail *et al.* (1982), for example, found that in the Flower Garden Banks area the form and the strength of the layer varied seasonally. The vertical placement of this layer varied as well, ranging from near-bottom to 42 meters above the bottom. It appeared that locally resuspended sediments only affected the near-bottom (within 3 m above the bottom) region. Some distinctions have been drawn between a "nepheloid layer," which is propagated over distance, and turbidity in the bottom boundary layer, which varies in thickness based on sediment type, bottom roughness, current velocity, fluid shear stress and degree of density stratification. The nepheloid layers may represent a major "capturing" and transporting mechanism of sedimented pollutants. McGrail and Carnes (1983) found that the nepheloid layer in the Gulf of Mexico reaches a maximum thickness of 30 m. The shear stress that maintains this layer is the result of varied processes including diurnal inertial oscillations and winter storms.

Offshore oil and gas exploration and production activities introduce particles into the waters adjacent to operating platforms mainly in the form of discharged drilling fluids and drill cuttings (see Neff *et al.*, Chapter 4; Neff, Chapter 10). Drill cuttings are particles of crushed rock originating in the formation rock being drilled and are dense, angular particles ranging from clay- to gravel-sized which are discharged continuously during drilling. Water-based drilling fluids or muds are colloidal dispersions of clays and polymers used to lubricate the drill bit. They are usually bulk-discharged with volumes ranging from 100 to 1000 bbl per discharge (500 to 2000 bbl/hr) (National Research Council, 1983). The major ingredients in drilling fluids are barite (barium sulfate), clay (usually bentonite), lignosulfonate, lignite and caustic.

Ranges of concentrations of total suspended solids in estuarine, coastal and continental shelf areas are summarized in Table 6.1. The relative contributions of sources of suspended particulates is highly site-specific and varies seasonally. Suspended particulate concentrations can be periodically much higher off the California coast than they are on the U.S. east coast due to the greater influence of pulsed riverine discharges off California (Kolpack, 1983). The Mississippi River's influence in the Gulf of Mexico is crucial in determining particulate concentration levels along the Louisiana and Texas coasts. The Yukon River in Alaska and the rivers delivering suspended sediment to the Beaufort Sea create important seasonal (spring/summer) maxima of suspended sediment concentrations (10-1000 ppm) in the region (Northern Technical Services, 1981).

Potential for pollutant-particle interactions (i.e., sorption) with subsequent sinking are greatest in areas of higher suspended particulate concentrations such

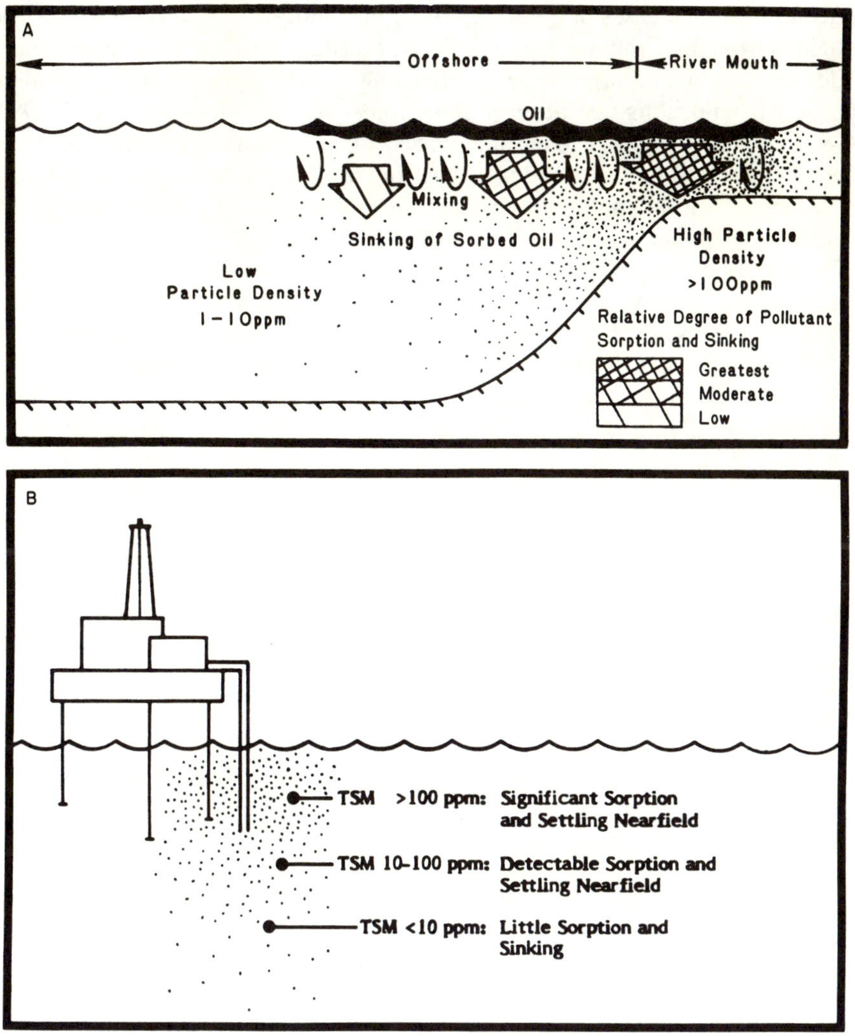

Figure 6.1. Schematic representation of how total suspended matter (TSM) concentrations promote sorption and sedimentation of oil from A, spills and B, platform discharges.

as those influenced by pulsed riverine discharges (Figure 6.1). Although high suspended particulate concentrations are present immediately adjacent to exploratory oil drilling platforms due to drilling mud discharges (up to 15,000 ppm or more; Ayers *et al.*, 1980), suspended concentrations decrease rapidly due to dilution (factors of 10^5 to 10^6 within one hour) and settling of particles. Components of drilling fluids either settle to the bottom or disperse by a factor of 10^6 to less than 1 ppm within 100 to 200 m of the point of discharge (National Research Council, 1983). The rate of attenuation of localized suspended particulate inputs depends on discharge rate, current speed and turbulent mixing.

Concentration decreases of 5 to 6 orders of magnitude occur at distances of 500 to 1000 m from high volume discharges (Ayers *et al.*, 1980) and at lesser distances for lower rates of discharge. Several other studies have addressed the particulate transport from drilling operations (Houghton *et al.*, 1980; Ray and Meek, 1980; EG&G, 1982; Northern Technical Services, 1983). Northern Technical Services (1983) determined that suspended levels from drilling discharges in the Beaufort Sea remained approximately three times the ambient (2 ppm) level 1900 m from the discharge studied.

The inputs of suspended solids from drilling platform inputs, while locally significant and potentially offering a significant transport mechanism of contaminants to the seabed, must be viewed in the light of their importance relative to riverine discharges. Kolpack (1983), for example, calculated that drilling fluid impacts have a small contribution (0.1%) to the overall water column suspended loadings in the Santa Barbara Channel. The main impact of these particulate inputs would be in the benthos immediately adjacent to the discharge point.

Particle Transport and Settling

Particles which enter coastal marine systems fall vertically through the water column and may be transported by: 1) tidal currents, 2) residual nontidal currents, 3) wave driven turbulence and 4) storm events. These forces interact with gravitational settling of a particle, the rate of which is dependent on its size and its density relative to sea water. The dynamics of fine-grained particle deposition are complex and are a function of numerous physical, chemical and biological processes. Fine particles often undergo several episodes of deposition, resuspension and transport other caused by short-term episodic events (e.g., storm-induced flow, ice scour, and turbidity currents) than to mean current flow. In estuaries and nearshore areas, tidal currents and longshore currents are the primary influences on particle settling, deposition and resuspension.

The composite particles of agglomerated primary particles dominate the volume and mass of suspended sediments. Particles may be agglomerated by electrochemical flocculation, bound by dissolved organic compounds or packaged by biological processes. Filter-feeding planktonic and benthic organisms package small particles as fecal or pseudofecal material which increases their settling velocity (Schubel and Kana, 1972) or, in the case of benthic suspension-feeders, sequester fine suspended material as agglomerated pellets (Rhoads, 1974). These suspended composite particles are common in the transitional zones between rivers and continental shelf waters (e.g., Manheim *et al.*, 1972).

Flocculation by electrochemical attraction between charged particles is another significant process which alters particle size and mass and hence settling velocity in areas of steep salinity gradients (i.e., estuaries). This type of particle association is of lesser importance in offshore areas. Direct collisions of primary particles with other particles and with other agglomerated particles are also responsible for the removal of particles from the water column (i.e., settling) in areas of high particle concentrations (estuaries, areas of discharge plumes from platforms or ocean dumped material). Schubel (1982) concludes, however, that in areas of lower particle concentrations such as the continental shelf and slope, biological processing accounts for most particle agglomeration and, hence, particle fluxes to the seabed.

Ultimately, a settling particle reaches the benthic boundary layer, the turbulent layer at the bottom where most momentum transfer occurs and where particles are incorporated into the bottom sediments. Whether or not a particle is deposited at a particular point depends on bottom characteristics (roughness, current velocity). Whether or not it remains on the bottom depends on the balance of gravitational and cohesive forces on the particle versus the shear stress trying to dislodge it.

Estimates of the rates of removal of particle-reactive pollutants from marine waters have been made by using thorium isotope ratio measurements in sea water. Li *et al.* (1979) showed that the time to remove half of ^{228}Th, and by inference other reactive, sorbed pollutants, by settling particles was 185 ± 35 days in the open ocean, 70 ± 10 days on the continental slope, 20 ± 2 days on the outer continental shelf and 17 ± 1 day in inner shelf waters. Similarly, Santschi *et al.* (1980) determined that half removal times in Narragansett Bay ranged from 1.5 to 15 days with fine-grained particle settling velocities of 1 to 11 meters per day. These calculations are of course subject to variations in suspended sediment concentrations. Increases in resuspension of bottom sediment can tend to reduce reactive pollutant removal time through increased rates of sorption, but pollutants can also be released from interstitial water and from desorption of particle-bound pollutants during resuspension.

CHEMICAL COMPOSITION OF RELEVANT POLLUTANT SOURCES

Offshore oil and gas exploration and production activities can introduce contaminants from a number of sources (Table 6.2). The relative importance of each of the sources in a given area differs depending on the exploration and production history and on the operational practices in a given area. The compositions of such sources will be briefly summarized here. More detailed information is presented in Chapter 4 and is readily available in other reports and publications (McAuliffe, 1969; National Academy of Sciences, 1975; Collins, 1975; Armstrong *et al.*, 1979; Jackson *et al.*, 1981; Middleditch, 1981a; Sauer, 1981; Lysyj, 1982; National Research Council, 1983, 1985).

The chemistry of petroleum is certainly well known in relation to questions regarding oil and gas development activities (e.g., Tissot and Welte, 1981). The chemistry of drilling fluids is variable but is well documented (National Research Council, 1983; Chapter 4). While the metal composition of drilling fluids has been studied little work has been done on characterizing the organic polymeric material in lignosulfonate drilling fluids.

The composition of formation waters or produced water is highly variable (Collins, 1975; Jackson *et al.*, 1981; Lysyj, 1982). Produced waters are for the most part an oily brine, brought to the surface along with produced hydrocarbons. The oil content of this water is usually reduced by gravity separation prior to discharge, but a variety of inorganic and organic constituents remain (Chapter

TABLE 6.2
Significant potential sources of contaminants from offshore oil and gas operations

Activity/source	*Petroleum hydrocarbons*	*Other organics*	*Trace metals*	*Solids*
Exploration/drilling muds and cuttings	*(**)[a]	**	***	****
Production/produced waters	***	***	**[b]	*
Exploration and production petroleum spillages	****	**	*	*

[a]Oil-based muds or diesel additives; also cuttings may contain higher layers of formation hydrocarbons.
[b]Very variable; may be more significant in some instances.
****Most important contaminant.
***Very important contaminant.
**Contaminant of lesser importance.
*Detectable contaminant.

4). Of the organic constituents, the most abundant constituents are the low molecular weight hydrocarbons (benzenes, toluenes, xylenes). These constituents are quite water soluble and, along with the other highly soluble compounds (e.g., phenols), are not readily sorbed to particles in the discharged brines or in ambient sea water. The relation between water solubility and the sorption coefficient (Figure 6.2) is critical. Those compounds having the lowest solubility, i.e., the polycyclic aromatics and the alkanes (not shown) are more likely to sorb onto particulates and be incorporated into sediment. However, the water soluble aromatics and other organics may be directly bioaccumulated by benthic animals should the produced water plume come in contact with the bottom. In shallow waters, such as in Trinity Bay, Texas (Armstrong *et al.*, 1979), sizable quantities of petroleum aromatics from produced waters can be both sorbed to sediments and bioaccumulated by benthic animals.

Each and all of these inputs must be evaluated in light of the normal, chronic input of pollutants to a given area and hence the resultant concentrations of pollutants in the sediments. Petroleum and anthropogenic hydrocarbons (see next section) are continually discharged into the coastal ocean. On a global basis offshore oil and gas activities account for a small part of the total hydrocarbon inputs (National Research Council, 1985; Chapter 4). However, these global budgets do not address site specific estimates of the relative importance of the various sources, pointing to a serious lack in existing data bases. In general, chronic, land-based (riverine or eolian) inputs of anthropogenic hydrocarbons account for the overwhelming percentage of hydrocarbons deposited in sediments. The Mississippi River and other river inputs largely influence the distribution of high molecular weight hydrocarbons in sediments of the northern Gulf of Mexico. In the North Atlantic, long-range fluxes of polycyclic aromatic hydrocarbons (PAH) by direct eolian transport or riverine delivery of PAH-laden

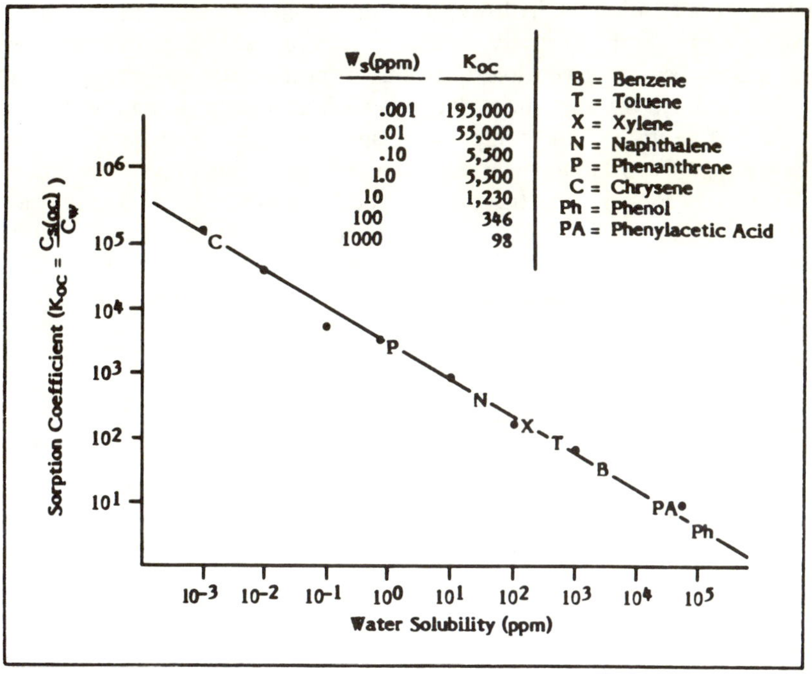

Figure 6.2. Equilibrium relationship of compound water solubility to sediment sorption coefficient.

urban particles account for most of the important anthropogenic hydrocarbon material in offshore areas (Gschwend and Hites, 1981). Notable exceptions are in areas of ocean disposal of municipal sludges or petroleum contaminated harbor dredged material, such as the New York Bight area, where petroleum-sourced PAH and other hydrocarbons dominate; areas near chronic sources of nonpoint source petroleum effluents (e.g., the Mississippi River delta); and areas near chronic petroleum laden ocean outfalls (e.g., Southern California Bight municipal outfalls; Eganhouse *et al.*, 1981).

TYPES AND SOURCES OF HYDROCARBONS IN SEDIMENTS

General Sources

Hydrocarbons are ubiquitous to marine sediments. These compounds may originate from: 1) biogenic sources—marine and terrestrial; 2) petrogenic sources—a) anthropogenic petroleum inputs from a variety of sources including municipal discharges, stormwater runoff, tanker washings, tanker accidents, and offshore activities (produced water, chronic spillages, drilling cuttings discharges, blowouts), b) natural petroleum sources (i.e., petroleum seeps); 3) pyrogenic (incomplete combustion) sources—from the anthropogenic combustion of oil, coal, wood, peat and from natural fires; and 4) diagenetic sources—the production or alteration of hydrocarbons in sediments mediated by time, temperature and microbial activity. These sources vary substantially in their hydrocarbon composition. Even hydrocarbons from a similar anthropogenic source may be modified through such processes such as predepositional weathering of spilled petroleum and the temperature and fuel type in incomplete reactions (Lee *et al.*, 1977). Offshore oil and gas exploration, production and transportation activities may contribute petrogenic hydrocarbon to sediments that may already contain, for example, pyrogenic polycyclic aromatic hydrocarbons (PAH) compounds, biogenic compounds of planktonic and terrigenous origin (alkanes, alkenes, polyolefins), and diagenetic saturated hydrocarbons and PAH compounds.

Chemical Composition of Hydrocarbons in Marine Sediments

Sources of hydrocarbons entering the marine environment are numerous and the number of individual hydrocarbon components quite large. Chemical and microbial alterations occur after introduction of a particular set of hydrocarbon compounds to the marine environment, a set originally attributable to a general type source, but subsequently modified.

Crude Oils

The chemical composition of crude oils from different producing regions and even from within a particular formation can vary tremendously. The chemical properties are also linked closely to environmental behavior and fate during spills (Koons, 1973). Crude oils contain thousands of different compounds formed during development. Hydrocarbons (i.e., compounds containing only carbon and hydrogen) are the most abundant compounds in crude oils, accounting for 50-98% of the total composition (Clark and Brown, 1977), although the majority of crude oils contain the higher relative amounts of hydrocarbons. While carbon (80-87%) and hydrogen (10-15%) are the main elements in petroleum, sulfur (0-10%), nitrogen (0-1%) and oxygen (0-5%) are important minor constituents present either in elemental form (i.e., sulfur) or as heterocyclic constituents and functional groups. Crude oils often contain wide concentrations of trace metals such as V, Ni, Fe, Al, Na, Ca, Cu and U. Although a wide range of chemical composition is one of the main tenets of petroleum geochemistry, Koons (1973) presented a composition for an average crude oil (Table 6.3).

Refined Petroleum Products

Many refined petroleum products are transported and are subject to introduction to the marine environment. These include gasoline, kerosine, jet fuels, fuel oils (No. 2 diesel, No. 4, No. 5, No. 6), bunker fuel oils, lubricating oils and mineral oils. As refining processes and terminologies differ worldwide, comparisons of compositions of refined products yield wide variation. An excellent discussion of the chemical properties of refined products is found in Clark and Brown (1977). This is an important class of input because diesel oil and mineral oil are important additives of drilling fluids.

TABLE 6.3
The "average" composition of crude oil (from Koons, 1973)

Component	*Percent content*
By molecular size	
Gasoline (C_5-C_{10})	30%
Kerosene (C_{10}–C_{12})	10%
Light distillate oil (C_{12}–C_{20})	15%
Heavy distillate oil (C_{20}–C_{40})	25%
Extremely heavy residuum oil	20%
By molecular type	
Saturated hydrocarbons (normal and branched alkanes)	30%
Naphthene hydrocarbons (cycloalkanes)	50%
Aromatic hydrocarbons	15%
Polar (NSO) compounds	5%

Biogenic Hydrocarbons

Hydrocarbons are synthesized by most marine plants and animals including microbiota (Han and Calvin, 1969; Davis, 1968), phytoplankton (Blumer *et al.*, 1971; Clark and Blumer, 1967), zooplankton (Blumer *et al.*, 1963, 1969; Blumer and Thomas, 1965a,b; Avignan and Blumer, 1968), benthic algae (Youngblood *et al.*, 1971; Youngblood and Blumer, 1973; Clark and Blumer, 1967) and fishes (Blumer *et al.*, 1969; Blumer and Thomas, 1965b). Organisms can both produce their own hydrocarbons or acquire them from food sources. Species of marine organisms synthesize limited numbers of hydrocarbon constituents over relatively narrow boiling ranges. Terrestrial plants (and *Sargassum*) produce C_{21} through C_{33} odd chain *n*-alkanes associated with the waxy coatings of leaves. These are major hydrocarbon components of most "unpolluted" coastal sediments.

Diagenetic Sources

Biogenic precursor molecules (e.g., terpenes, sterols, carotenoid pigments) may be altered after deposition in sediments by microbially mediated and chemical processes to yield a variety of chemical compounds. Diagenetic hydrocarbon constituents include: 1) aliphatic hydrocarbons, 2) cycloalkenes, 3) sterenes, 4) polycyclic aromatic hydrocarbons (PAH) and 5) pentacyclic triterpanes. One of the most significant sets of diagenetic produces are the PAH compounds, including some compounds which are also found in petroleum and other hydrocarbon sources as well (Wakeham *et al.*, 1981). These diagenetic compounds may constitute important components of recent sediment hydrocarbon assemblages. Perylene and retene are among those compounds formed in reducing sediments from higher plant precursors (Hites *et al.*, 1980; Aizenshtat, 1973).

Combustion Sources

Urban air particulate matter contains saturated and aromatic hydrocarbons formed during the incomplete combustion or pyrolysis of fossil fuels (wood, coal, oil) (Lee *et al.*, 1977). Polycyclic aromatic hydrocarbons formed during combustion processes are transported seaward via direct deposition on the sea surface or rainout over land followed by stormwater runoff. PAH compounds are therefore ubiquitous chemical components of marine systems throughout the world (Laflamme and Hites, 1978; Pancirov and Brown, 1977; Youngblood and Blumer, 1975; Windsor and Hites, 1979; Brown and Weiss, 1978; Boehm and Farrington, 1984).

PAH compounds from combustion sources are characterized by a lesser degree of alkylation than aromatics from petroleum. The degree of alkylation within a homologous series of aromatics (e.g., phenanthrenes) in a given PAH assemblage is dependent on the temperature of formation of the PAH. It is this principle which allows for the differentiation of combustion related inputs from fresh and weathered petroleum (Blumer, 1976; Lee *et al.*, 1977; Hites and Biemann, 1975; Youngblood and Blumer, 1975). Combustion sources contain relatively low quantities of two ringed aromatic families (e.g., naphthalenes). Therefore, the relative inputs of petroleum and combustion sources can be discerned from such plots of two- to five-ringed aromatics.

Other Sources

Anthropogenic hydrocarbons may be introduced through a variety of other sources (dredge spoil, sewage sludge, fly ash, industrial wastes) containing mixed inputs of hydrocarbon material (petroleum plus combustion material). In addition, the direct introduction of coal may be significant in certain areas. The saturated and aromatic hydrocarbon compositional nature of coal (Tripp *et al.*, 1981) is very similar to that from petroleum, both materials being formed through low temperature processes. Careful evaluation of PAH and organosulfur compositions of sediments can differentiate oil and coal (Hites *et al.*, 1980).

Summary

All of the sources discussed contribute to the hydrocarbon assemblage that one observes when one analyzes a sediment sample for "hydrocarbons." It is essential to differentiate petrogenic inputs from oil and gas development from other inputs. This is especially true for petrogenic PAH and heterocyclic aromatic compounds as these are the most important compounds related to long-term biological effects (Neff and Anderson, 1981). The analysis of other compound types (e.g., alkanes) may serve to diagnose the presence of petrogenic residues but these compounds generally are of no ecological consequence. The sources of PAH compounds in the marine environment and the different fate and bioavailability of pyrogenic and petrogenic PAH are summarized in Figure 6.3.

TRANSPORT OF PARTICULATE POLLUTANTS TO THE BENTHOS

Petroleum Transport to the Seabed

Several possible mechanisms of transport of oil to the benthos are shown in Figures 6.1 and 6.4. From case studies of oil spills several generalizations regarding transport of oil to the benthos can be made:

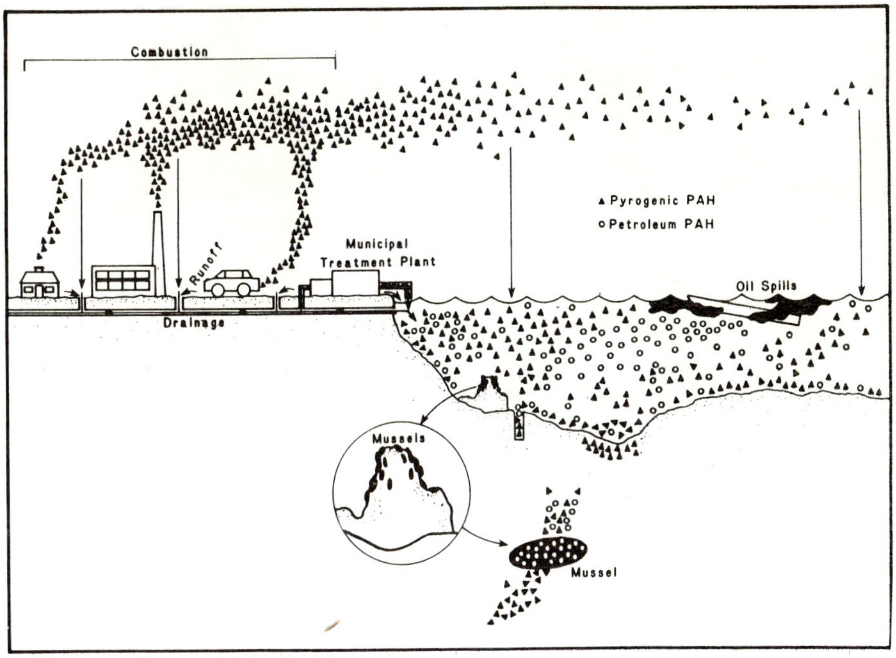

Figure 6.3. Pictorial representation of the various sources and fates of polycyclic aromatic hydrocarbons in the ocean (adapted from Farrington *et al.*, 1983).

1. The beaching (landfall) of oil from a spill followed by erosion or scouring of beach sediment is a major possible route of entry of oil into nearshore sediments (Boehm, 1983; Gundlach *et al.*, 1983).

2. Under conditions of low suspended particle concentrations in the water column (1 to 10 ppm) no significant transport of particle-sorbed oil to the seabed will occur. Under conditions of moderate suspended particle concentrations (10 to 100 ppm) significant quantities of oil may be sorbed to particles if sufficient mixing of oil and available particles occurs. Under extreme conditions of influx of particles (>100 ppm) and mixing of oil and particles in the water column, massive transport of sorbed oil to the offshore benthos can occur with possible severe offshore impact (see Figure 6.1).

3. The transport of biologically pelletized oil (zooplankton fecal pellets) to the bottom certainly has occurred and has resulted in analytically detectable oil residues in the bottom with some biological consequences (Johanssen *et al.*, 1980). In some spills this may represent the major transport path; however, this is probably quantitatively unimportant with regard to the overall fate of the mass of spilled oil.

4. The direct sinking of oil due to increased density of oil through weathering or physical fractionation of a spilled cargo has occurred (Grose *et al.*, 1979). The chances of direct sinking occurring is increased when dense oil (weathered crude or "heavy" oil) comes in contact with less dense (low salinity) sea water in areas

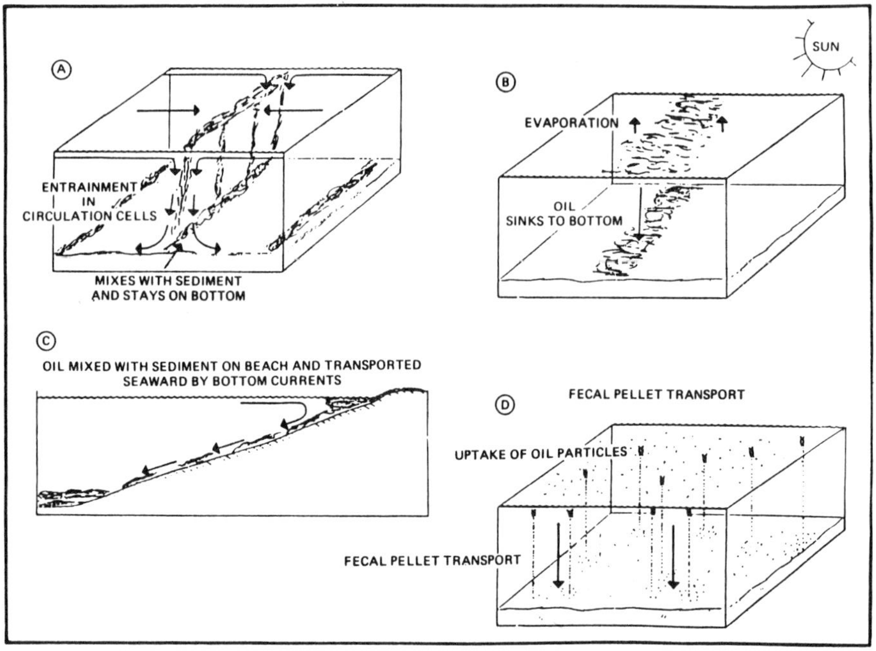

Figure 6.4. Hypothesized methods by which oil may be caused to sink and remain on the bottom.

of freshwater runoff or ice melting (Figure 6.5). Seawater densities can decrease to the point where weathered oil (density >1.01-1.02 g/cm^3) can sink. The sinking may proceed until a density discontinuity (pycnocline) is reached. At this pycnocline elevated levels of fine-grained suspended material may reside thus enhancing further oil-suspended matter interactions and possible sinking.

5. Sedimentation of petroleum hydrocarbon residues directly to the seafloor

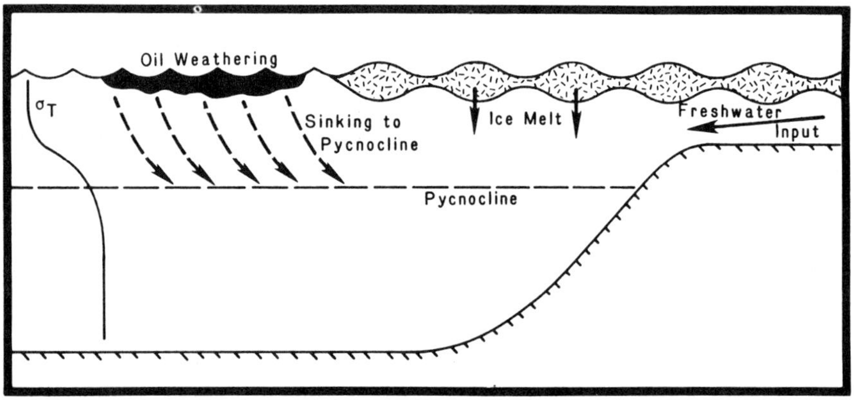

Figure 6.5. Transport of weathered oil to seabed in low density water.

may occur where discharge of organic-laden drilling fluids occurs. This process depends on the nature of the associated hydrocarbons. A light distillate such as diesel oil or a light fuel oil will largely partition into the aqueous phase or evaporate (Boehm, unpubl.). Heavier oil and hydrocarbons associated with the formation during drilling are more apt to remain associated with particles and settle with drilling mud particles.

The resulting concentration of petroleum or other associated contaminants (e.g., metals) in the sediments accompanying any of the possible transport paths

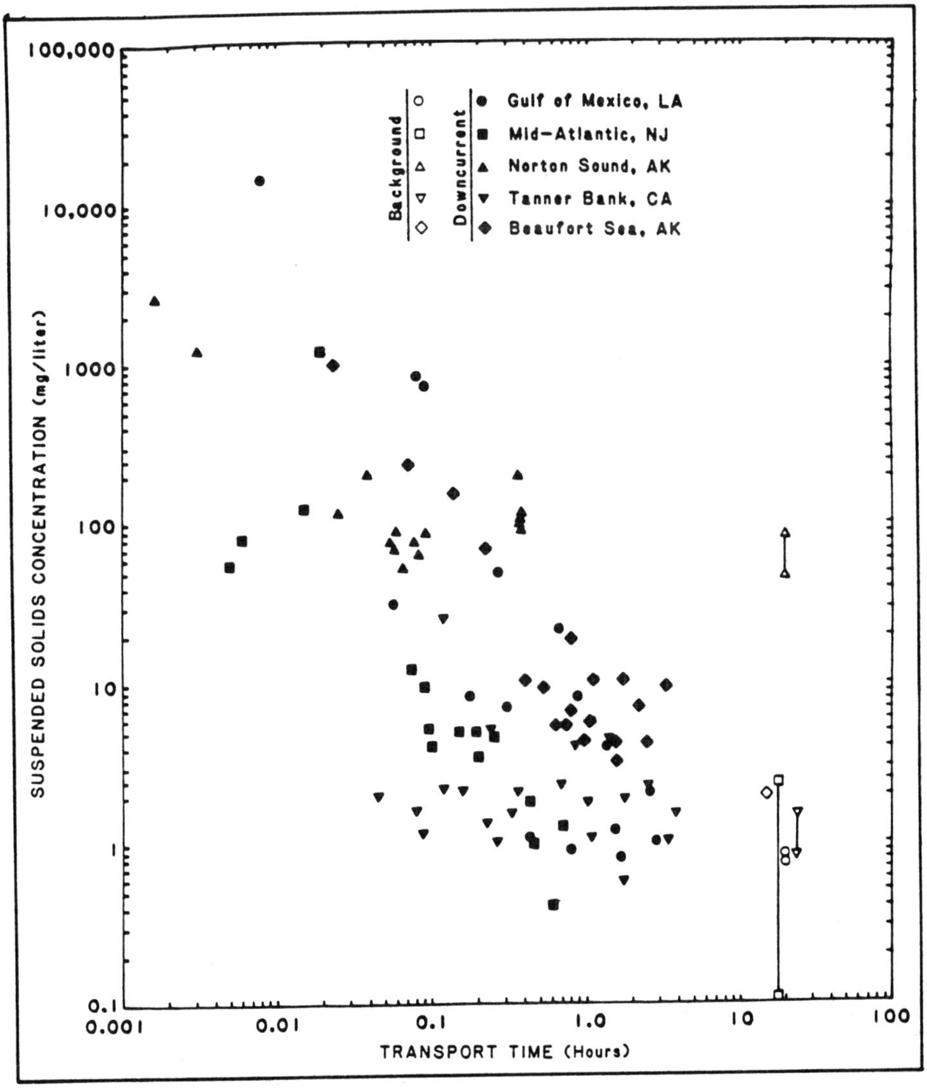

Figure 6.6. Relationship of suspended solids concentration in the water column following drilling fluid discharges to transport time of particulate plume (J. M. Neff, unpubl.).

depends on water depth (and hence dilution) and the offshore settling regime (i.e., currents and tidal energy). A rapid dilution of particulate materials and oil droplets introduced into the water column from point sources occurs. When drilling muds are discharged from platforms, dilutions of 10^4 of suspended solid materials can occur within 1.5 to 2.5 h after discharge and up to 10^5 within 6 h in most discharge scenarios (Figure 6.6). Dispersion time is important in assessing the potential impact of particle-bound pollutants. This time and the transport distance are related to the current velocity. Ayers *et al.* (1980) used a plot of transport distance and, hence, transport time to determine the concentration of barium in the plume of discharged drilling fluids. Within 5 min after discharge, concentrations were reduced to 0.001 to 0.01% of initial (discharge) concentration. Dispersion and hence dilution takes place due to turbulent entrainment, diffusion and settling. If a high concentration of hydrocarbons (>1%) were associated with the discharged muds and if all the oil remained with the solids (which is not the case; see below), and if the 500 bbl of mud were to be discharged in 30 m of water and in an area of low current velocity (4.5 cm/s), then a model developed by Sauer (1983) would predict that about 80% of the solids (initial concentration of solids in mud = 3.04×10^5 mg/l) would be deposited within 300 m of the platform in about a 60-m wide band. The resultant mud solids deposition, if evenly distributed in this area of seabed would result in approximately 0.5 mm of sedimentation (assuming a density of 3 g/cm^3). The resultant initial oil concentration in the top 1.0 cm of sea bottom would be approximately 0.4 mg/g sediment or 400 ppm. At increased depth and current velocity, this value would be much decreased. Futhermore, in reality, a significant quantity of oil associated with drilling fluids will partition into the water column. Laboratory partitioning studies of diesel oil additives to drilling fluids have indicated that a relatively small amount (<20%) of diesel hydrocarbons remain associated with rapidly settling particles, with most diesel remaining in the aqueous or fine particulate phases (Breteler *et al.*, 1985).

Accumulation of Particulate Pollutants in Sediments

Pollutants associated with settling particles from non-point source discharges (e.g., rivers) tend to accumulate in certain areas of the nearshore and continental shelf region. Pollutants associated with point source discharges (e.g., oil exploration and production rigs or dumping of dredged material) tend to settle differentially with a dense plume of cohesive or flocculated material settling close to the point of discharge in areas of moderate water depths (ten to several hundred meters). Discharges of drilling fluids from oil and gas exploration platforms results in a primary plume of rapidly sinking particles and a secondary plume (4-10% of total material) of fine-grained material generated by turbulent mixing of the primary plume. The secondary plume is transported as a diffuse cloud which drifts away from the source with the prevailing currents (Ayers *et al.*, 1980). A typical profile of suspended particle concentration contours after a bulk discharge of drilling fluids is shown in Figure 6.7. The bulk of this type of discharge settles rapidly over an area determined by water depth and current

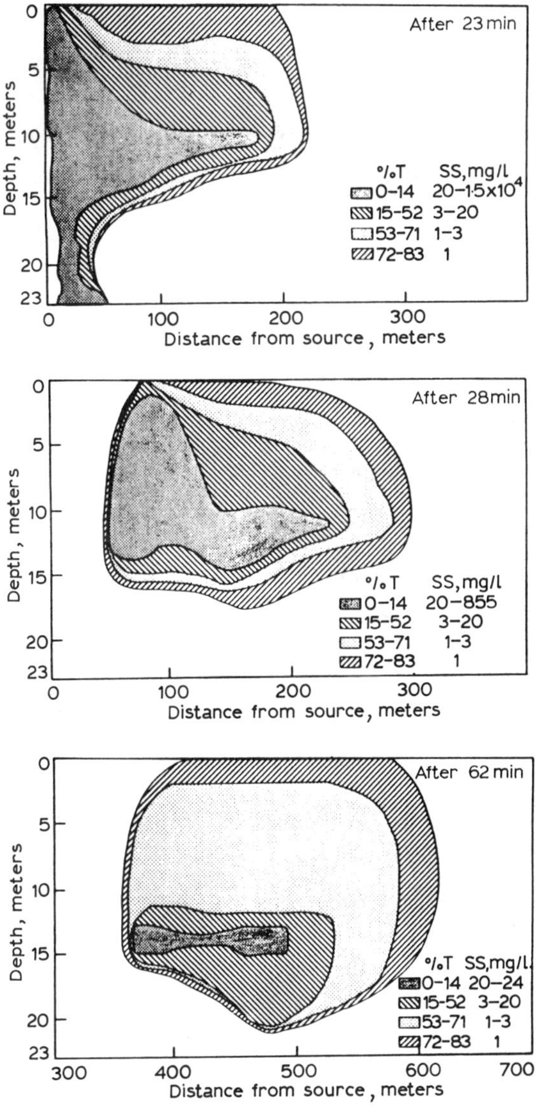

Figure 6.7. Time sequence illustrating fate of discharged drilling fluid particulates; side view of plume during 1000 bbl/hr discharge (from Ayers *et al.*, 1980).

speed. Secondary resuspension of material deposited from point discharges occurs with time in shallow to moderate depths (up to 200 m depth) (EG&G, 1982), with the resuspended material presumably being transported through a number of deposition and resuspension events to a final "depositional area."

Materials other than drilling muds, such as dredged material, sewage sludge or weathered petroleum in the form of tarballs have varying settling behavior. Dispersion and sinking of pollutant residues associated with these discharges

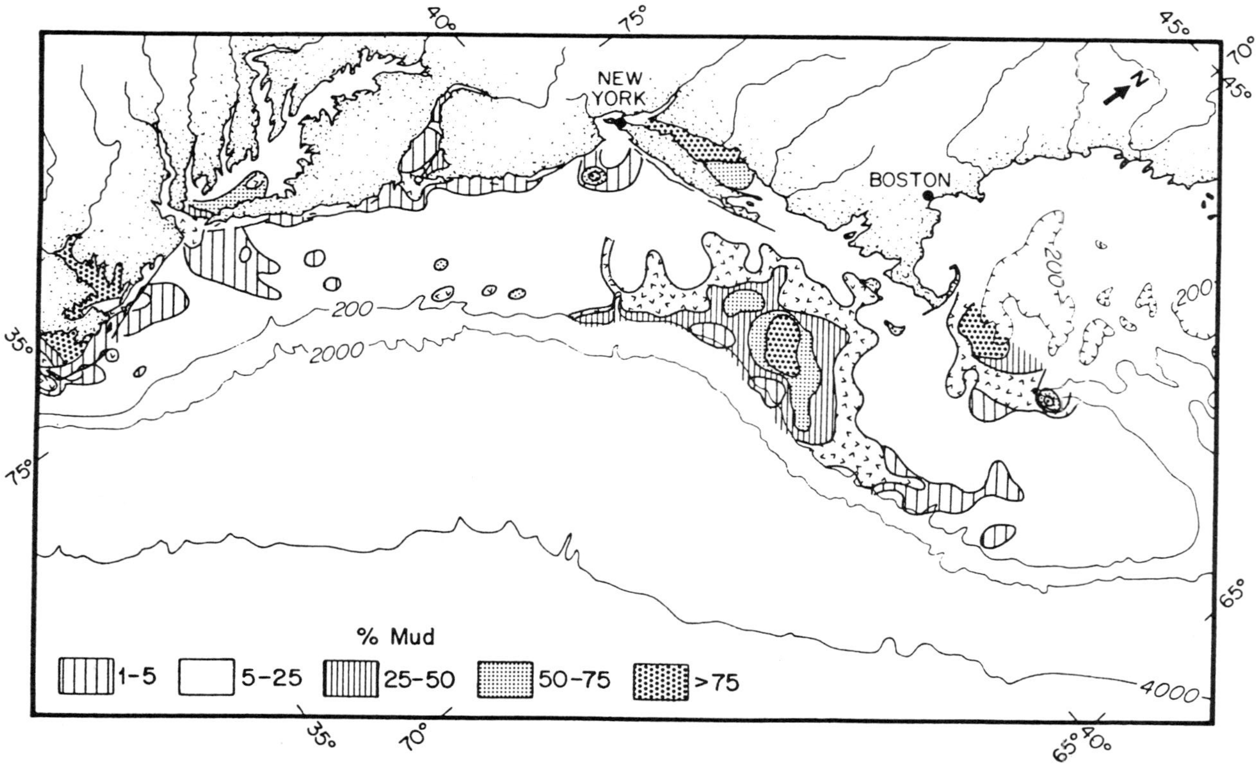

Figure 6.8. Distribution of mud (silt and clay) in surface sediments on the Middle Atlantic and New England continental shelf indicating the outer shelf "Mud Patch" off southern New England. The muddy deposits are thought to represent an area of deposition of fine sediments eroded from Georges Bank (from Milligan, 1972).

depends on their rate of discharge, postdischarge flocculation or coagulation and, hence, the final density of discharged particles. Dredged material discharges usually result in direct and rapid deposition of particle-associated pollutants. In the case of petroleum, oil can sink rapidly if oil mixes with high quantities of suspended sediments. Slower sinking can result from the gradual increasing density of petroleum residues as they weather or confront decreasing seawater densities.

Areas of deposition are topographically and dynamically suited for fine-grained sediment deposition. On the continental shelves these depositional areas may represent topographic lows such as the depositional basins off the California coast and elsewhere (Reed *et al.*, 1977), or other quiescent areas where net accumulation is occurring such as depressions between sand waves (Carmody *et al.*, 1973; Farrington and Tripp, 1977; Freeland *et al.*, 1979; Nittrouer *et al.*, 1979; Olsen *et al.*, 1980; Bothner *et al.*, 1981; Boehm, 1984) or mid-shelf areas, such as the "Mud Patch," west of Georges Bank (Figure 6.8), where the current dynamics allow sediment deposition (Twichell *et al.*, 1981). Other important areas of sediment transport and deposition (and potential pollutant deposition) on the continental shelves are the submarine canyons. Pollutants associated with fine grained particles tend to accumulate in and at the heads of canyons (e.g., Stanley and Freeland, 1978; Butman *et al.*, 1983; Boehm, 1984), although the hydrodynamic explanation for deposition at canyon heads is not clear. Depositional areas in water depths of up to several hundred meters are subjected to large storm induced resuspension events (Harris, 1976; Freeland *et al.*, 1979). Episodic events, such as turbidity currents occur off the California coast (e.g., Komar, 1969) and elsewhere, may be responsible for pollutant transport to the continental slope. Continual resuspension of particulates has been observed in the head of Lydonia and Oceanographer canyons (Butman *et al.*, 1983) and other east coast canyons as well.

Pollutant levels may be elevated or may actually increase in areas of low net accumulation (Olsen *et al.*, 1982). Mixing of sediments due to biological (bioturbation) or physical processes has been observed to exchange surface sediment with overlying particles. Aller *et al.* (1980) and Bothner *et al.* (1981) observed that even in areas of very low accumulation an excess of radioactive tracers in the sediments leads to the conclusion that "old sediment" is being exchanged for new, and perhaps pollutant-burdened, suspended particles.

The distributions of metal and organic constituents in sediments are related directly to total organic carbon content and inversely related to grain size on the continental shelves (e.g., Trefry, 1977, Boehm and Fiest, 1980) (e.g., Figure 6.9).

Levels of Accumulation of Pollutants

The net effects of deposition, resuspension and mixing account for the time averaged residual concentrations of metal or hydrocarbon pollutants in the bottom sediments. In the case of point source discharges from drilling platforms an interesting relationship appears to exist between net accumulation (Net =

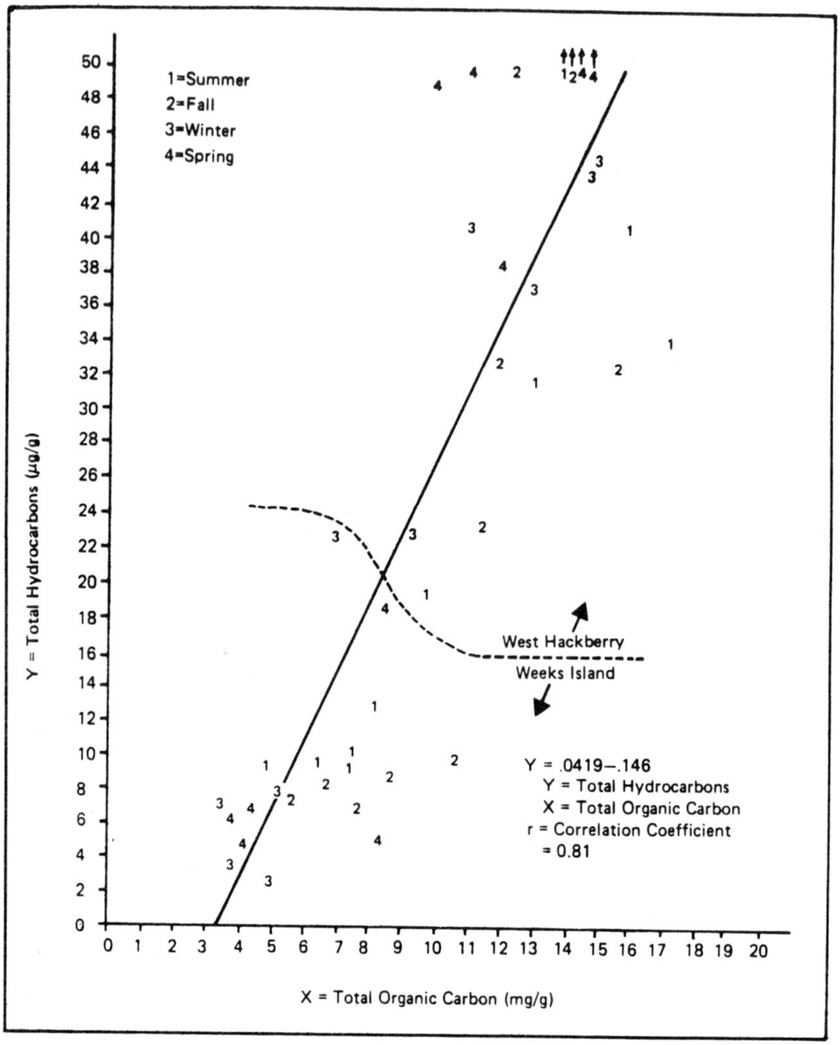

Figure 6.9. Total hydrocarbon concentration in surface sediment as a function of total organic carbon content on the Gulf of Mexico continental shelf (from Boehm and Fiest, 1980).

deposition - resuspension) and water depth. A study by Boothe and Presley (1983) of metal and hydrocarbon distributions around exploratory and production wells in the Gulf of Mexico suggests that sediments around wells in shallow water depths (<34 m) initially accumulate larger amounts of contaminants than do wells at deeper depths in the short term, but that contaminated sediments are transported out of the immediate area due to physical processes. In deeper waters (>100 m), deposited contaminants tend to reside in surrounding sediments for longer periods.

Bothner *et al.* (1983) determined that between 21 and 31% of the barite discharged from a drilling platform on Georges Bank was present in the sediments within 6 km of the drill site four weeks after drilling was completed. Their ability to detect drilling mud inputs to sandy sediments was enhanced by the analysis of the clay-sized fraction of the sediments. Concentrations of Al, Cr, Cu and Hg increased initially by a factor of two over background.

More net accumulation occurs around platforms at intermediate water depths (several hundred meters) due to the absence of significant resuspension of sediments even by large storm events. Sea surface discharges from wells drilled at greater water depths (about 1000 m) on the continental slope would tend to be dispersed due to the net effect of long settling time and current-induced transport, and discharged solids would not accumulate near the point of discharge. Instead the discharges might be spread over a much larger area of sea floor hence greatly diluting the net effect of deposition. However, drill cuttings from the first 2000 feet of drilled hole are discharged directly on the bottom. Thereafter the cuttings are circulated to the drilling platform and are discharged at the sea surface.

A number of studies have been undertaken recently to ascertain the levels and areal extent of contaminant distributions around exploratory and production oil and gas wells. The results of these studies are summarized in Table 6.4.

The degree of elevation of sediment metal concentration due to drilling fluid discharges around exploration platforms varies considerably with the rate of discharge, depth of the water and the distance from the point of discharge and current speed. It is not uncommon to find levels of barium in the sediments 10 to 100 times background levels within 500 to 1000 m of the discharge point. However, in areas where the sediment is reworked and resuspended by strong currents these levels decrease with time. While barium sulfate is heavier than other drilling fluid components, and therefore its fate may not represent the fate of all components, the large quantities of barium in drilling fluids make it a good tracer of drilling fluid distributions in sediments. Other metals show moderate to no elevation in the near-rig (within 125 m) environment due to dilution. In shallow environments, large amounts of discharged drilling muds with elevated levels of barium and, occasionally, chromium occur within several hundred meters of the discharge resulting in elevated contaminant and solids levels with possible burying of animals. These deposits would be subsequently dispersed over time due to physical factors such as resuspension and ice scouring.

Elevated hydrocarbon levels of several parts per million may also be short-lived. Water-based drilling fluids do not normally contain significant enough amounts of hydrocarbons to result in detectably elevated levels in surrounding sediments. However, the studies of sedimentation resulting from discharges of oil-based drilling muds, comprised of large quantities of diesel or other petroleum lubricants, have been carried out in the North Sea (e.g., COWI Consultants, 1982; Law and Blackman, 1981; Davies *et al.*, 1984). Due to the high petroleum content of these discharges, concentrations of petroleum hydrocarbons in surrounding sediments were elevated to varying extents depending on the nature of the added hydrocarbons and their affinity for drilling mud particles. Davies *et al.* (1984) summarized the results of studies conducted around nine North Sea

TABLE 6.4
Synopsis of studies on the fate of discharged drilling and production fluids in bottom sediments

Study	*Location*	*Water depth (m)*	*Well type*	*Metals*	*Hydrocarbons*	*Amount of material discharged*
Lytle and Lytle, 1979	Gulf of Mexico (Texas)	25	Exploration		No significant elevations	
Meek and Ray, 1980	Tanner Bank (Southern California)	55	Exploration	Deposition: 7–52 $g/m^2/day$ within 150 m of discharge point; 12% of discharge material 50–150 m from rig; other material initially deposited within 50 m; 70–80% of settled material transported out of area	Not measured	2850 bbl
Houghton *et al.*, 1980	Lower Cook Inlet (Alaska)	62	Exploration	Maximum of 3% cuttings in sediments by weight—100 m from discharge; constant resuspension	Not measured	
EG & G, 1982	Mid-Atlantic	120	Exploration	Increases in barium levels up to 6500 ppm (190 ppm background) initially; decreasing to 1000–2000 ppm one year later; evidence for sediment transport outside of the study region (3–2 km radius)	None detected	31,000 bbl
Grahl-Nielson *et al.*, 1980	North Sea	65	Exploration		Hydrocarbons detected 5 km^2; 10–1000 ppm (oil-based mud used); high of 1200 ppm	1250 m^3
Tillery and Thomas, 1980	Gulf of Mexico		Production Well (several platforms)	Unquantified gradients of metals (Ba, Cd, Cr, Cu, Pb, Zn) away from some platforms		

Crippen *et al.*, 1980	Beaufort Sea	7	Exploration	Within 45 m of discharge, elevated mercury levels 6.4 ppm over background 0.07 ppm; 1 ppm–150 m of site; Ba not analyzed; elevated levels of arsenic (12–23 ppm); cadmium (2–6 ppm); chromium (70 ppm); lead (470 ppm); zinc (1400 ppm); (highly contaminated barite used in drilling fluid)		
Armstrong *et al.*, 1979	Trinity Bay (Texas)	2–3	Production		Large hydrocarbon impact on sediments (100 ppm) due to low water depth and high suspended loadings in the water	
Northern Technical Services, 1983	Tern Island	6.7	Exploration	Elevated barium levels to 6000 ppm; other metals above background within 100 m of discharge point	Possible elevated levels of "oil and grease" (12 mg/l over 1 mg/l background	4000 bbl
Northern Technical Services, 1982	Beaufort Sea	Above ice disposal, 1–5 m water depth	Exploration	Up to 12,000 ppm Ba in sediments; commonly 1000–6000 ppm	No hydrocarbon increases attributable to drilling fluids	
Boothe and Presley, 1983	Gulf of Mexico	13–100	Production and Exploration	Up to 40,000 ppm; levels increase with water depth; deeper wells (100 m) contained 10–12% of total discharged Ba within a 500 m radius and 50–100% within a 3000 m radius; resuspension and transport effects shallower wells	Some evidence of hydrocarbon accumulation not quantified at a few platforms	600–10,000 m^3 cuttings discharged over times ranging from 2 months to 2 years
Bedinger, 1981	Gulf of Mexico (20 platforms)	6–75	Production		Some evidence of hydrocarbon accumulation (low incremental additions) around platforms	
Law and Blackman, 1981	North Sea		Exploration	Up to 800 ppm of Ba around rig	Up to 57 ppm (oil-based mud)	

(continued)

TABLE 6.4—*contd.*

Study	*Location*	*Water depth (m)*	*Well Type*	*Metals*	*Hydrocarbons*	*Amount of material discharged*
COWI Consultants, 1982	North Sea	37–42	Exploration	Up to 100 ppm of Ba within 500 m of discharge; elevated levels to 700 m away from rig	30–7000 ppm within 500 m of platform (7–95 time background) (one value of 7000 ppm; commonly 30 ppm)	500–1000 bbl per well
Pierce *et al.*, 1984	Gulf of Mexico (Flower Gardens)		Exploration		No petroleum in sediments attributable to discharges	27,000 liters
Ayers *et al.*, 1980	Gulf of Mexico		Exploration	90% of discharged fluid settled directly to the bottom		
Bothner *et al.*, 1983	Georges Bank	80, 140	Exploration	21–30% of discharged barium accounted for within 6 km of point of discharge (after 1 yr)		
Gettleson, 1980; Gettleson and Laird, 1980	Various		Exploration	3000 ppm barium within up to 100 m of rig (drilling fluids shunted to within 10 m of bottom) settlement of most solids within 1000 m of rig		
Anderson *et al.*, 1981	Buccaneer oil field	22	Production	Resuspension of fine-grained sediment does not allow for any significant deposition of metals; barium detected at elevated levels		
Middleditch, 1981a	Buccaneer oil field	20	Production		Low-level accumulation of *n*-alkanes noted around platforms; gradients away from platforms	

platforms employing oil-based muds. Hydrocarbon concentrations in sediments within 250 m of the platforms in some cases exceeded 1000 times the background concentration (i.e., as much as several thousand ppm), but the concentration gradient was very steep with background levels usually reached 2000 to 3000 m from the platform (Figure 6.10). It is interesting to note that Law and Blackman (1981) found a lack of correlation of barium and petroleum in sediments due to their differential postdischarge behavior.

In most massive offshore blowouts and spills (e.g., IXTOC-I, *Amoco Cadiz*), little direct sedimentation of petroleum hydrocarbons occurred offshore. Boehm and Fiest (1981) estimated that 1% of the spilled oil from the IXTOC-I blowout may have been sedimented in the immediate blowout region, accounting for perhaps a doubling of sedimentary hydrocarbon levels (50 to 100 ppm). Gundlach *et al.* (1983) estimated that 8% of the *Amoco Cadiz* oil was transported to offshore sediments resulting in 200 ppm levels. Maximum transport to the bottom occurs after stranding of the oil followed by its erosion. As previously stated, maximum sedimentation also occurs where oil mixes with high (>100 ppm) concentrations of suspended particles.

The oil spill during the Santa Barbara Channel blowout in 1969 resulted in large scale transport of oil to the sediments. Kolpack *et al.* (1971) estimated that resultant sediment oil concentrations were up to 1.4% (14 mg/g) of the sediment weight. The high levels of suspended particulate materials introduced to the area

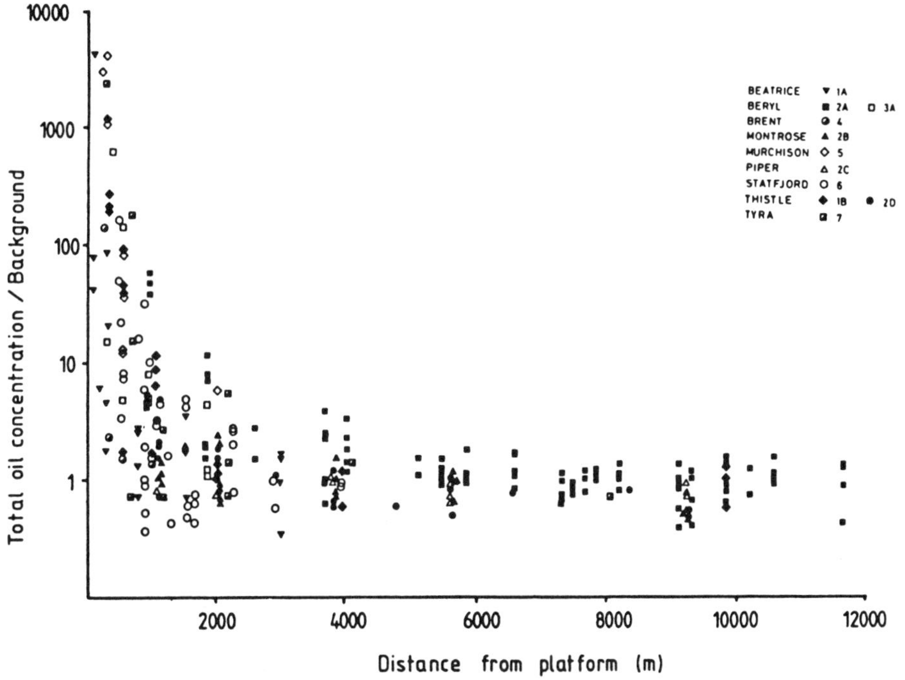

Figure 6.10. The relative concentration of total oil (related to background value) against distance from a production platform (from Davies *et al.*, 1984).

from runoff of the Santa Clara River provided a large source of sorbent material for the floating oil (Drake *et al.*, 1971).

In studies of the application of chemical dispersants to oil spills the following relevant points have come to light:

1. Effective chemical dispersion of oil reduces the affinity of oil for solid surfaces (i.e., particles) as long as the dispersant-oil micellar association persists (Mackay and Hossain, 1982).

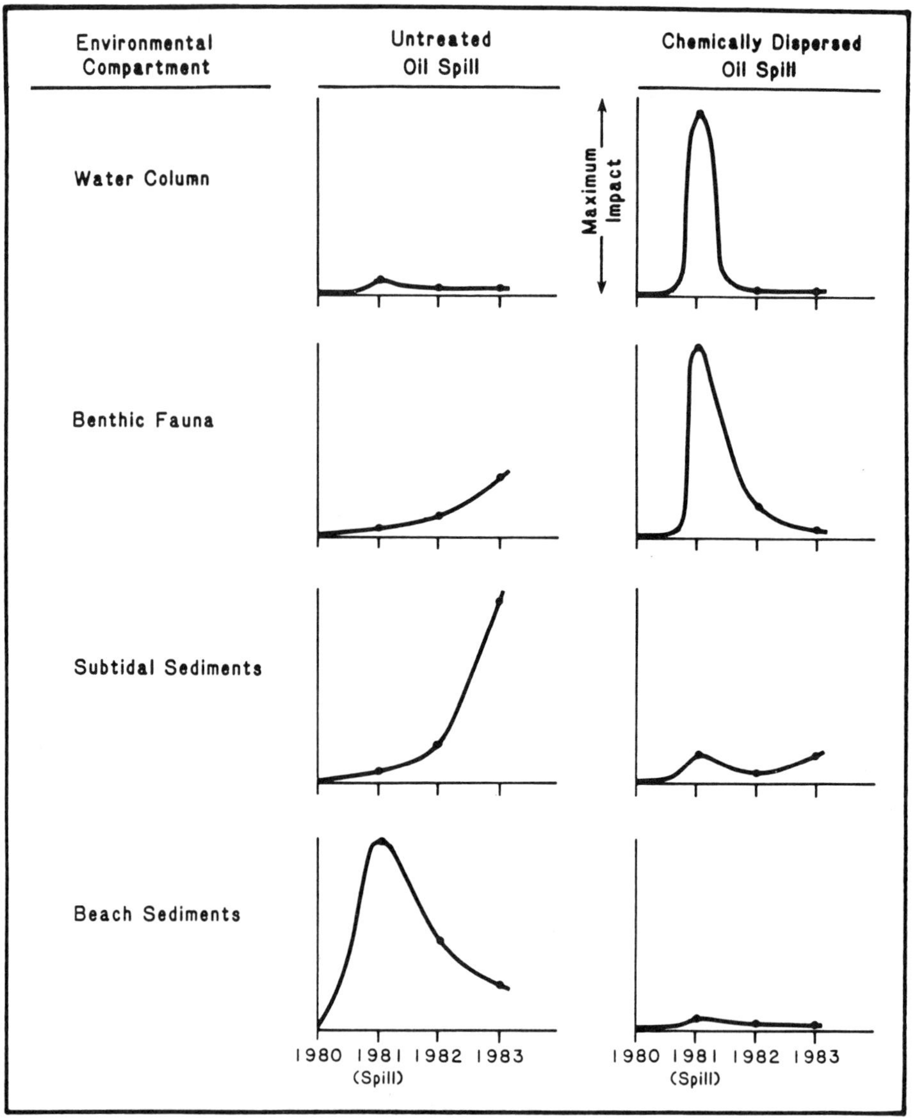

Figure 6.11. Summary of comparative fates of oil from the Baffin Island experimental oil spills (from Boehm *et al.*, 1985).

2. Effective chemical dispersion of oil greatly reduces the probability of beaching of an oil mass and hence erosion and offshore deposition.

3. Dispersion of oil may temporarily increase its bioavailability to exposed pelagic and benthic animals.

4. Studies (e.g., Gilfillan *et al.*, 1983; Page *et al.*, 1983; Boehm *et al.*, 1982a; Boehm, 1983) have suggested that the overall effect of dispersing oil prior to landfall (i.e., interaction with solid substrate) is to reduce the probability of any prolonged impact due to chronic exposure.

5. Boehm *et al.* (1985) have graphically summarized the relative impacts of chemically dispersed versus untreated, beached oil over the two years following the Baffin Island experimental oil discharges (Figure 6.11). After two years the large quantities of initially bioaccumulated oil from the chemical dispersion was nearly depurated. However, the chronic source of oil from introduction of the beached oil to subtidal sediments resulted in temporally increasing oil levels in sediments and animal tissues.

TRANSFORMATIONS OF DEPOSITED HYDROCARBONS AND METALS IN SEDIMENTS

Once hydrocarbons and trace metals are sedimented, their fate and, hence, their residence time and bioavailability is determined by physical, chemical and micro- and macrobiological processes.

Physical Processes

Physical mixing of surface sediment occurs due to bottom currents and storm-induced turbulence and can result in resuspension of sediments. This resuspension may affect a chemical exchange in the benthic boundary layer, wherein pollutants dissolved in interstitial waters are released while others may be sorbed or resorbed and sedimented. Redistribution of contaminants is also accomplished through mixing of surface sediment (5 to 15 cm within the sediment column) by bioturbation (Aller, 1977). This very important process results in: 1) irrigation of the sediment column resulting in possible dissolution of sediment-bound pollutants; 2) oxygenation of sediments (important to biodegradation; see Bartha and Atlas, Chapter 7); 3) changes in pollutant residence times in sediments as the residence time is a function of sedimentation rate, mixing rate and mixing depth; and 4) effects on the diagenesis of organic compounds and trace elements. Where sediments are not subjected to mixing due to wave and current-induced physical forces, the majority of sediment-to-water fluxes of solutes are mediated by bioturbation.

Overlying water mass movements can also influence organic and inorganic chemical distributions in sediments. For example, "wave pumping," or the oscillating hydrostatic pressure on the sea floor, can result in the movement of interstitial chemicals in surface sediments.

Chemical (Diagenetic) Processes

Chemical alterations that affect pollutants once sequestered in near-surface sediments include 1) dissolution of certain metals into interstitial waters, 2) precipitation of certain metals from soluble forms to insoluble sulfides and 3) dissolution of certain soluble nonpolar organic metabolites. In sediments, several factors favor the partitioning of pollutants into the aqueous, interstitial water phase. These include 1) a high solid to aqueous ratio causing saturation of interstitial water, 2) changes in redox conditions and 3) microbial activity.

Where the rate of bacterial decomposition of organic matter exceeds oxygen supply to the sediment column, changes in the Eh-pH conditions result which can make some elements more soluble (e.g., Mn) while others become less soluble (e.g., Cd) (Bender, 1976). Some metals such a Fe and Mn are more soluble in their divalent forms which prevail in anoxic zones of biologically mediated reactions (CO_2 to CH_4; NO_3^- to NH_3; $SO_4^=$ to H_2S). Once mobilized, divalent iron is precipitated as iron sulfide in abundance. The movement of these metals to the sediment-water interface through diffusion and compaction establishes concentration gradients which result in fluxes to the overlying water or reprecipitation in the oxic zone (e.g., Emerson *et al.*, 1979; Bischoff and Sayles, 1972). Remobilization is also enhanced by resuspension and bioturbation. Other metals such as Hg and Cd are immobilized in reducing sediments, but may be released through irrigation and, hence, oxygenation of the sediment column or during resuspension.

Organic compounds may enter interstitial waters and be returned to the overlying waters due to dissolution of soluble compounds, micellar solubilization in interstitial waters, or metabolic transformation followed by dissolution of a polar metabolite. Solubilities of organic compounds, in particular hydrocarbons, are an inverse function of molecular weight. One- and two-ringed aromatic hydrocarbons, if deposited, are readily lost from surface sediment layers due to dissolution and may be made bioavailable to animals exposed to these waterborne organics. Concentration gradients of organic compounds may also be established for organic compounds in sediments due to probable fluxes of compounds out of the sediments.

Biologically Mediated Transformations

Microbial processes are of great importance in determining the fate of some metallic and organic pollutants in surface sediments. The oxidation state of many trace metals (e.g., Cr, Mn) is biologically mediated. In addition, metals such as Hg, As, Se, S and the halogens may be biotransformed in volatile organic molecules (e.g., methylated species) under anoxic conditions.

Many studies have been published concerning the biodegradation of petroleum hydrocarbons (Chapter 7), and other organic pollutants. Given the availability of oxygen and nutrients, resident microbial populations will utilize hydrocarbons as substrates at varying rates. Sedimented oil was observed to be rapidly biodegraded in the *Amoco Cadiz* spill (Atlas *et al.*, 1981) and in the *Tsesis* spill (Boehm *et al.*, 1982b), while little biodegradation was evident from chemical results in the IXTOC-I blowout (Boehm and Fiest, 1982) and Baffin Island experiment spill (Boehm, 1983; Boehm *et al.*, 1985). Haines and Atlas (1982) determined that biodegradation of petroleum proceeds slowly in arctic environments with significant degradation occurring only after a year or more of environmental exposure. Petroleum hydrocarbons associated with sedimented drilling fluids would be expected to biodegrade with time given the proper nutrient and oxygen supplies, unless the compounds present inhibited degradation via their direct toxicity.

Once buried or mixed in the sediment below the oxic zone, which may be as little as several millimeters deep, no significant biodegradation will proceed due to limited oxygen availability (Winfrey *et al.*, 1982). Oxygenation by physical or biological processes would tend to accelerate biodegradation. There is evidence that bioturbation of marine sediments enhances oxygen irrigation and hence biodegradation of oiled sediments (Gordon *et al.*, 1978; Chapter 7). Studies examining the distributions of PAH in coastal and offshore sediments (e.g., Farrington *et al.*, 1983) suggest that PAH sources from petroleum are more readily degraded than associated PAH from pyrolytic inputs due to their availability to microbial populations.

Evidence exists (Boehm *et al.*, 1982a; Boehm, 1983) for the biodegradation of petroleum within the gut of arctic bivalves, owing probably to an indigenous, concentrated microbial population within the animal. These observations were made in an area where no chemical evidence of biodegradation was seen outside of the animals (i.e., in the sediments). It is not known whether this may represent a significant removal mechanism of oil from lightly contaminated substrates.

Weathering of Petroleum Hydrocarbons

The combined processes of evaporation, dissolution, microbial oxidation and photooxidation in addition to mediating physical processes result in an alteration of the chemical composition of petroleum. The composition is rapidly altered beyond the point where the oil can be definitively attributed to a particular source. The weathering of petroleum has been discussed at length in other reports (e.g., Jordan and Payne, 1980; Boehm, 1982b). As oil weathers, the following processes occur, thus changing the ultimate composition of petrogenic hydrocarbons reaching the sediments:

1. Loss of low boiling ($<C_{20}$) aromatic and saturated hydrocarbons through evaporation (Boehm and Fiest, 1982).
2. Loss of low boiling ($<C_{15}$) aromatic hydrocarbons through dissolution (Boehm *et al.*, 1982c).
3. An increased relative importance of unresolved naphthenic and naphthenoaromatic compounds (i.e., the unresolved complex mixture or UCM) (Farrington and Tripp, 1977).
4. An increased importance of highly branched aliphatic hydrocarbons (i.e, isoprenoids) relative to straight chain and singly methyl-branched molecules due to selective depletion of *n*-alkanes by biodegradation (Atlas *et al.*, 1981; Gundlach *et al.*, 1983).
5. An increased importance of alkylated (dimethyl to tetramethyl) phenanthrene and dibenzothiophene compounds relative to one- and two-ringed aromatics through combined weathering processes (Gundlach *et al.*, 1983; Overton *et al.*, 1981).

6. An increased importance of polycyclic aliphatic (e.g., pentacyclic triterpanes) compounds relative to all saturated components (Atlas *et al.*, 1981; Boehm *et al.*, 1982b).

The most abundant, persistent petroleum hydrocarbons that originate from spills or chronic discharges of crude oil, are the middle boiling range alkylated naphthalene, phenanthrene and dibenzothiophene compounds (Boehm, 1983; Boehm *et al.*, 1982b). Weathered petroleum and the benthic marine environment adjacent to weathering petroleum residues may contain oxidation products of hydrocarbons due to microbial oxidation.

FACTORS AFFECTING THE BIOAVAILABILITY OF SEDIMENT CONTAMINANTS

The transfer of hydrocarbons and trace metals associated with sediments to biological food webs is determined by their availability to the biota. The uptake of hydrocarbons by marine organisms can occur by three routes: 1) direct uptake from sea water, 2) uptake from sea water after partitioning into interstitial or boundary layer water or 3) direct uptake from sediment particles or from food. The efficiency of uptake from 1) or 2) is greater than from 3). Many studies (summarized in Neff and Anderson, 1981) have indicated that marine animals, especially bivalve molluscs, can acquire hydrocarbons directly from the aqueous phase both from the accommodated state (micellar droplets) and from the soluble state. There is less evidence for the direct uptake of hydrocarbons sorbed to sediment or in food (Neff and Anderson, 1981). Recent evidence (Boehm *et al.*, 1982a,b; Boehm, 1983) indicated that benthic detrital feeders (e.g., the bivalve *Macoma*) can acquire hydrocarbons from spilled oil via the water column and directly from sediments. Increases in sediment hydrocarbon levels in these studies due to offshore deposition of oil were paralleled by increased tissue levels. These acquired hydrocarbons were 1) not detectable in the water column and 2) indicative of a sorbed assemblage of hydrocarbons rather than a water soluble fraction.

A biogeochemical dilemma is encountered when one examines sediment and benthic animals from certain locations (e.g., New York Bight and Boston Harbor). Mussels from these areas contain a PAH mix, primarily consisting of compounds of petrogenic origin, whereas sediments from the same area contain predominantly pyrogenic PAH compounds (Farrington *et al.*, 1983) or a mix of petrogenic and pyrogenic compounds. Data for polychaetes yielded similar results (Farrington *et al.*, 1983). The authors suggest (Figure 6.3) that PAH from both petrogenic and pyrogenic origin enter the ecosystem. Pyrogenic PAH are more tightly sorbed to particles having been generated at high temperature as a part of soot. Petrogenic PAH enter the system in soluble, colloidal and loosely bound (sorbed) particulates. Thus, the petrogenic PAH are more readily available in all of these forms. Petrogenic PAH deposited in sediments are presumably more readily desorbed (with resuspension or bioturbation) or degraded by microorganisms thus accounting for a relative deficiency of petrogenic PAH in sediments.

In studying the bioconcentration of PAH in a benthic deposit-feeding clam (*Nucula*) in the New York Bight, Boehm (1982b) found that the ratio of levels of PAH in the tissues to those in the sediments increased as PAH molecular size increased. This generally directly followed the octanol-water partitioning coefficients of the aromatics thus indicating that the animals were acquiring pyrogenic PAH to a greater extent than petrogenic PAH. Polychaete worms on the other hand displayed equal "preference" for pyrogenic and petrogenic PAH and that the mode of uptake involved both sediment and water mediation. The inference here is that the uptake "situation" is highly variable and kinetically controlled. Animals can acquire hydrocarbons from both the sea water and sediment. Sediment input becomes very important where water column inputs are minor or when exposure times are long.

A number of geochemical factors affect the bioavailability of trace metals once deposited in sediments: 1) form of the metal (and hence solubility), 2) particle size distribution, 3) pH and Eh and 4) presence of "binding agents" such as Fe hydrous oxides and organic matter. Exposure to metals occurs in both food and particulate form and in solution. Concentrations of the free metal ion appear to control uptake from solution of Cd, Cu, Fe, Mn and Zn (Luoma, 1983). Thus, increasing salinity would decrease the free metal ion of Cd (Cd^{+2}) as chloride complexes are formed. Organic ligand availability apparently decreases bioavailability. Although the kinetics of bioavailability of metals from food and particles is much slower than from solution, the high concentrations of particulate metals increases the quantitative importance of this uptake "vector" (Luoma, 1983), although its exact significance is difficult to measure. Sediments are ingested by many animals. As with organics, the partitioning of metals into the organism from sediment depends on the binding strength of the metal and its equilibration time. Longer exposures to more slowly exchanging metals result in greater quantities of metal uptake (Luoma, 1983). Methylation of Hg, Sn and As greatly enhances their availability. Sediment particle size is important in bioavailability and uptake as it is known that the concentration of hydrocarbons and metals vary with the particle size fraction of a given sediment. Different benthic animals select different particle (sediment) sizes for food (Luoma, 1982).

One presently used method for predicting the maximal chemical uptake of sediment-associated metals and hydrocarbons considers a thermodynamic equilibrium between the lipids of benthic animals and the organic carbon reservoir in the sediments. This approach (Mackay, 1979; McFarland, 1983) implies that as the organic carbon content of the sediment decreases the maximal biotic pollutant concentration increases. Also, as the lipid content of the animal increases, the maximal biotic pollutant concentration increases:

$$\frac{\text{Animal Pollutant Concentration}}{\text{Animal Lipid Content}} \cong \frac{\text{Sediment Pollutant Concentration}}{\text{Sediment Organic Carbon Content}}$$

Several common hypotheses can be stated regarding the bioavailability of metals and organics:

1. Uptake from solution (interstitial or overlying waters) is more efficient

and rapid than uptake directly from sediments. Therefore processes favoring dissolution of certain metals (i.e., pH, Eh) and dissolution in general (bioturbation, resuspension) will tend to enhance bioavailability.

2. Metals and hydrocarbons are available directly from sediments, albeit more slowly than from solution. The availability is inversely proportional to the binding strength of the metal or hydrocarbon. Pyrogenic PAH are tightly sorbed and therefore less readily available. However, octanol-water partitioning data predicts that at equilibrium these larger PAH should be partitioned more strongly into animals.

3. Where exposure times are relatively short (days to weeks), kinetically regulated uptake mechanisms prevail (i.e., dissolution and aqueous phase uptake); where exposure times are relatively long (months to years), equilibrium based uptake mechanisms are most important. If an animal were exposed to contaminated sediments, presumably kinetic mechanisms would "eventually" be superseded by equilibrium mechanisms in determining the nature of metals and hydrocarbons in marine animals.

POSTDEPOSITIONAL TRANSPORT

The ultimate fate of contaminants associated with fine particles will depend on the exchange, transformation and burial processes described above, but also on the erosion, resuspension and advection of the particles from the site of initial deposition. Because contaminants are preferentially associated with finer sediment particles (e.g., clays) which may be cohesive, these transport processes are much more complex than if the particles were coarse silts or sands. Sediment cohesion results from 1) interparticle electrochemical attractions between clay mineral particles, 2) bonding by organic particles and 3) binding by mucal secretions of organisms and biogenic pelletization (Young, 1982). The shear stress on surface sediments (a function of current velocity, density of the particles and the fluid medium and particle diameter) determines whether the sediments are eroded from the bed. As the critical shear stress is reached, forces are sufficient to overcome cohesion, causing incipient motion and resuspension. Actually, though, the theoretical relationship which predicts erosion on the relationship between shear stress and surface sediment grain size properties (usually depicted in a Shields diagram; Madsen and Grant, 1976; Miller *et al.*, 1977) is oversimplified for practical application. Erodability is also influenced by the mix of sediment particles of different grain size, seabed roughness and biological effects (Rhoads *et al.*, 1978; Nowell *et al.*, 1981).

Bottom shear stress is affected by the combination of oscillatory currents caused by the passage of waves, tidal currents and residual nontidal currents resulting from geostrophic or meteorologically-forced flows. The absolute and relative importance of each of these current types varies widely among and within continental shelf environments. In most cases, geostrophic currents themselves are insufficient to erode bottom sediment, although they may add to the stress caused by the dominant waves, tides or storm-driven currents. Wave-induced oscillatory currents can cause large bottom stresses on shallow continental shelves. Storm swells can cause sediment resuspension in waters up to 100 m deep (Komar, 1969; Grant and Madsen, 1979; Cacchione and Drake, 1979; Butman and Moody, 1983). Tidal currents can also be very significant in certain environments, such as on Georges Bank (Emery and Uchupi, 1972; Knebel, 1981; Allen *et al.*, 1983). Intense flows driven by winds and the set-up of water against the coast during storms are also widely important on continental shelves (e.g., Butman *et al.*, 1979).

Particle Deposition

The particle settling velocity, the near-bed shear stress and the concentration of suspended particles affect deposition. Under conditions of low shear stress over smooth bottoms, a suspended particle that settles through the turbulent upper zone of the bottom boundary layer is trapped in the viscous sublayer flowing directly over the bed surface (McCave, 1970). Flow in this viscous sublayer is less turbulent and may be laminar. Deposition of fine particles which reach this layer is facilitated by the lack of turbulence. Models and observation of boundary layer deposition (e.g., Kline *et al.*, 1967) suggest that this sublayer periodically exchanges particles with the turbulent outerlayer.

The rate of collisions of particles and aggregates in the turbulent boundary layer may result in both the creation and destruction of aggregates (Einstein and Krone, 1962). The formation of large aggregates from smaller ones has been observed in the nepheloid layer of the Gulf of Mexico (Feely, 1976). However, shearing of aggregates by boundary layer flows may decrease deposition under turbulent regimes.

Bed Stability and Erosion

When the frictional fluid forces in the boundary layer (i.e., the bed shear stress) exceed the gravitational and cohesive forces, erosion commences. This point is known as the "critical shear stress" and may, under certain conditions, be defined by a "critical velocity" in the boundary layer. For noncohesive particles, gravity acting on the particles is the only force that must be overcome. The interparticle forces between cohesive sediments are much greater than the gravity forces on these fine-grained particles.

Many physical properties contribute to the "erosion resistance" of cohesive particles. These include: 1) water content (Postma, 1967), 2) clay mineralogy (Einselle *et al.*, 1974), 3) bed age (Young and Southard, 1978) accompanied by changes in water content and organic content, 4) organic content (Rhoads *et al.*, 1978; Young and Southard, 1978), 5) biogenic sediment modifications (Rhoads and Boyer, 1982; Eckman *et al.*, 1981) and 6) Bingham yield strength, or the force required to break bonds between aggregates of cohesive particles. The erosion resistance apparently increases with depth in the sediment. Boundary layer forces sufficient to erode the top several centimeters are often inadequate for eroding sediments deeper than this, of apparently higher cohesive strength (Partheniades, 1965).

Some potentially important pollutant "particulates" (e.g., tarballs, unaggregated fecal pellets) behave like noncohesive sediments and are therefore subject to different threshold conditions for their erosion. Particles of differing sizes and densities will undergo selective erosion and resuspension under the conditions prevalent in continental shelf environments wherein currents typically range from 10 to 50 cm/s. Particles, such as fecal pellets produced by marine invertebrates and organic detritus, will behave as noncohesive sediment and can be remobilized at lower threshold boundary layer currents than cohesive sediments (Fischer *et al.*, 1979; Nowell *et al.*, 1981).

The current velocity necessary for erosion of noncohesive (unconsolidated) sediment particles depends on the sediment particle size distribution and can be empirically depicted in a Hjulstrom diagram (Figure 6.12). The lowest velocities that may cause erosion of uncohesive particles are about 15 cm/s. Saila *et al.* (1972) indicated that during storms off the Rhode Island coast 25% of the larger waves were capable of producing bottom current speeds of 15 to 30 cm/s at 60-m depth.

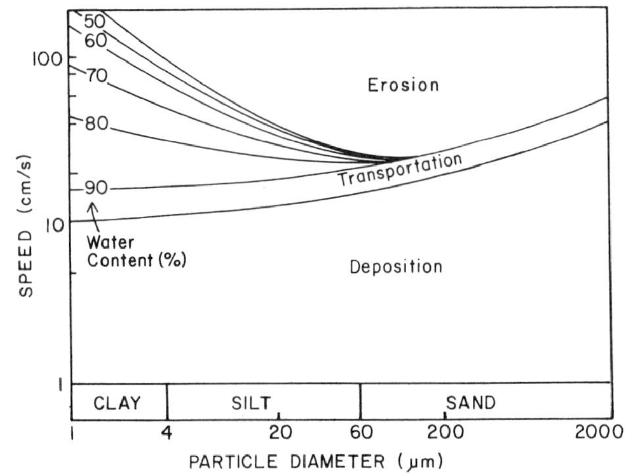

Figure 6.12. Relationship between current speed, particle diameter and sediment erosion, transport or deposition (after Kennett, 1982).

A number of studies have directly observed or measured the transport and resuspension of bottom sediments on the continental shelf under the influence of waves, currents and storms (e.g., Cacchione and Drake, 1982; Butman *et al.*, 1979; Butman and Moody, 1983; Lavelle *et al.*, 1978). On the other hand there have been few attempts to address the transport of drilling muds or cuttings as a function of bottom currents. In lower Cook Inlet, Alaska, maximum bottom currents reached 99 cm/s and were effective in eroding settled drilling mud solids (Dames & Moore, 1978). In 120 m of water in the Middle Atlantic Bight, EG&G (1982) found that bottom currents of 18 cm/s resulted in slow redistribution of settled drilling muds. At this depth the seabed was not influenced by wave-induced shear stresses. Bottom currents of 10 cm/s were able to resuspend flocs of drilling materials. Water depth is one of the most important parameters in predicting the postdepositional fate of particulate pollutants because it determines the strength of wave-induced bottom oscillatory currents which, working in consort with other currents, can cause large bottom stress resulting in sediment resuspension (Grant and Madsen, 1978).

In areas such as the Tanner Bank in Southern California (55 m depth), Meek and Ray (1980) calculated that even the largest particles in discharged drilling cuttings would be resuspended after initial sedimentation by an average steady bottom current of 20 cm/s plus an average wave-induced oscillatory current of 20 cm/s. The authors estimated that approximately 800 kg/day of deposited drilling cuttings in the vicinity of the drilling platform could be resuspended per meter of bottom width. This means that 6800 metric tons of deposited material could potentially have been resuspended and transported elsewhere during an 85-day period.

Factors other than water motion can profoundly affect sediment erosion. For example, in arctic marine environments, ice gouging or ice scour followed by sediment entrainment in sea ice are important processes of erosion and sediment transport in the shallow (2 to 5 m) water nearshore areas. In the Beaufort Sea those processes may extend considerable distances (2 to 16 km offshore). Ice gouging is reported to be capable of reworking sediments down to 20 to 50 cm in a period of 50 to 100 years (Barnes and Reimnitz, 1979).

Benthic organisms may also affect sediment erodibility. Biogenic activities of burrowing, deposit feeding and filter feeding affect the physical properties of sediments and, hence, their erodibility (Eckman *et al.*, 1981, Nowell *et al.*, 1981; Jumars and Nowell, 1984a,b; Grant *et al.*, 1982). Benthic organisms can affect erodibility and sediment transport via alteration of 1) fluid momentum impinging on the bed, 2) particle exposure to flow, 3) adhesion between particles and 4) particle momentum (Jumars and Nowell, 1984b). Organisms can increase cohesion through organic "glueing" (mucal binding through creation of tubes and mucal trails). Rhoads *et al.* (1975) estimated that critical erosion velocities increased 25 to 40% in flat beds 3 to 15 days after bacterial cultures were introduced into sediments. Sediments are also pelletized by deposit-feeding benthos. In some marine muds fecal pellets can form a large percentage of the bed surface deposit (Rhoads and Young, 1970). Decreased critical erosion velocities can be achieved where bioturbation increases the porosity and water content of sediments (Postma, 1967). Biogenic structures, such as tubes, affect the near-bottom turbulence field and, depending on their form and density, may either decrease or enhance erodibility (Rhoads and Young, 1970; Eckman *et al.*, 1981). In general, at low densities these features tend to promote erosion, while at high densities, microenvironments are formed with sheltered environments of reduced velocity and turbulence. The lee side of these features tend to accumulate suspended material.

Models

Mathematical modeling of sediment erosion processes is still in a developmental stage. Major advances have been made in sediment transport models (Grant and Madsen, 1982; Kachel, 1980; Van Rijn, 1981; Owen, 1977). Other recent advances in model development have been summarized by Nowell (1983). Observations of the velocity profile and turbulence in the benthic boundary layer have been difficult to obtain by direct measurement. These observations are critical in order to evaluate the bed shear stress which directly pertains to sediment erosion (Taylor and Dyer, 1977). Boundary layer models have been attempted to take into account nonplanar topography (Smith and McLean, 1977a), accelerating flow (Soulsby and Dyer, 1981; Grant and Madsen, 1982), wave-current interactions (Smith, 1977; Grant and Madsen, 1979), biogenic complications (Jumars *et al.*, 1981; Nowell *et al.*, 1981), stratification due to suspended sediments (Smith and McLean, 1977b; Soulsby and Dyer, 1981) and cohesion of fine sediments (Owen, 1977).

Young (1982) has pointed out that in order to apply boundary layer models to actual field predictions they must be coupled to regional circulation models, which are also in a stage of development (Allen, 1980; Winant, 1981; Weatherly and Martin, 1978). Recent observations of shelf flows (Mayer *et al.*, 1979; Cacchione and Drake, 1982; Larson *et al.*, 1981; Vincent *et al.*, 1981; Butman *et al.*, 1979) have contributed greatly to coupling theoretical models with actual field observations, although, to date, applications of models produce results that only roughly correspond to regional sedimentation patterns (Young, 1982). Until adequate methods are devised to establish a cohesive sediment bed in the laboratory, the question of comparisons of laboratory and field data will be difficult.

Contemporary studies of sediment transport—critical to understanding the fate of contaminants in continental shelf environments—involve physicists, geologists and biologists. An example is the High Energy Benthic Boundary Layer Experiment (HEBBLE), sponsored by the U.S. Office of Naval Research. HEBBLE developed and tested predictions about the response of cohesive sediments in a deep-sea (4820 m) location subject to episodic, swift currents. In a continental shelf environment, the Coastal Ocean Dynamics Experiment of the National Science Foundation, although primarily oriented to ocean physics, did assess the temporal and spatial variation in near-bottom flow, stress and sediment transport on the northern California shelf.

By using devices such as simple linear and circular flumes, empirical comparisons of laboratory and field data on critical erosion velocities have been made (Neumann *et al.*, 1970; Pierce *et al.*, 1970, Young and Southard, 1978). Young and Southard (1978) estimated τ_e, the erosional shear stress, and μ^*, the threshold shear velocity, through use of an *in situ* flume. Large variability in τ_e was observed between closely spaced field locations. Threshold shear velocity values obtained in the laboratory were found to be greater, by up to a factor of two, in laboratory flumes (Young and Southard, 1978). The authors noted that with time biological reworking decreased the τ_e values in the laboratory, making field and laboratory measurements more comparable. These temporal and spatial variations must be understood before laboratory and field measurements yield comparable results. Improvements in measurement techniques and instrumentation are needed to reduce experimental uncertainties.

Several theories have been advanced to describe the distribution of suspended sediment in the turbulent benthic boundary layer. These theoretical treatments (e.g., Hunt, 1954; Smith and Hopkins, 1972; Smith and McLean, 1977a,b, Smith, 1977) all assume that suspended sediment profiles can be described by sediment diffusion equations and Prandtl-Von Karman boundary layer equations for steady flow (Young, 1982). Agreement of Hunt's model and laboratory flume observations of suspended sand profiles is good.

ASSESSMENT STRATEGIES AND RESEARCH NEEDS

If one accepts the premise that long-term effects of oil and gas development in continental shelf environments are primarily important when 1) pollutant transport to the benthos occurs and 2) elevated pollutant levels associated with these activities persist, resulting in significant biological effects, then it follows that any desired assessment of these potential impacts hinges on the quantitative prediction of transport of pollutants to the benthos and the persistence of these pollutants in the sediments. Assessment studies should be designed to test the hypothesis: pollutants from a specific offshore oil and gas development activity will persist in the bioavailable zone of the sediment column.

Previous Assessment Studies

A number of assessments carried out in recent years have been directed at examining organic or metal distributions in offshore sediments surrounding: 1) exploratory drilling operations in shallow waters (Northern Technical Services, 1981), 2) exploratory drilling in deeper waters (EG&G, 1982; Boothe and Presley, 1983; Butman *et al.*, 1983), 3) production operations in shallow and intermediate depth waters (Bedinger, 1981; Ward *et al.*, 1979; Middleditch, 1981b; Boothe and Presley, 1983; Armstrong *et al.*, 1979; Anderson *et al.*, 1981) and 4) oil spills (Boehm, 1982c).

Central to the design of many of the studies has been the establishment of a series of stations surrounding existing exploratory or production platforms. Periodic "snapshots" of contaminant distributions in these sediments were then obtained through sediment sampling and analysis. Comparison of the distributions of metals and hydrocarbons around platforms and at presumed control or reference areas were the final outputs of the geochemical parts of the program. The results were often used to interpret biological data. The results from the Buccaneer Oil Field Study (Middleditch, 1981b), the Central Gulf of Mexico Platform Study (Bedinger, 1981), the Offshore Ecology Investigation (Ward *et al.*, 1979) primarily dealt with the time-averaged fate of discharged pollutants in sediments receiving inputs from production platforms. Linkages between actual discharged material and the inventory of contaminants in sediments were not directly established.

The general design of these studies is reviewed by Carney (Chapter 14), but, from a geochemical point-of-view, the designs suffered from the lack of measurement of several important parameters: amount and rate of materials discharged, ambient suspended particulate levels, amounts of particulate phase pollutants, the rate and areal extent of initial deposition, and pollutant/particle resuspension and transport (i.e., bottom current regime and inferred bed shear stress). The design of these studies seemed to assume that through a pattern or series of samples taken during or after possible pollutant inputs, one could infer a set of conclusions about dynamic fates and effects of various discharges. By the omission of these and other "process-related" parameters, the geochemical results of these studies are purely observational and do not support well predictive assessments of future offshore oil and gas-related activities.

Studies by EG&G (1982), Boothe and Presley (1983) and Armstrong *et al.* (1979) have been successful in illuminating important geochemically-based (transport and fate) issues through careful attention to: 1) measurement of process-related parameters, 2) assessment of temporal inputs and subsequent movement or burial of pollutant inputs and 3) application of background data and adequate data analytical techniques. The input of diagnostic metals (Boothe and Presley, 1983; EG&G, 1982) or diagnostic hydrocarbon components (Armstrong *et al.*, 1979; Boehm, 1982c) to the water column were evidence for the transport of these chemicals to the benthos, their persistence and associated biological effects.

Boothe and Presley (1983) focused on barium distributions in surface and near-surface sediments surrounding drilling platforms set in various water depths. The authors attempted to obtain integrated depositional "histories" (i.e., barium levels in the core) rather than focusing merely on recent deposition. The hypothesis suggested by the data, rather than directly tested in this study, was that long-term net deposition minus net erosion (i.e., accumulation) was related to water depth as influenced by bed shear stress (wave induced erosion) and settling as influenced by current dispersal. The EG&G (1982) study and the related study by Ayers *et al.* (1980) followed the depositional behavior of a plume of drilling fluid and resultant sediment accumulations as they varied with time. These temporal variations were evaluated in light of measured bottom current velocities.

A significant refinement to assessing the fate of drilling discharges was developed by Bothner *et al.* (1983) on Georges Bank. They improved the sensitivity of chemical measurements of barium in sediments by analyzing just the silt/clay-sized fraction of the sediment, thus improving the signal-to-noise ratio of their assessment of the amount of discharged drilling fluids associated with sediments. They were able to obtain a mass balance for barium released from a particular drilling platform. The Georges Bank monitoring program (Battelle/Woods Hole Oceanographic Institution, 1983; Bothner *et al.*, 1983) also used a regional grid of sampling stations in addition to two site-specific studies. The design of this program benefited from an extensive background knowledge of physical oceanographic and sediment transport processes in the area.

The study of "formation water" impacts on the shallow benthos of Trinity Bay (Armstrong *et al.*, 1979) was successful in that 1) a point-source, well-characterized impact was studied over an extended period of time, 2) there was analytical consistency between source characterization and environmental distribution measurements and 3) the data were evaluated in terms of a relation of specific toxic causal agents (naphthalenes not "petroleum hydrocarbons" as a general source) to specific biotal changes.

The studies which have been most useful in understanding the long-term fate and effects of drilling discharges have been designed taking into account: 1) the amount and chemical nature of the discharged material, 2) the water depth, current regime and depositional environments and 3) the postdepositional resuspension of deposited materials. These studies have led to hypotheses which can now be tested, but the ability to test these hypotheses is limited by a lack of data on 1) the full chemical characterization of drilling fluids, 2) the partitioning of metals and organics associated with drilling fluids into aqueous and particulate phases and 3) regional processes of sediment transport which control the long-term fate of deposited materials.

Ideally, such studies of the fate of contaminants which are themselves particulate or partition with particulate materials should integrate 1) sufficient reconnaissance to characterize the sedimentary regime in question; 2) statistically sound sampling of contaminant levels in sediments directed toward testing null hypotheses; 3) measurements of bottom sediment transport, or at least bottom currents; and 4) fate models including mass balance considerations.

Design of Future Studies

If one assumes that long-term effects on the benthos are directly linked to time-integrated contaminant exposure levels, then one could arrive at "predictive exposure" levels resulting from 1) drilling fluid discharge, 2) formation water discharge and 3) oil spill discharges or resulting contamination. This would necessitate predictions of the quantities of contaminants associated with these discharges in the sediments surrounding the point of discharge or area of impact. Studies of these three major discharge types should have laboratory and field verification components. The laboratory studies, which would involve the generation of new data and the evaluation of existing data, should focus on quantitatively defining the relevant processes presented in Table 6.5.

Field studies should be carefully designed to test hypotheses directly related to the long-term persistence of contaminants in sediments as influenced by physical transport processes, chemical transformations and bioaccumulation and depuration. Field studies should include the following:

1. Regional, spatial and temporal characterization of near-bottom stress, suspended and bottom sediments and physical oceanography.

2. Mass balance studies using actual discharges of drilling fluids and formation water to predict accumulation of contaminants in sediments. These studies should cover various water depths, be initiated prior to discharge and continue for a given amount of time after discharges cease. Environments should include those of low and high energy and with no previous oil and gas development and with continuing oil and gas development.

3. Use of field data to verify and refine depositional and erosional models (e.g., Brandsma *et al.*, 1980; Sauer, 1983).

4. Estimation of incremental addition of persistent contaminants to existing contaminants.

These field studies should include sampling of deposited material captured in particle interceptor traps and sectioned sediment cores, and should involve

TABLE 6.5
Study needs related to the long-term fate of sediment-associated contaminants resulting from offshore oil and gas development

Drilling fluids and cutting discharges

1. Flume studies on threshold movement of the settleable fraction of drilling fluid and drilling fluid ambient sediment mixtures.
2. Studies on threshold movement as a function of biotic activity (i.e., biological "packaging"; biogenic structures).
3. Studies on the vertical mixing of drilling fluids into the sediment column and then resuspension due to bioturbation.
4. Microscopic studies of the postdepositional "packaging" and other physical transformations of drilling fluid particles.
5. Transformation of heavy metals and organics from sorbed to dissolved state in oxic and anoxic sediments.
6. Partitioning of heavy metals and organics between aqueous and solid phases during mixing with sea water as a function of drilling fluid concentrations in sea water.
7. Studies on the association of metals and organics with different particle sizes within the drilling fluid mixture.
8. Characterization of organic compounds other than nonhydrocarbons in drilling fluids.
9. Quantification of formation-related hydrocarbons (shales) that may be introduced and deposited with drill cuttings, and the postdischarge partitioning of these formation hydrocarbons between sea water and particles.

Formation water discharges

1. The full characterization of organic compounds and metals in a variety of "typical" formation waters.
2. The partitioning of these compounds between particulate and aqueous phases, for different formation waters and particle concentrations.
3. Studies on the postdepositional fate of formation water contaminants, including biodegradation and partitioning into sea water.
4. Bioavailability and depuration by filter-feeding and detrital-feeding bivalves.

Oil spills and discharges

1. Studies on the association of petroleum hydrocarbons with different particle types and concentrations as a function of oil concentration and type (heavy and light crudes).
2. Mass balance studies to quantify the amounts of oil expected to be sedimented under differing suspended particulate concentrations.
3. Rates of hydrocarbon degradation as a function of 1) sediment oxygenation, 2) background hydrocarbon concentration, 3) background microbial populations abundance and 4) hydrocarbon molecular type.
4. Release of sedimented petroleum compounds into pore waters and overlying waters.
5. Bioavailability and depuration by filter-feeding and detrital-feeding bivalves.

fractionation of the sediments according to particle sizes. Scanning electron microscopy should be employed to examine the nature of the particles and the biogenic "packaging" of deposited particles. A sediment profiling camera (Rhoads and Cande, 1971; Rhoads and Germano, 1982) may be deployed as well in drilling fluid assessments to rapidly assess the extent of areal coverage of discharged and deposited muds and cuttings. Chemical measurements should focus on:

1. heavy metals - (fine fraction only in sandy sediments) as potential tracers of discharged material and as toxic components,
2. polynuclear aromatic compounds, especially medium molecular weight (2-3 rings) PAH and heterocyclic sulfur compounds, which are the most persistent of the petrogenic organics,
3. lignosulfonate polymers as tracers of the fine fraction of drilling fluids,
4. other polymers or characterized compounds in drilling fluids or formation waters (e.g., phenols, biocides), and
5. pentacyclic triterpanes and steranes as tracers of crude oil or shale from formation cuttings as differentiated from drilling mud additives such as diesel or mineral oils.

LITERATURE CITED

Aizenshtat, A. 1973. Perylene and its geochemical significance. Geochim. Cosmochim. Acta 37: 559-567.

Allen, J.S. 1980. Models of wind-driven currents on the continental shelf. Amer. Rev. Fluid Mech. 17: 389-433.

Aller, R.C. 1977. The influence of macrobenthos on chemical diagenesis of marine sediments. Ph.D. Thesis, Yale University, New Haven, Connecticut.

Aller, R.C., L.K. Benninger and J.K. Cochran. 1980. Tracking particle-associated processes in nearshore environments by use of $^{234}Th/^{238}U$ disequilibrium. Earth Planet. Sci. Lett. 47: 151-175.

Anderson, J.B., R.B. Wheeler and P.R. Schwarzer. 1981. Sedimentology and geochemistry of recent sediments. Pages 59-67 *in* B.S. Middleditch (ed.), Environmental Effects of Offshore Oil Production. The Buccaneer Gas and Oil Field Study. Plenum Press, New York.

Armstrong, H.W., K. Fucik, J.W. Anderson and J.M. Neff. 1979. Effects of oilfield brine effluent on sediments and benthic organisms in Trinity Bay, Texas. Mar. Environ. Res. 2: 55-69.

Atlas, R.M., P.D. Boehm and J.A. Calder. 1981. Chemical and biological weathering of oil from the *Amoco Cadiz* oil spillage, within the littoral zone. Estuar. Coastal Mar. Sci. 12: 589-608.

Avignan, J. and M. Blumer. 1968. On the origin of pristane in marine organisms. J. Lipid Res. 9: 350-352.

Ayers, R.C., T.C. Sauer, D.C. Stuebner and R.P. Meek. 1980. An environmental study to assess the impact of drilling fluids on water quality parameters during high rate, high volume discharges to the ocean. Page 351 *in* Research on Environmental Fate and Effects of Drilling Fluids and Cuttings, Volume I. American Petroleum Institute, Washington, D.C.

Barnes, P. and D. Fox. 1979. Sediment-laden first year sea ice, central Beaufort Sea, Alaska. Pages 1E-3E *in* P. Barnes and E. Reimnitz (eds.), Marine Environmental Problems in the Ice Covered Beaufort Sea Shelf and Coastal Regions. National Oceanic and Atmospheric Administration/Outer Continental Shelf Environmental Assessment Program, Annual Report, R.U. 205.

Barnes, P. and E. Reimnitz. 1979. Ice Gouge Obliteration and Sediment Redistribution Event: 1977-1978, Beaufort Sea, Alaska. U.S. Geological Survey, Open-File Report 78-848, 22 p.

Barnes, P., E. Reimnitz and D. Drake. 1977. Marine Environmental Problems in the Ice Covered Beaufort Sea Shelf and Coastal Regions. National Oceanic and Atmospheric Administration/Outer Continental Shelf Environmental Assessment Program, Annual Report, R.U. 205, 229 p.

Battelle/Woods Hole Oceanographic Institution. 1983. Georges Bank Infauna Monitoring Program. Final Rept., Year 1, Contract No. 14-12-001-29192, U.S. Dept. of Interior, Minerals Management Service, New York OCS Office, New York, 153 p.

Bedinger, C.A., Jr. (ed.). 1981. Ecological Investigations of Petroleum Production Platforms in the Central Gulf of Mexico. Volume I, Pollutant Fate and Effects Studies. Final Report to Bureau of Land Management, Contract No. AA551-CT8-17, New Orleans, Louisiana. Southwest Research Institute, San Antonio, Texas.

Bender, M.L. 1976. Sediment/sea interface exchange processes. Pages 14-17 *in* H.W. Windom and R.A. Duce (eds.), Marine Pollutant Transfer. D.C. Heath & Co., Lexington, Massachusetts.

Biggs, R.B., J.H. Sharp, T.M. Church and J.M. Tramontano. 1983. Optical properties, suspended sediment, and chemical associations with the turbidity maximum of the Delaware estuary. Can. J. Fish. Aquat. Sci. 40: 172-192.

Biscaye, P.E. and C.R. Olsen. 1976. Suspended particulate concentrations and compositions in the New York Bight. Pages 124-137 *in* M.G. Gross (ed.), Middle Atlantic Continental Shelf and the New York Bight. Amer. Soc. Limnol. Oceanogr. Special Symposium, Vol. 2: 1-441.

Bischoff, J.L. and F.L. Sayles. 1972. Pore fluid and mineralogical studies of recent marine sediments of 35° to 45°N, Gibraltar to mid-Atlantic ridge. Jour. Sed. Petrol. 42: 711-724.

Blumer, M. 1976. Polycyclic aromatic compounds in nature. Sci Am. 234: 34-45.

Blumer, M. and D.W. Thomas. 1965a. Phytodienes in zooplankton. Science 147: 1148-1149.

Blumer, M. and D.W. Thomas. 1965b. Zamene, isometric C_{19} monoolefins from marine zooplankton, fishes, and mammals. Science 148: 370-371.

Blumer, M., J. Gordon, J.C. Robertson and J. Sass. 1969. Phytol-derived C_{19} and di- and tri-olefinic hydrocarbons in marine zooplankton and fishes. Biochem. 8: 5067-4074.

Blumer, M., R.R.L. Guillard and T. Chase. 1971. Hydrocarbons of marine phytoplankton. Mar. Biol. 8: 183-189.

Blumer, M., M.M. Mullin and D.W. Thomas. 1963. Pristane in zooplankton. Science 140: 974.

Boehm, P.D. 1982a. Organic Pollutant Transformation and Bioaccumulation of Pollutants in the Benthos from Waste Disposal Associated Sediments. Annual Technical Report, NOAA Grant NA81RAD0020, National Oceanic and Atmospheric Administration/ Ocean Assessments Division, Rockville, Maryland.

Boehm, P.D. 1982b. Petroleum in the Marine Environment: Physical/Chemical Methods Background. Paper for National Academy of Sciences Publication Update. *In* Petroleum in the Marine Environment. National Academy of Sciences, Washington, D.C., 152 p.

Boehm, P.D. (ed.). 1982c. IXTOC Oil Spill Assessment. Final Report, Contract AA851-CTO-71. Bureau of Land Management, New Orleans, Louisiana.

Boehm, P.D. 1983. Long-term fate of crude oil in the Arctic nearshore environment - The BIOS experiments. Pages 280-291 *in* Proceedings of the Sixth Arctic Marine Oil Spill Program Technical Seminar. Environmental Protection Service, Ottawa, Canada.

Boehm, P.D. 1984. Aspects of hydrocarbon geochemistry of recent sediments in the Georges Bank Region. I. Saturated Hydrocarbons. Org. Geochem. 7: 11-23.

Boehm, P.D. and J.W. Farrington. 1984. Aspects of the polycyclic aromatic hydrocarbon geochemistry of recent sediments in the Georges Bank region. Env. Sci. Tech. 18: 840-845.

Boehm, P.D. and D.L. Fiest. 1980. Hydrocarbon Biogeochemistry - Biological and Chemical Survey of Texoma and Capline Sector Salt Dome Brine Disposal Sites off Louisiana. Final Report, National Oceanic and Atmospheric Administration Contract No. 03-78-D08-022, National Marine Fisheries Service, Galveston, Texas.

Boehm, P.D. and D.L. Fiest. 1981. Aspects of the transport of petroleum hydrocarbons to the offshore benthos during the IXTOC-I blowout in the Bay of Campeche. Pages 207-236 *in* Proceedings of a Symposium on Preliminary Results from the September 1979 *Researcher/Pierce* IXTOC-1 Cruise; (June 9-10, 1980, Key Biscayne, Florida). Office of Marine Pollution Assessment, National Oceanic and Atmospheric Administration, Rockville, Maryland.

Boehm, P.D. and D.L. Fiest. 1982. Subsurface distributions of petroleum from an offshore well blowout: The IXTOC-I blowout, Bay of Campeche. Env. Sci. Tech. 15: 67-74.

Boehm, P.D., D.L. Fiest, P. Hirtzer, L. Scott, R. Nordstrom and R. Engalhardt. 1982a. A biogeochemical assessment of the BIOS experimental spills: Transport pathways and fates of petroleum in benthic animals. Pages 581-618 *in* Proceedings of the Fifth Arctic Marine Oil Spill Program Technical Seminar. Environmental Protection Service, Ottawa, Canada.

Boehm, P.D., D.L. Fiest and A. Elskus. 1982b. Comparative weathering patterns of hydrocarbons from the *Amoco Cadiz* oil spill observed at a variety of coastal environments. *In* Proceedings, International on the *Amoco Cadiz* Fates and Effects of the Oil Spill. Brest, France. November 19-22, 1979. Centre National pour l'Exploration des Oceans, Brest.

Boehm, P.D., D.L. Fiest, D. Mackay and S. Paterson. 1982c. Physical chemical weathering of petroleum hydrocarbons from the Ixtoc-I blowout: Chemical measurements and a weathering model. Env. Sci. Tech. 16: 498-505.

Boehm, P.D., J.E. Barak, D.L. Fiest and A.A. Elskus. 1982d. A chemical investigation of the transport and fate of petroleum and hydrocarbons in littoral and benthic environments: The *Tsesis* oil spill. Mar. Env. Res. 6: 157-188.

Boehm, P.D., W. Steinhauer, A. Roquejo, O. Cobb, S. Duffy and J. Brown. 1985. Comparative fate of chemically dispersed and untreated oil in the Arctic: Baffin Island Oil Spill Studies 1980-1983. Pages 561-569 *in* Proceedings 1985 Oil Spill Conference. American Petroleum Institute, Washington, D.C.

Boothe, P.N. and B.J. Presley. 1983. Distribution and behavior of drilling fluid and cuttings around Gulf of Mexico drill sites. Draft Final Report. API Project No. 243. American Petroleum Institute, Washington, D.C.

Bothner, M.H., C.M. Parmenter and J.D. Milliman. 1981. Temporal and spatial variations in suspended matter in continental shelf and slope waters off the northeastern United States. Est. Coast. Shelf Sci. 13: 213-234.

Bothner, M.H., R.R. Rendigs, E. Campbell, M.W. Doughten, C.M. Parmenter, M.J. Pickering, R.G. Johnson and R.J. Gillison. 1983. The Georges Bank Monitoring Program: Analysis of Trace Metals in Bottom Sediments during the Second Year of Monitoring. Final Report to U.S. Minerals Management Service. U.S. Geological Survey, 88 p.

Bothner, M.H., E.C. Spiker, P.P. Johnson, R.R. Rendigs and J.P. Aruscavage. 1981. Geochemical evidence for modern sediment accumulation on the continental shelf off southern New England. J. Sed. Petrol. 51: 281-292.

Brandsma, M.G. and R.C. Sauer. 1983. The OOC Model: Prediction of short term fate of drilling fluids in the ocean. Part Two: Model results. *In* Proceedings of Minerals Management Service Workshop on Discharges Modeling, February 7-10, 1983, Santa Barbara, California.

Brandsma, M.G., L.R. Davis, R.C. Ayers and T.C. Sauer. 1980. Computer model to predict the short-term fate of drilling discharges in the marine environment. Page 588 *in* Research on Environmental Fate and Effects of Drilling Fluids and Cuttings, Volume I. American Petroleum Institute, Washington, D.C.

Breteler, R.J., P.D. Boehm, J.M. Neff and A.G. Requejo. 1985. Acute Toxicity of Drilling Muds Containing Hydrocarbon Additives and the Fate and Partitioning among Liquid, Suspended and Solid Phases. Final Report to American Petroleum Institute, Washington, D.C., 107 p.

Britch, R.P. 1976. Tidal Currents in Knik Arm, Cook Inlet, Alaska. M.S. Thesis, University of Alaska, Institute of Marine Resources, Fairbanks, Alaska.

Brooks, J., D. Wiesenburg, A. Schwab, R. Claude, E.L. Estes and R.F. Shokes. 1981. Surficial sediments and suspended particulate matter. Pages 69-115 *in* B.S. Middleditch (ed.), Environmental Effects of Offshore Oil Production. The Buccaneer Oil and Gas Field Study. Plenum Press, New York.

Brown, R.A. and F.T. Weiss. 1978. Fate and Effects of Polynuclear Aromatic Hydrocarbons in the Aquatic Environment. API, Environmental Affairs Department, Publication No. 4297. American Petroleum Institute, Washington, D.C., 23 p.

Burbank, D.C. 1974. Suspended Sediment Transport and Deposition in Alaskan Coastal Waters. University of Alaska, Institute of Marine Science, 222 pp.

Butman, B. and D.W. Folger. 1979. An instrument system for long-term sediment transport studies on the continental shelf. J. Geophys. Res. 84: 1215-1220.

Butman, B., M.A. Noble and D.W. Folger. 1979. Long-term observations of bottom current and bottom sediment movement on the mid-Atlantic continental shelf. J. Geophys. Res. 84: 1187-1205.

Butman, B. and J.A. Moody. 1983. Observations of bottom currents and sediment movement along the U.S. East Coast Continental Shelf during winter. Chapter 7 *in* B.A. McGregor (ed.), Environmental Geologic Studies on the United States Mid and North Atlantic Outer Continental Shelf Area, 1980-1983, Volume III. Final Report to U.S. Bureau of Land Management, New York.

Butman, B., M.A. Noble, J.A. Moody and M.H. Bothner. 1983. Lydonia Canyon Dynamics Experiment Progress Report, Preliminary Results. Chapter 8 *in* B.A. McGregor (ed.), Environmental Geologic Studies on the United States Mid and North Atlantic Outer Continental Shelf Area, 1980-1982. Rept. to U.S. Geological Survey.

Cacchione, D.A. and D.E. Drake. 1979. A new instrument system to investigate sediment dynamics on continental shelves. Mar. Geol. 30: 229-312.

Cacchione, D.A. and D.E. Drake. 1982. Measurements of storm generated bottom stresses on the continental shelf. J. Geophys. Res. 87: 1952-1960.

Carmody, D.J., J.B. Pearce and W.E. Yasso. 1973. Trace metals in the sediments of the New York Bight. Mar. Poll. Bull. 9: 132-135.

Clark, R.C., Jr. 1966. Occurrence of Normal Paraffin Hydrocarbons in Nature (Unpublished Manuscript). Woods Hole Oceanographic Institution Technical Report 66-34.

Clark, R.C., Jr. and D.W. Brown. 1977. Petroleum: Properties and analyses in biotic and abiotic systems. Pages 1-89 *in* D.C. Malins (ed.), Effects of Petroleum on Arctic and Subarctic Marine Environments and Organisms. Volume 1. Academic Press, New York.

Collins, A.G. 1975. Geochemistry of Oilfield Waters. Developments in Petroleum Science 1. Elsevier Publishing Co., New York, 946 p.

COWI Consultants. 1982. Environmental Impact of Discharge of Cuttings when Drilling with Low Toxic Oil-Based Mud at the Tyra Field, North Sea (unpublished report). Dansk Boreselskab A/S, November.

Crippen, R.W., S.L. Hood and G. Greene. 1980. Metal levels in sediment and benthos resulting from a drilling fluid discharge into the Beaufort Sea. Page 636 *in* Research on Environmental Fate and Effects of Drilling Fluids and Cuttings, Volume I. American Petroleum Institute, Washington, D.C.

Dames & Moore. 1978. Drilling Fluid Dispersion and Biological Effects Study for the Lower Cook Inlet C.O.S.T. Well. Report submitted to ARCO, Anchorage, Alaska.

Davies, J.M., J.M. Addy, R.A. Blackman, J.R. Blanchard, J.E. Ferbache, D.C. Moore, H.J. Somerville, A. Whitehead and T. Wilkinson. 1984. Environmental effects of the use of oil-based drilling muds in the North Sea. Mar. Pollut. Bull. 15: 363-370.

Davis, J.B. 1968. Paraffinic hydrocarbons in the sulfate-reducing bacterium *Desulfovibrio desulfuricans*. Chem. Geol. 3: 155-160.

Dayal, R., M.G. Heaton, M. Fuhrman and I.W. Duedall. 1981. A Geochemical and Sedimentological Study of the Dredged Material Deposit in the New York Bight. National Oceanic and Atmospheric Administration, Tech. Memo., OMPA-3, Dept. of Commerce, 174 p.

Drake, D.E. 1977. Suspended particulate matter in the New York Bight Apex, Fall 1973. J. Sediment. Petrol. 47: 209-228.

Drake, D.E., P. Fleischer and R.L. Kolpak. 1971. Transport and deposition of flood sediment, Santa Barbara Channel, California. Pages 181-217 *in* R.L. Kolpak (ed.), Biological and Oceanographical Survey of the Santa Barbara Channel Oil Spill 1969-1970. Volume II. Univ. of Southern California, Los Angeles, California.

Eckman, J.E., A.R.M. Nowell and P.A. Jumars. 1981. Sediment destabilization by animal tubes. J. Mar. Res. 72: 132-143.

EG&G. 1982. A Study of Environmental Effects of Exploratory Drilling on the Mid-Atlantic Outer Continental Shelf - Final Report of the Block 684 Monitoring Program. Prepared for Offshore Operators Committee under the supervision of Exxon Production Research Company, P.O. Box 2189, Houston, Texas.

Eganhouse, R.P., B.R.T. Simoneit and I.R. Kaplan. 1981. Extractable organic matter in urban stormwater runoff. 2. Molecular characterization. Environ. Sci. Technol. 15: 315-326.

Einselle, G., R. Overbect, H.U. Schwarz and G. Unsold. 1974. Mass physical properties, sliding and erodibility of experimentally deposited and differently consolidated clayey muds. Sedimentology 21: 339-372.

Einstein, H.A. and R.B. Krone. 1962. Experiments to determine modes of cohesive sediment transport in salt water. J. Geophys. Res. 67: 1451-1464.

Elder, D.L. and S.W. Fowler. 1977. Polychlorinated biphenyls: Penetration into the deep ocean by zooplankton fecal pellet transport. Science 197: 459-461.

Elmgren, R., S. Hansson, V. Larsson, B. Sundelin and P.D. Boehm. 1983. The *Tsesis* oil spill: Acute and long-term impact on the benthos. Mar. Biol. 73: 51-65.

Emerson, S., J. Orr, L. Jacobs and R. Jahnke. 1979. The oxidation rate of Mn^{+2} in an anoxic fjord and in estuarine sediments. Extended abstract *in* Proceedings of workshop: Processes Determining the Input, Behavior and Fate of Radionuclides and Trace Elements in the Ocean, with Particular Reference to Continental Shelf Environments. March 7-9, 1979, Sponsored by the U.S. Dept. of Energy, Washington, D.C.

Emery, K.O. and E. Uchupi. 1972. Western North Atlantic: Topography, Rocks, Structure, Water, Life, and Sediments. American Assoc. Petroleum Geologists, Tulsa, Oklahoma, 532 p.

Ewing, M. and E. Thorndike. 1965. Suspended matter in deep ocean water. Science 147: 1291-1294.

Farrington, J.W. and B.W. Tripp. 1977. Hydrocarbons in western North Atlantic surface sediments. Geochim. Cosmochim. Acta. 41: 1627-1641.

Farrington, J.W., J.M. Teal, B.W. Tripp, J.B. Livramento and A. McElroy. 1983. Biogeochemistry of Petroleum Components at the Sediment-Water Interface. Final Report, U.S. Dept. of Energy, Div. Biomedical-Environmental Programs under Contract DE-AC02-77EVO4256 A007, and U.S. Dept. of Interior, Bureau of Land Management.

Feely, R.A. 1976. Evidence of aggregate formation in a nepheloid layer and its possible role in the sedimentation of particulate matter. Mar. Geol. 20: M7-M13.

Fischer, J.S., J. Pickral and W.E. Odum. 1979. Organic detritus particles, initiation of motion criteria. Limnol. Oceanogr. 23: 529-532.

Freeland, G.L., D.J.P. Swift and R.A. Young. 1979. Mud deposits near the New York Bight dumpsites: Origin and behavior. Pages 73-95 *in* H.D. Palmer and M.G. Gross (eds.), Ocean Dumping and Marine Pollution: Geological Aspects of Waste Disposal. Dowden, Hutchinson and Ross, Inc., Stroudsburg, Pennsylvania.

Gettleson, D. 1980. Effects of oil and gas drilling operations on the marine environment. Pages 371-411 *in* E. Geyer (ed.), Marine Environmental Pollution. 1. Hydrocarbons. Elsevier Oceanography Series 27A, Elsevier Scientific Publishing Co., New York.

Gettleson, D.A. and C.E. Laird. 1980. Benthic barium levels in the vicinity of six drill sites in the Gulf of Mexico. Page 739 *in* Research on Environmental Fate and Effects of Drilling Fluids and Cuttings, Volume I. American Petroleum Institute, Washington, D.C.

Gilfillan, E.S., S.A. Hanson, D. Vallas, R. Gerber, D.S. Page, J. Foster, J. Hotham and S.D. Pratt. 1983. Effect of spills of dispersed and nondispersed oil on intertidal infaunal community structure. Pages 457-464 *in* Proceedings 1983 Oil Spill Conference. American Petroleum Institute, Washington, D.C.

Gordon, D.C., Jr., J. Dale and P.D. Keizer. 1978. Importance of sediment working by the deposit-feeding polychaete *Arenicola marina* on the weathering rate of sediment-bound oil. J. Fish. Res. Board Can. 35: 591-603.

Grahl-Nielson, O., S. Sunby, K. Westrheim and S. Wilhemlsen. 1980. Petroleum hydrocarbons in sediment resulting from drilling discharges from a production platform in the North Sea. Page 541 *in* Research on Environmental Fates and Effects of Drilling Fluids and Cuttings, Volume I. American Petroleum Institute, Washington, D.C.

Grant, W.D. and O.S. Madsen. 1979. Combined wave and current interaction with a rough bottom. J. Geophys. Res. 84: 1797-1808.

Grant, W.D. and O.S. Madsen. 1982. Moveable bed roughness in unsteady oscillatory flow. J. Geophys. Res. 87: 469-482

Grant, W.D., L.F. Boyer and L.P. Stanford. 1982. The effect of biological processes on the initiation of sediment motion in non-cohesive sediments. J. Mar. Res. 40: 659-677.

Grose, P.L., J.S. Mattson and H. Petersen. 1979. USNS *Potomac* Oil Spill Melville Bay, Greenland, 5 August 1977. National Oceanic and Atmospheric Administration, Washington, D.C.

Gschwend, P.M. and R.A. Hites. 1981. Fluxes of polycyclic aromatic hydrocarbons to marine and lacustrine sediments in the northeastern United States. Geochim. Cosmochim. Acta 45: 2359-2367.

Gundlach, E.R., P.D. Boehm, M. Marchand, R.M. Atlas, D.M. Ward and D.A. Wolfe. 1983. Fate of *Amoco Cadiz* oil. Science 221: 122-129.

Haines, J.R. and R.M. Atlas. 1982. *In situ* microbial degradation of Prudhoe Bay crude oil in Beaufort Sea sediments. Mar. Environ. Res. 7: 91-102.

Han, J. and M. Calvin. 1969. Hydrocarbon distribution of algae and bacteria and microbiological activity in sediments. Proc. National Acad. Sci. 64: 436-443.

Harris, W.H. 1976. Spatial and temporal variation in sedimentary grain-size facies and sediment heavy metal ratios in the New York Bight Apex. Pages 102-123 *in* M.G. Gross (ed.), Middle Atlantic Continental Shelf and the New York Bight. Amer. Soc. Limnol. Oceanogr. Special Symposium, Vol. 2: 1-441.

Hausknecht, K.A. 1980. Describe surficial sediments and suspended particulate matter. Volume V, Page 56 *in* W.B. Jackson and G.M. Faw (eds.), Biological/Chemical Survey of Texoma and Capline Sector Salt Dome Brine Disposal Sites off Louisiana, 1978-1979. National Oceanic and Atmospheric Administration. Technical Memorandum NMFS-SEFC-29. Available from National Technical Information Service, Springfield, Virginia.

Hites, R.A. and W.G. Biemann. 1975. Identification of specific organic compounds in a highly anoxic sediment by gas chromatographic-mass spectrometry and high resolution mass spectrometry. Pages 188-201 *in* R.P. Gibbs (ed.), Analytical Methods in Oceanography. Advances in Chemistry Series No. 147. American Chemical Society, Washington, D.C.

Hites, R.A., R.E. Laflamme and J.G. Windsor, Jr. 1980. Polycyclic aromatic hydrocarbons in marine/aquatic sediments. Pages 289-311 *in* L. Petrakis and F.T. Wiess (eds.), Petroleum in the Marine Environment. Advances in Chemistry Series No. 185. American Chemical Society, Washington, D.C.

Hjulstrom, F. 1939. Transportation of detritus by moving water. Pages 5-31 *in* P.D. Trask (ed.), Recent Marine Sediments. Amer. Assoc. of Petrol. Geol., Tulsa, Oklahoma.

Houghton, J.P., R.P. Britch, R.C. Miller, A.K. Runchal and C.P. Falls. 1980. Drilling fluid dispersion studies at the lower Cook Inlet, Alaska, C.O.S.T. Well. Page 285 *in* Research on Environmental Fate and Effects of Drilling Fluids and Cuttings, Volume I. American Petroleum Institute, Washington, D.C.

Hunt, J.N. 1954. The turbulent transport of suspended sediment in open channels. Proc. Roy. Soc. London, Ser. A, 224: 322-335.

Inman, D.L. 1949. Sorting of sediments in light of fluid mechanics. J. Sed. Petrol. 19: 51-70.

Jackson, G.F., E. Hume, M.J. Wade and M. Kirsch. 1981. Oil Content in Produced Brine on Ten Louisiana Production Platforms. Final Report Contract No. 68-03-2648. Environmental Protection Agency, Cincinnati, Ohio.

Johanssen, S., V. Larsson and P. Boehm. 1980. The *Tsesis* oil spill impact on the pelagic ecosystem. Mar. Pollut. Bull. 11: 284-293.

Jordan, R.E. and J.R. Payne. 1980. Fate and Weathering of Petroleum Spills in the Marine Environment: A Literature Review and Synopsis. Ann Arbor Science Publishers, Inc., Ann Arbor, Michigan, 174 p.

Jumars, P.A. and A.R. Nowell. 1984a. Effects of benthos on sediment transport: difficulties and functional grouping. Cont. Shelf Res. 3: 1115-130.

Jumars, P.A. and A.R. Nowell. 1984b. Fluid and sediment dynamic effects on marine benthic community structure. Amer. Zool. 24: 45-55.

Jumars, P.A., A.R.M. Nowell and R.F.L. Self. 1981. A simple model of flow-sediment-organism interaction. Mar. Geol. 42: 155-172.

Kachel, N.B. 1980. A Time-Dependent Model of Sediment Transport and Strata Formation on a Continental Shelf. Ph.D. Dissertation, Univ. of Washington, Seattle.

Karl, H.A. 1980. Influence of San Gabriel submarine canyon on narrow-shelf sediment dynamics, southern California. Mar. Geol. 34: 61-78.

Kennett, J.P. 1982. Marine Geology. Prentice Hall, Englewood Cliffs, New Jersey, 813 p.

Kline, S.J., W.C. Reynolds, F.A. Schraub and P.W. Runstadlen. 1967. The structure of turbulent boundary layers. J. Fluid Mech. 30: 741-773.

Knebel, H.J. 1981. Processes controlling the characteristics of the surficial sand sheet, U.S. Atlantic outer continental shelf. Mar. Geol. 42: 349-368.

Kolpack, R.L. 1983. Transport Processes in Central and Southern California Offshore Waters. Testimony for Environmental Protection Agency Region IX Public Hearing, August 11, 1983, Santa Barbara, California.

Kolpak, R.L., J.S. Mattson, H.B. Mark and T.C. Yu. 1971. Hydrocarbon content of Santa Barbara Channel sediments. Pages 276-295 *in* R.L. Kolpak (ed.), Biological and Oceanographical Survey of the Santa Barbara Channel Oil Spill 1969-1970. Volume II. Univ. of Southern California, Los Angeles, California.

Komar, P.D. 1969. The channelized flow of turbidity currents with application Monterey deep-sea fan channel. J. Geophys. Res. 74: 4544-4556.

Koons, C.B. 1973. Chemical composition: A control on the physical and chemical processes acting on petroleum in the marine environment. *In* Background Papers for a Workshop on Inputs, Fates, and Effects of Petroleum in the Marine Environment. Volume II. National Academy of Sciences, Washington, D.C.

Laflamme, R.E. and R.A. Hites. 1978. The global distribution of polycyclic aromatic hydrocarbons in recent sediments. Geochim. Cosmochim. Acta 42: 289-304.

Larson, L.H., Steinberg, R.W., N.C. Shi, M.D.H. Marselein and L. Thomas. 1981. Field investigations of the threshold of gradual motion by ocean waves and currents. Mar. Geol. 42: 105-132.

Lavelle, J.W., R.A. Young, D.J.P. Swift and T.L. Clarke. 1978. Nearbottom sediment concentration and fluid velocity measurements on the inner continental shelf, New York. J. Geophys. Res. 83: 6052-6062.

Law, R.J. and A.A. Blackman. 1981. Hydrocarbons in Water and Sediments from Oil-Producing Areas of the North Sea. ICES CM 1981/E-16 (unpublished manuscript).

Lee, M.L., G.P. Prado, J.B. Howard and R.A. Hites. 1977. Source identification of urban airborne polycyclic aromatic hydrocarbons by gas chromatographic mass spectrometry. Biomed. Mass. Spec. 4: 182-186.

Li, Y.-H., H.W. Feely and P.H. Santschi. 1979. ^{228}Th-^{228}Ra radioactive disequilibrium in the New York Bight and its implications for coastal pollution. Earth Planet. Sci. Lett. 42: 13-26.

Ludwick, J.C. and G.W. Domurat. 1982. A deterministic model of the vertical component of sediment motion in a turbulent fluid. Mar. Geol. 45: 1-15.

Luoma, S.N. 1982. Briefing document 4 - Transport of trace elements from particulates to biota. Page 227 *in* Pollutant Transfer by Particulates Workshop. January, 1982, U.S. Dept. of Commerce, Washington, D.C.

Luoma, S.N. 1983. Bioavailability of trace metals to aquatic organisms - A review. Science of the Total Environment 28: 1-22.

Lysyj, I. 1982. Chemical Composition of Produced Water at Some Offshore Oil Platforms. Final Report Contract No. 68-03-2648. Municipal Environmental Research Laboratory, Office of Research and Development, U.S. Environmental Protection Agency, Cincinnati, Ohio.

Lytle, T.F. and J.S. Lytle. 1979. Sediment hydrocarbons near an oil rig. Estuar. Coastal Mar. Sci. 9: 319-330.

Mackay, D. 1979. Finding frugacity feasible. Env. Sci. Tech. 13: 1218-1223.

Mackay, D., and K. Hossain. 1982. An exploratory study of naturally and chemically dispersed oil. Submitted to Environment Protection Service, Environment Canada, Ottawa.

Manheim, F.T., J.C. Hathaway and E. Uchupi. 1972. Suspended matter in surface waters of the northern Gulf of Mexico. Limnol. Oceanogr. 17: 17-27.

Manheim, F.T., R.H. Meade and G.C. Bond. 1970. Suspended matter in surface waters of the Atlantic continental margin from Cape Cod to the Florida Keys. Science 167: 371-376.

Mantz, P.A. 1977. Incipient transport of fine grains and flakes by fluids-extended Shields diagram. J. Hydr. Div., Proc. Amer. Soc. Civil Engineers 103: 601-615.

Mayer, D.A., D.V. Hansen and D.A. Ortman. 1979. Long-term current and temperature observations on the middle Atlantic shelf. J. Geophys. Res. 84: 1776-1792.

McAuliffe, C. 1969. Determination of dissolved hydrocarbons in subsurface brines. Chem. Geol. 4: 225-233.

McCave, I.N. 1970. Deposition of fine-grained suspended sediment from tidal currents. J. Geophys. Res. 75: 4151-4159.

McFarland, V.A. 1983. Estimating Bioaccumulation Potential of Chemicals in Sediment. Environment Effects of Dredging. Information Exchange Bulletin, D-83-4. U.S. Army Corps of Engineers, Waterways Experiment Station, Vicksburg, Mississippi.

McGrail, D.W. and M. Carnes. 1983. Shelf edge dynamics and the nepheloid layer in the northwestern Gulf of Mexico. Pages 251-264 *in* Critical Interface on Continental Margins. Society of Economic Paleontologists and Mineralogists Spec. Publ. No. 33, Tulsa, Oklahoma.

McGrail, D., R. Rezak and T. Bright. 1982. Environmental Studies at the Flower Gardens and Selected Banks: Northwestern Gulf of Mexico, 1979-1981. Final Report to Minerals Management Service, Contract No. AA851-CT0-25. Texas A&M University, College Station, 315 p.

Meade, R.H., P.L. Sachs, F.T. Manheim, J.C. Hathaway and D.W. Spencer. 1975. Sources of suspended matter in waters of the Middle Atlantic Bight. J. Sediment. Petrol. 45: 171-188.

Meek, R.P. and J.P. Ray. 1980. Induced sedimentation, accumulation and transport resulting from exploratory drilling discharges of drilling fluids and cuttings on the southern California outer continental shelf. Page 259 *in* Research on Environmental Fate and Effects of Drilling Fluids and Cuttings, Volume I. American Petroleum Institute, Washington, D.C.

Meyers, P.A. and J.G. Quinn. 1973. Association of hydrocarbons and mineral particles in saline solution. Nature 244: 1-5.

Middleditch, B.S. 1981a. Hydrocarbons and sulfur. Pages 13-50 *in* B.S. Middleditch (ed.), Environmental Effects of Offshore Oil Production. The Buccaneer Gas and Oil Field Study. Plenum Press, New York.

Middleditch, B.S. (ed.). 1981b. Environmental Effects of Offshore Oil Production. The Buccaneer Gas and Oil Field Study. Plenum Press, New York, 446 p.

Miller, M.C., I.N. McCave and P.D. Komas. 1977. Threshold of sediment motion under unidirectional currents. Sedimentology 24: 507-527.

Milliman, J.D. 1972. Marine geology. Chapter 10 *in* Coastal and Offshore Environmental Inventory. Cape Hatteras to Nantucket Shoals. Complement Volume. Marine Publication Series No. 3, Univ. of Rhode Island, Kingston, Rhode Island.

Milliman, J.D. 1980. Transfer of river-borne particulate material to the oceans. Pages 5-12 *in* River Inputs to Ocean Systems. Proc. of SCOR Workshop, 26-30 March, 1979. Rome, Italy. UNESCO, Paris, 384 p.

Milliman, J.D., M.H. Bothner and C.M. Parmenter. 1980. Sector in New England shelf and slope waters 1977-1978. Chapter 2 *in* J.M. Aaron (ed.), Environmental Geologic Studies in the Georges Bank Area. United States Northeastern Outer Continental Shelf 1975-1977. Final Report submitted to Bureau of Land Management, New York OCS Office.

Naidu, A.S. 1979. Sources, transport pathways, depositional sites and dynamics of sediments in the lagoon and shallow marine region, northern Arctic Alaska. National Oceanic and Atmospheric Administration/Outer Continental Shelf Assessment Program, Annual Report, 81 p.

National Academy of Sciences. 1975. Petroleum in the Marine Environment. National Academy of Sciences, Washington, D.C., 107 p.

National Research Council. 1983. Drilling Discharges in the Marine Environment. National Academy Press, Washington, D.C., 180 p.

Neff, J.M. and J.W. Anderson. 1981. Responses of Marine Animals to Petroleum and Specific Petroleum Hydrocarbons. Applied Science Publishers, Ltd., London, 177 p.

Neumann, A.C., C.D. Gebelein and T.P Scoffin. 1970. The composition, structure and erodibility of subtidal mats, Abaco, Bahama. J. Sed. Petrol. 40: 274-297.

Nittrouer, C.A., R.W. Sternberg, R. Carpenter and J.T. Bennett. 1979. The use of Pb-210 geochronology as a sedimentological tool: Application to the Washington continental shelf. Mar. Geol. 31: 297-316.

Northern Technical Services. 1981. Beaufort Sea Drilling Effluent Disposal Study. Prepared for the Reindeer Island Stratigraphic Test Well Participants. Submitted to SOHIO Alaska Petroleum Co., 329 p.

Northern Technical Services. 1982. Above-Ice Drilling Effluent Disposal Tests Sag Delta No. 7, Sag Delta No. 8 and Challenge Island No. 1, Wells Beaufort, Alaska. Submitted to SOHIO Alaska Petroleum Co., Anchorage, Alaska.

Northern Technical Services. 1983. Open-Water Drilling Effluent Disposal Study, Turn Island, Beaufort Sea, Alaska. Submitted to Shell Oil Company, Anchorage, Alaska.

Nowell, A.R.M. 1983. The benthic boundary layers and sediment transport. Rev. Geophys. Space Physics 21: 1181-1192.

Nowell, A.R.M., P.A. Jumars and J.E. Eckman. 1981. Effects of biological activity on the entrainment of marine sediments. Mar. Geol. 42: 133-153.

Olsen, C.R., P.E. Biscaye, H.J. Simpson, R.M. Trier, N. Kostyk, R.F. Bopp, L.H. Li and H.W. Feely. 1980. Reactor released radionuclides and fine grained sediment transport and accumulation patterns in Barnegat Bay, New Jersey and adjacent shelf waters. Estuar. Coastal Mar. Sci. 10: 119-142.

Olsen, C.R., N.H. Cutshall and T.A. Nelsen. 1982. Pollutant-particle dynamics in geochemical cycles. Briefing document 2 *in* Proceedings of a Pollutant Transfer by Particulates Workshop. Old Dominion University, Norfolk, Virginia. January 19-21, 1982. U.S. Dept. of Commerce, National Oceanic and Atmospheric Administration, Office of Marine Pollution Assessment, Rockville, Maryland.
Overton, E.B., J. McFall, S.W. Mascarella, C.F. Steele, S.A. Antoine, I.R. Politzer and J.L. Laseter. 1981. Petroleum residue sources identification after a fire and oil spill. Pages 541-546 *in* Proceedings in 1981 Oil Spill Conference. American Petroleum Institute, Washington, D.C.
Owen, M.W. 1977. Problems in the modeling of transport erosion, and deposition of cohesive sediments. Pages 515-537 *in* E.D. Goldberg, I.N. McCave, J.J. O'Brien and J.J. Steel (eds.), The Sea, Volume 6. John Wiley and Sons, New York.
Page, D.S., J.C. Foster, J.R. Hotham, E. Pendergast, S. Hebert, L. Gonzales, E.S. Gilfillan, S.A. Hanson, P. Gerber and D. Vallas. 1983. Long term fate of dispersed and undispersed crude oil in two nearshore test spills. Page 465 *in* Proceedings of 1983 Oil Spill Conference. American Petroleum Institute, Washington, D.C.
Pancirov, R.J. and R.A. Brown. 1977. Polynuclear aromatic hydrocarbons in marine tissues. Environ. Sci. Tech. 11: 989-992.
Partheniades, E. 1965. Erosion and deposition of cohesive soils. J. Hydr. Div., Proc. Amer. Soc. Civil Engineers 91: 105-138.
Pierce, R.H., D.C. Anne, F.I. Saksa and A. Weichert. 1984. Fate of organic compounds from spent drilling fluid discharged to the marine environment. Chapter 7, *in* Energy Wastes in the Ocean, Volume 4. Wiley-Interscience, New York.
Pierce, T.J., R.T. Jarman and C.M. deTurville. 1970. An experimental study of silt scouring. Proc. Inst. Civ. Engrs. 38: 231-243.
Pierre, J.W. and F.R. Siegel. 1979. Suspended particulate matter on the southern Argentine shelf. Mar. Geol. 29: 73-91.
Postma, H. 1967. Sediment transport and sedimentation in the estuarine environment. Pages 158-179 *in* G.H. Lauff (ed.), Estuaries. Amer. Assoc. Adv. Sci., Pub. No. 83.
Qinchun, X., Z. Liren and Z. Fugen. 1983. Features and transportation of suspended matter over the continental shelf off the Changjing estuary. Pages 400-412 *in* Proc. International Sympos., Sedimentation on the Continental Shelf with Special Reference to the East China Sea, Volume I. China Ocean Press, Beijing.
Ray, J.P. and R.P. Meek. 1980. Water column characterization of drilling fluids dispersion from an offshore exploratory well on Tanner Bank. Page 223 *in* Research on Environmental Fate and Effect of Drilling Fluids and Cuttings, Volume 1. American Petroleum Institute, Washington, D.C.
Reed, W.E., I.R. Kaplan, M. Sandstrom and P. Mankiewicz. 1977. Petroleum and anthropogenic influence on the composition of sediments from the southern depositional environmental California Bight. Pages 183-188 *in* Proceedings 1977 Oil Spill Conference. American Petroleum Institute, Washington, D.C.
Rhoads, D.C. 1974. Organism-sediment relations on the muddy sea floor. Oceanogr. Mar. Biol. Ann. Rev. 12: 263-300.
Rhoads, D.C. and L.F. Boyer. 1982. The effects of marine benthos on physical properties of sediments. A successional perspective. Pages 3-52 *in* D.L. McCall and M.J.S. Tevesz (eds.), Animal-Sediment Interactions. The Biogenic Alteration of Sediments. Plenum Press, New York.
Rhoads, D.C. and S. Cande. 1971. Sediment profile camera for *in situ* study of organism-sediment relations. Limnol. Oceanogr. 16: 110-114.
Rhoads, D.C. and J.D Germano. 1982. Characterization of organism-sediment relations using sediment profile imaging: An efficient method of remote ecological monitoring of the seafloor remote TM system. Mar. Ecol. Progr. Ser. 8: 115-126.
Rhoads, D.C. and D.K. Young. 1970. The influence of deposit feeding organisms on sediment stability and community trophic structure. J. Mar. Res. 28: 150-178.
Rhoads, D.C., K. Tenore and M. Browne. 1975. The role of resuspended bottom muds

in the food chain of shallow embayments. Pages 563-579 *in* L.E. Cronin (ed.), Estuarine Research, Volume 1. Academic Press, London.
Rhoads, D.C., J.Y. Yingst and W.J. Ullman. 1978. Seafloor stability in central Long Island Sound. Part I. Temporal changes in erodibility of fine-grained sediment. Pages 221-224 *in* M.L. Wiley (ed.), Estuarine Interactions. Academic Press, New York.
Rubenstein, N.I., E. Lores and N.R. Gregory. 1983. Accumulation of PCBs, mercury, and cadmium by *Nereis virens*, *Mercenaria mercenaria* and *Palaemonetes pugio* from contaminated harbor sediments. Aquatic Toxicology 3: 249-260.
Saila, S.B., S.D. Pratt and T.T. Polgar. 1972. Dredge Spoil Disposal in Rhode Island Sound. Mar. Techn. Rep. No. 2, Univ. of Rhode Island, 48 p.
Santschi, P.H., Y.H. Li, J.J. Bell, R.M. Trier and K. Kowtaluk. 1980. Pu in coastal marine environments. Earth Planet. Sci. Lett. 51: 248-265.
Sauer, T.C. 1981. Volatile liquid hydrocarbon characterization of underwater hydrocarbon vents and formation waters from offshore production operations. Environ. Sci. Technol. 15: 917.
Sauer, T.C. 1983. Modeling of the Fate of Drilling Fluid Discharges from California OCS Platforms. Testimony for Environmental Protection Agency Region IX Public Hearing, August 11, 1983, Santa Barbara, California.
Schubel, J. 1982. Brief Document 1, An eclectic look at fine particles in the coastal ocean. *In* Pollutant Transfer by Particulates Workshop. U.S. Dept. of Commerce, Washington, D.C.
Schubel, J.R. and T.W. Kana. 1972. Agglomeration of fine-grained suspended sediment in northern Chesapeake Bay. Power Technol. 6: 9-16.
Scruton, P.C. and D.G. Moore. 1953. Distribution of surface turbidity off Mississippi Delta. Amer. Assoc. Petrol. Geol. Bull. 37: 1067-1074.
Sharma, G.D. 1979. The Alaskan Shelf. Hydrographic, Sedimentary and Geochemical Environment. Springer-Verlag, Inc., New York, 498 p.
Shields, A. 1936. Anwendung der Ahnlichkeits mechanik und der Turbulenzforschung auf die Geschieb. Transl. by W.P. Ott and J.C. Van Vehelen, U.S. Dept. Agriculture, Soil Conservation Service Coop Laboratory, California Inst. Tech., Pasadena, California.
Smith, J.D. 1977. Modeling of sediment transport on continental shelves. Pages 539-578 *in* E.D. Goldberg, I.N. McCave, J.J. O'Brien and J.H. Steele (eds.), The Sea, Volume 6. John Wiley and Sons, New York.
Smith, J.D. and T.S. Hopkins. 1972. Sediment transport on the continental shelf off Washington and Oregon in light of recent current measurements. Pages 143-180 *in* D.J.P. Swift, D.B. Duane and O.H. Pilkey (eds.), Shelf Sediment Transport: Process and Pattern. Dowden, Hutchinson and Ross, Inc., Stroudsburg, Pennsylvania.
Smith, J.D. and S.R. McLean. 1977a. Spatially averaged flow over a wavy boundary. J. Geophys. Res. 82: 1735-1746.
Smith, J.D. and S.R. McLean. 1977b. Boundary layer adjustments to bottom topography and suspended sediment. Pages 123-151 *in* J.C.J. Nittoul (ed.), Bottom Turbulence. Elsevier, New York.
Soulsby, R.L. and K.R. Dyer. 1981. The form of the near-bed velocity profile in a tidally accelerating flow. J. Geophys. Res. 89: 8067-8974.
Stanley, D.J. and G.L. Freeland. 1978. The erosion-deposition boundary in the head of Hudson submarine canyon defined on the basis of submarine observations. Mar. Geol. 26: M37-M49.
Strakhov, N.M. 1967. Principles of Lithogenesis, Volume 1. Consultants Bureau, New York, 245 p.
Taylor, P.A. and K.R. Dyer. 1977. Theoretical models of flow near the bed and their implications for sediment transport. Pages 579-601 *in* E.D. Goldberg, I.N. McCave, J.T. O'Brien and J.M. Steele (eds.), The Sea, Volume 6. John Wiley and Sons, New York.
Tillery, J.B. and R.E. Thomas. 1980. Heavy metal contamination from petroleum

production platforms in the central Gulf of Mexico. Page 562 *in* Research on Environmental Fate and Effects of Drilling Fluids and Cuttings, Volume 1. American Petroleum Institute, Washington, D.C.

Tissot, B.P. and D.H. Welte. 1981. Petroleum Formation and Occurrence. Springer-Verlag, New York.

Trefry, J.H. 1977. The Transport of Heavy Metals by the Mississippi River and Their Fate in the Gulf of Mexico. Ph.D. Dissertation, Texas A&M University, College Station, 223 p.

Tripp, B.W., J.W. Farrington and J.M. Teal. 1981. Unburned coal as a source of hydrocarbons in surface sediments. Mar. Pollut. Bull. 12: 122-126.

Twichell, D.C., C.E. McClennen and B. Butman. 1981. Morphology and pressures associated with accumulation of the fine-grained sediment deposit on the southern New England Shelf. J. Sediment. Petrol. 51: 269-280.

Vincent, C.E., D.J.P. Swift and Halloral. 1981. Sediment transport in the New York Bight, North American continental shelf. Mar. Geol. 42: 369-398.

Wakeham, S.G., C. Schaffner and W. Giger. 1981. Diagenic polycyclic aromatic hydrocarbons in recent sediments: Structural information obtained by high performance liquid chromatography. Pages 353-363 *in* A.G. Douglas and J.R. Maxwell (eds.), Advances in Organic Geochemistry. Proc. 9th Intl. Mtg. on Organic Geochemistry, Newcastle-Upon-Tyne, England, September, 1979. Pergamon Press, New York.

Ward, C.H., M.R. Bender and D.J. Reish (eds.). 1979. The Offshore Ecology Investigation. Effects of Oil Drilling and Production in a Coastal Environment. Rice Univ. Studies 65: 1-589.

Weatherly, G.L. and P.J. Martin. 1978. On the structure and dynamics of the oceanic bottom boundary layer. J. Phys. Oceanog. 8: 557-570.

Winant, C.D. 1981. Coastal circulation and wind-induced currents. Amer. Rev. Fluid Mech. 12: 271-301.

Windsor, J.G., Jr. and R.A. Hites. 1979. Polycyclic aromatic hydrocarbons in Gulf of Maine sediments and Nova Scotia soils. Geochim. Cosmochim. Acta 43: 27-33.

Winfrey, M.R., E. Beck, P. Boehm and D.M. Ward. 1982. Impact of crude oil on sulfate reduction and methane production in sediments impacted by the *Amoco Cadiz* oil spill. Mar. Env. Res. 7: 175-194.

Young, R.A. 1982. Briefing document 3 - Mechanisms of erosion, deposition and transport of cohesive sediments in the boundary layer. Page 193 *in* Pollutant Transfer by Particulates Workshop, January, 1982. U.S. Dept. of Commerce, Washington, D.C.

Young, R.A. and J.B. Southard. 1978. Erosion of fine-grained marine sediments: Seafloor and laboratory experiments. Geol. Soc. Amer. Bull. 89: 663-672.

Youngblood, W.W. and M. Blumer. 1973. Alkanes and alkenes in marine benthic algae. Mar. Biol. 21: 163-172.

Youngblood, W.W., M. Blumer, R.L. Guillard and F. Fiore. 1971. Saturated and unsaturated hydrocarbons in marine benthic algae. Mar. Biol. 8: 190-201.

CHAPTER 7

TRANSPORT AND TRANSFORMATIONS OF PETROLEUM: BIOLOGICAL PROCESSES

Richard Bartha and Ronald M. Atlas

CONTENTS

INTRODUCTION

Although the limited nature of the global petroleum supply is now acutely perceived, petroleum continues to serve as the principal energy source of the United States and other industrialized nations. The production, transport, refining and end uses of petroleum hydrocarbons subject the marine environment to a considerable level of oil pollution. The biodegradation of petroleum pollutants, a major process mediating the fate of oil in the environment, has been the subject of several relatively recent reviews (Atlas, 1977, 1981, 1984; Atlas and Bartha, 1973a; Bartha and Atlas, 1977; Colwell and Walker, 1977; Crow *et al.*, 1974; Gutnick and Rosenberg, 1977; Hughes and McKenzie, 1975; Jordan and Payne, 1980; National Research Council, 1975, 1985).

Petroleums or crude oils are extremely complex mixtures of aliphatic, alicyclic and aromatic hydrocarbons and of some nonhydrocarbon compounds, such as naphthoic acids, phenols, thiols and heterocyclic nitrogen, sulfur and oxygen (NSO) compounds, as well as some metalloporphyrins (Atlas and Bartha, 1973a). The NSO compounds constitute the "resins;" the highly condensed and insoluble residue constitutes the ill-defined "asphaltene" fraction of the crude oils. Even when the most advanced techniques, such as computerized gas chromatography-mass spectrometry (GC-MS) analysis, are applied (Pancirov, 1974), hundreds of petroleum components remain unresolved and unidentified. According to their origin, crude oils vary greatly in composition. Furthermore, when spilled into the marine environment, petroleums are altered not only by biodegradation but also by evaporation, photooxidation, dissolution and emulsification. The effects of these abiotic processes (see Chapters 5 and 6) are difficult to separate from those of biodegradation. Compared to measuring the biodegradation of a single defined organic substrate, the monitoring of biodegradation or transformation of petroleum is a complex, demanding and relatively inaccurate procedure. It becomes necessary either to use relatively nonspecific monitoring techniques, such as CO_2 evolution, O_2 consumption and weight loss, or conversely, to follow the fate of individual hydrocarbons attempting to extrapolate from these results to the overall fate of the complex petroleum. Both approaches have obvious drawbacks.

No crude oil is completely biodegradable, even under the most favorable conditions. The proportion of nonvolatile components removable by biodegradation may vary, according to the nature of the petroleum, from as little as 11% to as much as 90% (Colwell and Walker, 1977). The "rate" of petroleum removal by biodegradation reflects the simultaneous or sequential removal of various components at various rates. This circumstance limits the validity of calculations of heterotrophic potential or microbial metabolic capacity for petroleum biodegradation. Nevertheless, all recent reviews agree that the nonvolatile components of most crude oils are removed from the marine environment predominantly by the biodegradation mechanism. A mass balance of spilled oil given by Mackay (1981), for a speculative but representative case, illustrates the importance of biodegradation as one of the principal self-purification mechanisms of the marine environment (Figure 7.1).

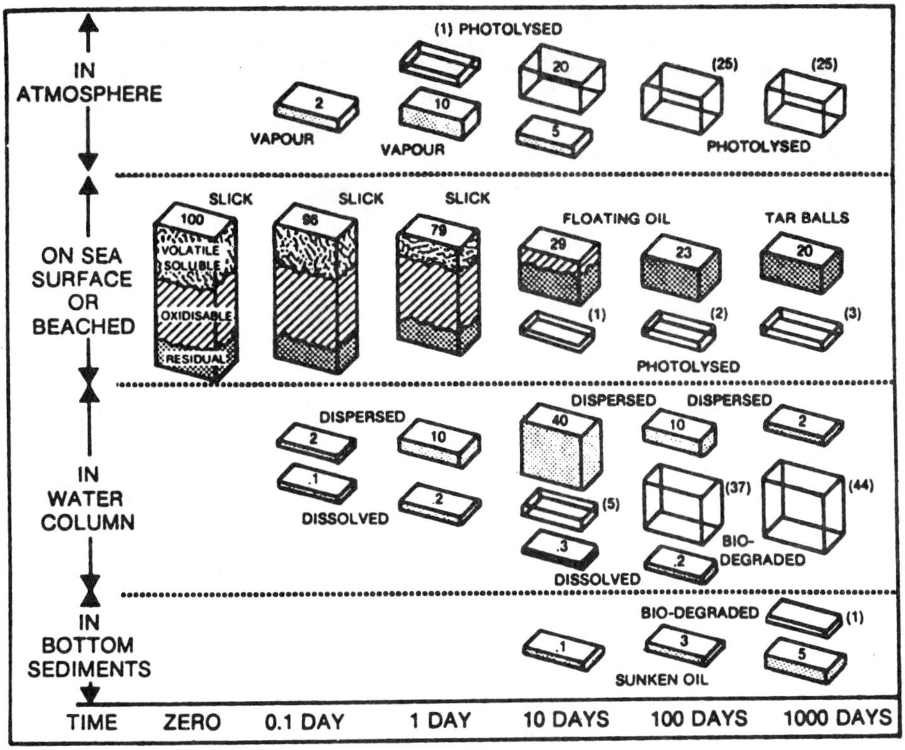

Figure 7.1. Speculative mass balance illustrating the distribution and conversion of an initial 100 volumes of oil at various times after a spill. Empty unshaded boxes represent oil converted to another chemical form (based on a similar diagram by J. N. Butler, Harvard University) (from Mackay, 1981).

HYDROCARBON-DEGRADING MARINE MICROORGANISMS

Identity and Distribution

The origin of the numerous hydrocarbon-utilizing microorganisms listed in books and reviews (Beerstecher, 1954; Davis, 1967; Fuhs, 1961; Nyns and Wiaux, 1969; Friede *et al.*, 1972) is often obscure. The genera listed in Table 7.1 are restricted to microorganisms capable of using hydrocarbons as their sole source of carbon and energy and reported in relatively recent papers (from 1970 to date) as originating from marine or brackish waters and sediments. Although, in some cases, characterization of the microorganisms was carried beyond the genus level, this information was omitted from the table as having little additional value. Undoubtedly, the varied hydrocarbon substrates and isolation procedures influenced the types and the diversity of the microorganisms isolated from marine samples. Nevertheless, the number of citations for each genus gives some indication of how prevalent the hydrocarbon-degrading representatives of the listed genera are in the marine environment. Based on this and on additional information gleaned from the papers (e.g., the frequency of occurrence in

TABLE 7.1
Hydrocarbon-degrading microorganisms isolated from marine and estuarine environments*

Bacteria	*References*	*Fungi*	*References*
Achromobacter	6,8,12,13,15	*Allescheria*	5
Acinetobacter	6,11,17	*Aspergillus*	5,7,17
Actinomyces	3	*Aureobasidium*	1,5,17
Aeromonas	5,17	*Botrytis*	5
Alcaligenes	4,6	*Candida*	1,10,17
Arthrobacter	5,8,14,15	*Cephalosporium*	7
Bacillus	3,5,8,12	*Cladosporium*	1,17
Beneckea	8	*Cunninghamella*	1
Coryneforms	3,5,8,15,17	*Debaromyces*	1
Erwinia	3	*Fusarium*	1
Flavobacterium	2,5,8,12,13	*Gonytrichum*	5
Klebsiella	3	*Hansenula*	17
Lactobacillus	3	*Helminthosporium*	5
Leucothrix	3	*Mucor*	1
Micrococcus	3,5,12	*Oidiodendrum*	5
Moraxella	3	*Paecylomyces*	5
Nocardia	2,3,8,13,17	*Phialophora*	5
Peptococcus	5	*Penicillium*	1,5,7,17
Pseudomonas	2,3,5,8,12,13,17	*Rhodosporidium*	1
Sarcina	12	*Rhodotorula*	1,17
Sphaerotilus	3	*Saccharomyces*	1
Spirillum	12	*Saccharomycopsis*	16
Streptomyces	3	*Scopulariopsis*	5
Vibrio	3,8,12,13,17	*Sporobolomyces*	1,17
Xanthomonas	3	*Torulopsis*	10
		Trichoderma	5
		Trichosporon	1

Algae	*References*
Prototheca	9,17

*Only microorganisms that were able to grow on hydrocarbons as the only sources of carbon and energy are listed. Isolations and identifications by one research group, if conducted essentially by the same techniques, are listed under a single reference number, even if reported in separate papers. Only reports published after 1969 are included. The taxonomic nomenclature was updated, where possible, to conform to the Eighth Edition of Bergey's Manual and current fungal nomenclature. Thus, some of the above designations differ from the originally published ones.
Key to the references: (1) Ahearn *et al.*, 1971; Ahearn and Meyers, 1971, Ahearn and Crow, 1980; (2) Atlas and Bartha, 1972a; Dean-Raymond and Bartha, 1975; (3) Austin *et al.*, 1977a,b; (4) Bertrand *et al.*, 1976; (5) Buckley *et al.*, 1976; (6) Byrom *et al.*, 1970; (7) Cerniglia and Perry, 1973; Perry and Cerniglia, 1973; Cerniglia *et al.*, 1980a; (8) Cundell and Traxler, 1973a,b, 1976; (9) Kockova-Kratochvilova and Havelkova, 1974; (10) LePetit *et al.*, 1970; (11) Makula *et al.*, 1975; (12) Mironov, 1970; Mironov and Lebed, 1972; (13) Mulkins-Phillips and Stewart, 1974a; (14) Reisfeld *et al.*, 1972; (15) Soli and Bens, 1972; (16) Stormer and Vinjansen, 1976; (17) Walker and Colwell, 1974a; Walker *et al.*, 1975a,b,c; Walker *et al.*, 1976a.

subsamples), the most prevalent marine bacterial hydrocarbon degraders belong, in decreasing order, to the genera *Pseudomonas*, *Achromobacter*, *Flavobacterium*, *Nocardia*, *Arthrobacter* and other coryneforms, *Vibrio*, *Bacillus*, *Micrococcus* and *Acinetobacter*.

The numerical taxonomy approach, applied recently to marine hydrocarbon-degrading bacteria by Austin *et al.* (1977a,b), resulted in the identification of some rather unexpected types as marine hydrocarbon degraders, including various representatives of the Enterobacteriaceae and the filamentous bacteria *Leucothrix* and *Sphaerotilus*. In addition to the approach used, the source (nearshore, polluted estuarine sediment) may have contributed to the isolation and identification of these unusual marine hydrocarbon degraders. The suggestion was advanced that some of the isolates may have acquired their hydrocarbon-degrading capabilities through plasmid transfer, an interesting but as yet not fully explored possibility.

Fungal hydrocarbon degraders have been isolated and identified by a few workers, and generalizations about the prevalence of certain forms are more precarious. It appears that among the yeast-like forms, *Candida*, *Rhodotorula*, *Aureobasidium* (=*Pullullaria pullullans*) and *Sporobolomyces* are the most frequent. Some members of the heterogeneous *Candida*, e.g., *C. lipolytica*, have recently been reclassified as *Saccharomycopsis* based on observation of the perfect stage (Stormer and Vinjansen, 1976). From the filamentous fungi, *Penicillium*, *Aspergillus* and *Cladosporium resinae* have been most frequently isolated. Both yeasts and filamentous fungi are predominantly associated with surface films and, presumably, contribute significantly to hydrocarbon biodegradation in cases of undisturbed surface slicks on quiescent waters. Bacteria seem to predominate in the biodegradation of dispersed or dissolved hydrocarbons in less-protected waters.

The achlorophyllous algae *Prototheca hydrocarbonea* and *Prototheca zopfii* are relatively recent additions to the list of marine hydrocarbon degraders (Kockova-Kratochvilova and Havelkova, 1974; Walker *et al.*, 1975c).

In addition to the microorganisms that are capable of using hydrocarbons for growth, there are undoubtedly many more that are capable of cometabolic hydrocarbon transformations (Perry, 1979; Cerniglia *et al.*, 1978, 1980a). In addition to bacteria and fungi, cometabolic transformations of hydrocarbons are performed by algae and even protozoa. Cometabolism refers to the gratuitous utilization of a substance that does not provide energy or nutrition to an organism while that organism is growing on another substrate as its source of carbon and energy. In effect, cometabolism reflects a metabolic error due to a breakdown in enzyme specificity. Cometabolism usually results in the formation of partially oxidized products rather than the complete degradation of the cometabolized substance.

In addition to cometabolism, microorganisms in natural environments can derive energy and carbon from multiple substrates. Almost all hydrocarbon degraders grow very well on nonhydrocarbon substrates. Masters and Zajic (1971) reported metabolism of *n*-heptadecane by phototrophically growing *Scenedesmus* (Chlorophycophyta) strains. Because cell yields increased in the presence of *n*-heptadecane, this is probably an example of myxotrophic growth (simultaneous autotrophic and heterotrophic metabolism) rather than of cometabolism. Cerniglia *et al.* (1980b) and Cerniglia and Gibson (1979) tested the ability of various Cyanobacteria and algae to transform naphthalene while growing phototrophically. From 18 cultures tested, all transformed naphthalene to 1-naphthol, although at relatively low rates. The naphthalene oxidizers were the Cyanobacteria *Oscillatoria* (two strains), *Microcoleus*, *Anabaena* (two strains), *Agmenellum*, *Coccochloris*, *Nostoc* and *Aphanocapsa*, the Chlorophycophyta *Chlorella* (two strains), *Dunaliella*, *Chlamydomonas* and *Ulva*, the Chrysophycophyta *Cylindrotheca* and *Amphora*, the Rhodophycophyta *Porphyridium* and the Phaeophycophyta *Petalonia*. One *Oscillatoria* strain also oxidized biphenyl (Cerniglia *et al.*, 1980c).

The marine ciliate protozoan *Parauronema acutum* transformed the polynuclear aromatic hydrocarbon derivatives 2-aminofluorene and 2-acetylaminofluorene to compounds that were active mutagens (Lindmark, 1981). Cometabolic transformations of hydrocarbons by bacteria, fungi, algae and protozoa will, undoubtedly, receive additional attention in the future, since they affect the solubility, toxicity and the ultimate fate of hydrocarbons in the marine environment.

The abundance of hydrocarbonoclastic microorganisms in the marine environment varies considerably according to environmental conditions and hydrocarbon pollution history. As with all viable counting procedures, the methodology employed by the various investigators greatly influences the density estimates obtained. Walker and Colwell (1976a) compared and optimized counting procedures for hydrocarbon-degrading microorganisms, but because standardized procedures have not been generally applied, the comparison of estimates from different publications has little meaning. Generally, pelagic environments with no history of oil pollution appear to have very low populations of hydrocarbon-degrading microorganisms. Less than 5% of 50-ml water samples taken from such areas contained any hydrocarbon degraders (ZoBell, 1969). Populations of hydrocarbon-degrading microorganisms are substantially higher along oceanic shipping lanes (Mironov, 1970) and in oil-polluted coastal areas (Polyakova, 1962; ZoBell and Prokop, 1966; Floodgate, 1976; Buckley *et al.*, 1976; LePetit *et al.*, 1977; Oppenheimer *et al.*, 1977). In Raritan Bay (Atlas and Bartha, 1973b), Chesapeake Bay (Colwell *et al.*, 1973; Walker and Colwell, 1974a), and Cook Inlet (Roubal and Atlas, 1978), positive correlations were found between the numbers of hydrocarbonoclastic microorganisms and oil pollution patterns. The enrichment of hydrocarbonoclastic microorganisms in response to hydrocarbon exposure was consistently observed in marine water and sediment samples incubated *in vitro* (Atlas and Bartha, 1972a; ZoBell, 1973; Traxler, 1973; Miget, 1973; Soli, 1973; Kator *et al.*, 1971; Kator, 1973; Perry and Cerniglia, 1973; Pritchard and Starr, 1973). Similar responses were observed in the field following experimental (Atlas and Schofield, 1975; Atlas *et al.*, 1976; Atlas and Busdosh, 1976; Horowitz and Atlas, 1977, 1978a; Kator and Herwig, 1977; Atlas, 1978a; Atlas *et al.*, 1978) or an accidental (Gunkel, 1968; Walker and Colwell, 1977; Colwell *et al.*, 1978; Stewart and Marks, 1978; Atlas and Bronner, 1981; Ward *et al.*, 1980) oil spills. Hydrocarbon pollution tends to enrich not only for hydrocarbon degraders but also for additional populations that utilize breakdown products but not intact hydrocarbons. Shifts occur also in proteolytic, amylolytic, and chitinolytic populations (Colwell *et al.*, 1978) that are not readily explainable at present.

Microbial Emulsification and Uptake of Hydrocarbons

Petroleum hydrocarbons have generally very low water solubility. Microorganisms utilizing these substrates typically position themselves at the hydrocarbon-water interface. This position assures maximal access to both the hydrocarbon substrate and water-soluble mineral nutrients. Interfacial tension minimizes the oil-water interface, imposing severe spatial constraints on the hydrocarbon-utilizing microbial populations. This spatial constraint is usually counteracted by production of surface-active emulsifying agents by the microorganisms. The increased surface of the finely dispersed oil provides additional interface area for microbial proliferation.

The ability to reduce the surface tension of a culture medium by the leakage of fatty acids and other metabolites from the microbial cell is a phenomenon not restricted to hydrocarbon degraders (LaRiviere, 1955). This phenomenon may be incidental for most microorganisms, but for hydrocarbon degraders it is an essential way to increase substrate availability. Some hydrocarbon degraders excrete copious emulsifying substances (Iguchi *et al.*, 1969; Abbott and Gledhill, 1971; Guire *et al.*, 1973). In the case of a marine *Arthrobacter* strain, the surface-active material consisted mainly of a mixture of long-chain fatty acids, presumably derived from the oxidation of petroleum hydrocarbons (Reisfeld *et al.*, 1972; Rosenberg *et al.*, 1975). Dispersants produced by pseudomonads and coryneforms have been studied by Zajic and co-workers (Zajic and Knetting, 1971, 1972; Zajic and Suplisson, 1972; Zajic *et al.*, 1974; Zajic and Panchal, 1976; Gerson and Zajic, 1977; Panchal and Zajic, 1977). These surface-active polymers were found to be high molecular weight, anthrone-positive polymers that were precipitated by 95% ethanol. A dispersant with similar characteristics is produced by two unidentified marine hydrocarbonoclastic bacteria (Floodgate, 1978).

It is still a matter of controversy whether a true solubilization of hydrocarbons in the aqueous medium is necessary as a precondition for their microbial uptake (Valenkar *et al.*, 1975), or whether liquid hydrocarbons, upon direct contact with the cell membrane, can be taken up directly without prior solubilization. This topic was reviewed in some detail by Gutnik and Rosenberg (1977), and currently most workers appear to accept direct uptake of liquid hydrocarbons as a major transport mechanism. There is no evidence of active transport mechanisms to hydrocarbons (Christensen, 1975). It is important to recognize that microbial dispersants promote direct microbial contact with the oil, as well as the solubilization of the oil.

Transport of hydrocarbons across the cell membrane or, at least, the intimate contact of hydrocarbons with the cell membrane, is a precondition to their metabolism. As it will become apparent in the following sections, the metabolism of hydrocarbons is initiated by membrane-bound oxidases requiring cofactors.

To date, no exoenzymes have been found that attack intact hydrocarbons, nor can any such finding be expected in view of our current understanding of hydrocarbon metabolism. Some recent reports indicate that certain hydrocarbonoclastic microorganisms are capable of accumulating and storing intracellularly large amounts of hydrocarbon substrates (Kennedy *et al.*, 1975; Cooney *et al.*, 1980; Crow *et al.*, 1980). Through this mechanism, microorganisms have a high potential for introducing undegraded hydrocarbons into marine and estuarine food webs.

The accumulation of hydrocarbons by microorganisms and subsequent transfer to higher trophic levels could have a long-term effect on consumer populations. In particular, bioaccumulation of hydrocarbons by the higher consumers could cause disease or other deleterious effects or could render fish and other economically important marine resources unfit for human consumption. The practical question concerns the concentration of hydrocarbon contaminants in these animals regardless of how the hydrocarbons accumulate; however, the mechanisms by which higher consumers accumulate hydrocarbons are of scientific interest. To examine the role of microorganisms in food web transfers of hydrocarbons, it would be practical to grow cultures of microorganisms in the presence of radiolabeled hydrocarbons and then to introduce these labeled organisms into microcosms containing higher consumers. Animals could later be assayed for the radiolabeled hydrocarbon tracers in order to indicate the importance of microbial accumulation of hydrocarbons in food web transfers.

MICROBIAL METABOLISM OF HYDROCARBONS

The literature on hydrocarbon metabolism by microorganisms is very extensive; general overviews of hydrocarbon metabolism dealing with all classes have been presented by Friede *et al.* (1972), Higgins and Gilbert (1978) and Chapman (1979).

Individual Hydrocarbons and Structural Classes

Aliphatic Hydrocarbons

The biodegradation of normal and branched alkanes was reviewed by McKenna (1971) and Ratledge (1978). Pirnik (1977) reviewed some specific problems connected with the biodegradation of methyl-branched alkanes. The most common type of primary metabolic attack by microorganisms on *n*-alkanes is mediated by mixed-function oxidases (monooxygenases) that, acting on the terminal carbon, convert the hydrocarbon molecule to a primary alcohol. Both cytochrome P-450 and rubredoxin systems mediated such oxidations, resulting in the same primary alcohol product (Figure 7.2). Both systems require molecular oxygen. *n*-Alkane oxidation via hydroperoxide intermediates, as proposed by Foster (1962) and reported in numerous reviews (e.g., Friede *et al.*, 1972), is not supported by evidence, and its validity is in doubt. A persistent controversy surrounds the proposed anaerobic oxidation of alkanes via dehydrogenation to alkenes and the subsequent addition of water across the double bond, leading to a secondary alcohol. Originally proposed by Senez and Azoulay (1961), this pathway has some recent support (Parekh *et al.*, 1977). However, consistent and abundant growth of microorganisms on alkanes under strictly anaerobic conditions has never been demonstrated, and the environmental significance of an anaerobic alkane oxidation pathway, if it exists, appears to be negligible.

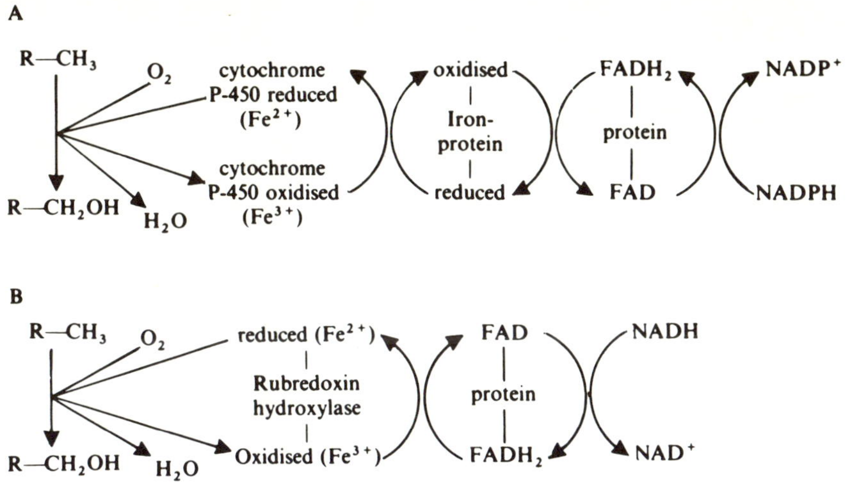

Figure 7.2. Mechanisms for oxidation of *n*-alkanes to primary alcohols. A, cytochrome P-450 system; B, rubredoxin system (from Ratledge, 1978).

Although in the great majority of cases the initial attack is directed at the terminal carbon atom of the hydrocarbon molecule, some microorganisms attack hydrocarbons subterminally, converting them to secondary alcohols (Markovetz, 1971). Oxidation continues to the keto and ester stage. The ester, most commonly a formate or acetate ester, is hydrolyzed yielding formic or acetic acid and a primary alcohol.

The primary alcohols, whether derived from terminal or subterminal oxidations, are further oxidized to aldehydes and fatty acids. The fatty acids are subsequently shortened by C_2 units by *beta*-oxidation. If *beta*-oxidation is hindered by branching, the fatty acid is attacked at the other terminal carbon by the process called *omega*-oxidation. The *omega*-terminus is progressively oxidized to an alcohol, aldehyde and carboxyl group. The resulting dicarboxylic acid is further degraded by *beta*-oxidation. The described pathways for *n*-alkane oxidation are schematically shown in Figure 7.3.

Branching of the aliphatic chain may interfere with *beta*-oxidation (Pirnik, 1977). Hydrocarbons with multiple methyl branches, such as pristane and phytane, are common in petroleum and are presumably derived from isoprenoid natural products. If the methyl branch is on the second carbon of the *iso*-alkane-derived fatty acid, *beta*-oxidation can take place, but the resulting fragment will be a propionyl- rather than an acetyl-SCoA. If the methyl branch is on the third

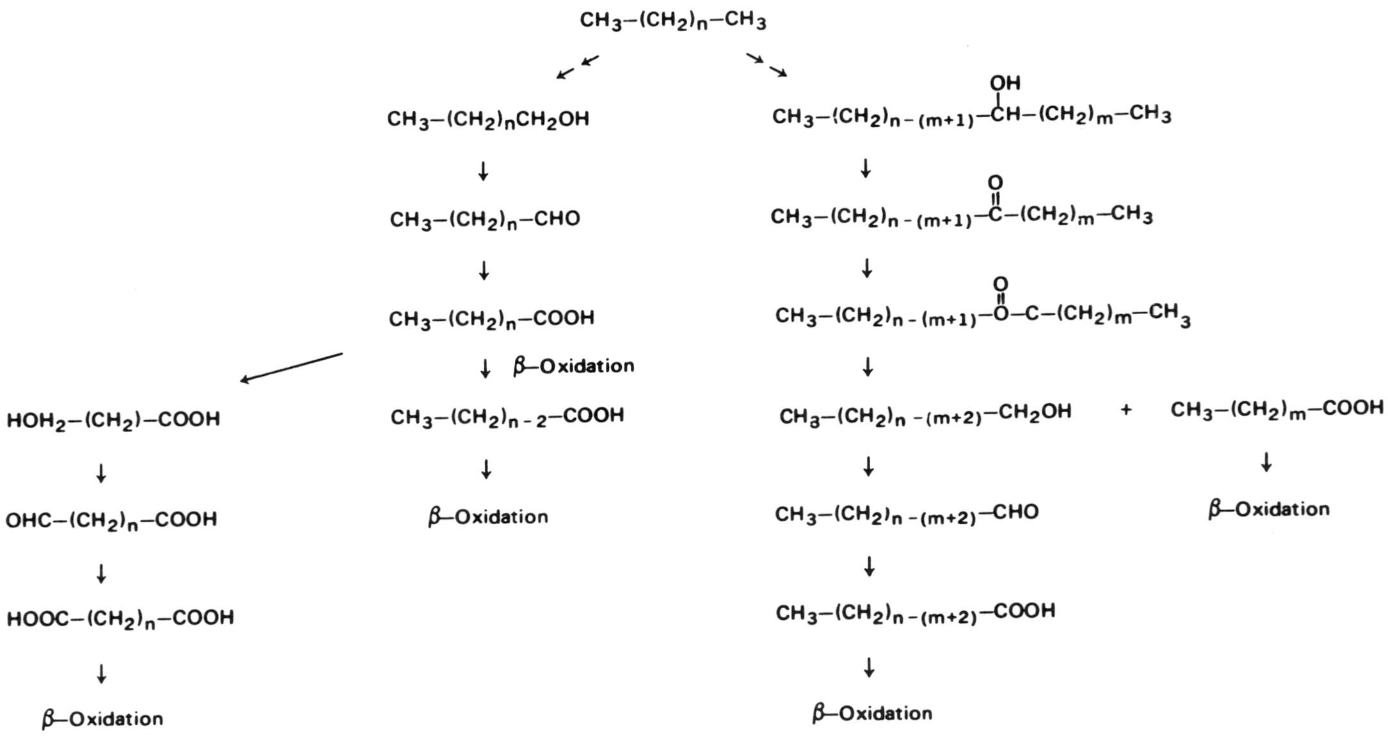

Figure 7.3. Degradative pathways of *n*-paraffins. The symbols *n* and *m* stand for a given number of CH_2 groups. Left, diterminal or *omega*-oxidation; center, monoterminal *beta*-oxidation, and right, subterminal oxidation (from Atlas and Bartha, 1973a).

carbon, the normal *beta*-oxidation sequence is blocked. Degradation may still be accomplished via *omega*-oxidation from the opposite terminus. Alternatively, the blockage posed by the methyl branch can be eliminated by the mechanism elucidated by Seubert and Fass (1964). This pathway, shown on Figure 7.4, essentially elongates the methyl branch by a carboxylation step and the resulting C_2 unit is released as acetic acid.

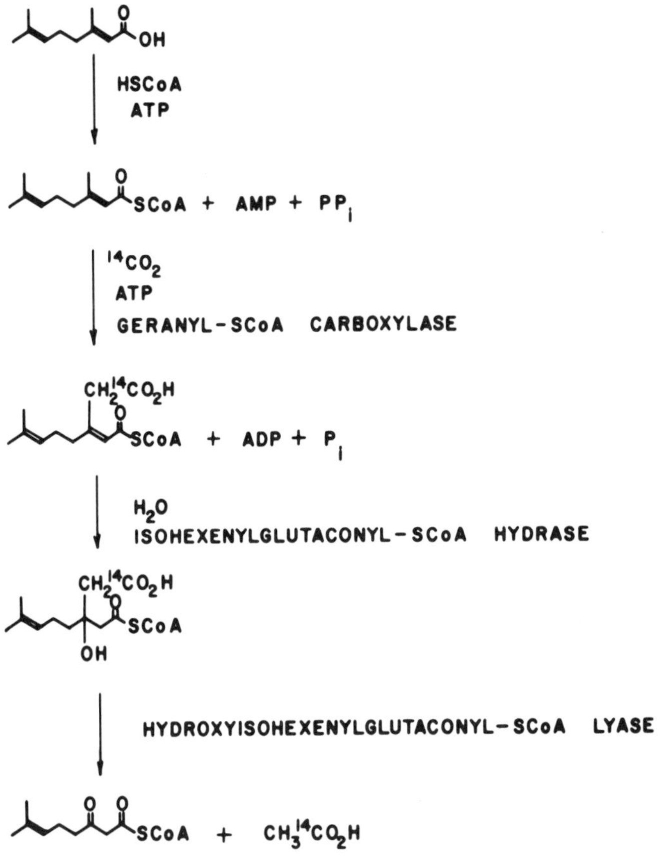

Figure 7.4. Oxidative demethylation of a 3-methyl branched acid, according to Seubert and Fass (1964). The illustrated metabolic fixation of radioactive carbon dioxide in free acetic acid provides evidence of the operation of such a pathway (from Pirnik, 1977).

Alkenes may be attacked, either as the alkanes at a saturated terminal carbon, or may be oxidized directly at the double bond with formation of an epoxy compound. It is not clear whether the epoxy intermediates are converted to diols enzymatically or by spontaneous hydration. Subsequently, one of the hydroxy groups is oxidized to carboxyl, resulting in cleavage to a fatty acid and a primary alcohol.

Alicyclic Hydrocarbons

The microbial metabolism of cycloalkanes has been the subject of two recent reviews (Perry, 1977; Trudgill, 1978). A curious fact connected with cycloalkane metabolism is the difficulty in isolating pure cultures capable of growth on unsubstituted cycloalkanes such as cyclopentane, cyclohexane or decalin (Pelz and Rehm, 1971; Beam and Perry, 1973, 1974a). Earlier positive reports have generally remained unconfirmed and probably resulted from insufficiently stringent experimental criteria. A recent isolate by Stirling *et al.* (1977), tentatively identified as a *Nocardia* strain, appears to possess all the enzymes necessary for cyclohexane utilization, but is auxotrophic for biotin and possibly for additional growth factors. It performs cyclohexane utilization at enhanced rates in dual culture with a *Pseudomonas* strain. This report, however, is a rarity, and cycloalkane metabolism in nature appears to occur primarily through cometabol-

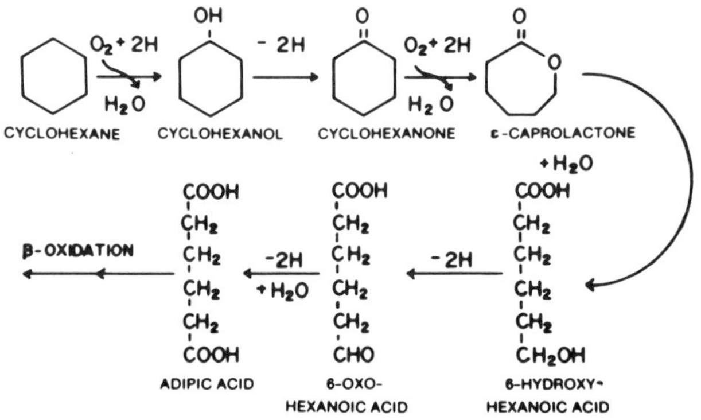

Figure 7.5. Microbial oxidation of cyclohexane as an example for metabolism of alicyclic hydrocarbons (from Atlas and Bartha, 1981).

ism followed by commensal utilization of the products by other microbial strains (Beam and Perry, 1974a; Perry, 1979).

Cooxidative, as well as metabolic degradation of cycloalkane, involves hydroxylation by an as yet not fully-characterized oxidase system to a corresponding cyclic alcohol (Figure 7.5). This product, in turn, is dehydrogenated to the corresponding cyclic ketone. The next step, catalyzed by mixed-function oxidase (monooxygenase) systems, converts the cyclic ketone to a lactone. Ring opening is catalyzed by a lactone hydrolase, or it may occur spontaneously, resulting in an *omega*-hydroxylated carboxylic acid. The hydroxy group is successively dehydrogenated, and the resulting dicarboxylic acid is metabolized further by *beta*-oxidation. It should be recognized that the described pathway was pieced together from cooxidative transformations of cycloalkanes, and from transformations by cycloalkanol and cycloalkanone utilizers that, in contrast to cycloalkane degraders, are isolated with ease. Because the whole metabolic sequence has not been demonstrated in a single organism on the enzymatic level,

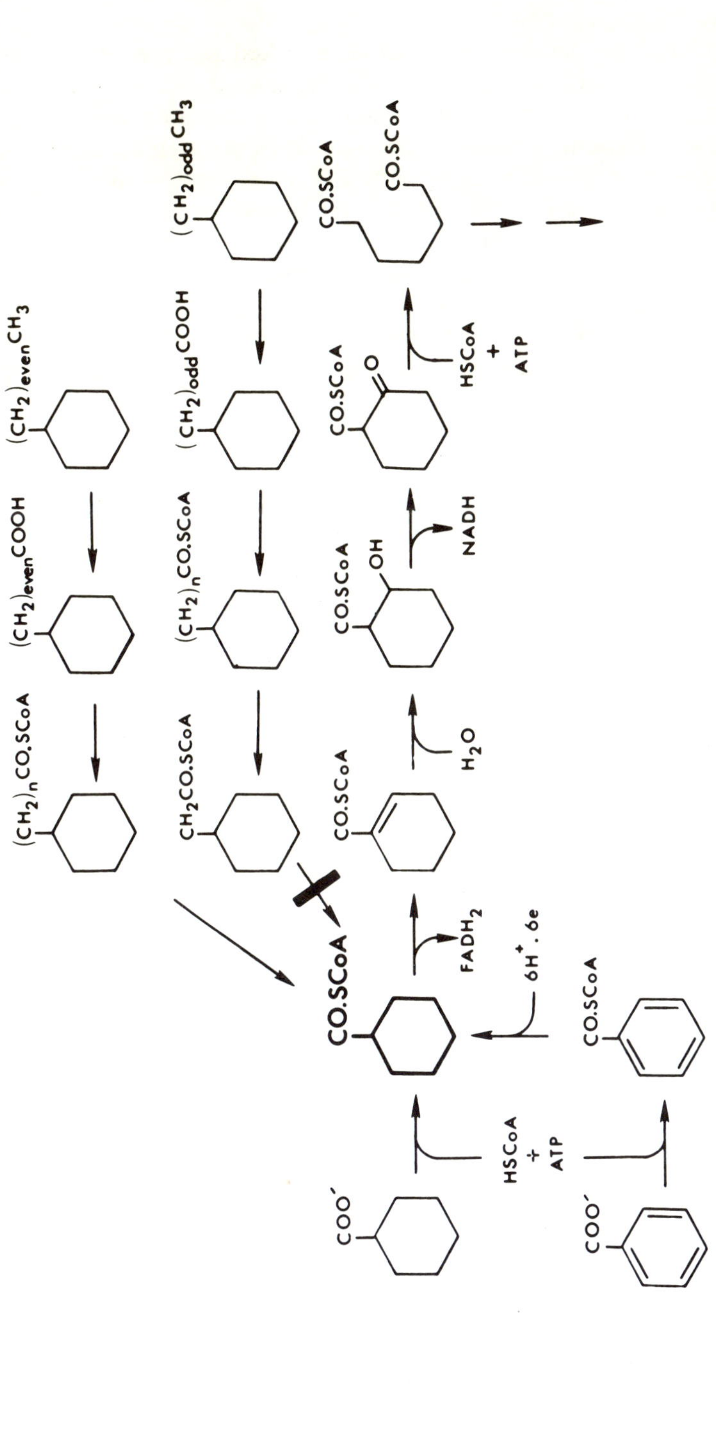

Figure 7.6. Microbial cleavage of cyclohexane carboxylate, benzoate and *n*-alkane-substituted cycloalkanes by *beta*-oxidation. Cycloalkanes with C_{odd} *n*-alkane side chains are probably integrated into this degradation route originally established for the anaerobic dissimilation of benzoate (from Trudgill, 1978).

a considerable amount of uncertainty remains concerning the above mechanism of cycloalkane biodegradation.

Alkyl-substituted cycloalkanes are generally attacked at the terminal (*omega*) alkyl carbon (Beam and Perry, 1974b). As in the case of alkanes, the terminal carbon is oxidized through the hydroxy- and oxo- stages to a carboxy group (Figure 7.6). The resulting cycloalkylcarboxylic acids are metabolized by *beta*-oxidation. An even number of carbons in the alkyl chain leads to cycloalkylacetic acid that is not readily utilized further. An odd number of carbons in the alkyl side chain leads to cycloalkylcarboxylate that can be metabolized further by dehydrogenation and partial aromatization. Hydroxylation is accomplished by addition of water; dehydrogenation leads to ring opening and the formation of a dicarboxylic acid. This pathway can operate under anaerobic conditions and was part of the same sequence that is involved in anaerobic benzoate utilization.

Aromatic and Condensed Polyaromatic Hydrocarbons

The microbial metabolism of aromatic hydrocarbons has been the subject of several recent reviews (Gibson, 1968, 1971, 1977; Hopper, 1978). As illustrated for the simplest aromatic hydrocarbon, benzene (Figure 7.7), initial bacterial oxidation occurs by dioxygenase attack. The postulated dioxetane product is first reduced to *cis*-1,2-dihydroxy-dihydrobenzene and is oxidized, in turn, to catechol, regenerating NADH in the process. In contrast, higher organisms, including eukaryotic microbial forms such as fungi and algae (Cerniglia *et al.*, 1978; Cerniglia *et al.*, 1980a, b), use a mixed-function oxidase (monooxygenase)

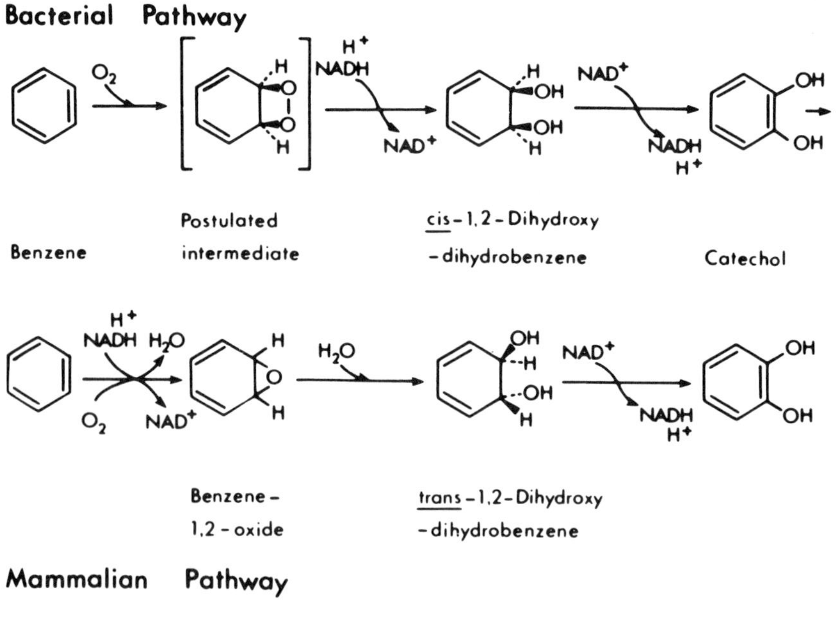

Figure 7.7. Schemes for conversion of benzene to catechol (from Hopper, 1978).

to produce benzene-1,2-oxide and, by addition of H_2O, *trans*-1,2-dihydroxy-dihydrobenzene. The latter is oxidized to catechol with regeneration of NADH. Although the first stable product, catechol, is identical in both cases, the differing mechanisms employed by prokaryotes and eucaryotes are of theoretical interest. Cyanobacteria appear to conform to the prokaryotic pattern (Cerniglia *et al.*, 1980c).

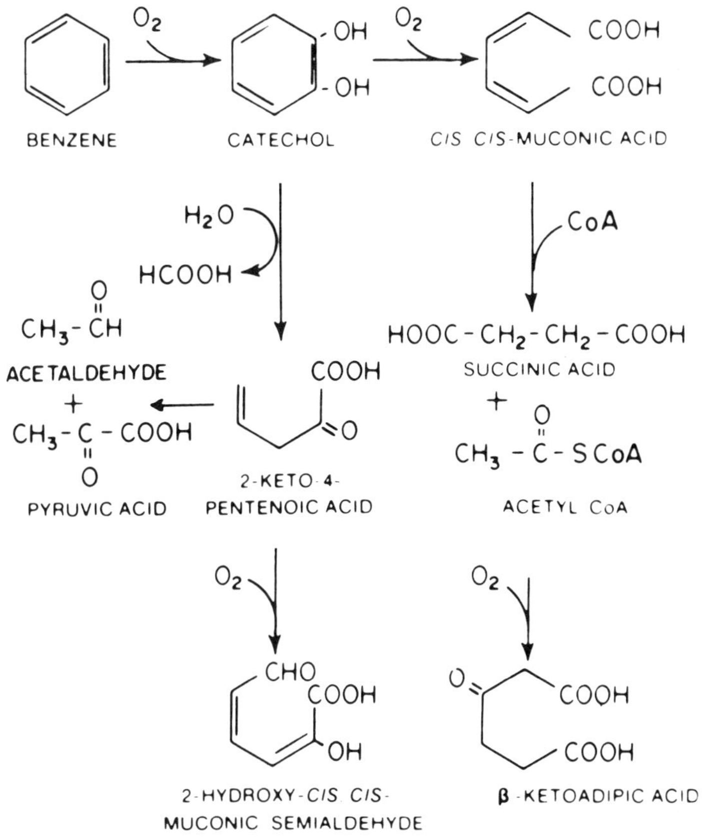

Figure 7.8. Microbial metabolism of the aromatic ring (simplified) by *meta* or *ortho* cleavage, as shown for benzene (from Atlas and Bartha, 1981).

The catechol ring is opened by either *ortho*- or *meta*-cleavage (Figure 7.8) yielding, in the first case, *cis*, *cis*-muconic acid, *beta*-ketoadipic acid and the succinate plus acetate fragments. In the second case the cleavage yields 2-hydroxy- *cis*, *cis*-muconic semialdehyde and subsequently the pyruvate plus 2-keto-4-pentenoic acid fragments are produced.

As reviewed by Cripps and Watkinson (1978), condensed polyaromatic hydrocarbons, having two or more fused aromatic rings, command special interest because some compounds in this group are potential carcinogens or may be transformed to carcinogens by microbial metabolism. Two- and three-ring condensed aromatic hydrocarbons such as naphthalene, anthracene, and phenanthrene are degraded by successive opening of the aromatic rings, essentially by the mechanism described for benzene (Figure 7.9). More highly-condensed polycyclic aromatic hydrocarbons such as benzo(a)pyrene and benzo(a)anthracene (Gibson, 1975, 1976) are cooxidized to dihydrodiols, and thus are activated to carcinogens. They are apparently not extensively degraded by pure cultures and are mineralized to CO_2 in the environment only at extremely

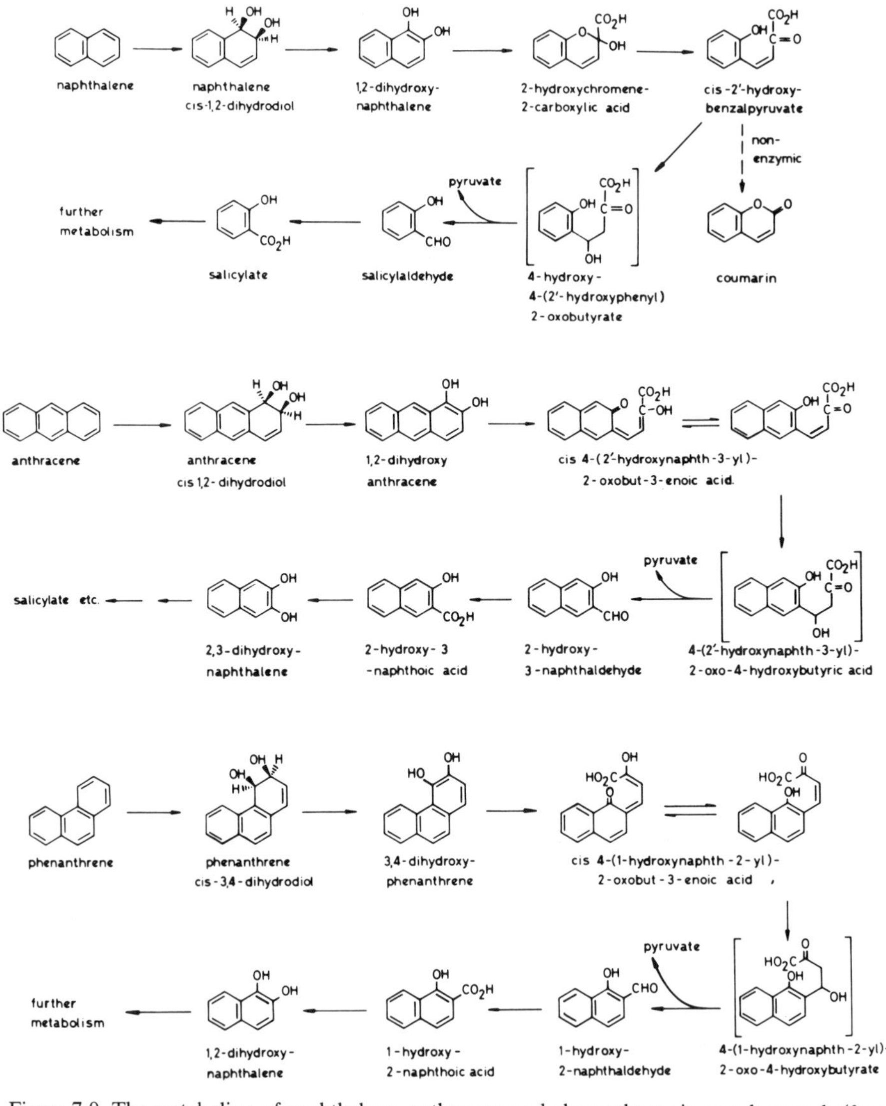

Figure 7.9. The metabolism of naphthalene, anthracene and phenanthrene in pseudomonads (from Cripps and Watkinson, 1978).

slow rates (Lee and Ryan, 1976; Lee, 1977; Herbes and Schwall, 1978). Recently, Wu and Wong (1981) reported microbial methyl hydroxylation of 7,12-dimethylbenzo(a)anthracene, resulting in carcinogenic activation.

Alkylaromatic hydrocarbons with short alkyl moieties, such as toluene, may be degraded by the mechanisms described for benzene (Figure 7.10). Alternatively, the initial attack may occur at the methyl group with a conversion, in several steps, to benzoic acid. Oxidative decarboxylation leads to catechol that is subject to ring cleavage. Phenylalkanes with long alkyl chains are regularly

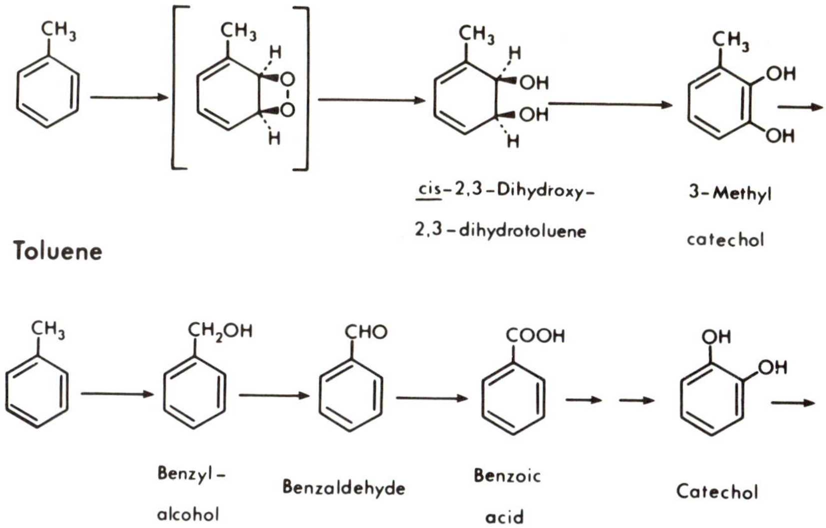

Figure 7.10. Pathways for toluene metabolism (from Hopper, 1978).

metabolized starting at the terminal carbon of the alkyl moiety (*omega*-oxidation). Successive *beta*-oxidation steps shorten the alkyl chain to benzoic acid (for odd carbon numbers) or to phenylacetic acid (for even carbon numbers). Benzoate is easily degraded as outlined above, but phenylacetic acid is more persistent and, in pure culture experiments, often accumulates as an end product.

Asphaltenes and Resins

These fractions of crude oils are defined by solubility and chromatographic elution characteristics. Asphaltenes are a heterogeneous and poorly-characterized assortment of compounds with high molecular weights and low volatility and solubility characteristics. The polar, often heterocyclic "NSO" compounds that are not hydrocarbons in the strict sense of the definition, make up the resin fraction of petroleum. Analytical techniques are, in general, inadequate to define the individual chemical structures of asphaltenes and are even less able to follow their fate in the environment. However, practical experience shows that these compounds are highly resistant to biodegradation. Tar that is high in asphaltic components is widely distributed and extremely persistent in the marine environment (Butler *et al.*, 1973). Tar and asphalt are used in wood preservation, roofing and pavements because of their hydrophobic character, persistence and antimicrobial properties. Biodegradation of these compounds, if any, is probably limited to small cometabolic changes that have little significance in terms of their overall environmental fate.

During "weathering" of a crude oil by photodegradation and biodegradation, the asphaltic fraction tends to increase in relative and, occasionally, also in absolute amount (Walker *et al.*, 1976b). The latter fact shows that, besides being

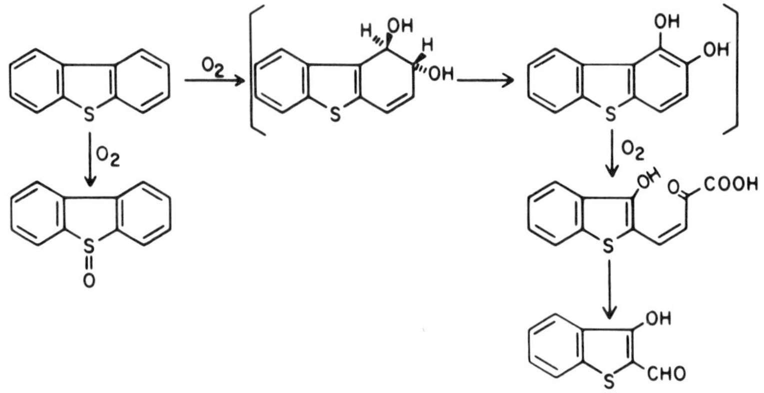

Figure 7.11. Pathways of dibenzothiophene oxidation by a *Beijerinckia* sp. showing aromatic ring cleavage and sulfur oxidation (from Laborde and Gibson, 1977).

highly resistant to degradation, asphaltic and "tar" material is actively formed from other components of the crude oil. Free radicals, that are formed in both photo- and biodegradation reactions and are capable of initiating condensation and polymerization processes, are responsible for this phenomenon.

The low molecular weight representatives of the resin (NSO) fraction, such as phenols, cresols, thiols, thiophenes, pyridines and pyrroles, have considerable toxicity towards microorganisms but, at least some of them, are likely to be biodegraded at low concentrations. Very little has been published in this area and available information is restricted to the condensed dibenzothiophenes (Yamada *et al.*, 1968; Nakatani *et al.*, 1968; Kodama *et al.*, 1970, 1973). Dibenzothiophene was converted to oxygenated products but the central thiophene ring remained intact. Hou and Laskin (1976) obtained similar results with *Pseudomonas aeruginosa* growing on *n*-alkanes in the presence of dibenzothiophene; one of the identified products was 4(2-(3-hydroxy)-thianaphthenyl)-2-hydroxy-3-butenoic acid. Cometabolism by *Beijerinckia* transformed dibenzothiophene as illustrated in Figure 7.11 (Laborde and Gibson, 1977).

Biodegradation of Petroleum Hydrocarbon Mixtures

The preceding discussion of biodegradation pathways dealt with individual classes of hydrocarbons. In case of a crude oil spill or discharges of formation waters, microorganisms are confronted with a large spectrum of potential hydrocarbon substrates simultaneously. In such situations, will all potential substrates be degraded simultaneously, or will biodegradation proceed in a predictable sequence? How will the presence of one substrate influence the biodegradability of another one? We are still far from being able to resolve the above questions completely, but some answers have begun to emerge.

Substrate Range

Hydrocarbons differ in their capacity to serve as microbial substrates. This is true in relation to a single microbial strain and also, in a broader sense, as related to the microbial community of a given marine environment. Based on the reviews listed in the Introduction, some generalization can be made concerning the suitability of various hydrocarbons as substrates but, in most cases, selective pressure was applied already at the time of isolation by supplying specific substrates. Therefore, particularly relevant are the studies by Soli and Bens (1972) and Soli (1973) who used a mixture of 28 hydrocarbons including *n*- and *iso*-alkanes, aromatic and alicyclic hydrocarbons. They picked colonies randomly from their seawater agar plates for subsequent substrate range evaluation. Their results are consistent with the following broadly based summary statements expressed, in part, in previous reviews (Bartha and Altas, 1977; Atlas, 1981).

1. *n*-Alkanes of the C_{10}-C_{22} range are the most readily and frequently utilized hydrocarbon substrates. The microbial degradation of low molecular weight gaseous C_1-C_4 alkanes is restricted to a few specialized species that have the necessary enzymes, and C_5-C_9 alkanes have solvent characteristics that are tolerated by the membranes of relatively few hydrocarbon degraders. The physical characteristics of *n*-alkanes above C_{22} are not favorable for biodegradation because at physiological temperatures they are solids with extremely low water solubility. Nevertheless, slow biodegradation of *n*-alkanes up to C_{44} in length has been demonstrated (Haines and Alexander, 1974).

2. *Iso*-alkanes are less readily utilized as compared to *n*-alkanes. Methyl branching in 3-position is a hindrance to *beta*-oxidation and relatively few alkane degraders possess mechanisms to bypass such blockage. Extensive branching, resulting in quaternary carbon atoms, may render an *iso*-alkane completely resistant to biodegradation.

3. Olefins tend to be more toxic and, at least under aerobic conditions, less readily utilizable than the corresponding alkanes. At least theoretically, olefins should be less stable under anaerobic conditions than alkanes because they can be hydroxylated without a need for oxygenases.

4. Monoaromatic hydrocarbons, because of their solvent properties, have considerable membrane toxicity but, in low concentrations, they are rapidly utilized by a considerable number of microorganisms. Condensed polyaromatics with two to four rings are somewhat less toxic and are biodegradable at rates that decrease with the level of condensation. Condensed polyaromatics with five and more rings fail to serve as growth substrates and are eliminated from the environment very slowly. The initial metabolic transformation steps, if any, are cometabolic.

5. Low molecular weight cycloalkanes, such as cyclohexane and decalin, exhibit a high degree of solvent-type membrane toxicity. They serve as growth substrates only in exceptional cases. At low concentrations, in mixed enrichments and in the environment, cycloalkanes are degraded at moderate rates. Initial cometabolic attack followed by commensal utilization of the products is the main mechanism of biodegradation. Highly condensed cycloalkanes are refractory.

6. Resins, when not highly condensed, may be subject to limited microbial metabolism. The highly condensed asphaltenes are virtually immune to biodegradation.

When hydrocarbons become available to a microbial community in a complex mixture such as petroleum, biodegradation of most petroleum compounds occurs simultaneously but at widely differing rates. Most rapid is the biodegradation of the *n*-alkanes. In the gas chromatographic profile of a crude oil, the *n*-alkanes form a characteristic peak series, and the reduction or elimination of these peaks is consistently observed early during biodegradation. The "envelope" part of the profile that represents the unresolved *iso*-alkane, cycloparaffin and aromatic components shows comparatively little change during the same time period. More sophisticated analytical approaches involving class separation by column chromatography followed by computerized GC-MS analysis (Walker *et al.*, 1975d), show, however, a slow reduction in the above components.

Diauxie or Sparing

In pure culture, a *Nocardia rhodochrous* strain (formerly *Brevibacterium erythrogenes*, Pirnik *et al.*, 1974) showed typical diauxie on *normal* and *iso*-alkanes, the presence of the former repressing the utilization of the latter. It is not clear whether such regulatory mechanisms play any substantial role in the case of complex microbial communities. The more rapid disappearance of *n*-alkanes appears to be associated with (a) the high numbers of microorganisms capable of degrading *n*-alkanes, and (b) the relatively high rates at which these structurally unhindered compounds can be metabolized. The *n*-alkane utilizers may repress utilization of other petroleum components in an ecological sense by effectively competing for limiting resources such as mineral nutrients and perhaps oxygen. After exhaustion of the *n*-alkanes, many microorganisms with narrow substrate ranges (Fredericks, 1966) would be eliminated and ecological succession would favor the hydrocarbon degraders that can utilize the more "difficult" substrates, even if they do so at lower rates. Some experimental evidence for such ecological succession was obtained with multi-stage continuous enrichments (Horowitz *et al.*, 1975).

Cometabolism

The complexity of the biodegradation sequence in a multi-substrate situation is increased by the phenomenon of cometabolism. As discussed earlier, some of the more "difficult" hydrocarbon substrates fail to support growth but may undergo limited cometabolic transformations by microorganisms growing on other, more easily utilizable hydrocarbons. The latter ones are often *n*-alkanes, and after these are exhausted, cometabolic changes that would pave the way for subsequent commensal utilization of the more difficult substrates may no longer occur.

Inhibitory Components

Inhibition of hydrocarbon-degrading microbial activity by the presence of toxic volatile components of some crude oils was reported by Atlas and Bartha (1972b) and Atlas (1975). No component of certain crude oils was biodegraded until bacteriostatic components of the crude were eliminated by evaporation. The time required for evaporation was dependent on water temperature. In yet another type of interaction between hydrocarbon substrates, Walker and Cooney (1975) reported that hexadecane oxidation by *Cladosporium resinae* was enhanced by the presence of *p*-xylene and toluene. The latter compounds were not metabolized, and the authors concluded that the basis of enhancement was an increase of hexadecane transport across the cell membrane caused by the presence of *p*-xylene and toluene.

Summary

The biodegradation of petroleum hydrocarbons varies greatly, leading to sequential disappearance of individual components and to successional changes in the degrading microbial community. The presence of one hydrocarbon substrate may influence the biodegradation of another in either a positive or a negative manner by several possible mechanisms. We start to recognize the range of substrate interactions that can occur, but we seldom know which ones actually do occur, and how significant each of them is in determining the overall fate of the polluting oil.

To examine the fate of individual hydrocarbons within varying crude oils, it would be possible to spike different oils with specific radiolabeled hydrocarbons and to measure the rate of $^{14}CO_2$ production. Such analyses should be compared with GC-MS analyses of the oil to determine whether overall rates of degradation of aromatic and aliphatic hydrocarbons correlate with the rates of degradation of these specific tracer compounds. These measurements would provide a basis for developing a predictive capability for estimating the persistence times of particular hydrocarbons within the context of different oil contamination situations.

PRODUCTS OF PETROLEUM BIODEGRADATION

Mineralization of Petroleum

The complete degradation of petroleum hydrocarbons, known as mineralization, produces carbon dioxide and water. Only a portion of the oil that is biodegraded, however, is converted to these end products. Part of the oil that is metabolized forms cellular biomass and intermediary products. Given a theoretical situation of 100 percent oil degradation, probably only 50-70% would be converted to carbon dioxide and water. Additionally, if an oil contains compounds other than hydrocarbons, e.g., pyridines and mercaptans, various other products, such as oxides of nitrogen and sulfur, would be formed during extensive oil biodegradation.

Biodegradation Intermediates and Their Effects

The microbial metabolism of hydrocarbons may result in the temporary accumulation of biodegradation intermediates. These may have an effect on the further biodegradation of hydrocarbons, on the microbial community, and on higher organisms. Unfortunately, our knowledge in this area is primarily based on *in vitro* studies conducted with pure cultures. It should be recognized that the extrapolation of these results to the marine environment is extremely tenuous.

Fatty acids are produced in the biodegradation of every hydrocarbon class, and fatty acids have considerable biological activity. In yeast cultures grown on *n*-hexadecane, Aida and Yamaguchi (1969) noticed a dialyzable inhibitory factor and identified lauric acid as the principal inhibitor. Atlas and Bartha (1973c) found that crude oil biodegradation by two marine bacteria was inhibited by a variety of accumulating fatty acids. LePetit and Tagger (1976) obtained similar results, acetate being the main identified inhibitory factor. Several marine *Pseudomonas* strains growing on naphthalenes and methylnaphthalenes were inhibited by salicyclic acid derivatives accumulating in the culture medium (D.D. Raymond and R. Bartha, unpublished results). It is suspected that effects such as the listed ones will tend to be minimal or absent under environmental conditions because of the degradation of fatty acids by a diverse microbial community and the dilution of any metabolites in large volumes of sea water.

In an interdisciplinary study conducted using large-scale marine microcosms, the fate of benzanthracene was followed over a 230-day period (Farrington *et al.*, 1982; Hinga *et al.*, 1980; Oviatt *et al.*, 1982). During this time 29% of this polynuclear aromatic hydrocarbon was mineralized to carbon dioxide, while the remaining detectable compound was evenly divided between parent compound and intermediate metabolic products. The hydrocarbon was rapidly removed from the water column and incorporated into the sediment. Once incorporated into the sediment, the degradation rate was very low. The incorporation of hydrocarbons or degradation products into the sediment had a long-lasting effect on the benthic organisms.

As discussed earlier, microbial metabolism of crude oil results in its dispersion. This not only increases the availability of the hydrocarbons to hydrocarbon degraders, but also increases the hydrocarbon exposure of other marine micro- and macroorganisms. Whether the biodegradation is spontaneous or artificially stimulated, the temporary increase in toxicity can be substantial: bacterially emulsified oil was 100-fold more toxic to sea urchin embryos than the intact crude oil (Rosenberg *et al.*, 1975). Dispersion of crude oils by hydrocarbonoclastic yeasts also increased toxicity to guppies (*Lebistes reticulatus*) (Cook *et al.*, 1973).

Somewhat paradoxical is the microbially mediated production of long-chain alkanes (waxes) during biodegradation of petroleum (R. Kallio, personal communication; Walker and Colwell, 1976b; Atlas *et al.*, 1981). These are produced only as a consequence of biodegradation and not of non-biological weathering. The mechanism of their formation is as yet unexplored. A head to head condensation of reactive biodegradation intermediates (e.g., free radicals) is considered a possible explanation for their appearance. They are unlikely to have inhibitory effects but may participate in the formation of tar globules.

Several aspects of intermediary metabolite formation have critical implications for the long-term effects of petroleum contaminants. Various microbial degradation products as well as photooxidation products probably play important roles in the formation of highly weathered residues. The formation of weathered residues leads to very long persistence times of hydrocarbon contaminants in marine ecosystems. The possible role of microorganisms in tar formation and the ability of microbes to eventually degrade highly weathered residues has not been sufficiently analyzed. The analytical complexity of examining weathered petroleum, with its high molecular weight and substituted components, makes this a very difficult problem to study. As new methods develop in analytical chemistry, it may be possible to approach the problems concerning microbes and weathered oil that are not presently possible to investigate.

In addition to questions relating to microorganisms and the formation and fate of weathered petroleum, it is important to consider the role of cometabolism in determining the fate of petroleum hydrocarbons. Cometabolism results in the formation of products that are not further degraded by the hydrocarbon-degrading populations. This tends to lead to the accumulation of high amounts of these intermediates in the environment. The oxygenated and partially-degraded hydrocarbons may be biologically active and some are potentially toxic or carcinogenic to higher organisms. The rates of cometabolic product formation and the persistence times of such products are important factors influencing the long-term effects of petroleum contaminants in marine ecosystems.

RATES OF PETROLEUM BIODEGRADATION

Factors Influencing Petroleum Biodegradation

The rate of petroleum biodegradation is the single most important parameter in the self-purification of the marine environment. Before trying to define the rates, the most important factors influencing them will be discussed. Some of the determining factors are inherent to the polluting oil, others are environmental and subject to variation. Some environmental parameters such as pH and salinity that are known to influence petroleum biodegradation in other environmental systems (Dibble and Bartha, 1979; Ward and Brock, 1978a) will not be discussed here. These parameters are in a generally favorable range and are remarkably uniform throughout the marine environment (Tait and DeSanto, 1972).

Organismic factors, such as the abundance and substrate range of microbes, were discussed in previous sections of this paper. Little is known about the rate of change of microbial communities in response to low levels of chronic pollution. It is clear that microbial communities in chronically-polluted areas are different than those in pristine areas. In areas contaminated by chronic inputs of hydrocarbons, there are elevated proportions of hydrocarbon utilizers in the microbial community compared to areas not subject to chronic petroleum contamination

(Atlas and Bartha, 1973b). It is also clear that elevated densities of hydrocarbon-degrading bacteria represent an increased potential for biodegradative removal of hydrocarbon contaminants.

Composition and Weathering of Petroleum

The biodegradable portion of various crude oils ranges from 11 to 90% (Colwell and Walker, 1977). A low percentage of biodegradation may result from a high amount of volatile components, since these will ordinarily evaporate much too fast from the water surface to allow significant biodegradation to take place. Because of this, the biodegradation percentage is often related to the "topped crude" (pre-evaporated oil from which the volatiles have been removed), rather than the intact one. Low biodegradation percentages can result also from high proportions of condensed polyaromatic, condensed cycloparaffinic and asphaltic petroleum components, because these compounds are either recalcitrant or are biodegraded at extremely slow rates.

Toxicity of certain petroleum components can delay or prevent the biodegradation of susceptible ones. Atlas and Bartha (1972b) and Atlas (1975) noted such action by volatile components of certain petroleums, but they did not characterize these components chemically. Toxicity may not be inherent to the crude oil but may be acquired upon exposure in the environment. Toxic and lipophilic substances such as pesticides (Seba and Corcoran, 1969; Hartung and Klinger, 1970), polychlorinated biphenyls (Sayler and Colwell, 1976), and mercury (Walker and Colwell, 1976c; Sayler and Colwell, 1976) can be concentrated in the oil slick 10^2-10^5 times above their ambient concentration in the water. Although heavy metal resistance may be genetically linked in plasmids coding for hydrocarbon-degrading enzymes (Chakrabarty and Friello, 1974; Walker and Colwell, 1974b), substantial mercury concentrations (85 ppm) in petroleum have prevented its biodegradation (Walker and Colwell, 1976c).

The weathering history of spilled petroleum greatly influences its availability for biodegradation. In addition to removal of volatile toxic components by evaporation, weathering can induce changes by photodegradation (Freegarde *et al.*, 1970; Burwood and Speers, 1974). Photooxidation occurs preferentially at tertiary carbons, thus removing some of the methyl branches that hinder biodegradation of *iso*-alkanes. However, Van der Linden (1978) reported that, at relatively high concentrations, photooxidation products of gas oil may become toxic to degrading microorganisms (see Payne *et al.*, Chapter 5, for a further discussion of photooxidation products). Agitation by wave action may lead to enhanced dispersal favorable to subsequent biodegradation but may also lead to formation of water-in-oil emulsions called "chocolate mousse" or "mousse" (Berridge *et al.*, 1968). The tendency to form mousse is in part an inherent property of some petroleums and in part is promoted by photooxidative (Burwood and Speers, 1974) and biodegradative (Berridge *et al.*, 1968) changes. Inside mousse aggregates the availability of oxygen and mineral nutrients is severely restricted and biodegradation is hindered. Mousse formation also reduces photooxidative and evaporative losses. Davis and Gibbs (1975) could not detect hydrocarbon loss from large mousse accumulations over a 2-year period. The formation of tarballs (Butler *et al.*, 1973) restricts biodegradation and weathering for similar reasons. Mousse and tarball formation have been suggested as major limiting factors of petroleum biodegradation in the IXTOC-I (Atlas *et al.*, 1980) and *Metula* (Colwell *et al.*, 1978) oil spills.

Temperature

As every other microbial activity, oil biodegradation is strongly temperature dependent. Hydrocarbon biodegradation can occur in the thermophilic range (Klug and Markovetz, 1967; Mateles *et al.*, 1967), up to 70°C. Although some thermophilic hydrocarbon-degrading bacteria (optimal growth at 55-65°C) have been isolated from estuarine sediments (Merkel *et al.*, 1978), it is unlikely that such microorganisms have a significant role in an environment where temperatures range from mesophilic to psychrophilic. Hydrocarbon biodegradation was reported at temperatures below 0°C (ZoBell and Agosti, 1972) because of arctic and subarctic oil exploration, a substantial interest developed in psychrophilic and psychrotrophic hydrocarbon degraders (Malins, 1977). Generally, at low water temperatures, the rate and extent of hydrocarbon biodegradation was severely restricted (Gunkel, 1968; ZoBell, 1969; Mulkins-Phillips and Stewart, 1974b). Somewhat surprisingly, Walker and Colwell (1974a) and Colwell *et al.* (1978) found slower but more extensive petroleum biodegradation at low (0 and 3°C, respectively) than at higher temperatures. The reduced toxicity of some hydrocarbon components at the lower temperatures was considered as a possible explanation for this unexpected finding. The more extensive albeit slower removal of hydrocarbons at low temperatures theoretically could lead to a slower but more complete recovery of the ecosystem. This question requires further investigation.

Some investigators determined the temperature-dependence of hydrocarbon biodegradation rates in terms of Q_{10} values. Gibbs *et al.* (1975) and Gibbs and Davis (1976) determined an average Q_{10} value of 2.7 over the 6-26°C range. Arhelger *et al.* (1977) compared the biodegradation of ^{14}C-dodecane at several subarctic and arctic locations and found 0.7 g/l/day at Port Valdez, 0.5 g/l/day in the Chukchi Sea and 0.001 g/l/day in the Arctic Ocean. Temperature and perhaps additional factors, such as the presence of different microbial populations, caused the decreasing rates of biodegradation observed at the increasingly polar locations.

Atlas and Bartha (1972b) and Atlas (1975) studied the biodegradation of various crude oils at low temperatures. In addition to the expected decline in biodegradation rates, they observed long lag periods with some crude oils prior to the start of biodegradation. Temperature-dependent evaporation of volatile toxic components was shown to explain this phenomenon. Similar results were found by Atlas (1975) for various crude oils. Based upon these results, Atlas concluded that biodegradation of oils containing low molecular weight components would likely be delayed in arctic ecosystems until the volatile fraction slowly evaporated. Although temperature clearly influences rates of petroleum biodegradation, other environmental factors, such as available concentrations of oxygen and mineral nutrients, often are the rate-limiting factors; oil degradation rates in arctic ecosystems are most likely directly limited by factors other than temperature.

Pressure

Most petroleum biodegradation is expected to take place in surface slicks or in the water column at shallow depths. However, some crude oils exceed the specific weight of water and others may do so at an advanced stage of weathering. Thus, hydrocarbons can enter the deep-sea environment and, consequently, the effect of hydrostatic pressure on oil biodegradation is of interest. The findings of Jannasch and Wirsen (1973) showed that high hydrostatic pressure in combination with low temperatures drastically reduces the rate of metabolic processes and prompted investigations in this direction using hydrocarbon substrates. A lack of biodegradation in deep-sea and anaerobic sediments may affect the benthic organisms living in such ecosystems. The long-term persistence of oil in such ecosystems could preclude ecological recovery for extended time periods.

Schwarz *et al.* (1974a,b, 1975) obtained an enrichment culture from sediment taken at 4940-m depth. This enrichment culture was incubated with radiolabeled *n*-tetradecane and *n*-hexadecane under *in situ* and ambient surface pressure and temperature conditions. At 20 and 25°C, 500-atm pressure delayed biodegradation only moderately. The same pressure at 4°C reduced metabolism by more than one order of magnitude as compared to a 1-atm control incubated at the same temperature. The authors concluded that the biodegradation of any petroleum residue that reaches the deep-sea environment will be exceedingly slow (Schwarz *et al.*, 1975).

Oxygen

As previously discussed, the initial attack on hydrocarbons is commonly performed by oxygenases, and thus the presence of molecular oxygen is critical for hydrocarbon biodegradation. There have been several reports of anaerobic conversion of hydrocarbons (Senez and Azoulay, 1961; Choteau *et al.*, 1962; Iizuka *et al.*, 1969; Traxler and Bernard, 1969; Parekh *et al.*, 1977). These were all *in vitro* studies with isolated cultures. Essentially, they demonstrated that *n*-alkanes, such as *n*-heptane or *n*-decane, can be dehydrogenated to the corresponding *n*-alkene with subsequent hydration across the double bond. Thus, a pathway appears to exist for the anaerobic utilization of alkanes, with sulfate or nitrate serving as electron sinks. However, it should be emphasized that consistent and abundant growth of microorganisms on hydrocarbons as the only sources of carbon and energy has not been demonstrated under strictly anaerobic conditions. For the utilization of hydrocarbons, which are fully reduced substrates, anaerobic conditions offer a rather marginal energy balance. For this or other yet undefined reasons, anaerobic hydrocarbon biodegradation in the environment is either undetectable, or orders of magnitude lower than aerobic hydrocarbon biodegradation (Bailey *et al.*, 1973; Ward and Brock, 1978b, Delaune *et al.*, 1980; Ward *et al.*, 1980). Table 7.2 shows the results of one of the most recent and sophisticated measurements, and shows that hydrocarbon

TABLE 7.2

Anaerobic hydrocarbon oxidation potential in reducing oil-polluted sediments following the *Amoco Cadiz* spill after 233 days of incubation (from Ward *et al.*, 1980)

	Percent of added ^{14}C recovered in oxidation products[a]					
	Aber Wrac'h mud flat			*Ile Grande mud flat*		
Compound	*0–5 cm aerobic*	*5–10 cm anaerobic*	*10–15 cm anaerobic*	*0–2 cm aerobic*	*2–7 cm anaerobic*	*7–12 cm anaerobic*
Hexadecane	33	3	+[b]	45	+	+
Heptadecane	25	1	0	34	0	+
Heptadecane	28	0	+	34	0	0
Benzene	26	0	2	45	0	0
Toluene (ring ^{14}C)	27	2	1.4	34	2.3	2.9
Toluene (methyl ^{14}C)	18	+	0	29	5	1.4
Naphthalene	72	0	0	78	0	0
Benzo(a)pyrene	3	0	0	0	0	0

[a] $^{14}CO_2$ + $^{14}CH_4$ produced by reduction of $^{14}CO_2$ on long-term incubation.
[b] + indicates that $^{14}CO_2$ was apparently present but was nonquantifiable because levels were very low.

biodegradation is restricted to the aerobic sediment, which in this study was the upper 5-cm sediment layer that was partially aerobic.

Because field measurements show that the presence of oxygen is critical for significant hydrocarbon biodegradation, the availability of oxygen must be considered as an important limiting factor for petroleum biodegradation in the marine environment. Oxygen limitation is highly unlikely in the case of surface slicks that are in direct contact with atmospheric oxygen. In the case of oil dispersed in the water column, oxygen limitation may occur. Depending on the temperature, sea water contains 6-11 mg dissolved oxygen per liter. The complete oxidation of hydrocarbons requires 3-4 weight equivalents in oxygen. Based on these figures, it can be calculated that the complete oxidation of 1 l of oil will exhaust the dissolved oxygen in 320,000 l of sea water (ZoBell, 1969). Whether or not oxygen actually becomes limiting depends on the oil concentration, the rate of its biodegradation, and oxygen replenishment by turbulent diffusion from the surface or photosynthesis. Temperature interacts with oxygen supply in two ways: elevated seawater temperatures reduce oxygen solubility and increase the rates of metabolic oxygen consumption. Thus, oxygen limitation in the water column is most likely to develop if the sea water is warm and calm, if the concentration of the dispersed oil is high, or if the degradation of other organic matter competes for dissolved oxygen. Aminot (1981) was able to measure the oxygen depletion in the water column underlying oil from the *Amoco Cadiz* spill and to use the oxygen deficit to calculate a rate of hydrocarbon oxidation of 0.2-0.4 mg/l.

Marine sediments, with the exception of their upper few centimeters or millimeters, tend to be anaerobic. The thickness of the upper oxygenated layer depends on the porosity, organic content and degree of the physical or biological (bioturbation) mixing of the sediment. The importance of bioturbation in providing oxygen for oil biodegradation activity has been documented by Gordon *et al.* (1978) who found that polychaetes were very important in determining the rates of weathering of sediment-bound oil. Petroleum that became incorporated into anaerobic marine sediments is essentially immune to biodegradation until some disturbance releases the oil or oxygenates the sediment. Haines and Atlas (1983) found that oil degradation in arctic sediments was not detectable for over one year, in part because of a lack of oxygen and bioturbation during this period.

Petroleum protected from biodegradation within anaerobic sediments can later be remobilized by disturbances to the sediment. Such introduction of undegraded hydrocarbons into ecosystems can cause sporadic and long-term introduction of potentially toxic compounds.

Mineral Nutrients

Petroleum as a microbial substrate supplies carbon and energy but little else. The nitrogen and sulfur of the NSO fraction are mostly in heterocyclic rings and generally insufficient and unavailable. Consequently, petroleum-degrading microorganisms must obtain their essential mineral nutrients from the sea water. Considering the composition of sea water (Tait and DeSanto, 1972) in relation to mineral nutrient requirements, phosphorus, nitrogen, and iron are likely to approach limiting concentrations, while other essential elements should be present in sufficient or excess concentration. *In vitro* experiments employing relatively high oil to sea water ratios have convincingly demonstrated the phosphorus and nitrogen limitation of petroleum biodegradation (Atlas and Bartha, 1972c). Iron limitation was confirmed in clear offshore sea water but not in sediment-rich coastal sea water (Dibble and Bartha, 1976). The need for phosphorus and nitrogen supplementation for optimal oil biodegradation activity in sea water was noted also by Bridie and Bos (1971), Reisfeld *et al.* (1972), Gibbs (1975), and LePetit and N'Guyen (1976). Based on the data of Atlas and Bartha (1972c) and Reisfeld *et al.* (1972), 1.5-2.5% N and 0.2% P (w/w) addition (calculated on basis of petroleum that was actually degraded) allowed maximal petroleum biodegradation in these *in vitro* experiments. Bridie and Bos (1971) found somewhat higher and Gibbs (1975) somewhat lower requirements, explainable by differing experimental conditions and petroleum:sea water ratios. A summary discussion of the relation of mineral nutrient requirements for biodegradation of oil pollutants was provided by Floodgate (1979) with the conclusion that the scarcity of mineral nutrients in sea water is often limiting for petroleum biodegradation not only *in vitro* but also under closely simulated environmental conditions.

Physical Form of the Oil

Oil in the marine environment can exist in several forms. Spilled oil may form a thin surface film or slick, may form a stable oil-water emulsion (mousse), may become associated with suspended particles in the water column, or may become entrained in the sediments as concentrated pockets of oil. The surface areas of these various physical forms of oil accessible to microorganisms, as well as the associated abiotic surroundings, have a marked influence on the fate of the petroleum hydrocarbons.

In a study of oil spilled from the IXTOC-I well blowout, Atlas *et al.* (1982) concluded that oil was protected within the stable mousse emulsion that formed. The limited surface area of the mousse and the flux of oxygen and mineral nutrients to the oil-degrading microorganisms at the oil-water interface undoubtedly limited the rate of degradation. Compared to the extremely limited degradation of oil in the mousse, Pfaender and Buckley (1980) found that oil from IXTOC-I was degraded within the water column; they estimated the rate of oil degradation to range from 0.01 to 44 ng/l/h.

In an *in situ* study of the fate of Prudhoe Bay crude oil in nearshore sediments of the Beaufort Sea, Haines and Atlas (1982) found that oil in arctic sediments formed discrete pockets or patches of concentrated oil; the oil in these sediments was degraded very slowly and only after 1-yr exposure was biodegradation evident. They concluded that several factors probably contributed to the slow rate of microbial degradation, including: limited populations of hydrocarbon-utilizing microorganisms; localized high oil concentrations; low temperatures; limiting nutrient concentrations (unfavorable C:N and C:P ratios); low oxygen tensions; and limited circulation of interstitial waters in fine-grained sediments. Abiotic weathering of the oil was also slow, with limited loss of low molecular weight aliphatic and aromatic hydrocarbons during 2-yr exposure. Significant features of the overall weathering process were: lack of initial loss of low molecular weight compounds; aliphatic compounds were not preferentially degraded over aromatic compounds; and C_{17} and lower molecular weight normal alkanes were preferentially degraded over higher molecular weight alkanes. Results of this study indicate that hydrocarbons will persist in a relatively unaltered state for several years if Beaufort Sea sediments are contaminated with petroleum.

Substrate Concentration

Studies have not been performed to determine whether oligotrophic microorganisms are capable of utilizing hydrocarbons at very low concentrations. This is important when considering low level discharges from oil and gas development and production activities. It is difficult to measure the lower concentration limits at which microorganisms utilize specific substrates, but by using radiolabeled hydrocarbon tracers it is possible to approximate the minimal hydrocarbon concentrations that would be metabolized by oligotrophic microorganisms. If oligotrophic microorganisms within the water column do not attack low levels of hydrocarbons, then the hydrocarbons are likely to become adsorbed onto sediment particles and accumulate in higher concentrations in areas of sediment deposition. In such areas where higher levels of hydrocarbons accumulate, microorganisms should degrade the degradable hydrocarbons assuming that other factors such as the lack of oxygen do not preclude hydrocarbon biodegradation. With respect to adsorption of hydrocarbons onto sediments, it is unknown whether this reduces the availability for biodegradation or whether microorganisms are able to attack adsorbed hydrocarbons. Clearly the nature of the sediment will affect the amount of adsorption and influence the availability of hydrocarbons for microbial attack. In a study aimed at examining microbial activities in areas of deposition of sediments with adsorped hydrocarbons, no evidence was found for hydrocarbon accumulation nor for altered microbial activities in areas predicted to be the sites of fine-grained sediment deposition from Cook Inlet that would have been contaminated with low levels of hydrocarbons from Cook Inlet oil production activities (Atlas *et al.*, 1983). This study left unresolved the rates at which microorganisms utilize adsorbed hydrocarbons. If microorganisms attack adsorbed hydrocarbons at decreased efficiencies, then increased amounts of hydrocarbons would be incorporated into detritus and into the food web.

Direct and Indirect Involvement of Animals in Oil Degradation

The role of marine vertebrates and invertebrates, including members of the zooplankton and the benthic community, in biodegradation of oil pollutants is two-fold. They have some role in direct metabolism of hydrocarbons (National Academy of Sciences, 1975) but compared to the extent of microbial metabolism of hydrocarbons, this direct role is of limited significance. The indirect roles of the protozoan members of the zooplankton as grazers on oil-degrading bacterial populations and the bioturbation of oiled sediments by members of the benthic community are more important.

Vertebrate and invertebrate animals rarely utilize petroleum hydrocarbons as nutrients. Nevertheless, they may consume petroleum hydrocarbons during grazing, filter-feeding or surface-feeding activities. Additional hydrocarbon exposure may take place through gills and, to a lesser extent, directly through the body surface. In all these cases, the metabolism of the hydrocarbons incorporated by the animal is not directed at deriving carbon and energy from them, but rather at their detoxification and excretion (see Capuzzo, Chapter 8).

Hydrocarbons in animals are primarily partitioned into fatty tissues. The excretion process in fish involves oxidation by cytochrome P-450-dependent monooxygenase systems followed by conjugation of the oxidation products. The resulting polar conjugates are excreted in urine (Lee *et al.*, 1972a). Metabolic transformations are less clear in invertebrates (Lee *et al.*, 1972b), but they are ultimately excreted in fecal pellets. Some *n*-alkane components of petroleum, once oxidized to fatty acids, may be extensively degraded in animals by the *beta*-oxidation mechanism.

The indirect effects of animals and protozoa on marine microbial oil-degrading populations appear to be significant. Studies have been conducted on the influence of grazing by protozoa on oil-degradation in freshwater ecosystems and these studies provide some insight into the probable indirect influence of these organisms on the rates of oil biodegradation in marine environments. Rogerson (unpublished data) surveyed a broad spectrum (67 species) of oil-degrading microbiota and, in general, determined that most species of actinomycetes and yeasts were either not palatable or did not support protozoan growth. However, with few exceptions eubacterial species were readily ingested and supported positive growth among the eight protozoan species tested. A detailed study (Rogerson and Berger, 1980, 1981a) of the growth responses of five freshwater and soil ciliate species (*Colpidium campylum*, *C. colpoda*, *Tetrahymena pyriformis*, *Colpoda cucullus*, and *Uroleptus* sp.), and a soil amoeba (*Naegleria gruberi*) on eight representative bacteria that utilized Norman Wells crude oil determined that the magnitude of the protozoan growth response was determined by the nature of the prey species. The protozoa could grow on most but not all of the hydrocarbon-grown bacteria. It was found that the protozoa feeding on the hydrocarbon-grown bacteria contained oil droplets in their food vacuoles, but that the presence of the oil in the medium or cells did not consistently raise or lower the growth rate of the protozoa.

The general resilience of protozoa to oil in laboratory experiments (Rogerson and Berger, 1981a,b) confirms the observations (Smith, 1968; Spooner, 1968; ZoBell, 1969; Andrews and Floodgate, 1974; Langlois, 1979) that ciliates and flagellates are frequently associated with oil spills. The predation by protozoa on hydrocarbon-degrading bacteria accelerates nutrient turnover (Stout, 1980). This should enhance the biological removal of oil, as nitrates and phosphates are often limiting factors in microbial degradation processes. This prediction has been confirmed by Rogerson and Berger (1982, 1983) in laboratory simulations of crude oil spills. Ratios of the n-C_{17}/pristane and n-C_{18}/phytane peak heights in bacterial cultures without ciliates incubated for up to two months were only slightly lower than sterile control values. However, pronounced decreases in these n-alkane/isoprenoid ratios were apparent in cultures containing both bacteria and ciliates. The predation by the ciliate *Colpidium colpoda* enhanced the bacterial degradation of crude oil, at least in this *in vitro* experiment.

The role of bioturbation by polychaetes in promoting oil biodegradation in sediments (Gordon *et al.*, 1978) was described earlier. Other burrowing benthic animals are expected to exert a similar positive effect on oil biodegradation.

Summary

These are a few examples of the complex considerations of the factors that influence the rates of oil biodegradation in various marine environments. Clearly, the fate of oil in the marine environment will depend on the particular set of abiotic parameters of a given habitat. The interactions of multiple factors will determine the overall rate of biodegradation. Factors such as favorable oxygen concentrations and a large surface area for microbial attack on a surface spill may be offset by low nutrient concentrations; similarly, the favorable nutrient concentrations of benthic sediments may be offset by the concentration of oil in anoxic pockets. While the rate-limiting factors have been elucidated, the interactions of these factors with respect to determining rates of oil biodegradation have not, and therefore, it remains difficult to predict the fate of oil in diverse marine habitats.

Measured Rates of Petroleum Biodegradation

The foregoing discussion indicates the number and complexity of the factors that influence the rate of petroleum biodegradation in sea water. In consideration of these factors, it is difficult to make generalized statements about rates, because these will vary tremendously according to the type of petroleum and the prevailing complex set of environmental conditions. Nevertheless, petroleum biodegradation rates are of central interest and need to be defined at least with the rather crude accuracy of an order of magnitude.

Petroleum is a multicomponent mixture, and each of its components degrades at a different rate, some fast, some slow, and some not at all. It is possible to measure an "overall" biodegradation rate (the average of the individual rates), but it would be misleading to use such data for construction of a linear regression. As the more biodegradable components are depleted and the more recalcitrant ones remain, the average rate will continuously decline.

Some rate measurements have been made with conventional techniques such as residual weight or quantitative gas chromatography (Coty and Leavitt, 1971; Atlas and Bartha, 1973d; Kanazawa, 1975; Dibble and Bartha, 1976; Walker *et al.*, 1975e), but most of the newer measurements rely on $^{14}CO_2$ production from labeled hydrocarbons (Robertson *et al.*, 1973; Caparello and LaRock, 1975; Seki, 1976; Walker and Colwell, 1976d; Arhelger *et al.*, 1977; Lee, 1977; Atlas, 1978b; Ward *et al.*, 1980; Atlas and Bronner, 1981; Atlas *et al.*, 1981; Traxler and Vandermeulen, 1981). The sophistication of this approach is increasing. The earlier measurements relied on a single radiolabeled *n*-alkane (usually hexadecane) as the spike (Robertson *et al.*, 1973; Caparello and LaRock, 1975; Seki, 1976); the later measurements included one or more representatives of each hydrocarbon class. For example, Walker and Colwell (1976c), who used hexadecane, naphthalene, toluene, and cyclohexane, found that these compounds were utilized at rates of decreasing order. Some authors (Ward *et al.*, 1980; Lee, 1977) included multi-ring condensed polynuclear aromatic compounds such as benzo(a)pyrene that, predictably, showed very slow biodegradation. For calculation of overall rates, the contributions from *n*-alkanes and the mono- and dicyclic aromatics are the most critical, the other hydrocarbon classes contribute relatively little to the total. The underlying assumption of the spiking approach is that the radiolabeled compound will biodegrade at a rate that is "representative" for the petroleum or certain classes of hydrocarbons in the petroleum. In a strict sense this is, of course, not true. One could argue with good justification that two petroleums of very different biodegradability will show identical rates, if spiked in the same manner. Conversely, one could obtain different rates for the same petroleum with dissimilar spiking. One may say that the radioactive spiking approach gives a good estimate of the hydrocarbon biodegradation potential of a marine habitat, but gives a poor prediction of how this potential will be realized on a particular petroleum. However, as compared to the disadvantages of other possible measurement approaches, radioactive spiking offers a highly sensitive, convenient, and moderately valid technique for assessment and prediction of petroleum biodegradation rates, and the use of this technique is likely to increase in the future.

Table 7.3 summarizes the orders of magnitude of hydrocarbon biodegradation rates obtained in different systems. Optimization of hydrocarbon biodegradation may involve temperature, aeration, mineral nutrients or all of the above. In

TABLE 7.3
The order of magnitude of hydrocarbon biodegradation rates*

System	*Order of magnitude ($g/m^3/day$)*	*References*
1) High density monocultures in fermenters, optimal conditions	10,000–100,000	6,8
2) Seawater community under partially optimized conditions	5–2,500	3,5,7,12
3) *In situ* marine enrichment (long incubation times)	0.5–60	1,2,3,4
4) *In situ* marine potential (short incubation times)	0.001–0.030	9,10,11

*The cited references served as basis for calculations; the above listed figures do not necessarily appear in the papers in this particular form. Key to the references: (1) Atlas and Bronner, 1981; (2) Atlas and Bartha, 1973d; (3) Atlas *et al.*, 1980; (4) Atlas *et al.*, 1981; (5) Caparello and La Rock, 1975; (6) Dibble and Bartha, 1976; (7) Coty and Leavitt, 1971; (8) Kanazawa, 1975; (9) Lee, 1977; (10) Robertson *et al.*, 1973; (11) Seki, 1976; (12) Walker *et al.*, 1976b.

some cases sea water alone, in others sea water-sediment mixtures, furnished the microbial community. Incubation times were typically long, ranging from several days to several weeks. *In situ* conditions combined with long incubation times provided ample opportunity for substantial population shifts. Short-term *in situ* petroleum degradation measurements are comparable to "heterotrophic potential" measurements in the classical sense (Wright and Hobbie, 1966). Obviously, the previous pollution history of the site and the prevailing environmental conditions strongly affect the outcome of *in situ* measurements.

A few comments on Table 7.3 are in order. Atlas *et al.* (1980) compared rates in a mineral supplements system with rates under *in situ* conditions and found petroleum biodegradation up to 300-fold higher in the nutrient-supplemented system. In polluted, nutrient-enriched sea water, Caparello and LaRock (1975) and Dibble and Bartha (1976) obtained remarkably similar results (2,500 and 2,000 $g/m^3/day$), even though the former authors used radioactive spiking and the latter ones used gas liquid chromatographic analysis.

Some highly atypical results have not been included in Table 7.3. Arhelger *et al.* (1977) reported *in situ* marine potential degradation up to 700 $g/m^3/day$. This is difficult to understand, since the same group using similar methods and environments previously reported rates around 0.001 $g/m^3/day$ (Robertson *et al.*, 1973), and the later paper does not comment on the apparent discrepancy. Traxler and Vandermeulen (1981) report in similar experiments 1.0-135 $g/m^3/day$ naphthalene biodegradation. Besides being unusually high, the corresponding *n*-hexadecane biodegradation rates were 0.5-1.7 $g/m^3/day$, a rather atypical ratio. Further work will be needed to reconcile these atypical findings with the data summarized in Table 7.3.

How do the rates summarized in Table 7.3 relate to the biodegradation rates of an oceanic oil spill under oxygenated conditions? The initial values are expected to be in the range of the *in situ* potential (case 4). On prolonged contact with the petroleum, population shifts occur, and biodegradation rates will approach the values of *in situ* enrichment. Stimulated biodegradation measures under favorable temperature conditions will raise the rates into the range of those specified for the optimized laboratory enrichments.

EFFECTS OF OIL POLLUTION CONTROL MEASURES ON PETROLEUM BIODEGRADATION

Responses to accidental marine oil spills, according to geographic environmental and economic considerations, range from no action to intense cleanup efforts. In the first case, natural physicochemical and biodegradative forces are relied upon to deal with the spilled oil. When these are considered to be too slow to prevent or ameliorate serious economic or environmental damage, the cleanup measures may affect the subsequent fate of the oil in the marine environment.

The ideal way to deal with a marine oil spill is to prevent its spreading by a floating barrier and to reclaim the floating oil layer with various collection devices. Unfortunately, much oil is usually lost before the containment barriers can be deployed, and these are effective only in protected bays with minimal wave and current action. These are not the conditions that commonly prevail during major marine disasters, but the technique is useful in harbors and at loading piers when spills occur due to equipment failure or human error.

Burning a floating slick is usually not feasible because of the rapid loss of low flashpoint petroleum components by evaporation. In any case, the combustion is incomplete, toxic compounds such as benzo(a)pyrene are formed, and air pollution is severe. Sinking of the oil with ground chalk or siliconized sand is a cosmetic measure that removes the surface slick and may protect birds from oiling. However, it may cause severe damage to benthic communities and prolong the effects of the spill by sequestering the oil in anaerobic sediments where, for lack of oxygen, little oil is degraded. Neither burning nor sinking appear to be environmentally sound cleanup procedures (Atlas and Bartha, 1973a).

Dispersal of an oil slick by detergents is an approved procedure if it becomes absolutely necessary to protect beaches and harbor installations from oiling. It is recognized, however, that the detergents, as well as the dispersal of the oil, may increase the damage to organisms in the water column and the benthos. As to the effect of this measure on microbial degradation of oil, dispersion should facilitate it unless the toxicity of the detergent is sufficient to counteract this favorable effect. Experimentally, Gatellier (1971), Robichaux and Myrick (1972), Gatellier *et al.* (1973), and Mulkins-Phillips and Stewart (1974c) found that most detergents enhanced petroleum biodegradation, but a lack of enhancement or even inhibition was caused by certain dispersant formulations. Atlas and Bartha

(1973e) tested several dispersants and oil herders and found that all enhanced the rate but not the total extent of petroleum biodegradation. Foght and Westlake (1982) found that dispersant addition differentially altered rates of paraffinic and aromatic hydrocarbon biodegradation; they found that *n*-alkane biodegradation was retarded and aromatic hydrocarbon biodegradation stimulated by the addition of Corexit 9527. In conclusion, while oil dispersants appear to be generally favorable to subsequent oil biodegradation activity, more work is needed to determine the actual effects of specific dispersants on *in situ* rates of oil biodegradation.

Potential of Stimulated Biodegradation for Oil Pollution Abatement

As stated earlier, biodegradation is the major mechanism for elimination of non-volatile hydrocarbons from the marine environment. It is logical to attempt to control this process and to exploit it for the cleanup of polluting oil. Attempts to do so were subject to a detailed review by Atlas (1977a). To increase the rate and extent of petroleum biodegradation, one may attempt to manipulate the parameters controlling the process, i.e., the types and numbers of petroleum degraders, the quality of the polluting oil and the prevailing environmental parameters. In practical terms, not much can be done about the composition of an accidental oil spill, nor about the prevailing water temperature and weather. Oxygen is not commonly limiting in the "slick" stage of an oil spill. Therefore, to date, attempts at stimulated oil biodegradation have been aimed either at the modification of petroleum-degrading microbial populations by "seeding" with selected or genetically engineered strains, or at relieving the mineral nutrient limitations of sea water by "fertilization" of the oil slicks.

The inoculation or "seeding" approach is appealing in its simplicity. Single isolated strains or mixed enrichments were used for this purpose by Miget *et al.* (1969), Kator *et al.* (1971, 1972), Bridie and Bos (1971), Atlas and Bartha (1972a), Reisfeld *et al.* (1972), and Horowitz and Atlas (1978b). A new development is the genetic engineering of hydrocarbon degraders for broad substrate range and high metabolic rates. The fact that some of the genes coding for hydrocarbon biodegradation pathways are plasmid-associated allowed the expansion of substrate range by plasmid transfer (Chakrabarty and Friello, 1974). The rapid expansion of this field was reviewed by Williams (1978).

A general criticism of the seeding approach is that the use of an allochthonous microbial population may not be necessary or effective in most cases, especially if applied without regard to the prevailing limitation by mineral nutrients (Atlas, 1977). Some prematurely commercialized inocula were completely ineffective in laboratory tests (Atlas and Bartha, 1973e).

The general limitation of petroleum biodegradation by the mineral nutrient deficiencies of sea water was discussed earlier. A few reports to the contrary (Kinney *et al.*, 1969; Arhelger *et al.*, 1977) are explainable by the fact that in these experiments hydrocarbons were present in true solution only, not suspended or as slicks. In an accidental spill, the addition of nitrogen, phosphorus, and sometimes iron definitely stimulates the biodegradation of floating or suspended oil. In a contained situation, e.g., the ballast tank of an oil tanker, mineral nutrients may be added as water-soluble salts (Rosenberg *et al.*, 1975). In the case of floating slicks, it becomes a problem to keep the mineral nutrients in contact with the slick. Kator *et al.* (1972) used paraffinized ammonium and phosphorus salts as fertilizer, but over 90% of the salts was released to the water phase during the first hour of incubation. Atlas and Bartha (1973d) used paraffinized urea (urea-paraffin clathrate) and octylphosphate. Subsequently, iron was added as ferric octoate (Dibble and Bartha, 1976). These “oleophilic fertilizers” remained associated with the floating oil and did not stimulate algal blooms. In a simulated field trial, they caused a 6- to 12-fold increase in petroleum biodegradation rates, depending upon whether or not an initial lag period during which oil toxicity prevented biodegradation was included in the calculation. Atlas and Schofield (1975), Atlas and Busdosh (1976) and Horowitz and Atlas (1977) subsequently tested the oleophilic fertilizer in Alaskan waters. Even in this hostile environment, consistent stimulation of oil biodegradation was achieved in both *in vitro* and in field tests. No algal blooms or toxicity to the tested invertebrates were noted in these experiments.

Olivieri *et al.* (1976) field-tested paraffinized $MgNH_4PO_4$ as a combined N and P fertilizer in Mediterranean waters and observed moderate stimulation of petroleum biodegradation. In a three-week period, 63% of the treated oil disappeared, as compared to 40% of the untreated control. Horowitz and Atlas (1977, 1978b) field-tested the oleophilic fertilizer developed by Atlas and Bartha (1973d) in Alaskan waters, both alone and in combination with freeze-dried bacterial inocula. The latter treatment approximately doubled the amount of petroleum that was biodegraded in 45 days.

In summary, oleophilic fertilizers, alone or in combination with microbial inocula, are capable of substantial acceleration of biodegradation, and appear to have no detrimental side effects. The approach appears to be cost-effective and could be applied under adverse weather conditions. Its drawbacks are the relative slowness of the process and possible storage problems or side effects of the microbial inocula. In the absence of commercial development of this technique, experience in actual spill incidents is lacking.

EFFECTS OF HYDROCARBONS ON MICROBIAL COMMUNITIES

In addition to considering the role of microorganisms in determining the fate of hydrocarbons in the environment, it is important to consider the potential impact of hydrocarbon contaminants on the functioning of microbial communities. Microorganisms carry out activities that are critical ecological functions. In a review of the effects of petroleum hydrocarbons on microbial metabolic activities, Pfaender and Buckley (1984) concluded that no overall generalizations can be made. Relatively few studies have been performed, and there is considerable variability in reported responses that depend upon the characteristics of the indigenous microbial community, environmental factors, the composition and concentration of the hydrocarbons in the contaminating oil and the previous history of exposure to oil (Table 7.4).

Hodson *et al.* (1977) as part of the CEPEX (Controlled Ecosystem Pollution Experiments) program found that glucose utilization was inhibited by the presence of all types of oil, although the refined petroleum products, No. 2 fuel oil and Bunker C oil, produced significantly greater degrees of inhibition than either of the crude oils. Griffiths *et al.* (1981) reported reduced rates of glucose and glutamate uptake by the indigenous microbial communities in arctic and subarctic waters exposed to crude oil. They concluded that exposure to petroleum hydrocarbons inhibits microbial metabolism and hence microbial productivity. In the Cook Inlet area of Alaska, they found that samples from areas near natural oil seeps or in shipping lanes all showed lesser effects on metabolism than were observed in samples from areas not previously exposed to oil. Their results indicate that pelagic populations can adjust to the presence of oil fairly rapidly and that prior exposure may select for a community adapted to oil.

Buckley (1980) found that the degree of inhibition of a synthetic crude oil on the turnover rate of amino acids by microbial communities from North Carolina salt marshes was proportional to the concentration of oil. The effect of oil on microbial communities was transitory, and recovery occurred within one month, again indicating that microbial communities have the potential for adaptation to low-level chronic inputs of hydrocarbons. The IXTOC-I well blowout in the Bay of Campeche appears to have stimulated microbial heterotrophic activity (Pfaender *et al.*, 1980). Only at a site very near the well head was any inhibition of metabolism observed, and then only the respiration of amino acids appeared to be inhibited. Alexander and Schwarz (1980) studied the effect of South Louisiana and Kuwait crude oil, at concentrations ranging from 0.001 to 0.1% v/v, on the heterotrophic uptake of glucose by marine microbial communities of the Gulf of Mexico and estuarine areas near Galveston Bay. They also found that oil had little or no effect on microbial metabolism. These results are probably indicative of the effects of low levels of hydrocarbons on microbial communities that have had previous exposure to petroleum hydrocarbons and which have adapted to the presence of hydrocarbon contaminants.

With respect to the effects of oil on denitrification and nitrogen fixation, Haines *et al.* (1981) found no significant effects of short-term exposure to crude oil upon benthic ecosystems in the Beaufort Sea and Cook Inlet, Alaska. However, long-term exposure resulted in significantly decreased rates of nitrogen fixation compared to unoiled controls in Cook Inlet. In the Beaufort Sea, however, prolonged exposure to Prudhoe Bay crude oil did not result in reduced rates of nitrogen fixation or denitrification.

Kator and Herwig (1977) reported that crude oil had no effect on chitin and cellulose decompositional processes. Walker *et al.* (1974, 1975f) found decreased microbial utilization of chitin, cellulose, lipids and protein when benthic communities were exposed to crude or refined oils. Buckley (1980) found that petroleum hydrocarbons dramatically decreased the utilization of *Spartina alterniflora*-derived crude fiber by salt marsh microbiota. The inhibition was attributed to one or a combination of the following reasons: 1) a shift in the metabolic

TABLE 7.4
Summary of petroleum effects on microbial community metabolic activity

Site and sample	*Oil type*	*Oil concentration*	*Parameters measured*	*Response*	*Reference*
Aquatic Communities:					
Saanich Inlet, British Columbia	No. 2 fuel oil	0–1200 μg/l	Glucose assimilation	0–85% inhibition	Hodson *et al.*, 1977
	Bunker C oil	0–1700 μg/l		0–80% inhibition	
	S. Louisiana oil	0–1500 μg/l		0–30% inhibition	
	Kuwait Crude	0–1200 μg/l		0–20% inhibition	
Cook Inlet, Alaska	Cook Inlet crude	1 l/ml	Glucose metabolism	22–92% inhibition	Griffiths *et al.*, 1981
	Cook Inlet crude	0.1% v/v	Glutamate uptake	+47 to −67%	
Beaufort Sea, Alaska	Prudhoe Bay Crude	1 μl/ml	Glucose metabolism	8–88% inhibition	
Straits of Georgia, British Columbia	Alberta Crude	1 cm layer on water sample	Heterotrophic production	Slight increase	Dietz *et al.*, 1976
			Heterotrophic potential	Slight decrease	
	Diesel fuel	1 cm layer on water sample	Heterotrophic production	No effect	
			Heterotrophic potential	No effect	
North Carolina salt marsh tidal creek water		0.5–4.0% v/v	Acute (30 min.) effect on amino acid turnover	0–80% inhibition	Buckley, 1980
North Carolina salt marsh tidal creek water		1.0% v/v	Long-term effect (35 d) amino acid turnover	No effect	
			Cellulose mineralization	90% inhibition	

Bay of Campeche Mexico	IXTOC-I Crude	10,000 g/l	Amino acid uptake	No effect	Pfaender *et al.*, 1980
			Amino acid respiration	90⁺% inhibition	
Bay of Campeche Mexico	IXTOC-I Crude	1.0% v/v	Amino acid uptake	Slight initial inhibition, then increase	Buckley *et al.* 1980
			Amino acid respiration	No effect	
	Fresh mousse	1.0% v/v	Amino acid uptake	300% stimulation	
			Amino acid respiration	600% stimulation	
	Weathered mousse	1.0% v/v	Amino acid uptake	100% stimulation	
			Amino acid respiration	800% stimulation	
Gulf of Mexico	South Louisiana Crude Kuwait Crude	0.001–0.1% v/v	Glucose uptake and mineralization	No effect 11 of 13 water samples	Alexander and Schwarz, 1980
Benthic Communities:					
Kasitsna Bay, Alaska long-term exposure	Cook Inlet Crude	50 ppt	Glucose uptake	38–60% inhibition	Griffiths *et al.*, 1981
			Glutamate uptake	21–73% inhibition	
Beaufort Sea long-term exposure	Prudhoe Bay Crude	50 ppt	Glucose uptake Glutamate uptake	No significant reduction until 13-month exposure, then glucose 62–85%, glutamate 47–71% inhibition	

(continued)

TABLE 7.4—*contd.*

Site and sample	*Oil type*	*Oil concentration*	*Parameters measured*	*Response*	*Reference*
Galveston Bay, Texas	South Louisiana Crude Kuwait Crude	0.001–0.1% v/v	Glucose uptake and mineralization	No effect on 10 of 15 samples	Alexander and Schwarz, 1980
Anaerobic sediment from Louisiana salt marsh	South Louisiana Crude	10% w/w	Methane production NO_3 reduction Fe, Mn, $SO_4^=$ reduction N-immobilization	No effect No effect No effect No effect	DeLaune *et al.*, 1979
Brittany coast sediment (beach estuary, marsh)	*Amoco Cadiz*	Spill	Sulfate reduction Methane production Acetate metabolism	No significant effect at any site	Winfrey *et al.*, 1982

activity of the microbial populations from the utilization of the *Spartina*-derived cellulose to the preferential degradation of the hydrocarbons; 2) a suppression of cellulose enzyme production in those cellulolytic bacteria which can utilize hydrocarbons or degradative products of the oil; and 3) the inhibition of bacterial colonization of the particulate organic matter by the oil either coating the particles and acting as a physical barrier, or by the oil interfering with the chemical detection of the particulate organic material by the bacteria.

Many of the same patterns observed in studies of water column microbiological characteristics have been reported for sediments. In general, where both water and sediment samples have been examined in the same environment, the inhibition of microbial metabolism noted in sediment is significantly less than that seen in water. These include the greater sensitivity of microbial communities from pristine arctic and subarctic areas to altered metabolism as a result of oil pollution and the significance of adaptation to prior oil exposure. Most of the metabolic effect studies performed to date have addressed aerobic, heterotrophic processes in the sediments. Little information is available regarding the impact of petroleum pollutants on anaerobic sediment processes. Winfrey *et al.* (1982) examined anaerobic microbial activities, methane-production and sulfate-reduction in sediments impacted by the *Amoco Cadiz* oil spill. They reported that methane production and sulfate reduction did not vary significantly in sediment samples from oiled and unoiled sites. It is not possible based upon this one study to dismiss the possible effects of hydrocarbons on critical anaerobic biogeochemical activities. This is clearly an area where additional research is needed.

In summary, there appears to be a significant amount of variation in both the type and magnitude of response observed when aquatic microbial communities are perturbed by oil. These effects vary with the type of oil, the concentration and the community that is exposed. Crude oil typically produces less inhibition than refined petroleum products when they are present at the same concentrations. Marine samples from arctic and subarctic environments appear to be affected to a greater degree and for a longer period of time than communities from temperate or tropical regions. It is evident that the effect of oil on microbial communities with a prior history of oil exposure is significantly less than with communities with no prior exposure.

It is also apparent that both crude and refined petroleum can impact microbial processes that are critical for proper ecological functioning. In ecosystems that are driven by detrital food webs, a negative impact on bacterial productivity could have a particularly severe long-term effect on the productivity of organisms at higher trophic levels. The degree to which oil can negatively affect microbial communities appears to be influenced by a number of factors. In many cases microbial communities appear to be able to adapt to hydrocarbon exposure and thus the long-term impact is mitigated. There is, however, little information concerning the mechanisms of oil inhibition, the long-term effects on important ecological functions of the microbial community and the effects of changes in the metabolic activities of the microbial community on other biota and the overall functioning of the ecosystem.

QUESTIONS FOR FUTURE RESEARCH

It is clear that most of the ocean's oil pollutants come from routine operations and not from major acute spillages. The majority of our information concerning oil biodegradation, however, deals with the fate of relatively high concentrations of oil in nonadapted ecosystems. More needs to be known about the fate of chronic low level hydrocarbon inputs. How much of the chronic oil input is biodegraded and how much ends up as tar or other persistent forms needs to be determined.

With respect to large acute inputs of oil pollutants, the major question following an oil spill is what will happen to the oil and what will be the ecological consequences of the spill. We do not yet have the capability of predicting with adequate precision the fate of oil and how long particular fractions will remain as environmental contaminants. In most major oil spills, such as the *Argo Merchant* and the IXTOC-I spills, we are even unable to develop accurate mass balances for where the oil has gone and how much remains as an environmental contaminant. It is clear that biodegradation is an effective means of environmental decontamination. In some ecosystems, however, biodegradation fails when certain environmental conditions, particularly the scarcity of essential nutrients and oxygen, restrict microbial growth and metabolism. Defining the physical distribution of oil and the critical environmental parameters of the various oiled habitats is a major part of predicting the biological fate of the oil.

The nature of the oil, in terms of both its physical form and chemical composition, has a major impact on its degradative fate. More needs to be determined about the interactions of components within the complex petroleum mixture, particularly with respect to sparing and cometabolism. At present, it is not clear for particular oils how the presence of one component will influence the persistence times of other components in the mixture. Additionally, it is not clear how physically resistant forms, such as tars, are produced. The role of microorganisms in the formation of tar and various oxidation products that resist further degradation must be investigated to understand the long-term persistence of petroleum hydrocarbons. Various oxidation products can be formed and can accumulate in the environment from chronic low-level hydrocarbon contaminants as well as from acute hydrocarbon spills. Such products can be toxic or carcinogenic and can have long-term effects.

The interactions of hydrocarbons with sediment, both suspended and deposited needs to be better understood to predict the availability of hydrocarbons for microbial degradation and hence the ability of microbes to degrade marine hydrocarbon contaminants. More also needs to be known about the benefits of dispersants for stimulating oil biodegradation. The data on dispersants and biodegradation are sparse and leave several critical gaps that must be answered before the benefits of dispersant use can be properly evaluated.

A further question is the movement of oil within the food web and the effects of low-level chronic oil releases on the normal roles of microorganisms within food webs. It is clear that some microorganisms sequester oil and that these microorganisms can be prey for organisms on higher trophic levels. The actual movement of oil up through the food web has not been adequately documented and thus the importance of this process is unclear. It is also clear that microorganisms are essential for ecological productivity and that in some cases hydrocarbons can alter essential microbial activities. More studies are needed to examine the effects of chronic low-level inputs of hydrocarbons into marine ecosystems on microbial communities and how changes in the microbial community are reflected within the communities of the organisms that constitute the higher levels of marine food webs.

Additionally, the lower concentration limits of hydrocarbons that can be degraded by microorganisms need to be established. Studies on oligotrophic hydrocarbon degraders have not been conducted. The ability of oligotrophic microbes to degrade low concentrations of hydrocarbons is extremely important when considering the fate of discharges from oil and gas exploratory and production operations. It is also necessary to consider the toxicity of metals and other components from oil and gas development effluents toward oligotrophic microorganisms.

Finally, the question is what to do when there is an oil spill to minimize persistence and thus long-term effects? Our knowledge of biodegradation indicates that an integrated approach to the oil pollution problem is desirable. Treatment methods should enhance rather than inhibit the natural rates of oil biodegradation. In some cases, it is possible to modify environmental parameters to enhance rates of hydrocarbon biodegradation, but such methods are rarely taken. How to translate effectively our scientific knowledge of hydrocarbon biodegradation into useful guidelines for industry and government stands out as a major challenge.

LITERATURE CITED

Abbott, B.J. and W.G. Gledhill. 1971. The extracellular accumulation of metabolic products by hydrocarbon-degrading microorganisms. Adv. Appl. Microbiol. 14: 249-388.

Ahearn, D.G. and S.A. Crow. 1980. Yeasts from the North Sea and *Amoco Cadiz* oil. Botanica Marina 23: 125-127.

Ahearn, D.G. and S.P. Meyers. 1971. The role of fungi in the decomposition of hydrocarbons in the marine environment. Pages 12-18 *in* Biodeterioration of Materials. Applied Science Publishers Ltd., London.

Ahearn, D.D., S.P. Meyers and P.G. Standard. 1971. The role of yeasts in the decomposition of oils in marine environments. Dev. Ind. Microbiol. 12: 126-134.

Aida, T. and K. Yamaguchi. 1969. Studies of the utilization of hydrocarbons by yeasts. IV. On the dialysis culture of *Mycotorula japonica* and "growth inhibitory factor" in the dialyzable material. Agric. Biol. Chem. (Japan) 13: 1244-1250.

Alexander, S.A. and J.R. Schwarz. 1980. Short-term effects of South Louisiana and Kuwait crude oils on glucose utilization by marine bacterial populations. Appl. Environ. Microbiol. 40: 341-345.

Aminot, A. 1981. Anomalies du systeme hydrobiologique cotier apres l'echouage de L'Amoco Cadiz: Considerations qualitatives et quantitatives sur la biodegradation in situ des hydrocarbures. Pages 223-242 *in Amoco Cadiz*: Fates and Effects of the Oil Spill. Centre National Pour L'Exploitation des Oceans, Paris, France.

Andrews, A.R. and G.D. Floodgate. 1974. Some observations on the interaction of marine protozoa and crude oil residues. Mar. Biol. 25: 7-12.
Arhelger, S.D., B.R. Robertson and D.K. Button. 1977. Arctic hydrocarbon biodegradation. Pages 270-275 *in* D.A. Wolfe (ed.), Fate and Effects of Petroleum Hydrocarbons in Marine Ecosystems and Organisms. Pergamon Press, Inc., New York.
Atlas, R.M. 1975. Effects of temperature and crude oil composition on petroleum biodegradation. Appl. Microbiol. 30: 396-403.
Atlas, R.M. 1977. Stimulated petroleum biodegradation. Crit. Rev. Microbiol. 5: 371-386.
Atlas, R.M. 1978a. An assessment of the biodegradation of petroleum in the Arctic. Pages 86-90 *in* M.W. Loutit and J.A.R. Miles (eds.), Microbial Ecology. Springer-Verlag, Berlin.
Atlas, R.M. 1978b. Measurement of hydrocarbon biodegradation potentials and enumeration of hydrocarbon utilizing microorganisms using ^{14}C radiolabelled spiked crude oil. Pages 196-204 *in* J.W. Costerton and R.R. Colwell (eds.), Native Aquatic Bacteria: Enumeration, Activity and Ecology. Publ. ASTM-STP 695. American Society for Testing and Materials, Philadelphia.
Atlas, R.M. 1981. Microbial degradation of petroleum hydrocarbons: An environmental perspective. Microbiol. Rev. 45: 180-209.
Atlas, R.M. (ed.). 1984. Petroleum Microbiology. Macmillan Publishing Co., Inc., New York.
Atlas, R.M. and R. Bartha. 1972a. Degradation and mineralization of petroleum by two bacteria isolated from coastal water. Biotechnol. Bioeng. 14: 297-308.
Atlas, R.M. and R. Bartha. 1972b. Biodegradation of petroleum in seawater at low temperatures. Can. J. Microbiol. 18: 1851-1855.
Atlas, R.M. and R. Bartha. 1972c. Degradation and mineralization of petroleum in seawater: Limitation by nitrogen and phosphorus. Biotechnol. Bioeng. 14: 309-317.
Atlas, R.M. and R. Bartha. 1973a. Fate and effects of oil pollution in the marine environment. Residue Rev. 49: 49-85.
Atlas, R.M. and R. Bartha. 1973b. Abundance, distribution and oil biodegradation potential of microorganisms in Raritan Bay. Environ. Pollut. 4: 291-300.
Atlas, R.M. and R. Bartha. 1973c. Inhibition by fatty acids of the biodegradation of petroleum. Antonie van Leeuwenhoek J. Microbiol. Serol. 39: 257-271.
Atlas, R.M. and R. Bartha. 1973d. Stimulated biodegradation of oil slicks using oleophilic fertilizers. Environ. Sci. Technol. 7: 538-541.
Atlas, R.M. and R. Bartha. 1973e. Effects of some commercial oil herders, dispersants and bacterial inocula on biodegradation of oil in seawater. Pages 283-289 *in* D.G. Ahearn and S.P. Meyers (eds.), The Microbial Degradation of Oil Pollutants. Publication No. LSU-SG-73-01. Center for Wetland Resources, Louisiana State University, Baton Rouge, Louisiana.
Atlas, R.M. and R. Bartha. 1981. Microbial Ecology - Fundamentals and Applications. Addison-Wesley, Reading, Massachusetts.
Atlas, R.M. and A. Bronner. 1981. Microbial hydrocarbon degradation within intertidal zones impacted by the *Amoco Cadiz* oil spillage. Pages 251-256 *in* Proceedings of the International Symposium on the *Amoco Cadiz*: Fates and Effects of the Oil Spill. Centre National Pour L'Exploitation des Oceans, Paris, France.
Atlas, R.M. and M. Busdosh. 1976. Microbial degradation of petroleum in the Arctic. Pages 79-86 *in* J.M. Sharpley and A.M. Kaplan (eds.), Proceedings of the Third International Biodegradation Symposium, Applied Science Publishers, Ltd., London.
Atlas, R.M. and E.A. Schofield. 1975. Petroleum biodegradation in the Arctic. Pages 183-198 *in* A.W. Bourquin, D.G. Ahearn and S.P. Meyers (eds.), Impact of the Use of Microorganisms on the Aquatic Environment. EPA 660/3-75-001. U.S. Environmental Protection Agency, Corvallis, Oregon.
Atlas, R.M., E.A. Schofield, F.A. Morelli and R.E. Cameron. 1976. Interactions of microorganisms and petroleum in the Arctic. Environ. Pollut. 10: 35-44.

Atlas, R.M., A. Horowitz and M. Busdosh. 1978. Prudhoe crude oil in Arctic marine ice, water and sediment ecosystems: Degradation and interactions with microbial and benthic communities. J. Fish. Res. Board Can. 35: 585-590.

Atlas, R.M., G. Roubal, A. Bronner and J. Haines. 1980. Microbial degradation of hydrocarbons in mousse from IXTOC-I. Pages 1-24 *in* Proceedings of Conference on *Researcher/Pierce* IXTOC-I Cruises. National Oceanographic and Atmospheric Administration, Atlantic Oceanographic and Meteorological Laboratories, Miami, Florida.

Atlas, R.M., P.D. Boehm and J.A. Calder. 1981. Chemical and biological weathering of oil from the *Amoco Cadiz* spillage within the littoral zone. Estuar. Coastal Shelf Sci. 12: 589-608.

Atlas, R.M., G. Roubal, A. Bronner and J. Haines. 1982. Biodegradation of hydrocarbons in mousse from the IXTOC-I well blowout. Pages 199-217 *in* L.H. Keith (ed.), Energy and Environmental Chemistry: Volume I - Fossil Fuels. Ann Arbor Science Publishers, Ann Arbor, Michigan.

Atlas, R.M., M.I. Venkatesan, I.R. Kaplan, R.A. Feely, R.P. Griffiths and R.Y. Morita. 1983. Distribution of hydrocarbons and microbial populations related to sedimentation processes in Lower Cook Inlet and Norton Sound, Alaska. Arctic 36: 251-261.

Austin, B., J.J. Calomiris, J.D. Walker and R.R. Colwell. 1977a. Numerical taxonomy and ecology of petroleum degrading bacteria. Appl. Environ. Microbiol. 34: 60-68.

Austin, B., R.R. Colwell, J.D. Walker and J.J. Calomiris. 1977b. The application of numerical taxonomy to the study of petroleum degrading bacteria isolated from the aquatic environment. Dev. Ind. Microbiol. 18: 685-696.

Bailey, N.J.L., A.M. Jobson and M.A. Rogers. 1973. Bacterial degradation of crude oil: Comparison of field and experimental data. Chem. Geol. 11: 203-221.

Bartha, R. and R.M. Atlas. 1977. The microbiology of aquatic oil spills. Adv. Appl. Microbiol. 22: 225-266.

Beam, H.W. and J.J. Perry. 1973. Cometabolism as a factor in microbial degradation of cycloparaffinic hydrocarbons. Arch. Mikrobiol. 91: 87-90.

Beam, H.W. and J.J. Perry. 1974a. Microbial degradation of cycloparaffinic hydrocarbons via cometabolism and commensalism. J. Gen. Microbiol. 82: 163-169.

Beam, H.W. and J.J. Perry. 1974b. Microbial degradation and assimilation of *n*-alkyl-substituted cycloparaffins. J. Bacteriol. 118: 394-399.

Beerstecher, E. 1954. Petroleum Microbiology. Elsevier Press, New York.

Berridge, S.A., M.T. Thew and A.G. Loriston-Clarke. 1968. Formation and stability of emulsions of water in crude petroleum and similar stocks. Pages 35-59 *in* P. Hepple (ed.), Scientific Aspects of Pollution of the Sea by Oil. Institute of Petroleum, London.

Bertrand, J.C., H.J.M. Doux and E. Azoulay. 1976. Metabolisme des hydrocarbures chez une bacterie marine. Biochimie. 58: 843-854.

Bridie, A.L. and J. Bos. 1971. Biological degradation of mineral oil in seawater. J. Inst. Pet. London 57: 270-277.

Buckley, E.N. 1980. Effects of Petroleum Hydrocarbons on Metabolic Activity and Community Composition of Suspended Heterotrophic Salt Marsh Bacteria. Ph.D. Dissertation, University of North Carolina, Chapel Hill, North Carolina.

Buckley, E.N., R.B. Jonas and F.K. Pfaender. 1976. Characterization of microbial isolates from an estuarine ecosystem: Relationship of hydrocarbon utilization to ambient hydrocarbon concentrations. Appl. Environ. Microbiol. 32: 232-237.

Buckley, E.N., F.K. Pfaender, K.L. Kylber and R.L. Ferguson. 1980. Response of the pelagic microbial community to oil from the Ixtoc I blowout: Model ecosystem studies. Pages 563-586 *in* D.K. Atwood (convenor), Proceedings of a Symposium on Preliminary Results from the September 1979 *Researcher/Pierce* IXTOC-I Cruise. National Oceanic and Atmospheric Administration, Boulder, Colorado.

Burwood, R. and G.C. Speers. 1974. Photooxidation as a factor in the environmental dispersal of crude oil. Estuarine Coastal Mar. Sci. 2: 117-135.

Butler, J.N., B.F. Morris and J. Sass. 1973. Pelagic Tar from Bermuda and the Sargasso Sea. Spec. Publ. No. 10, Bermuda Biological Station for Research, Bermuda.

Byrom, J.A., S. Beastall and S. Scotland. 1970. Bacterial degradation of crude oil. Mar. Pollut. Bull. 1: 25-26.

Caparello, D.M. and P.A. LaRock. 1975. A radioisotope assay for the quantification of hydrocarbon biodegradation potential in environmental samples. Microb. Ecol. 2: 28-42.

Cerniglia, C.E. and D.T. Gibson. 1979. Algal oxidation of aromatic hydrocarbons: Formation of 1-naphthol from naphthalene by *Agmenellum quadruplicatum*, strain PR-6. Biochem. Biophys. Res. Commun. 88: 50-58.

Cerniglia, C.E. and J.J. Perry. 1973. Crude oil degradation by microorganisms isolated from the marine environment. Z. Allg. Mikrobiol. 13: 299-306.

Cerniglia, C.E., R.L. Hebert, P.J. Szaniszlo and D.T. Gibson. 1978. Fungal transformation of naphthalene. Arch. Microbiol. 117: 135-143.

Cerniglia, C.E., R.H. Dodge and D.T. Gibson. 1980a. Studies on the fungal oxidation of aromatic hydrocarbons. Botanica Marina 23: 121-124.

Cerniglia, C.E., D.T. Gibson and C. Van Baalen. 1980b. Oxidation of naphthalene by cyanobacteria and microalgae. J. Gen. Microbiol. 116: 495-500.

Cerniglia, C.E., C. Van Baalen and D.T. Gibson. 1980c. Oxidation of biphenyl by the cyanobacterium *Oscillatoria* sp., strain JCM. Arch. Microbiol. 125: 203-207.

Chakrabarty, A.M. and D.A. Friello. 1974. Dissociation and interaction of individual components of a degradative plasmid aggregate in *Pseudomonas*. Proc. Natl. Acad. Sci. U.S.A. 75: 3109-3112.

Chapman, P.J. 1979. Degradation mechanisms. Pages 28-66 *in* A.W. Bourquin and P. H. Pritchard (eds.), Microbial Degradation of Pollutants in Marine Environments. EPA-66019-79-012. Environmental Protection Agency, Environmental Research Laboratory, Gulf Breeze, Florida.

Chouteau, J., E. Azoulay and J.C. Senez. 1962. Anaerobic formation of *n*-hept-1-ene from *n*-heptane by resting cells of *Pseudomonas aeruginosa*. Nature (London) 194: 576-578.

Christensen, H.N. 1975. Biological Transport, 2nd edition. W. A. Benjamin, Reading, Massachusetts.

Colwell, R.R. and J.D. Walker. 1977. Ecological aspects of microbial degradation of petroleum in the marine environment. Crit. Rev. Microbiol. 5: 423-445.

Colwell, R.R., J.D. Walker and J.D. Nelson, Jr. 1973. Microbial ecology and the problem of petroleum degradation in Chesapeake Bay. Pages 185-197 *in* D.G. Ahearn and S.P. Meyers (eds.), The Microbial Degradation of Oil Pollutants. Publication No. LSU-SG-73-01. Center for Wetland Resources, Louisiana State University, Baton Rouge, Louisiana.

Colwell, R.R., A.L. Mills, J.D. Walker, P. Garcia-Tello and V. Campose-P. 1978. Microbial ecology studies of the *Metula* spill in the Straits of Magellan. J. Fish. Res. Board Can. 35: 573-580.

Cook, W.L., J.K. Massey and D.G. Ahearn. 1973. The degradation of crude oil by yeasts and its effects on *Lesbistes reticulatus*. Pages 279-282 *in* D.G. Ahearn and S.P. Meyers, (eds.), The Microbial Degradation of Oil Pollutants. Publication No. LSU-SG-73-01. Center for Wetland Resources, Louisiana State University, Baton Rouge, Louisiana.

Cooney, J.J., C. Siporin and R.A. Smucker. 1980. Physiological and cytological responses to hydrocarbons by the hydrocarbon-using fungus *Cladosporium resinae*. Botanica Marina 28: 227-232.

Coty, V.F. and R.I. Leavitt. 1971. Microbial protein from hydrocarbons. Dev. Ind. Microbiol. 12: 61-71.

Cripps, R.E. and R.J. Watkinson. 1978. Polycyclic aromatic hydrocarbons: Metabolism and environmental aspects. Pages 113-134 *in* J.R. Watkinson (ed.), Developments in Biodegradation of Hydrocarbons. Applied Science Publishers, Ltd., London.

Crow, S.A., S.L. Bell and D.G. Ahearn. 1980. The uptake of aromatic and branched chain hydrocarbons by yeast. Botanica Marina 28: 117-120.

Crow, S.A., S.P. Meyers and D.G. Ahearn. 1974. Microbiological aspects of petroleum degradation in the aquatic environment. La Mer 12: 37-54.

Cundell, A.M. and R.W. Traxler. 1973a. The isolation and characterization of hydrocarbon-utilizing bacteria from Chedabucto Bay, Nova Scotia. Pages 421-426 *in* Proceedings of the Joint Conference on Prevention and Control of Oil Spills. American Petroleum Institute, Washington, D.C.

Cundell, A.M. and R.W. Traxler. 1973b. Microbial degradation of petroleum at low temperature. Mar. Pollut. Bull. 4: 125-127.

Cundell, A.M. and R.W. Traxler. 1976. Psychrophilic hydrocarbon-degrading bacteria from Narragansett Bay, Rhode Island, U.S.A. Mater. Org. 11: 1-17.

Davis, J.B. 1967. Petroleum Microbiology. Elsevier Publishing Co., New York.

Davis, S.J. and C.F. Gibbs. 1975. The effect of weathering on a crude oil residue exposed at sea. Water Res. 9: 275-285.

Dean-Raymond, D. and R. Bartha. 1975. Biodegradation of some polynuclear aromatic petroleum components by marine bacteria. Dev. Ind. Microbiol. 16: 97-110.

Delaune, R.D., W.H. Patrick, Jr. and R.J. Bursch. 1979. Effect of crude oil on a Louisiana *Spartina alterniflora* salt marsh. Environ. Pollut. 17: 21-31.

Delaune, R.D., G.A. Hambrick and W.H. Patrick, Jr. 1980. Degradation of hydrocarbons in oxidized and reduced sediments. Mar. Pollut. Bull. 11: 103-106.

Dibble, J.T. and R. Bartha. 1976. The effect of iron on the biodegradation of petroleum in seawater. Appl. Environ. Microbiol. 31: 544-550.

Dibble, J.T. and R. Bartha. 1979. Effect of environmental parameters on the biodegradation of oil sludge. Appl. Environ. Microbiol. 37: 729-739.

Dietz, A.S., T. Tuominen and L.J. Albright. 1976. Sublethal pollutant effects upon the native microflora of Georgia Strait. Pages 1083-1090 *in* J.M. Sharpley and A.M. Kaplan (eds.), Proceedings of the Third International Biodegradation Symposium. Applied Science Publishers, Ltd., London.

Farrington, J.W., B.W. Tripp, J.M. Teal, G. Mille, K. Tjessem, A.C. Davis, J.B. Livramento, N.A. Hayward and N.M. Frew. 1982. Biogeochemistry of aromatic hydrocarbons in the benthos of microcosms. Toxicol. Env. Chem. 5: 331-346.

Floodgate, G.D. 1976. Oil biodegradation in the oceans. Pages 87-92 *in* J.M. Sharpley and A.M. Kaplan (eds.), Proceedings of the Third International Biodegradation Symposium. Applied Science Publishers, Ltd., London.

Floodgate, G.D. 1978. The formation of emulsifying agents in hydrocarbonoclastic bacteria. Pages 82-85 *in* M.W. Loutit and J.A.R. Miles (eds.), Microbial Ecology. Springer-Verlag, Berlin.

Floodgate, G.D. 1979. Nutrient limitation. Pages 107-118 *in* A.W. Bourquin and P.H. Pritchard (eds.), Microbial Degradation of Pollutants in Marine Environments. EPA-66019-79-012. Environmental Protection Agency, Environmental Research Laboratory, Gulf Breeze, Florida.

Foght, J.M. and D.W.S. Westlake. 1982. Effect of the dispersant Corexit 9527. Can. J. Microbiol. 28: 117-122.

Foster, J.W. 1962. Hydrocarbons as substrates for microorganisms. Antoine van Leeuwenhoek J. Microbiol. Serol. 28: 241-274.

Fredricks, K.M. 1966. Adaptation of bacteria from one type of hydrocarbon to another. Nature (London) 209: 1047-1048.

Freegarde, M., C.G. Hatchard and C.A. Parker. 1970. Oil spilt at sea: Its identification, determination and ultimate fate. Lab Practice 20: 35-40.

Friede, J., P. Guire, R.K. Gholson, E. Gaudy and A.F. Gaudy. 1972. Assessment of Biodegradation Potential for Controlling Oil Spills on the High Seas. Rep. No. 4110 1/3. U.S. Coast Guard, Washington, D.C.

Fuhs, G.W. 1961. Der mikrobielle Abban von Kohlenwasserstoffen. Arch. Mikrobiol. 39: 374-422.

Gatellier, C.R. 1971. Les facteurs limitant la biodegradation des hydrocarbures dans l'epuration des eaux. Chim. Ind. (Paris) 104: 2283-2289.

Gatellier, C.R., J.L. Oudin, P. Fusey, J.C. Lacase and M.L. Priou. 1973. Experimental

ecosystems to measure fate of oil spills dispersed by surface active products. Pages 497-507 *in* Proceedings of Joint Conference on Prevention and Control of Oil Spills. American Petroleum Institute, Washington, D.C.
Gerson, D.F. and J.E. Zajic. 1977. Surfactant production from hydrocarbons by *Corynebacterium lepus*, sp. nov. and *Pseudomonas asphaltenicus* sp. nov. Dev. Ind. Microbiol. 19: 577-599.
Gibbs, C.F. 1975. Quantitative studies on marine biodegradation of oil. I. Nutrient limitation at 14°C. Proc. R. Soc. London Ser. B. 188: 61-82.
Gibbs, C.F. and S.J. Davis. 1976. The rate of microbial degradation of oil in a beach gravel column. Microb. Ecol. 3: 55-64.
Gibbs, C.F., K.B. Pugh and A.R. Andrews. 1975. Quantitative studies in marine biodegradation of oil. II. Effects of temperature. Proc. R. Soc. London Ser. B. 188: 83-94.
Gibson, D.T. 1968. Microbial degradation of aromatic compounds. Science 161: 1093-1097.
Gibson, D.T. 1971. The microbial oxidation of aromatic hydrocarbons. Crit. Rev. Microbiol. 1: 199-223.
Gibson, D.T. 1975. Oxidation of the carcinogens benzo(a)pyrene and benzo(a)anthracene to dihydrodiols by a bacterium. Science 189: 295-297.
Gibson, D.T. 1976. Microbial degradation of polycyclic aromatic hydrocarbons. Pages 57-66 *in* J.M. Sharpley and A.M. Kaplan (eds.), Proceedings of the Third International Biodegradation Symposium. Applied Science Publishers, Ltd., London.
Gibson, D.T. 1977. Biodegradation of aromatic petroleum hydrocarbons. Pages 36-46 *in* D.A. Wolfe (ed.), Fate and Effects of Petroleum Hydrocarbons in Marine Ecosystems and Organisms. Pergamon Press, Inc., New York.
Gordon, D.C., J. Dale and P.D. Keizer. 1978. Importance of sediment working by the deposit-feeding polychaete *Arenicola marina* on the weathering rate of sediment bound oil. J. Fish. Res. Bd. Can. 35: 591-603.
Griffiths, R.P., T.M. McNamara, B.A. Calwell and R.Y. Morita. 1981. Field observations on the acute effects of crude oil on glucose and glutamate uptake in samples collected from Arctic and subarctic waters. Appl. Environ. Microbiol. 41: 1400-1406.
Guire, P.E., J.D. Friede and R.K. Gholson. 1973. Production and characterization of emulsifying factors from hydrocarbonoclastic yeasts and bacteria. Pages 229-231 *in* D.G. Ahearn and S.P. Meyers (eds.), The Microbial Degradation of Oil Pollutants. Publication No. LSU-SG-73-01. Center for Wetland Resources, Louisiana State University, Baton Rouge, Louisiana.
Gunkel, W. 1968. Bacteriological investigations of oil-polluted sediments from the Cornish coast following the *Torrey Canyon* disaster. Helgol. Wiss. Meeresunters. 17: 151-158.
Gutnick, D.L. and E. Rosenberg. 1977. Oil tankers and pollution: A microbiological approach. Ann. Rev. Microbiol. 31: 379-396.
Haines, J.K. and M. Alexander. 1974. Microbial degradation of high-molecular weight alkanes. Appl. Microbiol. 28: 1084-1085.
Haines, J.R. and R.M. Atlas. 1982. *In situ* microbial degradation of Prudhoe Bay crude oil in Beaufort Sea sediment. Mar. Environ. Res. 7: 91-102.
Haines, J.R. and R.M. Atlas. 1983. Biodegradation of petroleum hydrocarbons in continental shelf regions of the Bering Sea. Oil and Petrochem. Poll. 1: 85-96.
Haines, J.R., R.M. Atlas, R.P. Griffiths and R.Y. Morita. 1981. Denitrification and nitrogen fixation in Alaskan continental shelf sediments. Appl. Env. Microbiol. 41: 412-421.
Hartung, R. and G.W. Klinger. 1970. Concentration of DDT by sedimented polluting oils. Env. Sci. Technol. 4: 407-410.
Herbes, S.E. and L.R. Schwall. 1978. Microbial transformation of polycyclic aromatic hydrocarbons in pristine and petroleum-contaminated sediments. Appl. Environ. Microbiol. 35: 306-316.
Higgins, I.J. and P.D. Gilbert. 1978. The biodegradation of hydrocarbons. Pages 80-

117 *in* K.W.A. Chater and H.J. Somerville (eds.), The Oil Industry and Microbial Ecosystems. Heyden and Son, Ltd., London.

Hinga, K.R., M.E. Pilson, R.F. Lee, J.W. Farrington, K. Tjessem and A.C. Davis. 1980. Biogeochemistry of benzanthracene in an enclosed marine ecosystem. Env. Sci. Technol. 14: 1136-1143.

Hodson, R.E., F. Azam and R.F. Lee. 1977. Effects of four oils on marine bacterial populations; controlled ecosystem pollution experiment. Bull. Mar. Sci. 27: 119-126.

Hopper, D.J. 1978. Microbial degradation of aromatic hydrocarbons. Pages 85-112 *in* J.R. Watkinson (ed.), Developments in Biodegradation of Hydrocarbons-1. Applied Science Publishers, Ltd., London.

Horowitz, A., D. Gutnick and E. Rosenberg. 1975. Sequential growth of bacteria on crude oil. Appl. Microbiol. 30: 10-19.

Horowitz, A. and R.M. Atlas. 1977. Continuous open flow-through system as a model for oil degradation in the Arctic Ocean. Appl. Environ. Microbiol. 33: 647-653.

Horowitz, A. and R.M. Atlas. 1978a. Crude oil degradation in the Arctic: Changes in bacterial populations and oil composition during one year exposure in a model system. Dev. Ind. Microbiol. 19: 517-522.

Horowitz, A. and R.M. Atlas. 1978b. Microbial seeding to enhance petroleum hydrocarbon biodegradation in aquatic Arctic ecosystems. Pages 15-20 *in* T.A. Oxley, D. Olsopp, and G. Becker (eds.), Biodeterioration. Pittman Publications, Ltd., London.

Hou, C.T. and A.I. Laskin. 1976. Microbial conversion of dibenzothiophene. Dev. Ind. Microbiol. 17: 351-362.

Hughes, D.E. and P. McKenzie. 1975. The microbial degradation of oil in the sea. Proc. R. Soc. London Ser. B. 189: 375-390.

Iguchi, T., I. Takeda and H. Oksawa. 1969. Emulsifying factors of hydrocarbon-assimilating yeasts. Agric. Biol. Chem. 33: 1657-1658.

Iizuka, H., M. Ilida and S. Fujita. 1969. Formation of *n*-decene-1 from *n*-decane by resting cells of *C. rugosa*. Z. Allg. Mikroabiol. 9:223-226.

Jannasch, H.W. and C.O. Wirsen. 1973. Deep sea microorganisms: *In situ* response to nutrient enrichment. Science 180: 641-643.

Jordan, R.E. and J.R. Payne. 1980. Fate and Weathering of Petroleum Spills in the Marine Environment. Ann Arbor Science Publishers, Ann Arbor, Michigan.

Kanazawa, M. 1975. Production of yeast from *n*-paraffins. Pages 438-453 *in* S.R. Tannenbaum and D.I.C. Wang (eds.), Single Cell Protein, Vol. 2. M.I.T. Press, Cambridge, Massachusetts.

Kator, H. 1973. Utilization of crude oil hydrocarbons by mixed cultures of marine bacteria. Pages 47-66 *in* D.G. Ahearn and S.P. Meyers (eds.), The Microbial Degradation of Oil Pollutants. Publication No. LSU-SG-73-01. Center for Wetland Resources, Louisiana State University, Baton Rouge, Louisiana.

Kator, H. and R. Herwig. 1977. Microbial responses after two experimental oil spills in an eastern coastal plain estuarine ecosystem. Pages 517-522 *in* Proceedings of the 1977 Oil Spill Conference. American Petroleum Institute, Washington, D.C.

Kator, H., C.H. Oppenheimer and R.J. Miget. 1971. Microbial degradation of a Louisiana crude oil in closed flasks and under simulated field conditions. Pages 287-296 *in* Joint Conference on Prevention and Control of Oil Spills. American Petroleum Institute, Washington, D.C.

Kator, H., R. Miget and C.H. Oppenheimer. 1972. Utilization of paraffin hydrocarbons in crude oil by mixed cultures of marine bacteria. *In* Symposium on Environmental Conservation, No. 13-14, Paper No. SPE-4206. Society of Petroleum Engineers of AIME, Dallas, Texas.

Kennedy, R.S., W.R. Finnerty, K. Suzdasana and R.A. Young. 1975. Microbial assimilation of hydrocarbons. I. The fine structure of a hydrocarbon oxidizing *Acinetobacter* sp. Arch. Microbiol. 102: 75-83.

Kinney, P.J., D.K. Button and D.M. Schell. 1969. Kinetics of dissipation and biodegradation of crude oil in Alaska's Cook Inlet. Pages 333-340 *in* Proceedings of

1969 Joint Conference on Prevention and Control of Oil Spills. American Petroleum Institute, Washington, D.C.
Klug, M.J. and A.J. Markovetz. 1967. Thermophilic bacteria isolated on *n*-tetradecane. Nature (London) 215: 1082-1083.
Kockova-Kratochvilova, A. and M. Havelkova. 1974. *Prototheca hydrocarbonea* n. sp.; Lebenszyklus, Metabolismus und Feinstruktur. Z. Allg. Mikrobiol. 14: 123-134.
Kodama, K., S. Nakatani, K. Umehara, K. Shimizu, Y. Minoda and K. Yamada. 1970. Microbial conversion of petro-sulfur compounds. III. Isolation and identification of products from dibenzothiophene. Agric. Biol. Chem. (Japan) 34: 1320-1324.
Kodama, K., K. Umehara, K. Shimizu, S. Nakatani, Y. Minoda and K. Yamada. 1973. Identification of microbial products from dibenzothiophene and its proposed oxidation pathway. Agric. Biol. Chem. (Japan) 37: 45-50.
Laborde, A.L. and D.T. Gibson. 1977. Metabolism of dibenzothiophene by a *Beijerinckia* species. Appl. Environ. Microbiol. 34: 783-790.
Langlois, G.A. 1979. The effect of petroleum hydrocarbons on microbial interactions in the marine environment. J. Protozool. 26: 22A.
LaRiviere, J.W.M. 1955. The production of surface active compounds by microorganisms and its possible significance in oil recovery. I. Some general observations on the change of surface tension in microbial cultures. Antonie van Leeuwenhoek J. Microbiol. Ser. 21: 1-8.
Lee, R.F. 1977. Fate of petroleum components in estuarine waters of the southeastern United States. Pages 611-616 *in* Proceedings of the 1977 Oil Spill Conference. American Petroleum Institute, Washington, D.C.
Lee, R.F. and C. Ryan. 1976. Biodegradation of petroleum hydrocarbons by marine microbes. Pages 119-126 *in* J.M. Sharpley and A.M. Kaplan (eds.), Proceedings of the Third International Biodegradation Symposium. Applied Science Publishers, Ltd., London.
Lee, R.F., R. Sauerheber and A.A. Benson. 1972a. Petroleum hydrocarbons: Uptake and discharge by the marine mussel, *Mytilus edulis*. Science 177: 344-346.
Lee, R.F., R. Sauerheber and G.H. Dobbs. 1972b. Uptake, metabolism and discharge of polycyclic aromatic hydrocarbons by marine fish. Mar. Biol. 17: 201-208.
LePetit, J., M.-H. N'Guyen and L. Deveze. 1970. Etude de l'intervention des levures dans la biodegradation en mer des hydrocarbures. Ann. Inst. Pasteur 118: 709-720.
LePetit, J. and M.-H. N'Guyen. 1976. Besoins en phosphore des bacteries metabolisant les hydrocarbures en mer. Can. J. Microbiol. 22: 1364-1373.
LePetit, J. and S. Tagger. 1976. Degradation des hydrocarbures en presence d'autres substances organiques par des bacteries isolees de l'eau de mer. Can. J. Microbiol. 22: 1654-1657.
LePetit, J., M.-H. N'Guyen and S. Tagger. 1977. Quelques donnees sur l'ecologie d'une zone marine littorales recevant les rejets d'une raffinerie de petrole. Environ. Pollut. 13: 41-56.
Lindmark, D.G. 1981. Activation of polynuclear aromatic hydrocarbons to mutagens by the marine ciliate *Paramecium acutum*. Appl. Environ. Microbiol. 41: 1238-1242
Mackay, D. 1981. Fate and behaviour of oil spills. Pages 7-27 *in* J.B. Sprague, J.H. Vandermeulen and P.G. Wells (eds.), Oil Dispersants in Canadian Seas - Research Appraisal and Recommendations. Environment Canada, Ottawa.
Makula, R.A., P.J. Lockwood and W.R. Finnerty. 1975. Comparative analysis of the lipids of *Acinetobacter* species grown on hexadecane. J. Bacteriol. 121: 250-258.
Malins, D.C. (ed.). 1977. Effects of Petroleum on Arctic and Subarctic Marine Environments and Organisms. Vol. 1. Nature and Fate of Petroleum. Academic Press, New York.
Markovetz, A.J. 1971. Subterminal oxidation of aliphatic hydrocarbons by microorganisms. Crit. Rev. Microbiol. 1: 225-237.
Masters, M.J. and J.E. Zajic. 1971. Myxotrophic growth of algae on hydrocarbon substrates. Dev. Ind. Microbiol. 12: 77-86.

Mateles, R.I., J.N. Baruah and S.R. Tannenbaum. 1967. Growth of a thermophilic bacteria on hydrocarbons: A new source of single cell protein. Science 157: 1322-1323.

McKenna, E.J. 1971. Microbial degradation of normal and branched alkanes. Pages 73-97 *in* Degradation of Synthetic Organic Molecules in the Biosphere. National Academy of Sciences, Washington, D.C.

Merkel, G.J., S.S. Stapleton and J.J. Perry. 1978. Isolation and peptidoglycan of gram-negative hydrocarbon-utilizing thermophilic bacteria. J. Gen. Microbiol. 109: 141-148.

Miget, R.J. 1973. Bacterial seeding to enhance biodegradation of oil slicks. Pages 291-309 *in* D.G. Ahearn and S.P. Meyers (eds.), The Microbial Degradation of Oil Pollutants. Publication No. LSU-SG-73-01. Center for Wetland Resources, Louisiana State University, Baton Rouge, Louisiana.

Miget, R.J., C.H. Oppenheimer, H.I. Kator and P.A. LaRock. 1969. Microbial degradation of normal paraffin hydrocarbons in crude oil. Pages 327-331 *in* Proceedings of the 1969 Oil Spill Conference. American Petroleum Institute, Washington, D.C.

Mironov, O.G. 1970. Role of microorganisms growing on oil in the self-purification and indication of oil pollution in the sea. Oceanology 10: 650-656.

Mironov, O.G. and A.A. Lebed. 1972. Hydrocarbon-oxidizing microorganisms in the North Atlantic. Hydrobiol. J. 8: 71-74.

Mulkins-Phillips, G.J. and J.E. Stewart. 1974a. Distribution of hydrocarbon-utilizing bacteria in northwestern Atlantic waters and coastal sediments. Can. J. Microbiol. 20: 955-962.

Mulkins-Phillips, G.J. and J.E. Stewart. 1974b. Effect of environmental parameters on bacterial degradation of Bunker C oil, crude oils and hydrocarbons. Appl. Microbiol. 28: 915-922.

Mulkins-Phillips, G.J. and J.E. Stewart. 1974c. Effect of four dispersants on biodegradation and growth of bacteria on crude oil. Appl. Microbiol. 28: 547-552.

Nakatani, S., T. Akasaki, K. Kodama, Y. Minoda and K. Yamada. 1968. Microbial conversion of petro-sulfur compounds. II. Culture conditions of dibenzothiophene-utilizing bacteria. Agric. Biol. Chem. (Japan) 32: 1205-1211.

National Research Council. 1975. Petroleum in the Marine Environment. National Academy of Sciences, Washington, D.C., 107 p.

National Research Council. 1985. Oil in the Sea. Inputs, Fates, and Effects. National Academy Press, Washington, D.C., 601 p.

Nyns, E.J. and A.R. Wiaux. 1969. Biology of hydrocarbons, 1966-1969. Agricultura (Luwain) 17: 3-56.

Olivieri, R., P. Bacchin, A. Robertiello, N. Oddo, L. Degen and A. Tonolo. 1976. Microbial degradation of oil spills enhanced by a slow-release fertilizer. Appl. Environ. Microbiol. 31: 629-634.

Oppenheimer, C.H., W. Gunkel and G. Gassman. 1977. Microorganisms and hydrocarbons in the North Sea during July-August 1975. Pages 593-610 *in* Proceedings of the 1977 Oil Spill Conference. American Petroleum Institute, Washington, D.C.

Oviatt, C., J. Frithsen, J. Gearing and P. Gearing. 1982. Low chronic additions of number 2 fuel oil: Chemical behavior, biological impact, and recovery in a simulated estuarine environment. Mar. Ecol. Prog. Ser. 9: 121-136.

Panchal, C.J. and J.E. Zajic. 1977. Isolation of emulsifying agents from species of *Corynebacterium*. Dev. Ind. Microbiol. 19: 569-576.

Pancirov, R.J. 1974. Compositional Data on API Reference Oils Used in Biological Studies: A #2 Fuel Oil, a Bunker C, Kuwait Crude Oil and South Louisiana Crude Oil. Report No. AID. IBA. 74. Esso Research, Linden, New Jersey.

Parekh, V.R., R.W. Traxler and J.M. Sobek. 1977. *n*-Alkane oxidation enzymes of a Pseudomonad. Appl. Environ. Microbiol. 33: 881-884.

Pelz, B.F. and H.J. Rehm. 1971. Resistance of decalin against microbiological oxidation. Arch. Mikrobiol. 77: 288-290.

Perry, J.J. 1977. Microbial metabolism of cyclic hydrocarbons and related compounds. Crit. Rev. Microbiol. 5: 387-412.

Perry, J.J. 1979. Microbial cooxidations involving hydrocarbons. Microbiol. Rev. 43: 59-72.

Perry, J.J. and C.E. Cerniglia. 1973. Studies on the degradation of petroleum by filamentous fungi. Pages 89-94 *in* D.G. Ahearn and S.P. Meyers (eds.), The Microbial Degradation of Oil Pollutants. Publication No. LSU-SG-73-01. Center for Wetland Resources, Louisiana State University, Baton Rouge, Louisiana.

Pfaender, F. and E.N. Buckley. 1980. Response of the pelagic community to oil from the IXTOC-I blowout: *In situ* studies. Pages 545-560 *in* Proceedings of Conference on *Researcher/Pierce* IXTOC-I Cruises. National Oceanic and Atmospheric Administration, Atlantic Oceanographic and Meteorological Laboratories, Miami, Florida.

Pfaender, F. and E.N. Buckley. 1984. Effects of petroleum on microbial communities. *In* R.M. Atlas (ed.), Petroleum Microbiology. Macmillan, New York.

Pfaender, F.K., E.N. Buckley and R.L. Ferguson. 1980. Response of the pelagic microbial community to oil from the IXTOC-I blowout: *in situ* studies. Pages 545-562 *in* D.K. Atwood (convenor), Proceedings from a Symposium on Preliminary Results from the September 1979 *Research/Pierce* IXTOC-I Cruise. National Oceanic and Atmospheric Administration, Boulder, Colorado.

Pirnik, M.P. 1977. Microbial oxidation of methyl branched alkanes. Crit. Rev. Microbiol. 5: 413-422.

Pirnik, M.P., R.M. Atlas and R. Bartha. 1974. Hydrocarbon metabolism by *Brevibacterium erythrogenes*: Normal and branched alkanes. J. Bacteriol. 119: 868-878.

Polyakova, I.N. 1962. Distribution of hydrocarbon-oxidizing microorganisms in water of Neva Bay. Mikrobiologiya 31: 1076-1081.

Pritchard, P.H. and T.J. Starr. 1973. Microbial degradation of oil and hydrocarbons in continuous culture. Pages 39-46 *in* D.G. Ahearn and S.P. Meyers (eds.), The Microbial Degradation of Oil Pollutants. Publication No. LSU-SG-73-01. Center for Wetland Resources, Louisiana State University, Baton Rouge, Louisiana.

Ratledge, C. 1978. Degradation of aliphatic hydrocarbons. Pages 1-46 *in* J.R. Watkinson (ed.), Developments in Biodegradation of Hydrocarbons-1. Applied Science Publishers, Ltd., London.

Reisfeld, A., E. Rosenberg and D. Gutnick. 1972. Microbial degradation of oil: Factors affecting oil dispersion in seawater by mixed and pure cultures. Appl. Microbiol. 24: 363-368.

Robichaux, T.J. and H.N. Myrick. 1972. Chemical enhancement of the biodegradation of crude oil pollutants. J. Petrol. Technol. 24: 16-20.

Robertson, B., S. Arhelger, P.J. Kinney and D.K. Button. 1973. Hydrocarbon biodegradation in Alaskan waters. Pages 171-184 *in* D.G. Ahearn and S.P. Meyers (eds.), The Microbial Degradation of Oil Pollutants. Publication No. LSU-SG-73-01. Center for Wetland Resources, Louisiana State University, Baton Rouge, Louisiana.

Rogerson, A. and J. Berger. 1980. The effect of crude oil on protozoan growth. J. Protozool. 27: 30A.

Rogerson, A. and J. Berger. 1981a. Effect of crude oil and petroleum-degrading microorganisms on the growth of freshwater and soil protozoa. J. Gen. Microbiol. 124: 53-59.

Rogerson, A. and J. Berger. 1981b. The toxicity of the dispersant Corexit 9527 and oil-dispersant mixtures to ciliate protozoa. Chemosphere 10: 33-39.

Rogerson, A. and J. Berger. 1982. The stimulatory role of ciliates in crude oil biodegradation. J. Protozool. 29: 487.

Rogerson, A. and J. Berger. 1983. Enhancement of the microbial degradation of crude oil by the ciliate *Colpidium colpoda*. J. Gen. Appl. Microbiol. 29: 41-50.

Rosenberg, E., E. Englander, A. Horowitz and D. Gutnik. 1975. Bacterial growth and dispersion of crude oil in an oil tanker during its ballast voyage. Pages 157-167 *in* A.W. Borquin, D.G. Ahearn and S.P. Meyers (eds.), Impact of the Use of Microorganisms on the Aquatic Environment. EPA-660-3-75001. Environmental Protection Agency, National Environmental Research Center, Corvallis, Oregon.

Roubal, G. and R.M. Atlas. 1978. Distribution of hydrocarbon-utilizing microorganisms and hydrocarbon biodegradation potentials in Alaska continental shelf areas. Appl. Environ. Microbiol. 35: 897-905.

Sayler, G.S. and R.R. Colwell. 1976. Partitioning of mercury and chlorinated biphenyl by oil, water and sediment. Env. Sci. Technol. 10: 1142-1145.

Schwarz, J.R., J.D. Walker and R.R. Colwell. 1974a. Hydrocarbon degradation at ambient and *in situ* pressure. Appl. Microbiol. 28: 982-986.

Schwarz, J.R., J.D. Walker and R.R. Colwell. 1974b. Growth of deep sea bacteria on hydrocarbons at ambient and *in situ* pressure. Dev. Ind. Microbiol. 15: 239-249.

Schwarz, J.R., J.D. Walker and R.R. Colwell. 1975. Deep-sea bacteria: Growth and utilization of *n*-hexadecane at *in situ* temperature and pressure. Can. J. Microbiol. 21: 682-687.

Seba, D.B. and E.F. Corcoran. 1969. Surface slicks as concentrators of pesticides in the marine environment. Pestic. Monit. J. 3: 190-193.

Seki, H. 1976. Method for estimating the decomposition of hexadecane in the marine environment. Appl. Environ. Microbiol. 31: 439-441.

Senez, J.C. and E. Azoulay. 1961. Dehydrogenation of paraffinic hydrocarbons by resting cells and cell free extracts of *Pseudomonas aeruginosa*. Biochim. Biophys. Acta 47: 307-316.

Seubert, W. and E. Fass. 1964. Untersuchungen über den bakteriellen Abban von Isoprenoiden. V: Der Mechanisms des Isoprenoidabbaues. Biochem. Z. 341: 35-44.

Smith, J.E. 1968. *Torrey Canyon*: Pollution and Marine Life. Cambridge University Press, Cambridge, England.

Soli, G. 1973. Marine hydrocarbonoclastic bacteria: Types and range of oil degradation. Pages 141-146 *in* D.G. Ahearn and S.P. Meyers (eds.), The Microbial Degradation of Oil Pollutants. Publication No. LSU-SG-73-01. Center for Wetland Resources, Louisiana State University, Baton Rouge, Louisiana.

Soli, G. and E.M. Bens. 1972. Bacteria which attack petroleum hydrocarbons in saline medium. Biotechnol. Bioeng. 14: 319-330.

Spooner, M. 1968. Comments during discussion. Page 67 *in* P. Hepple (ed.), Scientific Aspects of Pollution of the Sea by Oil. Institute of Petroleum, London.

Stewart, J.E. and L.J. Marks. 1978. Distribution and abundance of hydrocarbon-utilizing bacteria in sediments of Chedabucto Bay, Nova Scotia in 1976. J. Fish. Res. Board Can. 35: 581-584.

Stirling, L.A., R.J. Watkinson and I.J. Higgins. 1977. Microbial metabolism of alicyclic hydrocarbons: Isolation and properties of a cyclohexane-degrading bacterium. J. Gen. Microbiol. 99: 119-125.

Stormer, F.C. and A. Vinsjansen. 1976. Microbial degradation of Ekofisk oil in seawater by *Saccharomycopsis lipolytica*. Ambio. 5: 141-142.

Stout, J.D. 1980. The role of protozoa in nutrient cycling and energy flow. Adv. Microbial Ecol. 4: 1-50.

Tait, R.V. and R.S. DeSanto. 1972. Elements of Marine Ecology. Springer-Verlag, New York, 327 p.

Traxler, R.W. 1973. Bacterial degradation of petroleum materials in low temperature marine environments. Pages 163-170 *in* D.G. Ahearn and S.P. Meyers (eds.), The Microbial Degradation of Oil Pollutants. Publication No. LSU-SG-73-01. Center for Wetland Resources, Louisiana State University, Baton Rouge, Louisiana.

Traxler, R.W. and J.M. Bernard. 1969. The utilization of *n*-alkanes by *Pseudomonas aeruginosa* under conditions of anerobiosis. Int. Biodeterior. Bull. 5: 21-25.

Traxler, R.W. and J.H. Vandermeulen. 1981. Hydrocarbon-utilizing microbial potential in marsh, mudflat and sandy sediments from North Brittany. Pages 243-249 *in Amoco Cadiz*: Fates and Effects of the Oil Spill. Centre National pour L'Exploitation des Oceans, Paris.

Trudgill, P.W. 1978. Microbial degradation of alicyclic hydrocarbons. Pages 47-84 *in* J.R. Watkinson (ed.), Developments in Biodegradation of Hydrocarbons-1. Applied Science Publishers, Ltd., London.

Valenkar, S. K., S.M. Barneff, C.W. Houston and A.R. Thompson. 1975. Microbial

growth on hydrocarbons: Some experimental results. Biotechnol. Bioeng. 17: 241-251.
Van der Linden, A.C. 1978. Degradation of oil in the marine environment. Pages 165-200 *in* J.R. Watkinson (ed.), Developments in Biodegradation of Hydrocarbons-1. Applied Science Publishers, Ltd., London.
Walker, J.D. and R.R. Colwell. 1974a. Microbial degradation of model petroleum at low temperatures. Microb. Ecol. 1: 63-95.
Walker, J.D. and R.R. Colwell. 1974b. Mercury-resistant bacteria and petroleum degradation. Appl. Microbiol. 27: 285-287.
Walker, J.D. and R.R. Colwell. 1976a. Enumeration of petroleum-degrading microorganisms. Appl. Environ. Microbiol. 31: 198-207.
Walker, J.D. and R.R. Colwell. 1976b. Long-chain *n*-alkanes occurring during microbial degradation of petroleum. Can. J. Microbiol. 22: 886-891.
Walker, J.D. and R.R. Colwell. 1976c. Oil, chlorinated biphenyl, mercury and microorganism interactions. Environ. Sci. Technol. 10: 1145-1147.
Walker, J.D. and R.R. Colwell. 1976d. Measuring the potential activity of hydrocarbon-degrading bacteria. Appl. Environ. Microbiol. 31: 189-197.
Walker, J.D. and R.R. Colwell. 1977. Role of autochthonous bacteria in the removal of spilled oil from sediment. Environ. Pollut. 12: 51-56.
Walker, J.D. and J.J. Cooney. 1975. Effect of poorly metabolized hydrocarbons on substrate oxidation by *Cladosporium resinae*. J. Appl. Bacteriol. 39: 189-195.
Walker, J.D., P.A. Seesman and R.R. Colwell. 1974. Effects of petroleum on estuarine bacteria. Mar. Poll. Bull. 5: 186-188.
Walker, J.D., H.F. Austin and R.R. Colwell. 1975a. Utilization of mixed hydrocarbon substrate by petroleum-degrading microorganisms. J. Gen. Appl. Microbiol. 21: 27-39.
Walker, J.D., R.R. Colwell and L. Petrakis. 1975b. Degradation of petroleum by an alga, *Prototheca zopfii*. Appl. Microbiol. 30: 79-81.
Walker, J.D., R.R. Colwell, Z. Vaituzis and S.A. Meyer. 1975c. Petroleum-degrading achlorophyllous alga *Prototheca zopfii*. Nature 254: 423-424.
Walker, J.D., R.R. Colwell and L. Petrakis. 1975d. Microbial petroleum degradation: Application of computerized mass spectrometry. Can. J. Microbiol. 21: 1760-1767.
Walker, J.D., R.R. Colwell and L. Detrakis. 1975e. Evaluation of petroleum-degrading potential of bacteria from water and sediment. Appl. Microbiol. 30: 1036-1039.
Walker, J.D., P.A. Seesman and R.R. Colwell. 1975f. Effects of South Louisiana Crude oil and No. 2 fuel oil on growth of heterotrophic microorganisms, including proteolytic, lipolytic, chitinolytic and cellulolytic bacteria. Environ. Pollut. 9: 13-33.
Walker, J.D., R.R. Colwell and L. Petrakis. 1976a. Biodegradation of petroleum by Chesapeake Bay sediment bacteria. Can. J. Microbiol. 22: 423-428.
Walker, J.D., R.R Colwell and L. Petrakis. 1976b. Biodegradation rates of components of petroleum. Can. J. Microbiol. 22: 1209-1213.
Ward, D.M. and T.D. Brock. 1978a. Hydrocarbon biodegradation in hypersaline environments. Appl. Environ. Microbiol. 35: 353-359.
Ward, D.M. and T.D. Brock. 1978b. Anaerobic metabolism of hexadecane in marine sediments. Geomicrobiol. J. 1: 1-9.
Ward, D.M., R.M. Atlas, P.D. Boehm and J.A. Calder. 1980. Microbial biodegradation and the chemical evolution of *Amoco Cadiz* oil pollutants. Ambio 9: 277-283.
Williams, P.A. 1978. Microbial genetics relating to hydrocarbon degradation. Pages 135-164 *in* J.R. Watkinson (ed.), Developments in Biodegradation of Hydrocarbons-1. Applied Science Publishers, Ltd., London.
Winfrey, M.R., E. Beck, P. Boehm and D.M. Ward. 1982. Impact of crude oil on sulphate reduction and methane production in sediments impacted by the *Amoco Cadiz* oil spill. Mar. Environ. Res. 7: 175-194.
Wright, R.T. and J.E. Hobbie. 1966. The use of glucose and acetate by bacteria and algae in aquatic ecosystems. Ecology 47: 447-464.

Wu, J. and L.K. Wong. 1981. Microbial transformations of 7,12-dimethylbenz(a)anthracene. Appl. Environ. Microbiol. 41: 843-845.

Yamada, K., Y. Monoda, K. Komada, S. Nakatani and T. Akasaki. 1968. Microbial conversion of petrol-sulfur compounds. I. Isolation and identification of dibenzothiophene-utilizing bacteria. Agric. Biol. Chem. 32: 840-845.

Zajic, J.E. and E. Knetting. 1971. Flocculants from paraffinic hydrocarbons. Dev. Ind. Microbiol. 12: 87-98.

Zajic, J.E. and E. Knetting. 1972. Microbial emulsifer from Bunker C fuel oil. Chemosphere 1: 51-56.

Zajic, J.E. and C.J. Panchal. 1976. Bio-emulsifiers. CRC Crit. Rev. Microbiol. 5: 39-66.

Zajic, J.E. and B. Suplisson. 1972. Emulsification and degradation of "Bunker C" fuel oil by microorganisms. Biotechnol. Bioeng. 14: 331-343.

Zajic, J. E., B. Suplisson and B. Volesky. 1974. Bacterial degradation and emulsification of No. 6 fuel oil. Env. Sci. Technol. 8: 664-668.

ZoBell, C. E. 1969. Microbial modification of crude oil in the sea. Pages 317-326 *in* Proceedings of Joint Conference on Prevention and Control of Oil Spills. American Petroleum Institute, Washington, D.C.

ZoBell, C.E. 1973. Bacterial degradation of mineral oils at low temperatures. Pages 153-161 *in* D.G. Ahearn and S.P. Meyers (eds.), The Microbial Degradation of Oil Pollutants. Publication No. LSU-SG-73-01. Center for Wetland Resources, Louisiana State University, Baton Rouge, Louisiana.

ZoBell, C.E. and J. Agosti. 1972. Bacterial oxidation of mineral oils at sub-zero Celsius. American Society for Microbiology, Philadelphia, 72nd Annual Meeting, April 23-28, Abstr. E11.

ZoBell, C.E. and J.F. Prokop. 1966. Microbial oxidation of mineral oils in Barataria Bay bottom deposits. Z. Allg. Mikrobiol. 6: 143-162.

CHAPTER 8

BIOLOGICAL EFFECTS OF PETROLEUM HYDROCARBONS: ASSESSMENTS FROM EXPERIMENTAL RESULTS

Judith M. Capuzzo

CONTENTS

INTRODUCTION

Biological effects of petroleum hydrocarbons on marine organisms are dependent upon their bioavailability and persistence, the ability of the organism to accumulate and metabolize various hydrocarbons, and the interference of hydrocarbons with normal metabolic pathways that alter an organism's chances for survival and reproduction in the environment. In considering the long-term effects of offshore oil and gas development activities, it is important to ascertain what biological effects may result in subtle ecological changes and possible impairment of fisheries resources.

Long-term effects of accidental oil spills from tankers and barges have resulted in alterations in benthic community structure (Southward and Southward, 1978; Cabioch *et al.*, 1980; Sanders *et al.*, 1980; Elmgren *et al.*, 1981), inhibition of recruitment and alteration in energetics of benthic populations (Krebs and Burns, 1977; Gilfillan *et al.*, 1976; Gilfillan and Vandermeulen, 1978; Elmgren *et al.*, 1981) and loss of fisheries resources (Friha and Conan, 1981). With the exception of a blowout or oil spills during transport, however, long-term effects of offshore exploration and production phases will be related to low-level chronic discharges of drilling fluids and produced waters. Thus, predicting the impact of offshore oil and gas activities requires an understanding of the responses of marine biota to both chronic low-level discharges and accidental discharges of larger volumes. This can best be accomplished by comparing changes in offshore or coastal areas receiving chronic low-level discharges with observations made in experimental laboratory or field studies. Although experimental studies are constrained by a certain degree of artificiality, carefully conducted studies can contribute to our understanding of the range of potential toxic responses among populations of marine organisms.

The responses of marine organisms to discharges of petroleum hydrocarbons can be manifested at four levels of biological organization: 1) biochemical and cellular; 2) organismal, including the integration of physiological, biochemical, and behavioral responses; 3) population, including alterations in population dynamics; and 4) community, resulting in alterations in community structure and dynamics. The interrelationship of these responses are presented in Table 8.1.

Biological effects of contaminants can be manifested at biochemical, cellular and organismal levels of organization before disturbances are seen at the population level (Capuzzo, 1981a). All responses are not disruptive in nature and do not necessarily result in degeneration at the next level of biological organization. Only when the compensatory or adaptive mechanisms at one level begin to fail, do deleterious effects become apparent at the next level (Capuzzo, 1981a). An important aspect of comparing responses at various levels of biological organization is ascertaining the degree to which adaptive responses at each of the four levels can persist with increasing concentrations of petroleum hydrocarbons. The initial responses in each case are inductions of mechanisms to resist or reduce the toxicant impact, as by induction of toxicant metabolizing processes (at the biochemical level) or by selection of toxicant resistant forms (at the population level). Adaptive processes are capable of countering disruptive processes until the system reaches a toxicant threshold, at which point the adaptive potential is completely overridden by the degeneration imposed on the system by disruptive effects.

For predictive purposes, it is important to understand the early warning signs of stress at each level of organization before compensatory mechanisms are surpassed. From the biochemical level to the community level, the degree of system complexity, the number of compensatory mechanisms available, and the lag time to measure a response increase exponentially, thereby increasing the predictive difficulties at each level. Cairns (1983) argued that our ability to detect toxic effects at higher levels of biological organization is limited by the lack of reliable predictive tests at population, community and ecosystem levels. He further suggested that much effort is needed in these areas before we can adequately address environmental hazards as a result of input of any toxicant.

TABLE 8.1
Response levels of marine organisms to petroleum hydrocarbons

Level	*Types of responses*	*Effects at next level*
Biochemical–Cellular	Impairment of metabolic pathways	Disruption in energetics Reduction in energy stores
	Detoxification	Adaptation of organism
Organismal	Metabolic changes Behavioral changes Increased incidence of disease Reduction in growth and reproduction	Reduction in performance of populations
	Adjustment in rate functions Disease defense	Regulation and adaptation of populations
Population	Changes in population dynamics Adaptations of populations to stress	Effects on coexisting organisms and community No change at community level
Community	Changes in species composition	Deterioration of community
	Reduced energy flow	Reduced secondary production
	Ecosystem adaptation	No change in community stability

Experimental studies directed at determining effects on energy metabolism or effects that influence growth and reproduction would be most appropriate at linking effects at higher levels of organization. The focus of this chapter is to integrate our understanding of the effects of petroleum hydrocarbons on marine organisms and to develop the framework for experimental design from which predictions of effects at higher levels of organization can be made.

BIOASSAY DESIGN

In evaluating toxic responses of marine organisms to petroleum hydrocarbons, it is essential to evaluate bioassay design, the bioavailability and concentration of hydrocarbons, and the potential for bioaccumulation. Chemical analyses are an important part of experimental protocol as the composition, relative solubilities, and dispersion forming capabilities of specific oils may influence their toxicity (Anderson *et al.*, 1974a,b). Many earlier studies did not include chemical analyses and interpretations of bioassay results were based on nominal concentrations of oil (i.e., the amount of oil added at the beginning of the experiment without consideration given to loss of hydrocarbons through volatilization, degradation or other processes during exposure). Several methodologies have been developed for quantitatively estimating hydrocarbon concentrations. These include 1) ultraviolet (UV) spectrophotometry, 2) infrared spectrophotometry, and 3) gas chromatography.

The UV technique described by Neff and Anderson (1975a) was specifically designed for the determination of naphthalenes and alkylnaphthalenes in both water and animal tissue, as these components dominate the water soluble fraction of both crude and refined oils (Anderson *et al.*, 1974a). Although this technique is not suitable for quantification of the full range of hydrocarbons, it is useful for monitoring the consistency of preparations of water soluble fractions. The infrared technique recommended by the American Petroleum Institute (API, 1958) has been routinely used for the determination of total hydrocarbons in seawater samples (Anderson *et al.*, 1974a). More detailed analyses of hydrocarbon composition have been conducted using gas chromatography (Anderson *et al.*, 1974a; Bean *et al.*, 1980; Farrington *et al.*, 1982; Galloway *et al.*, 1983). Both packed column (Anderson *et al.*, 1974a) and glass capillary column (Farrington *et al.*, 1982) techniques have been employed, the latter technique providing better resolution of the individual components of a hydrocarbon mixture. Compounds are quantified and identified by comparison with peak heights and retention times of known standards. More detailed analyses of individual hydrocarbons are carried out by gas chromatography-mass spectroscopy.

Oil can be introduced to a bioassay system as: 1) water soluble fractions, 2) oil-in-water dispersions, 3) surface slicks, 4) oil-contaminated foods and 5) oil-contaminated sediments. Each type of experiment presents several advantages and disadvantages in predicting the fate and effects of petroleum hydrocarbons in the marine environment, as discussed below.

When oil is mixed with sea water, both large droplets and microdroplets (comprising a particulate phase) and homogeneous mixtures of the more soluble components (comprising the water soluble fraction) are formed. Water soluble fractions (WSF) are generally prepared by stirring set ratios of oil and sea water for various periods (hours to days). Stirring is followed by an equilibration period to allow separation of particulate and water soluble fractions (Anderson *et al.*, 1974a; Kauss and Hutchinson, 1975; Pulich *et al.*, 1974; Winters *et al.*, 1977b) and the aqueous phase is diluted to the desired concentration. Stirring conditions and the physical and chemical characteristics of the crude or refined oil in question will affect the distribution of individual hydrocarbons in the water soluble fraction (Anderson *et al.*, 1974a; Soto *et al.*, 1975a). If stirring conditions are carefully standardized, however, preparation of water soluble fractions may be highly reproducible. The composition of water soluble fractions is generally quite different from the composition of the original oil; for example, the relative concentrations of mono- and di-aromatics are much higher in water soluble fractions (Table 8.2).

Bioassays testing the toxicity of water soluble fractions and individual hydrocarbons have utilized both static and continuous-flow bioassay systems. Unless the water soluble fraction is replenished during the exposure period, hydrocarbon concentration decreases with time due to volatilization of aromatic compounds,

TABLE 8.2

Composition of reference oils and 10% water soluble fractions (Anderson *et al.*, 1974b; Bean *et al.*, 1980)

	South Louisiana crude oil		*Prudhoe Bay crude oil*		*No. 2 fuel oil*	
Compound	*Oil %*	*WSF (ppb)*	*Oil %*	*WSF (ppb)*	*Oil %*	*WSF (ppb)*
n-Paraffins:						
C_{14}	0.44	10.0	0.36	—	0.82	5.0
C_{15}	0.48	10.0	0.37	—	1.06	7.0
C_{16}	0.54	12.0	0.36	—	1.20	8.0
C_{17}	0.41	9.0	0.34	—	0.98	6.0
C_{18}	0.30	7.0	0.30	—	0.60	4.0
Total C_{12}–C_{24}	3.98	89.0	4.40	2.93	7.38	47.0
Aromatics:						
benzene	—	6750	—	36.1	—	550
toluene	—	4130	0.93	55.5	—	1040
alkylbenzenes	—	760	3.19	53.8	—	970
naphthalene	0.04	120	0.16	1.6	0.40	840
1-methyl-naphthalene	0.08	60	0.21	2.4	0.82	340
2-methyl-naphthalene	0.09	50	0.25	—	1.89	480
dimethyl-naphthalenes	0.36	60	0.80	1.4	3.11	240
trimethyl-naphthalenes	0.27	8	0.48	0.6	1.84	30
biphenyls	0.01	2	—	—	0.16	28
fluorenes	0.02	3	—	—	0.36	20
phenanthrenes	0.06	4	0.39	0.3	0.53	20
dibenzothiophenes	0.02	1	0.13	—	0.07	4
Total aromatics	0.94	11,948	6.54	151.7	9.18	4562
Total hydrocarbons	4.92	12,037	10.94	154.6	16.56	4609

such as naphthalene (Laughlin *et al.*, 1979). Several toxicant delivery systems have been designed to minimize volatilization and more stable exposure concentrations are therefore maintained (Benville and Korn, 1974; Maynard and Weber, 1981). In continuous-flow assay systems, continuous mixing of water soluble fractions and frequent replacement with fresh preparations can also result in fairly constant exposure concentrations (Moore *et al.*, 1980; Capuzzo and Lancaster, 1981). More sophisticated assay system designs have incorporated a "solubilizer" component for continuous production and delivery of water soluble fractions to exposure systems (Roubal *et al.*, 1977a; Krugel *et al.*, 1978; Benville *et al.*, 1981). Assays testing the toxicity of water soluble fractions have generally demonstrated the acute toxicity of high concentrations of these oil components but do not adequately represent the availability of such components in actual field conditions or address the potential long-term effects of chronic low-level input.

Oil-in-water dispersions (OWD) are more representative of actual field conditions following an oil spill but are more difficult to control because of the partitioning of hydrocarbons between aqueous and particulate phases. Dispersions can be prepared by various stirring or mixing procedures (Anderson *et al.*, 1974a; Gruenfeld and Frederick, 1977; Winters *et al.*, 1977b; Wong *et al.*, 1981). The composition of oil-in-water dispersions more closely resembles that of the parent oil, although only a small fraction of the oil is partitioned into microdroplets (Anderson *et al.*, 1974a).

Oil-in-water dispersions have been tested using both static and continuous-flow bioassay systems. In a static system, Anderson *et al.* (1974b) compared the toxicity of oil-in-water dispersions of a refined oil (No. 2 fuel oil) and two crude oils (South Louisiana and Kuwait) to several species of marine animals. These static assays were based on a single addition of oil at the beginning of the experiment; because of aeration and the instability of oil-in-water dispersions in static systems, up to 90% of the hydrocarbons were lost within 24 h. A continuous-flow system was designed by Vanderhorst *et al.* (1977) specifically for consistent introduction of oil-in-water dispersions. In this system oil and sea water are mixed, and undispersed oil is separated and removed by baffle systems before introduction of the oil-in-water dispersion to the exposure system. Simpler systems have been designed by Hyland *et al.* (1977), Clement *et al.* (1980) and Capuzzo and Lancaster (1981). The systems designed by Vanderhorst *et al.* (1977) and Capuzzo and Lancaster (1981) have also been used to compare the effects of naturally dispersed and chemically dispersed oil (Anderson *et al.*, 1981; Capuzzo and Lancaster, 1982).

Surface slicks in bioassay systems are generally made by pouring oil over the seawater surface in an exposure system and allowing partitioning of hydrocarbons without mixing. The toxicity of slick formations can be tested in static systems without replenishment of sea water in the exposure system (Clark and Finley, 1974a) or in continuous-flow systems where oil-contaminated sea water is slowly replaced or diluted with uncontaminated sea water (Bott *et al.*, 1976; Taylor and Karinen, 1977; Shaw *et al.*, 1977; Payne *et al.*, 1978) or uncontaminated sea water is allowed to fall through the slick and contaminated sea water is drained from the bottom of the exposure tank (Eisler, 1975; Rinkevich and Loya, 1979). Such assay conditions can be used to test the effects of various weathering processes in altering the toxicity of oil and can be used to model acute toxicity associated with the initial phases of slick formation.

The toxic effects of ingestion of oil-contaminated food have been evaluated by either incorporating oil or radiolabeled hydrocarbons in prepared diets (such as pelleted feeds or capsules; Corner *et al.*, 1973; Hardy *et al.*, 1974; Roubal *et al.*, 1977b; Whittle *et al.*, 1977; Varanasi *et al.*, 1979), pre-exposing food organisms (e.g., oysters, algae, brine shrimp) to oil before feeding them to experimental animals (Lee *et al.*, 1976; Malins and Roubal, 1982; Capuzzo and Lancaster, 1982) or feeding contaminated detritus to detrital-feeding organisms (Roesijadi *et al.*, 1978). In the latter two types of experiments, the hydrocarbon composition of oil-contaminated food will reflect the uptake, accumulation and metabolism of hydrocarbons by food organisms and bacteria and can be used to evaluate the transfer of hydrocarbons and metabolites between trophic levels.

Because oiled sediments may result in long-term exposures of benthic populations to petroleum hydrocarbons, bioassay systems have been designed to evaluate the effects of oil-contaminated sediments on marine organisms. A wide range of experimental chambers (e.g., beakers, aquaria, trays, raceways) can be employed in such systems and artificially oiled sediments or sediments collected from contaminated habitats can be utilized (Prouse and Gordon, 1976; Taylor and Karinen, 1977; Shaw *et al.*, 1977; Anderson *et al.*, 1977b; McCain *et al.*, 1978; Roesijadi and Anderson, 1979; Capuzzo, 1981b; Kalke *et al.*, 1982). Artificially oiled sediments are prepared through either introducing oil over the surface of clean sediments (Prouse and Gordon, 1976; Capuzzo, 1981b; Kalke *et al.*, 1982) or by the formation of oil-seawater-sediment slurries (Anderson *et al.*, 1977b; McCain *et al.*, 1978). Such systems can be used to evaluate the uptake, metabolism and depuration of hydrocarbon components in benthic systems.

Augenfeld *et al.* (1982) demonstrated that fine sediments retained polyaromatic hydrocarbons better than coarse sediments. Boehm *et al.* (1982) and Gearing *et al.* (1980) further demonstrated that newly sedimented oil could generally be found in the flocculent layer at the sediment surface. Gearing and Gearing (1983) found that for No. 2 fuel oil and individual hydrocarbons the residence time in the sediments and the percentage sedimented varied with the lipophilic nature and degradation potential of each hydrocarbon. Spies (Chapter 9) suggests that interstitial waters may also be an important reservoir for hydrocarbons. Thus, predicting the impact of oil-contaminated sediments on benthic populations requires a better understanding of the biogeochemistry of hydrocarbons within sediments and interstitial waters. Replicating the sediment flocculent layer and the partitioning of hydrocarbons to interstitial waters in laboratory exposures, however, is extremely difficult but probably represents the most realistic model of benthic exposures in the field.

Enclosed ecosystems are an additional bioassay design that have been used recently to evaluate the effects of oil on marine ecosystems (Davies *et al.*, 1980; Elmgren *et al.*, 1980; Grassle *et al.*, 1981; Gearing and Gearing, 1982a,b). Although enclosed ecosystems are limited by a degree of artificiality in predicting the responses of populations and communities to petroleum hydrocarbons, they provide a valuable link between laboratory and field investigations by defining the biological and geochemical parameters responsible for hydrocarbon transport and effects. Specifically, they are useful in establishing:

1) the relative sensitivities of various trophic levels;
2) the chemical interactions of hydrocarbon components with biota and expected trends in persistence and biogeochemical cycling of hydrocarbons within the ecosystem;
3) the potential disruption in energy flow as a result of the input of petroleum hydrocarbons; and
4) the recovery potential of the ecosystem after such inputs.

Davies *et al.* (1980) used plastic bag enclosures (304 m^3) in Loch Ewe on the west coast of Scotland to investigate the effects of the water soluble fractions of North Sea crude oil on pelagic ecosystems. Their studies included investigations of effects of oil on primary and secondary production and the role of microbial degradation in altering the composition of hydrocarbon mixtures. The Marine Ecosystems Research Laboratory (MERL) at the University of Rhode Island has been designed to consider both the water column and benthic components of enclosed ecosystems, and MERL enclosures (15 m^3) have been used to evaluate the long-term effects of water accommodated fractions of No. 2 fuel oil on benthic communities (Grassle *et al.*, 1981), trophic interactions (Elmgren *et al.*, 1980), biogeochemical cycling and metabolite production (Hinga *et al.*, 1980; Gearing and Gearing, 1982a,b; Lee *et al.*, 1982b). Skjoldal *et al.* (1982) used 10.7-m^3 enclosures in Lindaspollene, Norway to evaluate the effects of a spill of unweathered Ekofisk crude oil on the dynamics of plankton communities. Oil was added to the water surface in order to simulate a natural spill situation, including the production of photooxidized byproducts and effects on primary production and bacterial biomass.

Of all the experimental approaches discussed above, the use of enclosed ecosystems combined with a study of benthic processes and sediment geochemistry offer the most promise in delineating the long-term effects of hydrocarbon contamination resulting from offshore oil and gas development.

BIOACCUMULATION OF PETROLEUM HYDROCARBONS

Accumulation of petroleum hydrocarbons by marine biota is dependent on the biological availability of hydrocarbons in soluble and droplet forms, the length of exposure and the organism's capacity for metabolic transformations. Both chemical factors—such as solubility, adsorption-desorption kinetics and octanol/water partition coefficients (Veith *et al.*, 1979; Means *et al.*, 1979)—and biological factors—such as deposition in body lipid, surface:volume ratios, feeding habits and metabolism—affect the bioaccumulation of hydrocarbons in marine organisms (Neff, 1979; McLeese *et al.*, 1980).

Hydrocarbon adsorption on marine sediments is affected by the solubility of individual hydrocarbons and the grain size distribution and organic content of sediments (Paris *et al.*, 1978; Gearing *et al.*, 1980; Olsen *et al.*, 1982; Gearing and Gearing, 1983; Chapter 6), where the extent of adsorption appears to be inversely related to compound solubility and the size of the sorbent particles and directly related to organic content of sediments. Biological uptake of hydrocarbons from contaminated sediments may be attributable to desorption of hydrocarbon particles from sediment particles into interstitial waters (Chapter 9). Thus, bioavailability of sediment-adsorbed hydrocarbons would appear to be directly related to compound solubility and sediment grain size and inversely related to the organic content of contaminated sediments. Other factors such as hydrocarbon source—petrogenic versus pyrogenic—may also affect bioavailability. Farrington *et al.* (1983) suggested that polyaromatic hydrocarbons from petroleum sources are more available for biological uptake than those from pyrogenic sources due to the more tightly bound nature of the latter to pyrogenic particulates. Uptake of compounds may occur through adsorption onto body surfaces, water exchange at respiratory and feeding surfaces, and ingestion of food and detrital particles. Removal may occur through active mechanisms such as hydrocarbon metabolism and excretion of metabolized byproducts, and passive mechanisms such as diffusive exchange and the production of particulate products (eggs, molts and feces). The balance of these processes for individual hydrocarbons may result in selective uptake, accumulation and depuration.

Protophytes

Accumulation of various petroleum hydrocarbons has been observed in every phylogenetic group of eukaryotic organisms investigated to date (also see discussion in Chapter 7). Phytoplankton—both chrysophytes and chlorophytes—have the capacity to accumulate and metabolize both aliphatic and aromatic compounds. Thompson and Eglinton (1979) documented accumulation of petroleum derived aliphatic compounds in benthic diatoms collected from an oil-contaminated site, but only trace amounts of polyaromatic hydrocarbons were detected. Boutry *et al.* (1977) detected uptake and conversion of petroleum derived aliphatic compounds to fatty acids in the diatom *Chaetoceros simplex calcitrans*. Cerniglia *et al.* (1979, 1980) demonstrated that both eukaryotic and prokaryotic unicellular algae (blue-green algae) could accumulate and metabolize naphthalene under photoautotrophic conditions resulting in the production of several metabolites. Hinga *et al.* (1980) detected accumulation of ^{14}C-benz(a)anthracene in phytoplankton collected in the MERL microcosms. Although some metabolite production was detected, most of the label remained in the form of the parent compound. Macroalgae have also been shown to accumulate petroleum hydrocarbons, although details of metabolism to secondary metabolites are not well understood (Clark *et al.*, 1973, 1975; Burns and Teal, 1971).

Corals

Solbakken *et al.* (1984) examined the uptake and release of radiolabeled naphthalene and phenanthrene in 19 species of anthozoan corals and one species of hydrozoan coral. All species of coral rapidly accumulated both compounds, although differences in uptake rate were evident among the different species. Naphthalene was more rapidly eliminated than phenanthrene; only 2% of the initial concentration of naphthalene remained after 10 days, whereas release of phenanthrene by the corals took up to four months.

Zooplankton

Zooplankton can accumulate oil from droplets, soluble fractions and food, and can concentrate both aliphatic and aromatic compounds (Corner, 1978). Hydrocarbon accumulation in copepods appears to be related to the concentration of hydrocarbon components in sea water or food, to the lipid content of the exposed animal and indirectly to temperature (Corner *et al.*, 1976b; Harris *et al.*, 1977a). Because polar and boreal species of copepods store more lipids, they tend to accumulate more hydrocarbons than temperate or tropical forms. Such findings suggest that regional differences may occur in the importance of zooplankton transfer of hydrocarbons to higher trophic levels. Deposition of oil components in fecal pellets has also been observed and it can account for a significant proportion of the petroleum hydrocarbon budget of exposed zooplankton (Conover, 1971; Freegarde *et al.*, 1971; Harris *et al.*, 1977a).

Lee (1975) exposed several species of zooplankton, including copepods, euphausiids, amphipods, ctenophores and cnidarians, to radioisotopes of naphthalene, benzo(a)pyrene, 20-methyl-cholanthrene, and octadecane in addition to the water soluble fraction of No. 2 fuel oil and evaluated the extent of accumulation of petroleum derived aliphatic and aromatic compounds in marine zooplankton. With copepods, uptake of hydrocarbons was linear for the first three days of exposure with no further uptake observed for an additional seven day exposure. When copepods were transferred to uncontaminated sea water after three days of exposure, most of the radioactivity was eliminated, either through depuration or metabolism, although a hydrocarbon residue was still detectable after 28 days post exposure. Harris *et al.* (1977a) exposed the first naupliar stage of *Eurytemora affinis* to ^{14}C-naphthalene for 24 hours; after transferring the animals to uncontaminated sea water and rearing them to the adult stage (development time = 34 days), they found that 10% of the label could still be detected in the previously exposed animals.

Corner *et al.* (1976a) compared the uptake of radiolabeled naphthalene by *Calanus helgolandicus* through aqueous and dietary pathways and demonstrated that higher rates of accumulation could be detected through the dietary pathway. Depuration of naphthalene derived from aqueous fractions was similar to that observed by Lee (1975)—i.e., rapid elimination with only a small fraction of the label remaining after 10 days depuration. Naphthalene accumulated through the dietary pathway was not rapidly depurated and one-third of the label was still detectable in exposed animals.

The persistence of a detectable fraction of hydrocarbons in zooplankton during postexposure periods may be related to metabolite production. Sanborn and Malins (1977) demonstrated in the planktonic larval stages of *Cancer magister* and *Pandalus platyceros* that although radiolabeled naphthalene was rapidly depurated from tissues within 24-36 hours after exposure, metabolites were highly resistant to depuration. The persistence of hydrocarbon components in zooplankton biomass even after long periods of depuration could be a significant factor in the transfer of hydrocarbons or their metabolites to higher trophic levels and pose long-term consequences for aquatic food chains.

Benthos

Worms

Benthic animals may accumulate hydrocarbon components from contaminated sediments, although the distribution of hydrocarbons in exposed animals does not always reflect the distribution of such compounds in sediments. In both polychaete and sipunculan worms, petroleum hydrocarbons are rapidly accumulated from both sediment and water and some components may be rapidly depurated (Anderson *et al.*, 1977b; Rossi and Anderson, 1977; Gordon *et al.*, 1978; Lyes, 1979; Augenfeld *et al.*, 1982). Lee *et al.* (1981) observed significant accumulation of benzo(a)pyrene in *Nereis virens* collected from an oil-contaminated site in comparison with an uncontaminated site. Gordon *et al.* (1978) observed rapid accumulation of hydrocarbons in *Arenicola marina* exposed to sediments contaminated with Bunker C fuel oil. Both the sipunculan *Phascolosoma agassizii* and the polychaete *Neanthes arenaceodentata* accumulated naphthalenes from sediments during oil exposures but depurated these hydrocarbons to background levels within two weeks after transfer to clean sea water (Anderson *et al.*, 1977b; Rossi and Anderson, 1977). Rossi and Anderson (1977), however, observed differences between male and female *Neanthes* in their respective depuration rates of naphthalenes from the water soluble fraction of No. 2 fuel oil; ovigerous females maintained high naphthalene levels until eggs were released, suggesting that naphthalenes were accumulated in the lipid fraction of eggs. Early developmental stages contained high concentrations of naphthalenes; later stages, however, released accumulated hydrocarbons, coincident with the mobilization of lipid reserves from the yolk.

Augenfeld *et al.* (1982) exposed *Abarenicola pacifica* for 60 days to fine sediments contaminated with radiolabeled phenanthrene, chrysene and benzo(a)pyrene and found that uptake of each compound increased during the first two weeks of exposure to a maximum of four to six times sediment concentrations. After that time tissue concentration of chrysene remained fairly constant throughout the remainder of the exposure period, but concentrations of phenanthrene and benzo(a)pyrene decreased. In a later field experiment, Augenfeld *et al.* (1983) exposed the same species to fine sediments mixed with Prudhoe Bay crude oil; after 1 month exposure individual saturates and aromatics were below detection limits in worm tissues. Bioturbation resulted in a greater release of aromatic hydrocarbons from surface sediments.

Bivalves

A great volume of literature is available on the uptake and accumulation of hydrocarbons by marine bivalve molluscs either exposed to oil in laboratory studies or collected from oil-contaminated habitats. The uptake of petroleum hydrocarbons by marine bivalves is dependent on the bioavailability and composition of the hydrocarbon mixture, the duration of exposure and the nutritional status of the animal. Species differ in their rates of hydrocarbon uptake due to differences in filtration rates, lipid content and habitat (Stegeman and Teal, 1973; Clark and Finley, 1974b; Neff *et al.*, 1976b; Vandermeulen and Gordon, 1976; Augenfeld *et al.*, 1980).

Lee *et al.* (1972a) found that radiolabeled naphthalene and benzo(a)pyrene were accumulated by the mussel *Mytilus edulis* with the highest concentrations being detected in gill tissue. The authors hypothesized that the gills were the primary site of uptake due to the presence of a micellar layer that could absorb hydrophobic compounds such as hydrocarbons (Pasteels, 1968); once absorbed, hydrocarbons were transferred to other tissues. When transferred to clean sea water, mussels rapidly eliminated both hydrocarbons. Neff *et al.* (1976b) measured the accumulation and release of four polyaromatic compounds—phenanthrene, naphthalene, chrysene and benzo(a)pyrene—by the clam *Rangia cuneata*. Among the four compounds, phenanthrene was accumulated most rapidly and released most slowly. Although naphthalene was probably accumulated at a rapid rate, it was also depurated rapidly; thus little net uptake was observed. The authors suggested that hydrocarbon uptake and accumulation was most probably related to the solubility and octanol/water partition coefficients (K_{OW}) of specific hydrocarbons. Of the four compounds studied, naphthalene has the highest solubility and its K_{OW} favors rapid release when environmental concentrations decrease to background levels. Benzo(a)pyrene has a low solubility, is not readily available, but its K_{OW} would also favor slow release rates once it is absorbed. The uptake and release of the four compounds by *Rangia* are presented in Table 8.3.

TABLE 8.3
Bioaccumulation of aromatic hydrocarbons by *Rangia cuneata* (Neff *et al.*, 1976b)

Parameter	*Compound*			
	Naphthalene	*Phenanthrene*	*Chrysene*	*Benzo(a)pyrene*
Exposure concentration (ppm)	0.071	0.089	0.066	0.052
Tissue concentration after 24 h (ppm ± SD)	0.43 (±0.1)	2.85 (±1.1)	0.54 (±0.3)	0.45 (±0.1)
Bioaccumulation factor	6.1	32.0	8.2	8.7
Tissue concentration after 24-h depuration (ppm ± SD)	0.15 (±0.02)	2.47 (±1.2)	0.40 (±0.15)	0.38
% released in 24 h	66	13	26	16

Other studies of accumulation of aromatic compounds by marine bivalves have also shown the rapid accumulation and release of low molecular weight aromatics, such as naphthalene and alkyl-naphthalenes, and decreased rates of uptake and slow release of higher molecular weight aromatics such as anthracene, fluorene, benz(a)anthracene, and benzo(a)pyrene (Neff and Anderson, 1975b; Dunn and Stich, 1976; Lee *et al.*, 1972a).

Oysters (*Crassostrea virginica*) exposed to No. 2 fuel oil accumulated aromatic compounds to a higher degree than aliphatic compounds relative to the composition of No. 2 fuel oil, but also released aromatic compounds more rapidly upon transfer to uncontaminated sea water (Stegeman and Teal, 1973; Stegeman, 1974). DiSalvo *et al.* (1975), however, in transferring mussels *Mytilus edulis* from an uncontaminated site to an oil-contaminated site for 11 weeks observed that aliphatics were accumulated to a greater extent than aromatic compounds. When transferred to clean sea water, 90% of the hydrocarbon content was released after five weeks with similar concentrations of aliphatics and aromatics remaining. When mussels were transferred from a chronically-contaminated area to an uncontaminated site, approximately 50-65% of the aromatic hydrocarbon content was lost over a 10-week period but the aliphatic content changed little during that time. Analytical techniques used in this study, however, were based on thin-layer chromatography separations of extracted hydrocarbons, with quantification by densitometry after visualization by sulfuric acid charring, and specific compounds could not be differentiated.

Boehm and Quinn (1977) transferred quahogs *Mercenaria mercenaria* from chronically oil-contaminated stations to uncontaminated sea water and found that approximately 30% of the hydrocarbon content was depurated in 120 days. These investigators also used thin layer chromatography for separation of extracted hydrocarbons and differentiated "aromatic" compounds as those compounds that co-chromatographed with phenanthrene and dimethylnaphthalene standards and "aliphatic" compounds as those compounds that had a R_f greater than that of dimethylnaphthalene; this latter group included both aliphatic and naphthenic compounds. Analyses by packed column gas chromatography revealed that the dominant hydrocarbons in bivalve tissues were naphthenic, aromatic and naphtheno-aromatic compounds, and no qualitative differences in hydrocarbon distribution occurred over the duration of the experiment.

Differential rates of hydrocarbon uptake and release are evident among bivalve molluscs, related presumably to interspecific differences and exposure situations. Bieri and Stamoudis (1977) measured the uptake and release of specific hydrocarbons in *Crassostrea virginica* exposed to No. 2 fuel oil. Alkanes, branched alkanes, and olefins were accumulated and removed first; the alkylnaphthalenes (with up to five alkyl carbons), the biphenyls (with up to two alkyl carbons) and the fluorenes (with up to one methyl group) were accumulated and released next; and the highly substituted naphthalenes, biphenyls, and fluorenes, in addition to dibenzothiophenes and phenanthrenes, were accumulated and released last. The authors also observed that oysters continued to accumulate polyaromatic hydrocarbons after aqueous concentrations decreased to undetectable levels, presumably from hydrocarbons adsorbed to detrital particles.

Uptake and release of hydrocarbons were similar in *Mercenaria*, but total uptake was much lower.

Widdows *et al.* (1982) exposed *Mytilus edulis* to low concentrations (7-68 μg total hydrocarbons/1) of the water accommodated fraction of North Sea oil and monitored bioaccumulation and sublethal effects during both short-term (four weeks) and long-term (five months) exposures. Tissue hydrocarbon concentrations were determined by either UV spectrophotometry or gas chromatography after steam distillation. Uptake and accumulation of hydrocarbons appeared to be related to hydrocarbon concentration, the duration of exposure, the presence of algal particulate food, clearance rate of the animals and the solubility of individual hydrocarbons.

In two-week exposures to 7.7 or 68 μg total hydrocarbons/1, rapid uptake of aromatic hydrocarbons occurred in both the digestive gland and other tissues during the first seven days of exposure, with bioconcentration factors in the digestive gland of 850-910 and in other tissues of 180-260. During four-week exposures to 36 μg total hydrocarbons/1, initially alkanes were more rapidly accumulated than aromatics in the digestive gland, possibly related to the accumulation of particulate associated aliphatic compounds in the gut. Tissue concentrations of alkanes and aromatics were always higher in the digestive gland than in other tissues throughout the course of the experiment. In addition to quantitative differences between the digestive gland and other tissues, qualitative differences were also apparent: 1) nC_{10} - nC_{25} alkanes were detected in the digestive gland, whereas lower molecular weight alkanes were the dominant compounds in other tissues; and 2) alkylbenzenes and alkylnaphthalenes were the dominant aromatic compounds in both the digestive gland and other tissues, although their relative distribution did not reflect that of the water-accommodated fraction.

With long-term exposure to 30 μg total hydrocarbons/1, concentrations of aromatic hydrocarbons in the digestive gland increased rapidly during the early phase of exposure (up to 33 days), reached a steady state, and then increased further (after 100 days). Accumulation in other tissues, however, increased gradually throughout exposure and failed to reach a steady-state condition; these increases were not related to changes in lipid reserves during the course of the experiment. Similar observations have been made by Fossato and Canzonier (1976) with *Mytilus edulis* and by Clement *et al.* (1980) with *Macoma balthica*.

Farrington *et al.* (1982) measured the uptake of hydrocarbons by *Mytilus edulis* exposed to a spill of No. 2 fuel oil; the initial spill persisted for two days and the uptake and release of hydrocarbons by *Mytilus* were monitored during an 86-day postspill sampling period. More detailed analyses of hydrocarbons by glass capillary gas chromatography enabled these investigators to calculate the biological half-life of alkanes and aromatics. The biological half-life of each component was as follows: *n*-alkanes = 0.2-0.8 days; pristane = 1.5 days; C-2 (dimethyl or ethyl) naphthalenes = 0.9 days; C-3 naphthalenes = 1.5 days; phenanthrene = 2.1 days; methyl phenanthrenes = 1.7 days; unresolved complex mixtures = 2.8-3.9 days. The authors concluded that molecular weight and solubility of hydrocarbon components were the main controlling factors in the release of hydrocarbons by *Mytilus*, although molecular type and configuration were additional factors. These results are consistent with the findings of other investigators that initial rapid release of some hydrocarbons following exposure may occur. The persistence of other hydrocarbons, particularly those associated with the unresolved complex mixture (di- and tri-terpenoids, steranes, mono-aromatic diasteranes, etc.), needs to be further explored (Farrington *et al.*, 1982).

Bioavailability of sediment-bound hydrocarbons to bivalve molluscs appears to vary as a function of persistence of hydrocarbon components in sediments and the feeding behavior of individual species. Boehm *et al.* (1982) compared the uptake and depuration of hydrocarbon residues in the suspension-feeding bivalve *Mytilus edulis* and the deposit-feeding bivalve *Macoma balthica* following the *Tsesis* oil spill. They found that hydrocarbon residues in *Mytilus* were characteristic of "fresh" oil and depuration was complete within one year; whereas, hydrocarbon residues of *Macoma* were characteristic of weathered oil and depuration occurred slowly. Uptake of hydrocarbons by *Macoma* may have been complicated by continued re-exposure by newly-sedimented oil. Fucik *et al.* (1977) transferred *Rangia cuneata* to sediments near an oil-separator platform in Trinity Bay, Texas; although naphthalene concentrations in clams correlated with sediment concentrations, concentrations in clams were lower than those in sediments suggesting inefficient uptake.

Roesijadi *et al.* (1978) found that the efficiency of uptake of polyaromatic hydrocarbons by the deposit-feeding bivalve *Macoma inquinata* from sediments was less than uptake from sea water (Table 8.4). Augenfeld *et al.* (1982),

TABLE 8.4
Bioaccumulation of radiolabeled aromatic compounds by *Macoma inquinata* from oil-contaminated sediments and sea water (Roesijadi *et al.*, 1978)

Parameter	*Compound*			
	Phenanthrene	*Chrysene*	*Dimethylbenz(a)-anthracene*	*Benzo(a)-pyrene*
Sediment-net uptake (μ/g)	0.096	0.308	0.297	0.059
Sea water-net uptake (μg/g)	0.038	0.297	0.856	0.037
Sediment concentration at 7 days (μg/g)	0.49	8.37	4.63	0.64
Sea water concentration at 7 days (μg/g)	0.0037	0.00043	0.00063	0.00004
Bioaccumulation factor-sediment	0.20	0.04	0.06	0.09
Bioaccumulation factor-sea water	10.3	694	1349	861

however, found higher uptake rates by *Macoma* of radiolabeled chrysene and benzo(a)pyrene (magnification rates as high as 11.6 and 5.2 times sediment concentrations, respectively) and attributed the differences observed in the two studies to the persistence of each compound in the sediment preparations.

In controlled one-year field exposures, Anderson *et al.* (1983) compared tissue contamination and growth of the littleneck clam *Protothaca staminea* from contaminated sediments in which Prudhoe Bay crude oil had been mixed to a depth of either 3 or 10 cm. In sediments in which oil was added only to the upper 3 cm, hydrocarbon concentrations were reduced to background levels within one year; in sediments in which oil was mixed to 10 cm, measurable amounts of hydrocarbons persisted even after one year. Patterns of hydrocarbon loss from the sediments were related to the molecular weight and biodegradation of individual hydrocarbons, consistent with the findings of other investigators (Gearing and Gearing, 1983). The ratios for tissue concentrations to final sediment concentrations for phenanthrenes, alkylnapththalenes, and dibenzothiophenes reported by Anderson *et al.* (1983) ranged between 0.16 and 0.18, consistent with the findings of Roesijadi *et al.* (1978) for phenanthrene bioaccumulation in *Macoma*.

Crustaceans

Benthic crustaceans have been shown to accumulate oil from water, sediment and food (Anderson, 1975; Cox *et al.*, 1975; Burns, 1976; Neff *et al.*, 1976b; Sanborn and Malins, 1977, 1980). Miller *et al.* (1978) exposed the shrimp *Penaeus duorarum* to chrysene for 28 days followed by transfer to uncontaminated sea water for 10 days; although most of the chrysene was released during the depuration period, a measurable quantity persisted. Dillon (1982) examined the accumulation of dimethylnaphthalene from a contaminated food source by the grass shrimp *Palaemonetes pugio*; and although the compound was accumulated by the shrimp, it was lost rapidly upon removal of the contaminated food source. Lee *et al.* (1976) measured the accumulation of radiolabeled benzo(a)pyrene, methylcholanthrene and fluorene by the crab *Callinectes sapidus* and found maximum radioactivity after two days of exposure. Although uptake continued throughout exposure, uptake was balanced by rapid release of compounds and their metabolites. The gill was the primary site of uptake and the hepatopancreas was the major site of accumulation and metabolism. When blue crabs were fed shrimp or oysters containing radiolabeled compounds, only 10% of the activity of each compound was accumulated in tissues other than the stomach, and the remainder was lost as feces. Release of compounds and metabolites was similar to that observed for uptake from sea water.

Cox *et al.* (1975) exposed the shrimp *Penaeus aztecus*, the fiddler crab *Uca minax* and the crab *Sesarma cinereum* to No. 2 fuel oil for 38 days in a shrimp mariculture pond and measured the accumulation of naphthalenes. The alkylnaphthalenes were accumulated to a greater extent than naphthalene and, following transfer to uncontaminated sea water, all compounds were depurated within 10 days. Burns (1976), however, noted that *Uca* collected from an oil-contaminated site had measurable hydrocarbon concentrations even four years after a spill of No. 2 fuel oil had occurred, presumably due to continued uptake of hydrocarbons from contaminated sediments.

Sanborn and Malins (1980) exposed adult spot shrimp *Pandalus platyceros* to water soluble fractions of Prudhoe Bay crude oil and found that shrimp accumulated a wide range of low molecular weight aromatic compounds, primarily the C_2-C_5 alkyl-substituted benzenes and the C_1-C_3 alkyl-substituted naphthalenes. Bioconcentration factors for the alkylbenzenes were directly related to the degree of alkylation but no evidence of such a relationship was observed with the accumulation of naphthalenes.

Fish have also been shown to rapidly accumulate and release petroleum hydrocarbons. Anderson *et al.* (1974b) measured the uptake and release of naphthalene and 1-methylnaphthalene by the sheepshead minnow *Cyprinodon variegatus* after a four-hour exposure to 1 mg/l of each compound. Although the compounds were accumulated to 60 μg/g and 210 μg/g, respectively, 90% of the compounds were released within 29 h. Lee *et al.* (1972b) exposed three species of fish—mudsucker *Gillichthys mirabilis*, sculpin *Oligocottus maculosus* and sand dab *Citharichthys stigmaeus*—to radiolabeled naphthalene and benzo(a)pyrene and found the main route of uptake to be through the gills and the highest accumulation to be in the liver and gall bladder. After 24-h depuration, 90% of the naphthalene was released primarily as metabolites; the release of benzo(a)pyrene, however, was much slower. Other investigators have observed accumulation of hydrocarbons in liver, gall bladder, spleen, brain, kidney and muscle (Statham *et al.*, 1976; DiMichele and Taylor, 1978; Melancon and Lech, 1978; Miller *et al.*, 1978).

Melancon and Lech (1978) measured the uptake, distribution and release of radiolabeled naphthalene and 2-methylnaphthalene in the trout *Salmo gairdneri*. Length of exposure (8 hours versus 26 days) had a significant effect on both uptake of hydrocarbons and release of hydrocarbons and metabolites. In the long-term exposure, 2-methylnaphthalene and its metabolites were released more rapidly than naphthalene, but metabolites in general were released less rapidly than parent compounds. The highly substituted alkylbenzenes and alkylnaphthalenes were accumulated more rapidly than the less substituted forms in both coho salmon *Oncorhynchus kisutch* and starry flounder *Platychthys stellatus* (Roubal *et al.*, 1977b, 1978), although there were interspecific differences in both rates of uptake and release.

From the limited amount of data available, uptake of hydrocarbons by fish from ingestion of oil-contaminated food appears to be quite small (Corner *et al.*, 1976b; Dixit and Anderson, 1977; Whittle *et al.*, 1977), although accumulation of various components has been observed (Roubal *et al.*, 1977b; Collier *et al.*, 1978; Solbakken and Palmork, 1980; Nava and Engelhardt, 1980).

McCain *et al.* (1978) measured the uptake of aromatic hydrocarbons from Prudhoe Bay oil-contaminated sediments by the English sole *Parophrys vetulus*. Concentrations of most aromatic compounds in fish tissues—such as tetramethylbenzene and 2- methylnaphthalene—were within the same order of magnitude as sediment concentrations, although some aromatic compounds—such as phenanthrene, fluorene, and trimethylnaphthalene—were not accumulated in fish tissue. Hydrocarbons were readily accumulated in skin, muscle and liver, but after 27 days measurable quantities could only be detected in the liver. In similar experiments Varanasi and Gmur (1981a) exposed English sole to oil-contaminated sediments spiked with radiolabeled benzo(a)pyrene and naphthalene and found that naphthalene was more readily accumulated and released than benzo(a)pyrene. They also found that a large percentage of the naphthalene accumulated in the liver was in the form of the parent compound and not metabolites unlike the accumulation of benzo(a)pyrene. They concluded that benzo(a)pyrene was metabolized and eliminated through the bile, whereas a large proportion of naphthalene might be excreted prior to metabolism through the gill, skin and mucus, as suggested by Thomas and Rice (1981) and Varanasi *et al.* (1978).

Neff (1979) proposed a two-compartment model to partially explain the differences in accumulation and depuration of petroleum hydrocarbons in animals from acutely and chronically oil-contaminated sites. Uptake and storage of hydrocarbons in depot lipids, such as energy reserves or gonadal reserves, may be retained until those reserves are mobilized for nutritional or reproductive needs. Hydrocarbons associated with more labile hydrophobic compartments, such as membrane lipids and cellular macromolecules, however, may be more rapidly released when ambient levels decrease. A two-phase depuration—initial rapid release of hydrocarbons followed by a slower, more gradual release—could be explained by these differences in lipid compartments. Hydrocarbon metabolism, however, also plays an important role in regulating accumulation and release of petroleum hydrocarbons.

Relationship between Bioaccumulation and Toxicity

The relationships between accumulated hydrocarbons and toxic effects and between reversibility of effects and release of hydrocarbons are difficult to establish. Few studies have focused on both aspects simultaneously. Anderson (1977) concluded that there was little agreement between sublethal responses of marine organisms to petroleum hydrocarbons and the level of hydrocarbon contamination in tissues. Gilfillan *et al.* (1977), however, reported a negative correlation between energetic measurements, such as carbon flux, and the concentration of aromatic hydrocarbons in tissues of the soft-shelled clam *Mya arenaria*. Widdows *et al.* (1982) found a similar relationship between various energetic parameters and aromatic hydrocarbon tissue concentrations in *Mytilus edulis*. No correlation in either study was observed between effects and tissue concentrations of total hydrocarbons or aliphatic compounds.

Anderson *et al.* (1980) investigated the relationship between mortality and accumulated hydrocarbons in the mysid *Neomysis awatschensis* and two species of shrimp *Hippolyte clarkii* and *Pandalus davae* after exposure to Prudhoe Bay crude oil. Accumulation of di- and tri-aromatic compounds in the tissues of these crustaceans was not cumulative and could not be used to explain toxic effects. The authors suggested that further analysis of mono-aromatic hydrocarbons and metabolites might provide a better clue as to the relationship between body burden and toxic effects.

Capuzzo *et al.* (1984) suggested that different sublethal effects might be related to exposure, uptake, and metabolism of different hydrocarbons. In studies with larval stages of the American lobster *Homarus americanus*, suppression of respiratory activity during and after exposure to South Louisiana crude oil correlated with accumulated benzene, thiophene, toluene and alkylbenzenes; whereas alterations in lipid metabolism occurred only during exposure and might have been related to the detoxication of higher molecular weight aromatic hydrocarbons.

Malins and Hodgins (1981), in their review of the effects of petroleum hydrocarbons on marine fish, concluded that although aromatic hydrocarbons are accumulated in fish tissues, subsequent metabolism and excretion of metabolized byproducts reduce body burdens and increased hydrocarbon concentrations are not always detected. They further suggested that a clearer relationship might exist between toxic effects and accumulated metabolites. Thus, it is apparent that the sublethal effects of oil exposure may be modified by the ability of the organism to accumulate and metabolize various hydrocarbons. In evaluating long-term effects of hydrocarbon exposure, it is important to understand not only factors that influence bioaccumulation, but also the metabolic capacity for alteration and consequences of accumulated hydrocarbons.

BIOTRANSFORMATION OF PETROLEUM HYDROCARBONS

Petroleum hydrocarbons, including polyaromatic components, and other lipid soluble foreign compounds are metabolized by many marine vertebrate and invertebrate animals (Bend and James, 1978; Stegeman, 1981), as well as bacteria, microalgae and fungi. Metabolism of these compounds will affect their disposition or enhance their removal, but may also result in their transformation to potentially more toxic derivatives within an organism or within the ecosystem. Understanding the balance between detoxication and toxication is critical to determining the metabolic fate of transformed hydrocarbons. Transformation mechanisms are discussed in detail in Chapter 7 and therefore will only be highlighted here.

Microbes

Bacteria are capable of degrading a wide array of hydrocarbons, including the utilization of monocyclic aromatic compounds such as benzene as a carbon source (Gibson, 1976, 1977) and the partial or complete degradation of polyaromatic compounds such as naphthalene (Jerina *et al.*, 1971). An important distinction of bacterial degradation of polyaromatic hydrocarbons is the formation of *cis*-dihydrodiols through a dioxetane intermediate (Gibson, 1976, 1977). Higher molecular weight aromatic compounds may also be partially degraded to phenolic or acidic metabolites (Dean-Raymond and Bartha, 1975; Gibson *et al.*, 1975; Gibson, 1977).

Lee and Takahashi (1977) studied the rates of polyaromatic hydrocarbon degradation by microbial populations in enclosed ecosystems exposed to No. 2

fuel oil. Degradation rates of naphthalene and methylnaphthalene were generally low prior to oil addition, but were accelerated three days after addition. Higher molecular weight aromatic compounds such as benzo(a)pyrene and fluorene, however, showed little or no degradation (Table 8.5). Davies *et al.* (1980) investigated the microbial degradation of benzo(a)pyrene, hexadecane, and naphthalene in pelagic enclosures exposed to the water soluble fraction of North Sea crude oil. Naphthalene was degraded extremely rapidly (rates as high as μg/l/day) but other substrates were only slowly degraded. Bacterial populations from estuarine habitats have an increased capacity for hydrocarbon degradation in comparison with coastal and open ocean populations (Lee, 1977) and hydrocarbons are more rapidly degraded under aerobic conditions (R.F. Lee *et al.*, 1978; Delaune *et al.*, 1980; Ward *et al.*, 1980).

TABLE 8.5
Microbial degradation of aromatic hydrocarbons in a controlled ecosystem enclosure (Lee and Takahashi, 1977)

Compound	*Depth (m)*	*Incubation time (hours)*	*Time after oil addition (days)*	*Degradation rate (μg/l/day ± SD)*
Benzo(a)pyrene	5–10	48	0	0
		24	3	1.0 ± 0.7
Fluorene	5–10	48	0	0
		48	3	0
Methyl-naphthalene	0–5	24	0	10 ± 6
		24	3	26 ± 4
Naphthalene	0–5	24	0	10 ± 3
		16	3	250 ± 7
	5–10	10	4	100 ± 5
		10	4	500 ± 11

Fungi, unlike bacteria but like most vertebrates and invertebrates studied to date, possess a cytochrome P-450 mediated mixed-function oxygenase (MFO) enzyme system, where a *trans*-dihydrodiol is formed through an arene oxide intermediate (Duppel *et al.*, 1973; Ferris *et al.*, 1973, 1976; Cerniglia and Gibson, 1978). In the fungus *Cunninghamella elegans*, naphthalene was degraded to a dihydrodiol form and hydrocarbon degrading activity could be induced when the fungus was cultured in medium containing naphthalene, 3-methylcholanthrene or phenobarbital. Cerniglia *et al.* (1979) demonstrated that blue-green algae were capable of producing both *cis*- and *trans*-dihydrodiol metabolites of naphthalene and, thus, are intermediate between bacteria and higher organisms.

Animals

Bend and James (1978), Lee (1981) and Stegeman (1981) have recently reviewed the details of hydrocarbon metabolism in vertebrates and invertebrates.

Lipophilic compounds (both foreign compounds and some naturally occurring organic compounds such as steroids) may be converted by reduction, oxidation, hydrolysis or conjugation to more water soluble metabolites, facilitating their elimination from biological systems (Lee, 1981). This is accomplished through cytochrome P-450 mediated mixed function oxygenases that oxidize such lipophilic compounds by hydroxylation, O-dealkylation, N-dealkylation or epoxidation (Figure 8.1). The mixed-function oxygenase system in mammals, fish,

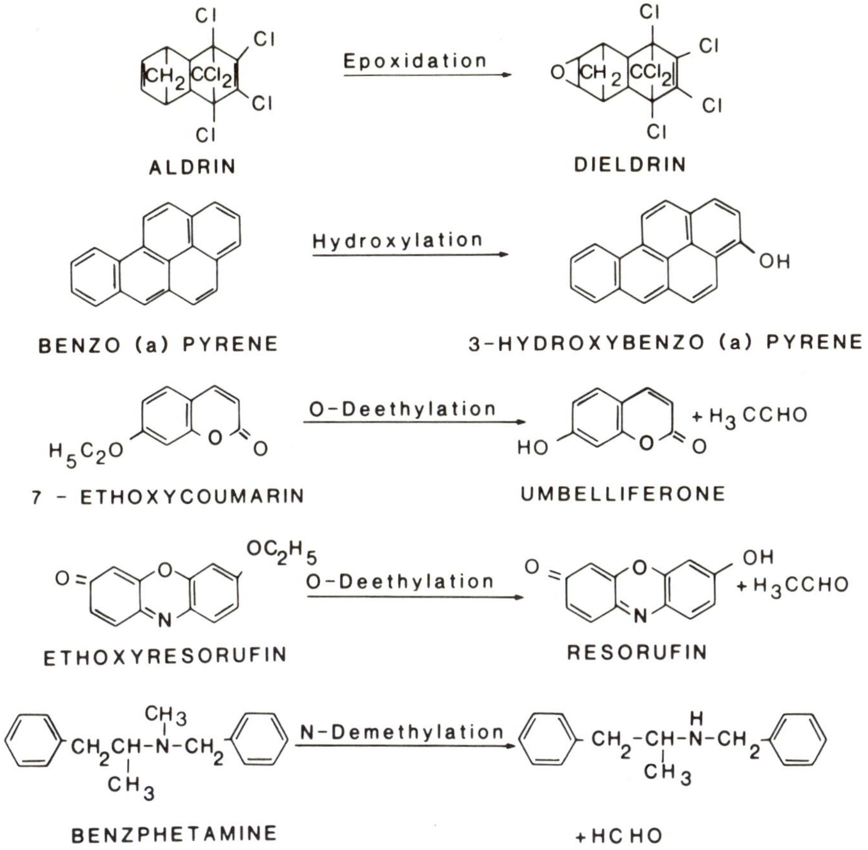

Figure 8.1. Mixed-function oxygenase reactions (Lee, 1981).

insects and probably most other invertebrates is a multicomponent system consisting of a phospholipid, cytochrome P-450 (with a spectrum of maximum absorbance at 450 nm), and NADPH cytochrome P-450 reductase (Lu, 1976; Nakatsugawa and Morelli, 1976; Philpot *et al.*, 1976; Singer and Lee, 1977; Lee *et al.*, 1979; Singer and Lee, 1977; Lee, 1981). The metabolism of benzo(a)pyrene is presented as an example in Figure 8.2.

Secondary metabolites can be formed through other metabolic reactions such as epoxide hydrolase, glutathione-S-transferase and other conjugating enzymes.

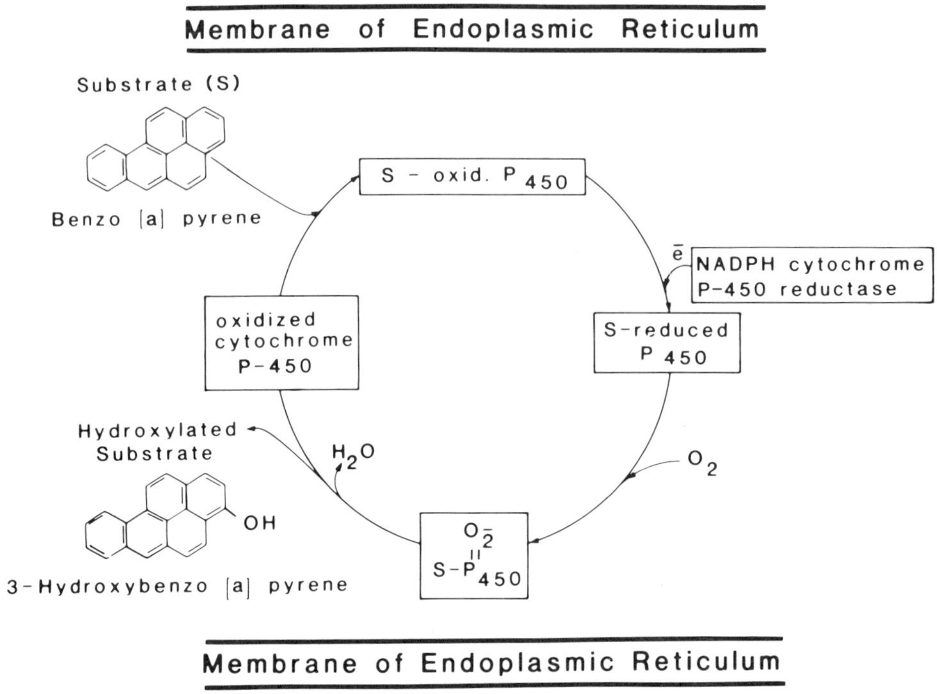

Figure 8.2. Microsomal electron transport reactions involved in the metabolism of benzo(a)pyrene (Lee, 1981).

The consequences of metabolite production are generally 1) the conversion of toxic, hydrophobic molecules to more soluble compounds that may be readily excreted and 2) a reduction of toxic effects. Metabolism of some compounds can result in the formation of metabolites that are more toxic than the parent compounds and that may possess carcinogenic, mutagenic and teratogenic potential because of their binding to cellular macromolecules (Figure 8.3). Oxidative metabolism of polyaromatic hydrocarbons is accomplished through highly electrophilic arene oxides, some of which may bind covalently to macromolecules such as DNA, RNA or proteins and result in mutagenic or toxic effects (Stegeman, 1981; Neff, 1985). Metabolic activation by mixed-function oxygenases and epoxide hydrolases is also a prerequisite for carcinogenesis by many foreign organic compounds including polyaromatic hydrocarbons (Jerina, 1983).

Activity of the mixed-function oxygenase system may be altered by both endogenous and exogenous factors. In both vertebrate and invertebrate systems, increases in cytochrome P-450 content and mixed-function oxygenase activity have been observed with exposure to exogenous inducing agents, such as polyaromatic hydrocarbons (Lee, 1981; Stegeman, 1981; Spies *et al.*, 1982). The patterns of induction, however, vary considerably among different phylogenetic groups, particularly in the response to phenobarbital-like and methylcholanthrene-like inducers (Stegeman, 1981). Induction usually results in only a small increase in cytochrome P-450 content, accompanied by a larger increase

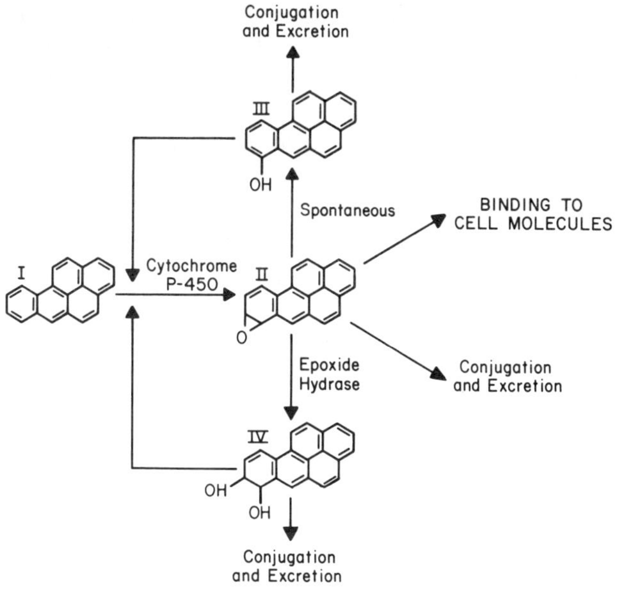

Figure 8.3. Metabolism of benzo(a)pyrene by fish; I is metabolized by cytochrome P-450 to an epoxide (II) which may rearrange to a phenol (III) or be hydrated to a diol (IV). Epoxidation may occur at any of several sites on the molecule and compounds III and IV may again serve as substrates (from Stegeman, 1977).

in mixed-function oxygenase activity (James and Bend, 1980); little change in the activity of secondary enzymes (such as epoxide hydrolase or glutathione-S-transferase), however, may occur. Induction of activity may at least partially reflect changes in the relative proportions of various cytochrome P-450 isozymes that may be responsible for metabolism of specific substrates (Williams and Buhler, 1982; Klotz, 1983; Klotz *et al.*, 1983). Endogenous factors such as sex, developmental or reproductive status and nutrition may also influence activity of the mixed-function oxygenase system (Stegeman, 1981). The interaction of mixed-function oxygenase activity and steroid hormone synthesis and action is also apparent (Lee, 1981; Stegeman, 1981).

Lee (1975) examined the potential of marine zooplankton to metabolize aromatic hydrocarbons. Ctenophores and cnidarians accumulated and excreted aromatic compounds such as naphthalene and benzo(a)pyrene but showed no capacity for metabolism. Planktonic crustaceans including copepods and amphipods, however, could metabolize aromatic compounds. When the copepod *Euchaeta japonica* was exposed to radiolabeled naphthalene, as much as 88% of the compound was converted to metabolites within 24 hours. In similar experiments *Calanus plumchrus* exposed to radiolabeled benzo(a)pyrene retained metabolites for several days postexposure. Lindmark (1981) demonstrated that marine ciliates were unable to metabolize benzo(a)pyrene and benz(a)anthracene but could metabolize amino polyaromatic hydrocarbons, such as 2-aminofluorene and 2-acetylaminofluorene.

Corner *et al.* (1976a) demonstrated that naphthalene accumulated by the copepod *Calanus helgolandicus* through a dietary pathway could also be metabolized and during a short-term exposure (24 hours) metabolites were rapidly excreted. Harris *et al.* (1977b) also demonstrated the conversion of naphthalene to metabolites during a four to six day exposure but the metabolites were retained and not rapidly excreted. Sanborn and Malins (1977) observed an accumulation of metabolites of naphthalene in larval *Pandalus platyceros* and Varanasi and Malins (1977) suggested that the toxic effects of low levels of naphthalene to these larvae might be related to their inability to release toxic metabolites.

The pathways responsible for hydrocarbon metabolism in planktonic crustaceans are similar to those observed in other invertebrates and vertebrates (Walters *et al.*, 1979). Sanborn and Malins (1980) reported that larval and adult *Pandalus* have the capacity to convert naphthalene to conjugated and nonconjugated structures, such as glucuronide, sulfate, dihydrodiol and phenolic derivatives. Similar metabolites have been detected in other species of benthic crustaceans (Burns, 1976; Lee *et al.*, 1976; Corner *et al.*, 1973; Meyer and Bakke, 1977).

Lee *et al.* (1976) found that benzo(a)pyrene was metabolized to polar metabolites in the hepatopancreas of the blue crab *Callinectes sapidus* and Conner and Singer (1981) isolated cytochrome P-450 from the microsomal fraction of hepatopancreas taken from the blue crab. The hepatopancreas was not the site of maximum benzo(a)pyrene hydroxylase activity as higher activity was detected in the green gland and the pyloric stomach (Singer and Lee, 1977). Crustaceans are also capable of metabolizing phenanthrene, benz(a)anthracene and chrysene to various phenol and diol derivatives (Singer *et al.*, 1980) although considerable interspecific and seasonal differences in cytochrome P-450 activity have been detected among several species of crabs (Lee *et al.*, 1982a).

Bivalve molluscs have generally been considered to possess extremely low mixed-function oxygenase activity. Vandermeulen and Penrose (1978) failed to detect production of phenolic metabolites of benzo(a)pyrene in whole extracts of *Mya arenaria*, *Mytilus edulis* and *Ostrea edulis*. Anderson (1978), however, detected the production of quinone metabolites in the microsomal fraction of digestive gland extracts in *Mercenaria mercenaria* and *Crassostrea virginica*. Moore *et al.* (1980) reported the presence of a xenobiotic detoxication/toxication system involving NADPH reductase, glucose-6-phosphate dehydrogenase, and aldrin epoxidation in *Mytilus edulis*. They further demonstrated that some elements of the system could be induced with exposure to aromatic hydrocarbons, resulting in the production of diol epoxide metabolites. The authors suggested, however, that because of the low level of enzymatic activity of components of the detoxication system, hydrocarbon metabolism did not play a major role in the removal of aromatic hydrocarbons from the tissues of bivalve molluscs.

Lee and Singer (1980) detected mixed-function oxygenase activity in the polychaete worms *Nereis virens* and *Capitella* sp. Higher levels of cytochrome P-450 and benzo(a)pyrene monooxygenase activity were found in *Nereis* after worms were fed food contaminated with benz(a)anthracene. Natural populations of *Nereis virens* collected from an oil-contaminated site had higher levels of cytochrome P-450 and mixed-function oxygenase activity than polychaetes collected from an uncontaminated site (Lee *et al.*, 1981). Although *Capitella* had low levels of mixed function oxygenase activity at the beginning of exposure to oil-contaminated sediments, long-term exposure of several generations resulted in increased activity in later generations. Other studies have also shown that *Capitella* may enhance microbial decomposition of hydrocarbons in subsurface sediments through bioturbation (Lee *et al.*, 1979). In later studies *Nereis succinea* fed oil-contaminated food showed reductions in growth and increases in mixed-function oxygenase activity (Lee *et al.*, 1981).

A large volume of literature exists on the characterization and induction of cytochrome P-450 mixed-function oxygenases in marine fish (see reviews by Bend and James, 1978; Stegeman, 1981). Polyaromatic hydrocarbons are active inducers of enzymatic activity in both hepatic and extrahepatic tissues of marine fish, but induction may be modified by temperature, route of uptake of aromatic compounds and other environmental factors. There are wide interspecific and intraspecific differences among marine fish in the induction of benzo(a)pyrene monooxygenase activity and cytochrome P-450 content by aromatic hydrocarbons (Pedersen *et al.*, 1976; Bend and James, 1978; Stegeman, 1981), presumably related to high variability in enzyme activities in control fish. Metabolites produced by fish exposed to polyaromatic hydrocarbons are primarily *trans*-diols. There is considerable variation in the distribution of metabolites produced by different species of fish, although the benzo-ring diols of benzo(a)pyrene account for 40-50% of the metabolites produced by marine teleosts (Stegeman, 1981). In general metabolism of benzo(a)pyrene in hepatic preparations of teleost fish results in formation of high percentages of benzo-ring (7,8- and 9,10-dihydrodiols) but little K-region (4,5-dihydrodiol) metabolites (Figure 8.4).

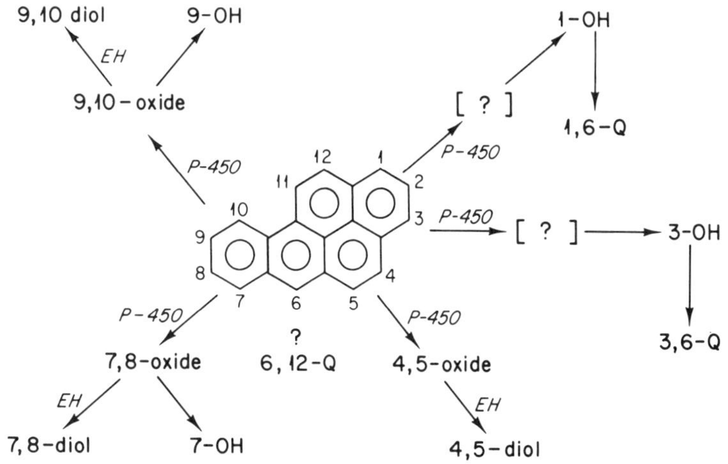

Figure 8.4. Pathways by which the major primary metabolites of benzo(a)pyrene may be formed *in vitro* by aquatic species. Not all metabolites indicated are formed by each species. Whether 6,12-quinone originates from initial metabolism at the 6-carbon or at the 11,12 position is not clear; 1,6- and 3,6-quinone might also be derived from 6-OH-BP if this product is formed. P-450—cytochrome P-450; EH—epoxide hydrolase (from Stegeman, 1981).

Metabolites and their Effects

Metabolism on the benzo-ring is associated with the production of mutagenic and carcinogenic derivatives (7,8-diol-9,10-epoxides). Although further metabolism may result in inactivation and excretion of metabolites, there appears to be a tendency among induced animals to form highly reactive metabolites that are resistant to inactivation. This may be related to the greater capacity of induced forms of cytochrome P-450 to form carcinogenic and mutagenic metabolites (Stegeman, 1981). The induction of cytochrome P-450 mixed-function oxygenases can, therefore, have very different consequences for species of marine fish. Much less is known concerning the carcinogenic and mutagenic potential of metabolite production in the invertebrates.

The liver is the major site of hydrocarbon metabolism in marine fish, although extrahepatic tissues may also possess high activity (Stegeman, 1981). Binder and Stegeman (1980) demonstrated the presence of mixed-function oxygenases in embryonic stages of *Fundulus heteroclitus* before the development of a functional liver. The gall bladder is the major site of metabolite excretion in marine fish (Lee *et al.*, 1972b; Melancon and Lech, 1978; Collier *et al.*, 1978; Varanasi and Gmur, 1981a). Metabolites excreted in bile by marine fish are primarily conjugated derivatives of oxygenated forms, whereas excretion of other metabolites such as glucuronides or unconjugated derivatives appears to vary between species and type of compound being metabolized (Varanasi and Gmur, 1981a). Biliary metabolites of naphthalene, dimethylnaphthalene, phenanthrene and benzo(a)pyrene include conjugates of dihydrodiols, phenols and quinones (Melancon and Lech, 1978; Solbakken *et al.*, 1980; Varanasi and Gmur, 1981a). Metabolites may also be lost through routes other than biliary excretion (Lee *et al.*, 1972b; Bend and James, 1978). Metabolite production may balance hydrocarbon uptake so that little uptake is observed (Lu *et al.*, 1977; Malins and Hodgins, 1981) or may result in retention of metabolites in various tissues (Varanasi and Gmur, 1981b).

A summary of mixed function oxygenase activity in marine animals is presented in Table 8.6. The activities of epoxide hydrolase and glutathione transferase with benzopyrene-4,5-oxide as a substrate are reported in Table 8.7. Comparison of Tables 8.6 and 8.7 suggests that the phylogenetic distributions of *in vitro* activities of epoxide hydrolase and glutathione transferase are not consistent with those of monooxygenase activities (Stegeman, 1981).

Persistence of metabolites in the tissues of marine organisms may induce toxic effects, particularly when metabolites bind to cellular macromolecules such as DNA, RNA or protein. Varanasi and her coworkers demonstrated the metabolic activation of benzo(a)pyrene and subsequent binding of metabolites *in vitro* to DNA from deproteinized salmon sperm when preparations were incubated with liver homogenates from untreated, 3-methylcholanthrene- and benzo(a)pyrene-treated starry flounder *Platichthys stellatus*, English sole *Parophrys vetulus* and coho salmon *Oncorhynchus kisutch* (Varanasi and Gmur, 1980; Varanasi *et al.*, 1980). Further studies demonstrated that similar processes occurred *in vivo* in fish exposed to benzo(a)pyrene (Varanasi and Gmur, 1981a,b; Varanasi *et al.*, 1981). The distribution of unmetabolized and metabolized benzo(a)pyrene in English sole is presented in Table 8.8.

TABLE 8.6
Cytochrome P-450 and mixed-oxygenase activity in marine invertebrates and fish (Lee, 1981)

Species	*Tissue*	*MFO*[a]	*Cytochrome P-450*[b]	*Reference*
Invertebrates:				
Balanus eburneus	digestive gland	43 ± 1	0.11 ± 0.01	Stegeman and Kaplan, 1981
	intestine	16 ± 4	0.09 ± 0.03	
Callinectes sapidus	stomach	37 ± 13	0.14	Singer *et al.*, 1980
	hepatopancreas	—	0.064 ± 0.009	Singer, unpub.
			0.18 ± 0.08	James *et al.*, 1979b
Homarus americanus	hepatopancreas	—	0.043–0.13	Elmamlouk *et al.*, 1974
Nereis virens	intestine	21 ± 4	0.069 ± 0.024	Lee and Singer, 1980
Panulirus argus	hepatopancreas	—	0.91 ± 0.39	James *et al.*, 1979b
Fish:				
Archosargus probatocephalus	liver	—	0.28 ± 0.09	James and Bend, 1980
Fundulus heteroclitus	liver	267	0.37	Stegeman, 1979
Raja erinacea	liver	—	0.24 ± 0.08	James and Bend, 1980
Salmo gairdneri	liver	14.4 ± 4.7	0.24 ± 0.02	Stegeman and Chevion, 1980

[a]fluorescence equivalent to pmoles of 3-hydroxyB(a)P/mg microsomal protein/60 minutes ± SD.
[b]nmoles/mg microsomal protein, ± SD.

A wide range of morphological, cytological and developmental abnormalities have been observed in marine animals exposed to petroleum or petroleum components (Table 8.9). The relationship between these abnormalities and metabolite production has not been clearly established, although the relationship is strongly suggested (Malins, 1982). Biochemical changes indicative of altered energy metabolism and balance were observed among populations of *Pleuronectes platessa* collected from oil-contaminated estuaries following the spill from the *Amoco Cadiz* (Neff, 1983), coincident with histopathological disorders (Haensly *et al.*, 1982; Stott *et al.*, 1983). Malins *et al.* (1983) reported a high incidence of organic free radicals in liver microsomes associated with idiopathic liver lesions from English sole collected from sediments with a high concentration of aromatic hydrocarbons. The organic free radicals appeared to be an XP· type, presumably derived from the metabolism of a polyaromatic hydrocarbon linked

TABLE 8.7
Microsomal epoxide hydrolase and cytosolic glutathione transferase activities with benzo-pyrene-4,5-oxide (James *et al.*, 1979a; Stegeman, 1981).

Species	*Epoxide hydrolase (nmol/min/mg)*	*Glutathion transferase (nmol/min/mg)*
Invertebrates:		
Callinectes sapidus	3.9 ± 2.6	0.2 ± 0.1
Homarus americanus	6.3 ± 5.3	3.3 ± 1.8
Mya arenaria	2.1	0.3
Mytilus edulis	0.05	0.3
Panulirus argus	18.8 ± 6.1	0.4 ± 0.2
Fish:		
Archosargus probatocephalus	3.0 ± 0.9	38.5 ± 23.1
Dasyatis sabina	0.23 ± 0.21	1.1 ± 0.2
Pseudopleuronectes americanus	3.6 ± 3.0	5.3 ± 2.5
Raja erinacea	0.09 ± 0.5	23.7 ± 6.5
Stenotomus chrysops	0.9–1.2	

Nanomoles BP-4,5-oxide hydrated or conjugated ± SD.

TABLE 8.8
Distribution of unmetabolized and metabolized benzo(a)pyrene in liver and bile of English sole *Parophrys vetulus* (Varanasi and Gmur, 1981b)

Form of b(a)p	*% in liver*	*% in bile*[a]
Unmetabolized	2	2 ± 3
Ethyl acetate soluble metabolites	16	14 ± 10
Glucuronides	<5	43 ± 3
Sulfates	<5	13 ± 3
Other water soluble metabolites	45	28 ± 2
Bound to cellular constituents	34	

[a] mean of 3 replicates ±1 SD.

to a macromolecule such as protein. The exact relationship between free radicals derived from biotransformations and the development of lesions is not understood.

Monitoring Aspects

Measurement of induction of mixed-function oxygenase systems has been proposed as a possible monitoring tool for predicting the effects of exposure to oil or other organic contaminants on natural populations of marine animals

TABLE 8.9
Morphological, cytological and developmental abnormalities in marine organisms exposed to petroleum hydrocarbons (adapted from Malins, 1982, with additions)

Species	*Hydrocarbon*	*Concentration (exposure)*	*Effect*	*Reference*
Invertebrates:				
Homarus americanus	crude oil	100–1000 ppb (15 d)	distended chromatophores	Forns, 1977
Mercenaria mercenaria	phenol	100 ppb (24 h)	cytological damage-gill, digestive gland	Fries and Tripp, 1977
Mytilus edulis	aromatics	30 ppb (34–182 d)	functional and structural changes in lysosomes	Low *et al.*, 1981 Moore and Clarke, 1982
Paracentrotus lividus	benzo(a) pyrene	4.5 ppb (3 h)	abnormal cleavage in embryos	Ceas, 1974
Fish:				
Clupea harengus	benzene	0.9 ppb (24 h)	delayed embryonic development	Eldridge *et al.*, 1977
	crude oil	<1 μg/g	morphological abnormalities	Smith and Cameron, 1979
Fundus heteroclitus	naphthalene	200 ppb (15 d)	ischemia, lateral line necrosis	DiMichele and Taylor 1978
Hypomesus pretiosus	crude oil	54–113 ppb (3 h/d for 15–21 d)	necrotic neurons in embryos	Hawkes and Stehr, 1982
Menidia beryllinia	crude oil	5–100 mg/l 5–50% WSF (21–30 d)	gill and olfactory hyperplasia, pancreatic atrophy and necrosis	Solangi and Overstreet, 1982
Menidia menidia	crude oil	140 ppb (168 h)	hyperplasia	Gardner, 1975
Mugil cephalus	crude oil	exp. spill 4–5 μg/g (56 d)	fin erosion	Minchew and Yarbrough, 1977
Parophrys vetulus	crude oil	0.2–1.0% sediments (0.75–4 mo.)	hepatocellular lipid vacuolization, weight loss	McCain and Malins, 1982

(continued)

TABLE 8.9—*contd.*

Species	*Hydrocarbon*	*Concentration (exposure)*	*Effect*	*Reference*
Pleuronectes platessa	crude oil	*Amoco Cadiz*	fin necrosis, hyperplasia, decreased hepatocellular vacuolation	Haensly *et al.*, 1982
			delayed or suppressed ovarian development	Stott *et al.*, 1983
Pseudo-pleuronectes americanus	crude oil	sediments 1 1/45 kg (4–5 mos.)	liver hypertrophy	Fletcher *et al.*, 1982
Salmo gairdneri	crude oil	200 μl/l OWD	gill damage, chloride cell abnormalities	Engelhardt *et al.*, 1981
Trinectes maculatus	crude oil	5 100 mg/l 5 50% WSF (38–60 d)	necrosis of olfactory, pancreatic atrophy and necrosis	Solangi and Overstreet, 1982

(Payne, 1976; Kurelec *et al.*, 1979; Stegeman, 1980). This will require differentiating the catalytic functions of multiple forms of cytochrome P-450 and understanding the basis for seasonal fluctuations or sex-linked differences in mixed-function oxygenase activity. The significance of chronic induction is uncertain, however, as it may imply a functional adaptation for detoxication (Spies *et al.*, 1982) or an increased potential for pathological damage (Neff, 1983). Before we can effectively use this as a monitoring tool, we must link our understanding of induced changes in the mixed-function oxygenase system with long-term consequences in reproduction (due to possible altered metabolism of endogenous compounds such as steroid hormones) and disease (due to possible increased production of carcinogenic and mutagenic metabolites).

COMPARATIVE ASPECTS OF ACUTE TOXICITY

Acute toxic effects of petroleum hydrocarbons to marine organisms are not simply dose dependent but are clearly dependent on the bioavailability of toxic components. Toxic responses may result when hydrophobic components bind to lipophilic sites within the cell and interfere with metabolic processes or when metabolites bind to cellular macromolecules and alter cellular or subcellular structure (Neff, 1979). The differential toxicity of crude and refined oils appears to be related to the relative availability and persistence of specific aromatic components (Anderson, 1979).

The toxicity of individual hydrocarbons is related to their solubility. Thus, alkyl substituted benzenes and naphthalenes are more toxic than unsubstituted forms and highly insoluble compounds (chrysene, benzo(a)pyrene and benz(a)anthracene) have extremely low acute toxicities (Figure 8.5).

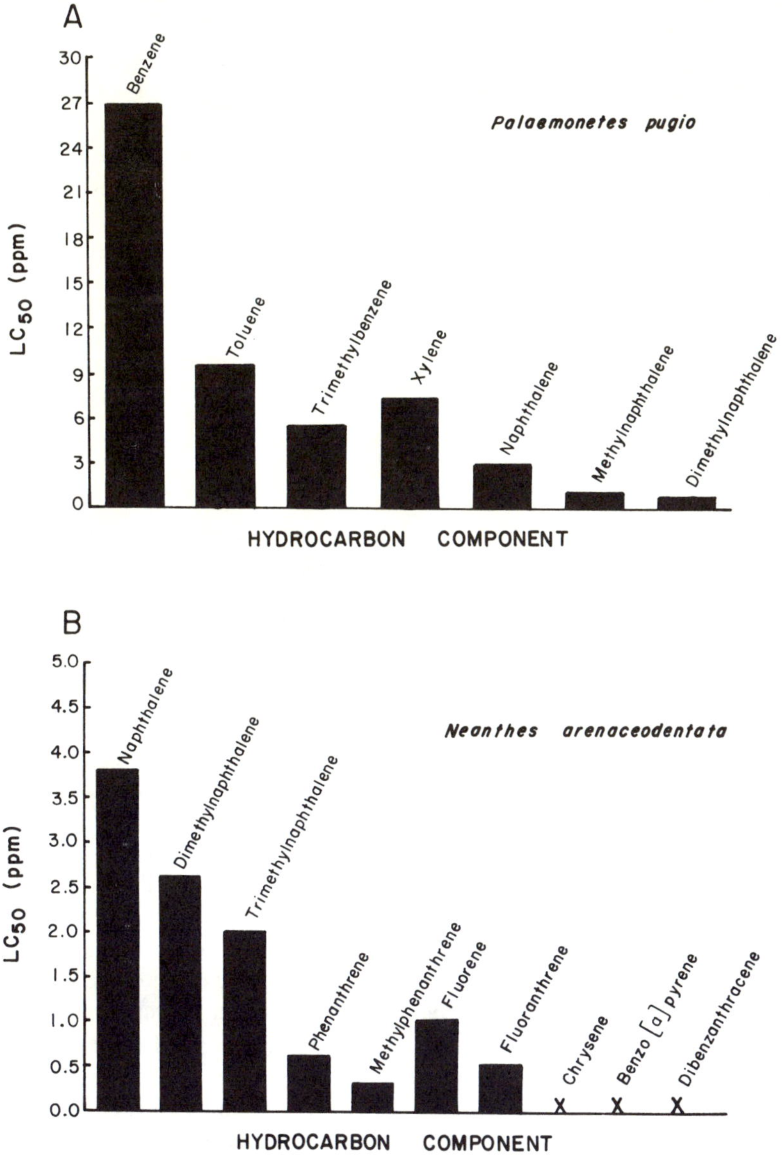

Figure 8.5. 96-hour LC_{50} values for A, *Palaemonetes pugio* and B, *Neanthes arenaceodentata* exposed to aromatic hydrocarbons (data from Neff *et al.*, 1976a).

Phytoplankton

The acute toxicity of petroleum hydrocarbons to marine phytoplankton is generally assessed by comparing growth rates and photosynthetic activities of single species cultures and natural phytoplankton assemblages with uncontaminated controls. Many studies have documented a reduction in photosynthetic activity and growth of both natural phytoplankton assemblages and single species cultures as a result of oil exposure (Gordon and Prouse, 1973; Shiels *et al.*, 1973; Lacaze, 1974; Pulich *et al.*, 1974; Soto *et al.*, 1975c; Parsons *et al.*, 1976; Vandermeulen and Ahearn, 1976; Lee and Takahashi, 1977; Brooks *et al.*, 1977; Hsiao, 1978; Hsiao *et al.*, 1978; Kusk, 1978; Federle *et al.*, 1979; Karydis, 1979). Acute toxicity appears to be related to the specific composition of the water soluble fraction of crude and refined oils (Nuzzi, 1973; Winters *et al.*, 1976, 1977a). Variation in toxic effects among natural assemblages have been observed seasonally (Gordon and Prouse, 1973; Fontaine *et al.*, 1975), possibly as a result of the differential sensitivity of individual species of phytoplankton and the composition of phytoplankton assemblages (Pulich *et al.*, 1974). Exposure to low levels of hydrocarbons (5-100 μg/1) appears to result in an enhancement of phytoplankton growth (Dunstan *et al.*, 1975; Parsons *et al.*, 1976; Prouse *et al.*, 1976); this will be discussed further in the following section.

Macroalgae

Reduced photosynthetic rates have also been observed among macroalgae as a result of exposure to high concentrations of oil (Clendenning and North, 1959; Schramm, 1972; Shiels *et al.*, 1973), whereas exposure to low concentrations appear to enhance photosynthetic rates (Shiels *et al.*, 1973). Clendenning and North (1959) found a two day delayed response of reduced photosynthesis by the giant kelp *Macrocystis pyrifera* exposed to a diesel oil dispersion that they suggested might be due to delayed penetration of hydrocarbons into plant tissues. Mitchell *et al.* (1970) further observed that *Macrocystis* secreted a mucus in response to oil exposure that minimized the contact between plant tissues and hydrocarbon components. Macroalgal gametes appear to be particularly sensitive to oil exposure, as demonstrated by Steele (1977) with the brown alga *Fucus edentatus*.

The mechanisms responsible for disruption in growth and impairment of photosynthesis are not well understood. Van Overbeek and Blondeau (1954) suggested that hydrocarbon molecules could disrupt the plasma membrane of plant cells by displacing certain lipid components and thus altering membrane permeability. Furthermore, they suggested that photosynthesis could be impaired as a result of hydrocarbons dissolving the lipid phase of the grana of chloroplasts and increasing the distance between individual chlorophyll molecules. Toxic effects may also result from disruption in mitochondrial membranes and inhibition of energy metabolism (Baker, 1970). The assumption that disruption in growth and impairment of photosynthesis are a result of these mechanisms is supported by the findings of Boney and Corner (1959), Vandermeulen and Ahearn (1976) and Hutchinson *et al.* (1979).

Animals

The acute toxicity of petroleum hydrocarbons to marine animals is assessed by the measurement of an LC_{50} value—i.e., the concentration of a hydrocarbon mixture or specific hydrocarbon that results in 50% mortality of the test organism during a designated exposure period. A large volume of literature exists on the acute toxicity of crude and refined oils and specific hydrocarbons to marine animals, although comparison of LC_{50} values among different species is difficult because of differences in bioassay protocol, failure to measure hydrocarbon concentrations in exposure systems, and failure to consider loss of hydrocarbons through volatilization and degradation. Using comparative data sets generated under similar laboratory protocols, however, one can begin to compare the differential sensitivity of various phylogenetic groups and developmental stages to hydrocarbon exposure. The LC_{50} values reported in this section should not be considered as absolute values, as bioassays were conducted under different conditions and hydrocarbon concentrations are often reported as the measured concentration at the beginning of the exposure period, and thus may underestimate actual toxic concentrations. The data do serve, however, as comparative reference data for assessing differences in phylogenetic, developmental and biogeographic sensitivity.

The toxicity of petroleum hydrocarbons to marine zooplankton has been extensively reviewed by Corner (1978). Both holoplanktonic and meroplanktonic forms appear to be sensitive to hydrocarbon exposure and acute LC_{50} values range from approximately 0.02 to 10.0 mg/1 (based on measured initial hydrocarbon concentrations; National Research Council, 1985) with a few higher values reported for exposure to oil-water-dispersions of some crude oils. Although Corner (1978) reported some values that extend beyond this range, many early results were based on nominal (not measured) hydrocarbon concentrations. As evident in Table 8.10, there appears to be no differential sensitivity of various phylogenetic groups, although considerable variation in interspecific and stage specific sensitivity may occur.

Developmental Stages

Rosenthal and Alderice (1976) suggested that the most sensitive stages in the life cycle of marine fish (and presumably other multicellular organisms as well) to pollutant exposure are the development of gonadal tissue, the development of early embryonic (pre-gastrulation) stages, and the larval transition to exogenous food sources and metamorphosis. Several investigators have demonstrated that early embryonic and larval stages are more sensitive than later larval stages (Ceas, 1974; Wells and Sprague, 1976; Donahue *et al.*, 1977; Lonning, 1977; Linden, 1978; Cucci and Epifanio, 1979; Sharp *et al.*, 1979; Vashchenko, 1980; Capuzzo and Lancaster, 1981). Sharp *et al.* (1979) found that the early embryonic stages of *Fundulus heteroclitus* were more sensitive than later embryonic and larval stages to the water soluble fraction of No. 2 fuel oil, possibly as a result of reduced membrane permeability to hydrocarbons (Sharp *et al.*, 1979) or increased capability for detoxication (Binder and Stegeman, 1980) among later stages.

TABLE 8.10

Toxicity of petroleum hydrocarbons to marine zooplankton (National Research Council 1985). WSF = water soluble fraction; OWD = oil-in-water dispersion

Phylogenetic group	*Oil type*	*Exposure time (h)*	LC_{50} *(ppm)*	*Reference*
Protozoa	crude-WSF	90	1.7	Lanier and Light, 1978
Ctenophora	No. 2-WSF	24	0.6	Lee, 1975
Mollusca				
Bivalve				
embryo	crude-WSF	48	0.2–12	Byrne and Calder, 1977
	No. 2-WSF	48	0.4	
	No. 6-WSF	48	1.0	
larvae	crude-WSF	48	0.25–>25	
	No. 2-WSF	48	1.3	
	No. 6-WSF	48	3.2	
larvae	crude-WSF	240	0.05–2.1	
	No. 2-WSF	240	0.5	
	No. 6-WSF	240	1.6	
Pteropods	No. 2-OWD	48	<0.2	Winters *et al.*, 1977b
Crustacea				
Barnacles	No. 2-WSF	1	2.6	Blundo, 1978
Copepods	crude-OWD	96	13.5–73	Lee *et al.*, 1980 Sekarah and Foy, 1978
	No. 6-OWD	96	5.1	Hollister *et al.*, 1980
	No. 2-WSF	72	1.0–2.5	Lee and Nicol, 1977
Amphipods	crude-OWD	48	0.8	Linden, 1976a
	No. 2-OWD	48	0.3	
	No. 4-OWD	48	6.2	
Decapods				
Shrimp larvae	crude-WSF	96	0.5–8.5	Mecklenburg *et al.*, 1977 Broderson *et al.*, 1977 Rice *et al.*, 1976
postlarvae	crude-OWD	96	>1000	Anderson *et al.*, 1974b
	No. 2-OWD	96	1.7–9.4	Anderson *et al.*, 1974b Tatem *et al.*, 1978
	crude-WSF	96	>19.8	Anderson *et al.*, 1974b
	No. 2-WSF	96	1.2–6.6	Tatem *et al.*, 1978 Neff *et al.*, 1976a
	No. 6-WSF	96	1.9	Anderson *et al.*, 1974b Tatem *et al.*, 1978
Lobster larvae	crude-OWD	96	0.9–4.9	Wells and Sprague, 1976
		720	0.14	
Crab larvae	crude-WSF	96	1.3–>10.8	Mecklenburg *et al.*, 1977 Broderson *et al.*, 1977 Rice *et al.*, 1976 Cucci and Epifanio, 1979
Mysids	crude-OWD	96	0.05–0.2	Wells, 1976
Teleostei eggs	No. 2-WSF	96	1.5	Sharp *et al.*, 1979
	No. 6-WSF	144	0.02–0.2	Kuhnhold, 1978
larvae	crude-OWD	96	43	Vuorinen and Axell, 1980
	No. 2-WSF	96	1.5	Sharp *et al.*, 1979

Although early developmental stages (eggs, embryonic and larval stages) are generally considered to be more sensitive than later developmental stages, there is no consistent trend in the comparison of sensitivities of juvenile and adult organisms. Neff *et al.* (1976a) compared the sensitivity of different life stages of four species of crustaceans to water soluble fractions of No. 2 fuel oil (Figure 8.6). Although larvae of *Palaemonetes pugio* were clearly more sensitive than postlarvae and adults, postlarval and early juvenile stages of some species were more tolerant than adults. In similar experiments with the polychaete *Neanthes arenaceodentata*, Rossi and Anderson (1976) compared the sensitivity of 4-, 18-, 32-, and 40-segment juveniles in addition to 60-segment mature adult worms.

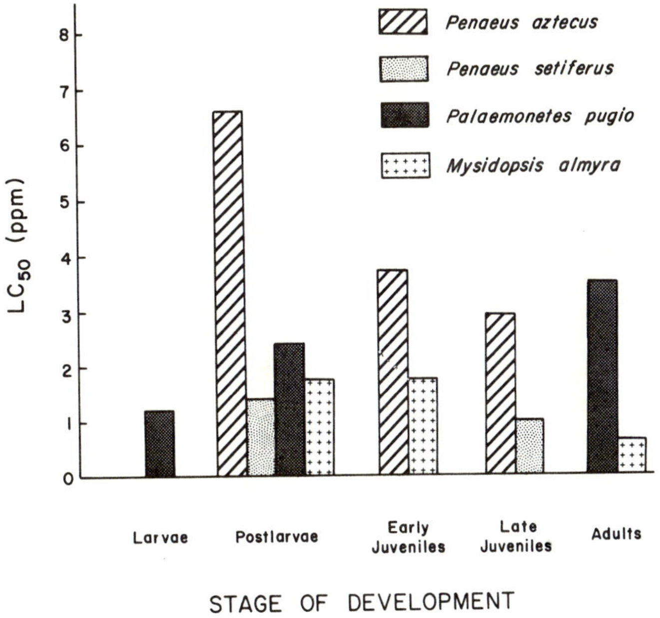

Figure 8.6. Comparative sensitivity (96-hour LC_{50} values) of various developmental stages of marine crustaceans to water soluble fractions of No. 2 fuel oil (data from Neff *et al.*, 1976a).

During 96-hour exposures, toxicity increased with increased size and age and mature gravid females were more tolerant than mature males. The authors suggested that the rich lipid stores of early juvenile stages and gravid female adult worms were used to sequester hydrocarbons and prevent uptake by other tissues, thus reducing toxic effects. Under conditions of chronic exposure, however, juveniles were found to be more sensitive than either adults or larvae (Rossi and Anderson, 1978).

Extrinsic Factors

Toxic responses by marine animals to petroleum hydrocarbons may be modified by other environmental factors, such as salinity and temperature; intrinsic factors, such as the level of feeding and reproductive activity; and habitat. Moles *et al.* (1979) and Levitan and Taylor (1979) found salinity to be an important factor in the survival of migrating salmonids and estuarine killifish, respectively, exposed to aromatic hydrocarbons. Korn *et al.* (1979) and Thomas and Rice (1979) found that salmonids were more sensitive to toluene exposure at low exposure temperatures, presumably due to both the greater persistence of toluene at low temperatures and decreased rates of hydrocarbon metabolism. Fletcher *et al.* (1981) found that winter flounder *Pseudopleuronectes americanus* were more sensitive to oil-contaminated sediments during summer exposures than during winter exposures due to lack of feeding activity and reduced sediment reworking during the winter months.

Jackson *et al.* (1981) found that ghost crabs *Ocypode quadrata* collected during reproductive season were more sensitive to petroleum hydrocarbons than individuals collected during other times of the year. The authors suggested that this response might be a result of lowered energy reserves in the reproductively active animals, but such responses might also result from interactions between hormonal changes associated with reproduction and hydrocarbon detoxication reactions.

Rice *et al.* (1976) compared the sensitivities of 27 species of fish and invertebrates from polar seas to water soluble fractions of Cook Inlet crude oil and No. 2 fuel oil (Figure 8.7). No. 2 fuel oil was slightly more toxic than Cook Inlet crude oil to most species tested and fish were consistently among the most sensitive species. Some invertebrates—specifically, intertidal invertebrates including gastropod and bivalve molluscs, echinoderms and crustaceans—were more resistant than others suggesting either increased tolerance or effective avoidance responses (e.g., mucous production) of intertidal invertebrates. Comparison of the data observed for polar species (Rice *et al.*, 1976, 1977a) with those observed for temperate species (Anderson *et al.*, 1974a,b; Neff *et al.*, 1976a; Rossi *et al.*, 1976) suggests that the polar species are more sensitive than their temperate water counterparts. This trend, however, may simply be due to the greater persistence and availability of aromatic hydrocarbons due to the lower rate of weathering at low experimental temperatures (Jordan and Payne, 1980; Chapters 5 and 7).

Toxicity Index

Anderson *et al.* (1980) proposed the use of a toxicity index that incorporates both exposure time and LC_{50} values to compare mortality data.

The index is calculated by the equation

$$X^a Y = C,$$

where, X = concentration in ppm;
Y = time to LC_{50} in days;
a = arbitrary constant; and
C = toxicity index.

In applying this index to mortality data collected for three species of crus-

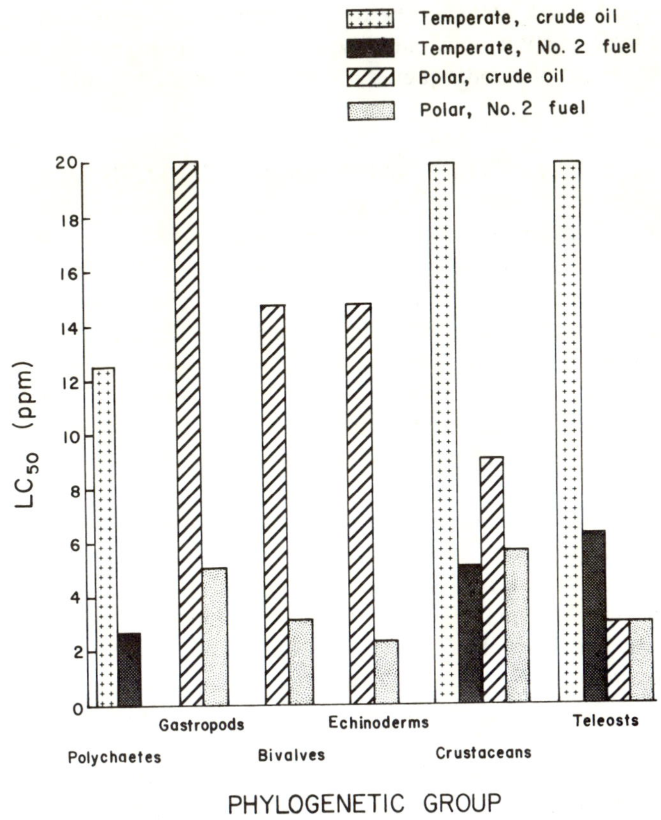

Figure 8.7. Comparative toxicity (96-hour LC_{50} values) of petroleum hydrocarbons to fish and invertebrates from polar and temperate seas (data from Anderson *et al.*, 1974b; Neff *et al.*, 1976a; Rice *et al.*, 1976; Rossi *et al.*, 1976).

taceans—*Neomysis awatschensis*, *Hippolyte clarkii* and *Pandalus danae*—exposed to an oil-water-dispersion of Prudhoe Bay crude oil, Anderson *et al.* (1980) observed a linear relationship between mortality and the product of time and concentration and suggested that this relationship could be used in a predictive manner to compare acute lethal effects with sublethal responses on an "equal exposure" basis. With the data in Table 8.10 and Figure 8.7 converted to a time-exposure basis, the acute toxicity index is equivalent to 0.003-2.35 ppm-days for planktonic organisms, 0.05-2.25 for polar species of crustaceans and fish, and 0.23-4.95 for temperate species of crustaceans and fish. Although these values are within the range one would expect sublethal responses to occur, the wide range of values would appear to limit the usefulness of this index as a predictor of sublethal impact.

The acute toxicity of petroleum hydrocarbons to marine organisms varies considerably among species, developmental stage and routes of exposure. Although precise interspecific comparison is difficult because of differences in bioassay protocol, the differential sensitivity of various phylogenetic groups, various life cycle stages and species from different biogeographical regions may be related to hydrocarbon bioavailability, capacity for hydrocarbon biotransformation and the metabolic consequences of hydrocarbon exposure.

SUBLETHAL AND CHRONIC EFFECTS

Data gathered from several recent oil spills have demonstrated that the medium and higher molecular weight aromatic compounds, such as the alkylated phenanthrenes and alkylated dibenzothiophenes, are among the most persistent petroleum hydrocarbons in both animal tissues and sediments (Grahl-Nielsen *et al.*, 1978; Roesijadi *et al.*, 1978; Teal *et al.*, 1978; Boehm *et al.*, 1981, 1982). Thus, although short-term sublethal stress may be the result of exposure to a wide range of hydrocarbons, long-term chronic stress is most probably the result of exposure to medium and higher molecular weight aromatic compounds (see Chapter 6).

Sublethal effects of petroleum hydrocarbons on marine organisms may be manifested at all levels of biological organization. The level of impact will be dependent on the duration of exposure to toxic concentrations and the compensatory mechanisms available for recovery following exposure.

At the subcellular level, effects can be manifested in changes in energy metabolism, alterations in cellular structure and function, and enhancement of chromosome mutation. The mutagenic potential of metabolites of polyaromatic hydrocarbons has been well documented in mammalian systems and may have similar effects on fish and invertebrates. Chromosomal aberrations and increased incidence of sister chromatid exchange have been observed in fish with exposure to benzo(a)pyrene (Hooftman and Vink, 1981). Cod and pollock eggs collected after the *Argo Merchant* oil spill showed a wide range of developmental abnormalities including greater cytological deterioration, abnormal differentiation and greater incidence of mitotic abnormalities (Longwell, 1977). Other investigators have also reported a wide range of developmental abnormalities in early life history stages as a result of oil exposure (Kuhnhold, 1974; Linden, 1976b; Linden *et al.*, 1980; Rabalais *et al.*, 1981). The implications of these responses on the survival and fitness of populations remain largely unexplored.

Alterations in growth and energy metabolism have been observed in both plants and animals as a result of oil exposure. Effects of low levels of hydrocarbons (5-100 μg/1) appear to enhance the growth of phytoplankton (Dunstan *et al.*, 1975; Parsons *et al.*, 1976; Prouse *et al.*, 1976), presumably due to either utilization of hydrocarbon components as metabolic substrates by plant cells (Soto *et al.*, 1975a,c) or the presence of growth regulating compounds, such as auxins, in the oil (Gordon and Prouse, 1973). Growth stimulation of macroalgae has also been observed with exposure of algal sporelings to polyaromatic hydrocarbons (Boney and Corner, 1962; Boney, 1974).

The mechanisms responsible for disruption in growth and impairment of photosynthesis of phytoplankton at higher exposure concentrations are not well understood. Hutchinson *et al.* (1979) suggested a correlation between hydrocarbon toxicity and ion "leakage" from exposed cells and Vandermeulen and Ahearn (1976) found alterations in the ATP/ADP balance with exposure of phytoplankton to naphthalene. Soto *et al.* (1975b) found changes in the chemical composition of *Chlamydomonas angulosa* with exposure to naphthalene with significant increases in both lipid and carbohydrate content and decreases in protein content being observed during exposure. Biochemical composition reverted back to that of controls upon transfer of cultures to uncontaminated sea water. Davavin *et al.* (1975) in their studies of macroalgae suggested that oil exposure could result in inhibition of biosynthetic pathways and polymerization of DNA and RNA.

In mesocosm experiments testing the effects of unweathered Ekofisk crude oil, Skjoldal *et al.* (1982) observed reductions in primary production, coincident with the accumulation of polar compounds in the water column presumably derived from photooxidation; increases in bacterial numbers, coincident with high rates of phosphate uptake and mineralization of naphthalene and hexadecane; and reduced grazing pressure, as a result of a reduction in ciliate predators. Although a significant alteration in phytoplankton species composition was also evident during the eight-week experiment, the enclosed ecosystems continued to be autotrophically dominated throughout exposure.

Exposure of marine zooplankton to petroleum hydrocarbons may result in alterations in feeding, growth and reproduction. Berdugo *et al.* (1977) observed reduced egg production rates of *Eurytemora affinis* with short-term exposure to the water soluble fraction of a high aromatic heating oil. Chronic long-term exposure of *Eurytemora* to aromatic hydrocarbons resulted in long-term changes in reproductive effort as evidenced by reductions in life span, the number of nauplii produced, mean brood size and egg production rates (Ott *et al.*, 1978). The toxic effects might not have been due to specific inhibition of reproductive processes, but related to reduced feeding rates of exposed copepods. Cowles and Remillard (1983) observed decreased ingestion rates and egg viability in the copepod *Centropages hamatus* exposed to 10-80 μg/1 of South Louisiana crude oil, although egg production rates appeared to be unaffected. The authors suggested that biosynthetic pathways involved in oogenesis could be influenced by low levels of hydrocarbons. Cowles (1983) further documented alterations in swimming activity and food perception in *Centropages* during oil exposure.

Berman and Heinle (1980) found that the feeding behavior of copepods in MERL enclosures was altered both qualitatively and quantitatively with exposure to sublethal concentrations of No. 2 fuel oil, resulting in a reduction in particle retention. Vargo (1981) found that physiological changes of copepod populations in the same enclosure experiments correlated with subsequent changes in zooplankton abundance and species composition of the zooplankton community. Both MERL experiments were carried out at oil concentrations ranging from 52 to 265 μg/1 total hydrocarbons, with average values of 90 and 190 μg/1. Although both phytoplankton and zooplankton may be affected by oil exposure, effects may be only temporary as exposed populations are rapidly replaced by high reproductive rates and immigration of populations from unaffected areas. Because of the sporadic and seasonal abundance of meroplanktonic forms, exposure of larval stages of benthic or pelagic species to petroleum hydrocarbons, however, may have greater consequences at the population level as a result of reduced recruitment.

Effects of petroleum hydrocarbons on reproductive and developmental processes can include interference of hydrocarbon components with hormone synthesis (Truscott *et al.*, 1983) and normal sex pheromone responses (Derenbach *et al.*, 1980; Derenbach and Gerek, 1980), loss of gametes (Renzoni, 1975; Steele, 1977), impairment of gonad development (Rinkevich and Loya, 1979), transfer of hydrocarbons from gonads to early developmental stages (Rossi and Anderson, 1977; Koster and Van den Beggelaar, 1980), reduced hatching success (Anderson *et al.*, 1977a; Ernst *et al.*, 1977; Kuhnhold, 1978; Sharp *et al.*, 1979), and increased incidence of developmental abnormalities (Hawkes and Stehr, 1982; Smith and Cameron, 1979).

Responses of planktonic larval stages to hydrocarbon exposure have been manifested in delayed development (Wells, 1972; Katz, 1973; Wells and Sprague, 1976; Byrne and Calder, 1977; Caldwell *et al.*, 1977; Winters *et al.*, 1977b; Laughlin *et al.*, 1978; Cucci and Epifanio, 1979; Laughlin and Neff, 1979), reduced feeding (Wells and Sprague, 1976; Johns and Pechenik, 1980), reduced growth (Struhsaker *et al.*, 1974; Renzoni, 1975; Linden, 1976b; Tatem, 1977; Johns and Pechenik, 1980; Laughlin and Neff, 1980), inhibition of molting in larval crustaceans (Winters *et al.*, 1977b; Mecklenburg *et al.*, 1977: Cucci and Epifanio, 1979; Laughlin and Neff, 1979), morphogenic abnormalities (Kuhnhold, 1974; Linden, 1976b; Linden *et al.*, 1980), inhibition of yolk utilization (Linden *et al.*, 1980), and the presence of abnormal intermediate larval stages (Wells and Sprague, 1976; Cucci and Epifanio, 1979; Laughlin and Neff, 1979).

Successful development and metamorphosis of planktonic larval stages of marine organisms are dependent on the balance and efficient utilization of energy reserves (Holland, 1978), with lipid reserves being either of primary or secondary importance in the energetics of larval development and metamorphosis. Sharp *et al.* (1979) suggested that hydrocarbon exposure of embryonic and larval stages of fish might result in the shunting of energy reserves away from critical differentiation and morphogenic processes to be used for metabolic maintenance.

The interference of lipophilic components of petroleum hydrocarbons with lipid metabolism is a possible mechanism for disruption in developmental potential of planktonic larval stages (Capuzzo *et al.*, 1984). Exposure of larval stages of the American lobster *Homarus americanus* to 250 μg/1 total hydrocarbons for 96 hours resulted in reduced respiration rates, delayed molting, reduced growth and disruptions in the normal patterns of lipid storage, utilization and synthesis.

Alterations in growth and energetics of adult fish and benthic invertebrates have also been observed with exposure to petroleum hydrocarbons. A wide range of metabolic changes have been observed with petroleum exposure including alterations in lysosome stability (Moore *et al.*, 1978; Moore 1979; Viarengo and Moore, 1982), ratios of free amino acids (Roesijadi and Anderson, 1979; Augenfeld *et. al.*, 1980), freezing tolerance (Aarset and Zachariassen, 1982), and glucose metabolism (Riley and Mix, 1981) in bivalves; acid phosphatase and cathepsin D activities (Drewa *et al.*, 1977), plasma copper levels (Dillon, 1981;

1983), hemolymph alkalosis (Sabourin, 1982) and serum protein and lipid levels (Payne *et al.*, 1983) in crustaceans; hepatic lipogenesis (Stegeman and Sabo, 1976; Sabo and Stegeman, 1977), plasma chloride levels (Payne *et al.*, 1978), and smoltification (Folmar *et al.*, 1982) in fish. Such findings suggest that exposure to sublethal levels of petroleum hydrocarbons may significantly impair metabolic functions including acid-base balance, osmoregulation, energy mobilization and oxygen transport.

The most important physiological changes associated with petroleum exposure are those that may adversely affect an organism's growth and survival and, thus, its potential ability to contribute to the population gene pool. Alterations in growth potential may take place as a result of changes in feeding behavior, respiratory metabolism or digestive efficiencies. Animals from polar regions may be particularly sensitive to impairment of energetics because of the sporadic seasonal abundance of food and the dependence on long-term energy reserves and slow recovery rates as a result of reduced fecundity, dispersal and growth rates of many polar species (Dunbar, 1968; Clarke, 1979).

Reductions in physiological measurements (such as respiration rates, carbon turnover rates, and scope-for-growth indices) have correlated with reduced growth rates measured for bivalve populations from oil-contaminated habitats (Gilfillan *et al.*, 1976; Gilfillan and Vandermeulen, 1978). Gilfillan *et al.* (1977) and Gilfillan (1980) suggested that alterations in energetics and growth of bivalve populations might be related to tissue burdens of aromatic compounds. Widdows *et al.* (1982) further demonstrated that with long-term exposure of *Mytilus* to oil concentrations as low as 30 μg/l total hydrocarbons, a negative correlation existed between both cellular and physiological stress indices (lysosomal latency and scope-for-growth) and tissue concentrations of aromatic hydrocarbons.

Roesijadi and Anderson (1979) and Augenfeld *et al.* (1980) found changes in condition index of *Macoma inquinata* and *Protothaca staminea*, respectively, with exposure to sediments contaminated with Prudhoe Bay crude oil. *Macoma*, a deposit-feeding bivalve, was more sensitive than *Protothaca*, a suspension-feeding bivalve, presumably as a result of differences in feeding habits. Anderson *et al.* (1983) observed reduced growth of *Protothaca* with exposure to oil-contaminated sediments in a one-year field experiment with initial sediment concentrations of Prudhoe Bay crude oil equivalent to 68-80 μg/g total hydrocarbons.

Edwards (1978) observed changes in respiration rate, growth rate, and net carbon turnover in the sand shrimp *Crangon crangon* with chronic exposure to the water soluble fraction of North Sea Brent Field crude oil. Acute exposure (24-48 h) to low concentrations (5% WSF = 1 ppm) resulted in a reduction in respiration rate, but increases and subsequent decreases in respiration rate were observed at higher concentrations (10% WSF). Similar changes in respiratory activity have been observed by W.Y. Lee *et al.* (1978) in the shrimp *Lucifer faxoni*, by Anderson *et al.* (1974b) in the shrimps *Penaeus aztecus* and *Palaemonetes pugio*, by Percy (1977) in the amphipod *Onisimus* (*Boekisimus*) *affinis*, and by Capuzzo *et al.* (1984) in the lobster *Homarus americanus*. Anderson (1977) suggested that these changes in respiratory activity might be related to a metabolic response to the naphthalene component of water soluble fractions. Cantelmo *et al.* (1982), however, found similar changes in respiration rate with exposure of the blue crab *Callinectes sapidus* to sublethal concentrations of benzene and dimethylnaphthalene. Respiratory responses have been observed at concentrations as low as 30-250 μg/1 total hydrocarbons (Baden and Hagermann, 1981; Widdows *et al.*, 1982; Capuzzo *et al.*, 1984) and may be related to changes in oxygen transport, damage to gill membranes or alterations in energy metabolism.

Physiological changes in marine fish include alterations in heart beat (Wang and Nicol, 1977; Anderson *et al.*, 1977a; Linden, 1978), coughing rate (Rice *et al.*, 1977b; Barnett and Toews, 1978); and respiration rate (Anderson *et al.*, 1974b; Barnett and Toews, 1978; Thomas and Rice, 1979). Thomas and Rice (1979) found that exposure of pink salmon to low levels of No. 2 fuel oil, Cook Inlet crude oil, naphthalene or toluene resulted in increased oxygen consumption rates immediately after exposure; respiration rates subsequently declined to normal values. They suggested that increased respiration might be the result of increased oxygen demand during detoxication processes. Decreased respiration rates were observed in several species of salmonids exposed to crude oil concentrations in the range of 0.3-1.6 mg/1 (Barnett and Toews, 1978: Duval, 1979; Fink and Duval, 1980). Altered growth rates and impairment of gonad development have also been observed in fish exposed to petroleum hydrocarbons (McCain *et al.*, 1978; Moles *et al.*, 1981).

Sublethal changes in energetics as a result of petroleum exposure may also result in greater susceptibility to other environmental stresses, such as disease, as a result of the high energy demand of tissue repair. Several studies have shown a direct correlation between hydrocarbon stress and increased incidence of fish diseases such as fin erosion (Minchew and Yarbrough, 1977; Giles *et al.*, 1978), and other histopathological conditions (McCain *et al.*, 1978; Sindermann, 1982).

Behavioral responses of an organism to pollutant stress may serve as a mechanism for detection of adverse pollutant concentrations, followed by the triggering of adaptive mechanisms such as altered feeding behavior or inducing an avoidance response. At an extreme level of stress, adaptive behaviors are overridden and an organism's ability to respond to environmental stimuli may become impaired, temporarily until the stress is removed or permanently if chemosensory mechanisms are irreversibly damaged. Blumer (1969) was the first to suggest that exposure to petroleum hydrocarbons could interfere with chemoreception, resulting in altered behavior patterns. Behavioral aberrations associated with oil exposure include altered sexual mating behavior (Kittredge *et al.*, 1974), feeding behavior (Jacobsen and Boylan, 1973; Atema and Stein, 1974; Berdugo *et al.*, 1977; Berman and Heinle, 1980; Pearson *et al.*, 1981; Busdosh, 1981; Cowles, 1983), burrowing behavior of benthic invertebrates (Prouse and Gordon 1976; Gordon *et al.*, 1978; Augenfeld, 1980; Olla *et al.*, 1983), swimming activity (Berge *et al.*, 1983) and schooling behavior of fishes (Gardner, 1975). Neurophysiological studies with chemoreceptors in the lateral flagella of antennules of the lobster *Homarus americanus* provided evidence that oil added to mussel extracts elicited nerve firing responses distinct from those elicited by mussel extract alone (Atema *et al.*, 1979). Avoidance responses of fish to low levels of petroleum hydrocarbons have also been documented (Rice, 1973; Maynard and Weber, 1981; Weber *et al.*, 1981; Hellstrom and Doving, 1983). Behavioral responses can be observed at concentrations as low as 0.1-0.4 μg/1 and thus may serve as extremely sensitive indicators.

Recovery from oil exposure at the organismal level is dependent on the duration of exposure and the bioavailability of specific hydrocarbons throughout exposure. Behavioral and physiological responses observed during short-term exposure (a few hours to several days) may be restored to control levels following transfer to uncontaminated sea water (Berge *et al.*, 1983; Cowles, 1983; Capuzzo *et al.*, 1984), although recovery does not always occur immediately upon transfer (Baden and Hagermann, 1981; Capuzzo *et al.*, 1984). Capuzzo *et al.* (1984) in their study of short-term exposure of larval stages of *Homarus americanus* to 250 μg/1 of South Louisiana crude oil found that recovery in terms of normal respiratory activity was not immediate upon transfer to uncontaminated seawater; values for respiration rates and ammonia excretion rates remained at a depressed level up to one week following exposure and correlated with tissue uptake of benzene, thiophenes, alkylbenzenes and toluene. Other sublethal responses that were evident during exposure (depressed O:N ratios and alterations in lipid storage patterns) were restored to control levels soon after exposure.

Recovery from long-term oil exposure is somewhat more complicated. Widdows *et al.* (1982) in their study of long-term exposure of *Mytilus edulis* to 30 μg/1 of North Sea oil found no evidence of 11 gradual recovery or acclimation to exposure conditions but a gradual deterioration in physiological condition, as evidenced by physiological and cellular stress indices. With chronic exposure of larval and juvenile crustaceans to water soluble or water-accommodated fractions of crude or refined oils, increased tolerance or increased acclimation to exposure conditions has generally been observed during exposure, resulting in little difference in growth or physiological parameters between oil-exposed and control animals at the end of a long-term experiment (Wells and Sprague, 1976; Laughlin *et al.*, 1978; Cucci and Epifanio, 1979; Capuzzo, 1981b). This may at least in part be due to an increased capacity for detoxication among older, post-metamorphic animals, although there is a paucity of data available to support this hypothesis at the present time. Benthic crustaceans and fish exposed to oil-contaminated sediments, however, continue to exhibit sublethal responses throughout exposure; this suggests that differences in exposure conditions, particularly in the persistence and availability of specific hydrocarbons in interstitial waters (Chapter 9), may elicit long-term responses (McCain *et al.*, 1978; Capuzzo, 1981b).

The integration of physiological and behavioral disturbances may result in alterations at the population and community levels. As illustrated in Figure 8.8, impairment of behavioral, developmental and physiological processes may occur at concentrations significantly lower than acutely toxic levels; such responses may alter the long-term survival of affected populations. Populations of plaice collected from oil-contaminated estuaries following a spill from the *Amoco Cadiz*

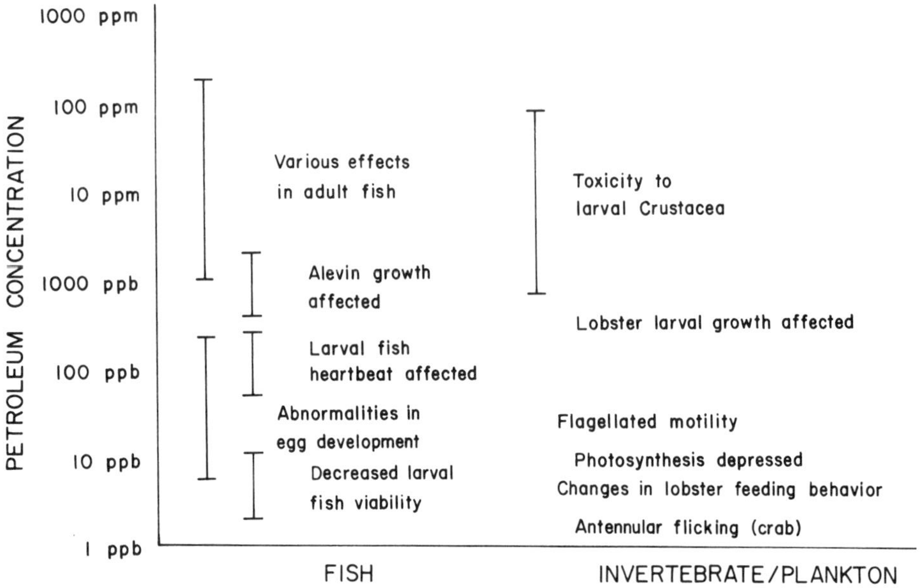

Figure 8.8. Comparison of lethal and sublethal effects of petroleum hydrocarbons on fish and invertebrates (from Vandermeulen and Capuzzo, 1983).

exhibited reduced growth rates and fecundity in addition to histopathological aberrations (Conan, 1982). Hauschildt-Lillge (1982) exposed the oligochaete *Lumbricillus lineatus* through five generations (15 months) to water soluble fractions of Arabian Light crude oil. Although this species was highly resistant to short-term oil exposure (Giere and Hauschildt, 1979), long-term exposure resulted in significant alterations in reproductive success. During the first generation, reduced egg fertility and hatching success were compensated by increased coccoon production at all exposure concentrations except 100% WSF. Embryonic development was not altered by exposure, but reduced growth and prolonged maturation and generation times were evident. Subsequent generations also exhibited reduced reproductive potential, but no effects on maturation or generation times were observed. In the 20% and 100% WSF exposures, coccoon production continued to decline and ceased with the F4 generation. In the 50% WSF exposure, however, productivity in terms of coccoon production increased among succeeding generations but hatching success decreased and the incidence of developmental abnormalities increased with each successive generation.

Krebs and Burns (1977) studied populations of the fiddler crab *Uca pugnax* for seven years after a spill of No. 2 fuel oil from the barge *Florida*; they observed long-term reductions in recruitment, population density, female:male ratios of adult crabs, behavioral changes and high overwintering mortality. Recovery of crab populations was correlated with the disappearance of naphthalenes and alkylated naphthalenes from contaminated sediments. Similar patterns of long-term changes in recruitment and density of benthic fauna have been observed with other oil spills (Gilfillan and Vandermeulen, 1978; Thomas, 1978; Cabioch *et al.*, 1980; Sanders *et al.*, 1980; Beslier *et al.*, 1980; Glémarec and Hussenot,

1981; Elmgren *et al.*, 1983). Mesocosm experiments have further documented that the major long-term changes in marine ecosystems as a result of petroleum inputs occur in the benthos (Grassle *et al.*, 1981; Oviatt *et al.*, 1982).

Although mesocosm experiments are discussed in more detail by Spies (Chapter 9), various aspects of these experiments will be discussed here as they relate to disturbance and recovery of populations of marine organisms and marine communities following oil exposure. In a series of three experiments conducted in the MERL mesocosms from 1977 to 1979 (Grassle *et al.*, 1981; Oviatt *et al.*, 1982), the impacts and recovery of a controlled marine ecosystem were monitored in response to chronic inputs of No. 2 fuel oil. For the first two experiments (1977 and 1978) oil-in-water dispersions of No. 2 fuel oil were added on a semi-continuous basis for 24- and 17-week periods, respectively; the third experiment was designed to evaluate the recovery potential of the ecosystem and no oil was added. In the 1977 experiment water column hydrocarbon levels were maintained at approximately 190 μg/l total hydrocarbons and toward the end of the exposure period 151 μg total hydrocarbons/g dry sediment were detected in the upper 3 cm of sediment with concentrations as high as 527 μg/g in the surface flocculent layer (Grassle *et al.*, 1981; Oviatt *et al.*, 1982). Benthic macrofaunal and meiofaunal populations were significantly reduced after the 24-week exposure. Benthic protozoa and diatoms, however, increased, presumably as a result of reduced predation. Time-series analyses indicate that the greatest degree of benthic impact occurred during the summer months (Oviatt *et al.*, 1982). During the 1978 experiments, water column hydrocarbon concentrations were maintained at 90 μg/l total hydrocarbons. Approximately 50% of the added oil was recovered in the surface flocculent layer of the sediments. Similar effects on macrofaunal abundance were evident in the second experiment although effects on meiofauna were less dramatic.

In the recovery experiment effects on the benthos were evident one year after the cessation of oil experiments, although there were no measurable impacts on the water column. Sediment concentrations declined rapidly initially, but a residual (10-20%) remained one year later. Differences in degradation rates of specific hydrocarbons were apparent throughout the exposure periods with higher loss rates occurring during the summer months, coincident with the highest degree of benthic impact (Grassle *et al.*, 1981; Oviatt *et al.*, 1982).

Elmgren and Frithsen (1982) compared the effects of oil exposure on the MERL mesocosm with changes observed in the water column and benthos following the *Tsesis* oil spill. Responses of various components of the ecosystem were similar in both situations—increased phytoplankton abundance and reduced zooplankton, macrofaunal and meiofaunal abundances. Although losses of zooplankton biomass were evident, observations at the *Tsesis* spill site suggest that this reduction is transitory as the recovery time for zooplankton was relatively short. Recovery rates for the meiofaunal component of the benthos may also be rapid, but recovery of the macrofaunal component occurs slowly (Grassle *et al.*, 1981).

Alterations in benthic biomass as a result of oil exposure can also result in changes in physical and chemical features of the benthos. Kalke *et al.* (1982)

demonstrated in benthic colonization experiments that oil exposure resulted in a reduction in the depth of the oxygenated layer by approximately one-half; this could result in a reduction in subsurface benthic production and changes in such processes as nutrient regeneration. Oviatt *et al.* (1982) also reported reductions in benthic respiration and nutrient flux during the recovery phase of the MERL mesocosm experiments.

Although alteration of benthic communities is an important impact of chronic hydrocarbon inputs in many marine habitats, it is difficult to ascertain whether differential recovery rates exist among different communities (offshore versus shallow) due to the lack of comparable data on hydrocarbon inputs, degree of sediment contamination, and rates of recovery for macrofauna and meiofauna. Recovery rates from disturbance will be dependent on the initial degree of impact, the rate of removal and persistence of hydrocarbons from the impacted area, and the recolonization rates of indigenous fauna. More detailed characterizations of the relative sensitivities and recovery rates of different marine communities must await further field verification.

Long-term effects of oil contamination on fisheries may be both direct, through the loss of reproductive or recruitment success, or indirect, through the disruption in food chain dynamics. Both types of impacts are difficult to assess because of our lack of knowledge of natural variability in reproductive effort and recruitment and flexibility in prey selection by commercially important species of fish and shellfish. Spaulding *et al.* (1983) developed a multicomponent fisheries assessment model based on an oil spill fates model, a shelf hydrodynamics model, an ichthyoplankton model and a fishery population model. The model is useful in predicting the spatial and temporal interactions of spill conditions with the spawning and development of early life history stages. Such an approach can be modified to include the impact of chronic exposure conditions on reproductive effort in addition to larval development and recruitment success of demersal species. Indirect effects will be more difficult to estimate and quantify, however, for alterations in food chain dynamics to have a significant impact on demersal fishery stocks, an extensive area of the benthos would have to be degraded.

CONCLUSIONS AND RECOMMENDATIONS

Although a large volume of literature exists on the effects of petroleum hydrocarbons on marine organisms, derived from laboratory studies, the majority of studies have been carried out at concentrations higher than environmentally realistic. Although these studies have contributed substantially to our understanding of potential for long-term consequences of petroleum discharges in the marine environment, a better understanding of the balance between adaptive and disruptive responses of organisms to chronic low-level discharges is needed in order to predict the consequences of offshore oil and gas development. Despite this information gap, several important generalizations can be derived from laboratory studies that have been conducted during the past decade. Long-term effects of petroleum hydrocarbons to marine organisms are related to the persistence and bioavailability of specific hydrocarbons, the ability of organisms to metabolize various hydrocarbons, the fate of metabolized products, and the interference of hydrocarbons with metabolic processes that may alter an organism's chances for survival and reproduction in the environment.

The uptake, accumulation and toxicity of individual hydrocarbons are dependent on solubility and the partitioning between hydrophobic and hydrophilic compartments. Uptake of hydrocarbons has been demonstrated from aqueous, dietary and sedimentary pathways. Retention of hydrocarbons in lipophilic cellular compartments may result in disruptions in membrane functions or alterations in energetic processes and impairment of an organism's adaptive capacity within its natural habitat. Capacity for metabolism of lipophilic compounds may influence the disposition or removal of aromatic hydrocarbons by marine organisms. Furthermore, if metabolites are retained and bind to cellular macromolecules, detoxication/toxication reactions may result in a wide range of cytological, morphological and developmental abnormalities. The cause and effect relationship of xenobiotic metabolism and cellular and subcellular damage has not, however, been clearly established.

The differential sensitivity of various phylogenetic groups, various life cycle stages and species from different biogeographical regions appears to be related to hydrocarbon bioavailability, capacity for hydrocarbon biotransformation and the metabolic consequences of hydrocarbon exposure. The increased sensitivity of early developmental stages and seasonal differences in the responses of adult animals may be related to stage-specific or seasonal dependency on particular metabolic processes (e.g., storage and mobilization of lipid reserves, synthesis of steroid hormones, etc.). Animals from polar regions may be particularly sensitive to impairment of energy storage because of the sporadic seasonal abundance of food and the dependence on long-term energy reserves and recovery rates may be slow as a result of reduced fecundity and dispersal and growth rates of many polar species in comparison with temperate species.

Sublethal effects include impairment of feeding, growth, development, energetics and recruitment that may result in alterations in both reproductive and developmental success and changes in community structure and dynamics. It is difficult to ascertain, however, the relationship between chronic responses of organisms to petroleum hydrocarbons and large-scale alterations in the functioning of marine ecosystems and harvesting of fishery resources. The sensitivity of early developmental stages, the impairment of reproductive processes, and the long-term effects on populations suggest that chronic exposure conditions may certainly alter the dynamics of benthic populations, including populations of demersal fish.

In assessing long-term impact of offshore oil and gas development activities, it is important to understand the conditions under which hydrocarbons persist in benthic environments and the sublethal effects that lead to reduced growth, delayed development and reduced reproductive effort and result in population decline and the loss of that population's function in marine communities. Such a task requires a combined laboratory, mesocosm, and field approach that addresses: 1) the physical processes—specifically, flow characteristics—that influence the partitioning of hydrocarbons between sediments and interstitial waters; 2) the chemical processes that influence hydrocarbon persistence and degradation rates in sediments and interstitial waters; and 3) the long-term biological effects that alter population stability and function and the consequences of such effects on resource utilization. The first two aspects are important in establishing realistic exposure scenarios—both in time and space—and the third is important in linking sublethal stress indices with predictions of population and community effects.

Although a wide range of sublethal stress indices have been proposed for monitoring the responses of organisms to pollutants (McIntyre and Pearce, 1980), few have been linked to the survival potential of the individual or the reproductive potential of a population. Particularly sensitive responses that may show a relationship with population effects include biochemical responses that relate either to energy metabolism and membrane function (such as lysosome stability) or detoxication (such as induction of mixed-function oxygenase activity), and physiological responses (such as scope-for-growth or hormonal changes) that influence the energy available for growth and reproduction or other aspects of reproductive and developmental processes. These indices can be integrated by the functional relationship of metabolic function and energy turnover for growth and reproduction. No single index can provide the predictive capability to evaluate population changes, and future studies should be directed at defining the relationship of multiple responses. For example, laboratory studies directed at defining relationship between mixed-function oxygenase induction in marine animals and the long-term consequences in regards to reproduction and disease would greatly strengthen the use of mixed-function oxygenase induction as a monitoring tool.

In designing a monitoring program for determining long-term impact of hydrocarbon contamination resulting from offshore oil and gas development on marine ecosystems, efforts should be directed at:

1) determining the persistence and degradation rates of hydrocarbons, particularly the medium and higher molecular weight aromatic compounds, within the sediments in the vicinity of discharges from platforms or coastal processing facilities and the flux rates of these hydrocarbons between sediments, interstitial waters and biota;

2) relating hydrocarbon content in sediments, interstitial waters and biota to changes in benthic biomass, the structure of benthic communities and recruitment of benthic populations; and

3) using demersal fish and shellfish populations with limited or no migratory behavior as models, relate hydrocarbon content (both parent compounds and metabolites) of sediments and interstitial waters with hydrocarbon content of tissues, activity of detoxication enzymes, seasonal alterations in energy reserves, recruitment of juvenile stages, reproductive condition, and incidence of disease and histopathologic conditions. Such information should provide an analysis of potential long-term change at both the organismal and population levels before irreversible damage occurs at the community and ecosystem levels.

LITERATURE CITED

Aarset, A.V. and K.E. Zachariassen. 1982. Effects of oil pollution on the freezing tolerance and solute concentration of the blue mussel *Mytilus edulis*. Mar. Biol. 72: 45-51.

American Petroleum Institute. 1958. Determination of Volatile and Non-Volatile Oily Material. Infrared Spectrometric Method, No. 733-48. American Petroleum Institute, Washington, D.C.

Anderson, J.W. (ed.). 1975. Laboratory Studies on the Effects of Oil on Marine Organisms: An Overview. American Petroleum Institute Publication No. 4349, Washington, D.C.

Anderson, J.W. 1977. Responses to sublethal levels of petroleum hydrocarbons, Are they sensitive indicators and do they correlate with tissue contamination? Pages 95-114 *in* D.A. Wolfe (ed.), Fate and Effects of Petroleum Hydrocarbons in Marine Ecosystems and Organisms. Pergamon Press, New York.

Anderson, J.W. 1979. An assessment of knowledge concerning the fate and effects of petroleum hydrocarbons in the marine environment. Pages 3-21 *in* W.B. Vernberg, A. Calabrese, F.P. Thurberg and F.J. Vernberg (eds.), Marine Pollution: Functional Responses. Academic Press, New York.

Anderson, J.W., D.B. Dixit, G.S. Ward and R.S. Foster. 1977a. Effects of petroleum hydrocarbons on the rate of heart beat and hatching success of estuarine fish embryos. Pages 241-258 *in* F.J. Vernberg, A. Calabrese, F.P. Thurberg and W.B. Vernberg (eds.), Physiological Responses of Marine Biota to Pollutants. Academic Press, New York.

Anderson, J.W., S.L. Kiesser, R.M. Bean, R.G. Riley and B.L. Thomas. 1981. Toxicity of chemically dispersed oil to shrimp exposed to constant and decreasing concentrations in a flowing system. Pages 69-75 *in* Proceedings of 1981 Oil Spill Conference (Prevention, Behavior, Control, Cleanup). American Petroleum Institute, Washington, D.C.

Anderson, J.W., S.L. Kiesser and J.W. Blaylock. 1980. The cumulative effect of petroleum hydrocarbons on marine crustaceans during constant exposure. Rapp. P.-V. Réun. Cons. Int. Explor. Mer 179: 62-70.

Anderson, J.W., L.J. Moore, J.W. Blaylock, D.L. Woodruff and S.L. Kiesser. 1977b. Bioavailability of sediment-sorbed naphthalenes to the sipunculid worm, *Phascolosoma agassizii*. Pages 276-285 *in* D.A. Wolfe (ed.), Fate and Effects of Petroleum Hydrocarbons in Marine Ecosystems and Organisms. Pergamon Press, New York.

Anderson, J.W., J.M. Neff, B.A. Cox, H.E. Tatem and G.M. Hightower. 1974a. Characteristics of dispersions and water-soluble extracts of crude and refined oils and their toxicity to estuarine crustaceans and fish. Mar. Biol. 27: 75-88.

Anderson, J.W., J.M. Neff, B.A. Cox, H.E. Tatem and G.M. Hightower. 1974b. The effects of oil on estuarine animals: Toxicity, uptake and depuration, respiration. Pages 285-310 *in* F.J. Vernberg and W.B. Vernberg (eds.), Pollution and Physiology of Marine Organisms. Academic Press, New York.

Anderson, J.W., R.G. Riley, S.L. Kiesser, B.L. Thomas and G.W. Fellingham. 1983. Natural weathering of oil in marine sediments: Tissue contamination and growth of the littleneck clam *Protothaca staminea*. Can. J. Fish. Aquat. Sci. 40 (Suppl. No. 2): 70-77.

Anderson, R.S. 1978. Benzo(a)pyrene Metabolism in the American Oyster *Crassostrea virginica*. Ecological Research Series EPA-600/3-78-009. U.S. Environmental Protection Agency, Washington, D.C.

Atema, J. and L.S. Stein. 1974. Effects of crude oil on the feeding behavior of the lobster *Homarus americanus*. Environ. Pollut. 6: 77-86.

Atema, J., E.B. Karnofsky and S. Oleszko-Szuts. 1979. Lobster behavior and chemoreception: Sublethal effects of No. 2 fuel oil. Pages 122-134 *in* F.S. Jacoff (ed.), Advances in Marine Environmental Research. Environmental Research Laboratory,

Office of Research and Development, U.S. Environmental Protection Agency, Narragansett, Rhode Island.
Augenfeld, J.M. 1980. Effects of Prudhoe Bay crude oil contamination on sediment working rates of *Abarenicola pacifica*. Mar. Environ. Res. 3: 307-313.
Augenfeld, J.M., J.W. Anderson, S.L. Kiesser, G.W. Fellingham, R.G. Riley and B.L. Thomas. 1983. Exposure of *Abarenicola pacifica* to oiled sediment: Effects on glycogen content and alterations in sediment-bound hydrocarbons. Pages 443-449 *in* Proceedings of 1983 Oil Spill Conference (Prevention, Behavior, Control, Cleanup). American Petroleum Institute, Washington, D.C.
Augenfeld, J., J.W. Anderson, D.L. Woodruff and J.L. Webster. 1980. Effects of Prudhoe Bay crude oil-contaminated sediments on *Protothaca staminea* (Mollusca: Pelecypoda): Hydrocarbon content, condition index, free amino acid level. Mar. Environ. Res. 4: 135-143.
Augenfeld, J.M., R.G. Riley, B.L. Thomas and J.W. Anderson. 1982. The fate of polyaromatic hydrocarbons in an intertidal sediment exposure system: I. Bioavailability to *Macoma inquinata* (Mollusca: Pelecypoda) and *Abarenicola pacifica* (Annelida: Polychaeta). Mar. Environ. Res. 7: 31-50.
Baden, S. and L. Hagerman. 1981. Ventilatory responses of the shrimp *Palaemon adspersus* to sublethal concentrations of crude oil extract. Mar. Biol. 63: 129-133.
Baker, J.M. 1970. The effects of oil on plants. Environ. Pollut. 1: 27-44.
Barnett, J. and D. Toews. 1978. The effect of crude oil and the dispersant, Oilsperse 43, on respiration and coughing rates in Atlantic salmon (*Salmo salar*). Can. J. Zool. 56: 307-310.
Bean, R.M., J.W. Blaylock and R.G. Riley. 1980. Application of trace analytical techniques to a study of hydrocarbon composition upon dispersion of petroleum in a flowing seawater system. Pages 235-246 *in* L. Petrakis and F.T. Weiss (eds.), Petroleum in the Marine Environment. Advances in Chemistry Series, No. 185. American Chemical Society, Washington, D.C.
Bend, J.R. and M.O. James. 1978. Xenobiotic metabolism in marine and freshwater species. Pages 128-188 *in* D.C. Malins and J.R. Sargent (eds.), Biochemical and Biophysical Perspectives in Marine Biology, Volume 4. Academic Press, New York.
Benville, P.E., Jr. and S. Korn. 1974. A simple apparatus for metering volatile liquids into water. J. Fish. Res. Board Can. 31: 367-368.
Benville, P.E., Jr., T.G. Yocom, P. Nunes and J.M. O'Neill. 1981. Simple, continuous-flow systems for dissolving the water-soluble components of crude oil into seawater for acute or chronic exposure of marine organisms. Water Res. 5: 1197-1204.
Berdugo, V., R.P. Harris and S.C. O'Hara. 1977. The effect of petroleum hydrocarbons on reproduction of an estuarine planktonic copepod in laboratory cultures. Mar. Pollut. Bull. 8: 138-143.
Berge, J.A., K.I. Johannessen and L.-O. Reiersen. 1983. Effects of water soluble fraction of North Sea crude oil on the swimming activity of the sand goby *Pomatoschistus minutus* (Pallas). J. Exp. Mar. Biol. Ecol. 68: 159-167.
Berman, M.S. and D.R. Heinle. 1980. Modification of the feeding behavior of marine copepods by sublethal concentrations of water-accommodated fuel oil. Mar. Biol. 56: 59-64.
Beslier, A., J.L. Birrien, L. Cabioch, C. Larsonneur and J. LeBorgne. 1980. La pollution des Baies de Morlaix et de Lannion par les hydrocarbures de l'*Amoco Cadiz*: Repartition sur les fonds et evolution. Helgoländer Meeresunters. 33: 209-224.
Bieri, R.H. and V.C. Stamoudis. 1977. The fate of petroleum hydrocarbons from a No. 2 fuel oil spill in a seminatural estuarine environment. Pages 332-344 *in* D.A. Wolfe (ed.), Fate and Effects of Petroleum Hydrocarbons in Ecosystems and Organisms. Pergamon Press, New York.
Binder, R.L. and J.J. Stegeman. 1980. Induction of aryl hydrocarbon hydroxylase activity in embryos of an estuarine fish. Biochem. Pharmacol. 29: 949-951.
Blumer, M. 1969. Oil pollution of the ocean. Pages 5-13 *in* D.P. Hoult (ed.), Oil on the Sea. Plenum Press, New York.

Blundo, R. 1978. The toxic effects of the water-soluble fraction of No. 2 fuel oil and of three aromatic hydrocarbons on the behavior and survival of barnacle larvae. Contrib. Mar. Sci. 21: 35-37.

Boehm, P.D. and J.G. Quinn. 1977. The persistence of chronically accumulated hydrocarbons in the hard-shell clam *Mercenaria mercenaria*. Mar. Biol. 44: 227-233.

Boehm, P.D., J.E. Barak, D.L. Fiest and A.A. Elskus. 1982. A chemical investigation of the transport and fate of petroleum hydrocarbons in littoral and benthic environments, the *Tsesis* oil spill. Mar. Env. Res. 6: 157-188.

Boehm, P.D., D.L. Fiest and A. Elskus. 1981. Comparative weathering patterns of hydrocarbons from the *Amoco Cadiz* oil spill observed at a variety of coastal environments. Pages 159-173 *in* Proceedings of International Symposium on the *Amoco Cadiz*: Fates and Effects of the Oil Spill. CNEXO, Brest, France.

Boney, A.P. 1974. Aromatic hydrocarbons and the growth of marine algae. Mar. Pollut. Bull. 5: 185-186.

Boney, A.D. and E.D.S. Corner. 1959. Application of toxic agents in the study of ecological resistance of intertidal red algae. J. Mar. Biol. Ass., U.K. 38: 267-275.

Boney, A.D. and E.D.S. Corner. 1962. On the effects of some carcinogenic hydrocarbons on the growth of sporelings of marine red algae. J. Mar. Biol. Ass., U.K. 42: 579-585.

Bott, T.L., K. Rogenmuser and P. Thorne. 1976. Effect of No. 2 fuel oil, Nigerian crude oil, and used crankcase oil on the metabolism of benthic algal communities. Pages 373-393 *in* Sources, Effects and Sinks of Hydrocarbons in the Aquatic Environment. American Institute of Biological Sciences, Arlington, Virginia.

Boutry, L.-C., M. Bordes, A. Feurier, M. Barbier and A. Saliot. 1977. La diatomée marine *Chaetoceros simplex calcitrans* Paulsen et son environment. IV. Relations avec le milieu de culture: étude des hydrocarbures. J. Exp. Mar. Biol. Ecol. 28: 41-51.

Broderson, C.C., S.D. Rice, J.W. Short, T.A. Mecklenburg and J.F. Karinen. 1977. Sensitivity of larval and adult Alaskan shrimp and crabs to acute exposures of the water-soluble fraction of Cook Inlet crude oil. Pages 575-578 *in* Proceedings of 1977 Oil Spill Conference (Prevention, Behavior, Control, Cleanup). American Petroleum Institute, Washington, D.C.

Brooks, J.M., G.A. Fryxell, D.F. Reid and W.M. Sackett. 1977. Gulf underwater flare experiment (GUFEX): Effects of hydrocarbons on phytoplankton. Pages 45-75 *in* C.S. Giam (ed.), Pollutant Effects on Marine Organisms. Lexington Books, D.C. Heath, Lexington, Massachusetts.

Burns, K.A. 1976. Hydrocarbon metabolism in the intertidal fiddler crab *Uca pugnax*. Mar. Biol. 36: 5-11.

Burns, K.A. and J.M. Teal. 1971. Hydrocarbon Incorporation into the Salt Marsh Ecosystem from the West Falmouth Oil Spill. Woods Hole Oceanographic Institution, Ref. 71-69. Woods Hole, Massachusetts, 23 p.

Busdosh, M. 1981. Long-term effects of the water soluble fraction of Prudhoe Bay crude oil on survival, movement and food search success of the arctic amphipod *Boeckosimus* (*Onisimus*) *affinis*. Mar. Environ. Res. 5: 167-180.

Byrne, C.J. and J.A. Calder. 1977. Effect of the water-soluble fractions of crude, refined, and waste oils on the embryonic and larval stages of the Quahog clam *Mercenaria* sp. Mar. Biol. 40: 225-231.

Cabioch, L., J.C. Dauvin, J. Mora Bermudez and C. Rodriguez Babio. 1980. Effets de la marée noire de l'*Amoco Cadiz* sur le benthos sublittoral du nord de la Bretagne. Helgoländer Meeresunters. 33: 192-208.

Cairns, J. 1983. Are single species toxicity tests alone adequate for estimating environmental hazard? Hydrobiologia 100: 47-57.

Caldwell, R.S., E.M. Calderone and M.H. Mallon. 1977. Effects of seawater-soluble fraction of Cook Inlet crude oil and its major aromatic components on larval stages of the Dungeness crab, *Cancer magister* Dana. Pages 210-220 *in* D.A. Wolfe (ed.), Fate and Effects of Petroleum Hydrocarbons in Marine Ecosystems and Organisms. Pergamon Press, New York.

Cantelmo, A., L. Mantel, R. Lazell, F. Hospod, E. Flynn, S. Goldberg and M. Katz.

1982. The effects of benzene and dimethylnaphthalene on physiological processes in juveniles of the blue crab, *Callinectes sapidus*. Pages 349-389 *in* W.B. Vernberg, A. Calabrese, F.P. Thurberg and F.J. Vernberg (eds.), Physiological Mechanisms of Marine Pollutant Toxicity. Academic Press, New York.

Capuzzo, J.M. 1981a. Predicting pollution effects in the marine environment. Oceanus 24: 25-33.

Capuzzo, J.M. 1981b. Crude Oil Effects to Developmental Stages of the American Lobster. Woods Hole Oceanographic Institution, Ref. 81-75. Woods Hole, Massachusetts, 102 p.

Capuzzo, J.M. and B.A. Lancaster. 1981. Physiological effects of South Louisiana crude oil on larvae of the American lobster (*Homarus americanus*). Pages 405-423 *in* F.J. Vernberg, A. Calabrese, F.P. Thurberg and W. B. Vernberg (eds.), Biological Monitoring of Marine Pollutants. Academic Press, New York.

Capuzzo, J.M. and B.A. Lancaster. 1982. Physiological effects of petroleum hydrocarbons on larval lobsters: hydrocarbon accumulation and interference with lipid metabolism. Pages 477-501, *In* W.B. Vernberg, A. Calabrese, F.P. Thurberg and F.J. Vernberg (eds.), Physiological Mechanisms of Marine Pollutant Toxicity. Academic Press, New York.

Capuzzo, J.M., B.A. Lancaster and G. Sasaki. 1984. The effects of petroleum hydrocarbons on lipid metabolism and energetics of larval development and metamorphosis in the American lobster (*Homarus americanus)*. Mar. Environ. Res. 14: 201-228.

Ceas, M.P. 1974. Effects of 3-4-benzopyrene on sea urchin egg development. Acta Embryologiae Experimentalis 3: 267-272.

Cerniglia, C.E. and D.T. Gibson. 1978. Metabolism of naphthalene by cell extracts of *Cunninghamella elegans*. Arch. Biochem. Biophys. 186: 121-127.

Cerniglia, C.E., D.T. Gibson and C. Van Baalen. 1979. Algal oxidation of aromatic hydrocarbons; formation of 1-naphthol from naphthalene by *Agmenellum quadruplicatum*, strain PR-6. Biochem. Biophys. Res. Commun. 88: 50-58.

Cerniglia, C.E., D.T. Gibson and C. Van Baalen. 1980. Oxidation of naphthalene by cyanobacteria and microalgae. J. Gen. Microbiol. 116: 495-500.

Clark, R.C. and J.S. Finley. 1974a. Acute effects of outboard effluent on two marine shellfish. Environ. Sci. Technol. 8: 1009-1014.

Clark, R.C. and J.S. Finley. 1974b. Tidal aquarium for laboratory studies of environmental effects on marine organisms. Prog. Fish Cultur. 36: 134-137.

Clark, R.C., J.S. Finley, B.G. Patten and E.E. DeNike. 1975. Long-term chemical and biological effects of a persistent oil spill following the grounding of the *General M. C. Meigs*. Pages 479-487 *in* Proceedings of 1975 Oil Spill Conference (Prevention, Behavior, Control, Cleanup). American Petroleum Institute, Washington, D.C.

Clark, R.C., J.S. Finley, B.G. Patten, D.F. Stefoni and E.E. DeNike. 1973. Interagency investigations of a persistent oil spill on the Washington coast. Pages 793-808 *in* Proceedings of 1973 Oil Spill Conference (Prevention and Control of Oil Spills). American Petroleum Institute, Washington, D.C.

Clarke, A. 1979. Living in cold water: K-strategies in antarctic benthos. Mar. Biol. 55: 111-119.

Clement, L.E., M.S. Stekoll and D.G. Shaw. 1980. Accumulation, fractionation and release of oil by the intertidal clam *Macoma balthica*. Mar. Biol. 57: 41-50.

Clendenning, K.A. and W.J. North. 1959. Effects of wastes on the giant kelp, *Macrocystis pyrifera*. Pages 82-91 *in* E.A. Pearson (ed.), Proceedings of the First International Conference on Waste Disposal in the Marine Environment. Pergamon Press, New York.

Collier, T.K., L.C. Thomas and D.C. Malins. 1978. Influence of environmental temperature on disposition of dietary naphthalene in coho salmon (*Oncorhynchus kisutch*) isolation and identification of individual metabolites. Comp. Biochem. Physiol. 61C: 23-28.

Conan, G. 1982. The long-term effects of the *Amoco Cadiz* oil spill. Phil. Trans. R. Soc.

London B 297: 323-333. Reprinted in R.B. Clark (ed.). 1982. The Long-Term Effects of Oil Pollution in Marine Populations, Communities and Ecosystems. The Royal Society, London, 259 p.

Conner, J.W. and S.C. Singer. 1981. Purification scheme for cytochrome P-450 of the blue crab, *Callinectes sapidus* Rathbun. Aquatic Toxicol. 1: 271-278.

Conover, R.J. 1971. Some relations between zooplankton and bunker C oil on Chedabucto Bay following the wreck of the tanker *Arrow*. J. Fish. Res. Board Can. 28: 1327-1330.

Corner, E.D.S. 1978. Pollution studies with marine plankton. Part 1. Petroleum hydrocarbons and related compounds. Adv. Mar. Biol. 15: 289-380.

Corner, E.D.S., R.P. Harris, C.C. Kilvington and S.C.M. O'Hara. 1976a. Petroleum compounds in the marine food web: short-term experiments on the fate of naphthalene in *Calanus*. J. Mar. Biol. Ass., U.K. 56: 121-133.

Corner, E.D.S., R.P. Harris, K.J. Whittle and P.R. Mackie. 1976b. Hydrocarbons in marine zooplankton and fish. Pages 71-106 *in* A.P.M. Lockwood (ed.), Effects of Pollutants on Aquatic Organisms. Cambridge University Press, Cambridge, England.

Corner, E.D.S., C.C. Kilvington and S.C.M. O'Hara. 1973. Qualitative studies on the metabolism of naphthalene in *Maia squinado* (Herbst). J. Mar. Biol. Ass., U.K. 53: 819-832.

Cowles, T.J. 1983. Effects of exposure to sublethal concentrations of crude oil on the copepod *Centropages hamatus*. II. Activity patterns. Mar. Biol. 78: 53-57.

Cowles, T.J. and J.F. Remillard. 1983. Effects of exposure to sublethal concentrations of crude oil on the copepod *Centropages hamatus*. I. Feeding and egg production. Mar. Biol. 78: 45-51.

Cox, B.A., J.W. Anderson and J.C. Parker. 1975. An experimental oil spill: the distribution of aromatic hydrocarbons in the water, sediment, and animal tissues within a shrimp pond. Pages 607-612 *in* Proceedings of 1975 Oil Spill Conference (Prevention, Behavior, Control, Cleanup). American Petroleum Institute, Washington, D.C.

Cucci, T.L. and C.E. Epifanio. 1979. Long-term effects of water-soluble fractions of Kuwait oil on the larval and juvenile development of the mud crab *Eurypanopeus depressus*. Mar. Biol. 55: 215-220.

Davavin, I.A., O.G. Mironov and I.M. Tsimbal. 1975. Influence of oil on nucleic acids of algae. Mar. Pollut. Bull. 6: 13-14.

Davies, J.M., I.C. Baird, L.C. Massie, S.J. Hay and A.P. Ward. 1980. Some effects of oil-derived hydrocarbons in an enclosed ecosystem and a consideration of their implications for monitoring. Rapp. P.-V. Réun. Cons. Int. Explor. Mer. 179: 201-211.

Dean-Raymond, D. and R. Bartha. 1975. Biodegradation of some polynuclear aromatic petroleum components by marine bacteria. Dev. Ind. Microbiol. 16: 97-110.

Delaune, R.D., G.A. Hambrick, III and W.H. Patrick, Jr. 1980. Degradation of hydrocarbons in oxidized and reduced sediments. Mar. Pollut. Bull. 11: 103-106.

Derenbach, J.B. and M.V. Gerek. 1980. Interference of petroleum hydrocarbons with the sex pheromone reaction of *Fucus vesiculosus* (L.). J. Exp. Mar. Biol. Ecol. 44: 61-65.

Derenbach, J.B., W. Boland, E. Folster and D.G. Muller. 1980. Interference tests with the pheromone system of the brown alga *Cutleria multifida*. Mar. Ecol. Prog. Ser. 3: 357-361.

Dillon, T.M. 1981. Effects of dimethylnaphthalene and fluctuating temperatures on estuarine shrimp. Pages 79-85 *in* Proceedings of 1981 Oil Spill Conference (Prevention, Behavior, Control, Cleanup). American Petroleum Institute, Washington, D.C.

Dillon, T.M. 1982. Dietary accumulation of dimethylnaphthalene by the grass shrimp *Palaemonetes pugio* under stable and fluctuating temperatures. Bull. Environ. Contam. Toxicol. 28: 149-153.

Dillon, T.M. 1983. Oxygen consumption in the shrimp, *Palaemonetes pugio*, exposed to fluctuating temperatures and food contaminated with the diaromatic petroleum hydrocarbon, dimethylnaphthalene. Estuar. Coast. Shelf Sci. 16: 403-413.

DiMichele, L. and M.H. Taylor. 1978. Histopathological and physiological responses of

Fundulus heteroclitus L. to naphthalene exposure. J. Fish. Res. Board Can. 35: 1060-1066.

DiSalvo, L.H., H.E. Guard and L. Hunter. 1975. Tissue hydrocarbon burden of mussels as a potential monitor of environmental hydrocarbon insult. Environ. Sci. Technol. 9: 247-251.

Dixit, D. and J.W. Anderson. 1977. Distribution of naphthalenes within exposed *Fundulus similis* and correlations with stress behavior. Pages 633-636 *in* Proceedings of 1977 Oil Spill Conference (Prevention, Behavior, Control, Cleanup). American Petroleum Institute, Washington, D.C.

Donahue, W.H., R.T. Wang, M. Welch and J.A.C. Nicol. 1977. Effects of water-soluble components of petroleum oils and aromatic hydrocarbons on barnacle larvae. Environ. Pollut. 13: 187-202

Drewa, G., Z. Zbytniewski and F. Pautsch. 1977. The effect of detergent "Solo" and crude oil on the activities of Cathepsin D and acid phosphatase in hemolymph of *Crangon crangon* L. Polish Arch. Hydrobiol. 24: 279-284.

Dunbar, M.J. 1968. Ecological Development of Polar Regions. Prentice-Hall, Englewood Cliffs, New Jersey, 199 p.

Dunn, B.P. and H.F. Stich. 1976. Release of the carcinogen benzo(a)pyrene from environmentally contaminated mussels. Bull. Environ. Contam. Toxicol. 1: 398-401.

Dunstan, W.M., L.P. Atkinson and J. Natoli. 1975. Stimulation and inhibition of phytoplankton growth by low molecular weight hydrocarbons. Mar. Biol. 31: 305-310.

Duppel, W., J.M. Lebeault, and M.J. Coon. 1973. Properties of yeast cytochrome P-450-containing enzyme system which catalyzes the hydroxylation of fatty acids, alkanes and drugs. Eur. J. Biochem. 36: 583-592.

Duval, W.S. 1979. The sub-lethal effects of hydrocarbons on the bioenergetics and productivity of selected marine fauna. Pages 83-86 *in* Proceedings of Arctic Marine Oil Spill Program Technical Seminar, Edmonton, Alberta. Fisheries and Environment, Canada Environmental Research Branch, Ottawa.

Edwards, R.R.C. 1978. Effects of water-soluble oil fractions on metabolism, growth and carbon budget of the shrimp *Crangon crangon*. Mar. Biol. 46: 259-265.

Eisler, R. 1975. Toxic, sublethal, and latent effects of petroleum on Red Sea macrofauna. Pages 535-540 *in* Proceedings of 1975 Oil Spill Conference (Prevention, Behavior, Control, Cleanup). American Petroleum Institute, Washington, D.C.

Eldridge, M.B., T. Echeverra and J.A Whipple. 1977. Energetics of Pacific herring (*Clupea harengus pallasi*) embryos and larvae exposed to low concentrations of benzene, a monoaromatic component of crude oil. Trans. Am. Fish. Soc. 106: 452-461.

Elmamlouk, T.H., T. Gessner and A.C. Brownie. 1974. Occurrence of cytochrome P-450 in hepatopancreas of *Homarus americanus*. Comp. Biochem. Physiol. 48B: 419-425.

Elmgren, R. and J.B. Frithsen. 1982. The use of experimental ecosystems for evaluating the environmental impact of pollutants: A comparison of an oil spill in the Baltic Sea and two long-term, low-level oil addition experiments in mesocosms. Pages 153-165 *in* G.D. Grice and M.R. Reeve (eds.), Marine Mesocosms: Biological and Chemical Research in Experimental Ecosystems. Springer-Verlag, New York.

Elmgren, R., J.F. Grassle, J.P. Grassle, D.R. Heinle, G. Langlois, S.L. Vargo and G.A. Vargo. 1980. Trophic interactions in experimental marine ecosystems perturbed by oil. Pages 779-800 *in* J.P. Geisy (ed.), Microcosms in Ecological Research. Department of Energy Symposium Series, Conf. 781101. Augusta, Georgia.

Elmgren, R., S. Hanson, U. Larsson and B. Sundelin. 1981. The *Tsesis* oil spill: acute and long-term impact on the benthic ecosystem. I.C.E.S. Contribution C.M. 1981/E, 24 p.

Elmgren, R., S. Hanson, U. Larsson, B. Sundelin and P.D. Boehm. 1983. The *Tsesis* oil spill: Acute and long-term impact on the benthos. Mar. Biol. 73: 51-65.

Engelhardt, F.R., M.P. Wong and M.E. Duey. 1981. Hydromineral balance and gill morphology in rainbow trout *Salmo gairdneri*, acclimated to fresh and sea water as affected by petroleum exposure. Aquat. Toxicol. 1: 175-186.

Ernst, V.V., J.M. Neff and J.W. Anderson. 1977. The effects of water-soluble fraction of No. 2 fuel oil on the early development of the estuarine fish, *Fundulus grandis* Baird and Girard. Environ. Pollut. 14: 25-35.

Farrington, J.W., A.C. Davis, N.M. Frew and K.S. Rabin. 1982. No. 2 fuel oil compounds in *Mytilus edulis*: Retention and release after an oil spill. Mar. Biol. 66: 15-26.

Farrington, J.W., E.D. Goldberg, R.W. Risebrough, J.H. Martin and V.T. Bowen. 1983. U.S. "Mussel Watch" 1976-1978: An overview of the trace-metal, DDE, PCB, hydrocarbon, and artificial radionuclide data. Environ. Sci. Technol. 17: 490-496.

Federle, T., W. Vestal, J. Robie, G.R. Hater and M.C. Miller. 1979. Effects of Prudhoe Bay crude on primary production and zooplankton in Arctic tundra thaw ponds. Mar. Environ. Res. 2: 2-18.

Ferris, J.P., M.J. Fasco, F.L. Stylianopoulou, D.M. Jerina, J.W. Daly and A.M. Jeffrey. 1973. Monooxygenase activity in *Cunninghamella bainerii*. Evidence for a fungal system similar to liver microsomes. Arch. Biochem. Biophys. 156: 97.103.

Ferris, J.P., L.H. MacDonald, M.A. Patrie and M.A. Martin. 1976. Aryl hydrocarbon hydroxylase activity in the fungus *Cunninghamella bainerii*. Evidence of the presence of cytochrome P-450. Arch. Biochem. Biophys. 175: 443-452.

Fink, R.P. and W.S. Duval. 1980. The sublethal and lethal effects of the water soluble fraction of Prudhoe Bay crude oil on juvenile coho salmon (*Oncorhynchus kisutch*). Pages 158-181 *in* Proceedings of Arctic Marine Oil Spill Program Technical Seminar. Environment Canada, Edmonton, Alberta.

Fletcher, G.L., J.W. Kiceniuk, and U.P. Williams. 1981. Effects of oiled sediments on mortality, feeding, and growth of winter flounder *Pseudopleuronectes americanus*. Mar. Ecol. Prog. Ser. 4: 91-96.

Fletcher, G.L., M.J. King, J.W. Kiceniuk and R.F. Addison. 1982. Liver hypertrophy in winter flounder following exposure to experimentally oiled sediments. Comp. Biochem. Physiol. 73C: 457-462.

Folmar, L.C., W.W. Dickhoff, W.S. Zaugg and H.O. Hodgins. 1982. The effects of Aroclor 1254 and No. 2 fuel oil on smoltification and sea-water adaptation of coho salmon (*Oncorhynchus kisutch*). Aquat. Toxicol. 2: 291-299.

Fontaine, M., J.C. Lacaze, X. LePemp and O. Villedon DeNaide. 1975. Des interactions entres pollutions thermiques et pollutions par hydrocarbures. Pages 115-122 *in* Les Journees et sur les Pollutions Marines. Commission Internationale pour l'Exploration Scientifique de la Mer Mediterranée, Monaco.

Forns, J.M. 1977. The effects of crude oil on larvae of lobster *Homarus americanus*. Pages 569-573 *in* Proceedings of 1977 Oil Spill Conference (Prevention, Behavior, Control, Cleanup). American Petroleum Institute, Washington, D.C.

Fossato, V.U. and W.J. Canzonier. 1976. Hydrocarbon uptake and loss by the mussel *Mytilus edulis*. Mar. Biol. 36: 243-250.

Freegarde, M., C.G. Hatchard and C.A. Parker. 1971. Oil at sea: its identification, determination and ultimate fate. Laboratory Practice 20: 35-40.

Fries, C.R. and M.R. Tripp. 1977. Cytological damage in *Mercenaria mercenaria* exposed to phenol. Pages 174-181 *in* D.A. Wolfe (ed.), Fate and Effects of Petroleum Hydrocarbons in Marine Ecosystems and Organisms. Pergamon Press, New York.

Friha, M. and G. Conan. 1981. Long-term impact of hydrocarbon pollution from the *Amoco Cadiz* on the mortality of plaice (*Pleuronectes platessa*) in the Aber Benoit estuary. I.C.E.S. Contribution C.M. 1981/E, 55 p.

Fucik, K.W., H.W. Armstrong and J.M. Neff. 1977. The uptake of naphthalenes by the clam, *Rangia cuneata*, in the vicinity of an oil separator platform in Trinity Bay, Texas. Pages 637-640 *in* Proceedings of 1977 Oil Spill Conference (Prevention, Behavior, Control, Cleanup). American Petroleum Institute, Washington, D.C.

Galloway, W.B., J.L. Lake, D.K. Phelps, P.F. Rogerson, V.T. Bowen, J.W. Farrington, E.D. Goldberg, J.L. Laseter, G.C. Lawler, J.H. Martin and R.W. Risebrough. 1983. The Mussel Watch: Intercomparison of trace level constituent determinations. Environ. Toxicol. Chem. 2: 395-410.

Gardner, G.R. 1975. Chemically induced lesions in estuarine or marine teleosts. Pages

657-693 *in* W.E. Ribelin and G. Migaki (eds.), The Pathology of Fishes. University of Wiscosin Press, Madison, Wisconsin.

Gearing, J.N. and P.J. Gearing. 1983. Suspended load and solubility affect sedimentation of petroleum hydrocarbons in controlled estuarine ecosystems. Can. J. Fish. Aquat. Sci. 40 (Suppl. No. 2): 54-62.

Gearing, P.J. and J.N. Gearing. 1982a. Behavior of No. 2 fuel oil in the water column of controlled ecosystems. Mar. Env. Res. 6: 115-132.

Gearing, P.J. and J.N. Gearing. 1982b. Transport of No. 2 fuel oil between water column, surface microlayer, and atmosphere in controlled ecosystems. Mar. Environ. Res. 6: 133-144.

Gearing, P.J., J.N. Gearing, R.J. Pruell, T.L. Wade and J.G. Quinn. 1980. Partitioning of No. 2 fuel oil in controlled estuarine ecosystems: Sediments and suspended particulate matter. Environ. Sci. Technol. 14: 1129-1136.

Gibson, D.T. 1976. Microbial degradation of carcinogenic hydrocarbons and related compounds. Pages 224-238 *in* Sources, Effects and Sinks of Hydrocarbons in the Aquatic Environment. American Institute of Biological Sciences, Washington, D.C.

Gibson, D.T. 1977. Biodegradation of aromatic petroleum hydrocarbons. Pages 36-46 *in* D.A. Wolfe (ed.), Fate and Effects of Petroleum Hydrocarbons in Marine Ecosystems and Organisms. Pergamon Press, New York.

Gibson, D.T., V. Mahdevan, D.M. Jerina, H. Yagi and H.J.C. Yeh. 1975. Oxidation of the carcinogens benzo(a)pyrene and benzo(a)anthracene to dihydrodiols by a bacterium. Science 189: 295-297.

Giere, O. and D. Hauschildt. 1979. Experimental studies on the life cycle and production of the littoral oligochaete *Lumbricillus lineatus* and its response to oil pollution. Pages 113-122 *in* E. Naylor and R.G. Hartnoll (eds.), Cyclic Phenomena in Marine Plants and Animals. Pergamon Press, Oxford, England.

Giles, R.C., L.R. Brown and C.D. Minchew. 1978. Bacteriological aspects of fin erosion in mullet exposed to crude oil. J. Fish Biol. 13: 113-117.

Gilfillan, E.S. 1980. The use of scope-for-growth measurements in monitoring petroleum pollution. Rapp. P.-V. Réun. Cons. Int. Explor. Mer 179: 71-75.

Gilfillan, E.S. and J.H. Vandermeulen. 1978. Alterations in growth and physiology of soft shell clams, *Mya arenaria*: chronically oiled with Bunker C from Chedabucto Bay, Nova Scotia, 1970-76. J. Fish. Res. Board Can. 35: 630-636.

Gilfillan, E.S., D. Mayo, S. Hanson, D. Donovan and L.C. Jiang. 1976. Reduction in carbon flux in *Mya arenaria* caused by a spill of No.6 fuel oil. Mar. Biol. 37: 115-123.

Gilfillan, E.S., D.W. Mayo, D.S. Page, D. Donovan and S. Hanson. 1977. Effects of varying concentrations of petroleum hydrocarbons in sediments on carbon flux in *Mya arenaria*. Pages 299-314 *in* F.J. Vernberg, A. Calabrese, F.P. Thurberg and W.B. Vernberg (eds.), Physiological Responses of Marine Biota to Pollutants. Academic Press, New York.

Glémarec, M. and E. Hussenot. 1981. Définition d'une succession écologique en milieu meuble anormalement enrichie en matieres organiques a la suite de la catastrophe de l'*Amoco Cadiz*. Pages 499-512 *in* Proceedings of International Symposium on the *Amoco Cadiz*, Fate and Effects of the Oil Spill. CNEXO, Brest, France.

Gordon, D.C. and N.J. Prouse. 1973. The effects of three oils on marine phytoplankton photosynthesis. Mar. Biol. 22: 329-333.

Gordon, D.C., J. Dade and P.D. Keizer. 1978. Importance of sediment working by the deposit feeding polychaete, *Arenicola marina*, on the weathering rate of sediment-bound oil. J. Fish. Res. Board Can. 35: 591-603.

Grahl-Nielsen, O., J.T. Staveland and S. Wilhelmsen. 1978. Aromatic hydrocarbons in benthic organisms from coastal areas polluted by Iranian crude oil. J. Fish. Res. Bd. Canada 35: 615-623.

Grassle, J.F., R. Elmgren and J.P. Grassle. 1981. Response of benthic communities in MERL experimental ecosystems to low level, chronic additions of No. 2 fuel oil. Mar. Environ. Res. 4: 279-297.

Gruenfeld, M. and R. Frederick. 1977. The ultrasonic dispersion, source identification, and quantitative analysis of petroleum oils in water. Rapp. P.-V. Réun. Cons. Int. Explor. Mer. 171: 33-38.

Haensly, W.E., J.M. Neff, J.R. Sharp, A.C. Morris, M.F. Bedgood and P.D. Boehm. 1982. Histopathology of *Pleuronectes platessa* L. from Aber Wrac'h and Aber Benoit, Brittany, France: Long-term effects of the *Amoco Cadiz* crude oil spill. J. Fish Diseases 5: 365-391.

Hardy, R., P.R. Mackie, K.J. Whittle and A.D. McIntyre. 1974. Discrimination in the assimilation of n-alkanes in fish. Nature 252: 577-578.

Harris, R.P., V. Berdugo, E.D.S. Corner, C.C. Kilvington and S.C.M. O'Hara. 1977a. Factors affecting the retention of a petroleum hydrocarbon by marine planktonic copepods. Pages 286-304 *in* D.A. Wolfe (ed.), Fate and Effects of Petroleum Hydrocarbons in Marine Ecosystems and Organisms. Pergamon Press, New York.

Harris, R.P., V. Berdugo, S.C.M. O'Hara and E.D.S. Corner. 1977b. Accumulation of ^{14}C-1-naphthalene by an oceanic and an estuarine copepod during long-term exposure to low-level concentrations. Mar. Biol. 42: 187-195.

Hauschildt-Lillge, D. 1982. Long-term effects of petroleum hydrocarbons on the life cycle and productivity of the littoral oligochaete *Lumbricillus lineatus*. Neth. J. Sea Res. 16: 502-510.

Hawkes, J.M. and C.M. Stehr. 1982. Cytopathology of the brain and retina of embryonic surf smelt (*Hypomesus pretiosus*) exposed to crude oil. Environ. Res. 27: 164-178.

Hellstrom, T. and K.B. Doving. 1983. Perception of diesel oil by cod (*Gadus morhua*). Aquat. Toxicol. 4: 303-315.

Hinga, K.R., M.E.Q. Pilson, R.F. Lee, J.W. Farrington, K.Tjessem and A.C. Davis. 1980. Biogeochemistry of benzanthracene in an enclosed marine ecosystem. Environ. Sci. Technol. 14: 1136-1143.

Holland, D.L. 1978. Lipid reserves and energy metabolism in the larvae of benthic marine invertebrates. Pages 85-123 *in* D.C. Malins and J.R. Sargent (eds.), Biochemical and Biophysical Perspectives in Marine Biology, Volume 4. Academic Press, New York.

Hollister, T.A., G.S. Ward and P.R. Parrish. 1980. Acute toxicity of a No. 6 fuel oil to marine organisms. Bull. Environ. Contam. Toxicol. 24: 656-661.

Hooftman, R.N. and G.J. Vink. 1981. Cytogenetic effects on the eastern mud minnow, *Umbra pygmaea*, exposed to ethyl methane-sulphonate, benzo(a)pyrene and river water. Ecotoxicol. Environ. Safety 5: 261-269.

Hsiao, S.I.C. 1978. Effects of crude oils on the growth of Arctic marine phytoplankton. Environ. Pollut. 17: 93-107.

Hsiao, S.I.C., D. Kittle and M.G. Foy. 1978. Effects of crude oils and the oil dispersant Corexit on primary production of Arctic marine phytoplankton and seaweed. Environ. Pollut. 15: 209-221.

Hutchinson, T.C., J.A. Hellebust, D. Mackay, D. Tamm and P.B. Kauss. 1979. Relation of hydrocarbon solubility to toxicity and cellular membrane effects. Pages 541-547 *in* Proceedings of 1979 Oil Spill Conference (Prevention, Behavior, Control, Cleanup). American Petroleum Institute, Washington, D.C.

Hyland, J.L., P.F. Rogerson and G.R. Gardner. 1977. A continuous flow bioassay system for the exposure of marine organisms to oil. Pages 547-550 *in* Proceedings of 1977 Oil Spill Conference (Prevention, Behavior, Control, Cleanup). American Petroleum Institute, Washington, D.C.

Jackson, L., T. Bidleman, and W. Vernberg. 1981. Influence of reproductive activity on toxicity of petroleum hydrocarbons to ghost crabs. Mar. Poll. Bull. 12: 63-65.

Jacobson, S.M. and D.B. Boylan. 1973. Effect of seawater soluble fraction of kerosene on chemotaxis in a marine snail, *Nassarius obsoletus*. Nature 241: 213-215.

James, M.O. and J.R. Bend. 1980. Polycyclic aromatic hydrocarbon induction of cytochrome P-450 independent mixed-function oxidases in marine fish. Toxicol. Appl. Pharmacol. 54: 113-117.

James, M.O., E.R. Bowen, P.M. Dansette and J.R. Bend. 1979a. Epoxide hydrase and

glutathione S-transferase activities with selected alkene and arene oxides in several marine species. Chem. Biol. Interact. 25: 321-344.
James, M.O., M.A.Q. Khan and J.R. Bend. 1979b. Hepatic microsomal mixed function oxidase activities in several species common to coastal Florida. Comp. Biochem. Physiol. 62C: 155-164.
Jerina, D.M. 1983. Metabolism of aromatic hydrocarbons by the cytochrome P-450 system and epoxide hydrolase. Drug Metabolism and Disposition 11: 1-4.
Jerina, D.M., J.W Daly, A.M. Jeffrey and D.T. Gibson. 1971. *Cis*-1, 2-dihydroxy-1, 2-dihydronaphthalene: A bacterial metabolite from naphthalene. Arch. Biochem. Biophys. 142: 394-396
Johns, D.M. and J.A. Pechenik. 1980. Influence of the water accommodated fraction of No. 2 fuel oil on energetics of *Cancer irroratus* larvae. Mar. Biol. 55: 247-254.
Jordan, R.E. and J.R. Payne. 1980. Fate and Weathering of Petroleum Spills in the Marine Environment. A Literature Review and Synopsis. Ann Arbor Science, Ann Arbor, Michigan, 174 p.
Kalke, R.D., T.A. Duke, and R.W. Flint. 1982. Weathered IXTOC I oil effects on estuarine benthos. Estuar. Coast. Shelf Sci. 15: 75-84.
Karydis, M. 1979. Short term effects of hydrocarbons on the photosynthesis and respiration of some phytoplankton species. Bot. Mar. 22: 281-285.
Katz, L.M. 1973. The effects of water-soluble fractions of crude oil on larvae of the decapod crustacean *Neopanope texana* (Say). Environ. Pollut. 5: 199-204.
Kauss, P.B. and T.C. Hutchinson. 1975. Studies on the susceptibility of *Ankistrodesmus* species to crude oil components. Verh. Internat. Verein Limnol. 19: 2155-2164.
Kittredge, J.S., F.T. Takahashi and F.O. Sarinana. 1974. Bioassays indicative of some sublethal effects of oil pollution. Pages 891-897 *in* Proceedings of Marine Bioassays Workshop. Marine Technology Society, Washington, D.C.
Klotz, A.V. 1983. Purification and Characterization of the Hepatic Microsomal Monooxygenase System from the Coastal Marine Fish *Stenotomus chrysops*. Ph.D. Thesis, Massachusetts Institute of Technology/Woods Hole Oceanographic Institution, WHOI-83-43, Woods Hole, Massachusetts 270 p.
Klotz, A.V., J.J. Stegeman and C. Walsh. 1983. An aryl hydrocarbon hydroxylating hepatic cytochrome P-450 from the marine fish *Stenotomus chrysops*. Arch. Biochem. Biophys. 226: 578-592.
Korn, S., D.A. Moles and S.D. Rice. 1979. Effects of temperature on the median lethal limit of pink salmon and shrimp exposed to toluene, naphthalene, and Cook Inlet crude oil. Bull. Environ. Contam. Toxicol. 21: 521-525.
Koster, A.S. and J.A.M. Van den Beggelaar. 1980. Abnormal development of *Dentalium* due to the *Amoco Cadiz* oil spill. Mar. Pollut. Bull. 11: 166-169.
Krebs, C.T. and K.A. Burns. 1977. Long term effects of an oil spill on populations of the salt-marsh crab *Uca pugnax*. Science 197: 484-487.
Krugel, S., D. Jenkins and S.A. Klein. 1978. Apparatus for the continuous dissolution of poorly water-soluble compounds for bioassays. Water Res. 12: 269-272.
Kuhnhold, W.W. 1974. Investigations on the toxicity of seawater extracts of three crude oils on eggs of cod (*Gadus morhua* L.). Ber. dt. wiss. Kommn. Meeresforsch. 23: 165-180.
Kuhnhold, W.W. 1978. Effects of the water soluble fraction of a Venezuelan heavy fuel oil (No. 6) on cod eggs and larvae. Pages 126-130 *in* The Wake of the *Argo Merchant*, Proceedings of Symposium, January 1978. Center for Ocean Management Studies, University of Rhode Island, Kingston.
Kurelec, B., Z. Matijasevic, M. Rijavic, M. Alacevic, S. Britvic, W.E.G. Muller and R.K. John. 1979. Induction of benzo(a)pyrene monoxygenase in fish and the *Salmonella* test as a tool for detecting mutagenic/carcinogenic xenobiotics in the aquatic environment. Bull. Environ. Contam. Toxicol. 21: 799-807.
Kusk, K.O. 1978. Effects of crude oil and aromatic hydrocarbons on the photosynthesis of the diatom *Nitzschia pilea*. Physiol. Plant 43: 1-6.

Lacaze, J.C. 1974. Ecotoxicology of crude oils and the use of experimental marine ecosystems. Mar. Poll. Bull. 5: 153-156.

Lanier, J.J. and M. Light. 1978. Ciliates as bioindicators of oil pollution. Pages 651-676 *in* Proceedings of Conference on Assessment of Ecological Impacts of Oil Spills. National Technical Information Service No. AD-A072 859. American Institute of Biological Sciences, Washington, D.C.

Laughlin, R.B., Jr. and J.M. Neff. 1979. Interactive effects of salinity, temperature and polycyclic aromatic hydrocarbons on the survival and development rate of larvae of the mud crab *Rhithropanopeus harrisii*. Mar. Biol. 53: 281-291.

Laughlin, R.B., Jr. and J.M. Neff. 1980. Influence of temperature, salinity and phenanthrene (a petroleum derived polycyclic aromatic hydrocarbon) on the respiration of larval mud crabs, *Rhithropanopeus harrisii*. Estuar. Coast. Mar. Sci. 10: 655-669.

Laughlin, R.B., Jr., O. Linden and J.M. Neff. 1979. A study on the effects of salinity and temperature on the disappearance of aromatic hydrocarbons from the water-soluble fraction of No. 2 fuel oil. Chemosphere 10: 741-749.

Laughlin, R.B., Jr., L.G.L. Young and J.M. Neff. 1978. A long-term study of the effect of water-soluble fractions of No. 2 fuel oil on the survival, development rate, and growth of the mud crab *Rhithropanopeus harrisii*. Mar. Biol. 47: 87-95.

Lee, R.F. 1975. Fate of petroleum hydrocarbons in marine zooplankton. Pages 549-553 *in* Proceedings of 1975 Oil Spill Conference (Prevention, Behavior, Control, Cleanup). American Petroleum Institute, Washington, D.C.

Lee, R.F. 1977. Accumulation and turnover of petroleum hydrocarbons in marine organisms. Pages 60-70 *in* D.A. Wolfe (ed.), Fate and Effects of Petroleum Hydrocarbons in Marine Ecosystems and Organisms. Pergamon Press, New York, 446 p.

Lee, R.F. 1981. Mixed function oxygenases (MFO) in marine invertebrates. Mar. Biol. Letts. 2: 87-105.

Lee, R.F. and S.C. Singer. 1980. Detoxifying enzyme system in marine polychaetes: increases in activity after exposure to aromatic hydrocarbons. Rapp. P.-V. Réun. Cons. Int. Explor. Mer. 179: 29-32.

Lee, R.F. and M. Takahashi. 1977. The fate and effect of petroleum in controlled ecosystem enclosures. Rapp. P.-V. Réun. Cons. Int. Explor. Mer. 171: 150-156.

Lee, R.F., J.W. Corner, D. Page, L.E. Ray and C.S. Giam. 1982a. Cytochrome P-450 dependent mixed function oxygenase systems in marsh crabs. Pages 145-159 *in* W.B. Vernberg, A. Calabrese, F.P. Thurberg and F.J. Vernberg (eds.), Physiological Mechanisms of Marine Pollutant Toxicity. Academic Press, New York.

Lee, R.F., K. Hinga and G. Almquist. 1982b. Fate of radiolabelled polycyclic aromatic hydrocarbons and pentachlorophenol in enclosed marine ecosystems. Pages 123-135 *in* G.D. Grice and M.R. Reeve (eds.), Marine Mesocosms: Biological and Chemical Research in Experimental Ecosystems. Springer-Verlag, New York.

Lee, R.F., W.S. Gardner, J.W. Anderson, J.W. Blaylock and J. Barwell-Clarke. 1978. Fate of polycyclic aromatic hydrocarbons in controlled ecosystem enclosures. Environ. Sci. Technol. 12: 832-838.

Lee, R.F., C. Ryan and M.L. Neuhauser. 1976. Fate of petroleum hydrocarbons taken up from food and water by the blue crab *Callinectes sapidus*. Mar. Biol. 37: 363-370.

Lee, R.F., R. Sauerheber and A.A. Benson. 1972a. Petroleum hydrocarbons: uptake and discharge by the marine mussel *Mytilus edulis*. Science 177: 344-346.

Lee, R.F., R. Sauerheber and G.H. Dobbs. 1972b. Uptake, metabolism and discharge of polycyclic aromatic hydrocarbons by marine fish. Mar. Biol. 17: 201-208.

Lee, R.F., S.C. Singer, K.R. Tenore, W.S. Gardner and R.M. Philpot. 1979. Detoxification system in polychaete worms, importance in the degradation of sediment hydrocarbons. Pages 23-37 *in* W.B. Vernberg, F.P. Thurberg, A. Calabrese and F.J. Vernberg (eds.), Marine Pollution: Functional Responses. Academic Press, New York.

Lee, R.F., J. Stolzenbach, S. Singer and K.R. Tenore. 1981. Effects of crude oil on growth and mixed-function oxygenase activity in polychaetes, *Nereis* sp. Pages 323-334

in F.S. Vernberg, A. Calabrese, F.P. Thurberg and W.B. Vernberg (eds.), Biological Monitoring of Marine Organisms. Academic Press, New York.
Lee, W.Y. and J.A.C. Nicol. 1977. The effects of the water soluble fractions of No. 2 fuel oil on the survival and behavior of coastal and oceanic zooplankton. Environ. Pollut. 12: 279-292.
Lee, W.Y., A. Norris and D. Boatwright. 1980. Mexican oil spill: a toxicity study of oil accommodated in seawater on invertebrates. Mar. Pollut. Bull. 11: 231-234.
Lee, W.Y., K. Winters and J.A.C. Nicol. 1978. The biological effects of the water-soluble fractions of a No. 2 fuel oil to the planktonic shrimp, *Lucifer faxoni*. Environ. Pollut. 15: 167-183.
Levitan, W.M. and M.H. Taylor. 1979. Physiology of salinity-dependent naphthalene toxicity in *Fundulus heteroclitus*. J. Fish. Res. Board Can. 36: 615-620.
Linden, O. 1976a. Effects of oil on the amphipod *Gammarus oceanicus*. Environ. Pollut. 10: 239-250.
Linden, O. 1976b. The influence of crude oil and mixtures of crude oil/dispersants on the ontogenic development of the Baltic herring, *Clupea harengus membras* L. Ambio 5: 136-140.
Linden, O. 1978. Biological effects of oil on early development of the Baltic herring *Clupea harengus membras*. Mar. Biol. 45: 273-283.
Linden, O., R. Laughlin, Jr., J.R. Sharp and J.M. Neff. 1980. The combined effect of salinity, temperature and oil on the growth pattern of embryos of the killifish *Fundulus heteroclitus* Walbaum. Mar. Environ. Res. 3: 129-144.
Lindmark, D.G. 1981. Activation of polynuclear aromatic hydrocarbons to mutagens by the marine ciliate *Parauronema acutum*. Appl. Environ. Microbiol. 41: 1238-1242.
Longwell, A.C. 1977. A genetic look at fish eggs and oil. Oceanus 20: 45-58.
Lonning, S. 1977. The effects of crude Ekofisk oil and oil products on marine fish larvae. Astarte 10: 37-47.
Lowe, D.M., M.N. Moore and K.R. Clarke. 1981. Effects of oil on digestive cells in mussels: Quantitative alterations in cellular and lysosomal structure. Aquat. Toxicol. 1: 175-186.
Lu, A.Y.H. 1976. Liver microsomal drug-metabolizing enzyme system: Functional components and their properties. Fed. Proc. 35: 2460-2463.
Lu, P., R.L. Metcalf, N. Plummer and D. Mandel. 1977. The environmental fate of three carcinogens: benzo(a)pyrene, benzidine, and vinyl chloride evaluated in laboratory model ecosystem. Arch. Environ. Contam. Toxicol. 6: 129-142.
Lyes, M.C. 1979. Bioavailability of a hydrocarbon from water and sediment to the marine worm *Arenicola marina*. Mar. Biol. 55: 121-127.
Malins, D.C. 1982. Alterations in the cellular and subcellular structure of marine teleosts and invertebrates exposed to petroleum in the laboratory and field: A critical review. Can. J. Fish. Aquat. Sci. 39: 877-889.
Malins, D.C. and H.O. Hodgins. 1981. Petroleum and marine fishes: A review of uptake, disposition, and effects. Environ. Sci. Technol. 15: 1272-1280.
Malins, D.C. and W.T. Roubal. 1982. Aryl sulfate formation in sea urchins (*Strongylocentrotus droebachiensis*) ingesting marine algae (*Fucus distichus*) containing 2,6-dimethylnaphthalene. Environ. Res. 27: 290-297.
Malins, D.C., M.S. Myers and W.T. Roubal. 1983. Organic free radicals associated with idiopathic liver lesions of English sole (*Parophrys vetulus*) from polluted marine environments. Environ. Sci. Technol. 17: 679-685.
Maynard, D.J. and D.D. Weber. 1981. Avoidance reactions of juvenile coho salmon (*Oncorhynchus kisutch*) to monocyclic aromatics. Can. J. Fis. Aquat. Sci. 38: 772-778.
McCain, B.B. and D.C. Malins. 1982. Effects of petroleum hydrocarbons on selected demersal fish and crustaceans. Pages 315-326 *in* G.F. Mayer (ed.), Ecological Stress and New York Bight: Science and Management. Estuarine Research Federation, Columbia, South Carolina.
McCain, B.B., H.O. Hodgins, W.D. Gronlund, J.W. Hawkes, D.W. Brown, M.S. Myers

and J.H. Vandermeulen. 1978. Bioavailability of crude oil from experimentally oiled sediments to English sole (*Parophrys vetulus*), and pathological consequences. J. Fish. Res. Board Can. 35: 657-664.

McIntyre, A.D. and J.B. Pearce (eds.). 1980. Biological Effects of Marine Pollution and the Problems of Monitoring. Rapp. P.-V. Réun. Cons. Int. Explor. Mer. 179: 1-346.

McLeese, D.W., C.D. Metcalfe and D.S. Pezzack. 1980. Uptake of PCBs from sediment by *Nereis virens* and *Crangon septemspinosa*. Arch. Environ. Contam. Toxicol. 9: 507-518.

Means, J.C., J.J. Hassett, S.G. Wood and W.L. Banwart. 1979. Sorption properties of energy-related pollutants and sediments. Pages 327-340 *in* P.W. Jones and P. Leber (eds.), Polynuclear Aromatic Hydrocarbons. Ann Arbor Science, Ann Arbor, Michigan.

Mecklenburg, T.A., S.D. Rice and J.F. Karinen. 1977. Molting and survival of king crab (*Paralithoides camtschatica*) and coonstripe shrimp (*Pandalus hypsinotus*) larvae exposed to Cook Inlet crude oil water-soluble fraction. Pages 221-228 *in* D.A. Wolfe (ed.), Fate and Effects of Petroleum Hydrocarbons in Marine Ecosystems and Organisms. Pergamon Press, New York.

Melancon, M.J., Jr. and J.J. Lech. 1978. Distribution and elimination of naphthalene and 2-methylnaphthalene in rainbow trout during short- and long-term exposures. Arch. Environ. Contam. Toxicol. 7: 207-220.

Meyer, T. and T. Bakke. 1977. The metabolism of biphenyl. V. Phenolic metabolites in some marine organisms. Acta. Pharmacol. Toxicol. 40: 201-208.

Miller, D.L., J.P. Corliss, R.N. Farragut and H.C. Thompson, Jr. 1978. Accumulation and elimination of the polynuclear aromatic hydrocarbon chrysene by mangrove snapper (*Lutjanus griseus*) and pink shrimp (*Penaeus duorarum*). Unpublished manuscript. Southeast Fisheries Center, Miami Laboratory, National Marine Fisheries Service, Miami, Florida.

Minchew, C.D. and J.D Yarbrough. 1977. The occurrence of fin rot in mullet (*Mugil cephalus*) associated with crude oil contamination of an estuarine pond-ecosystem. J. Fish Biol. 10: 319-323.

Mitchell, C.T., E.K. Anderson, L.G. Jones and W.J. North. 1970. What oil does to ecology. J. Water Pollut. Control Fed. 42: 812-818.

Moles, A., S. Bates, S.D. Rice and S. Korn. 1981. Reduced growths of coho salmon fry exposed to two petroleum components, toluene and naphthalene in fresh water. Trans. Am. Fish. Soc. 110: 430-436.

Moles, A., S.D. Rice and S. Korn. 1979. Sensitivity of Alaskan freshwater and anadromous fishes to Prudhoe Bay crude oil and benzene. Trans. Am. Fish. Soc. 108: 408-414.

Moore, M.N. 1979. Cellular responses to polycyclic aromatic hydrocarbons and phenobarbitol in *Mytilus edulis*. Mar. Env. Res. 2: 255-263.

Moore, M.N. and R. Clarke. 1982. Use of microstereology and quantitative cytochemistry to determine the effects of crude oil-derived aromatic hydrocarbons on lysosmal structure and function in a marine bivalve mollusc, *Mytilus edulis*. Histochem. J. 14: 713-718.

Moore, M.N., D.R. Livingstone, P. Donkin, B.L. Bayne, J. Widdows and D.M. Lowe. 1980. Mixed function oxygenase and xenobiotic detoxification/toxification systems in bivalve molluscs. Helgoländer Meeresunters. 33: 278-291.

Moore, M.N., D.M. Lowe and P.E.M. Fieth. 1978. Lysosomal responses to experimentally injected anthracene in the digestive cells of *Mytilus edulis*. Mar. Biol. 48: 297-302.

Nakatsugawa, T. and M.A. Morelli. 1976. Microsomal oxidation and insecticide metabolism. Pages 61-114 *in* C.F. Wilkinson (ed.), Insecticide Biochemistry and Physiology. Plenum Press, New York.

National Research Council. 1985. Oil in the Sea. Input, Fates, and Effects. National Academy Press, Washington, D.C., 601 p.

Nava, M.E. and F.R. Engelhardt. 1980. Compartmentalization of ingested labelled petroleum in tissues and bile of the American eel (*Anguilla rostrata*). Bull. Environ. Contam. Toxicol. 24: 879-885.
Neff, J.M. 1979. Polycyclic Aromatic Hydrocarbons in the Aquatic Environment. Applied Sci. Publ., London. 266 p.
Neff, J.M. 1985. The use of biochemical measurements to detect pollutant-mediated damage to fish. Pages 155-183 *in* R.D. Cardwell, R. Purdy and R.C. Bahner (eds.), Aquatic Toxicology and Hazard Assessment. Seventh Symposium. Special Technical Publ. No. 854. American Society for Testing and Materials, Philadelphia, Pennsylvania.
Neff, J.M. and J.W. Anderson. 1975a. An ultraviolet spectrophotometric method for the determination of naphthalene and alkylnaphthalenes in the tissues of oil-contaminated marine animals. Bull. Environ. Contam. Toxicol. 14: 122-128.
Neff, J.M.and J.W. Anderson. 1975b. Accumulation, release and distribution of benzo(a)pyrene C in the clam *Rangia cuneata*. Pages 469-471 *in* Proceedings of 1975 Oil Spill Conference (Prevention, Behavior, Control, Cleanup). American Petroleum Institute, Washington, D.C.
Neff, J.M., J.W. Anderson, B.A. Cox, R.B. Laughlin, Jr., S.S. Rossi and H.E. Tatum. 1976a. Effects of petroleum on survival, respiration and growth of marine animals. Pages 515-539 *in* Sources, Effects and Sinks of Hydrocarbons in the Aquatic Environment. American Institute of Biological Sciences, Washington, D.C.
Neff, J.M., B.A. Cox, D. Dixit and J.W. Anderson. 1976b. Accumulation and release of petroleum-derived aromatic hydrocarbons by four species of marine animals. Mar. Biol. 38: 279-289.
Nuzzi, R. 1973. Effects of water soluble extracts of oil on phytoplankton. Pages 809-813 *in* Proceedings of 1973 Oil Spill Conference. American Petroleum Institute, Washington, D.C.
Olla, B.L., A.J. Bejda and W.H. Pearson. 1983. Effects of oiled sediment on the burrowing behavior of the hard clam *Mercenaria mercenaria*. Mar. Environ. Res. 9: 183-193.
Olsen, C.R., N.H. Cutshall and J.L. Larsen. 1982. Pollutant-particle associations and dynamics in coastal marine environments: A review. Mar. Chem. 11: 501-533.
Ott, F.S., R.P. Harris and S.C.M. O'Hara. 1978. Acute and sublethal toxicity of naphthalene and three methylated derivatives to the estuarine copepod, *Eurytemora affinis*. Mar. Environ. Res. 1: 49-58.
Oviatt, C., J. Frithsen, J. Gearing and P. Gearing. 1982. Low chronic additions of No. 2 fuel oil: Chemical behavior, biological impact and recovery in a simulated estuarine environment. Mar. Ecol. Prog. Ser. 9: 121-136.
Paris, D.F., W.C. Steen and G.L. Baughman. 1978. Role of physico-chemical properties of Aroclors 1016 and 1242 in determining their fate and transport in aquatic environments. Chemosphere 4: 319-325.
Parsons, T.R., W.K.W. Li and R. Waters. 1976. Some preliminary observations on the enhancement of phytoplankton growth by low levels of mineral hydrocarbons. Hydrobiol. 51: 85.
Pasteels, J.J. 1968. Pinocytose et athrocytose par l'epithelium branchial de *Mytilus edulis*. Z. Zellforsch. 92: 239-259.
Payne, J F. 1976. Field evaluation of benzopyrene hydroxylase induction as a monitor for marine petroleum pollution. Science 191: 945-946.
Payne, J.F., J. Kiceniuk, R. Misra, G.L. Fletcher and R.J. Thompson. 1983. Sublethal effects of petroleum hydrocarbons on adult American lobsters (*Homarus americanus*). Can. J. Fish. Aquat. Sci. 40: 705-717.
Payne, J.F., J.W. Kiceniuk and W.R. Squires. 1978. Pathological changes in a marine fish after a 6-month exposure to petroleum. J. Fish. Res. Board Can. 35: 665-667.
Pearson, W.H., S.E. Miller, J.W. Blaylock and B.L. Olla. 1981. Detection of the water soluble fraction of crude oil by the blue crab *Callinectes sapidus*. Mar. Environ. Res. 5: 3-11.

Pedersen, M.G., W.K. Hershberger, P.K. Zachariah and M.R. Juchau. 1976. Hepatic biotransformation of environmental xenobiotics in six strains of rainbow trout (*Salmo gairdneri*). J. Fish. Res. Bd. Can. 33: 666-675.

Percy, J.A. 1977. Effects of dispersed crude oil upon the respiratory metabolism of an Arctic marine amphipod *Onisimus* (*Boekisimus*) *affinis*. Pages 192-200 *in* D.A. Wolfe (ed.), Fate and Effects of Petroleum Hydrocarbons in Marine Ecosystems and Organisms. Pergamon Press, New York.

Philpot, R.M., M.O. James and J.R. Bend. 1976. Metabolism of benzo(a)pyrene and other xenobiotics by microsomal mixed function oxidases in marine species. Pages 184-199 *in* Sources, Effects and Sinks of Hydrocarbons in the Marine Environment. American Institute of Biological Sciences, Washington, D.C.

Prouse, N.J. and D.C. Gordon, Jr. 1976. Interactions between the deposit feeding polychaete *Arenicola marina* and oiled sediment. Pages 407-422 *in* Sources, Effects and Sinks of Hydrocarbons in the Aquatic Environment. American Institute of Biological Sciences, Washington, D.C.

Prouse, N.J., D.C. Gordon and P.D. Kiezer. 1976. Effects of low concentrations of oil accommodated in sea water on the growth of unialgal marine phytoplankton cultures. J. Fish. Res. Board Can. 33: 810-818.

Pulich, W.M., Jr., K. Winters and C. Van Baalen. 1974. The effects of a No. 2 fuel oil and two crude oils on the growth and photosynthesis of microalgae. Mar. Biol. 28: 87-94.

Rabalais, S.C., C.R. Arnold and N.S. Wohlschlag. 1981. The effects of IXTOC I oil on the eggs and larvae of red drum (*Sciaenops ocellata*). Texas J. Sci. 33: 33-38.

Renzoni, A. 1975. Toxicity of three oils to bivalve gametes and larvae. Mar. Pollut. Bull. 6: 125-128.

Rice, S.D. 1973. Toxicity and avoidance tests with Prudhoe Bay oil and pink salmon fry. Pages 667-670 *in* Proceedings of 1973 Oil Spill Conference. American Petroleum Institute, Washington, D.C.

Rice, S.D., J.W. Short and J.F. Karinen. 1976. Toxicity of Cook Inlet crude oil and No. 2 fuel oil to several Alaskan marine fishes and invertebrates. Pages 394-406 *in* Sources, Effects and Sinks of Hydrocarbons in the Aquatic Environment. American Institute of Biological Sciences, Washington, D.C.

Rice, S.D., J.W. Short and J.F. Karinen. 1977a. Comparative oil toxicity and comparative animal sensitivity. Pages 78-94 *in* D.A. Wolfe (ed.), Fate and Effects of Petroleum Hydrocarbons in Marine Ecosystems and Organisms. Pergamon Press, New York.

Rice, S.D., R.E. Thomas and J.W. Short. 1977b. Effect of petroleum hydrocarbons on breathing and coughing rates and hydrocarbon uptake-depuration in pink salmon fry. Pages 259-277 *in* F.J. Vernberg, A. Calabrese, F.P. Thurberg and W.B. Vernberg (eds.), Physiological Responses of Marine Biota to Pollutants. Academic Press, New York.

Riley, R.T. and M.C. Mix. 1981. The effects of naphthalene on glucose metabolism in the European flat oyster *Ostrea edulis*. Comp. Biochem. Physiol. 70C: 13-20.

Rinkevich, B. and Y. Loya. 1979. Laboratory experiments on the effects of crude oil on the Red Sea coral *Stylophora pistillata*. Mar. Pollut. Bull. 10: 328-330.

Roesijadi, G. and J.W. Anderson. 1979. Condition index and free amino acid content of *Macoma inquinata* exposed to oil-contaminated marine sediments. Pages 69-83 *in* W.B. Vernberg, A. Calabrese, F.P. Thurberg and F.J. Vernberg (eds.), Marine Pollution: Functional Responses. Academic Press, New York.

Roesijadi, G., J.W. Anderson and J.W. Blaylock. 1978. Uptake of hydrocarbons from marine sediments contaminated with Prudhoe Bay crude oil: influence of feeding type of test species and availability of polycyclic aromatic hydrocarbons. J. Fish. Res. Board Can. 35: 608-614.

Rosenthal, H. and D.F. Alderdice. 1976. Sublethal effects of environmental stressors, natural and pollutional, on marine fish eggs and larvae. J. Fish. Res. Board Can. 33: 2047-2065.

Rossi, S.S. and J.W. Anderson. 1976. Toxicity of water-soluble fractions on No. 2 fuel oil and South Louisiana crude oil to selected stages in the life history of the polychaete, *Neanthes arenaceodentata*. Bull. Environ. Contam. Toxicol. 16: 18-24.
Rossi, S.S. and J.W. Anderson. 1977. Accumulation and release of fuel oil-derived diaromatic hydrocarbons to the polychaete *Neanthes arenaceodentata*. Mar. Biol. 39: 51-55.
Rossi, S.S. and J.W. Anderson. 1978. Petroleum hydrocarbon resistance in the marine worm *Neanthes arenaceodentata* (Polychaeta: Annelida) induced by chronic exposure to No. 2 fuel oil. Bull. Environ. Contam. Toxicol. 20: 515-521.
Rossi, S.S., J.W. Anderson and G.S. Ward. 1976. Toxicity of water-soluble fractions of four test oils for the polychaetous annelids, *Neanthes arenaceodentata* and *Capitella capitata*. Environ. Pollut. 10: 9-18.
Roubal, W.T., D.H. Bovee, T.K. Collier and S.I. Stranahan. 1977a. Flow-through system for chronic exposure of aquatic organisms to seawater-soluble hydrocarbons from crude oil, construction and applications. Pages 551-556 *in* Proceedings of 1977 Oil Spill Conference (Prevention, Behavior, Control, Cleanup). American Petroleum Institute, Washington, D.C.
Roubal, W.T., T.K. Collier and D.C. Malins. 1977b. Accumulation and metabolism of carbon-14 labeled benzene, naphthalene, and anthracene by young coho salmon (*Oncorhynchus kisutch*). Arch. Environ. Contam. Toxicol. 5: 513-529.
Roubal, W.T., S.I. Stranahan and D.C. Malins. 1978. The accumulation of low molecular weight aromatic hydrocarbons of crude oil by coho salmon (*Oncorhynchus kisutch*) and starry flounder (*Platichthys stellatus*). Arch. Environ. Contam. Toxicol. 7: 237-244.
Sabo, D.J. and J.J. Stegeman. 1977. Some metabolic effects of petroleum hydrocarbons in marine fish. Pages 279-287 *in* F.J. Vernberg, A. Calabrese, F.P. Thurberg and W.B. Vernberg (eds.), Physiological Responses of Marine Biota to Pollutants. Academic Press, New York.
Sabourin, T. 1982. Respiratory and circulatory responses of the blue crab to naphthalene and the effect of acclimation salinity. Aquat. Toxicol. 2: 301-318.
Sanborn, H.R. and D.C. Malins. 1977. Toxicity and metabolism of naphthalene: A study with marine larval invertebrates. Proc. Soc. Exp. Biol. Med. 154: 151-154.
Sanborn, H.R. and D.C. Malins. 1980. The disposition of aromatic hydrocarbons in adult spot shrimp (*Pandalus platyceros*) and the formation of metabolites of naphthalene in adult and larval spot shrimp. Xenobiotica 10: 193-200.
Sanders, H.L., J.F. Grassle, G.R. Hampson, L.S. Morse, S. Garner-Price and C.C. Jones. 1980. Anatomy of an oil spill: long-term effects from the grounding of the barge *Florida* off West Falmouth, Massachusetts. J. Mar. Res. 38: 265.380.
Schramm, W. 1972. Investigations on the influence of oil pollution on marine algae. 1. The effect of crude oil films on the CO_2 gas exchange outside the water. Mar. Biol. 14: 189-198.
Sekerah, A. and M. Foy. 1978. Acute lethal toxicity of Corexit 9527/Prudhoe Bay crude oil mixtures to selected Arctic invertebrates. Spill Tech. Newsletter 3: 37-41.
Sharp, J.R., K.W. Fucik and J.M. Neff. 1979. Physiological bases of differential sensitivity of fish embryonic stages to oil pollution. Pages 85-108 *in* W.B. Vernberg, A. Calabrese, F.P. Thurberg and F.J. Vernberg (eds.), Marine Pollution: Functional Responses. Academic Press, New York.
Shaw, D.G., A.J. Paul and E.R. Smith. 1977. Responses of the clam *Macoma balthica* to Prudhoe Bay crude oil. Pages 493-494 *in* Proceedings of 1977 Oil Spill Conference (Prevention, Behavior, Control, Cleanup). American Petroleum Institute, Washington, D.C.
Shiels, W.E., J.J. Goering and D.W. Hood. 1973. Crude oil phytotoxicity studies. Pages 413-446 *in* D.W. Hood, W.E. Shiels and E.J. Kelly (eds.), Environmental Studies of Port Valdez. University of Alaska Institute of Marine Sciences, Occasional Publ. No. 3.
Sindermann, C.J. 1982. Implications of oil pollution in production of disease in marine

organisms. Phil. Trans. R. Soc. London B 297: 385-399. Reprinted in R.B. Clark (ed.). 1982. The Long-Term Effects of Oil Pollution on Marine Populations, Communities and Ecosystems. The Royal Society, London.

Singer, S.C. and R.F. Lee. 1977. Mixed function oxygenase activity in the blue crab, *Callinectes sapidus*: Tissue distribution and correlation with changes during molting and development. Biol. Bull. 153: 377-386.

Singer, S.C., P.E. March, F. Gonsoulin and R.F. Lee. 1980. Mixed function oxygenase activity in the blue crab, *Callinectes sapidus*: characterization of enzyme activity from stomach tissue. Comp. Biochem. Physiol. 65C: 129-134.

Skjoldal, H.R., T. Dale, H. Haldorsen, B. Pengerud, T.F. Thingstad, K. Tjessem and A. Aaberg. 1982. Oil pollution and plankton dynamics. 1. Controlled ecosystem experiment during the 1980 spring bloom in Lindaspollene, Norway. Neth. J. Sea Res. 16: 511-523.

Smith, R.L. and J.A. Cameron. 1979. Effect of water soluble fraction of Prudhoe Bay crude oil on embryonic development of Pacific herring. Trans. Am. Fish. Soc. 108: 70-75.

Solangi, M.A. and R.M. Overstreet. 1982. Histopathological changes in two estuarine fishes, *Menidia beryllina* (Cope) and *Trinectes maculatus* (Bloch and Schneider), exposed to crude oil and its water soluble fractions. J. Fish Diseases 5: 13-35.

Solbakken, J.E. and K.H. Palmork. 1980. Distribution of radioactivity in the chondrichthyes *Squalus acanthias* and the osteichthyes *Salmo gairdneri* following intragastric administration of (9-^{14}C) phenanthrene. Bull. Environ. Contam. Toxicol. 25: 902-908.

Solbakken, J.E., A.H. Knap, T.D. Sleeter, C.E. Searle and K.H. Palmork. 1984. Investigation into the fate of ^{14}C-labelled xenobiotics (naphthalene, phenanthrene, 2,4,5,2′,4′,5′-hexachlorobiphenyl, octachlorostyrene) in Bermudian corals. Mar. Ecol. Prog. Ser. 16: 149-154.

Solbakken, J.E., K.H. Palmork, T. Neppelberg and R.R. Scheline. 1980. Urinary and biliary metabolites of phenanthrene in the coalfish (*Pollachius virens*). Acta Pharmacol. Toxicol. 46: 127.

Soto, C., J.A. Hellebust and T.C. Hutchinson. 1975a. Effect of naphthalene and aqueous crude oil extracts on the green flagellate *Chlamydomonas angulosa*. II. Photosynthesis and the uptake and release of naphthalene. Can. J. Bot. 53: 118-126.

Soto, C., J.A. Hellebust and T.C. Hutchinson. 1975b. The effects of aqueous extracts of crude oil and naphthalene on the physiology and morphology of a freshwater green alga. Verhandlungen der Internatiolen Vereinigung fur theoretische und angewandte Limnologie 19: 2145-2154.

Soto, C., J.A. Hellebust, T.C. Hutchinson, and T. Sawa. 1975c. Effect of naphthalene and aqueous crude oil extracts on the green flagellate, *Chlamydomonas angulosa* I. Growth. Can. J. Bot. 53: 109-117.

Southward, A.J. and E.C. Southward. 1978. Recolonization of rocky shores in Cornwall after use of toxic dispersants to clean up the *Torrey Canyon* spill. J. Fish. Res. Board Can. 35: 682-706.

Spaulding, M.L., S.B. Saila, E. Lorda, H. Walker, E. Anderson and J.C. Swanson. 1983. Oil-spill fishery impact assessment model: Application to selected Georges Bank fish species. Estuar. Coast. Shelf Sci. 16: 511-541.

Spies, R.B., J.S. Felton and L. Dillard. 1982. Hepatic mixed-function oxidases in California flatfishes are increased in contaminated environments and by oil and PCB ingestion. Mar. Biol. 70: 117-127.

Statham, C.N., M.J. Melancon, Jr. and J.J. Lech. 1976. Bioconcentration of xenobiotics in trout bile: A proposed monitoring aid for some waterborne chemicals. Science 193: 680-681.

Steele, R.L. 1977. Effects of certain petroleum products on reproduction and growth of zygotes and juvenile stages of the alga *Fucus edentatus* de la Pyl (Phaeophyceae: Fucales). Pages 138-142 *in* D.A. Wolfe (ed.), Fate and Effects of Petroleum Hydrocarbons in Marine Ecosystems and Organisms. Pergamon Press, New York.

Stegeman, J.J. 1974. Hydrocarbons in shellfish chronically exposed to low levels of fuel oil. Pages 329-348 *in* F.J. Vernberg and W.B. Vernberg (eds.), Pollution and Physiology of Marine Organisms. Academic Press, New York.

Stegeman, J.J. 1977. Fate and effects of oil in marine animals. Oceanus 20: 59-66.

Stegeman, J.J. 1979. Temperature influence on basal activity and induction of mixed-function oxygenase activity in *Fundulus hetreroclitus*. J. Fish. Res. Board Can. 36: 1400-1405.

Stegeman, J.J. 1980. Mixed function oxygenase studies in monitoring for effects of organic pollution. Rapp. P.-V. Réun. Cons. Int. Explor. Mer 179: 33-38.

Stegeman, J.J. 1981. Polynuclear aromatic hydrocarbons and their metabolism in the marine environment. Pages 1-59 *in* H.V. Gelboin and P.O.P. Ts'O (eds.), Polycyclic Hydrocarbons and Cancer, Volume 3. Academic Press, New York.

Stegeman, J.J. and M. Chevion. 1980. Sex differences in cytochrome P-450 and mixed-function oxygenase activity in gonadally mature trout. Biochem. Pharmacol. 29: 553-558.

Stegeman, J.J. and H.B. Kaplan. 1981. Mixed-function oxygenase activity and benzo(a)pyrene metabolism in the barnacle *Balanus eburneus* (Crustacea, Cirripedia). Comp. Biochem. Physiol. 68C: 55-61.

Stegeman, J.J. and D.J. Sabo. 1976. Aspects of the effects of petroleum hydrocarbons on intermediary metabolism and xenobiotic metabolism in marine fish. Pages 423-431 *in* Sources, Effects and Sinks of Hydrocarbons in the Aquatic Environment. American Institute of Biological Sciences, Washington, D.C.

Stegeman, J.J. and J.M. Teal. 1973. Accumulation, release and retention of petroleum hydrocarbons by the oyster *Crassostrea virginica*. Mar. Biol. 22: 37-44.

Stott, G.G., W.E. Haensly, J.M. Neff and J.R. Sharp. 1983. Histopathologic survey of ovaries of plaice *Pleuronectes platessa* L. from Aber Wrac'h and Aber Benoit, Brittany, France: long-term effects of the *Amoco Cadiz* crude oil spill. J. Fish Diseases 6: 429-437.

Struhsaker, J.W., M.B. Eldridge and T. Echeverria. 1974. Effects of benzene (a water-soluble component of crude oil) on eggs and larvae of Pacific herring and northern anchovy. Pages 253-284 *in* F.J. Vernberg and W.B. Vernberg (eds.), Pollution and Physiology of Marine Organisms. Academic Press, New York.

Tatem, H.E. 1977. Accumulation of naphthalenes by grass shrimp: Effects on respiration, hatching, and larval growth. Pages 201-209 *in* D.A. Wolfe (ed.), Fate and Effects of Petroleum Hydrocarbons in Marine Ecosystems and Organisms. Pergamon Press, New York.

Tatem, H.E., B.A. Cox and J.W. Anderson. 1978. The toxicity of oils and petroleum hydrocarbons to estuarine crustaceans. Estuar. Coastal Mar. Sci. 6: 365-373.

Taylor, T.L. and J.F. Karinen. 1977. Response of the clam, *Macoma balthica* (Linnaeus), exposed to Prudhoe Bay crude oil as unmixed oil, water-soluble fraction, and oil-contaminated sediment in the laboratory. Pages 229-237 *in* D.A. Wolfe (ed.), Fate and Effects of Petroleum Hydrocarbons in Marine Ecosystems and Organisms. Pergamon Press, New York.

Teal, J.M., K. Burns and J.W. Farrington. 1978. Analyses of aromatic hydrocarbons in intertidal sediments resulting from two spills of No. 2 fuel oil in Buzzards Bay, Massachusetts. J. Fish. Res. Bd. Can. 35: 510-520.

Thomas, M.L.H. 1978. Comparison of oiled and unoiled intertidal communities in Chedabucto Bay, Nova Scotia. J. Fish. Res. Bd. Can. 35: 707-716.

Thomas, R.E. and S.D. Rice. 1979. The effect of exposure temperatures on oxygen consumption and opercular breathing rates of pink salmon fry exposed to toluene, naphthalene, and water soluble fractions of Cook Inlet crude oil and No. 2 fuel oil. Pages 39-52 *in* W.B. Vernberg, A. Calabrese, F.P. Thurberg and F.J. Vernberg (eds.), Marine Pollution: Functional Responses. Academic Press, New York.

Thomas, R.E. and S.D. Rice. 1981. Excretion of aromatic hydrocarbons and their metabolites by freshwater and saltwater Dolly Varden char. Pages 425-448 *in* F.J.

Vernberg, F.P. Thurberg, A. Calabrese and W.B. Vernberg (eds.), Biological Monitoring of Marine Pollutants. Academic Press, New York.

Thompson, S. and G. Eglinton. 1979. The presence of pollutant hydrocarbons in estuarine epipelic diatom populations. II. Diatom slimes. Estuar. Coastal Mar. Sci. 8: 75-86.

Truscott, B., J.M. Walsh, M.P. Burton, J.F. Payne and D.R. Idler. 1983. Effect of acute exposure to crude petroleum on some reproductive hormones in salmon and flounder. Comp. Biochem. Physiol. 75C: 121-130.

Vanderhorst, J.R., C.I. Gibson, L.J. Moore and P. Wilkinson. 1977. Continuous-flow apparatus for use in petroleum bioassay. Bull. Environ. Contam. Toxicol. 17: 577-584.

Vandermeulen, J.H. and J.M. Capuzzo. 1983. Understanding sublethal pollutant effects in the marine environment. Paper No. 9 *in* M.A. Champ and M. Trainor (eds.), Ocean Waste Management: Policy and Strategies. Background Papers of Symposium, May 2-6, 1983, University of Rhode Island, Kingston.

Vandermeulen, J.H. and T.P. Ahearn. 1976. Effect of petroleum hydrocarbons on algal physiology: review and progress report. Pages 107-126 *in* A.P.M. Lockwood (ed.), Effects of Pollutants on Aquatic Organisms. Cambridge University Press, Cambridge, England.

Vandermeulen, J.H. and D.C. Gordon, Jr. 1976. Re-entry of 5 year old stranded bunker C fuel oil from a low-energy beach into the water, sediments, and biota of Chedabucto Bay, Nova Scotia. J. Fish. Res. Board Can. 33: 2002-2010.

Vandermeulen, J.H. and W.R. Penrose. 1978. Absence of aryl hydrocarbon hydroxylase (AHH) in three marine bivalves. J. Fish. Res. Board Can. 35: 643-647.

Van Overbeek, J. and R. Blondeau. 1954. Mode of action of phytotoxic oils. Weeds 3: 55-65.

Varanasi, U. and D.J. Gmur. 1980. Metabolic activation and covalent binding of benzo(a)pyrene to deoxyribonucleic acid catalyzed by liver enzymes of marine fish. Biochem. Pharm. 29: 753-761.

Varanasi, U. and D.J. Gmur. 1981a. Hydrocarbons and metabolites in English sole (*Parophrys vetulus*) exposed simultaneously to (^{3}H)benzo(a)pyrene and (^{14}C)naphthalene in oil-contaminated sediment. Aquat. Toxicol. 1: 49-68.

Varanasi, U. and D.J. Gmur. 1981b. *In vivo* metabolism of naphthalene and benzo(a)pyrene by flatfish. Pages 367-376 *in* M. Cooke and A.J. Dennis (eds.), Chemical Analysis and Biological Fate: Polynuclear Aromatic Hydrocarbons. Battelle Press, Columbus, Ohio.

Varanasi, U. and D.C. Malins. 1977. Metabolism of petroleum hydrocarbons: accumulation and biotransformation in marine organisms. Pages 175-200 *in* D.C. Malins (ed.), Effects of Petroleum on Arctic and Subarctic Marine Environments. Volume II. Biological Effects. Academic Press, New York.

Varanasi, U., D.J. Gmur, and M.M. Krahn. 1980. Metabolism and subsequent binding of benzo(a)pyrene to DNA in pleuronectid and salmonid fish. Pages 455-470 *in* A. Bjorseth and J. Dennis (eds.), Polynuclear Aromatic Hydrocarbons: Biological Effects. Battelle Press, Columbus, Ohio.

Varanasi, U., D.J. Gmur and P.A. Treseler. 1979. Influence of time and mode of exposure on biotransportation of naphthalene by juvenile starry flounder (*Platichthys stellatus*) and rock sole (*Lepidopsetta bilineata*). Arch. Environ. Contam. Toxicol. 8: 673-692.

Varanasi, U., J.E. Stein and T. Hom. 1981. Covalent binding of benzo(a)pyrene to DNA in fish liver. Biochem. Biophys. Res. Commun. 103: 780-787.

Varanasi, U., M. Uhler and S.I. Stranahan. 1978. Uptake and release of naphthalene and its metabolites in skin and epidermal mucus of salmonids. Toxicol. Appl. Pharmacol. 44: 277-289.

Vargo, S.L. 1981. The effects of chronic low concentrations of No. 2 fuel oil on the physiology of a temperate estuarine zooplankton community in the MERL microcosms. Pages 295-322 *in* F.J. Vernberg, A. Calabrese, F.P. Thurberg and W.B. Vernberg (eds.), Biological Monitoring of Marine Pollutants. Academic Press, New York.

Vashchenko, M.A. 1980. Effects of oil pollution on the development of sex cells in sea urchins. Helgoländer Meeresunters. 33: 297-300.

Veith, D.G., D.L. Defoe and B.V. Bergstedt. 1979. Measuring and estimating the bioconcentration factor of chemicals in fish. J. Fish. Res. Board Can. 36: 1040.

Viarengo, A. and M.N. Moore. 1982. Effects of aromatic hydrocarbons on the metabolism of the digestive gland of the mussel *Mytilus edulis* L. Comp. Biochem. Physiol. 71C: 21-25.

Vuorinen, P. and M.B. Axell. 1980. Effects of the water soluble fraction of crude oil on herring eggs and pike fry. I.C.E.S., C.M. 1980/E:30, 10 p.

Walters, J.M., R.B. Cain, I.J. Higgin and E.D.S. Corner. 1979. Cell-free benzo(a)pyrene hydroxylase activity in marine zooplankton. J. Mar. Biol. Ass. U.K. 59: 553-564.

Wang, R.T. and J.A.C. Nicol. 1977. Effects of fuel oil on sea catfish: Feeding activity and cardiac responses. Bull. Environ. Contam. Toxicol. 18: 170-176.

Ward, D.M., R.M. Atlas, P.D. Boehm and J.A. Calder. 1980. Microbial biodegradation and the chemical evolution of *Amoco Cadiz* oil pollutants. Ambio 9: 277-283.

Weber D.D., D.J. Maynard, W.D. Gronlund and V. Konchin. 1981. Avoidance reactions of migrating adult salmon to petroleum hydrocarbons. Can. J. Fish. Aquat. Sci. 38: 779-781.

Wells, P.G. 1972. Influence of Venezuelan crude oil on lobster larvae. Mar. Pollut. Bull. 3: 105-106.

Wells, P.G. 1976. Effects of Venezuelan crude oil on young stages of the American lobster (*Homarus americanus*). Ph.D. Thesis, University of Guelph, Guelph, Ontario.

Wells, P.G. and J.B. Sprague. 1976. Effects of crude oil on American lobster (*Homarus americanus*) larvae in the laboratory. J. Fish. Res. Board Can. 33: 1604-1614.

Whittle, K.J., J. Murray, P.R. Mackie, R. Hardy and J. Farmer. 1977. Fate of hydrocarbons in fish. Rapp. P.-V. Réun. Cons. Int. Explor. Mer. 171: 139-142.

Widdows, J., T. Bakke, B.L. Bayne, P. Donkin, D.R. Livingstone, D.M. Lowe, M.N. Moore, S.V. Evans and S.L. Moore. 1982. Responses of *Mytilus edulis* on exposure to the water accommodated fraction of North Sea oil. Mar. Biol. 67: 15-31.

Williams, D.E. and D.R. Buhler. 1982. Purification of cytochromes P-448 from B-naphthoflavone treated rainbow trout. Biochim. Biophys. Acta 717: 398-404.

Winters, K., J.C. Batterton and C. Van Baalen. 1977a. Phenalen-1-one: Occurrence in a fuel oil and toxicity to microalgae. Environ. Sci. Tech. 11: 270-272.

Winters, K., C. Van Baalen and J.A.C. Nicol. 1977b. Water soluble extractives from petroleum oil: Chemical characterization and effects on microalgae and marine animals. Rapp. P.-V. Réun. Cons. Int. Explor. Mer. 171: 166-174.

Winters, K., R. O'Donnell, J.C. Batterton and C. Van Baalen. 1976. Water soluble components of four fuel oils: Chemical characterization and effects on growth of microalgae. Mar. Biol. 36: 269-276.

Wong, C.K., F.R. Engelhardt and J.R. Strickler. 1981. Survival and fecundity of *Daphnia pulex* on exposure to particulate oil. Bull. Environ. Contam. Toxicol. 26: 606-612.

CHAPTER 9

THE BIOLOGICAL EFFECTS OF PETROLEUM HYDROCARBONS IN THE SEA: ASSESSMENTS FROM THE FIELD AND MICROCOSMS

Robert B. Spies

CONTENTS

INTRODUCTION

Given our meager understanding of how marine ecosystems function and the causes of their variability, the question of whether chronic low-level petroleum contamination poses a serious threat to life in the sea is a particularly difficult one to answer completely (Walton, 1981). To the extent which we do not understand why things change in ecosystems, our answer will be incomplete—a situation which engenders various degrees of concern and motivates further research (Hardy *et al.*, 1977; Hedgpeth, 1978; Sanders *et al.*,1980). Realizing our limitations and the variable circumstances of petroleum contamination, the aim of this review is to evaluate the results of field and microcosm studies and the extent these results allow us to predict the outcome of continued and expanded offshore petroleum extraction, transportation and related sources of contamination. Emphasis will be placed on sublittoral studies for which 1) there are extensive biological and chemical analyses, 2) effects have been suggested at very low concentrations, 3) no effects have been claimed in areas of extensive contamination, or 4) insights into critical processes or effects are evident.

While the orientation of the review is toward understanding contaminated whole ecosystems, thc high concentrations and persistence of petroleum in sediments as well as the fixed character of their infaunal populations naturally steer most studies toward the benthos, with the notable exception of microcosm experiments.

The effects of petroleum on benthic communities, as well as other ecosystem components (plankton, nekton and epibenthos), will be discussed first with particular attention given to thresholds of toxicity, stimulation phenomena, the most sensitive organisms and circumstances which produce the most severe and lasting effects. Second, a comparison of several accidental spills, sites of chronic contamination, microcosm studies and offshore platforms will be made with a discussion of what factors could be responsible for the variety of outcomes. Finally, strategies that promise to advance our understanding of the outcome of chronic contamination will be discussed. Emphasis will be on finding a common denominator for comparing various outcomes.

THE EFFECTS OF PETROLEUM ON ECOSYSTEM COMPONENTS

Plankton and Nekton

By way of introduction, it should be said that because of turbulent mixing in the sea pollution biologists consider the impact of petroleum on plankton as an extremely difficult problem to assess in field studies. Many also consider the probable long-term impacts relatively minor (also because of turbulent mixing) compared to those on the benthos except under special circumstances such as chronic contamination of a water body with restricted flushing or if a massive spill were to take place during the spawning of important fish stocks with pelagic eggs and larvae. Although problems should not be ignored because they are difficult to study, the preponderance of evidence indicates that turbulent mixing and other processes combine to produce relatively short residence times of petroleum in the water column (e.g. Grahl-Nielsen *et al.*, 1979; Chapter 5) and that chronic effects on plankton are in most circumstances unlikely. There are probable exceptions to this, of course, in shallow estuaries, such as San Francisco Bay where the suspended particulates have hydrocarbon concentrations up to 6188 μg/g and the standing inventory of petroleum in the bay has been estimated to be at least 13 metric tons (DiSalvo and Guard, 1975). Possibly after some oil spills, sediment-associated oil may also become redistributed by water movement after the initial spill concentrations have subsided (Wolfe, 1978; Sanders *et al.*, 1980). Still there have been several accidental and some experimental spills where contamination of the plankton has been observed and transitory effects documented. Metabolic activity of bacteria appears to be quite sensitive to the presence of very low concentrations of hydrocarbons. Also the rather dramatic changes that can occur in enclosed water columns of microcosm experiments draw our attention to petroleum's effects (or implied effects) on plankton, especially in exposure systems where trophic linkage might have a role in the suggested effects of low-level petroleum exposure on benthic populations (Grassle *et al.*, 1981; Ritacco and Sastry, 1983).

Bacterioplankton

There are three major aspects of bacterial response to petroleum contamination: 1) alteration of species composition or community metabolism by selective encouragement of hydrocarbon degraders, biochemical induction of hydrocarbon-degrading enzymes in microbes or toxic inhibition of enzymes that degrade the usual sorts of available organic matter; 2) bacterial response may provide a major pathway for degradation of petroleum; and 3) numbers of bacteria in the water often increase dramatically in response to petroleum and may influence predator abundance (e.g., ciliates and flagellates) as well as compete for nutrients with algae. The first two aspects dominate the literature, while the third remains relatively unexplored (Skjoldal *et al.*, 1982), probably because marine biologists have been slow to appreciate the role of bacterial production in food webs and ecological studies often start with a premise of hydrocarbon toxicity.

The release of petroleum in sufficient concentrations to the sea usually results in selectively increased mineralization rates for petroleum hydrocarbons (Lee and Anderson, 1977; Lee *et al.*, 1978; Davies *et al.*, 1980; Skjoldal *et al.*, 1982). There may be a lag time in response, especially in pristine areas. In Saanich Inlet, British Columbia, 12 hours passed after addition of No. 2 fuel oil to the CEPEX (Controlled Ecosystem Pollution Experiments) enclosures before naphthalene mineralization rates measurably increased (Lee *et al.*, 1978). In the Loch Ewe (Scotland) experimental microcosms, peak mineralization rates for naphthalene did not occur until 8 to 10 days after addition of North Sea crude oil to an approximate concentration of 120 μg/l (Davies *et al.*, 1980). Also, in controlled ecosystems in a Norwegian fjord the potential for naphthalene mineralization also peaked eight days after addition of Ekofisk oil (Skjoldal *et al.*, 1982).

Degradation rate increases are selective. In the CEPEX enclosures the higher molecular weight aromatics fluorene, benzanthracene and benzo(a)pyrene were not measurably degraded (Lee and Takahashi, 1977). In the Loch Ewe experiments benzo(a)pyrene was not degraded, but the hexadecane mineralization rate did increase (Davies *et al.*, 1980). In a comparison of hydrocarbon degradation rates in amended water samples taken from two estuarine rivers in the southeastern U.S. and the Gulf Stream, the riverine waters had higher rates, and low molecular weight aromatics and alkanes were degraded relatively faster than high molecular weight aromatics (Lee, 1977).

Little information is available on threshold concentrations needed to elicit a response in hydrocarbon degraders. One month's exposure to 10 μg/l of No. 2 fuel oil in the CEPEX enclosures did not affect oxidation potential of hydrocarbons (Hodson *et al.*, 1977). On the other hand, Gunkel *et al.* (1980) have found large increases in hydrocarbon degraders in the vicinity of the North Sea oil fields, where water concentrations of the more persistent and soluble aromatics (naphthalenes, phenanthrenes, and dibenzothiophenes) were found to be on the order of 0.1 μg/l (Grahl-Nielsen *et al.*, 1979). This area did experience a major spill in 1977, and it is conceivable that elevated numbers of hydrocarbon degraders linger in the area.

Petroleum can inhibit bacterial heterotrophic activity, although some caution should be exercised in interpreting experimental results where toxic effects may not be easily distinguished from adaptive shifts in the metabolism of the bacterial community. Hodson *et al.* (1977) found that oil concentrations in CEPEX enclosures greater than 300 μg/l inhibited heterotrophic glucose utilization in a CEPEX experiment, while a concentration of 80 μg/l of Bunker "C" oil stimulated glucose utilization. Griffiths *et al.* (1981) studied the effects of Prudhoe Bay and Cook Inlet crude oils on glucose utilization rates in 215 arctic and subarctic water samples. Ten ml water samples were dosed with 10 μl of oil by surface addition. Mean glucose uptake rates were reduced from 37 to 58%, and, according to the authors, substrate kinetics suggested inhibition. The data also showed that near the Cook Inlet oil fields the reduction in glucose mineralization rates was less severe than in more pristine areas.

In contrast to these studies of cold water more pristine areas, Alexander and Schwarz (1980) studied the effects of similar concentrations of South Louisiana and Kuwait crude oils on glucose utilization by water samples from the Gulf of Mexico and found quite different results. Eleven of the 13 water samples showed no inhibition, and inhibition occurred only at the highest oil concentration tested (0.1%) in the other two. The authors attributed the general lack of effects to the relatively low toxicity of the oils. It also seems probable, however, that because their samples were taken in the vicinity of the Houston Ship Channel and the Mississippi River that the flora in the area could be adapted to chronic hydrocarbon exposure (as suggested by Griffiths *et al.*, 1981). I am unaware of comparable work for offshore southern California, which in some ways may be intermediate between Canadian and Alaskan waters and the Gulf of Mexico.

There is abundant evidence that petroleum contamination results in increases in the proportions of hydrocarbon degraders in the plankton and in many cases the absolute numbers of bacterioplankton. In CEPEX enclosures to which No. 2 fuel oil was added, a ten-fold increase in bacterial cells and cell clumps was observed in the first month of exposure (Lee and Takahashi, 1977). The cell exudates from phytoplankton may have influenced this result. In the Loch Ewe microcosms, addition of 100 μg of North Sea crude resulted in larger microbial populations as evidenced by increased amino acid mineralization rates (Davies *et al.*, 1980). Hagström (1977) measured a ten-fold increase in bacterial numbers following oil addition to a microcosm. In an experimental spill of weathered and unweathered Louisiana crude oil in a southeastern Virginia tidal salt marsh, densities of hydrocarbon-degrading bacteria increased several orders of magnitude, but the mean levels of chitonolytic, cellulytic and other heterotrophic bacteria and fungi did not differ from unoiled controls (Kator and Herwig, 1977).

The increase in hydrocarbon-degrading bacteria or total bacterioplankton following oil spills has been documented in several cases: for the *Arrow* spill in Chedabucto Bay, Nova Scotia (Cundell and Traxler, 1973); for the *Metula* spill in the Straits of Magellan (Colwell *et al.*, 1978); for the *Tsesis* oil spill in the Baltic Sea (Johansson *et al.*, 1980); and for the *Amoco Cadiz* spill off France's Brittany coast (Atlas and Bronner, 1981). Increases of oil-degrading bacteria have also been apparent around offshore oil platforms in Alaska (Kinney *et al.*, 1969) and in the North Sea (Gunkel *et al.*, 1980).

Phytoplankton

While the focus of this review is on field and microcosm studies, phytoplankton are a special case where phenomena seen in the field, and especially in microcosms, cannot be appreciated without reference to laboratory studies. Since the phytoplankton are the major primary producers in the oceans and certainly also in microcosms, it is important to venture into the laboratory literature to begin to understand the possible causes of the sometimes dramatic shifts in phytoplankton populations in oil effects experiments (see also Chapters 7 and 8). The factors affecting their rates of carbon fixation and other characteristics (e.g., size) will have important repercussions for the entire food web in long-term petroleum exposures. Several systematic reviews of petroleum's effects on phytoplankton and algae are available (O'Brien and Dixon, 1975; Vandermeulen and Ahearn, 1976; Connell and Miller, 1981); and therefore, the treatment here is more topical than it is exhaustive, focusing on inhibition and stimulation, interspecific sensitivities, influence of oil type, threshold phenomena and trophic and nutrient interactions.

Both inhibition and stimulation of phytoplankton have been measured in laboratory cultures. Most commonly, inhibition of growth rate has been shown (Galtsoff *et al.*, 1935; Mironov and Lanskaya, 1969; Mommearts-Billiet, 1973; Nuzzi, 1973; Pulich *et al.*, 1974; Soto *et al.*, 1975a,b; and Batterton *et al.*, 1978a,b). As expected this inhibition can be reflected in a depressed rate of carbon fixation (Kauss *et al.*, 1973). There are also many instances of phytoplankton and algal growth stimulation (Prouse *et al.*, 1976; Boney and Corner, 1962; Boney, 1974; Dunstan *et al.*, 1975; Parsons *et al.*, 1976). Oil exposure has had no measurable effects on several species (Hsaio, 1978).

The sensitivity of algae varies among major groups and sometimes within a species. There are several reports that green algae are more sensitive than diatoms, blue-green algae or flagellates (Winters *et al.*, 1976, 1977a; Batterton *et al.*, 1978a). The flagellates are probably the least sensitive to inhibition as a group (Dunstan *et al.*, 1975; Hsaio, 1978; Mahoney and Haskins, 1980). Such generalizations are based on relatively few species and may not be sustained with further study. Some idea of the possible variability in the sensitivity of closely related forms comes from a comparison of the toxicity of Nigerian crude oil to three strains of the cosmopolitan diatom *Skeletonema costatum*. In these strains the LC_{50}s ranged three-fold from 0.5 to 1.5 μg/l (Mahoney and Haskins, 1980).

As with other marine organisms, phytoplankton are more sensitive to certain types of petroleum. Fuel oil, particularly No. 2, fuel oil is more toxic to algae than crude oils (Mommearts-Billiet, 1973; Pulich *et al.*, 1974; Batterton *et al.*, 1978a). The general tendency has been to ascribe differences in oil toxicity to the naphthalene content of aqueous oil extracts. The elimination of toxicity of such extracts to algae by volatile stripping (Batterton *et al.*, 1978a) is consistent with this view; however, the presence of highly toxic aromatic amines (e.g. *p*-toluidine, Batterton *et al.*, 1978b) or toxic oxygenated compounds (pyroles, quinolines and idoles, Winters *et al.*, 1977b) may to some extent be responsible for the greater toxicity of some refined oils. Perhaps the toxicity of an aromatic amine, *p*-toluidine, to blue-green algae (Batterton *et al.*, 1978b) is related to the ability of some blue-green algae to metabolize aromatic compounds (Cerniglia *et al.*, 1979, 1980a,b) possibly converting them into potentially highly cytotoxic compounds (e.g., Kato *et al.*, 1983). Hutchinson *et al.* (1979) has shown that hydrocarbon toxicity to two species of microalgae is inversely related to water solubility for 38 compounds, mostly aromatic hydrocarbons.

With varying species sensitivity to oil and differences between oils, it is not surprising that determining threshold concentrations is difficult. Again, based on a somewhat scanty literature, the threshold for effects in the most sensitive conditions appears to be 20 to 50 μg/l for fuel oil extracts and considerably higher and more variable for crude oils. Fixation of $^{14}CO_2$ by natural phytoplankton populations was found to be inhibited by a No. 2 fuel oil at 50 μg/l (Gordon and Prouse, 1973). The growth of the diatom *Thalassiosira pseudonana* was affected at 40 μg/l of No. 2 fuel oil (Pulich *et al.*, 1974). In the CEPEX experiments the diatom *Ceratulina bergonii* declined after addition of No. 2 fuel oil at 40 μg/l (Lee and Takahashi, 1977), but whether this was a direct toxic effect or some sort of secondary effect, for example nutrient competition by microbes or the lack of vertical turbulence, is not known. Similarly reduced photosynthesis was measured in the Loch Ewe microcosms (Davies *et al.*, 1980).

Stimulatory effects of oil at low concentrations have also been reported. In both CEPEX and MERL (Marine Ecosystems Research Laboratory) microcosms, blooms of microflagellates followed additions of No. 2 fuel oil at 20 and 40 μg/l, respectively (Parsons *et al.*, 1976; Elmgren *et al.*, 1980). The pigmented flagellate that bloomed in the CEPEX exposures, *Chrysochromula kappa*, was also stimulated in culture by fuel oil concentrations of 50 μg/l (Parsons *et al.*, 1976). The dinoflagellate *Dunaliella tertiolecta* is stimulated by emulsified crude oil at concentrations of 106 μg/l (Prouse *et al.*, 1976). This same species as well as the coccolithophorid *Circosphaera carterae* were stimulated by concentrations of benzene, toluene and xylene at concentrations of 100 μg/l and below. No. 2 fuel oil with volatiles removed lacked the stimulatory effect on the former species (Dunstan *et al.*, 1975).

At slightly higher concentrations, Vargo *et al.*, (1982) found that 100-200 μg/l of No. 2 fuel oil increased diversity and standing crop of phytoplankton in the MERL microcosm experiments. In one study extremely high concentrations (up to 10,000 mg/l nominal concentrations) were found to have no effect on the growth of the flagellate *Chlamydomonas pulsatille* (Hsaio, 1978). Nutrient stress may affect toxicity thresholds. Batch-cultured marine diatoms deficient in phosphorus were found to be more susceptible to toxicity than those that were not so limited (Stoll and Guillard, 1974).

One aspect of laboratory studies that probably deserves more attention is the role that microorganisms might play in indirectly affecting growth conditions of microalgae. Since effects have been recorded on algae in the general range of 50 to 100 μg/l of petroleum and this is the range in which bacterial enhancement might be expected, the potential exists for nutrient competition and perhaps other phenomena that may not be representative of the conditions in nature where advection is a greater factor.

In accidental spills, controlled field experiments and microcosm experiments, the results of oil exposure are even less assignable to direct toxicity or growth stimulation due to possible multiple interactions. So, for instance, the microflagellate blooms in MERL microcosms at low oil concentrations have been ascribed to decreased grazing pressure by zooplankton, rather than direct stimulatory effects (Elmgren *et al.*, 1980; Vargo *et al.*, 1982). Zooplankton abundances in both oiled and unoiled treatments were very similar during the time of rapid increase of phytoplankton, but Elmgren *et al.* (1980) estimate that benthic filter feeders in the control tanks prevented the bloom from developing there. Exposure of laboratory cultures of the flagellate *Chrysochromula kappa* to 50 μg/l of No. 2 fuel oil induced growth; therefore, the bloom of this species during the CEPEX oil exposures could be due to direct growth enhancement (Parsons *et al.*, 1976; Lee and Takahasi, 1977).

Another explanation derives from the observations of Takahashi *et al.* (1975) that the decline of dominant phytoplankton (often centric diatoms) followed by blooms of flagellates is a natural successional sequence as nutrient depletion occurs. Menzel (1980) offers an explanation that may be related: the response to stress in a dynamic community is to favor growth of smaller organisms. Could it be that the addition of oil and the resultant stimulation of bacteria depletes nutrients and this causes the observed successions seen after some spills and in many microcosm experiments? The increased surface area-to-volume ratio of the smaller cells could make them more efficient at absorbing nutrients. The surfaces for microbial growth provided by tank walls would be expected to accelerate this phenomenon in microcosms. It is interesting that nitrogen limitation occurred in the oil-treated MERL microcosms (Elmgren *et al.*, 1980), but that in the Loch Ewe microcosms receiving North Sea crude oil (initial concentration, 100 μg/l) and nutrients, two species of the diatom genus *Leptocylindricus* remained dominant, although flagellates were continually present in small numbers.

The exact cause of the flagellate blooms is not fully understood. Besides direct stimulation, it is possible that: 1) flagellates compete well for nutrients at very low concentrations; 2) that many may be heterotrophs (at least nonpigmented forms) feeding directly on increased bacteria (Haas and Webb, 1979; Davis, 1982; Fenchel, 1982; Azam *et al.*, 1983); 3) the observed blooms result from relaxed grazing pressure (Elmgren *et al.*, 1980); or 4) diatoms may slowly sink in quiet waters allowing motile flagellates to increase (Eppley and Weiler,

1979). Carefully controlled laboratory experiments with phytoplankton cultures, mindful of bacterial activity, could differentiate between the first two possibilities.

Whatever the cause, phytoplankton stimulation has been noted in several oil pollution incidents. Increased productivity was indicated after the *Bravo* spill (Lannergren, 1978) and the *Tsesis* spill (Johansson, 1980). During the *Torrey Canyon* spill there were some indications of increased numbers of flagellates (including *Chrysochromula kappa*) (Smith, 1970). Also, a shelfwide plankton bloom was coincident with widespread spillage of Bunker "C" oil from the *Kurdistan* off Nova Scotia, although cause and effect are quite uncertain here and a bloom is expected at that time of year (O'Boyle, 1980).

In an experimental spill in a southeastern salt marsh, weathered and unweathered South Louisiana crude oils were introduced into several enclosures. In the exposure with fresh crudes, dissolved aromatic hydrocarbons reached a maximum concentration of 6.7 μg/l in 31 h, while in the enclosure exposed to weathered oil a maximum concentration of 7.7 μg/l was attained in only 6 h (Bieri *et al.*, 1977). Both oil enclosures experienced initial depressions of carbon fixation rates and ATP concentrations in the water column followed by a stimulatory effect only in the unweathered treatment. Primary productivity values returned to control levels after approximately a week (Bender *et al.*, 1977).

Another series of experimental spills was carried out in estuarine ponds and impoundments in Mississippi at nominal concentrations of 1.45 mg/l of Empire crude oil. This exposure resulted in a 44 to 65% depression of primary productivity and a 30 to 50% reduction in planktonic respiration after the spill. There was an apparent recovery but depressed values were still evident two weeks after the spill. In the impoundment spills with Arabian, Nigerian and Empire mix crudes, some depression of radiocarbon uptake was evident at first, followed by a significant enhancement in the Nigerian and Arabian crude-exposed impoundments (De La Cruz, 1982).

Zooplankton

As with the remainder of this review, results of laboratory tests of effects on single species will not be covered in any depth, rather emphasis is on field and microcosm studies. For thorough summaries of the effects of laboratory oil exposures, several reviews are available that include a treatment of the zooplankton literature (FAO, 1977; Anderson, 1979; Connell and Miller, 1981; Chapter 8), and one review focuses specifically on zooplankton (Corner, 1978).

To briefly summarize, lethal oil concentrations (96-h LC_{50}s) are in the range of 0.05 to 9.4 mg/l (Chapter 8). A whole array of sublethal effects have been seen below 1 mg/l and these include feeding and other behaviors as well as reproduction and development, which seem to be particularly sensitive. Connell and Miller (1981) have also suggested that effects can be seen with concentrations as low as 10 μg/l.

In the 1975 CEPEX experiments, as microflagellates increased and diatoms decreased following treatment with No. 2 fuel oil, the microzooplankton, mainly the tintinnid protozoan *Helicostomella subulata* and rotifers, increased (Lee *et al.*, 1978). An addition of naphthalenes to the CEPEX systems in 1976 in concentrations of 40 μg/l had somewhat different results. Naupliar copepods decreased, microflagellate increases were not seen and rotifer populations were similar in control and oiled exposures. The ctenophore *Pleurobranchia pileus* decreased when nominal concentrations of 160 μg/l naphthalenes (decreasing to 0 after 20 days) were introduced (Lee and Anderson, 1977).

In the Loch Ewe microcosms the addition of seawater-mixed North Sea crude oil to a concentration of 100 μg/l resulted in a number of effects on zooplankton (Davies *et al.*, 1980). Mainly, the calanoid copepod population declined severely by 15 days. Not only were adults affected, but development of nauplii and eggs, laid just after the oil was added, was severely reduced, although egg production by the three main copepod genera, *Acartia*, *Pseudocalanus* and *Temora*, was similar in treatment and control enclosures. Carnivorous zooplankton was also reduced in the oil enclosure, although it is uncertain how much was due directly to oil and how much was an indirect effect of reduced prey populations. The authors concluded that the main effect of oil was a continual instability of the copepod populations.

In the MERL microcosms a variety of zooplankton effects have been described after the addition of No. 2 fuel oil in various experiments. Respiration and excretion rates of zooplankton were affected by oil exposure, showing stimulation at low concentrations and inhibition at higher concentrations. The inhibition was irreversible at 181 μg/l and stimulation was seen as low as 91 μg/l. Oxygen-to-nitrogen ratios of copepods decreased at the lower concentration, but increased at the higher concentration (Vargo, 1981), which may reflect nitrogen limitation of these systems. In these same MERL experiments, zooplankton biomass and total zooplankton both decreased at these exposures relative to controls, while total numbers of zooplankton species were very similar during a year of recovery. The dominant zooplankters, two copepod species, showed somewhat different responses. At 190 μg/l *Acartia clausi* decreased 69% relative to the control but in the following year increased 2% relative to the control at a lower exposure (90 μg/l). The other copepod, *Acartia tonsa*, was much denser than controls in both exposure levels and in the following recovery year (Oviatt *et al.*, 1982).

The results of the oil exposures in microcosms on zooplankton should not be accepted uncritically as representing long-term effects in the sea. While there undoubtedly was some cumulative sublethal toxicity, many of the fluctuations are probably due to trophic interactions, and it is not yet clear how wall effects, patchiness and lack of horizontal mixing and, in the case of CEPEX, vertical mixing, may have combined with oil effects to produce the observed changes. In other words to what extent did enclosure effects influence trophic interactions? Also, it seems unlikely that such concentrations of oil would persist for long in the sea except under conditions of persistent contamination of a water body with restricted circulation.

During some accidental spills, effects on zooplankton have been apparent, and in about as many cases no effects have been apparent where there was a reasonable effort made to study zooplankton. Where zooplankton has been less well studied, naturally few effects have been found. Zooplankton was surveyed following the *Torrey Canyon*, Santa Barbara and *Argo Merchant* oil spills, but the studies were less than definitive and little could be ascribed to oil effects. In the *Torrey Canyon* spill, no quantitative data on plankton were taken but qualitative tows under oil slicks showed copepods in normal numbers and in apparent healthy condition (Smith, 1970). In the Santa Barbara oil spill, the plankton studies did not start until three to four months after the spill; and although a series of 11 stations was sampled over a year, there were little comparative data and no effects could be attributed to the oil (McGinnis, 1971). In the *Argo Merchant* spill, oil contamination in the guts and fecal pellets of zooplankton was observed and fish eggs had some evidence of oil contact (Kuhnhöld, 1978). Longwell (1977) claimed some mortality of fish eggs and malformed pollock embryos were due to the spill. No effects were seen in the *Kurdistan* spill on zooplankton and ichthyoplankton that could be related to oil (O'Boyle, 1980), but again the sampling carried out could not be expected to detect anything but a very drastic effect.

For several spills fairly thorough sampling was carried out, but still the definitive study on spill effects on zooplankton has not been done. For the *Bravo* spill in the North Sea, the chemical data accompanying the biological samples was quite good. No acute effects on the zooplankton were detected, although levels of hydrocarbons up to 256 μg/l of oil were measured in the water (Bjørke, 1977).

In the massive *Amoco Cadiz* spill of Arabian light crude, fairly well-documented effects were seen on zooplankton. There were immediate effects on zooplankton which persisted in the offshore areas for 15 days and in the inshore areas for 30 days. High mortality was documented as well as a widespread effect on zooplankton metabolism, as indicated by amylase-to-protein and trypsin-to-protein ratios (Samain *et al.*, 1981). Copepods were also observed with oil in their guts (Mackie *et al.*, 1978). The respiration of zooplankton was also apparently elevated (Hendrickson *et al.*, 1978 as cited in Wells, 1981).

In the *Tsesis* oil spill in the Baltic there were reduced numbers of zooplankton found in the immediate vicinity of the tanker wreck when compared to a distant control station. No changes in species composition were apparent, but 50% of the net zooplankton within the affected areas was visibly contaminated in the first few weeks and this decreased to about 20% after three weeks (Johansson *et al.*, 1980).

The *Arrow* spill of No. 6 fuel oil in Chedabucto Bay, Nova Scotia resulted in a fine dispersion of oil that was turbulently mixed down to 10-m depth. It was estimated that up to 10% of the oil in the water column was ingested by or associated with zooplankton. The feces collected contained up to 7% oil (Conover, 1971).

Very little has been reported on zooplankton in areas of chronic pollution which could be attributed solely to petroleum hydrocarbons. There is a large amount of data on the zooplankton of contaminated estuaries and coastal areas, but there appears to be no way to separate petroleum effects from those of other contaminants. In the Buccaneer Oil Field study, Middleditch *et al.* (1979) measured alkanes in net zooplankton and found concentrations ranging from 0.25 to 3.58 μg/kg. In most instances the alkane distributions appeared to be dominated by biogenic compounds, although some samples had petroleum characteristics.

Nekton and Motile Epibenthos

This discussion is limited to fishes since the evidence of effects or lack thereof on other motile invertebrates is extremely poor. Since it is so notoriously difficult to do accurate fish population studies in the sea, and the question of petroleum avoidance by fishes is a relatively open one, most of our reliable information from the field is based on fish kills or observed sublethal effects on individual fish. Such effects have been noted for spills, but I am unaware of reported effects (not including contamination or mixed-function oxidase (MFO) activity) on fishes attributable to petroleum in a chronically contaminated offshore area, although some finrot apparently occurs in fishes from the Buccaneer Oil Field (Middleditch, 1981). The argument could also be made that properly designed studies have not been carried out on the effects of oil on fish populations (Carney, Chapter 14). Rather contrasting views of the potential problem of oil pollution effects on fishes are presented by Malins and Hodgins (1981) and Payne (1982). A recent review of Rice (1981) treats the large literature of laboratory toxicity studies as well as the limited field research. Also, Sindermann (1982) has recently reviewed the relationships between oil pollution and disease in fish. One laboratory study worth mentioning here is the work of Kuhnhöld (1978) that shows effects of oil on flounder embryonic development at concentrations of 100 ppb.

Extensive fish kills have been associated with only a few spills. In the *Florida* spill large numbers of scup (*Stenotomus chrysops*) and tomcod (*Microgadus tomcod*) washed up on Silver Beach, North Falmouth (Hampson and Sanders, 1969). After the *Amoco Cadiz* spill large numbers of fish were reported washed up on the beach at Brest (O'Sullivan, 1978). Unfortunately the doses of oil causing these kills are not known. Sublethal effects of petroleum on fishes have been noted in a few more instances, but still the field data are quite sparse. On the other hand, the large populations of fishes that are associated with offshore platforms have been well documented and are common knowledge (see, for example, Simpson, 1977). A number of possible negative scenarios have been suggested in connection with this phenomenon, such as biomagnification of pollutants with subsequent return to man or sequestering of populations away from areas without platforms. The data are sparse or nonexistent to settle these concerns.

There is an interesting possible indirect effect of oil on fish that has been postulated as a result of the *Tsesis* oil spill, where the reproduction of herring was greatly reduced. Apparently amphipods, which normally grazed the fungal growths on the benthic eggs of these fish, were reduced to the point where the fungus was not controlled, causing high mortality to eggs (Nellbring *et al.*, 1980).

The best-documented effect of spilled petroleum on fish comes from the work of Haensly *et al.* (1982) in the aftermath of the *Amoco Cadiz* spill. A large number of histopathological abnormalities was described in plaice (*Pleuronectes platessa* L.) from Aber W'rach and Aber Benoit up to two years after these confined estuaries were originally heavily contaminated. The predominant conditions were fin and tail necrosis, hyperplasia and hypertrophy of gill lamellar mucous cells, gastric gland regeneration, increased hepatocellular vacuolization (lipid), increased concentration of hepatic macrophage centers and lateral trunk muscle fiber degeneration. Other conditions occurred with lesser frequency. None of the hydrocarbons identified in muscle or livers could be correlated with *Amoco Cadiz* oil. The lack of detectable aromatic hydrocarbons in tissue was probably the result of the activity of mixed-function oxidases (MFO), an aromatic hydrocarbon-hydroxylating enzyme system, and conjugating enzymes most active in the liver. While the circumstantial evidence for a oil-related effect was quite definitive, the MFO activities of affected and non-affected fish could have been informative.

Stegeman and Sabo (1976) and Sabo and Stegeman (1977) reported that *Fundulus heteroclitus* taken from Wild Harbor marsh following the *Florida* spill had a lower rate of net lipogenesis than those from an uncontaminated marsh. Wild Harbor fish were apparently metabolizing petroleum hydrocarbons entering their tissue, as evidenced by elevated MFO (Burns, 1976; Stegeman, 1978). Parent hydrocarbons were not in their tissues five years after the spill, although the marsh remained contaminated (Burns and Teal, 1979).

Although not without limitations, the measurement of MFO activities of P-450 enzymes in field populations can be a sensitive and useful indicator of petroleum exposure. Since aromatic hydrocarbons are often metabolized to undetectable levels following exposure (e.g., McCain *et al.*, 1978), these measurements are probably a more reliable indicator of exposure of fish to aromatic hydrocarbons than measurements of these compounds in tissues. Field induction of hepatic MFO activity has been apparent after small spills (Stegeman, 1978; Walton *et al.*, 1978). Under conditions of chronic contamination, induced MFO activity has been apparent in the North Sea associated with areas where offshore platforms have disposed of oil-based drilling muds (Bell *et al.*, 1983). Also around the natural petroleum seeps in the Santa Barbara Channel, two species of sanddabs (*Citharichthys*) have elevated levels of hepatic aryl hydrocarbon hydroxylase (AHH), a particular MFO activity (Spies *et al.*, 1982).

Since MFO activity can also be induced in fish by PCBs and probably some other xenobiotic compounds (Gruger *et al.*, 1977) and can vary with sex (Stegeman, 1980), reproductive state (Walton *et al.*, 1978) and season (Spies *et al.*, 1982; Walton *et al.*, 1983; Lindström-Sappä, 1985), a good understanding of these sources of variability for each species is essential before field data can be interpreted or the number of fish needed to detect significant induction estimated.

One isozyme that is highly inducible by petroleum appears to be either a 54×10^3 dalton protein in some marine species (Stegeman *et al.*, 1981; Spies *et al.*, 1982; Klotz *et al.*, 1983) or a 57×10^3 dalton protein in trout (Elcombe *et al.*, 1979; Stegeman *et al.*, 1981). These isozymes may be quite inactive in pristine environments and have large increases in contaminated environments. They can be detected by gel electrophoresis (Stegeman *et al.*, 1981), and antibodies to one inducible form in scup *Stenotomus chrysops* (P450E) have been made (Kloepper-Sams *et al.*, 1986) and applied to detecting apparent induction in deep sea fish by chlorinated hydrocarbons (Stegeman *et al.*, 1986). The use of inhibitors of the activity of induced P-450 isozymes, e.g., P-450E which is induced by PAH-type inducers in fish and is inhibited by 7,8-benzoflavone (Klotz *et al.*, 1983), are also useful probes in field studies.

Beyond the monitoring application of P-450 enzymes, current research may reveal how they could play a role in altering hormone balance (Ungvary *et al.*, 1981; Forlin and Haux, 1985) or the viability of gametes (Spies *et al.*, 1985). Should such causal relationships exist, then MFO activity will assume more importance than just a sensitive indicator of contamination. The potential for population-level effects mediated through chronically-elevated MFO must be regarded as a serious possibility deserving further investigation.

An additional tool for detecting exposure of fish to oil, specifically polynuclear aromatic hydrocarbons, is to monitor their bile using a combination of high performance liquid chromatography and fluorescence detection (Krahn *et al.*, 1984).

Excellent reviews of hydrocarbon metabolism in fish (Stegeman, 1981) and invertebrates (Lee, 1981) are available (see also Chapter 8).

Benthos

The integration and reflection of pollutant effects in benthic communities is a paradigm in marine environmental research, particularly for invertebrate macroinfauna. Consequently, benthic community studies are usually the dominant component in field studies of petroleum pollution. The literature is quite large, and here again effort is more selective than exhaustive. In this section I will be dealing mainly with results from refereed journals. The reports of large offshore investigations will be treated separately.

Sediment Microbes

The response of benthic microflora to oil has not been studied in microcosms. CEPEX had no native sediments, and in the MERL and Loch Ewe microcosms sediment microbes were not directly studied. Oviatt *et al.* (1982) made measurements of benthic respiration and nutrient flux in MERL tanks: fluxes of oxygen, ammonia, nitrite, nitrate, phosphorus, dissolved organic phosphorus and silicate were all appreciably lower in the oil recovery microcosms (tanks that had received chronic oil inputs for the previous two years at 90 to 190 μg/l of water-accommodated No. 2 fuel oil and then received no fresh contamination).

A number of experimental and simulated spills have indicated that in most circumstances an increase in hydrocarbon degraders and degradation potential is to be expected; other heterotrophs may or may not be positively influenced, and various microbial functions can be affected. In an experimental spill of oil in an intertidal salt marsh in southeastern Virginia, the numbers of hydrocarbon degraders increased several orders of magnitude within several days and remained high for a year. Other heterotrophs did not differ between treatments (Kator and Herwig, 1977). In an experimental spill of a heavy (No. 5) fuel oil in a Georgia salt marsh, the sediments showed a significant reduction in CO_2 production and a significant increase in total adenylates. Assayed degradation potentials for aromatic hydrocarbons increased (e.g., for phenanthrene) to seven times the control value (Lee *et al.*, 1981). In a simulated spill using sediments from Kasitsna Bay, Alaska, samples were treated with 50 ppt Cook Inlet crude oil. Many were amended with various sorts of organic matter, and they were all returned to the field. After a year changes in a variety of processes were evident and varied with the type of organic material added. In non-amended sediments, oil significantly depressed glucose uptake, CO_2 evolution, N_2 fixation, denitrification and phosphatase activity. There were significant increases in respiration and methane evolution (Griffiths *et al.*, 1981). Unfortunately data on pore water chemistry were not available for the small incubation bottles that were set out in the field. One would expect to see some of the reported changes accompanying anaerobiosis.

Most microbial studies of accidental spills occur well after the fact. So for the *Metula* spill in the remote Straits of Magellan, Colwell *et al.* (1978) found evidence of increased activity of petroleum degraders two years after the spill and did a "mini" spill to determine microbial responses to fresh oil. This resulted in an immediate and dramatic increase in aerobic heterotrophs and petroleum degraders relative to other functional types (e.g., chitin degraders). In a re-examination of the Chedabucto Bay spill after six years, only two of 76 sediment samples had large populations of petroleum degraders (Stewart and Marks, 1978). In one of the few cases where microbial work was occurring at the time of a major spill, Juge (1970) described elevated bacterial numbers in sediments at the distal end of a sewage-affected gradient following the Santa Barbara spill.

Larger proportions of petroleum degraders in the sediment flora also occur in areas of chronic contamination. This has been described for such contrasting sites as Chesapeake Bay (Walker and Colwell, 1973), various northwestern Atlantic coastal sites (Mulkins-Phillips and Stewart, 1974), areas of the Mediterranean (Azoulay *et al.*, 1983), and oil fields in the North Sea, where up to 100% of sediment microbes are estimated to be petroleum degraders (Gunkel *et al.*, 1980).

On the beaches around Coal Oil Point, near Santa Barbara, California an area chronically-contaminated by natural submarine petroleum seepage, the hydrocarbon-utilizing potential of incubated sediment samples was higher than in other less-contaminated areas (Caparello and LaRock, 1975). In this same area, benthic studies of active sublittoral seepage areas have been carried out (Spies *et al.*, 1980; Davis and Spies, 1980). In these sediments, there is a gradient of increasing sediment ATP with increasing amounts of fresh oil, that appears to be mainly associated with the <150-μm sediment fraction, suggesting increasing microbial biomass (Davis and Spies, 1980). Also, often associated with areas of intense seepage are mats of the filamentous, mixotrophic, sulfide-reducer, *Beggiatoa* sp., thus it is apparent that the carbon, sulfur and nitrogen cycles (since this species probably also fixes N_2 and assimilates nitrate, Nelson *et al.*, 1982) are probably affected by oil degradation in this environment. Degradation rates of several radiolabeled hydrocarbons measured in sediment slurries were greater in areas of seepage, as well as sulfate reduction rates, measured in intact cores. Further, in areas of seepage radiolabel was cellularly incorporated, while in a nearby non-seepage area such fixation did not occur (Montagna *et al.*, in press) The measurement of various naturally-occurring isotopes in components of this system, mainly the infaunal invertebrates, has suggested that petroleum degradation and chemoautotrophy are tightly coupled processes in sediments and have a key role in passing petroleum carbon and energy into the food chain (Spies and DesMarais, 1983).

Deveraux and Sizemore (1982) have isolated 62 bacterial strains on hydrocarbon media from sediment samples taken from various parts of Galveston Bay. For all samples, plasmids (extranuclear DNA that can carry hydrocarbon-degrading genes) were detected in 21% of the strains isolated on crude oil, while plasmids occurred in 12% of those strains isolated on media with polynuclear aromatic hydrocarbons. Plasmids do not appear to be important in the strains isolated from a previously-uncontaminated site experiencing a recent spill, but they do appear to be more important in a chronically-contaminated area of the bay, where plasmid incidence was highest in strains isolated on aromatic hydrocarbons. From these results it would be tempting to postulate two levels of response, an immediate and dramatic shift to favor degradation of labile hydrocarbons, followed by an eventual selection under conditions of chronic contamination for a flora capable of degrading the more refractory components of petroleum. This would seem worth investigating further.

Although I have only touched briefly on the subject of biodegradation rates of hydrocarbons (see reviews by Walker and Colwell, 1976; Colwell and Walker, 1977; Bartha and Atlas, 1977; Atlas, 1981; Chapter 7), it seems apparent that there are few accurate *in situ* measurements. Most rates have been measured on samples taken to the laboratory and made into slurries and measured out of context of the original chemical environment (see Walker and Colwell, 1976 for a discussion of methodology), although some studies have been done where the weathering process has been measured *in situ* in containers put into the environment. More research and method development is needed here. The implications to other benthic organisms of microbial population shifts in response to oil is another area needing further work.

Meiobenthos

There have been few field and microcosm studies of petroleum effects on meiobenthic communities. In the MERL microcosms, where the exposure tanks received a 6-month dose of water-dispersed No. 2 fuel oil averaging 190 μg/l, the meiofauna showed an irregular decline during and extending beyond the period of oil addition. Total meiofaunal density was almost always higher in the control tanks. However, density declines were also evident in the control tanks over the same period. Whatever processes were responsible for those declines, they were apparently either accelerated by the addition of oil or additive with oil toxicity.

Harpacticoid copepods declined drastically in both treatments, but were the lowest in the oil exposures $3\frac{1}{2}$ months after the start of the exposures. The harpacticoid population had not recovered when the experiment terminated 4 months later. About a month after the end of the oil additions, the populations in the control tanks reached approximately the same low level. The ostracods had very low densities in both treatments from the start, but the oiled tanks did not experience the mid-summer blooms seen in the control tanks. Foraminifera showed immediate population increases in response to oil and remained higher than the control tanks until the oil additions stopped (Grassle *et al.*, 1981).

The ciliates were consistently more abundant in the MERL oil exposures, an outcome consistent with increases of planktonic ciliates in the CEPEX enclosures (Lee and Takahashi, 1977). Lanier (1978) had also shown stimulation of ciliates in experimental spills on saltwater ponds. This was consistent with the earlier findings of Andrews and Floodgate (1974) that ciliates are often associated with oil in sea water, apparently feeding on the bacteria associated with oil.

The remainder of our information on meiofaunal community response to oil comes mainly from several spill studies. In an experimental spill in Louisiana, a South Louisiana crude oil was applied to a littoral *Spartina alterniflora* marsh with a dose of 2 l/m^2. Meiofauna was sampled before the experiment and then on days 2, 5, 10, 20, 30, 60, 95 and 144. No mortality could be measured in any group of meiofauna, although a 2-cm thick layer of oil formed on the sediment surface. Some nematodes and copepods showed significant increases on two sampling dates. After 144 days copepod densities did decrease, indicating a possible effect on this group (Fleeger and Chandler, 1983).

The effects of a marine diesel oil spill on the meiofauna of a beach in Hong Kong was studied by Wormwald (1976). Harpacticoid copepods were apparently most drastically affected, being only 2% of their density in a control area. Recovery of this group did not occur until sediment oil content dropped below 1000 to 6000 ppm. Overall, the nematodes were much less drastically affected. There were some indications that the meiofauna penetrated farther into the sediments as oil concentrations dropped. Anaerobic conditions persisted in the affected sediments after the spill. The recovery of the meiofauna was well underway in 15 months.

A somewhat similar meiofaunal response was seen in a littoral beach in South Africa after oil washed ashore from the collision of the *Venpet* and *Venoil*. Harpacticoid copepod populations were clearly depressed, but had recovered in 6 months. Nematodes appeared to be largely unaffected (Fricke *et al.*, 1981). Rather large nematode populations have been reported from beaches affected by oil from the *Amoco Cadiz* (Chasse, 1978). They also occur in large numbers in bacterial mats overlying active submarine petroleum seepages (Spies *et al.*, 1980; Montagna and Spies, 1985).

Meiofaunal studies were also carried out in the wake of the *Tsesis* spill (Elmgren *et al.*, 1983). At the most heavily oiled station (20), total meiofauna, excluding nematodes but including turbellaria, kinorhynchs, harpacticoids, ostracods and temporary meiofauna decreased relative to densities found three years previously at the same station and when compared to a contemporary control station (15). Again, nematodes appeared to be relatively less affected. Interpretation of the effects is complicated by natural fluctuations (e.g., a sharp decline in meiofauna at the control station the summer following the spill) and the lack of detectable petroleum hydrocarbons in sediments from the impacted area. Apparently most of the hydrocarbons in the benthos were associated with the difficult-to-sample flocculent layer. However, evidence of benthic contamination was apparent in the tissues of the bivalve *Macoma balthica*, which feeds in the flocculent layer.

A spill of light Arabian crude and Bunker C fuel oil from the grounding of the *Monte Urquiola* resulted in about 30,000 tons of oil washing ashore in northern Spain. Giere (1979) made a study of the meiofauna of the beaches, both exposed and protected, and compared them to an uncontaminated reference beach. In the most heavily impacted areas (at Mera), the meiofauna was almost totally obliterated by a thick layer of oil. In areas of moderate contamination, only a few nematodes survived. In contrast the reference beach (Corme) had a variety of harpacticoid copepods and nematodes. A year later all locations showed signs of recovery and the moderately-oiled sites had almost fully recovered.

Meiofauna associated with *Beggiatoa* mats found on natural oil seeps near Isla Vista, California were recently studied (Montagna and Spies, 1985). The *Beggiatoa* mats occur directly on areas of active seepage. The hydrocarbon concentrations in the sand below these mats may be 50% or greater with dissolved pore water concentrations of 1 ppm, while outside the mat areas where lower seepage rates prevail total hydrocarbons are in the range of 3000 to 10,000 ppm and the pore water concentration are 45 to 100 ppb (Stuermer *et al.*, 1982). Meiofauna densities are half as large in the more heavily-oiled *Beggiatoa* mats compared to the sites adjacent to the mats. In the mats nematodes comprise 96% of the meiofauna and harpacticoid copepods 1%, but adjacent to the mats nematodes comprise 70% and harpacticoids 19% of the populations. However, the density of harpacticoids adjacent to the bacterial mats is unusually high (474 individuals/10 cm^2) for an area where total hydrocarbons exceed several thousand ppm.

Macroinfauna

Extensive investigation of macroinfaunal communities of microcosms has only been done in the MERL tanks. The effects of oil exposure were measured in these systems during three consecutive years, 1977-1979. In 1977 periodic doses of dispersions of No. 2 fuel oil for 25 weeks resulted in a time-averaged mean of 190 μg/l in the water and after 20 weeks the sediment concentration was 109 μg/g (dry). During 1978 the dose was more evenly applied and resulted in a time-averaged mean water concentration of 90 μg/l. The exposure lasted four months and then the populations were studied for an additional year after the oil dose was complete. Results of the macroinfaunal studies during the 1977 experiment are reported by Grassle *et al.* (1981) and Elmgren *et al.* (1980). The results of the second and third years' efforts are included in an overview paper (Oviatt *et al.*, 1982).

In the 1977 experiment, although species diversity was not affected (see Smith *et al.*, 1979, for a discussion of this phenomena), the densities of the infauna declined during the nine-month experimental period. The main difference between treatments was a weakly developed summer density maximum in the control tanks which contrasted with a steady decline in the oiled tanks. Mean densities in both tanks converged by September. Large crabs in two of the controls could have conceivably had an influence on infaunal density.

The most abundant species at the onset of the 1977 experiment was *Mediomastus ambiseta*, which showed a pattern of fluctuation somewhat similar to that of the total infaunal density. In the oil treatments there was a brief increase at the start of the experiment followed by a steady decline and by late June, a complete decline. In the control tanks there was a density peak in April followed by a complete decline by August. At the end of the experiment the number and biomass of larger animals (not sampled during the experiment) were significantly higher in the unoiled tanks. The only consistent increases seen in the tanks, other than the meiofauna groups already mentioned, were in the polychaete *Chaetozone* sp. and in total polychaete larvae. The ampeliscid amphipod *Ampelisca abdita* was severely reduced in the oiled treatments. The same general effects described for 1977 in some detail by Elmgren *et al.* (1980) and Grassle *et al.*, (1981) were also apparent from the data for subsequent years (Oviatt *et al.*, 1982). It should be noted that in 1977 (and presumably to some extent in the other years) that total inorganic nitrogen was severely limited in the oiled tanks (never higher than 1.2 μg atoms/l) as contrasted with the controls (mean values above 4 μg atoms/l for most of the summer) (Elmgren *et al.*, 1980). This limitation seems to have been reflected in the zooplankton oxygen-to-nitrogen ratio, which increased with higher oil concentrations (Vargo, 1981).

There are a number of experimental spills and oiled-sediment tray experiments where effects on the macrobenthos were measured. In an experimental spill in a marsh tidal stream system in the York River, Virginia, the effects of fresh and weathered South Louisiana crude oil were determined on enclosed portions of the marsh. The enclosures were open below the water surface for water exchange. Each enclosure was about 800 m^2, consisting of some intertidal areas and open water, but mostly marsh. Each oil exposure received 3 bbl (570 l) of oil. Dominant macrofauna populations were followed for 39 weeks. Nereid polychaete, chironomid and amphipod populations were all depressed during most of this time in the oil treatments (Bender *et al.*, 1977). The oligochaete, *Peloscolex* sp., however, was stimulated in the oil treatments. Generally, weathered and unweathered oil had similar effects on the benthos, an interesting result considering other evidence that it is the volatile fractions of the oil that are most toxic. Because the weathering of the oil took place in outdoor tanks before the onset of the experiment, the production of toxic compounds from photooxidation seems a distinct possibility. Although concentrations of water-accommodated aromatic compounds peaked shortly after the spills at 6 to 7 ppb and some 76 h after the spill in the fish *Fundulus heteroclitus*, there was no indication of how persistent hydrocarbons were in the sediments of the enclosures (Bieri *et al.*, 1977). Shaw *et al.* (1976) exposed *Macoma balthica* for up to 44 days to oiled sediment in the field without any apparent mortality, but did see an effect when oil was later applied directly to the sediment surface.

In a series of experiments to test the effect of oiled sediments on the recruitment of benthic organisms, trays of sediment were set out in the intertidal zone of Sequim Bay, Washington (Anderson *et al.*, 1978). Initial oil concentrations in the separate treatments were: 5000 ppm (I), 6000 ppm (II) and 700 ppm (III). The higher concentrations were associated with coarse sand while some finer sediment was mixed into III. By 100 days, concentrations in I and II had decreased by 82 to 88% but only by 21% in III. A variety of different invertebrates was collected from these and control sediment trays in four sets of samples over a year. Because of apparent tray-related effects on variability, crustacean data were discounted and data on five species, two bivalves and three polychaetes, were compared. None of the results of this comparison indicated an inhibition of recruitment of these species, although between-replicate variability was quite high.

The effects of weathered oil from the IXTOC-I on sediments colonized by benthic organisms was studied near Port Aransas, Texas (Kalke *et al.*, 1982). Clean sand was colonized in the laboratory seawater system and on the seafloor for a period of eight weeks. Sixty grams of oil were added to some replicates and these were maintained, along with untreated replicates in the laboratory for an additional four weeks. At the end of the experiment there were no apparent effects of the oil on the laboratory-colonized sediments. In contrast, the oil-treated sediments colonized in the field showed reduction in nearly all faunal parameters (e.g., species, number, biomass, density, diversity), although only total density and density of some species were significantly different. A cluster analysis of the fauna suggested that field-colonized trays treated with weathered oil were different from untreated trays. Oil treatments also had a reduction in the depth of the redox potential discontinuity. It is not known whether the observed effects were due directly to toxicity or to effects from the probably higher oxygen demand of the oiled sediments. It is also not known whether enclosing the sediments in containers might exacerbate the effects of the oil in relation to what might occur in unenclosed sediments.

The number of studies on marine benthos following accidental spills is quite large and space does not permit, nor is it my purpose to attempt a comprehensive treatment of this literature. I have decided to focus mainly on three spills in which substantial amounts of oil reached the benthos: the Chevron Main Pass Block 41 spill that occurred in offshore Louisiana in 1970, the *Florida* or West Falmouth oil spill that occurred in Buzzards Bay in 1969 and washed into Wild Harbor marsh and boat basin and the *Tsesis* oil spill that occurred in the Baltic Sea in 1977.

The Chevron Main Pass Block oil spill occurred in 1970 from a platform in the Gulf of Mexico 11 miles east of the Mississippi River Delta (McAuliffe *et al.*, 1975). At least 65,000 bbl of oil were spilled over a three-week period. Approximately 2000 bbl of dispersants were applied. The greatest concentrations of oil in the water at the platform and one mile away were: dissolved hydrocarbons, 0.2 to 0.001 ppm; oil-in-water emulsion, 71 to 1 ppm and dispersant, 1 to 3 ppm to not detected (<0.2 ppm). Most of the hydrocarbon input to the sediments appeared to be within 5 miles of the platform where sediment samples (taken with the macrofauna samples) had C_{12}-C_{23} hydrocarbons from 25 to 105 mg/l (mean 31 mg/l). The hydrocarbons showed evidence of rapid weathering, losing most of the alkanes within 40 days. Benthic samples were taken over a large area, including inshore bays, close to the Mississippi Delta in Chandeleur and Breton Sounds and clustered around the platform. There were three replicate suction dredge samples taken at each station and pooled to give a total sample area of 0.3 m^2. Unfortunately, pooling replicates eliminated the possibility of calculating within-station variability. These were collected in a bag with 1.0-mm mesh and then later sieved through a 1.2-mm sieve screen. Many of the smaller organisms were undoubtedly lost using this procedure. Samples were taken in both 1970 and 1971. Shannon-Weaver (Shannon-Wiener) indices and crustacean-polychaete ratios were calculated for each station (see Chapter 14 on use of derived values).

One-way analysis of variance was used to compare the various biological parameters with distance from the platform and stepwise multiple regression analysis was used to determine correlations of physical and chemical parameters with various biological measures. In an effort to compensate for the skewing effect that large number of bivalves (*Abra aequalis* and *Mulinia lateralis*) had on diversity, the Shannon-Weaver statistic (H') was calculated with and without (corrected) these large values. It was apparent that lesser numbers of species were characteristic of samples taken close to the delta and in the inshore bays.

The analysis of variance for the various biological parameters as a function of distance from the platform showed significant correlations for numbers of species, H' and H' (corrected) and the crustacean/polychaete ratio, which would indicate a negative influence of the platform or the spill on these biological parameters. In contrast to this analysis, Sharp and Appan (1982) claim that there was no relation between H' and the distance from the spill. However, these analyses were carried out with samples up to 40 miles away and included a wide variety of environments. Because the platform is only 11 miles from the delta, the significant trends may well be more a measure of the effects of riverine outflow on the surrounding marine communities.

In the multiple regression analysis of 1970 data, 13 significant relationships were found between the biological parameters and days after the spill, distance from the platform, silt content, sand content and organic matter. Distance from the platform correlated with the same parameters in the stepwise multiple regression as revealed by the analysis of variance, but several additional significant relationships emerged: between silt content and H', between sand content, H', and crustacean/polychaete ratio and between sediment organic matter content, the numbers of species and numbers of individuals (adjusted for high densities of the two bivalves). As might be expected, these data suggest that sediment composition and organic matter are important factors to infaunal distribution in the study area. Perhaps the inclusion of salinity and depth data would have revealed further relationships.

If there was a platform-related effect in 1970, the correlation analysis of sediment-hydrocarbon content (C_{12}-C_{23} hydrocarbons by gas chromatography and C_{12}+ hydrocarbons gravimetrically) with previously mentioned biological parameters did not show it. The only significant regression was a positive one between H′ and C_{12}-C_{23} hydrocarbons and it may have been spurious.

In conclusion, there appeared to be a depression of numbers of individuals, numbers of species and perhaps the crustacean/polychaete ratio within 5 miles of the platform in 1970, but no direct relationship with sediment hydrocarbons could be found. The cause of this effect remains unclear. Beyond 10 miles natural forces dominated faunal distributions.

The most detailed analyses of benthic communities in the wake of a spill were those carried out after the grounding of the barge *Florida* in Buzzards Bay, Massachusetts in September 1969. About 65 to 70 thousand liters of No. 2 fuel oil were spilled. The oil became emulsified in water as it was driven toward the Wild Harbor area by strong winds. Four miles of coastline were contaminated. The oil had its greatest effect and persisted the longest in the harbor boat basin and marsh.

The studies carried out by scientists from Woods Hole Oceanographic Institution provide a unique view of the long-term effects of No. 2 fuel oil in the muddy sediments of a contaminated water body. The benthic fauna from a series of 13 stations stretching from the offshore areas in Buzzards Bay into the Wild River Harbor were described from immediately after the spill up to April 1973. These results, subjected to an extraordinary number of analyses, are reported by Sanders *et al.* (1980). Data on the fourth and fifth years after the spill (1973, 1974) are reported by Michael *et al.* (1975).

A control station in Sippiwissett marsh four kilometers to the south was established and sampled five months after the spill and three additional times in 1971 in order to provide a comparison to the heavily-oiled parts of Wild River Harbor. This station was sampled six times again in 1974 for later comparisons.

The benthic studies of the affected area were based largely on samples from six stations that varied from lightly- to very heavily-oiled. Three of the less affected stations were offshore in Buzzards Bay (stations 5, 20, 35). Of the more heavily affected stations, two were near the mouth of the harbor (stations 9 and 10) and one was well back in the harbor (station 31). Paired samples for fauna were taken by a Van Veen grab (1/125 m^2) or cores (1/128 m^2) and washed through a 0.297-mm sieve. Although more than 413 sample sets were taken in the first $3\frac{1}{2}$ years, the detailed analyses by Sanders *et al.* (1980) are based on 142 of these. Another sample in each set was taken for hydrocarbon analysis and granulometry.

A large number of analyses was performed on the benthic data in order to understand how these communities fluctuated after the spill and the role played by opportunistic and subdominant species in these communities. Among the methods used were coefficient of variation, index of constancy, Shannon-Wiener diversity, a measure of evenness, rarefraction curves and an agglomerative clustering technique. Since only two replicates were taken at each station, none of the data is reported with error estimates. It is apparent that some of the data represent single samples.

The immediate effect of the spill was a large mortality of the fauna, with many worms, crustaceans and fish washing up on local beaches (Hampson and Sanders, 1969). Marsh grasses contaminated with oil also died.

As an introduction to interpreting the more subtle biological effects, it should be noted that the stations experienced different amounts of initial oil exposure; the oil also weathered differently at various stations. There was persistent and relatively undegraded oil in the harbor (Teal *et al.*, 1978) and a transitory dose at offshore stations. Background concentrations of sediment hydrocarbons in the area varied from approximately 20 to 100 ppm. The most heavily affected station (31) experienced about 1100 ppm during 1969. This increased during the first year to well over 2000 ppm and stayed high through 1971. At two of the offshore stations (9 and 10) elevated levels, generally 150 to 300 ppm, were present during the first year. There was also a seaward migration of oil out of the harbor in the late winter of 1969 and spring of 1970 and again in later years.

The faunal changes for the most part could be linked to the duration and severity of oil dose and the extent of weathering. There was an immediate severe and prolonged reduction in species richness in the harbor followed by a spreading bloom of the opportunistic polychaete *Capitella capitata*. There was a gradual recovery of species numbers in the harbor through 1970 as *Capitella capitata* populations crashed, although the lack of ampeliscid amphipods as late as 1973 indicated a lingering effect of the oil. As late as 1973 and 1974 the numbers of species in Sippiwissett marsh samples were nearly always higher than those of Wild Harbor (Michael *et al.*, 1975).

At the offshore stations, apparently coincident with the seaward spread of petroleum in the winter of 1969-1970, animal density and numbers of species declined and did not increase again until late in the summer of 1970. Apparently the opportunistic polychaete *Mediomastus ambiseta* played a role in the offshore areas similar to that of *Capitella* spp. at the inshore stations, increasing in numbers as the rest of the fauna declined. Although less pronounced than at more inshore areas near the mouth of Wild Harbor, the farthest offshore areas sampled (Stations 5, 20 and 35) also experienced a decline in numbers of species and a bloom of *Mediomastus ambiseta*.

Further clues to the more subtle effects of the oil were provided by abundance of ampeliscid amphipods, that proved to be particularly susceptible to the fuel oil. In the harbor there were a few ampeliscids alive before the full impact of the spill was felt but then they soon disappeared from the harbor and had not reappeared as late as June 1973. At stations 9 and 10 there were more alive than dead amphipods just after the spill, but then from October 1969 to August 1970 all amphipods collected from this area were dead. After August 1970, 70% of all amphipods found there were alive.

The remainder of the analyses of benthic community data focused on subtle changes in the communities along the offshore-onshore gradient and between Wild Harbor and Sippiwissett marsh using a variety of statistical techniques. These measurements basically all traced the changing roles of various species, particularly opportunists, focusing on the influence that dominants and sub-dominants had at the various stations on overall community diversity and stability.

The measures of faunal fluctuation, coefficient of variation and constancy, tended to show similar patterns. In the offshore areas, where petroleum exposure was less, faunal variation was slight, but the variation increased in the inshore areas. The greater inshore variation was apparent whether comparisons were made on the basis of the whole faunas in each area or on the basis of shared species.

The diversity measures H′ (the Shannon-Wiener Statistic) and evenness were not consistently useful, being most sensitive to the large numbers of opportunistic species and relatively insensitive to dramatic decreases in the fauna. It was concluded that species richness was a much more important and useful component of diversity than was evenness.

The long-term effects on the benthos were concluded to have been greatest in the enclosed harbor area and were obviously related to persistent high concentrations of oil that degraded only slowly in these largely anaerobic sediments. An exact estimate of faunal recovery time in the most heavily affected areas is difficult as it depends on an interpretation of when successional phenomena merged with normal seasonal changes. This judgment, in turn, depends on how good a control area the Sippiwissett marsh was for the Wild Harbor marsh, the precision was for the various faunal parameters with only one or two replicate samples, and the extent to which a similar onshore-offshore transect in an unaffected area would show comparable changes. The available evidence certainly indicates that recovery of the Wild Harbor marsh was well along five years after the spill, but probably not complete. The stations farthest offshore had recovered within a year with the shallower nearshore stations taking perhaps two or three years longer to reach a similar stage. One phenomenon that is suggested by the *Capitella capitata* data is population oscillations continuing for several years after the initial toxic effects had passed.

The population of fiddler crabs *Uca pugnax* in Wild Harbor marsh was also reduced relative to Sippewissett marsh for at least seven years (Krebs and Burns, 1977). Behavioral effects, abnormal burrow shapes and reduced female-to-male ratios were seen at Wild Harbor. Crab density was correlated with hydrocarbon concentrations within the marsh as was the density of newly settled juveniles. As had been noted in other areas, particularly in more sheltered areas of the eastern U.S., substantial spills of fuel oils usually result in measurable biological effects for several years (Dow and Hurst, 1975; Michael *et al.*, 1975; Hampson and Moull, 1978). The patterns of faunal changes in this spill are consistent with effects described for other sources of organic materials in the marine environment as described by Pearson and Rosenberg (1978).

The third case of effects on the macrobenthos is from the *Tsesis* spill (Elmgren *et al.*, 1983). This again was a fuel oil spill and resulted in the contamination of approximately 30 km^2 of the Baltic Sea near the Swedish coast. The spill was about 1000 metric tons and occurred in October, 1977. Faunal changes were measured mainly by comparing the most heavily affected area (Station 20) and an area of lesser effect (Station 21) using prespill data and data collected for about three years after the spill. This was done against a general background of increasing eutrophication in the Himmerfjörd due to a sewage plant. As mentioned in the meiofauna section, measurable fuel oil hydrocarbons were not found in the sediment samples, but concentrations of fuel oil hydrocarbons up to 2000 μg/g were found in *Macoma balthica* in the most heavily contaminated areas (Elmgren *et al.*, 1983). This and other available evidence suggests that the hydrocarbons were mainly associated with the flocculent layer at the sediment-water interface. From sediment trap data it was estimated that at least 5 tons of the oil reached the benthos associated with fine particulate matter (Johansson *et al.*, 1980).

The immediate impact of the spill was a reduction of macrofauna abundance at Station 20, mainly due to the near total disappearance of the two amphipods *Pontoporeia affinis* and *P. femorata* and the polychaete *Harmothoe sarsi*. At the less heavily affected station, the amphipods were reduced but not the polychaetes. Recovery to prespill levels at station 20 did not occur until November, 1979, with some indications of relapse in the amphipod biomass through 1981. There was also a significant increase in abnormal embryos of *P. affinis* at station 20 over a distant reference station in February and March, 1978. However, because of the tendency of the amphipod populations to vary naturally, the depression of the *P. femorata* population at station 21 cannot be considered unambiguously to be an oil effect. Both *Macoma balthica* and, to a lesser extent the priapulid *Halicryptus spinulosus*, increased following the spill. The significant increase of *M. balthica* is consistent with the view that this species is oil tolerant and often thrives in contaminated sediments (Shaw *et al.*, 1976; Taylor and Karinen, 1977).

There are four general situations of chronic hydrocarbon input that have been studied in detail. These are in areas of refinery input, offshore oil platforms, produced water discharges and natural submarine seepages. Because of the complicating and possibly dominant influence of sulfide and ammonia toxicity (e.g. DeGraeve *et al.*, 1980) as well as appreciable concentrations of metals, refinery inputs will not be considered here. The interested reader is referred to Baker (1973), Wharfe (1975), Leppäkoski and Lindstrom (1978), McLusky (1982), and Dicks and Hartley (1982). The papers by Leppäkoski and Lindstrom (1978) and McLusky (1982) have good descriptions of faunal change in relation to changing refinery effluents.

The discussion of the effect of offshore platforms on benthos will be deferred since the benthic studies provide about the only meaningful discussion of effects directly associated with offshore activity. Therefore I shall discuss effects of chronic discharge associated with separator platforms (see also Chapter 10) and natural submarine seeps.

Armstrong *et al.* (1979) studied the effects of a produced water (brine) discharge of 4 to 10 x 10^3 bbl/day in Trinity Bay, Texas. The water was only 2.5 m deep in the area of the platform. The fauna and sediment chemistry were correlated along a series of radiating transects. The fauna was found to be severely depressed up to 150 m from the platform. Naphthalenes were the dominant compounds in the sediments and had a concentration of 6 to 10 ppm at 150 m. Decreases in numbers of individuals were seen as far as 600 to 1200 m from the platform and corresponded to sediment naphthalene concentrations of 4 to 8 ppm. In general there was an inverse correlation between naphthalene content and faunal abundance. The polychaetes *Streblospio benedicti* and *Polydora ciliata* appeared to be more resistant than other species.

The benthic ecology of a natural submarine petroleum seepage area in the Santa Barbara Channel has been investigated in a series of studies by my laboratory. We have compared infaunal communities within the Isla Vista petroleum seep with those of a nearby area without seepage. Based on the results of the community structure studies we formulated hypotheses of organic enrichment, adaptation, environmental toxicity and sublethal effects that have guided continuing investigations.

The Isla Vista seep is one of a group of seeps around Coal Oil Point that release an estimated 50 to 100 bbl/d of petroleum into the Pacific Ocean (Allen *et al.*, 1970). The ever-present oil slicks from the seeps often wash ashore on the local beaches but more frequently appear to move westward as they weather and disperse. There is a long history of seepage in this area, probably at least several tens of thousands of years (Simoneit and Kaplan, 1980). Consequently total hydrocarbon concentrations of sediments can exceed 1000 ppm (Reed *et al.*, 1977). Since dense and diverse communities of benthic organisms are found all along the mainland shelf of the Santa Barbara Channel and they appear to be similar to those in other areas of California (Jones, 1969; Barnard and Hartman, 1959), my opinion has been that total hydrocarbon concentrations of shelf sediments of southern California are not particularly biologically meaningful. The total hydrocarbon load of sediments is often dominated by highly-weathered asphaltic compounds, and there is little reason to believe these compounds are biologically reactive. Low molecular weight components, total alkanes and, perhaps, total aromatics are probably more important measures. With these considerations we established a comparison station for the benthic ecology studies about 1 km to the east of the petroleum seep, where total hydrocarbons are over 2000 ppm, but where the hexane and toluene fractions of the whole sediment extracts are approximately 20 times less than at the seep study site. Further, dissolved hydrocarbons, mainly mono- and diaromatic compounds, in the interstitial water of the seep site are 45 to 117 ppb, while those of the comparison site are 0.2 to 5 ppb (Stuermer *et al.*, 1982). Also, in sieving more than 150 benthic samples from each station over several years we have not seen evidence of fresh oil droplets in the comparison station sediments, while those of the seep station have fresh oil droplets in every core and we have to clean the sticky oil from our sieves after washing the samples.

The petroleum arrives in the surface sediments from the underlying Monterey Shale (Miocene) and apparently is similar to the production oil pumped from similar formations farther offshore except that the normal alkane series is not a dominant feature (Spies *et al.*, 1980). The seep oil is, however, as toxic or more toxic than other oils in the area, as measured by its effects on starfish embryo growth (Spies and Davis, 1982).

The communities of both areas are representative of an inshore facies of the *Nothria-Tellina* community (Jones, 1969). Of the 320 species identified in several years of sampling, 72% were found in both areas. It is difficult to attach much importance to the other 28%, since nearly all of these were undersampled rare species. Two differences that seem significant are the small numbers of phoxocephalid amphipods and the large number of oligochaetes at the Isla Vista Seep, otherwise more than 60% of the species contribute to the differences in density between the two areas (Spies and Davis, 1979).

The seep station had consistently larger densities of total infauna than the comparison area over a 28-month period, but numbers of species were only slightly higher. Although small scale fluctuations of Shannon-Wiener diversity were apparent, both areas had, on the average, similar values for H′. Evenness was fairly constant, and the differences seen were not consistent. Measures of skewness and kurtosis of the dominance-diversity curves were also quite similar despite the density differences between stations and pronounced seasonal fluctuations (Davis and Spies, 1980).

These results and our observations of mats of the mixotrophic, sulfide-oxidizing, filamentous bacteria *Beggiatoa* sp. on the areas of particularly active seepage led us to hypothesize that microbially-mediated organic enrichment was occurring as a result of the seepage. We also formulated several other hypotheses to guide further process-oriented research on this system (Spies *et al.*, 1980).

A several-fold increase of sediment ATP content, especially in the <150-μm fraction, in areas of heavy seepage strongly suggested that large microbial populations were associated with fresh oil (Spies *et al.*, 1980). Further evidence of organic enrichment (i.e., the utilization of petroleum carbon and energy in the benthic food web) was obtained by examining naturally occurring isotope ratios of carbon and sulfur in the tissues of benthic macrofauna, in *Beggiatoa* sp. and in interstitial H_2S. Shifts of $\delta^{13}C$ in the tissues of 12 infaunal species toward isotopically lighter carbon at the petroleum seep relative to the comparison area indicated an apparent effect of the petroleum on carbon flow. Measurements of ^{14}C content and $\delta^{34}S$ of a deep-feeding maldanid polychaete, *Praxillella affinis pacifica*, with a $\delta^{13}C$ shift of -2.73‰ at the petroleum seep, indicated that 14% more of its carbon was of fossil origin than at the comparison station. It was also estimated that chemoautrophically-fixed carbon contributed 13% more to the carbon of this species at the seep station than at the comparison station. Further isotopic evidence was presented that this was energetically mediated through the sulfur cycle (rapid H_2S production and utilization) linked to microbial hydrocarbon degradation. Allowing a few more assumptions, this led to a calculation that the petroleum carbon source had a mean $\delta^{13}C$ of -35‰, indicative of very light liquid hydrocarbons and gases (Spies and DesMarais, 1983). This seems to support the original judgment that fresh oil would be a biologically more meaningful criterion for establishing station locations.

A recently completed study of rates of benthic metabolism at the comparison site and two seepage sites indicated that both hydrocarbon degradation and sulfate reduction rates are higher in seep sediments. Oxygen flux was greater in the seepage areas when averaged over three days, but considerable temporal and spatial variability precluded a conclusion of significant site differences (Montagna *et al.*, in press).

Several different tests of the adaptation hypothesis have been made. In one of these, adult starfish, *Patiria miniata*, from the Isla Vista Seep and other less contaminated areas were spawned and their developing embryos exposed to water soluble fractions of seep oil (WSF, 100% ≅ 10 ppm hydrocarbons). This experiment was repeated three times, and no consistent differences in the growth responses of the starfish embryos to petroleum could be detected (Spies and Davis, 1982).

We have also compared uptake from sea water and metabolic conversion of ^{14}C-naphthalene in starfish from different populations. There were some indications of greater naphthalene accumulation by the digestive glands of seep starfish, but rates of metabolism were not different. Hepatic MFO activity, as measured by the aryl hydrocarbon hydroxylase (AHH) assay, was quite low (only about 1% of sanddab activity) and not different in seep and Monterey Bay starfish (Spies and Ireland, unpubl.). The elevated activity of MFO in two species of sanddabs in the seep area was discussed earlier. These enzymes can be viewed as a useful ontogenetic adaptation to chronic contamination of these fish to petroleum hydrocarbons.

There are several other aspects of these seep studies that should be mentioned. We have seen a smaller gonad index in adult *Patiria miniata* collected from the oil seep area relative to a rocky reef area some 5 miles to the northwest (Spies *et al.*, 1980). We do not know if this effect is due to petroleum hydrocarbons or to some other factor since we analyzed reproductive state and hydrocarbon content of gonads and digestive glands and found no relationships. It should be noted, however, that aromatic hydrocarbons were in extremely low concentrations in these tissues (they would only be detected in part per trillion concentrations using ion monitoring techniques in mass spectrometry). There could be some effect of hydrocarbon metabolites on reproductive potential of these species and this would not have been detected using our approach. We have measured activities of aryl hydrocarbon hydroxylase in the range of 1-5 picomoles 3-OH BAP/mg protein/min in this species and the activities are not different in contaminated and uncontaminated environments.

Reproductive effects in this environment had been previously studied by Straughan (1976). Studies of three abalone species (*Haliotis*) and *Mytilus californianus* revealed no consistent differences between Coal Oil Point populations and those in comparison areas. It was apparent, however, that natural variability between several sites was considerable and care should be taken in interpreting such studies based on distant comparison sites.

To summarize the seep studies, the phenomenon of organic enrichment mediated by hydrocarbon and sulfide-oxidizing microbes must be considered likely under the right conditions of chronic oil contamination and these processes can be the predominant ecological effect of oil. Short-term adaptations to petroleum in fishes do occur with the development of appropriate hydrocarbon-degrading enzymes. No evidence has yet been found of special long- or short-term adaptations in local invertebrates. Sublethal effects of seeping petroleum on reproduction deserve further study, especially since concentrations of interstitial petroleum hydrocarbons (45 to 100 μg/l) are in the range for which long-term effects of oil have been claimed in the MERL microcosms (Grassle *et al.*, 1981; Oviatt *et al.*, 1982, Ritacco and Sastry, 1983).

THE EFFECTS OF OFFSHORE PLATFORMS ON MARINE COMMUNITIES

"No one is certain at what concentrations in the ocean petroleum is dangerous to organisms, but there is no question that it causes scientific anxiety" (Sanders *et al.*, 1980, p.142).

Whether low-level chronic petroleum contamination that might result from the more-or-less routine operation of offshore platforms and small spills has effects or whether such effects could be detected with some confidence are concerns central to the purposes of this review. The appropriate questions are: Can effects be measured among the many natural controlling factors of animal and plant distribution? and Are the effects significant?

If there are indeed measurable effects from offshore petroleum activities, we would expect *a priori* that effects will be difficult to demonstrate, given that conditions producing the most pronounced effects documented for oil pollution are not generally present in offshore areas: shallow water, restricted dispersion, high concentrations of suspended particulates, fine-grained anaerobic sediments and spillage of distilled fuel and diesel oils.

I will discuss results and different interpretations of several major studies in the Gulf of Mexico and the North Sea. The scope of some of these studies have been quite large and voluminous data have been gathered (at great cost), and it is beyond the modest dimensions of this effort to review them in detail. Carney (Chapter 14) discusses design problems encountered in these studies. I will, however, attempt to relate aspects of these studies that are critical to their usefulness in understanding the effects of petroleum in the offshore environment.

The Offshore Ecology Investigation

The most widely known and discussed study is the Offshore Ecology Investigation (OEI) conducted by the Gulf Universities Research Consortium (GURC). It was carried out in eight consecutive seasonal samplings in 1972-1974 as a coordinated effort involving 23 principal investigators. Its purpose was to determine if 25 years of oil production involving more than 1500 offshore platforms has had a cumulative effect on the marine ecosystems of coastal Louisiana. Despite the fact that such extensive oil production activity made this area of the United States a worst case for possible effects, its location just west of the Mississippi River delta makes it a difficult location in which to attempt to identify causative ecological factors besides salinity, turbidity, high organic loadings, sediment type and periodic hypoxia. This study also did not have clearly-defined criteria for assessing what was a long-term impact (see Carney, Chapter 14).

Problems of high natural variation were apparently recognized by the designers of the study (Sharp, 1979), but as often happens the magnitude and strength of these natural controlling factors were effectively underestimated in the study design. In retrospect, some sampling reconnaissance, especially for petroleum hydrocarbons and habitat types, could have helped formulate a more effective study design.

The OEI sampling scheme was to evaluate both localized platform effects as well as the overall "health" of the ecosystems in the whole study area. Sampling was carried out at platform and control sites within Timbalier Bay and directly offshore, a transect connecting these two areas (to evaluate onshore-offshore fluxes), and four onshore-offshore transects to the east (towards the Mississippi Delta) and which were considered "upstream" in the prevailing currents from platform activities (but apparently were not always "upstream") (Menzies *et al.*, 1979).

Investigations were made in the areas of shelf currents and hydrography (Oetking *et al.*, 1979), bay hydrography (Price, 1979), turbidity (Griffin, 1979), sediment characterization (Jones and Williams, 1979), nutrients (Burchfield *et al.*, 1979), metals (Montalvo and Brady, 1979), organic and inorganic carbon (Brent *et al.*, 1979), hydrocarbons (Laseter and Ledet, 1979), microbes and hydrocarbons (Oppenheimer *et al.*, 1979), phytoplankton (Fucik and El-Sayed, 1979), planktonic copepods (Marum, 1979), benthic algae (Humm and Bert, 1979), benthic molluscs and crustaceans (Farrell, 1979), platform algae (Bert and Humm, 1979), foraminifera (Ostrum, 1979), littoral polychaetes (Kritzler, 1979; Lewis and Fish, 1979), pelagic, epipelagic and infaunal invertebrates (Waller, 1979), fish (Perry, 1979), temporal changes in offshore macrofauna (Thompson, 1979), fouling communities of platforms (George and Thomas, 1979) and recent geological history (Morgan, 1979). There were no measurements made of sediment hydrocarbons.

The original final reports of the senior investigators were not generally available, however, a consensus report (Menzies *et al.*, 1979) was widely distributed. A senior investigator (Oppenheimer, 1977) and an oil industry representative (Mertens, 1978) also published brief versions of these conclusions. The conclusions were that the study showed that there were no effects and: 1) natural phenomena in this environment have a much larger impact than petroleum-related activities; 2) petroleum-related contaminants could not be tied to platform sources and were in such low concentrations to not present a known hazard to marine life; and 3) that there is every indication of "good ecological health."

On the basis of the reports originally submitted to the GURC Committee Sanders (1981) vigorously attacked the study design and these conclusions. The following major points of criticism were made:

1. *The control stations were not adequate.* The current and hydrodynamic regime would allow relatively rapid and wide distribution of contaminants from the platforms. Further, there is chemical evidence that dissolved low molecular weight hydrocarbons ($<C_6$), although themselves probably not toxic, but indicative of petroleum contamination, are several times higher along this part of the Gulf of Mexico than areas farther offshore (Sackett and Brooks, 1974). The obvious conclusion from this is that there cannot be laboratory-type (petroleum-hydrocarbon-free) controls in the study area.

2. *Estimates of microbial degradation rates in the field were based on optimized*

laboratory conditions. Oppenheimer *et al.* (1974) proposed that microbial degradation of petroleum in the study area was sufficient to prevent a buildup of hydrocarbons. This has been criticized because rates measured in the laboratory were made on well-aerated sediment slurries supplied with sufficient nutrients and such optimal conditions often do not occur in the marine benthos (see Chapter 7).

3. *There were insufficient data on petroleum hydrocabons in sediments*. Although data on hydrocarbons were published eventually (Laseter and Ledet, 1979), the data were not sufficiently dense to support the level of biological effort. There was a lack of detailed information on concentrations and characteristics of sediment hydrocarbons for correlations to be made or discounted with the biological data. Apparently, the study design failed to recognize the sediments as the most probable long-term reservoir of hydrocarbons or the potential of such measurements along with biological data to support conclusions.

4. *The benthic fauna was not indicative of a "healthy ecosystem."* Sanders (1981) attacked the conclusion of "good ecological health" on two levels. First, he compared the densities of organisms in the OEI study area with those found in shallow marine habitats elsewhere in the world (locations were not stated, but assumed to include his study areas in Buzzards Bay, Massachusetts) and found the OEI densities significantly lower in many cases. Second, he interpreted the presence of large numbers of two opportunistic species, *Mulinia lateralis* and *Spiochaetopterus oculatus*, as indicative of polluted conditions in the study area and cites a series of studies where these species are found under heavily polluted conditions, particularly where organic enrichment phenomena deplete bottom water oxygen in poorly circulating bays and harbors. Presumably, these species can also indicate naturally stressful conditions, which are quite evident in the study area.

5. *Spatial and temporal replicates were pooled in calculating benthic diversity*. In Farrell's OEI benthic studies (1974a,b) all the replicates over several sampling periods from a single station were pooled to calculate a diversity value for each station. Needless to say, this was a very unusual procedure in a benthic study and produced an artificially high diversity value.

6. *The amount of data collected at the platform and control sites was not sufficient to show effects*. More data collected around the platforms would have allowed better conclusions of effect or no effect to be drawn.

A part of the published reports of the GURC/OEI effort was an evaluation of the studies by a second group of scientists (Bender *et al.*, 1979). The following points on study design and conclusions were made:

1. The probable uniform contamination of the whole study area by hydrocarbons renders the traditional study design of "affected and control sites" inappropriate. Therefore, the original consensus report conclusions of no effect were not warranted. The available chemical data did not demonstrate the platforms are significant sources of pollutants with the possible exception of unweathered oil, but the data were not dense enough to draw conclusions with statistical significance.

2. In Timbalier Bay the fauna has not shown a pattern consistent with an

adverse effect of oil, rather salinity and turbidity are the factors that control animal distribution.

3. In conclusion they summarized that there is no evidence in the benthic data that there is enviornmental stress from oil drilling and production activities.

After reading the 1979 OEI report, the appraisal by Bender *et al.* (1979), and comments by Sanders (1981), several things concerning study design and conclusions seem apparent:

1. Despite its long history of intensive offshore petroleum development, the southeastern Louisiana coast is a difficult place to conduct such a study. There are very little, if any, predevelopment data for comparison. An appropriate control or comparison may not exist. The study area is subject to high turbidity, fluctuating salinities and periodic hypoxia and anoxia, and the flora and fauna are under stress from these conditions. Because natural variability in the ecology of this area is potentially great, only very extensive and clear cut effects would be expected to be detected. The Mississippi River is not only a fluctuating source of highly turbid fresh water but probably also a major source of anthropogenic pollutants including hydrocarbons of which the designers of the study were apparently aware but did little to document before embarking on the study. The failure to realistically assess the contributions of hydrocarbons to the study area by the Mississippi River coupled with insufficient support of the general organic chemistry effort in the study were crippling defects that precluded much resolution in correlating possible biological effects with oil production.

2. The study had only a vague notion of what it was attempting to find. Therefore, the study design could not be precise (see the extensive comments on this point in Chapter 14).

3. While the presence of opportunistic species such as *Mulinia lateralis* and *Spiochaetopterus oculatus* are often indicative of stressful environments and we might expect them in Timbalier Bay, even in the absence of petroleum production, their presence alone is not definitive evidence of petroleum-related stress.

4. As hydrocarbon components and different oils vary so widely in their toxicity, the comparison of two widely separated geographic areas (e.g., New England and Gulf of Mexico) exposed to different types of petroleum pollution and having different exposure histories on the basis of total sediment hydrocarbons may have little biological meaning.

The Central Gulf Platform Study

This study was carried out in 1978 and 1979 on the Louisiana shelf and included areas that were in the earlier OEI studies. Four primary platform sites and four control sites were sampled in three consecutive seasons in the 1978-1979 study period. A scheme of sampling from 100 to 2000 m away from the platforms was adopted to indicate local as well as regional effects of petroleum development (Bedinger *et al.*, 1981). Again, there was a large variety of studies including hydrography, water column hydrocarbons, metals and hydrocarbons in sediments, numbers and activities of microbes including hydrocarbon degraders, and populations of meiofauna, macroinfauna, macroepifauna and platform-associated fouling organisms and fish.

The microbial data were synthesized with the following specific objectives: to compare predominant microbial types of platform and control sites, to determine the effects of temperature and nutrients on hydrocarbon oxidation rates and to determine process rates of major microbially-mediated processes in sediments (Brown *et al.*, 1981). The objectives of the biological data synthesis were: to compare communities at control and platform sites with an emphasis on indicator species, and to attempt to correlate biological parameters with various physical and chemical conditions (particularly contamination) related to offshore platforms.

Although not focusing on Timbalier Bay where the OEI studies showed salinities were often below 20‰, this study faced the same problems—differentiating a possible effect of offshore platform activity on an ecosystem so subjected to great natural stress, especially from turbid fresh water, and also, possibly, anthropogenic contaminants from the Mississippi River. These problems were recognized from the beginning, and the study introduction included a particularly lengthy discussion of the periodic hypoxia seen in bottom waters over a portion of the study area in offshore Louisiana and its contributing causes (Bedinger *et al.*, 1981). The designers of the study, however, apparently believed by taking an approach emphasizing fate and effects and including more data than the OEI on hydrocarbons and other contaminants, particularly in sediments near platforms, that the study objectives could be achieved.

The objectives of much of the microbial research were compromised by the logistics of the sampling cruises. Apparently, there were not enough facilities aboard the ship to carry out many of the planned measurements of microbial rate processes (e.g., nitrogen fixation). Consequently, all the sediment samples were frozen immediately after collection. Although freezing may not affect bacterial enumeration in sediment samples (Stewart and Marks, 1978), the effect of freezing on rate processes was not determined. In their conclusions the authors dwelt on the enumeration data. They found that microbial populations sometimes differed significantly between platform and control sites and other times did not, but the differences were not consistent between the cruises. The influence of the Mississippi River seemed evident in the data, particularly during periods following high outflow. This effect may have been somewhat exaggerated by sampling just the top 2 cm of sediment, which seems more likely to reflect recent riverine influence. The sediment surface layers are probably under the strong influence of currents and waves on the shallow Louisiana shelf. Then the authors concluded that there was no difference found between oil-degrading potential of control and platform sites and that there was no evidence of adverse effect of low concentrations of oil on microbial activity. However, no data were presented to indicate whether or not freezing the samples would have produced different results and changed these conclusions.

The benthic sampling on the three cruises resulted in the collection of 560 8-cm^2 cores for meiofauna, 840 0.09-m^2 Smith-McIntyre grabs for macroinfauna and 40 9-m otter trawls for macroepifauna and demersal fish (Baker *et al.*, 1981). Meiofaunal species diversity during two cruises was higher at the four primary platform sites than at the control sites, but the reverse was true during the third cruise. Data clustering techniques identified three meiofaunal groups that tended to occur in similar sites and correlated with distance from shore, depth, salinity, dissolved oxygen, temperature and presence of hypoxic conditions. Species diversity of macroinfauna was higher at all primary platform sites during all cruises. Clustering identified four macrofaunal groups that correlated with distance from shore, depth, salinity, temperature, percent sand, percent silt, total organic carbon and presence of hypoxic bottom conditions. Species diversity of macroepifauna and demersal fish were also highest at the primary platform sites than at the control sites during all three cruises. Two groups clustered that correlated with distance from shore, depth and presence of hypoxic bottom conditions.

The remainder of the data treatment involved extensive regression of a very large set of physical and chemical parameters against summary ecological statistics (diversity, evenness, numbers of species and numbers of individuals) and then against abundances of common species. This extensive statistical exercise resulted in many significant correlations. Some of these correlations resulted in intuitively understandable relationships, such as a significant inverse relationship between percent silt and numbers of species. Others such as the inverse correlation between propane concentrations in the water column and cyatholamid nematode abundance are obscure, and many of the correlations are undoubtedly spurious.

The density of the macroinfauna ranged from 45 to 9338/m^2 in the samples from this study. Although this compares favorably with densities found in other parts of the Gulf of Mexico and the U.S. Atlantic seaboard, such comparisons are probably best made on the basis of median densities and should also include some measure of species richness.

Two factors that may well have controlled faunal communities in the study area were a major tropical storm, Debra, that cut short the second cruise, and the hypoxia that covered a portion of the inner shelf west of the Mississippi River. Hypoxic conditions (<2 ppm dissolved oxygen) occurred during the late spring and summer cruises.

The organic chemical analyses in this study were reported by Nulton *et al.* (1981). They found low molecular weight hydrocarbons from several fold to approximately ten times open ocean values. Stratification of the water column tended to be reflected in higher near bottom concentrations. Sea water collected near two secondary platform sites in the second cruise had methane concentrations up to 24 μg/l.

High molecular weight hydrocarbons were always detected in sediment samples. Values ranged from a mean high of 87.4 μg/g at one platform site (S6) to a mean low of 5.7 μg/g at another platform site (S19). Samples from one primary platform site and four secondary platform sites were determined to contain high total hydrocarbon content relative to the controls. At only one platform site (S11) did total hydrocarbon content decrease with distance from the platform. The average values for control sites (35 μg/g) exceeded those of 11 of the 20 platform sites. In gas chromatographic analyses of sediment extracts, the unresolved complex mixture (UCM) accounted for an average of 96.4% of the total hydrocarbons indicating that a highly weathered fraction predominated.

For the unsaturated hydrocarbon fraction of sediments, one platform site (P1) had high concentrations (average of 34.9 μg/g) and one of these samples had 364 μg/g.

Most of the measurements of tissue hydrocarbons were done on fish and macroepifauna. No unresolved complex mixtures of hydrocarbons were found in these tissues. The unsaturated fractions in 8 of 19 demersal fish and 10 of 31 macroepifauna contained aromatic hydrocarbons. The range of concentrations were from 10 to 220 ppb with most <70 ppb. The most commonly occurring compounds were methylnaphthalene and 1,3 dimethylnaphthalene.

Nulton *et al.* (1981) concluded that there were no significant relationships between the age of the platform and environmental effects in the surrounding area. Also, there was no relationship between the amount of oil and gas production at a platform and the amount of environmental contamination. The platform with perhaps the greatest level of detectable contamination was P1, with a produced water discharge of 20,000 bbl/d. Assuming that this discharge had the allowable limit of 30 ppm hydrocarbons, this would result in the discharge of approximately 25 gallons of oil/day. The high amount of sand in the discharge probably also aided in transporting hydrocarbons to the benthos. Finally, the authors concluded that the Mississippi River is probably the principal source of hydrocarbons to the study area. This appears to be conjecture as no data were included to document this claim.

There are several comments that should be made about the Central Gulf Platform Study.

1. Although in the introduction to the study (Bedinger *et al.*, 1981) the influence of the Mississippi River was recognized as a principal controlling force of ecological variability and a source of adventitious hydrocarbon pollution, nothing was done in the study to document the river-derived loading (in terms of quantity and composition) of the study area which would have had been especially helpful for understanding the sources of organic contaminants. An imaginative chemical oceanographer might have found a few good conservative tracers, or perhaps they have already been described. Such tracers would have been very helpful in sorting out contributions of platforms and terrestrial sources to hydrocarbons found in the study area.

2. In the study conclusions it is strongly implied that hydrocarbons are having a chronic sublethal effect on the fauna of the study area. This was made on the basis of a review of the hydrocarbon toxicity literature and the concentrations of hydrocarbons found in the study area. The extrapolation of laboratory toxicity data to the field is fraught with difficulties. For example, the comparison was made mainly on the basis of total hydrocarbons. Since most of the toxic effects described for marine organisms are from relatively low molecular weight aromatic hydrocarbons and the sediment hydrocarbons found in this study were dominated by highly-weathered mixtures, this comparison is misleading. Until some evidence is obtained of the toxicity of highly-weathered petroleum compounds mixtures which characteristically show up in gas chromatographic analyses as an unresolved complex mixture, such comparisons should not be made. However, low molecular weight aromatic hydrocarbons in the range of 100 ppb found in

this study in some fish may be causing sublethal effects and should be a cause for concern and further investigation, particularly in regard to sources.

The Buccaneer Gas and Oil Field Study

The third large environmental study in the Gulf of Mexico was carried out at the Buccaneer Gas and Oil Field on the Texas continental shelf. This study was done between 1976 and 1980, and a book on the study has been published (Middleditch, 1981). The field consists of 18 structures of which 14 are satellite platforms, two are production platforms and two are crew quarters. The main source of hydrocarbon contamination comes from discharge of about 600 bbl/day of produced water which contain about 3 ppm of extractable hydrocarbons (based on one sample analysis (Middleditch, 1981, p.27)). There are, of course, many other miscellaneous contaminants and sources of disturbance in the area related to the oil field (e.g., sewage discharge, drill cuttings and muds, structure-related current changes, etc.) that are beyond the scope of this chapter (see Dicks, 1982 for a discussion of many of these).

The main study was preceded by a year-long pilot study which concluded that there was sufficient contamination by hydrocarbons to warrant a large multidisciplinary effort (Harper *et al.*, 1976). The field had been in production since the early 1960s, and it was felt that any ecological effects of chronic contamination would be evident by this time. The study included a wide range of efforts: hydrocarbon chemistry, sediment geochemistry, characterization of suspended sediments and particulates, organic carbon and carbon isotopes of sediments, macroinfauna and meiofauna, fouling communities, birds, bacteria, fishes, crustaceans, ecosystems modeling, hydrography, discharge modeling and contaminant transport and dispersion. The approach was characterized as interdisciplinary rather than multidisciplinary; however, it was apparent the study was driven from several different points, rather than having the results obtained guide the continuing efforts.

The study was intended to avoid some of the rather unproductive far-field designs used in the OEI and Central Gulf Platform studies, concentrating instead on near-field effects with a closer link between biological and chemical analyses. Still the results available early in the study indicated that effects from platforms were measurable within only about 100 meters and this should have been enough to suggest a change of the study strategy (see comments by Dicks, 1981). This situation of limited areal effects was thought to be due to bottom characteristics and hydrography. Apparently the surface sediments during certain seasons are in nearly continual motion over a somewhat more consolidated basement, and hydrocarbons reaching the benthos are soon dispersed. Little evidence of persistent and chronic accumulation of sediment hydrocarbons was found beyond the immediate areas of the discharges and even sediments beneath platforms could show wide daily fluctuations in hydrocarbon concentrations.

For some reason most of the reports of hydrocarbon analyses of sediments and tissues were limited to the alkane fractions, although the sampling scheme clearly indicated that the aromatic fractions were analyzed. Since many of the predominant alkanes are readily metabolized by microbes and the alkanes are less toxic than the aromatics, this seems to be a rather unproductive approach if the study intent truly was "to determine specific pollutants, their quantity and effects."

Besides some low-level contamination of sediments and some organisms (Middleditch *et al.*, 1977, 1978, 1979), the ecological effects found were generally quite limited. The fouling communities near some discharge pipes were severely inhibited within about a meter. A more widespread effect on the benthos, reduced numbers of individuals and species around the platforms was apparent; however, there were also areas well away from the platform that had large amounts of clay and showed similar effects. It was therefore unclear whether the near platform effects were due to contamination or related to sediment variability. Obviously, a reconnaissance of the study area prior to establishing the final study design would have suggested a stratified sampling design. No specific regressions of sediment-associated hydrocarbons and faunal parameters were reported. Whether this was a futile exercise because of the previously mentioned problems with sediment mobility was not discussed.

The Ekofisk Oilfield Study

One of the best documented cases of platform-related effects was carried out over the space of several years in the North Sea (Addy *et al.*, 1978). The initial studies were done in 1973, about the time that production started, and then again in 1975 and 1977. Sediment samples for benthic fauna, hydrocarbons and granulometry were taken along five radiating transect lines that originated near two platforms and an oil storage tank and extended out 6000 m. Benthic sampling was done with a 0.1-m^2 Day grab, and for faunal analysis the contents were washed through a 1.0-mm screen. Ten replicates were taken at most stations.

A rather uniform community was found through the area in 1973 (Dicks, 1975), but by 1975 changes in the benthic community structure were evident at six of the stations closest to the platforms. By 1977 the number of affected stations had increased to 14, and the changes were measurable in the benthic community up to 2.5 km from the platforms. The effect was greatest in 1977 and was evident as a reduction in total numbers of individuals and species, however Shannon-Wiener diversity was higher at near platform sites. The reduction of density in the most common species, the polychaete *Myriochele oculata*, was rather dramatic near the platform, but the decrease was accompanied by increases in another species, *Chaetozone setosa*. Other species showing decreases near the platforms were the polychaete *Owenia fusiformis* and the ophiuroid *Amphiura filiformis*, while others increasing near the platforms were the polychaete *Pholoe minuta* and the bivalve *Arctica islandica*.

The biological data were regressed against distance from the platforms with the following results: population density (r=0.78), number of species (r=0.7), density of *Myriochele oculata* (r=0.78) and Shannon-Wiener diversity (r=0.62). Shannon-Wiener diversity was the only measure that increased with distance from the platform.

The chemical data showed concentrations of total organic-extractable material in sediments up to 400 μg/g. Total saturated hydrocarbons reached a concentration material at one station of 100 μg/g, but were otherwise less than 50 μg/g. Unsaturated compounds occurred in concentrations of up to 25 μg/g. The unresolved complex mixtures, so characteristic of sediment petroleum hydrocarbon residues, were relatively quite higher at one station close to the platforms. The nC_{18}/nC_{29} ratio, as an indicator of the relative amount of undegraded oil, was highest only around the one most northerly platform at Ekofisk.

The correlation of the hydrocarbon data with biological parameters revealed some interesting relationships. There were negative relationships between *Myriochele oculata* density and total organic-extractable material and UCM (for both r=-0.32, P=0.06). The opposite was true for the density of *Chaetozone setosa* where total organic-extractable material (r=0.45, P=0.01) and the UCM (r= 0.35, P=0.045) were positively correlated.

The authors were cautious in their interpretation of these data and suggest that there may be causes other than oil contamination for these changes, but "that a relationship may well exist in the field between the oil content of sediments and the community structure of the seabed fauna" (p. 532).

In their analyses of the granulometric data, the authors suggest that the sediments are rather uniform throughout the area but that there are slight differences in the sediments. They suggested that it would be difficult to assess the significance of larger proportions of fines near the platforms without further research.

Since it seemed that a variation of 2.5 to 7.7% in proportion of fine material (silt and clay) could be biologically significant, I regressed percent fines against the density of the most common species, *Myriochele oculata*. To plot these values I approximated the density for *M. oculata* from the published figures so the density values are not exact. However, in Figure 9.1 it can be seen that there is a strong relationship between these variables (r=-0.72). The fit between these

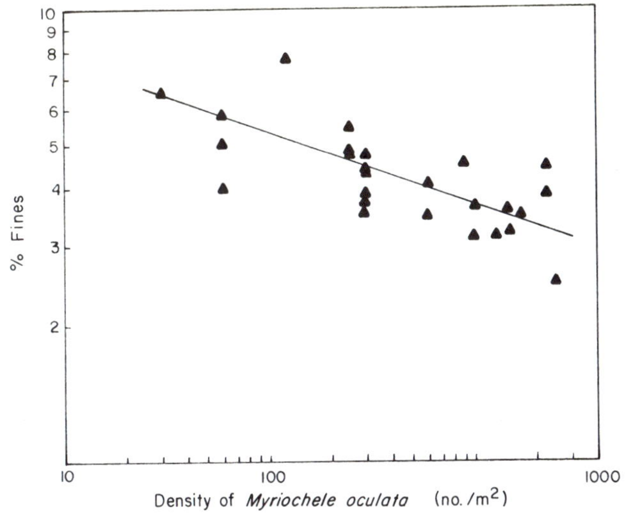

Figure 9.1. Relationship between percent fines in sediments and density of *Myriochele oculata* in the 1977 Ekofisk oilfield monitoring data (based on data in Addy *et al.*, 1978).

parameters is much better than any reported for the biological data and hydrocarbon concentrations. This relationship in conjunction with a significant relationship between *M. oculata* and *C. setosa* density suggest that it may be some parameter associated with fine sediments, quite probably total organic carbon (not measured), that produced these changes. The observed changes were undoubtedly a result of the presence of the platforms, but factors other than hydrocarbons are probably responsible. The physical effect of the platforms on hydrography and, in turn, on sediment deposition or perhaps sewage discharges or release of mussel pseudofeces from the platforms may be related to these changes.

The Forties Oilfield Study

An environmental monitoring study of another oilfield in the British sector of the North Sea has recently been published (Hartley and Ferbrache, 1983). The Forties Oilfield study is somewhat similar to that of the previously described Ekofisk Oilfield study in that preproduction (1975) benthic studies were available and benthic surveys were done again in 1978 and 1981. In this field there have been a total of 80 wells drilled from four platforms and, unlike other North Sea locations, only water-based drilling muds were used. Also there have been no spills in the field over one ton. The largest discharges have been the muds, cuttings and treated produced water. There were 24 benthic sampling stations arranged in two crossing 10-km long transects that include one of each of the four platforms on each arm. The transects run north-south and east-west. The field slopes to the west and the depths range from approximately 100 m in the east to 125 m in the west. Therefore, the study of this field has provided information on the potential impacts of relatively deep-water production. Six replicate samples were taken at each station, five for infauna and one for sediment parameters, during each of the three surveys.

Comparing the results of the three surveys, the authors found that the fauna remains rich and diverse, but that there have been notable increases of two polychaete species, *Capitomastus* sp. and *Chaetozone setosa*. These are reminiscent of changes seen elsewhere, most notably at the Ekofisk Oilfield. The benthic fauna changed as a function of physical changes in the sediments and appeared, with one possible exception, to be unrelated to hydrocarbon content. No platform-related gradients were found in the benthic fauna, although one set of cores from near a platform in 1982 had an elevated hydrocarbon content associated with some faunal change.

Gravimetric analyses for "aliphatic" and "aromatic" hydrocarbons in 1982 showed only one sample with 91 ppm total hydrocarbons, but otherwise were less than 20 ppm. The contribution of biogenic hydrocarbons to these concentrations were not indicated. Such values are not necessarily indicative of long-term accumulation.

PREDICTING ECOSYSTEM EFFECTS OF CONTAMINATION

It seems clear that petroleum contamination can result in little apparent effect, stimulation or sublethal and lethal effects on marine organisms. Effects on organisms can lead directly and indirectly to population changes, for example the invasion of biologically accommodated communities by opportunists. In turn, population shifts alone or acting in concert with natural forces may set up oscillations in community composition lasting several years after the direct effect of petroleum has apparently subsided (Sanders *et al.*, 1980). Chronic contamination can also result in stimulated communities having a stable dominance-diversity structure (Davis and Spies, 1980).

With a better mechanistic understanding of how doses are delivered, which hydrocarbons are biologically active and the sensitivity of different organisms, we could predict the outcome of each case of contamination. Beyond a physical/chemical/degradative model to approximate the doses to organisms, a dose-dependent physiological model would be required for each species that would include several hundred compounds each potentially decaying at a different rate. Further extensive laboratory toxicity testing of single hydrocarbons and mixtures would be required to derive the appropriate structure-activity relationships for such a model, although toxic effects of hydrocarbons to microalgae appear to be inversely related to water solubility (Hutchinson *et al.*, 1979). Such a mechanistic approach would then require a further layer of complexity—predictive ecological modeling, a branch of marine science still in gestation. And so the rewards of this kind of approach seem to be beyond our reach presently.

The alternative to such a mechanistic approach is an empirical comparison of well-documented cases of contamination to identify factors that are most important in producing effects, the approach I have attempted here. So I will summarize circumstances that have resulted in the greatest ecological changes and then hazard some guesses as to which are the most important factors. Finally, I will speculate as to the physical state and concentrations of hydrocarbons in water and sediments that may cause different outcomes. By doing this I hope to begin to reconcile such findings that on one hand, several thousand ppm of crude oil in sediments do not inhibit the settlement of larval polychaetes in sediments (Anderson *et al.*, 1978), stimulate the development of a dense infaunal community in a natural petroleum seep (Spies and Davis, 1979) and have some stimulatory effects on salt marsh meiofauna (Fleeger and Chandler, 1983), but on the other hand, smaller concentrations of refined oils have been identified with lethal effects (Sanders *et al.*, 1980; Elmgren *et al.*, 1983; Grassle *et al.*, 1981).

Circumstances that have been associated with the greatest ecological damage include restricted dispersion or initial very high concentrations of oil in the water from a massive spill, incorporation of fresh petroleum into sediments or the overlying flocculent layer, contamination from refined oils, large numbers of species existing in biologically accommodated communities and chronic contamination, especially when fresh petroleum is being continuously introduced. Many of these conditions occurred during the *Florida* spill (Sanders *et al.*, 1980), the *Tsesis* spill (Elmgren *et al.*, 1983), the Hong Kong Picnic Beach spill (Wormwald, 1976), the Sears Point pipeline leakage (Dow and Hurst, 1975) and the MERL microcosms (Grassle *et al.*, 1981; Oviatt *et al.*, 1982). These circumstances are also implicated in stimulatory responses, with the exception that crude oils are more often involved in stimulation. Stimulatory effects may be felt longer than the immediate toxic effects of refined oils.

As numerous authors have commented and is common knowledge among biologists doing oil pollution research, refined oils, especially No. 2 fuel oil, are much more toxic than crude oils and even moderate spills in inshore areas are likely to produce measurable biological damage. This appears to be the single most important factor associated with differing amounts of biological damage. Certainly the transport of large amounts of hydrocarbons to sediments under conditions which favor persistence of fresh petroleum, either through continual input or slow degradation and removal rates, is a second factor consistently associated with ecological change, both toxic and stimulatory. Related to this is the third factor, restricted dispersion. Release of oil to a confined body of water, cove, bay, harbor or estuary, is very often associated with ecological change. The converse of this is that open ocean spills are less likely to have measurable impacts unless they come ashore. It is of course very difficult to generalize about oil effects, and there are certainly exceptions to many of these. It is the circumstances of each case that control the outcome, and environments, oils, weather, and other related factors are so variable that predictions are difficult (Kerr, 1977). Returning somewhat to a mechanistic analysis, it must be the concentration of toxic or stimulatory compounds, the time of exposure, the sensitivity of species and the stability of the community that determine the effects of oil contamination.

In a speculative way, I would like to propose a common denominator for comparing dissimilar cases of contamination. Realizing that direct contact, ingestion and metabolism may also play roles, I hypothesize that different outcomes are a function of the concentrations of aromatic hydrocarbons in the interstitial waters or water at the sediment-water interface, or at least the outcomes correlate with these factors. I chose this for several reasons. First, aromatic hydrocarbons are considered the most toxic components of oil. Second, they are also the most soluble components and they can diffuse throughout the sediments, whereas the available concentrations of other hydrocarbons are severely limited by their solubility. Third, concentration of dissolved aromatic hydrocarbons provides a better measure of exposure for benthic organisms since solid phase measurements, e.g., total hydrocarbons, do not take into account how the oil is distributed within the sediment or what the sizes are of the discrete oil droplets. Fourth, since fresh oil is generally either more inhibitory or stimulatory than weathered oil and aromatics diffuse out of the solid phase of oil during weathering, the proposed measure is also a good index of the extent of weathering.

Although the hypothesis that pore water and near-bottom water hydrocarbon concentrations control ecological changes may be true for fresh oil in sediments, what of the potential toxicity of the more persistent aromatic hydrocarbons associated with weathered oil that leach out or are degraded only slowly and whose dissolved-phase concentrations will always be very low? Because these persist and can bioaccumulate, are they not a special case? I would argue that these compounds may not be a cause of community change. Although the inverse relationship between toxicity and water solubility that applies to microalgae

(Hutchinson *et al.*, 1979) may apply as well as to the infauna, this is offset by extremely low concentrations of very high molecular weight hydrocarbons in the interstitial water.

Unfortunately, few measurements have been made of interstitial water hydrocarbon content. In our studies of natural petroleum seeps, we have found a distribution of hydrocarbons in the interstitial water very similar to that of a laboratory-prepared water soluble fraction, dominated by mono- and diaromatic hydrocarbons. A sample taken from areas of very intense seepage had total interstitial hydrocarbons concentrations of 1.3 ppm. These limited areas of heavy seepage were depauperate in numbers of individuals and species (Spies *et al.*, 1980). However, in areas where the infauna was dense and diverse the concentrations were 45 to 117 ppb. Values for comparison station were 0.2 to 5 ppb (Stuermer *et al.*, 1982). These interstitial concentrations were attained where total sediment hydrocarbons ranged from approximately 2000 to 10,000 ppm and much of this was highly weathered asphalt-like material. I am unaware of similar data from areas of chronic contamination; although in the case of the MERL tanks and the Trinity Bay separator platform, values might be able to be extrapolated from water or sediment concentrations. Over one long period of oil addition to the MERL tanks the time-averaged water concentration was 190 μg/l. One might guess that interstitial water concentration could be several times larger—perhaps up to 1 mg/l (ppm).

In the case of the shallow water separator platform discharging hydrocarbon-laden brine, sharp changes in community composition were seen where 2 to 4 ppm naphthalenes occurred in the sediments (Armstrong *et al.*, 1979). Allowing for the weight of the sediment and some absorption to the sediment particles, the interstitial water concentrations of naphthalenes were perhaps 0.5 ppm. While these are merely guesses, they indicate that perhaps continuous concentrations of aromatics in interstitial water in the range of 0.5 to 1.0 ppm may be a common denominator for inducing toxic effects on the benthos.

I would suggest that interstitial water concentrations in the range of 0.02 to 0.10 ppm are stimulatory. The evidence indicates that the stimulation of benthic communities is mediated through microbes (Spies and DesMarais, 1983) and that the main ecological changes occurring along spatial and temporal concentration gradients are consistent with the theory of organic enrichment (Pearson and Rosenberg, 1978).

Interstitial water concentrations below 0.01 ppm are suggested to have no measurable effects, probably being below the concentrations required to stimulate microbes. However, since microbes are surface-active and may relate closer to the solid phase of oil, this may be an oversimplification.

SUMMARY, CRITICAL EVALUATION AND RECOMMENDATIONS

This section summarizes in a general way the results of different approaches to studying ecosystem effects of oil contamination, the limitations of each approach, gaps in our knowledge and recommendations for future study.

In microcosms oil can cause relatively rapid changes in bacterioplankton which are followed, under some conditions, by shifts in phytoplankton and zooplankton and eventually the benthos. All these occur with concentrations of No. 2 fuel oil as low as 90 μg/l; plankton effects will occur as low as 40 μg/l. The lower limit of measurable effects is probably 10 μg/l. It is obvious from the microcosm literature that a mechanistic appreciation of changes induced by enclosure and pollutant addition is not yet available. Therefore, population declines in many species seen, for instance, in the MERL experimental tanks are probably in some degree due to trophic linkage (beginning with microbial responses) as well as to cumulative sublethal toxicity.

Several questions about the MERL microcosm experiments should be considered if the results are to be extrapolated to actual sites of contamination. First, is the continual nitrogen limitation seen in the oil treatment tanks a real factor in chronically contaminated environments? And second, if nitrogen limitation is more severe because of enclosure, does it play a role in the eventual decline of benthic organisms? Third, what is the cause of the decline in the benthos of control microcosms? Are the responsible factors interactive with oil in the exposure tanks? Without clear answers to these questions, extrapolation of the results of such microcosms to the natural environment must only be conditional. Perhaps the true value of microcosms is captured in a recent quote, "Large experimental ecosystems have probably taught us more about the general ecological interactions in such systems than about subtle long-term effects of pollutants" (Steele, 1979).

Spilled oil can have transitory effects on plankton at the site of contamination that may last up to several weeks, but generally the effects are measurable for only several days, if at all. Advection and dilution preclude the measurement of these effects with any precision for an extended period of time, with the possible exception of some chronically contaminated bays, estuaries or harbors. Even there the presence of other pollutants would preclude any definite conclusions.

Oil spill effects on the benthos have recently been summarized by Dauvin (1982) for the *Amoco Cadiz* spill and his observations seem to largely agree with what has been described for well-studied spills. Dauvin recognized three stages:

1. A brief period of mortality, especially for amphipods, following exposure to fresh oil.
2. A medium-term change where some species populations continue at levels normally expected, while a few populations of opportunistic polychaete species proliferate greatly.
3. An eventual return to the original community state.

The effects are, of course, dose-dependent and can be less severe (even not measurable) or more severe according to the conditions outlined previously in this review. Dauvin suggests that in oligotrophic areas the appearance of opportunistic species is more fleeting, whereas in eutrophic bays and estuaries decrease in browsers leads to the accumulation of organic material and blooms of detritivorous polychaetes. While there may be a significant effect of reduced grazing on population changes (an argument made for MERL mesocosms as well), it is also true that oil itself is a largely degradable organic material, which, through microbial utilization, can contribute to the diet of detritus or deposit-feeding infauna (e.g., Spies and DesMarais, 1983), a result that had been previously predicted (Stanley *et al.*, 1978).

Studies of the effects of oil using experimental spills have the advantages of reality, time for planning and logistics, and the possibility of taking sufficient prespill data. The disadvantages include the difficulty of obtaining authorization for deliberate spills and the possibility that there are nonlinear scaling factors when trying to project the results obtained in a small area to a much larger accidental spill. The logistical disadvantages of studying large accidental spills are well known. It has also been difficult to relate doses to effects except under exceptionally favorable conditions.

As there are temporal gradients in the benthic effects that follow a spill, in areas of chronic petroleum contamination there are spatial gradients where zones of faunal suppression, stimulation and finally a return to a normal community may exist, again, in general concordance with the theory of organic enrichment (Pearson and Rosenberg, 1978). The study of areas of chronic contamination, away from the influence of other anthropogenic pollutants seems to hold great promise for isolating petroleum's ecological effects. We have studied offshore natural petroleum seeps for several other reasons as well: 1) they are a surrogate for the "worst case" offshore chronic contamination; 2) they are accessible and can be revisited at relatively little expense to refine experimental approaches; 3) there is no doubt that the observed effects are really representative of what occurs in the ocean; 4) quasi-stable gradients exist within them; and 5) they are amenable to manipulative field experimentation.

The shortcomings of such studies can include: 1) a laboratory-type "control" is not likely to exist; 2) often there is no precontamination data; and 3) there is the possibility of long-term adaptation of eucaryotic populations and every indication of accommodation of the microbial biocoenosis to oil. This could exist to the extent that it is difficult to confidently transfer results to transient spill incidents in pristine environments. Continued research in biological accommodation on the population and community level would seem to promise further insight into how well results can be transferred as well as provide a basis for assessing the seriousness of chronic petroleum exposure itself. So far little evidence has been found for long-term adaptation (genetic shifts) in multicellular organisms, but this deserves further study.

Measurable ecological changes do occur around offshore platforms but except for artificial reef effects or changes brought about by a cuttings pile, such changes are subtle and may not be detectable without a great sampling effort. Changes in the proportions of fine sediments, or correlative factors (e.g., total organic carbon), may be responsible. A mild form of organic enrichment is suspected. Beyond some contamination of organisms by petroleum, there is little convincing evidence of significant effects from petroleum around offshore platforms.

One could well argue that either 1) this is because there are no measurable effects in these environments or 2) the studies carried out so far have not been optimally designed to detect effects (see Carney, Chapter 14). At any rate there is presently a pervasive feeling among pollution biologists that classical ecological surveys by bottom-sampling cannot detect subtle population effects due to pollution. Perhaps the best hope for further research may lie in the area of relating tissue accumulations or some correlative factor (i.e., MFO activity) to subtle physiological changes (growth, reproduction) most closely related to population changes. This leads us to Lewis' dilemma, as outlined in Chapter 1.

The disadvantages of studying platform effects include: 1) the cost and effort needed to measure a small effect against a background of often great natural variability and stress; 2) the difficulty of assigning cause and effect to small changes observed; and 3) often, in large oil fields, appropriate control areas may not exist or predevelopment data are not available.

Turning now to areas that require more research, it seems apparent that priority should be given to searching for a common chemical denominator with which to compare different responses of the benthos to sediment contamination. The hypothesized correlation between porewater hydrocarbons and effects seems worth examining. There are a number of other questions that revolve around the relationships between sediment sources, actual doses and effects. It would seem worthwhile to investigate the relationship between hydrocarbon water solubility and toxicity for marine animals, as they are inversely correlated for microalgae (Hutchinson *et al.*, 1979). In this connection, it would also be important to know the relative decay of different hydrocarbon components of porewater, as there is some evidence, counter to intuition, that soluble hydrocarbons can decay from solution at similar rates, independent of their molecular weights or boiling points (Stuermer *et al.*, 1982). It would also seem important to determine whether the chemical differences producing different toxic responses between No. 2 fuel oil and crude oils are more of a quantitative or qualitative nature. The occurrence of phenols and analines in high concentrations in fuel oil extracts certainly suggests a qualitative difference in toxicity (Winters *et al.*, 1977b). Finally, investigation of the bioavailability and toxicity of highly-weathered complex hydrocarbon mixtures should help resolve the anxiety over the seriousness of long-term, low-level contamination of sediments by oil, since this is the form in which most chronic contamination occurs.

In the design of studies for detection of change around offshore platforms Americans could learn much from the published reports of the North Sea oil fields (Addy *et al.*, 1978; Hartley and Ferbrache, 1983). With some possible refinements, these seem to be good models for study designs in offshore areas, e.g., Alaska, Santa Maria Basin, western Santa Barbara Channel and Georges Bank, provided that a complicated sedimentary environment, in which a stratified sampling scheme would be more appropriate (see Chapter 14), does not exist. The American indulgence in rather aimless data gathering on the continental shelves over the last decade has told us precious little about either petroleum effects or controlling factors in marine ecosystems.

Some of the resources used to compile data might be better spent answering well-defined questions about the function and controlling factors in continental shelf ecosystems, for "Our difficulties of the moment must always be dealt with somehow; but our permanent difficulties are difficulties of every moment" (Eliot, 1949, p.5). Monitoring programs and basic process research could be funded together. In this way, more productive approaches and some progress might result from better interaction of basic and applied oceanography.

LITERATURE CITED

Addy, J.M., D. Levell and J.P. Hartley. 1978. Biological monitoring of sediment in Ekofisk Oilfield. Pages 515-539 *in* Proceedings Conference on Assessment of the Ecological Impact of Oil Spills. American Institute of Biological Sciences, Washington, D.C.

Alexander, S.K. and J.R. Schwarz. 1980. Short-term effects of South Louisiana and Kuwait crude oils on glucose utilization by marine bacterial populations. Appl. Environ. Microbiol. 40: 341-345.

Allen, A.A., R.S. Schlueter and P.G. Mikoloj. 1970. Natural oil seepage at Coal Oil Point, Santa Barbara, California. Science 170: 974-977.

Anderson, J.W. 1979. An assessment of knowledge concerning the fate and effects of petroleum hydrocarbon in the marine environment. Pages 3-22 *in* W.A. Vernberg, F.S. Verberg, A. Calabrese and F.P. Thurberg (eds.), Marine Pollution: Functional Responses. Academic Press, New York.

Anderson, J.W., R.G. Riley and R.M. Bean. 1978. Recruitment of benthic animals as a function of petroleum hydrocarbon concentrations in the sediment. J. Fish. Res. Board Can. 35: 776-790.

Andrews, A.R. and G.D. Floodgate. 1974. Some observations on the interactions of marine protozoa and crude oil residues. Mar. Biol. 25: 7-12.

Armstrong, H.W., K. Fucik, J.W. Anderson and J.M. Neff. 1979. Effects of oilfield brine effluent on sediments and benthic organisms in Trinity Bay, Texas. Mar. Environ. Res. 2: 55-69.

Atlas, R.M. 1981. Microbial degradation of petroleum hydrocarbons: An environmental perspective. Microbiol. Rev. 45: 180-209.

Atlas, R.M. and A. Bronner. 1981. Microbial hydrocarbon degradation within intertidal zones impacted by the *Amoco Cadiz* oil spillage. Pages 251-256 *in Amoco Cadiz*, Fates and Effects of the Oil Spill. Le Centre National Pour L'Exploitation des Oceans, Paris.

Azam, F., T. Fenchel, J.G. Field, J.S. Gray, L.A. Meyer-Reil and F. Thingstad. 1983. The ecological role of water column microbes in the sea. Mar. Ecol. Prog. Ser. 10: 257-263.

Azoulay, E., M. Colin, J. Dubreuil, H. Dou, G. Mille and G. Giusti. 1983. Relationship between hydrocarbons and bacterial activity in Mediterranean sediments: Part 2 - Hydrocarbon degrading activity of bacteria from sediments. Mar. Environ. Res. 9: 19-36.

Baker, J. 1973. Biological effects of refinery effluents. Pages 715-723 *in* Proceedings Joint Conference on Prevention and Control of Oil Spills. American Petroleum Institute, Washington, D.C.

Baker, J.H., K.T. Kimball, W.D. Jobe, J. Janousek, C.L. Howard and P.R. Chase. 1981. Part 6. Benthic biology. Pages 1-317 *in* C.A. Bedinger, Jr., (ed.), Ecological Investigations of Petroleum Production Platforms in the Central Gulf of Mexico, Vol. 1. Pollutant Fate and Effects Studies. Southwest Research Institute, San Antonio, Texas.

Barnard, J.L. and O. Hartman. 1959. The sea bottom off Santa Barbara, California: Biomass and community structure. Pacific Nat. 1: 1-16.

Bartha, R. and R.M. Atlas. 1977. The microbiology of aquatic oil spills. Adv. Appl. Microbiol. 22: 225-266.

Batterton, J.C., K. Winters and C. Van Baalen. 1978a. Sensitivity of three microalgae to crude oils and fuel oils. Mar. Environ. Research. 1: 31-41.

Batterton, J.C., K. Winters and C. Van Baalen. 1978b. Analines: Selective toxicity to blue green algae. Science 199: 1068-1070.

Bedinger, C.A., Jr., R.E. Childers, J.W. Cooper, K.T. Kimball and A. Kwok. 1981. Part 1. Background, program organization and study plan. Pages 1-53 *in* C.A. Bedinger (ed.), Ecological Investigations of Petroleum Production Platforms in the Central Gulf of Mexico. Vol. 1. Pollutant Fate and Effects Studies. Southwest Research Institute, San Antonio, Texas.

Bell, J.S., C. Houghton and J.M. Davies. 1983. A comparison of the levels of hepatic MFO in fish caught close to and distant from some North Sea oilfields. *In* Second International Symposium on Responses of Marine Organisms to Pollutants, April, 1983 (Abstract).

Bender, M.E., E.A. Shearls, R.P. Ayres, C.H. Hershner and R.J. Huggett. 1977. Ecological effects of experimental oil spill on eastern coastal plain estuarine ecosystems. Pages 505-509 *in* Proceedings 1977 Oil Spill Conference. American Petroleum Institute, Washington, D.C.

Bender, M.E., D.J. Reish and C.H. Ward. 1979. Independent appraisal. Re-examination of the Offshore Ecology Investigation. Pages 35-116 *in* C.H. Ward, M.E. Bender and D.J. Reish (eds.), The Offshore Ecology Investigation: Effects of Oil Drilling and Production in a coastal Environment. Rice University Studies 65: 1-589.

Bert, T.M. and H.J. Humm. 1979. Checklist of the marine algae on the offshore oil platforms of Louisiana. Pages 43-44 *in* C.H. Ward, M.E. Bender and D.J. Reish (eds.), The Offshore Ecology Investigations: Effects of Oil Drilling and Production in a Coastal Environment. Rice University Studies 65: 1-589.

Bieri, R.H., V.C. Stamoudis and M.K. Cueman. 1977. Chemical investigations of two experimental oil spills in an estuarine ecosystem. Pages 511-515 *in* Proceedings 1977 Oil Spill Conference. American Petroleum Institute, Washington, D.C.

Bjørke, H., E. Ellingsen and S.A. Iversen. 1977. Zooplankton, fish, eggs and larvae. Pages 1-8 *in* The Ekofisk Blow-Out, Compiled Norwegian Contributions, *ICES*, C.M. 1977/E, 55, Section 10: 1-8.

Boney, A.D. 1974. Aromatic hydrocarbons and the growth of marine algae. Mar. Pollut. Bull. 5: 185-186.

Boney, A.D. and E.D.S. Corner. 1962. On the effects of some carcinogenic hydrocarbons on the growth of sporelings of marine red algae. J. Mar. Biol. Ass., U.K. 42: 579-585.

Brent, C.R., H.P. Williams, W.A. Bergin, J.L. Tyvoll and T.E. Meyers. 1979. Organic carbon, inorganic carbon, and related variables in offshore oil production areas of the northern Gulf of Mexico. Pages 245-264 *in* C.H. Ward, M.E. Bender and D.J. Reish (eds.), The Offshore Ecology Investigation: Effects of Oil Drilling and Production in a Coastal Environment. Rice University Studies 65: 1-589.

Brown, L.R., J.D. Walker, G.W. Childers and R.W. Landers, Jr. 1981. Part 5. Microbiology and microbiological processes. Pages 122-215 *in* C.A. Bedinger, Jr., (ed.), Ecological Investigations of Petroleum Production in the Central Gulf of Mexico. Vol. 1. Pollutant Fate and Effects Studies. Southwest Research Institute, San Antonio, Texas.

Burchfield, H.P., R.J. Wheeler and W. Subra. 1979. Nutrient concentrations in Timbalier Bay, a Louisiana oil patch. Pages 223-234 *in* C.H. Ward, M.E. Bender and D.J. Reish (eds.), The Offshore Ecology Investigation: Effects of Oil Drilling and Production in a Coastal Environment. Rice University Studies 65: 1-589.

Burns, K.A. 1976. Microsomal mixed function oxidases in an estuarine fish, *Fundulus heteroclitus*, and their induction as a result of environmental contamination. Comp. Biochem. Physiol. 533: 443-446.

Burns, K.A. and J.M. Teal. 1979. The West Falmouth oil spill: Hydrocarbons in the salt marsh ecosystem. Estuar. Coastal Mar. Sci. 8: 349-360.

Caparello, D.M. and P.A. LaRock. 1975. A radioisotope assay for quantification of hydrocarbon biodegradation potential in environmental samples. Microb. Ecol. 2: 28-42.

Cerniglia C.E., D.T. Gibson and C. Van Baalen. 1979. Algal oxidation of aromatic compounds: Formation of 1-napthol from naphthalene by *Agmenellum quadruplicatum*, strain PR-6. Biochem. Biophys. Res. Comm. 88: 50-58.

Cerniglia, C.E., D.T. Gibson and C. Van Baalen. 1980a. Oxidation of naphthalene by cyanobacteria and microalgae. J. Gen. Microbiol. 116: 495-500.

Cerniglia, C.E., C. Van Baalen and D.T. Gibson. 1980b. Oxidation of biphenyl by the cyanobacterium, *Oscillatonia* sp. strain JCM. Arc. Microbiol. 125: 203-207.

Chasse, C. 1978. The ecological impact on and near shores by the *Amoco Cadiz* oil spill. Mar. Pollut. Bull. 9: 298-301.

Colwell, R.R. and J.D. Walker. 1977. Ecological aspects of microbial degradation of petroleum in the marine environment. Crit. Rev. Microbiol. 5: 423-445.

Colwell, R.R., A.L. Mills, J.D. Walker, P. Garcia-Tello and V. Campose-P. 1978. Microbial ecology studies of the *Metula* spill in the Straits of Magellan. J. Fish. Res. Bd. Canada. 35: 573-580.

Connell, D.W. and G.J. Miller. 1981. Petroleum hydrocarbons in aquatic ecosystems - behavior and effects of sublethal concentrations: Part 2. CRC Critical Reviews in Environmental Control 11: 105-162.

Conover, R.S. 1971. Some relations between zooplankton and Bunker C oil in Chedabucto Bay following the wreck of the tanker *Arrow*. J. Fish. Res. Board Can. 28: 1327-1330.

Corner, E.D.S. 1978. Pollution studies with marine plankton. Part 1, Petroleum hydrocarbons and related compounds. Adv. Mar. Biol. 15: 289-380.

Cundell, A.M. and R.W. Traxler. 1973. The isolation and characterization of hydrocarbon-utilizing bacteria from Chedabucto Bay, Nova Scotia. Pages 421-426 *in* Proceedings Joint Conference Prevention and Control of Oil Spills. American Petroleum Institute, Washington, D.C.

Dauvin, J.C. 1982. Impact of *Amoco Cadiz* oil spill on the muddy fine sand *Abra alba* and *Melinna palmata* community from the Bay of Morlaix. Estuar. Coastal Shelf Sci. 14: 517-531.

Davies, J.M., I.C. Baird, L.C. Massie, S.J. Hay and A.P. Ward. 1980. Some effects of oil derived hydrocarbons in an enclosed ecosystem and a consideration of their implications for monitoring. Rapp. P.-V. Réun. Cons. Int. Explor. Mer 179: 201-211.

Davis, P.G. 1982. Bacterivorous flagellates in marine waters. Ph.D. Thesis, University of Rhode Island, 180 p.

Davis, P.H. and R.B. Spies. 1980. Infaunal benthos of a natural petroleum seep: Study of community structure. Mar. Biol. 59: 31-41.

DeGraeve, G.M., R.L. Overcast and H.L. Bergman. 1980. Toxicity of underground coal gasification condensor water and selected constituents to aquatic biota. Arch. Environ. Toxicol. 9: 543-555.

De La Cruz, A.A. 1982. Effects of oil on phytoplankton metabolism in natural and experimental estuarine ponds. Mar. Environ. Res. 7: 257-263.

Deveraux, R. and R.K. Sizemore. 1982. Plasmid incidence in marine bacteria isolated from petroleum polluted sites on different petroleum hydrocarbons. Mar. Poll. Bull. 13: 198-202.

Dicks, B.M. 1975. Offshore biological monitoring. Pages 325-440 *in* J.M. Baker (ed.), Marine Ecology and Oil Pollution. Applied Science Publishers, Barking, England.

Dicks, B. 1981. Offshore oil impact (a book review). Mar. Poll. Bull. 12: 368-369.

Dicks, B.M. 1982. Monitoring the biological effects of North Sea platforms. Mar. Pollut. Bull. 13: 221-227.

Dicks, B.M. and J.P. Hartley. 1982. The effects of repeated small oil spillages and chronic discharges. Phil. Trans. R. Soc. Lond. B 297: 285-307. Reprinted in R.B. Clark (ed.). 1982. The Long-Term Effects of Oil Pollution in Marine Populations, Communities and Ecosystems. The Royal Society, London.

DiSalvo, L.H. and H.E. Guard. 1975. Hydrocarbons associated with particulate matter in San Francisco Bay waters. Pages 169-173 *in* Proceedings 1975 Conference on Prevention and Control of Oil Pollution. American Petroleum Institute, Washington, D.C.

Dow, R.L. and J.W. Hurst, Jr. 1975. The ecological, chemical and histopathological evaluation of an oil spill site. Part 1. Ecological Studies. Mar. Pollut. Bull. 6: 164-166.
Dunstan, W.M., L.P. Atkinson and J. Natoli. 1975. Stimulation and inhibition of phytoplankton growth by low molecular weight hydrocarbons. Mar. Biol. 31: 305-310.
Elcombe, C.R., R.B. Franklin and J.J. Leach. 1979. Induction of microsomal hemoprotein(s) P-450 in the rat and rainbow trout by polyhalogenated biphenyls. Ann. N.Y. Acad. Sci. 320: 193-203.
Eliot, T.S. 1949. Christianity and culture. Harcourt, Brace, New York.
Elmgren, R., G.A. Vargo, J.F. Grassle, J.P. Grassle, D.R. Heinle, G. Longelis and S.L. Vargo. 1980. Trophic interactions in experimental marine ecosystems perturbed by oil. Pages 779-800 *in* J.P. Giesy (ed.), Microcosms in Ecological Research. U.S. Dept. of Energy, Washington, D.C.
Elmgren, R., S. Hanson, U. Larsson, B. Sundelin and P.D. Boehm. 1983. The "Tsesis" Oil Spill: Acute and long-term impact on the benthos. Mar. Biol. 73: 51-65.
Eppley, R.W. and C.S. Weiler. 1979. The dominance of nanoplankton as an indicator of marine pollution: A critique. Oceanol. Acta. 2: 241-245.
FAO (Food and Agricultural Organization of the United Nations). 1977. Impact of Oil on the Marine Environment. Reports and Studies No. 6 IMCO/FAO/UNESCO/ WMO/ WHO/IAEA/UW. Joint Group of Experts on Scientific Aspects of Marine Pollution (GESADP).
Farrell, D. 1974a. Benthic communities in the vicinity of producing oil wells on the shallow Louisiana continental Shelf. *In* Abstracts, Summaries, and Conclusions from the Offshore Ecology Investigation, 1972-1974, Gulf Universities Research Consortium (GURC), Galveston, Texas.
Farrell, D. 1974b. Benthic communities in the vicinity of producing oil wells in Timbalier Bay, Louisiana. Pages 14-95 *in* Abstracts, Summaries, and Conclusions from the Offshore Ecology Investigation, 1972-1974, Gulf Universities Research Consortium (GURC), Galveston, Texas.
Farrell, D.H. 1979. Benthic molluscan and crustacean communities in Louisiana. Pages 40-43 *in* C.H. Ward, M.E. Bender and D.J. Reish (eds.), The Offshore Ecology Investigation: Effects of Oil Drilling and Production in a Coastal Environment. Rice University Studies 65: 1-589.
Fenchel, T. 1982. Ecology of heterotrophic microflagellates. IV. Quantitative occurrence and importance as bacterial consumers. Mar. Ecol. Prog. Ser. 9: 35-42.
Fleeger, J.W. and G.T. Chandler. 1983. Meiofauna responses to an experimental oil spill in a Louisiana salt marsh. Mar. Ecol. Prog. Ser. 11: 257-264.
Forlin, L. and C. Haux. 1985. Increased excretion in the bile of 17β-[^{3}H] estradiol-derived radioactivity in rainbow trout treated with β-naphthoflavone. Aquat. Toxicol. 6: 197-208.
Fricke, A.H., H. F-K. O. Hennig and M.J. Orren. 1981. Relationships between oil pollution and psammolittoral meiofauna density of two South African beaches. Mar. Environ. Res. 5: 59-77.
Fucik, K.W. and S.Z. El-Sayed. 1979. Effect of oil production and drilling operations on the ecology of phytoplankton in the OEI study area. Pages 325-353 *in* C.H. Ward, M.E. Bender and D.J. Reish (eds.), The Offshore Ecology Investigation: Effects of Oil Drilling and Production in a Coastal Environment. Rice University Studies 65: 1-589.
Galtsoff, P.S., H.F. Prytherch, R.O. Smith and V. Koehring. 1935. Effects of crude oil pollution on oysters in Louisiana Waters. Bull. Bureau of Fisheries, Washington, D.C. 18: 143-210.
George, R.Y. and P.J. Thomas. 1979. Biofouling community dynamics in Louisiana shelf oil platforms in the Gulf of Mexico. Pages 553-574, *in* C.H. Ward, M.E. Bender and D.J. Reish (eds.), The Offshore Ecology Investigation: Effects of Oil Drilling and Production in a Coastal Environment. Rice University Studies 65: 1-589.
Giere, O. 1979. The impact of oil pollution on intertidal meiofauna. Field studies after the La Coruna Spill, May 1976. Cahiers de Biol. Mar. 20: 231-251.

Gordon, D.C., Jr. and N.J. Prouse. 1973. The effects of three oils on marine phytoplankton photosynthesis. Mar. Biol. 22: 329-333.

Grahl-Nielsen, O., K. Westrheim and S. Wilhelmsen. 1979. Petroleum hydrocarbons in the North Sea. Pages 629-632 *in* Proceedings 1979 Oil Spill Conference. American Petroleum Institute, Washington, D.C.

Grassle, J.F., R. Elmgren and J.P. Grassle. 1981. Response of benthic communities in MERL experimental ecosystems to low level, chronic additions of No. 2 fuel oil. Mar. Environ. Res. 4: 279-297.

Griffin, G.M. 1979. Evaluation of the effects of oil production platforms on the turbidity of Louisiana Shelf Waters. Pages 159-179 *in* C.H. Ward, M.E. Bender and D.J. Reish (eds.), The Offshore Ecology Investigation: Effects of Oil Drilling and Production in a Coastal Environment. Rice University Studies 65: 1-589.

Griffiths, R.P., B.A. Caldwell, W.A. Brosch and R.Y. Morita. 1981. The long-term effects of crude oil on uptake and respiration of glucose and glutamate in arctic and subarctic marine sediments. Appl. Envir. Microbiol. 42: 792-801.

Gruger, E.H., Jr., M.M. Wekel, P.T. Numoto and D.R. Craddock. 1977. Induction of hepatic aryl hydrocarbon hydroxylase in salmon exposed to petroleum dissolved in seawater and to petroleum and polychlorinated biphenyls, separate and together, in food. Bull. Envir. Contam. Toxicol. 17: 512-520.

Gunkel, W., G. Gassman, C.H. Oppenheimer and I. Dundas. 1980. Preliminary results of baseline studies of hydrocarbons and bacteria in the North Sea, 1975, 1976 and 1977. Pages 223-247 *in* Resistencia a los Antibioticos y Microbiologia Marina, VI Congresso Nacional Microbiologia, 1977, Santiago de Compostila, Spain.

Haas, L.W. and K.L. Webb. 1979. Nutritional mode of several non-pigmented microflagellates from the York River Estuary, Virginia. J. Exp. Mar. Biol. Ecol. 39: 125-134.

Haensly, W.E., J.M. Neff, J.R. Sharp, A.C. Morris, M.F. Bedgood and P.D. Boehm. 1982. Histopathology of *Pleuronectes platessa* L. from Aber Wrac'h and Aber Benoit, Brittany, France; Long-term effect of the *Amoco Cadiz* oil spill. J. Fish. Dis. 5: 365-391.

Hagström, A. 1977. The fate of oil in a model ecosystem. Ambio 6: 229-231.

Hampson, G.G. and H.L. Sanders. 1969. Local oil spill. Oceanus 15: 8-11.

Hampson, G.R. and E.T. Moull. 1978. No. 2 Fuel oil spill in Bourne, Massachusetts. Immediate assessment of the effects of marine invertebrates and a 3-year study of growth and recovery of a salt marsh. J. Fish. Res. Board Can. 35: 731-744.

Hardy, R., P.R. Mackie and K.J. Whittle. 1977. Hydrocarbons and petroleum in the marine ecosystem - A review. Rapp P.-V. Réun. Cons. Int. Explor. Mer 171: 17-26.

Harper, D.E., Jr., R.J. Scrudato and C.S. Giam. 1976. Pilot study of the Buccaneer Oil Field (Benthos and Sediments). A Preliminary Environmental Assessment of the Buccaneer Oil and Gas Field. Unpubl. report, Texas A&M University, 63 p.

Hartley, J.P. and J. Ferbrache. 1983. Biological monitoring of the Forties oilfield (North Sea). Pages 407-414 *in* Proceedings 1983 Oil Spill Conference. American Petroleum Institute, Washington, D.C.

Hedgpeth, J.W. 1978. As blind men see the elephant: The dilemma of marine ecosystem research. Pages 3-15 *in* M.L. Wiley (ed.), Estuarine Interactions. Academic Press, New York.

Hendrickson, P., H.J. Hirche and U. Junghans. 1978. Preliminary results from Polumar III (April 13-18, 1978). (Suroit III). ATP, primary production and respiration rates in seston: Respiration rates in zooplankton (*Amoco Cadiz* spill). Act. Colloques Cent. Natn. Exploit. Oceans, No. 6: 209-224.

Hodson, R.E., F. Azam and R.F. Lee. 1977. Effects of four oils on marine bacterial populations: Controlled ecosystem pollution experiment. Bull. Mar. Sci. 27: 119-126.

Hsaio, S.I.C. 1978. Effects of crude oil on the growth of Arctic marine phytoplankton. Environ. Pollut. 17: 93-107.

Humm, H.J. and T.M. Bert. 1979. The benthic marine algae of Timbalier Bay, Louisiana.

Pages 37-40 *in* C.H. Ward, M.E. Bender and D.J. Reish (eds.), The Offshore Ecology Investigation: Effects of Oil Drilling and Production in a Coastal Environment. Rice University Studies 65: 1-589.

Hutchinson, T.C., J.A. Hellebust, D. McKay and P. Kauss. 1979. The relationship of hydrocarbon solubility to toxicity in algae and cellular membrane effects. Pages 541-547 *in* Proceedings 1979 Oil Spill Conference. American Petroleum Institute, Washington, D.C.

Johansson, S. 1980. Impact of oil on the pelagic ecosystem. Pages 61-80 *in* J.J. Kineman, R. Elmgren and S. Hansen (eds.), The Tsesis Oil Spill. National Oceanic and Atmospheric Administration, Washington, D.C.

Johansson, S., U. Larsson and P. Boehm. 1980. The *Tsesis* oil spill, impact on the pelagic ecosystem. Mar. Pollut. Bull. 11: 284-293.

Jones, G.F. 1969. The benthic macrofauna of the mainland shelf of southern California. Allan Hancock Monogr. Mar. Biol. 4: 1-219.

Jones, J.I. and S.E. Williams. 1979. The distribution and origin of bottom sediments in Timbalier Bay, Louisiana and the adjacent offshore area. Pages 201-222 *in* C.H. Ward, M.E. Bender and D.J. Reish (eds.), The Offshore Ecology Investigation: Effects of Oil Drilling and Production in a Coastal Environment. Rice University Studies 65: 1-589.

Juge, D.M. 1970. A study of the bacterial population of bottom sediments in the Santa Barbara Channel after the oil spill. Pages 179-222 *in* Biological and Oceanographical Survey of the Santa Barbara Channel Oil Spill, 1969-1970. Vol. 1. Biology and Bacteriology. Sea Grant Publ. No. 2, Allan Hancock Foundation, University of Southern California, Los Angeles, California.

Kalke, R.D., T.A. Duke and R.W. Flint. 1982. Weathered IXTOC I oil effects on estuarine benthos. Estuar. Coastal Shelf Sci. 15: 75-84.

Kato, R., T. Kamataki and Y. Yamazo. 1983. N-hydroxylation of carcinogenic and mutagenic aromatic amines. Envir. Health Persp. 49: 21-25.

Kator, H. and R. Herwig. 1977. Microbial responses after two experimental oil spills in an eastern coastal plain estuarine ecosystem. Pages 517-522. Proceedings 1977 Oil Spill Conference. American Petroleum Institute, Washington, D.C.

Kauss, P.B., T.C. Hutchinson, C. Soto, J. Hellebrest and M. Griffiths. 1973. The toxicity of crude oil and its components to freshwater algae. *In* Conference on Prevention and Control of Oil Spills. American Petroleum Institute, Washington, D.C.

Kerr, R. 1977. Oil in the ocean: Circumstances control its impact. Science 198: 1134-1136.

Kinney, P.J., D.K. Button and D.M. Schell. 1969. Kinetics of dissipation and biodegradation of crude oil in Alaska's Cook Inlet. Pages 333-340 *in* Joint Conference on Prevention and Control of Oil Spills. American Petroleum Institute, Washington, D.C.

Kloepper-Sams, P.J., S.S. Park, H.V. Gelboin and J.J. Stegeman. 1986. Immunochemical analysis of cytochrome P-450 isozymes in fish with polyclonal and monoclonal antibodies against cytochrome P-450E from scup. Fed. Proc. 45: 320.

Klotz, A.V., J.J. Stegeman and C. Walsh. 1983. An aryl hydrocarbon hydroxylating hepatic cytochrome P-450 from the marine fish *Stenotomus chrysops*. Arch. Biochem. Biophys. 226: 578-592.

Krahn, M.M., M.S. Meyers, D.G. Burrows and D.C. Malins. 1984. Determination of metabolites of xenobiotics in bile of fish from polluted waterways. Xenobiotica 14: 633-646.

Krebs, C.T. and K.A. Burns. 1977. Long-term effects of an oil spill on populations of the salt-marsh crab *Uca pugnax*. Science 197: 484.

Kritzler, H. 1979. Oil production and ecology of the littoral Polychaeta of Timbalier Bay. Pages 47-49 *in* C.H. Ward, M.E. Bender and D.J. Reish (eds.), The Offshore Ecology Investigation: Effects of Oil Drilling and Production in a Coastal Environment. Rice University Studies 65: 1-589.

Kuhnhöld, W.W. 1978. Impact of the *Argo Merchant* oil spill on macrobenthic and pelagic organisms. Pages 153-179 *in* Proceedings Conference Assessment Ecological Impacts of Oil Spills, June 1978, Keystone, Colorado. American Institute of Biological Sciences, Washington, D.C.

Lanier, J.J. 1978. Ciliates as bioindicators of oil pollution. Pages 651-663 *in* Proceedings Conference on Assessment of Ecological Impacts of Oil Spills, June 1978, Keystone, Colorado. American Institute of Biological Sciences, Washington, D.C.

Lannergren, C. 1978. Net and nanoplankton: Effects of an oil spill in the North Sea. Bot. Mar. 21: 353-356.

Laseter, J.L. and E.J. Ledet. 1979. Hydrocarbons and free fatty acids associated with the air/water interface, sediments, and beaches of the Timbalier Bay offshore Louisiana area. Pages 265-277 *in* C.H. Ward, M.E. Bender and D.J. Reish (eds.), The Offshore Ecology Investigation: Effects of Oil Drilling and Production in a Coastal Environment. Rice University Studies 65: 1-589.

Lee, R.F. 1977. Fate of petroleum components in estuarine waters of the southeastern United States. Pages 611-616 *in* Proceedings 1977 Oil Spill Conference. American Petroleum Institute, Washington, D.C.

Lee, R.F. 1981. Mixed-function oxygenases (MFO) in marine invertebrates. Mar. Biol. Lett. 2: 87-105.

Lee, R.F. and J.W. Anderson. 1977. Fate and effect of naphthalenes in controlled ecosystem enclosures. Bull. Mar. Sci. 27: 127-134.

Lee, R.F., B. Dornsief, F. Gonsoulin, K. Tenore and R. Hanson. 1981. Fate and effects of a heavy fuel oil spill on a Georgia salt marsh. Mar. Environ. Res. 5: 125-143.

Lee, R.F. and M. Takahashi. 1977. The fate and effect of petroleum in controlled ecosystem enclosures. Rapp. P.-V. Réun. Cons. Int. Explor. Mer 171: 150-156.

Lee, R.F., M. Takahashi and J. Beers. 1978. Short term effects of oil on plankton in controlled ecosystems. Pages 635-650 *in* Proceedings Conference on Assessment of Ecological Impacts of Oil Spills, June 1978, Keystone, Colorado. American Institute of Biological Sciences, Washington, D.C.

Leppäkoski, E.S. and L.S. Lindstrom. 1978. Recovery of benthic macrofauna from chronic pollution in the sea area off a refinery plant, southwest Finland. J. Fish. Res. Board Can. 35: 766-775.

Lewis, P.L. and A.G. Fish. 1979. The ecology of the littoral marine polychaetes of Timbalier Bay. Pages 511-528 *in* C.H. Ward, M.E. Bender and D.J. Reish (eds.), The Offshore Ecology Investigation: Effects of Oil Drilling and Production in a Coastal Environment. Rice University Studies 65: 1-589.

Lindström-Sappä, P. 1985. Seasonal variation of the xenobiotic metabolizing enzyme activities in the liver of male and female vendance (*Coregonus albula* L.). Aquat. Toxicol. 6: 323-331.

Longwell, A.C. 1977. A genetic look at fish eggs and oil. Oceanus 20: 46-58.

Mackie, P.R., R. Hardy, E.I. Butler, R.M. Holligan and M.F. Spooner. 1978. Early samples of oil in water and some analyses of zooplankton. Mar. Pollut. Bull. 9: 296-297.

Mahoney, B.M. and H.H. Haskins. 1980. The effects of petroleum hydrocarbons on the growth of phytoplankton recognized as food forms for the eastern oyster, *Crassostrea virginica* Gmelin. Environ. Pollut. 22: 123-132.

Malins, D.C. and H.O. Hodgins. 1981. Petroleum and marine fishes: A review of uptake, disposition and effects. Environ. Sci. Technol. 11: 1272-1280.

Marum, J.P. 1979. Significance of distribution patterns of planktonic copepods in Louisiana coastal waters and relationships to oil drilling and production. Pages 355-377 *in* C.H. Ward, M.E. Bender and D.J. Reish (eds.), The Offshore Ecology Investigation: Effects of Oil Drilling and Production in a Coastal Environment. Rice University Studies 65: 1-589.

McAuliffe, C.D., A.E. Smalley, R.D. Groover, W.M. Welsh, W.S. Pickle and G.E.

Jones. 1975. Chevron Main Pass Block 41 Oil Spill: Chemical and biological investigations. Pages 555-566 *in* Proceedings 1975 Joint Conference on Prevention and Control of Oil Spills. American Petroleum Institute, Washington, D.C.

McCain, B.B., H.O. Hodgins, W.D. Gronlund, J.W. Hawkes, D.W. Brown, M.S. Meyers and J.H. Vandemeulen. 1978. Bioavailability of crude oil from experimentally oiled sediments to English sole (*Parophrys vetulus*), and pathological consequences. J. Fish. Res. Board Can. 35: 657-664.

McGinnis, D.R. 1971. Observations on the zooplankton of the eastern Santa Barbara Channel from May, 1969, to March, 1970. Pages 49-53 *in* Biological and Oceanographical Survey of the Santa Barbara Channel Oil Spill, 1969-1970, Volume I. Biology and Bacteriology. Sea Grant Publ. No. 2, Allan Hancock Foundation, University of Southern California, Los Angeles, California.

McLusky, D.S. 1982. The impact of petrochemical effluent in the fauna of an intertidal estuarine mudflat. Estuar. Coastal Shelf Sci. 4: 489-499.

Menzel, D.W. 1980. Applying results derived from experimental microcosms to the study of natural pelagic marine ecosystems. Pages 742-752 *in* J.P. Giesy (ed.), Microcosms in Ecological Research. U.S. Dept. of Energy, Washington, D.C.

Menzies, R.J., J.P. Morgan, C.H. Oppenheimer, S.Z. El-Sayed and J.M. Sharp. 1979. Design of the Offshore Ecology Investigation. Pages 19-32 *in* C.H. Ward, M.E. Bender and D.J. Reish (eds.), The Offshore Ecology Investigation: Effects of Oil Drilling and Production in a Coastal Environment. Rice University Studies 65: 1-589.

Mertens, E.W. 1978. The impact of oil on marine life: A summary of the field studies. Pages 508-514 *in* Sources, Effects and Sinks of Hydrocarbons in the Aquatic Environment. American Institute of Biological Sciences, Washington, D.C.

Michael, A.D., C.R. Van Raalte and L.S. Brown. 1975. Long-term effects of an oil spill at West Falmouth, Massachusetts. Pages 573-582 *in* Proceedings Conference on Prevention and Control of Oil Pollution. American Petroleum Institute, Washington, D.C.

Middleditch, B.S. (ed.). 1981. Environmental Effects of Offshore Oil Production. The Buccaneer Gas and Oil Field Study. Plenum Press, New York.

Middleditch, B.S., B. Basile and E.S. Chang. 1977. Environmental effects of offshore oil production: Alkanes in the region of the Buccaneer Oilfield. J. Chromat. 142: 777-785.

Middleditch, B.S., B. Basile and E.S. Chang. 1978. Discharge of alkanes during offshore oil production in the Buccaneer Oilfield. Bull. Environ. Contam. Toxicol. 20: 59-65.

Middleditch, B.S., E.S. Chang and B. Basile. 1979. Alkanes in plankton from the Buccaneer Oilfield. Bull. Environ. Contam. Toxicol. 21: 421-427.

Mironov, O.G. and L.A. Lanskaya. 1969. Growth of marine microscopic algae in seawater contaminated with hydrocarbons. Biologiva Morya 17: 31-38 (in Russian).

Mommearts-Billiet, F. 1973. Growth and toxicity tests on the marine nano-planktonic algae *Platymonas tetrathele* G.S. West in the presence of crude oil and emulsifiers. Environ. Poll. 4: 261-282.

Montagna, P.A. and R.B. Spies. 1985. Meiofauna and chlorophyll associated with *Beggiatoa* mats of a natural submarine petroleum seep. Mar. Environ. Res. 16: 231-242.

Montagna, P.A., J.E. Bauer, M.C. Prieto, D. Hardin and R.B. Spies. In press. Benthic metabolism in a natural coastal petroleum seep. Mar. Ecol. Prog. Ser.

Montalvo, J.G., Jr. and V. Brady. 1979. Concentrations of Hg, Pb, Zn, Cd and As in Timbalier Bay and the Louisiana oil patch. Pages 235-243 *in* C.H. Ward, M.E. Bender and D.J. Reish (eds.), The Offshore Ecology Investigation: Effects of Oil Drilling and Production in a Coastal Environment. Rice University Studies 65: 1-589.

Morgan, J.P. 1979. Recent geological history of the Timbalier Bay area and adjacent continental shelf. Pages 575-589 *in* C.H. Ward, M.E. Bender and D.J Reish (eds.), The Offshore Ecology Investigation: Effects of Oil Drilling and Production in a Coastal Environment. Rice University Studies 65: 1-589.

Mulkins-Phillips, G.J. and J.E. Stewart. 1974. Distribution of hydrocarbon-utilizing bacteria in northwestern Atlantic waters and coastal sediments. Can. J. Microbiol. 20: 955-962.

Nellbring, S., S. Hanson, G. Aneer and L. Westin. 1980. Impact of oil on local fish fauna. Pages 193-20 *in* J.J. Kineman, R. Elmgren and S. Hanson (eds.), The *Tsesis* Oil Spill. National Oceanic and Atmospheric Administration, Washington, D.C.

Nelson, D.C., J.B. Waterbury and H.W. Jannasch. 1982. Nitrogen fixation and nitrate utilization by marine and freshwater *Beggiatoa*. Arch. Microbiol. 133: 172-177.

Nulton, C.P., C.F. Bohstedt, D.E. Johnson and S.J. Martin. 1981. Part 3. Organic chemical analyses. Pages 133-183 *in* C.A. Bedinger, Jr. (ed.), Ecological Investigation of Petroleum Production in the Central Gulf of Mexico. Vol. 1, Pollutant Fate and Effects Studies. Southwest Research Institute, San Antonio, Texas.

Nuzzi, R. 1973. Effects of water soluble extracts of oil on phytoplankton. Pages 809-813 *in* Proceedings 1973 Conference on Prevention and Control of Oil Spills. American Petroleum Institute, Washington, D.C.

O'Brien, P.Y. and P.S. Dixon. 1975. The effects of oils and oil components on algae. A review. Br. Phycol. J. 11: 115-142.

O'Boyle, R.N. 1980. Distribution of oil, chlorophyll, and larval fish on the Scotian Shelf during April and May 1979 following the *Kurdistan* spill. *In* J.H. Vandermeulen (ed.), Scientific Studies during the Kurdistan Tanker Incident: Proceedings of a Workshop. Bedford Inst. Oceanogr. Report Series, B1-R-80-3.

Oetking, P., B. Black, R. Watson and C. Merks. 1979. Physical studies of the near-shore continental shelf of south central Louisiana: Currents and hydrography. Pages 119-143 *in* C.H. Ward, M.E. Bender and D.J. Reish, (eds.), The Offshore Ecology Investigation: Effects of Oil Drilling and Production in Coastal Environment. Rice University Studies 65: 1-589.

Oppenheimer, C.H. 1977. The offshore ecology investigation 1972-1974. Rapp. P.-V. Réun. Cons. Int. Explor. Mer. 171: 147.

Oppenheimer, C.H., R. Miget and H. Kator. 1974. Hydrocarbons in seawater and organisms and microbiological investigations. Gulf Universities Research Consortium, Appendix to Rept. No. 138, The Offshore Ecology Investigation, Final Project Planning Council Concensus Report.

Oppenheimer, C.H., R. Miget and H. Kator. 1979. Ecological relationships between marine microorganisms and hydrocarbons in the OEI study area, Louisiana. Pages 287-324 *in* C.W. Ward, M.E. Bender and D.J. Reish (eds.), The Offshore Ecology Investigations: Effects of Oil Drilling and Production in a Coastal Environment. Rice University Studies 65: 1-589.

Ostrum, C.L. 1979. The distribution of recent formanifera in Timbalier Bay, Louisiana. Pages 44-47 *in* C.H. Ward, M.E. Bender and D.J. Reish (eds.), The Offshore Ecology Investigation: Effects of Oil Drilling and Production in a Coastal Environment. Rice University Studies 65: 1-589.

O'Sullivan, A.J. 1978. The *Amoco Cadiz* oil spill. Mar. Pollut. Bull. 9: 123-128.

Oviatt, C.A., J. Frithsen, J. Gearing and P. Gearing. 1982. Low chronic additions of No. 2 fuel oil: Chemical behavior, biological impact and recovery in a simulated estuarine environment. Mar. Ecol. Progr. Ser. 9: 121-136.

Parsons, K.R., W.K.W. Li and R. Waters. 1976. Some preliminary observations on the enhancement of phytoplankton growth by low levels of mineral hydrocarbons. Hydrobiol. 51: 85-89.

Payne, J.F. 1982. Critique of petroleum and marine fishes: A review of uptake, disposition and effects. Environ. Sci. Technol. 16: 370-372.

Pearson, T.H. and R. Rosenberg. 1978. Macrobenthic succession in relation to organic enrichment and pollution of the marine environment. Oceanogr. Mar. Biol. Ann. Rev. 16: 229-311.

Perry, A. 1979. Fish of Timbalier Bay and offshore Louisiana environments collected by trawling. Pages 537-545 *in* C.H. Ward, M.E. Bender and D.J. Reish (eds.), The Offshore Ecology Investigation: Effects of Oil Drilling and Production in a Coastal Environment. Rice University Studies 65: 1-589.

Price, K.C. 1979. Onshore hydrography of Timbalier Bay, Louisiana. Pages 145-158 *in* C.H. Ward, M.E. Bender and D.J. Reish (eds.), The Offshore Ecology Investigation: Effects of Oil Drilling and Production in a Coastal Environment. Rice University Studies 65: 1-589.

Prouse, N.J., D.C. Gordon, Jr. and P.D. Keizer. 1976. Effects of low concentrations of oil accommodated in sea water on the growth of unialgal marine phytoplankton cultures. J. Fish. Res. Board Can. 33: 810-818.

Pulich, W.M., Jr., K. Winters and C. Van Baalen. 1974. The effects of a No. 2 fuel oil and two crude oils on the growth and photosynthesis of microalgae. Mar. Biol. 28: 87-94.

Reed, W.E., I.R. Kaplan, M. Sandstrom and P. Mankiewicz. 1977. Petroleum and anthropogenic influence on the composition of sediments from the southern California Bight. Pages 183-188 *in* Proceedings 1977 Oil Spill Conference. American Petroleum Institute, Washington, D.C.

Rice, S.D. 1981. Review: Effects of oil on fish. Unpubl. background paper for 1981 National Academy of Sciences Workshop, Clearwater Beach, Florida.

Ritacco, P.J. and A.W. Sastry. 1983. The metabolic, biochemical and reproductive cycles of *Nucula annulata* as affected by hydrocarbon contamination in mesocosms. Fourth International Ocean Disposal Symposium, Plymouth, England, 11-15 April, 1983 (Abstract).

Sabo, D.J. and J.J. Stegeman. 1977. Some metabolic effects of petroleum hydrocarbons on marine fish. Pages 279-287 *in* F.J. Vernberg, A. Calabrese, F.P. Thurberg and W. Vernberg, (eds.), Physiological Responses of Marine Biota to Pollutants. Academic Press, New York.

Sackett, W.M. and J.M. Brooks. 1974. Use of low molecular-weight hydrocarbon concentrations as indicators of marine pollution. Pages 171-173 *in* Marine Pollution Monitoring, Proceedings of a Symposium and Workshop, May, 1974. U.S. Bureau of Standards, Washington, D.C.

Samain, J.F., J. Moal, J.R. LeCoz, J.Y. Daniel and A. Coum. 1981. Impact de L'*Amoco Cadiz* sur l'ecophysiologie du zooplankton: Une nouvelle possibilite de surveillance ecologique. Pages 481-498 *in Amoco Cadiz*, Fate and Effects of the Oil Spill. Centre Nat. L'Exploit. des Oceans, Paris.

Sanders, H.L. 1981. Environmental effects of oil in the marine environment. Pages 117-146 *in* Safety and Offshore Oil: Background Papers of the Committee on Assessment of Safety of OCS Activities. National Research Council, National Academy Press, Washington, D.C.

Sanders, H.L., J.F. Grassle, G.R. Hampson, L.S. Morse, S. Garner-Price and C.C. Jones. 1980. Anatomy of an oil spill: Long-term effects from the grounding of the barge *Florida* off West Falmouth, Massachusetts. J. Mar. Res. 38: 265-380.

Sharp, J.M. 1979. The cumulative effects of petroleum drilling and production in coastal and near-shore areas. Pages 3-15 *in* C.H. Ward, M.E. Bender and D.J. Reish (eds.), The Offshore Ecology Investigation: Effects of Oil Drilling and Production in a Coastal Environment. Rice University Studies 65: 1-589.

Sharp, J.M. and S.G. Appan. 1982. The cumulative ecological effects of normal offshore petroleum operations contrasted with those resulting from continental shelf oil spills. Phil. Trans. R. Soc. Lond. B 297: 309-322. Reprinted in R.B. Clark (ed.). 1982. The Long-Term Effects of Oil Pollution in Marine Populations, Communities and Ecosystems. The Royal Society, London.

Shaw, D.G., A.J. Paul, L.M. Cheek and H.M. Feder. 1976. *Macoma balthica*: An indicator of oil pollution. Mar. Poll. Bull. 7: 29-31.

Simoneit, B.R.T. and I.R. Kaplan. 1980. Triterpenoids as molecular indicators of paleo seepage in recent sediments of the Southern California Bight. Mar. Environ. Res. 3: 113-128.

Simpson, R.A. 1977. The biology of two offshore oil platforms. IMER, 76-13, University of California, Institute Marine Research, La Jolla, 14 p.

Sindermann, C.J. 1982. Implications of oil pollution in production of disease in marine organisms. Phil. Trans. R. Soc. London B 297: 385-399. Reprinted in R.B. Clark (ed.). 1982. The Long-Term Effects of Oil Pollution in Marine Populations, Communities and Ecosystems. The Royal Society, London.

Skjoldal, H.R., T. Dale, H. Haldorsen, B. Pengerud, T.F. Thingstad, K. Tjessem and A. Aaberg. 1982. Oil pollution and plankton dynamics 1. Controlled ecosystem experiment during the 1980 spring bloom in Lindaspollene, Norway. Netherlands J. Sea Res. 16: 511-523.

Smith, J.E. 1970. "Torrey Canyon" Pollution and Marine Life. A report by the Plymouth Laboratory of the Marine Biological Association of the United Kingdom. Cambridge University Press, U.K.

Smith, W., V.R. Gibson, L.S. Brown-Leger and J.F. Grassle. 1979. Diversity as an indicator of pollution: Cautionary results from microcosm experiments. Pages 269-277 *in* Ecological Diversity in Theory and Practice. International Cooperative Publishing House, Fairland, Maryland.

Soto, C., J.A. Hellebust, T.C. Hutchinson and T. Sawa. 1975a. Effect of naphthalene and aqueous crude oil extracts on the green flagellate *Chlamydomonas angulosa*. I. Growth. Canad. J. Bot. 53: 109-117.

Soto, C., J.A. Hellebust and T.C. Hutchinson. 1975b. Effect of naphthalene and aqueous crude oil extracts on the green flagellate *Chlamydomonas angulosa*. II. Photosynthesis and the uptake and release of naphthalene. Canad. J. Bot. 53: 118-126.

Spies, R.B., P.H. Davis and D.H. Stuermer. 1980. Ecology of a submarine petroleum seep off the California Coast. Pages 208-263 *in* R. Geyer (ed.), Environmental Pollution. 1. Hydrocarbons. Elsevier, Amsterdam.

Spies, R.B., J.S. Felton and L. Dillard. 1982. Hepatic mixed-function oxidases in California flatfishes are increased in contaminated environments and by oil and PCB ingestion. Mar. Biol. 70: 117-127.

Spies, R.B., D.W. Rice, Jr., P.A. Montagna and R.R. Ireland. 1985. Reproductive success, xenobiotic contaminants and hepatic mixed-function oxidase (MFO) activity in *Platichthys stellatus* populations from San Francisco Bay. Mar. Environ. Res. 17: 117-121.

Spies, R.B. and P.H. Davis. 1979. The infaunal benthos of a natural oil seep in the Santa Barbara Channel. Mar. Biol. 50: 227-237.

Spies, R.B. and P.H. Davis. 1982. Toxicity of Santa Barbara seep oil to starfish embryos: Part 3 - Influence of parental exposure and the effects of other crude oils. Mar. Environ. Res. 6: 3-11.

Spies, R.B. and D.J. DesMarais. 1983. Natural isotope study of trophic enrichment of marine benthic communities by petroleum seepage. Mar. Biol. 73: 67-71.

Stanley, S.O., T.H. Pearson and C.M. Brown. 1978. Marine microbial ecosystems and the degradation of organic pollutants. Pages 60-79 *in* K.W.A. Chater and H.J. Somerville (eds.), The Oil Industry and Microbial Ecosystems. Heyden and Sons, Ltd., London.

Steele, J.H. 1979. The uses of experimental ecosystems. Philos. Trans. R. Soc. London B, Biol. Sci. 286: 583-595.

Stegeman, J.J. 1978. Influence of environmental contamination on cytochrome P-450 mixed-function oxygenases: Implications for recovery in the Wild Harbor Marsh. J. Fish Res. Bd. Canada 35: 668-674.

Stegeman, J.J. 1980. Mixed-function oxygenase studies in monitoring for effects of organic pollution. Rapp. P.-V. Réun. Cons. Int. Explor. Mer 179: 33-38.

Stegeman, J.J. 1981. Polynuclear aromatic hydrocarbons and their metabolism in the

marine environment. Pages 1-59 *in* H.V. Gelboin and P.O.P. Ts'O (eds.), Polycyclic Hydrocarbons and Cancer, Volume 3. Academic Press, New York.
Stegeman, J.J., A.V. Klotz, B.R. Woodin and A.M. Pajor. 1981. Induction of hepatic cytochrome P-450 in fish and the indication of environmental induction in scup (*Stenotomus chrysops*). Aquat. Toxicol. 1: 197-212.
Stegeman, J.J., P.J. Kloepper-Sams and J.N Farrington. 1986. Monoxygenase induction and chlorobiphenyls in the deep-sea fish *Coryphaonoides armatus*. Science 231: 1287-1289.
Stegeman, J.J. and D.J. Sabo. 1976. Aspects of the effects of petroleum hydrocarbons on intermediary metabolism and xenobiotic metabolism in marine fish. Pages 424-436 *in* Sources, Effects and Sinks of Hydrocarbon in the Aquatic Environment. American Institute of Biological Sciences, Washington, D.C.
Stewart, J.E. and L.J. Marks. 1978. Distribution and abundance of hydrocarbon-utilizing bacteria in sediments of Chedabucto Bay, Nova Scotia, in 1976. J. Fish. Res. Board Can. 35: 581-584.
Straughan, D. 1976. Sublethal effects of natural chronic exposure to petroleum in the marine environment. A.P.I. Publ. 4280. American Petroleum Institute, Washington, D.C., 120 p.
Stoll, D.R. and R.R.L. Guillard. 1974. Synergistic effect on napthalene, toxicity and phosphate deficiency in a marine diatom. *In* 37th Annual Meeting, American Society of Limnology and Oceanography (Abstract).
Stuermer, D.H., R.B. Spies, P.H. Davis, D.J. Ng, C.J. Morris and S. Neal. 1982. The hydrocarbons in the Isla Vista marine seep environment. Mar. Chem. 11: 413-426.
Stuermer, D.H., R.B. Spies, and P.H. Davis. 1982. Toxicity of Santa Barbara seep oil to starfish embryos. Part 1. Hydrocarbon composition of test solutions and field samples. Mar. Environ. Res. 5: 275-286.
Takahashi, M., W.H. Thomas, D.L.R. Seibert, J. Beers, P. Koeller and T.R. Parson. 1975. The replication of biological events in enclosed water columns. Arc. V. Hydrobiol. 76: 5-23.
Taylor, T.L. and J.F. Karinen. 1977. Response of the clam, *Macoma balthica*, exposed to Prudhoe Bay crude oil as unmixed oil, water-soluble fraction and oil-contaminated sediment in the laboratory. Pages 229-232 *in* D.A. Wolfe (ed.), Fate and Effects of Petroleum Hydrocarbons in Marine Ecosystems and Organisms. Pergamon Press, New York.
Teal, J.M., K. Burns and J. Farrington. 1978. Analyses of aromatic hydrocarbons in intertidal sediments resulting from two spills of No. 2 fuel oil in Buzzards Bay, Massachusetts. J. Fish. Res. Board Can. 35: 510-520.
Thompson, J.R. 1979. A study of the temporal changes in offshore macrofauna in the northern Gulf of Mexico during the development of the offshore oil industry. Pages 547-551 *in* C.H. Ward, M.E. Bender and D.J. Reish (eds.), The Offshore Ecology Investigation: Effects of Oil Drilling and Production in a Coastal Environment. Rice University Studies 65: 1-589.
Ungvary, G., B. Varga, E. Horvath, E. Tatrai and G. Folly. 1981. Study on the role of maternal sex hormone production and metabolism in the embryotoxicity of para-xylene. Toxicology 14: 263-268.
Vandermeulen, J.H. and T.P. Ahearn. 1976. Effect of petroleum hydrocarbons on algal physiology; review and progress report. Pages 107-126 *in* A.P.M. Lockwood (ed.), Effects of Pollutants on Aquatic Organisms. Cambridge University Press, Cambridge, U.K.
Vargo, S.L. 1981. The effects of chronic low concentrations of No. 2 fuel oil on the physiology of a temperate estuarine zooplankton community in the MERL microcosms. Pages 295-322 *in* F.J. Vernberg, F.P. Calabrese, A. Thurberg and W.B. Verberg (eds.), Biological Monitoring of Marine Pollutants. Academic Press, New York.
Vargo, G.A., M. Hutchins and G. Almquist. 1982. The effect of low, chronic levels of

No. 2 fuel oil in natural phytoplankton assemblages in microcosms: 1. Species composition and seasonal succession. Mar. Environ. Res. 6: 245-264.

Walker, J.D. and R.R. Colwell. 1973. Microbial ecology of petroleum utilization in Chesapeake Bay. Pages 685-690 *in* Proceedings of Joint Conference Prevention and Control of Oil Spills. American Petroleum Institute, Washington, D.C.

Walker, J.D. and R.R. Colwell. 1976. Petroleum: Degradation by estuarine microorganisms. Pages 177-204 *in* J.M. Sharpley and A.M. Kaplen, (ed.), Third International Biodegradation Symposium: Sessions I, XIII, XXV, XVII, XIX. Applied Science Publishers, Ltd., London.

Waller, R.S. 1979. Pelagic, epibenthic and infaunal invertebrates of Timbalier Bay and offshore environment. Pages 529-536 *in* C.H. Ward, M.E. Bender and D.J. Reish (eds.), The Offshore Ecology Investigation. Effects of Oil Drilling and Production in a Coastal Environment. Rice University Studies 65: 1-589.

Walton, D.G., W.R. Penrose and J.M. Green. 1978. The petroleum-inducible mixed function oxidase of canner (*Tautoglabrus adsperus* Walbaum 1792): Some characteristics relevant to hydrocarbon monitoring. J. Fish. Res. Board. Can. 35: 1547-1552.

Walton, D.G., L.L. Fancey, J.M. Green, K.W. Kiceniuk and W.R. Penrose. 1983. Seasonal changes in aryl hydrocarbon hydroxylase activity of a marine fish *Tautoglabrus adsperus* (Walbaum) with and without petroleum exposure. Comp. Biochem. Physiol. 76C: 247-253.

Walton, S. 1981. Academy looks again at petroleum in the marine environment. Bioscience 31: 93-96.

Wells, P.G. 1981. Petroleum hydrocarbons and marine zooplankton. Unpubl. background paper for 1981 National Academy of Sciences Workshop, Clearwater Beach, Florida.

Wharfe, J.R. 1975. A study of the intertidal macrofauna around the B.P. Refinery (Kent) Limited. Environ. Poll. 9: 1-12.

Winters, K., R. O'Donnell, J.C. Batterton and C. Van Baalen. 1976. Water soluble components of four fuel oils: Chemical characterization and effects on growth of microalgae. Mar. Biol. 36: 269-276.

Winters, K., J.C. Batterton, R. O'Donnell and C. Van Baalen. 1977a. Fuel oils: Chemical characterization and toxicity to microalgae. Pages 167-184 *in* C.S. Giam (ed.), Pollutant Effects on Marine Organisms. Lexington Books, Lexington, Massachusetts.

Winters, K., C. Van Baalen and J.A.C. Nicol. 1977b. Water soluble extractives from petroleum oils: Chemical characterization and effect on microalgae and marine annuals. Rapp. P.-V. Reun. Cons. Int. Explor. Mar. 171: 166-174.

Wolfe, D.A. 1978. The *Amoco Cadiz* oil spill, a summary of observations made by U.S. Scientists 23 March - 10 May, 1978. Mar. Pollut. Bull. 9: 28-242.

Wormwald, A.P. 1976. Effects of spill of marine diesel oil on the meiofauna of a sandy beach at Picnic Bay, Hong Kong. Environ. Pollut. 11: 117-130.

CHAPTER 10

BIOLOGICAL EFFECTS OF DRILLING FLUIDS, DRILL CUTTINGS AND PRODUCED WATERS

Jerry M. Neff

CONTENTS

INTRODUCTION

The first offshore oil wells were drilled in the 1890s from piers extending from the southern California coast. The first offshore oil field in the Gulf of Mexico was developed in the late 1930s and the first producing well out of sight of land was completed 12 miles off the Louisiana coast in 1947. By the end of 1982, approximately 27,000 wells had been drilled in U.S. coastal waters (American Petroleum Institute, 1982). In 1983, there was a total of 4056 offshore platforms in operation in U.S. waters, 3600 of them off Louisiana (Essertier, 1984). A total of 1320 offshore wells was completed in 1982, and it is expected that 1485 wells will be completed offshore per year by 1985 (Offshore, 1983). Oil and gas production from the Federally controlled U.S. outer continental shelf (OCS) currently accounts for 8% of total domestic oil production and 24% of domestic gas production. The U.S Geological Survey has estimated that as much as 33.8% of the nation's undiscovered recoverable oil and 28.1% of natural gas may lie beneath U.S. coastal and outer continental shelf waters (Kash, 1983).

This constantly accelerating pace of exploration for and development of oil and gas resources in U.S. coastal, outer continental shelf and continental slope waters has led to a growing concern that such activities may cause serious long-term damage to the marine environment and the living resources it supports. The purpose of this review is to critically evaluate our current state of knowledge about the biological impacts of operational discharges resulting from offshore oil and gas exploration, development, and production.

During well drilling and during production of oil and gas offshore, a wide variety of liquid, solid and gaseous wastes are produced on the platform, some of which are discharged to the ocean. The major discharges associated with exploratory and development drilling are drill cuttings and drilling fluids. From 200 to about 1000 metric tons of drilling fluid solids and a similar amount of drill cuttings may be discharged intermittently to the ocean during drilling of an offshore well. During the production of oil or gas, connate or fossil water from the reservoir may be pumped as well. Some of this produced water is discharged to the ocean or coastal waters following treatment. During production of oil or gas, an offshore platform may generate from 0 to 1.5 million liters of produced water per day.

Water-based drilling fluids of the types most frequently used on the U.S. continental shelf are specially formulated mixtures of clays and/or polymers, weighting agents, lignosulfonates and other materials suspended in water. Barium, chromium, zinc and lead may be present at substantially higher concentration in drilling fluids than in natural marine sediments. Produced water destined for ocean discharge may contain up to about 48 ppm petroleum hydrocarbons, and elevated concentrations of barium, beryllium, cadmium, chromium, copper, iron, lead, nickel, silver and zinc. It may also contain small amounts of the natural radionuclides, ^{226}Ra and ^{228}Ra, and up to several hundred ppm of nonvolatile dissolved organic material of unknown composition. Details of the composition of these operational discharges are considered in detail in Chapter 4.

FATE OF OPERATIONAL DISCHARGES

Drilling Fluid and Cuttings

A water-based drilling fluid is a slurry of solid particles of different sizes and densities in water (see Chapter 4). Drilling fluid additives may be water soluble, colloidal or particulate. Clay, silt and cuttings have a density of about 2.6 g/cc. Silt and unflocculated clays settle in calm sea water at estimated rates of about 1.4 x 10^{-2} to 5.8 x 10^{-5} cm/s (Smedes *et al.*, 1981). Much of the clay in drilling mud, however, tends to flocculate upon contact with sea water, resulting in more rapid settling of this fraction. Barite, despite its fine grain size (<64 μm), may settle more rapidly because of its high density. Because of this physical/chemical heterogeneity, drilling fluids and cuttings undergo rapid and substantial fractionation and dispersion upon discharge to the ocean.

According to a dispersion/dilution model developed by Brandsma *et al.* (1980), drilling mud discharged from a submerged discharge pipe can be viewed as going through three distinct phases: convective descent of the jet of material, dynamic collapse and passive diffusion. In the convective descent phase, the plume descends rapidly, entraining low-density particles and bending toward the direction of current flow. The larger and/or denser solids in the drilling fluid continue their descent until they hit the bottom, while the lighter smaller particles and soluble materials undergo dynamic collapse when the plume encounters a level of neutral density. The lighter plume then undergoes further dilution by passive diffusion and convective mixing of the ambient medium. The upper plume generally contains less than 10% of the drilling fluid solids. The remaining 90% settles directly to the bottom. Critical determinants of the impacts of discharged drilling fluids and cuttings on water column biota are the rate and extent of these dispersion/dilution processes.

Several field studies have shown that drilling fluids discharged to the ocean are diluted rapidly to very low concentrations, usually within 1000 to 2000 m downcurrent from the discharge pipe and within 2 to 3 h of discharge (Ayers *et al.*, 1980a,b; Ecomar, Inc., 1978, 1983; Houghton *et al.*, 1980; Northern Technical Services, 1983). Quite frequently, dilutions of 1000-fold or more are encountered within 1 to 3 m of the discharge.

The effects of a material like drilling fluid on water column organisms will depend not only on its inherent toxicity but also on actual exposure concentrations and durations in the field. Thus, a graphical plot of drilling fluid concentration versus transport time (distance from rig at which sample is taken divided by current speed) in the field provides a good basis for predicting the impact of drilling fluids on water column organisms (Figure 10.1). In five field studies performed in different geographic regions of the U.S. outer continental shelf, drilling fluids were diluted to "background" or near background concentrations (based on suspended solids concentrations) within 0.1 to 4 h. Suspended solids concentrations in drilling mud plumes fell below 1000 mg/l in less than 1.5 min and below 10 mg/l within one hour of discharge. Other markers of the drilling mud plume (Ba, Cr, percent transmittance) give similar results. Using particulate barium as a tracer and ultraclean analytical techniques, drilling mud plumes have been traced to more than 3 km from the point of discharge (Trocine and Trefry, 1983). The estimated dilution of drilling fluid solids at this distance was one-billion-fold. At a current speed of 10-20 cm/s the drilling fluid plume would require 4 to 8 h to travel this far and reach this dilution.

The distance from an exploratory platform to which drilling fluid solids are

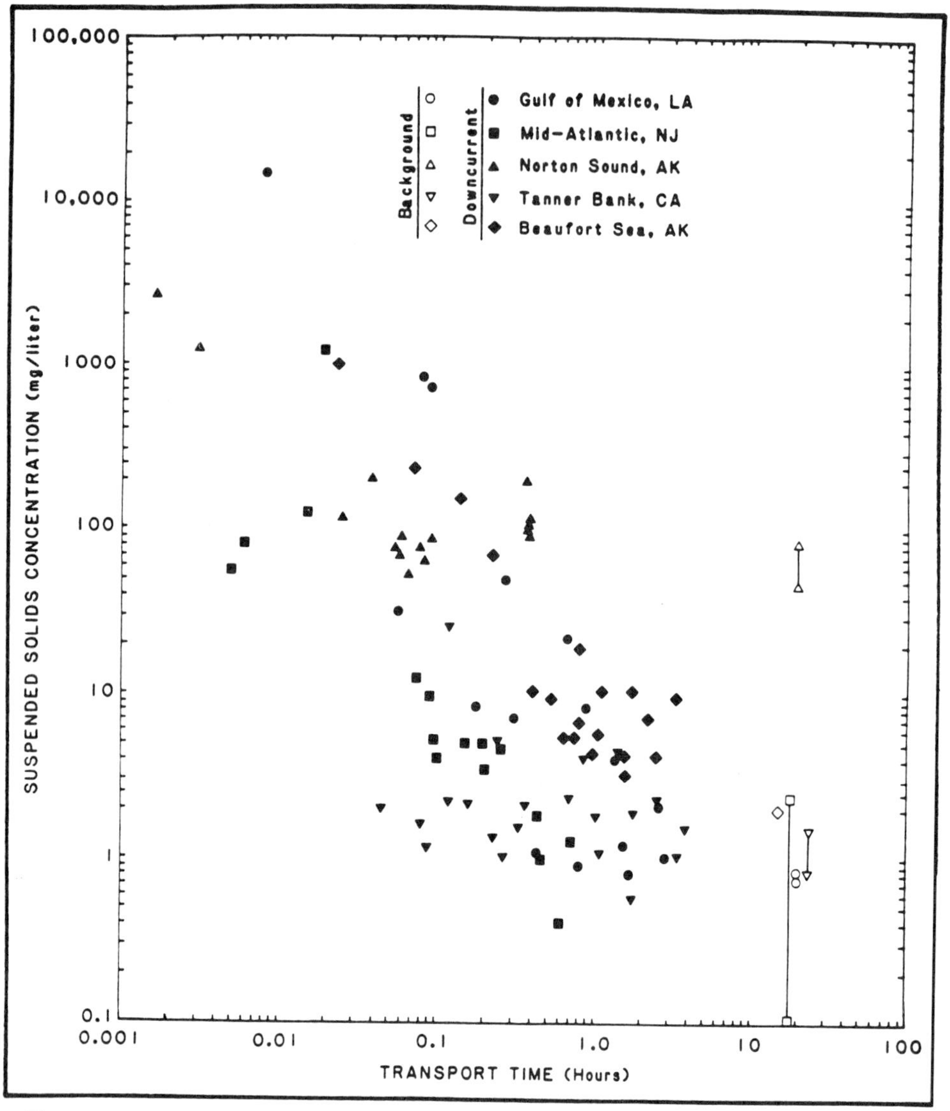

Figure 10.1. Graphic presentation of results of five drilling fluid dispersion studies performed in U.S. outer continental shelf waters. Concentration of suspended solids is used as an indicator of drilling fluid solids concentration and is plotted against transport time (distance from rig divided by current speed). Undiluted drilling fluids contained from 200,000 to 1,400,000 ppm suspended solids before discharge. Data from Ayers *et al.*, 1980b; EG&G, Environmental Consultants, 1982; Ecomar, 1978, 1983; Northern Technical Services, 1981.

dispersed and their concentration in bottom sediments depends on the types and quantities of drilling fluids discharged, hydrographic conditions at the time of discharge and height above the bottom at which discharges are made (Gettleson and Laird, 1980). Because barite (barium sulfate) is a major ingredient of many drilling fluids used on the U.S. outer continental shelf and is both very dense and insoluble in sea water, barium frequently is used as a marker for the settleable fraction of drilling fluid. In several investigations performed to date, barium concentration in bottom sediments was highest near the rig and decreased markedly with distance from the rig (Dames and Moore, Inc., 1978; Crippen *et al.*, 1980; Gettleson and Laird, 1980; Meek and Ray, 1980; Trocine *et al.*, 1981; Northern Technical Services, 1981, 1982, 1983; Bothner *et al.*, 1982, 1983; EG&G, Environmental Consultants, 1982; Boothe and Presley, 1983). Barium concentrations may reach concentrations 10 to 20 times above background in sediments near the discharge. Concentrations of barium in surficial sediments of 5000 mg/kg have been reported near an exploratory rig site (Trefry *et al.*, 1983; Trocine and Trefry, 1983), compared to a normal background of 200-300 mg Ba/kg in sediments from the area. Barium concentrations in excess of 40,000 ppm above background have been reported in surficial sediments within about 100 m of the discharge from a multiple-well development platform in the Gulf of Mexico (Petrazzuolo, 1983). Usually the increment in barium concentration is restricted primarily to the upper few centimeters of the sediments. In most cases, there is a steep gradient of decreasing barium concentration in surficial sediments with lateral distance to background concentrations 1000 to 1500 m downcurrent of the discharge point.

Other drilling mud-associated metals are much less elevated than barium in bottom sediments near the rig. Visible accumulations on the bottom of drilling discharges, primarily drill cuttings, have been reported near drilling rigs in the Gulf of Mexico (Zingula, 1975), offshore Southern California (Bascom *et al.*, 1976), on the mid-Atlantic outer continental shelf (EG&G, Environmental Consultants, 1982), and the Beaufort Sea (Northern Technical Services, 1981), but not on Georges Bank (Battelle/Woods Hole Oceanographic Institute, 1983, 1984) or Cook Inlet, Alaska (Dames and Moore, Inc., 1978). These cuttings piles may be as much as a few meters high and 100-200 m in diameter. In nondepositional and high-energy environments, accumulations of drilling fluid and cuttings solids are dispersed from their deposition site by current-induced resuspension, bed transport and bioturbation (National Research Council, 1983). Bothner *et al.* (1983) estimated the half-time for washout of barite from sediments near an exploratory rig on Georges Bank to be about 0.4 years.

Produced Water

Dilution of produced water upon discharge to the ocean is very rapid, the actual rate being dependent upon such factors as total dissolved solids concentration of the produced water, current speed, vertical convective mixing of the water column and water depth. Based on a model developed by the Massachusetts Institute of Technology, it was estimated that saturated brine (about 320 parts per thousand (ppt) salinity) from the Bryan Mound Strategic Petroleum Reserve salt dome would be diluted to within 5 ppt of ambient seawater salinity within 30 m of the discharge point (Federal Energy Administration, 1977). In the Buccaneer gas and oil field, concentrations of total hydrocarbons in the water column rarely exceeded 30 g/l (Middleditch, 1981b), with a maximum of 43 g/l, whereas produced water from the production platform contained about 20 mg/l total resolved petroleum hydrocarbons (see Chapter 4). Unfortunately, Middleditch monitored primarily alkanes, instead of aromatics. There was a relatively large and variable background of biogenic alkanes in the water which made detection and interpretation of dilution gradients difficult. In Trinity Bay, Texas, a shallow estuary, total resolved hydrocarbon concentrations in produced water were diluted by 2400-fold within 15 m downcurrent of the discharge pipe located 1 m above the bottom in 2-3 m of water (Armstrong *et al.*, 1979). Crude oil tanker ballast water (with a hydrocarbon composition similar to that of produced water) discharged to Valdez Harbor from the Valdez ballast treatment facility (Lysyj *et al.*, 1981) or to the Red Sea at Yanbu, Saudi Arabia from crude oil tankers (Neff *et al.*, 1983), was diluted 500-fold or more within 150 m of the discharge and 1000-3000 fold within 500-1000 m of the discharge.

Concentrations of low molecular weight volatile hydrocarbons (C_1-C_{14}) in the water column of the northwestern Gulf of Mexico are two orders of magnitude higher than in unpolluted open ocean waters (Brooks *et al.*, 1977; Sauer, 1980). These hydrocarbons are thought to have been derived from underwater venting of waste gas and discharge of produced water.

Less information is available about dilution of heavy metals in produced water. Middleditch (1984) reported that elevated levels of heavy metals generally are not observed in the water column near produced water discharges. Dilution of most produced waters by 10-100 fold will reduce heavy metals to near ambient concentrations (see Chapter 4). Montalvo and Brady (1979) measured concentrations of mercury, lead, zinc, cadmium and arsenic in the water column of Timbalier Bay and offshore Louisiana. There were no statistically significant differences in metals concentrations in the water column near oil platforms and at nearby reference stations, with two exceptions. There was a gradient of decreasing zinc concentration in the water with distance from a group of offshore platforms from about 8 g/l near a platform to about 1.5 g/l 1800 m away at mid-water depth. The authors suggested that the zinc may have been derived from sacrificial anodes on the submerged structure of the platform. In the bay, water samples obtained near a workover rig contained slightly elevated concentrations of lead (mean, 8 g/l compared to background of 1 g/l).

Where suspended sediment concentrations are high, as in the northwestern Gulf of Mexico, dissolved and colloidal hydrocarbons and metals from produced water tend to become adsorbed to suspended particles and sediment to the bottom (see Boehm, Chapter 6). In Trinity Bay, Texas, sediments 15 m downcurrent from a produced water discharge, contained high concentrations of C_{10}-C_{28} alkanes and aromatic hydrocarbons from C_3-benzenes to C_3-phenanthrenes (Armstrong *et al.*, 1979). A gradient of decreasing sediment naphthalenes concentrations extended away from the discharge in all directions for up to 5000 m of the discharge. In deeper water, more typical of offshore drilling activities, elevated levels of hydrocarbons are restricted to a much smaller area of the bottom or are not detected at all. In the Buccaneer Field, located in about 20 m of water, elevated levels of *n*-alkanes were detected in surficial sediments within a radius of about 15-20 m of the discharge (Middleditch, 1981a). However, sediment resuspension and transport resulted in rapid changes in sediment hydrocarbon concentrations on almost a daily basis.

Elevated levels of barium, cadmium, chromium, copper, lead, strontium, and zinc have been detected in surficial sediments in the vicinity of production platforms in the northwestern Gulf of Mexico (Tillery and Thomas, 1980; Wheeler *et al.*, 1980). These metals may be derived from discharge of drilling fluids and produced water and by corrosion or leaching of submerged rig structures, antifouling paints and sacrificial anodes. The magnitude of elevation in the concentration of metals other than barium in sediments usually is small.

The radioisotopes of radium do not readily adsorb to organic and mineral particles (Hanan, 1981; Landa and Reid, 1983). Landa and Reid (1983) showed that sorption of ^{226}Ra from produced water by natural sediments was rapid and was inversely related to produced water and receiving water salinity. Retention of ^{226}Ra in the sediment involved precipitation reactions with sulfates and ion exchange reactions with clays. Little ^{226}Ra was complexed with the organic fraction of sediment. In coastal Louisiana marshes which had received large volumes of produced water effluents for many years, radium isotope enrichment in the marsh sediment was only 1.5-2 times above background (Hanan, 1981; Landa and Reid, 1983). Thus most of the radium in produced water remains with the soluble phase of the effluent and will be dispersed and diluted in the water column.

TOXICITY OF DRILLING FLUID AND PRODUCED WATER INGREDIENTS

Drilling Fluids

Of the major drilling fluid ingredients, only chrome or ferrochrome lignosulfonate and sodium hydroxide can be considered moderately toxic. Chesser and McKenzie (1975) reported a 96-h LC_{50} (concentration causing 50% mortality in 96 h) of 465 mg/l for a chrome-treated lignosulfonate and the white shrimp *Penaeus setiferus*. The 144-h LC_{50}s of ferrochrome lignosulfonate to larvae of Dungeness crabs *Cancer magister* and dock shrimp *Pandalus danae* were 210 and 120 mg/l, respectively (Carls and Rice, 1980). Concentrations between 50 and 150 mg/l inhibited swimming in the larvae. Sea scallops *Placopecten magellanicus* which had been exposed for 20 days to 500 mg/l solutions of ferrochrome lignosulfonate showed structural histological damage to delicate gill ctenidia and an inhibition of ciliary activity in the gills (Morse *et al.*, 1982). Results presented by Krone and Biggs (1980) seem to indicate that reef corals are quite sensitive to drilling fluids containing chrome lignosulfonate; however, the conclusions are equivocal because of poor experimental design and inadequate replication.

Little information is available about the toxicity of sodium hydroxide to marine organisms. Toxic effects of this material to marine organisms are due exclusively to pH elevation. Because of the high buffer capacity of sea water and the rapid dilution expected for this soluble material following ocean disposal of drilling fluid, pH changes great enough to cause damage to the most delicate organisms (i.e., a pH increase from 7.8 to 8.5 caused abnormal development in starfish embryos; Chaffee and Spies, 1982), will be restricted to the immediate vicinity of the discharge and will be of very short duration.

The two major ingredients in most water-based drilling fluids, bentonite clay and barite, are practically inert toxicologically. They may, however, cause physical damage through abrasion or clogging, or they may change the texture and grain size of sediments where they settle, rendering the substrate less suitable for habitation by some benthic species and more suitable for others. The acute toxicity of these natural mineral materials to marine organisms usually is greater than 7000 mg/l (96-h LC_{50}; National Research Council, 1983).

Swimming activity in larvae of Dungeness crabs and dock shrimp was inhibited following exposure for 24 to 119 h to 400-4280 mg/l suspensions of barite or bentonite in sea water (Carls and Rice, 1980). Shrimp *Palaemonetes pugio*, exposed to a substrate of particulate barite for up to 106 days, ingested the barite (Brannon and Rao, 1979; Conklin *et al.*, 1980). Although this did not affect survival of the shrimp, several sublethal responses were observed. Barite ingestion caused damage to the epithelium of the posterior midgut, possibly in part by abrasion.

A 5-mm layer of barite interfered with recruitment of planktonic larvae of polychaetes and molluscs to natural sandy sediments in aquaria receiving unfiltered natural sea water (Tagatz and Tobia, 1978; Tagatz *et al.*, 1980; Cantelmo *et al.*, 1979). Lesser amounts of barite on or mixed with the sediment had little effect on recruitment of benthic meiofauna or macrofauna to the substrate. The effects of barite on recruitment probably were due primarily to a change in the texture of the sandy sediment. Three species of reef corals were able to clear their upper surfaces of heavy layers of deposited barite or bentonite clay and appeared to suffer no serious adverse effects from such drastic exposure (Thompson and Bright, 1977).

Lignite is a low grade soft coal and is added as a finely-ground powder to drilling fluids. Its acute toxicity to marine animals is in excess of 8000 mg/l (National Research Council, 1983).

Several metals, in the forms in which they occur in used drilling fluids, apparently have low toxicities to marine organisms. Chromium is the most abundant metal in many drilling fluids. In drilling mud, it is almost exclusively in the trivalent form and is tightly associated with lignosulfonate-clay complexes. In this form, chromium is much less toxic than ionic hexavalent chromium in solution. Slightly soluble hexavalent chromium salts, such as calcium chromate and zinc chromate, are carcinogenic in mammals following tracheal instillation and intramuscular or intrapleural injection (Norseth, 1981). Several chromate salts also show evidence of genetic toxicity and mutagenicity in several *in vitro* tests (Petrilli and DeFlora, 1982; Bianchi *et al.*, 1983). Trivalent chromium salts are neither carcinogenic nor mutagenic. Chromates ingested by marine animals or humans in water or food are reduced to the trivalent state by thermostable reducing agents in gastric juice and saliva (Petrilli and DeFlora, 1982), and, therefore, do not represent an important carcinogenic hazard when administered by this route.

Most other metals detected at elevated levels in some drilling fluids are present primarily as impurities in barite and bentonite or are derived from pipe dope or corrosion of drill pipe. These metals are present as insoluble metal sulfides or metal granules. Insoluble metal salts or metals tightly adsorbed to sediments are not remobilized readily into the water column and have a very limited bioavailability to marine organisms (Neff *et al.*, 1978; Amiard *et al.*, 1983; Hunt and Smith, 1983; Luoma, 1983).

There is growing evidence that diesel fuel may contribute significantly to the toxicity of those drilling fluids that contain it. Neff *et al.* (1981) were the first to suggest that volatile aromatic hydrocarbons, such as those found in diesel fuel, contributed significantly to the toxicity of some drilling fluids. Subsequently, Conklin *et al.* (1983) reported a statistically significant inverse relationship ($r = -0.58$, $P < 0.05$) between the 96-h LC_{50} of 18 controversial drilling mud samples from a drilling operation in Mobile Bay, Alabama to molting grass shrimp *Palaemonetes pugio* and the concentration in the muds of petroleum hydrocarbons identified as being derived from a No. 2 fuel oil (diesel). The drilling muds contained 170-8040 μl petroleum/l whole mud and had acute toxicities of 14,560-360 ppm drilling mud added, respectively.

In an ongoing research program, the Environmental Protection Agency (EPA), Gulf Breeze, Florida is sponsoring several projects to assess the acute toxicity and sublethal effects in marine organisms of 11 used offshore drilling muds. These drilling muds were obtained from offshore drilling sites in the Gulf of Mexico and were supplied to EPA by the Petroleum Equipment Suppliers Association. The mean 96-h LC_{50} values for bioassays performed to date with liquid phase, suspended phase and suspended whole mud preparations of the drilling muds and opossum shrimp *Mysidopsis bahia* were 176,500, 25,145 and 649 ppm mud added, respectively (T. Duke, EPA, pers. comm.). Several of the muds contained petroleum hydrocarbons identified as diesel fuel. There was a statistically significant inverse relationship between the 96-h LC_{50}s to opossum shrimp and the concentration in the drilling muds of petroleum hydrocarbons ($r = -0.73$, $P < 0.05$). There was no correlation between drilling mud toxicity and concentration of chromium in the drilling muds ($r = -0.5$, $P > 0.2$). The drilling muds contained 100-9430 ppm petroleum hydrocarbons, and 42-1345 ppm total chromium. In these studies, the drilling fluids tested differed from one another in many chemical and physical characteristics in addition to hydrocarbon content. Therefore, it was not possible to quantify precisely the contribution of diesel fuel to the total toxicity of the drilling fluids or to determine if there was a hydrocarbon concentration at which diesel fuel in a drilling fluid was not toxic.

The toxicity of crude and refined petroleums to marine organisms has been studied extensively (see reviews of Baker, 1976; Malins, 1977; Rice *et al.*, 1977; Neff and Anderson, 1981; Chapter 8). Acute toxicities of different crude and refined petroleums to different species of marine organisms are extremely variable. Most 96-h LC_{50}s fall in the range of 1-1000 mg oil/l. Some very sensitive larvae and early life stages of marine animals may have LC_{50}s of about 0.1-1 mg/l. Sublethal responses to oil, especially behavioral modifications, have been reported following acute exposure to petroleum concentrations in the low μg/l (parts per billion) range.

High aromatic diesel fuels, similar to those sometimes added to drilling fluids, are the most or among the most toxic petroleum products to marine organisms. Most if not all of the petroleum hydrocarbons in a used diesel-treated drilling mud, however, will be tightly adsorbed to the bentonite clay fraction of the mud, most of which remains in the surface plume following discharge. For this reason, drilling muds containing even a few percent diesel oil usually do not produce a visible oil sheen or slick when discharged to the ocean.

Petroleum hydrocarbons adsorbed to organic and inorganic particles are much less bioavailable to marine organisms than hydrocarbons in solution or dispersion in the water column (Anderson, 1982; Neff and Breteler, 1983).

Because of the low bioavailability of sediment-adsorbed hydrocarbons, most benthic marine animals are able to tolerate relatively high concentrations of sediment hydrocarbons. Chronic exposure to sediments containing 500-1200 ppm crude oil initially resulted in weight loss and hepatocellular vacuolization in English sole *Parophrys ventulus* (McCain *et al.*, 1978), reduced condition index and altered tissue-free amino acid ratios in clams, *Protothaca staminea* and *Macoma inquinata* (Roesijadi and Anderson, 1979; Augenfeld *et al.*, 1980) and reduced feeding rate by the polychaete *Abarenicola pacifica* (Augenfeld, 1980). When three experimental ecosystems at the Marine Ecosystems Research Laboratory at the University of Rhode Island were dosed with 190 μg/l (ppb) No. 2 fuel oil in the water column for 25 weeks, 109 mg/kg (ppm) petroleum hydrocarbons accumulated in the bottom sediments (Grassle *et al.*, 1980; Oviatt *et al.*, 1982). There was a highly significant decline in the numbers of macrofaunal and meiofaunal individuals in the benthos of the oiled tanks as compared to the control. The greater impact of No. 2 fuel oil than crude oil on benthic infaunal animals may be due to the higher concentration of toxic aromatic hydrocarbons in the former or to damage to pelagic larvae caused by the presence of petroleum hydrocarbons in the water column in the experimental ecosystem tanks.

The fate of diesel fuel which is discharged to the ocean is unknown. Breteler *et al.* (unpublished) have shown in laboratory studies that when a chrome lignosulfonate drilling fluid containing 5% diesel fuel is added to sea water in a mud/sea water ratio of 1/1000, 60-90% of the hydrocarbons (including most of the volatile and toxic mono-aromatics) remain with the soluble or colloidal (aqueous) phase. Another 10-30% (including most of the naphthalenes) remains with the fine-grained suspended particulate phase. Only 2-10% of the hydrocarbons remain with the rapidly-settling phase and accumulate on the bottom. The settleable fraction prepared from drilling mud containing 0.5 or 5% mineral oil (Mentor 28) or 0.5% low sulfur diesel fuel was no more toxic to benthic animals than the drilling fluid without hydrocarbon additive. The settleable fraction prepared from drilling fluid containing higher concentrations of diesel fuel was more toxic. It is probable that dilution of the aqueous and suspended particulate phases of drilling mud containing hydrocarbon additives will be sufficiently rapid in offshore waters that no adverse effects of the hydrocarbons will occur in water column organisms. Some impacts in the benthos could occur if large amounts of hydrocarbon-laden drilling fluid solids accumulated in a particular location.

A number of minor drilling mud ingredients (based on amount added per unit volume of drilling mud) show a moderate-to-high toxicity to marine organisms (National Research Council, 1983). Formaldehyde is moderately toxic. Aldacide (a formaldehyde polymer, paraformaldehyde) is recommended for use in amounts up to 300 g/bbl (about 1,500 ppm paraformaldehyde). Its rate of loss from discharged used drilling mud by dilution, biodegradation and evaporation would probably be sufficient to maintain its ambient concentration well below toxic levels. Paraformaldehyde depolymerizes to formaldehyde, the active biocide, upon contact with water. Although formaldehyde is a suspect carcinogen when administered to mammals via the inhalation route, it is extremely unlikely that traces of formaldehyde in solution would be carcinogenic to marine organisms.

Another material, used in small amounts in some drilling muds, which could show significant toxicity to marine animals is detergent (surfactant). Detergents are used to aid dispersion in the aqueous phase of the mud of poorly soluble mud components such as aluminum stearate, oil and gilsonite. Detergents of the type used in drilling fluids have acute toxicities to marine organisms in the range of 0.4 to 40 mg/l (Wildish, 1972; Abel, 1974).

Produced Water

The chemical properties of produced water that could cause harmful effects in marine organisms and ecosystems include elevated salinity, altered ion ratios, low dissolved oxygen, petroleum hydrocarbons and other organics, and heavy metals. Since most produced waters have dissolved solids concentrations higher than that of sea water (30-39 ppt), discharge of such waters to the ocean results in a localized elevation in seawater salinity. The concentration ratios of several major ions (particularly calcium and magnesium) in produced water may be markedly different than those of sea water, leading to adverse reactions in organisms in the receiving waters. Effects of elevated salinity and altered ion ratios on marine animals have been little studied. Some shallow seas and coastal lagoons, particularly in arid regions, periodically experience salinities between 50 and 100 ppt and are populated by a wide variety of salinity-tolerant species (Kinne, 1971). Species richness in such hypersaline environments usually is much less than in coastal waters that have more typical salinity regimes.

As part of the U.S. Strategic Petroleum Reserve Program, the toxicity of brines from Texas and Louisiana salt domes to marine organisms was evaluated (NOAA, Marine Assessment Division, 1978). Several species of phytoplankton, eggs and larvae of spotted sea trout *Cynoscion nebulosus* and white shrimp *Penaeus setiferus*, and juvenile and adult polychaete worms *Neanthes arenaceodentata* and *Nereis limbata* were able to tolerate additions of brine to sea water sufficient to elevate the salinity of the ambient medium by 4 to 8 ppt (1-3% brine in sea water of 30 ppt salinity). Hypersaline medium prepared with salt dome brine was not significantly more toxic than hypersaline medium prepared with artificial sea salts (Instant Ocean). Unless the volume and turnover rate of the receiving water are very small, mixing and dilution of discharged produced water with the receiving water is so rapid that significant elevations of ambient salinity are unlikely. No gradient of salinity was detected around the produced water discharge pipe of a separator platform in shallow Trinity Bay, Texas where ambient salinities varied seasonally between 1 and 15 ppt (Armstrong *et al.*, 1979). Therefore, elevated salinity or altered ionic ratios of produced water effluents are unlikely to contribute significantly to any impacts of produced water in the marine environment.

The toxicity of petroleum hydrocarbons to marine animals was discussed briefly earlier in this chapter and in more detail by Capuzzo (Chapter 8). The produced water effluent from an oil/water separator has a hydrocarbon composition similar to that of the water soluble fraction or water-accommodated fraction of crude oil. These fractions have been used extensively in laboratory marine biological effects studies. The toxicity of a crude oil water soluble fraction to marine organisms is related to the concentrations in it of light aromatic hydrocarbons (benzenes, naphthalenes, and phenanthrenes) (Anderson *et al.*, 1974; Neff *et al.*, 1976). The acute toxicity of aromatic hydrocarbons to marine plants (Hutchinson *et al.*, 1980) and animals (Neff, 1979) increases as molecular weight increases (and aqueous solubility decreases). However, only aromatic hydrocarbons in the molecular weight range from benzene (MW 78.1) to fluoranthene and pyrene (MW 202) are acutely toxic to aquatic organisms. Higher molecular weight aromatics, apparently because solubility falls below the aqueous concentration required to cause a response (Gehrs, 1978), are not acutely toxic to aquatic organisms. However, some of these, such as benz(a)anthracene and benzo(a)pyrene may produce chronic effects such as cancer. Middleditch (1981a) reported concentrations of benzo(a)pyrene up to 5 μg/l (mean 1.2 μg/l) in produced water from the Buccaneer field. Unfortunately, other polycyclic aromatic hydrocarbons and sulfur heterocyclics apparently were not quantified in the produced water.

The total organic load of produced water may be as high as 500 mg/l (Lysyj, 1982), and more than 90% of this is dissolved nonvolatile organic material which has not been adequately characterized chemically. Small amounts of phenols, amino acids, fatty acids, other organic acids, alcohols, and naphthenic and humic substances have been identified in produced water (Collins, 1975; Lysyj, 1982). The toxicity of these materials to marine organisms is not known.

Deck drainage, which may contain a variety of chemicals such as detergents, solvents and metals, is processed through the oil/water separator before discharge to the ocean. In addition, a wide variety of chemicals may be added to the process stream of the oil/water separator and ultimately appear in the effluent water (Middleditch, 1984). These may include biocides, coagulants, corrosion inhibitors, cleaners, dispersants, emulsion breakers, paraffin control agents, reverse emulsion breakers, and scale inhibitors. The concentrations of these materials in produced water effluent and their toxicity to marine organisms are poorly understood.

The biocide, acrolein, which was used in the Buccaneer field, has an acute toxicity to brown shrimp *Penaeus aztecus* of 0.1 mg/l (48-h LC_{50}) (Folmar, 1977). Two other biocides, glutaraldehyde and alkyldimethylbenzyl chloride, and a surfactant were added to produced water in the Buccaneer field at different times. Most biocides are used at concentrations up to about 20 mg/l and are scavenged from the effluent stream with sodium bisulfite (Middleditch, 1983). Concentrations in the final effluent rarely exceed about 2 mg/l. Laboratory bioassays (Zein-Eldin and Keney, 1979; Rose and Ward, 1981) and field studies with caged fish (Workman and Jones, 1979) indicate that produced water is more toxic when biocides are being used than when they are not.

High concentrations of elemental sulfur were detected in produced water from the Buccaneer field (Middleditch, 1981b). Elemental sulfur probably is unique to produced water from reservoirs associated with intrusive salt domes and is not a usual ingredient of produced water (Collins, 1975). Sulfur may be reduced to highly toxic hydrogen sulfide under reducing conditions. This could happen in anoxic basins, but is unlikely in the water column.

Several metals that are present at elevated concentrations in some produced waters are toxic to very toxic to marine animals (LC_{50} less than 10 mg/l). These include arsenic, beryllium, cadmium, chromium, copper, lead, mercury, nickel, silver and zinc. As discussed above, concentrations of most of these metals will be reduced to near ambient levels when the produced water is diluted 10- to 100-fold following discharge to the sea. The critical factors which determine whether these metals will have an adverse impact on organisms in the receiving water are the physical and chemical forms of the metals (Jenne and Luoma, 1977; Breteler and Neff, 1983; Luoma, 1983). Relatively little is known about the chemical forms of metals in produced water (Collins, 1975). Three factors which strongly influence metal speciation and the types of metal ion complexes in produced and receiving waters are the ionic strength and composition of the medium, its pH and Eh. The solubility of most metals decreases as the ionic strength of the medium increases. However, the concentration of several metals in high ionic strength brines is higher than their theoretical solubility (e.g., Ba and Fe) due to complex, poorly understood ion interactions (Ostroff, 1965). These metals may be in the form of metal ion complexes or colloidal suspensions.

Most produced waters have a neutral to slightly acidic pH (see Chapter 4), and Eh usually is low (Collins, 1975). Connate waters usually are anoxic and have a negative Eh. Collins (1969) reported that produced water from the Anadarko Basin, Texas-Oklahoma, had an Eh ranging from -270 to -300 mV. The Eh is a measure of the oxidation-reduction or redox potential of the medium. It is expressed in volts and at equilibrium is related to the proportion of oxidized and reduced species present. A positive Eh value indicates an oxidized state and a negative Eh indicates a reduced state. Therefore, in produced water, the more reduced form of metal ions and compounds will be favored over the more oxidized forms. At moderately low pH and Eh, several metals will be in forms that are moderately soluble and bioavailable. Under strongly reducing conditions, most metals will be in the form of insoluble metal sulfides, and will be non-available and nontoxic. During and after discharge to the ocean, produced water will be oxygenated, resulting in an increase in the redox potential to the oxidized state. Reduced species of some metals will be oxidized, increasing or decreasing their solubility and apparent bioavailability. Soluble polysulfide complexes may be oxidized to sulfates. Ferrous iron will be oxidized to colloidal

Fe hydrous oxides, resulting in adsorption and coprecipitation of several other metals (Patrick *et al.*, 1977; Gambrell *et al.*, 1980). Some of these oxidations may be inhibited by the presence of organic compounds in the produced water or receiving water.

Bascom (1983) has suggested that inorganic metals in the sea are not a hazard to marine animals or to persons who eat those animals. This conclusion was based on the observation that metals discharged to the ocean quickly precipitate or adsorb to particulate matter and settle to the bottom in a form that is not very bioavailable or toxic to marine organisms. More information must be gained about the chemical forms of metals in produced waters and the chemical transformations that take place when produced water is discharged to the oceans before definitive conclusions can be made about the impacts of these metals in the marine environment.

ACUTE TOXICITY OF USED DRILLING FLUIDS AND PRODUCED WATERS

Bioassay Protocols

A used drilling fluid, especially a treated mud from a deep hole, is an extremely heterogeneous mixture. It contains water soluble materials, clay-sized particles of moderate density that sediment slowly in sea water, and larger or denser particles that sediment rapidly. In addition, montmorillonite and attapulgite clays in muds flocculate upon contact with sea water, forming larger particles which tend to settle more rapidly than dispersed clay. These fractions tend to separate rapidly when the drilling fluid is added to sea water in a bioassay aquarium or when it is discharged from an offshore drilling rig. This makes it extremely difficult to design a bioassay protocol in which test organisms are exposed uniformly and reproducibly to a drilling mud-sea water mixture of known concentration or that at least roughly simulates the kind of exposure an organism might encounter in the vicinity of the drilling mud discharge from an offshore rig. Because of the complexity of the chemical and physical processes that take place when a used drilling fluid is discharged to the ocean, none of the bioassay protocols used to date is completely satisfactory.

The simplest approach has been to add whole drilling mud to sea water on a volume:volume basis to establish several exposure concentrations. Test organisms are exposed to these mixtures, which are aerated, mixed or left unmixed during the bioassay (McLeay, 1976; Houghton *et al.*, 1980; Tornberg *et al.*, 1980).

Another approach is to evaluate the toxicity of different drilling fluid fractions or drilling fluid-sea water mixtures that roughly simulate the types of exposure organisms in different marine habitats might encounter. These 87 bioassay protocols are similar to those recommended for evaluation of the environmental impact of dredged material (EPA/COE, 1977). In the EPA/COE dredge material bioassay protocols, which have been adopted with minor modifications for EPA Region I, II, and IX bioassays for compliance with NPDES (National Pollutant Discharge Elimination System) permits for discharge of drilling muds on the mid- and north Atlantic and California outer continental shelf, one part drilling mud is mixed with four parts sea water, and the phases are allowed to separate for one hour. The supernatant is called the suspended particulate phase, and the sedimented fraction is designated the solid phase. A liquid phase is prepared by centrifuging and filtering the suspended particulate phase. An initial dilution of one part drilling mud with nine parts sea water and a 20-h settling period has been used by several investigators (Gerber *et al.*, 1980; Neff *et al.*, 1980, 1981; Bookhout *et al.*, 1984; Chaffee and Spies, 1982). The liquid and suspended particulate phases simulate the types of mud fractions a pelagic or planktonic organism might encounter in the water column, while the solid phase simulates the type of exposure a benthic organism might encounter.

Performance of acute lethal bioassays with produced water presents many fewer technical difficulties. Bioassay media are prepared by volumetric additions of produced water to receiving water or filtered sea water according to standard protocols (EPA 1975, 1978). Exposure concentrations may be measured in terms of nominal volume of produced water added per volume of dilution water (Rose and Ward, 1981), and/or, in the case of high salinity brines, as ppt salinity (Andreasen and Spears, 1983). Care must be taken in collecting, transporting and storing samples of produced water to prevent excessive loss of volatile hydrocarbons. The effluent produced water from the oil/water separator is likely to contain bacteria unless the system has been heavily treated with biocides. Upon storage of such samples in sealed containers, significant amounts of highly toxic hydrogen sulfide may be generated by reduction of sulfur or sulfate by sulfate-reducing bacteria. These chemical changes in produced water during storage will substantially affect its acute toxicity and therefore should be controlled or at least monitored.

Drilling Fluids

A rather extensive body of information is available concerning the acute lethal toxicity and the acute-chronic sublethal effects of used drilling fluids, of the types used for exploration and development drilling in U.S. coastal and offshore waters, to marine organisms. According to the National Research Council (1983), the acute toxicity of at least 70 used water-based drilling fluids has been evaluated in 400 bioassays with at least 62 species of marine animals from the Atlantic and Pacific Oceans, the Gulf of Mexico and the Beaufort Sea (Table 10.1). Petrazzuolo (1983) lists 415 bioassays with 70 species of marine organisms and 68 different samples of drilling fluid. Five major marine phyla were represented among the bioassay animals, including Chordata (12 species of fish), Arthropoda (30 species of crustaceans), Mollusca (12 species of molluscs), Annelida (6 species of polychaetes) and Echinodermata (1 species of sea urchin). Larvae and other early life stages (considered to be more sensitive than adults to pollutant stress) were included. To date, only one species of marine plants, the diatom *Skeletonema costatum*, has been tested. Because liquid and suspended particulate phase preparations of lignosulfonate drilling fluids are highly colored even at low concentrations, algal bioassays are neither feasible nor realistic.

Although bioassay methods and conditions varied considerably, the results

TABLE 10.1

Summary of results of acute lethal bioassays with drilling fluids and marine/estuarine organisms. Most median lethal concentration (LC_{50}) values are based on 96-h bioassays and results are expressed as parts per million (mg/l or μl/l) mud added (from National Research Council, 1983)

Organism	*Number of species tested*	*Number of fluids tested*[b]	*Number of bioassays*	*Number of LC_{50} values (ppm)*					
				Not determinable	*100*	*100–999*	*1000–9999*	*10,000–99,999*	*>100,000*
Phytoplankton	1	9	12	5	0	0	7	0	0
Invertebrates									
Crustaceans									
Copepods	2	17	39	4	2	11	15	7	0
Isopods	2	4	6	0	0	0	0	1	5
Amphipods	4	8	19	0	0	0	0	5	14
Mysids[a]	5	18	41	1	0	1	0	21	18
Shrimp[a]	10	40	76	0	0	12	15	31	18
Crabs[a]	6	18	35	1	0	0	5	16	13
Lobsters[a]	1	2	7	0	0	0	1	3	3
Molluscs									
Gastropods	5	5	10	0	0	0	0	2	8
Bivalves[a]	7	14	33	0	0	0	1	15	17
Echinoderms									
Sea Urchins[a]	1	2	4	0	0	0	0	1	3
Polychaetes	6	14	28	0	0	0	0	9	19
Finfish[a]	12	32	90	0	0	0	3	52	35
Totals	62	72	400	11	2	24	47	163	153
Percentage, as a fraction of the total number of drilling fluid bioassays				2.8	0.5	6.0	11.8	40.8	38.2

[a]Includes results for embryonic, larval and early life stages.

[b]In many cases, the same drilling fluid was used for bioassays with several species. In a few cases, more than one investigator evaluated the toxicity of a single drilling fluid.

were quite consistent. Nearly 80% of the 400 LC_{50} values were higher than 10,000 ppm drilling fluid added. Only two LC_{50} values were below 100 ppm; these were for the copepod *Acartia tonsa* exposed to drilling fluids, heavily treated with chromate and diesel fuel, from Mobile Bay, Alabama (Gilbert, 1981). The NPDES permit for this well prohibited ocean disposal of drilling fluids, and the fluids were formulated without consideration of their marine toxicity. In the ongoing EPA drilling mud bioassay program, an unfractionated whole drilling mud bioassay with a drilling mud sample from the Gulf of Mexico containing 9430 μl/l (ppm) diesel fuel and 814 mg/l chromium had a 96-h LC_{50} of 26 ppm for juvenile mysid shrimp *Mysidopsis bahia* (T. Duke, personal communication).

When the results of acute lethal bioassays with drilling fluids are normalized to account for those species which have been tested more frequently (Petrazzuolo, 1983), 44% of LC_{50}s are in excess of 100,000 ppm, and 46% are between 10,000 and 100,000 ppm drilling fluid added. Six percent of LC_{50} values fall between 1000 and 10,000 ppm and 1-2% are in the 100-1000 ppm range. Thus, 90% of bioassays performed to date give LC_{50} values in excess of 10,000 ppm, defined as practically nontoxic by IMCO/FAO/UNESCO/WMO (1969).

The most sensitive species tested included the estuarine copepod *Acartia tonsa*, the oceanic copepod *Centropages typicus*, larvae of dock shrimp *Pandalus danae*, pink salmon fry *Oncorhynchus gorbuscha*, larvae of lobster *Homarus americanus*, juvenile ocean scallops *Placopecten magellanicus*, and mysid shrimp (*Mysidopsis*, *Neomysis*, *Acanthomysis* and *Mysis*). In most cases, larval and/or juvenile life stages were more sensitive than adults.

Whenever comparisons were made (Gerber *et al.*, 1980; Carls and Rice, 1980; Neff, 1980, 1982b; Marine Bioassay Labs, 1982; Ayers *et al.*, 1983; T. Duke, EPA, personal communication), the sensitive species were more sensitive to whole mud suspensions and suspended particulate phase preparations than to liquid phase preparations, indicating that suspended particulates in the drilling fluids contributed to the toxicity of the drilling fluids. Liquid phase preparations of drilling fluids also were toxic. Thus, the toxicity of drilling fluids is due to a combination of the chemical toxicity of water-accommodated mud ingredients and chemicals associated with the particulate phase, and physical irritation and damage to delicate gill and other body structures from mud particles.

The toxicity of different types of drilling fluids varies. However, information about the types and compositions of drilling fluids used in bioassays is incomplete. Lightly-treated muds and spud muds (used during the initial phases of drilling) have a toxicity, in most cases, not markedly different from that of suspended clay (McFarland and Peddicord, 1980). Their soluble fractions usually are nontoxic. Muds that have been treated heavily with chrome lignosulfonate-dichromate mixtures, surfactants and/or diesel fuel are the most toxic. Both the soluble and particulate phases of such muds have a similar toxicity.

When NPDES permits were granted for offshore drilling on the mid-Atlantic outer continental shelf in 1978, the Offshore Operators Committee Task Force on Environmental Science, and the EPA Region II, developed a list of eight general or generic drilling mud types that included virtually all types of water-based muds commonly used on the U.S. outer continental shelf. The generic mud concept has been incorporated into NPDES permits issued by EPA Regions I, II, III, IX, and X and is under consideration for future permits in other regions (Ayers *et al.*, 1983). Bioassays performed according to the EPA Region II protocols are conducted on field samples of muds representative of the eight generic drilling fluid types. Operators then may be allowed to discharge drilling fluids of the eight generic types without conducting additional bioassays. If specialty additives are used, bioassay and approval of the EPA Regional Administrator is required before the muds containing them are discharged to the ocean.

For Region II NPDES permits, liquid and suspended particulate phase bioassays were performed with opossum shrimp *Mysidopsis bahia* and solid phase bioassays were performed with hard shell clams *Mercenaria mercenaria* (Ayers *et al.*, 1983). The 96-h LC_{50} values reported by two laboratories for the liquid and suspended particulate phase preparations of the eight generic drilling muds ranged from 11,600 to 200,000 ppm and 5000 to 200,000 ppm drilling mud added, respectively. In only one of the bioassays was the 96-h LC_{50} lower than 10,000 ppm mud added; 97% of LC_{50}s were above 10,000 ppm, and 75% of the LC_{50} values were above 100,000 ppm mud added. The drilling fluid giving the lowest LC_{50} value was a potassium/polymer mud (Mud No. 1). The solid phase of only one drilling fluid (Mud No. 2, a lignosulfonate seawater mud) produced mortalities statistically significantly higher than controls. Thus, the generic muds, can be ranked as among the least toxic drilling fluids evaluated to date.

Ranking of different major marine taxa according to their relative sensitivities to drilling fluids, based on results of bioassays performed to date, as Petrazzuolo (1983) has attempted, should be done with caution. The sample size of total species evaluated (at least 62 species) is sufficiently large to provide a reasonable indication of the range of sensitivity that might be expected among marine organisms as a whole. However, the number of representatives of each major taxon is not large enough to indicate the full range of sensitivity of each major taxonomic group.

Produced Water

Relatively little information is available about the acute lethal toxicity of produced water to marine organisms. The results of recent bioassays are summarized in Tables 10.2 and 10.3. Only 11 species, all but two from the western Gulf of Mexico, of crustaceans and teleost fish have been evaluated in 54 bioassays. More than 88% of LC_{50} values were above 10,000 μl/l (ppm) produced water in sea water. The lowest LC_{50} values were obtained with produced water that had been treated with the biocides, glutaraldehyde and alkyldimethyl benzyl chloride, which were not scavenged before the produced water was discharged to the ocean (Zein-Eldin and Keney, 1979). Barnacles *Balanus tintinnabulum* and crested blennys *Hypleurochilus geminatus* were more sensitive to unaerated than aerated produced water from the Buccaneer field (Rose and Ward, 1981). The authors attributed this difference to the oxygen demand of the produced water itself or the ammonium bisulfite added to the produced water to scavenge the acrolein biocide. However, the greater toxicity of unaerated produced water may also be due in part to the presence of volatile low molecular weight aromatic hydrocarbons (benzene-xylenes, present at concentrations of 10-13 mg/l in produced water from the Buccaneer field, see Chapter 4) in it. These would be stripped from bioassay media rapidly by aeration. Andreasen and Spears (1983) showed that produced water added to natural sea water was more toxic than concentrated artificial sea water to sheepshead minnows *Cyprinodon variegatus* (96-h LC_{50}, 52 and 66 ppt salinity, respectively). The produced water evaluated

TABLE 10.2

Summary of results of acute lethal bioassays with produced water and marine/estuarine animals. Most median lethal concentration (LC_{50}) values are based on 96-h bioassays and results are expressed as parts per million (μl/l) produced water added to sea water

Organism	*Number of species tested*	*Number of bioassays*	*Number of LC_{50} values (ppm)*		
			1000–9999	*10,000–99,999*	*>100,000*
Crustaceans	4	37	5	20	12
Teleost fish	7	17	1	0	16
Totals	11	54	6	20	28
Percentage, as a fraction of all produced water bioassays			11.1	37.0	51.9

by these investigators contained 17.9 ppm petroleum hydrocarbons, including 2.4 ppm light aromatics. The total hydrocarbon concentration of produced water at the LC_{50} concentration was 11 mg/l. By comparison, the 96-h LC_{50} of the water soluble fraction of southern Louisiana crude oil to *C. variegatus* was greater than 19.7 mg/l (Anderson *et al.*, 1974). These results seem to indicate that, although petroleum hydrocarbons contributed to the toxicity of produced water, other properties of the produced water also contributed to its toxicity.

The most sensitive organism evaluated by Rose and Ward (1981) was larval brown shrimp *Penaeus aztecus*. The LC_{50} ranged from 160,000 ppm at 3 h to 8000 ppm at 48 h. Rose and Ward (1981) applied a conservative application factor of 0.01 to these values and compared the resulting limiting permissible concentrations of produced water (the estimated highest concentration causing no adverse biological impacts) to the estimated concentrations of produced water in the ocean at different times after discharge, based on a conservative use of the dispersion model developed for vertically distributed pollutants in the Buccaneer Field (Smedes *et al.*, 1981). The limiting permissible concentration of produced water was approximately four orders of magnitude greater than the estimated environmental concentration of formation water at all times (Figure 10.2). The two curves showed no tendency toward convergence at some time greater than 48 h. The authors concluded that discharge of formation water does not represent a potential hazard to water column organisms drifting or swimming through the waste water plume.

TABLE 10.3
Results of acute lethal bioassays with produced water and estuarine and marine animals (primarily from the Gulf of Mexico). LC_{50} values are given as μl/l produced water added to sea water

Species/life stage	*Test conditions*	*Source of produced water*	*96-h LC_{50}*	*Reference*
Crustaceans				
Brown Shrimp				
Penaeus aztecus				
Larvae	Static, aerated, 28°C	Buccaneer Field, Texas	8000–10,000[a]	Rose and Ward, 1981
Subadult	Static, aerated, 18–25°C	Buccaneer Field, Texas	60,000–183,000	Rose and Ward, 1981
Subadult	Flow-through, aerated, 25–29°C	Buccaneer Field, Texas	44,000	Rose and Ward, 1981
Adult	Static, aerated, 18–25°C	Buccaneer Field, Texas	78,000–178,000	Rose and Ward, 1981
Adult	Static, not aerated, 22°C	Aransas Bay, Texas	500,000	Andreasen and Spears, 1983
White Shrimp				
Penaeus setiferus				
Subadult	Static, aerated, 18–25°C	Buccaneer Field, Texas	56,000–133,000	Rose and Ward, 1981
Subadult	Flow through, aerated, 22°C	Buccaneer Field, Texas	62,000	Rose and Ward, 1981
Juvenile	Treated with biocides[b]	Buccaneer Field, Texas	1750–6500	Zein-Eldin and Keney, 1978
	Untreated	Buccaneer Field, Texas	>100,000	Zein-Eldin and Keney, 1978
Adult	Static, aerated, 18–25°C	Buccaneer Field, Texas	37,000–92,000	Rose and Ward, 1981
Adult	Static, not aerated, 22°C	Aransas Bay, Texas	500,000	Andreasen and Spears, 1983
Barnacle				
Balanus tintinnabulum	Static, aerated, 18–25°C	Buccaneer Field, Texas	33,000–154,000	Rose and Ward, 1981
Brine Shrimp				
Artemia salina	Static, not aerated, 18°C	Buccaneer Field, Texas	8000	Rose and Ward, 1981
	Flow-through, aerated, 25–29°C	Buccaneer Field, Texas	51,000	Rose and Ward, 1981
	Static, aerated, 20°C	California + NW Gulf of Mexico	33,300–>1,000,000	Palmer, 1978

Teleost Fish				
Crested Blenny *Hypleurochilus geminatus*	Static, aerated, 22–25°C	Buccaneer Field, Texas	158,000–408,000	Rose and Ward, 1981
Killifish	Static, not aerated, 24°C	Buccaneer Field, Texas	7000	Rose and Ward, 1981
Fundulus heteroclitus	Static, aerated, 20°C	California + NW Gulf of Mexico	22,700–>1,000,000	Palmer, 1978
Sheepshead Minnow *Cyprinodon variegatus*	Static, not aerated, 22°C	Aransas Bay, Texas	600,000	Andreasen and Spears, 1983
Pinfish *Lagodon rhomboides*	Static, not aerated, 22°C	Aransas Bay, Texas	500,000	Andreasen and Spears, 1983
Atlantic Croaker *Micropogonias undulatus*	Static, not aerated, 22°C	Aransas Bay, Texas	350,000	Andreasen and Spears, 1983
White Mullet *Mugil curema*	Static, not aerated, 22°C	Aransas Bay, Texas	500,000	Andreasen and Spears, 1983
Stickleback *Gasterosteus aculeatus*	Not known	Platform Harry, California	750,000–1,000,000	U.S. Dept. Interior, 1975

[a]48-h bioassays.
[b]Biocides included glutaraldehyde (K-31) and alkyldimethyl benzyl chloride (KC-14).

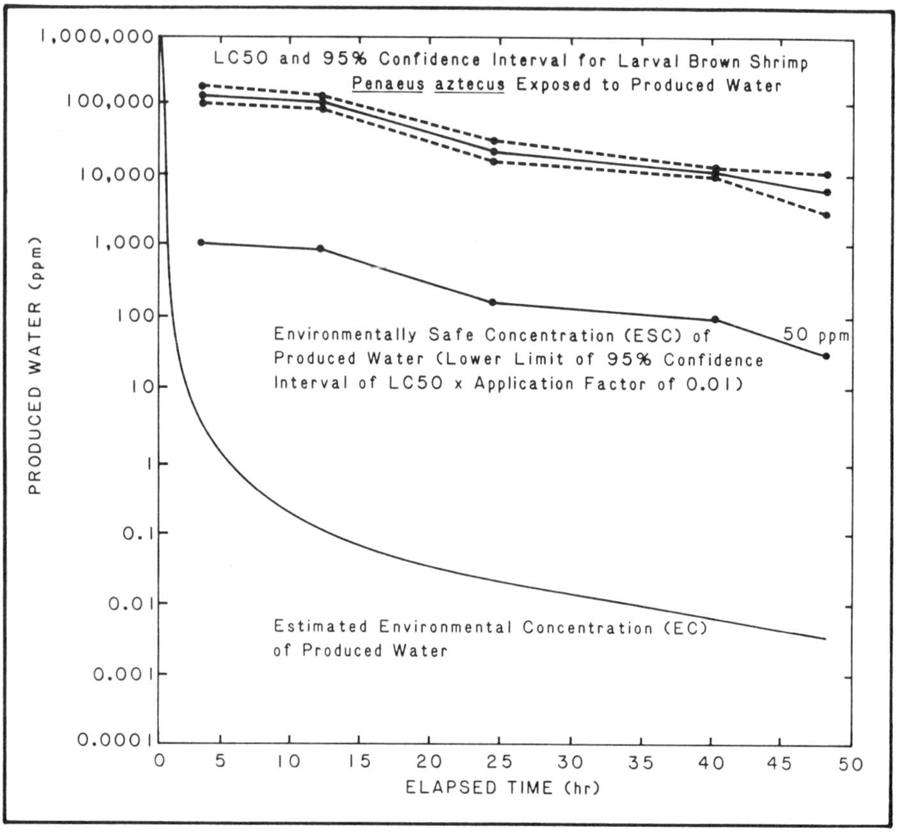

Figure 10.2. LC_{50} values, their 95% confidence intervals and estimated environmentally safe concentrations of produced water from the Buccaneer oil field (estimated at 1% of lower 95% confidence interval) for larval brown shrimp after different lengths of exposure compared to estimated dilution curve for produced water discharged from the Buccaneer platform (from Rose and Ward, 1981).

Conclusions

Acute lethal bioassays are useful primarily for ranking and comparing the relative toxicity of different chemicals or mixtures and for comparing the sensitivities of different species or life stages to a particular pollutant. The joint IMCO/FAO/UNESCO/WMO Joint Group of Experts on the Scientific Aspects of Marine Pollution (1969) has classified different grades or degrees of toxicity of chemicals to marine animals according to LC_{50} values. These toxicity grades and corresponding LC_{50} values as interpreted by Sprague (1973) are: very toxic, 1 ppm; toxic, 1-100 ppm; moderately toxic, 100-1000 ppm; slightly toxic, 1000-10,000; practically nontoxic, 10,000 ppm. By this classification, 90% of the 72 drilling fluids tested to date on 62 species of marine organisms can be considered practically nontoxic. Less than 1% are ranked toxic. Only one of the generic muds can be considered slightly toxic. The rest are practically nontoxic. Larval, juvenile and molting crustaceans are more sensitive to drilling muds than are most other life stages and species. The most toxic drilling muds are those that contain high concentrations of hexavalent chromium, diesel oil, surfactant, etc.

More than 88% of bioassays with produced water gave results indicating the produced water was practically nontoxic. Eleven percent of the results were in the slightly toxic range, and these were for produced water which had been treated with biocides or was not aerated during the bioassay. Brown shrimp larvae were the most sensitive organisms tested to date. However, only 11 species of crustaceans and fish have been evaluated. Although results to date indicate that produced water is likely to have little or no adverse impact on water column organisms, more information on the sensitivity of planktonic and nektonic organisms of a wider variety of taxa and geographic areas is needed to more fully assess the potential impacts of produced water discharges on receiving waters. For both drilling fluids and produced water, results of carefully-designed experiments to assess the relative importance of such factors as hydrocarbons, other organics, solids, metals and pH alteration in the toxicity of these complex effluents would be most useful in assessing potential long-term impacts on the marine environment.

CHRONIC AND SUBLETHAL EFFECTS OF DRILLING FLUIDS AND PRODUCED WATER

Studies of chronic and/or sublethal effects often are better predictors than acute lethal bioassays of the potential environmental impact of a pollutant, because the former may utilize exposure conditions that more closely simulate those that organisms might encounter in their natural environment. Chronic or sublethal effects studies may include exposures during particularly sensitive life stages and processes (such as embryo-larval development, molting in crustaceans and reproduction) and thereby provide better insights into the true toxicity and potential long-term effects of the pollutant. In addition, they may provide useful insights into the nature and mechanisms of pollutant toxicity and may suggest the types of responses to look for in field studies of the impacts of oil drilling and production-related discharges in the marine environment.

Drilling Fluids

Investigations of the chronic and/or sublethal effects of drilling fluids have been performed with at least 40 species of marine animals, including 10 species of corals, 6 species of molluscs, 15 species of crustaceans, 1 species of polychaete worm, 5 species of echinoderms, and 3 species of teleost fish. Results of these investigations are summarized in Table 10.4.

Responses to sublethal concentrations of drilling muds which have been measured to date include altered burrowing behavior and chemosensory responses in lobsters; alterations in patterns of embryological or larval development in lobsters, crabs, sand dollars, starfish and fish; depressed feeding in larval lobsters, larval cancer crabs, and juvenile hake; decreased food assimilation and growth efficiency in opossum shrimp; depressed growth and skeletal depo-

TABLE 10.4

Summary of chronic and/or sublethal responses of marine animals to water-based chrome or ferrochrome lignosulfonate-type drilling fluids. The lowest exposure concentrations eliciting a response are given

Animals	*Nature and length of exposure*	*Responses*	*References*
Reef Corals (ten species)	2 mm–12 cm layer for up to 10 h; 1–1000 ppm suspension for 4–42 days	Polyp retraction, mucus production, expulsion of zooxanthellae, depressed respiration, NH_3 excretion and uptake, photosynthesis, calcification and growth, altered biochemical composition, bacterial infection, death	Dodge, 1982; Hudson and Robbin, 1980; Kendall *et al.*, 1983; Krone and Biggs, 1980; Szmant-Froelich *et al.*, 1982; Thompson, 1980; Thompson and Bright, 1977, 1980; Thompson *et al.*, 1980.
Bivalve Molluscs (six species)	50–33,000 ppm suspension for 3–100 days	Depressed filtration, byssus thread formation, NH_3 excretion, shell growth, condition index, increased respiration, altered free amino acid ratios, behavior	Gerber *et al.*, 1980, 1981; Gilbert, 1981, 1982; Houghton *et al.*, 1980; Neff, 1980; Powell *et al.*, 1982; Rubinstein *et al.*, 1980; Olla *et al.*, 1982
Crustaceans (fifteen species)	7.7–100,000 ppm suspension for 5 min-42 days; 1–7 mm layer for up to 4 days	Decreased chemosensory response, inhibition of feeding, altered behavior in larvae and juveniles, cessation of swimming in larvae, increased duration of larval and juvenile development, decreased or increased enzyme activity, gill histopathology, decreased long-term larval and juvenile survival	Atema *et al.*, 1982b; Bookhout *et al.*, 1984; Capuzzo and Derby, 1982; Carls and Rice, 1980; Carr *et al.*, 1980; Conklin *et al.*, 1980; Gerber *et al.*, 1980, 1981; Gilbert, 1981; Houghton *et al.*, 1980; Neff, 1980; Olla *et al.*, 1982
Polychaete Worms (one species)	10 ppm suspension, 100 days	33% mortality	Rubinstein *et al.*, 1980
Echinoderms (five species)	10–100,000 ppm suspensions, 2 days-duration of larval development	Depressed fertilization, decreased development rate, increased incidence of development anomalies	Chaffee and Spies, 1982; Crawford, 1983; Crawford and Gates, 1981
Teleost Fish (three species)	3800–30,000 ppm for duration of embryo development; 7–17 ml/l for 3–12 days	Decreased development rate, depressed embryonic heart beat, developmental anomalies, gill histopathology, feeding and avoidance behavior, growth	Crawford and Gates, 1981; Houghton *et al.*, 1980; Olla *et al.*, 1982; Sharp *et al.*, 1984

sition in corals, starfish embryos, scallops, oysters and mussels; altered respiration and nitrogen excretion rates in corals, mussels and lobsters; changes in tissue enzyme activity in crustaceans; gill histopathology in shrimp and salmon fry; altered tissue-free amino acid ratios in corals and oysters; polyp retraction and mucus hypersecretion in corals; and avoidance behavior in sea scallops, sand shrimp and hake.

All the drilling fluids evaluated were of the chrome or ferrochrome lignosulfonate-type, which is the drilling fluid type used most frequently for exploratory drilling on the continental shelf of the United States. Several of the drilling muds tested, including the most toxic ones, are known to have contained high concentrations of diesel fuel or other petroleum material. These include the 18 muds from Mobile Bay, Alabama (Conklin *et al.*, 1983), the July and August muds from the Jay Field, Florida (Atema *et al.*, 1982b; Cappuzo and Derby, 1982), and the medium weight mud from the Gulf of Mexico used by Neff (1980) and Gerber *et al.* (1980, 1981). Diesel fuels, including No. 2 fuel oil, are known to be quite toxic to marine organisms (Malins, 1977; Neff and Anderson, 1981), and undoubtedly contribute significantly to the toxicity of those muds containing them.

In most of the chronic/sublethal effects studies performed to date, exposure conditions did not closely simulate those the organisms might encounter in their natural habitat. In nearly all cases, either exposure concentrations were much higher or the duration of exposure (particularly to unfractionated mud) was much longer than would ever occur in the field. In some cases, benthic animals were exposed to whole unfractionated drilling mud layered on or mixed with natural uncontaminated sediments. Unless drilling mud is shunted directly to the bottom, considerable fractionation will occur as it descends through the water column. Soluble fractions of the mud not tightly adsorbed to clay particles, including the more soluble and toxic aromatic fractions of diesel oil, may not reach the bottom at all. Several field studies, discussed above, have shown that drilling fluids are diluted to very low concentrations within 2-6 h of discharge to the ocean (see Chapter 4).

Significant biological responses were recorded at nominal drilling mud concentrations ranging from 1 to 160,000 ppm and as a 1-mm to 12-cm layer on natural sediment. In several cases, deleterious sublethal responses in marine animals were observed at drilling fluid concentrations only slightly lower than those that were acutely toxic. For instance, the 144-h LC_{50} was 1.4 to 3 times higher than the concentration causing cessation of swimming in 50% of test animals in 144 h for Stage I larvae of six species of marine crustaceans from the Pacific coast of Alaska exposed to suspensions and liquid phase preparations of a used chrome lignosulfonate drilling mud from Cook Inlet, Alaska (Carls and Rice, 1980). The behavioral responses did not occur until 4-24 h after the start of exposure, while mortality was delayed for 48-72 h. Because of the relatively low toxicity of the drilling muds and the delayed behavioral or lethal responses in the crustacean larvae, the authors suggested that planktonic larvae might be affected only within a few meters of the drilling mud discharge.

In several other species, however, significant sublethal responses were recorded at drilling mud concentrations orders of magnitude lower than acutely lethal concentrations. In this category are the American lobster *Homarus americanus* and several of the molluscs, particularly the ocean scallop *Placopecten magellanicus*. Most of the drilling muds used in these studies were heavily treated with chromate and/or diesel oil.

In addition, reef corals appear to be quite sensitive to chrome lignosulfonate-type drilling fluids. Exposure to drilling mud elicits partial or complete polyp retraction in the coral, accompanied in many cases by a hypersecretion of mucus (Thompson, 1980; Thompson and Bright, 1980; Thompson *et al.*, 1980). These are defensive reactions which, if they persist for long due to continued pollutant insult, may lead to decreased nutrient assimilation and production (Szmant-Froelich *et al.*, 1982), altered biochemical composition (Kendall *et al.*, 1983; Parker *et al.*, 1984), depressed respiration and nitrogen excretion (Krone and Biggs, 1980; Szmant-Froelich *et al.*, 1982), partial or complete inhibition of growth and calcium carbonate skeleton deposition (Hudson and Robbin, 1980; Szmant-Froelich *et al.*, 1982; Dodge, 1982; Kendall *et al.*, 1983), extrusion of zooxanthellae (Thompson, 1980; Kendall *et al.*, 1983), bacterial infection (Parker *et al.*, 1984) and eventually death. These responses are elicited by chronic exposure to drilling mud concentrations of 100 ppm or less, although there are large interspecies differences in sensitivity to drilling fluids. Equivalent concentrations of suspended kaolin clay produce lesser responses, indicating that the responses observed are not due to turbidity increases alone (Thompson, 1980; Kendall *et al.*, 1983).

The 96-h LC_{50} values for five used drilling fluids containing 0 to 4% diesel fuel and up to 1.3 ppm phenols ranged from 73.8 to greater than 500 ppm for larvae of American lobsters *Homarus americanus* (Capuzzo and Derby, 1982; Derby and Capuzzo, 1984). The most toxic drilling fluids were those with the highest concentrations of diesel fuel. At exposure concentrations of 50 to 100 ppm of the most toxic muds, there were reductions in growth rates, molting frequencies, respiration rates, feeding rates and growth efficiencies in the larvae. In stage IV postlarvae, continuous exposure for 36 days to 7.7 ppm suspensions of diesel fuel-treated drilling fluid caused a decrease in the rate of molting and growth and altered feeding behavior (Atema *et al.*, 1982b).

In adult lobsters, 10 and 100 ppm suspensions in sea water of whole used drilling fluid heavily treated with chromate and diesel fuel, altered neural activity of walking leg chemosensory neurons, as measured by extracellular neurophysiological recording techniques (Derby and Atema, 1981). Similar chemosensory responses were observed in lobsters exposed to 0.1 to 1.0 ppm of the water-accommodated fraction of a diesel fuel (Atema *et al.*, 1982a), indicating that responses to the drilling fluids may have been due to the 2-4% diesel fuel they contained (a nominal 0.2-4.0 ppm diesel fuel would have been present in the 10-100 ppm suspensions of drilling fluid). An increase or decrease in the number of action potentials in response to a standard food cue, as a result of pollutant exposure, may or may not be reflected as a change in feeding behavior in the intact animal, making the results of such studies difficult to interpret. Atema *et al.* (1982b) also described a variety of behavioral responses, including alterations in burrow construction, tail flipping and locomotion, during exposure to 1-7 mm layers on natural sediments of whole unfractionated drilling mud, as well as pure bentonite clay or barite. Some of these behavioral responses probably were due to the strong preference of juvenile lobsters for coarse substrates as compared to substrates containing significant amounts of silt- and clay-sized material (Pottle and Elner, 1982).

Gilbert (1981) reported a statistically significant inhibition of shell growth in 2-day-old embryos of ocean scallops *Placopecten magellanicus* during exposure to seawater suspensions of several used drilling fluids from Mobile Bay and the Jay Field, Florida at nominal concentrations ranging from 100 to 10,000 ppm. In subsequent experiments with 11 used drilling muds containing 0 to 1% diesel fuel from the Gulf of Mexico, exposure for 48 h to liquid and suspended particulate phase preparations at concentrations ranging from 64 to greater than 3000 ppm resulted in inhibition of shell formation in day-old embryos of hard shell clams *Mercenaria mercenaria* (Gilbert, 1982). The same drilling fluids at nominal concentrations as low as 10 to 100 ppm interfered with fertilization or caused abnormal embryonic development in sand dollars *Echinarachnius parma* and sea urchins *Lytechinus variegatus*, *L. pictus*, and *Strongylocentrotus purpuratus* (Crawford, 1983; Crawford and Gates, 1981). When juvenile scallops *Placopecten magellanicus* and mussels *Mytilus edulis* were exposed for 30-40 days to 50-250 ppm suspensions of used drilling fluids, rates of shell growth were inhibited significantly (Gerber *et al.*, 1981).

Embryos of the bat starfish *Patiria miniata* were exposed for 48 h to 13 samples of the liquid phase of used ferrochrome lignosulfonate drilling muds collected at different depths while drilling a slant hole from the Hondo platform in the Santa Barbara Channel, California (Chaffee and Spies, 1982). Three of the muds caused significant reductions in growth rate at concentrations as low as 500 ppm (v/v) mud added. Some other muds caused significant growth enhancement at concentrations up to 15,000 ppm. Degree of inhibition of embryo growth was correlated with increasing chromium concentration in the muds. Embryos exposed to 5000-15,000 ppm of the different drilling mud liquid phases had increased incidences of developmental anomalies. All embryos exposed to 25,000 ppm or higher concentrations of the mud developed anomalously or died. The authors concluded that adverse effects on water column organisms during ocean discharge of such drilling muds could occur only within short distances of the discharge pipes of offshore rigs. Such effects would be transitory.

Olla *et al.* (1982) performed three experiments in which great care was taken to simulate as closely as possible the fractionation that takes place as drilling fluids are dispersed in the water column. They investigated the effects of the settleable fraction of a 12.6 lb/gal lignosulfonate drilling fluid from Galveston County, Texas (115,000 ppm Ba, 380 ppm Cr) on behavior of a model demersal/benthic community consisting of juvenile red hake *Urophycis chuss*, ocean scallops *Placopecten magellanicus*, and sand shrimp *Crangon septemspinosa*.

To obtain a settleable fraction, the investigators allowed drilling fluid to settle through a 6.1-m (20-ft) vertical column of sea water for one hour. The lower three meters of water were collected. This was repeated three times to simulate settling through 18 m (59 ft) of water. From 2.5 to 6 liters (containing 10-15% drilling fluid solids) of the settleable fraction was added near the surface and allowed to settle naturally to the bottom of 360-l aquaria in three experiments. In the first two experiments, there was a single addition of drilling fluid; in the third, there were four additions at three-day intervals. The amount of drilling fluid solids added to each treatment aquarium ranged from 0.59 kg solids/m^2 of sediment in the first experiment to 4.7 kg solids/m^2 in the third. If it is assumed that all the barium was retained in the settleable fraction, the nominal increment in barium concentration in the upper 2 cm of sediment due to these treatments ranged from 1265 to 10,350 ppm.

The scallops, shrimp and hake were not affected behaviorally by the presence of drilling fluid settleable solids layered on or mixed with the sediments; however, introduction of the settleable fraction into the treatment aquaria provoked immediate and strong, but short-lived, reactions in three species. There was increased swimming activity in scallops and shrimp. Shrimp and fish first tried to avoid the plume of settling drilling fluid and then congregated near the water surface after they were enveloped by the plume. These behavior patterns persisted for 2-3 h after introduction of the drilling fluid and did not change for the four successive drilling fluid introductions in the third experiment. Normal behavior resumed after drilling fluid solids settled to the bottom and the water cleared. During the 11-12 day duration of experiments 2 and 3, hake from treatment aquaria consumed fewer shrimp but did not grow more slowly than fish from control aquaria. Thus, in these more environmentally realistic experiments, benthic animals showed only minimal responses to drilling fluid settleable fraction and responses were primarily to the suspended particulate plume.

Microcosm Studies

Laboratory microcosms have been used to study the effects of used whole drilling fluids in suspension or layered on or mixed with natural sandy sediments on recruitment of planktonic larvae to the benthos. Natural unfiltered sea water is pumped directly from the ocean or estuary through aquaria which contain clean natural sediments or sediments containing the contaminant under investigation. After several weeks, animals settling in the substrate of the aquaria are counted and identified. Alternatively, trays containing clean or contaminated sediments are placed on the bottom in the ocean or estuary and the types and numbers of organisms settling in the substrate are monitored. Flint *et al.* (1982) and Tagatz and Deans (1983) compared these two techniques for assessing potential impacts of contaminants on the benthos. There were several significant differences in the characteristics of the communities developing in the two systems and natural benthic communities. Flint *et al.* (1982) concluded that the field-colonized sediment trays were superior to laboratory-colonized aquaria for simulating impacts of disturbance on the benthos, whereas Tagatz and Deans (1983) concluded that the laboratory microcosms were sufficiently similar to the field situation to allow application of the results of the laboratory studies to nature.

In several experiments with drilling fluids, relatively high concentrations of whole unfractionated drilling fluid layered on or mixed with sandy sediments were required to significantly affect recruitment of benthic invertebrates (Rubinstein *et al.*, 1980; Tagatz *et al.*, 1978, 1980, 1982). Recruitment of some species was depressed severely at all levels of applied drilling fluids. Other species occurred preferentially in sediments contaminated with drilling fluids. Included in the latter category were certain species of bacteria and microeucaryotes (ciliates, nematodes, etc.) (Smith *et al.*, 1982). Some of these effects could be due to changes in sediment texture owing to the presence of drilling mud.

When marine aquaria were supplied with unfiltered natural sea water containing 50 ppm used chrome lignosulfonate drilling mud for eight weeks, numbers of tunicates, molluscs and polychaetes settling in the sandy substrate or on the walls of the aquaria were significantly decreased in comparison to numbers settling in control aquaria (Tagatz *et al.*, 1982). Several differences in community parameters were detected. These differences could have been due to physical and/or chemical interference of suspended drilling muds with survival and/or settlement of planktonic larvae or to accumulation of drilling muds in the sediments over time (which was noted but not quantified), altering sediment texture.

Interpretation of Laboratory Biological Effects Studies with Drilling Fluids in Relation to Field Observations

The results summarized here show that concentrations lower than 1000-10,000 ppm of most used offshore drilling fluids are unlikely to cause any acute damage to marine organisms. The most toxic drilling fluids tested produced chronic and/or sublethal effects at concentrations as low as 10-100 ppm. Dilution of such muds by 104 would render them nontoxic. Based on five field dispersion studies discussed above, discharged drilling fluids are diluted to below 1000 ppm within minutes and to less than 10 ppm within three hours (Figure 10.1). This time span is substantially less than the duration of exposure employed for all acute lethal bioassays and most chronic/sublethal effects studies. In addition, during dilution in the water column, drilling fluids undergo substantial fractionation. Thus, water column organisms never will be exposed to whole unfractionated drilling fluids or even to the lighter liquid or suspended particulate fractions of drilling fluids for long enough to elicit lethal or sublethal responses.

There could, however, be localized effects on benthic organisms and communities where drilling muds and cuttings settle out and accumulate on the bottom. The areal extent of accumulation of sufficient drilling mud and cuttings to cause serious damage to the benthos through chemical toxicity, change in sediment texture or outright burial is likely to be small, perhaps a few hundred meters in diameter near an exploratory rig and somewhat larger near a development platform. The rate of recovery of the impacted area will depend on the extent of damage and the rate of removal of drilling mud and cuttings from the area through resuspension and bed transport. Sandy bottom communities in high energy environments, usually are tolerant to sedimentation and may recover rapidly (Rhoads *et al.*, 1978; Boesch and Rosenberg, 1981), whereas complex coral reef communities or some deep-sea communities are very sensitive to burial and may require many years to recover from impacts of mud and cuttings accumulation (Grassle, 1977; Jumars, 1981; Pearson, 1981).

Produced Water

Practically no information is available about the chronic and/or sublethal responses of marine organisms to produced water. In recent reviews, the chronic and/or sublethal effects of produced water have been inferred from the very substantial published information about the chronic and sublethal effects of petroleum hydrocarbons and heavy metals to marine organisms (Koons *et al.*, 1977; Menzie, 1982; Middleditch, 1984). Such extrapolations must be made with caution, since they do not take into account possible synergistic or antagonistic interactions of the different chemical and physical characteristics of produced water which might influence its toxicity.

Early studies by Lunz (1950) indicated that the pumping rate of oysters *Crassostrea virginica* was not affected by produced water concentrations of 50,000 ppm. Trays of oysters were placed at several distances from produced water outfalls in five estuarine lakes or bays in Louisiana (Menzel, 1950; Menzel and Hopkins, 1951, 1953). In only one case, the Lake Barre field, where produced water was being discharged at an average rate of 9.5 million l/day, oysters placed 8 m from the outfall suffered heavy mortalities. Oysters located as far as 23 m from the outfall suffered some mortalities. Growth rate of oysters was depressed between 23 and 46 m from the outfall. Attempts have been made to culture penaeid shrimp (Thompson *et al.*, 1972) and oysters (Ogle *et al.*, 1977, 1978) in cages suspended in the water column below offshore production platforms from which produced water was being discharged. No adverse effects of the produced water were observed. However, oysters in the offshore suspension culture grew more slowly and had a higher incidence of fungal infection than oysters from inshore reference sites. These differences were attributed to the higher level of nutrients in the inshore area and the intolerance of the fungal parasite for low salinity water.

On the Buccaneer platform, there was a decreased abundance of fouling organisms, and particularly barnacles *Balanus tintinnabulum* from the surface to a depth of about 3 m on a platform leg immediately below the produced water discharge located 1 m above the water surface (Howard *et al.*, 1980). The abundance of barnacles increased below 1-m water depth and condition index of barnacles collected between 1-2 m below the water surface of the impacted platform leg was not significantly different from that of nearby reference barnacles collected from the same depth interval (Boland, 1980).

Even though the few chronic and sublethal effects studies performed to date indicate that whole produced water is not very toxic, further research definitely is warranted to assess possible subtle effects of chronic produced water discharges on water column and benthic organisms. Particularly important would be experimental studies of possible biological effects of produced water discharges to shallow coastal waters and salt marshes.

Bioavailability of Contaminants from Drilling Fluids and Produced Water

The components in both drilling fluids and produced water of major environmental concern are petroleum hydrocarbons and heavy metals. One of the important questions relating to these materials is whether marine animals can accumulate them in their tissues from water or sediment to concentrations high enough to be toxic to the animals themselves or to higher trophic levels, including human consumers of fishery products.

In water-based drilling fluids containing added diesel fuel, a majority of petroleum hydrocarbons will be sorbed to the bentonite clay fraction of the drilling mud and will be dispersed in the water column with this slowly-settling fraction of the mud (Breteler *et al.*, 1983). Most of the hydrocarbons eventually may desorb from the clay and evaporate to the atmosphere or be degraded by bacteria, or they may be deposited with the clay on the bottom. Similarly, petroleum hydrocarbons in discharged produced water may evaporate, or sorb to suspended particles and be deposited in bottom sediments. Under the extreme conditions of Trinity Bay, Texas, light aliphatic and aromatic hydrocarbons from produced water apparently were carried away and diluted rapidly by dispersion and evaporation, while higher molecular weight hydrocarbons accumulated in bottom sediments near the discharge site (Armstrong *et al.*, 1979).

Hydrocarbons in solution are much more bioavailable to marine organisms than those which are sorbed to bottom sediments or detritus (Rossi, 1977; Roesijadi *et al.*, 1978a,b; McCain *et al.*, 1978; Lyes, 1979; Neff, 1979, 1982a; Augenfeld *et al.*, 1982; Anderson, 1982). The bioaccumulation factor (concentration in tissues divided by concentration in sediments) for aromatic hydrocarbons associated with sediments and detritus usually is less than one, but may be as high as 11. By comparison, bioaccumulation factors from water for aromatic hydrocarbons frequently are in the range of 100 to several thousand (Neff *et al.*, 1976; Roesijadi *et al.*, 1978a).

There have been no published laboratory investigations to date of the uptake of petroleum hydrocarbons from diesel-treated drilling fluids. The only experimental study of uptake of petroleum hydrocarbons from produced water is that of Fucik *et al.* (1977). Clams *Rangia cuneata* placed in trays on the bottom near the produced water outfall of a separator platform in Trinity Bay, Texas accumulated aromatic hydrocarbons to high concentrations in their tissues. When placed in clean sea water in the laboratory, the clams released the accumulated hydrocarbons rapidly. Barnacles, shrimp, some benthic organisms and several species of fish from the vicinity of the Buccaneer platform contained slightly elevated levels of *n*-alkanes, some of which were identified as petrogenic and may have been derived in part from produced water (Middleditch, 1981b).

Elevated concentrations of barium and less frequently chromium, zinc, cadmium, lead and mercury, presumably derived in part from discharge drilling fluids and/or produced water, have been reported in the water or bottom sediments or both in the immediate vicinity of offshore exploratory and production platforms (Ecomar, Inc., 1978; Crippen *et al.*, 1980; Gettleson and Laird, 1980; Tillery and Thomas, 1980; Trocine *et al.*, 1981; Wheeler *et al.*, 1980; Anderson *et al.*, 1981a; Northern Technical Services, 1981; Bothner *et al.*, 1982, 1983; EG&G Environmental Consultants, 1982; Trocine and Trefrey, 1983).

There have been several laboratory investigations of the bioaccumulation of some metals from drilling fluids and drilling fluid ingredients (Brannon and Rao, 1979; Liss *et al.*, 1980; McCulloch *et al.*, 1980; Page *et al.*, 1980; Rubinstein *et al.*, 1980; Tornberg *et al.*, 1980; Espey Huston & Associates, Inc., 1981; Gerber *et al.*, 1981; Carr *et al.*, 1982). A statistically significant accumulation of barium and chromium and an indication of a slight accumulation of copper, cadmium and lead have been demonstrated. In all cases, the magnitude of metal bioaccumulation has been small. The bioavailability of sediment-adsorbed metals generally is very low (Jenne and Luoma, 1977; Bryan, 1983; Luoma, 1983). Recent evidence indicates that the bioavailability of sediment-adsorbed metals and nonpolar organics, such as petroleum aromatics, is inversely related to sediment organic carbon content and directly related to sediment grain size and pollutant concentration (Breteler and Neff, 1983; Neff and Breteler, 1983). Organic carbon concentration in sediments near offshore oil platforms often is elevated due to organic detritus generated by the fouling community and associated fauna of the rig structure (Behrens, 1981). This may result in a decreased bioavailability of sedimented contaminants from rig effluents.

There have been no published reports to date of laboratory studies of the bioaccumulation of metals from produced water by marine organisms. Of particular interest are the natural radionuclides ^{226}Ra and ^{228}Ra. Radium is considered a bone-seeking element and tends to accumulate preferentially in the calcified exoskeleton of marine invertebrates (van der Borght, 1963; Moore *et al.*, 1973) and bones of fish (Holtzman, 1969). In marine and freshwater fish from uncontaminated areas, the concentrations of ^{226}Ra plus its daughters, ^{210}Pb and ^{210}Po, were approximately 20 pCi/100 g bone ash and 2.4 pCi/100 g wet muscle tissue (Holtzman, 1969). Soft tissues of clams, oysters and squid contained about 20 pCi/100 g wet weight. Concentrations of ^{226}Ra in sea water are generally under 1 pCi/l, indicating significant bioaccumulation factors for this radionuclide from the ambient medium. In a freshwater stream contaminated with uranium mill wastes, Anderson *et al.* (1963), reported the following bioaccumulation factors for ^{226}Ra: attached filamentous algae, 500-1000; fish skeleton, 100; fish muscle, 3. These results point to a need for research on the bioaccumulation and food chain transfer of ^{226}Ra and its daughters from produced water in marine and coastal marine ecosystems.

FIELD STUDIES

Laboratory effects studies like those described above provide the basis for making preliminary decisions about whether discharges of drilling fluids, drill cuttings and produced water to the ocean should be permitted (Beller, 1983). They also provide the basis for regulating how these discharges should take place to prevent damage to the environment. If, based on laboratory studies and site-specific environmental considerations, it is concluded that serious damage is unlikely to result from such discharges, field monitoring programs may then be recommended or required. Monitoring programs may be of two types: compliance monitoring, to ensure that discharges are being performed as prescribed in the permit; and surveillance monitoring, to verify that adverse effects do not occur and to detect them if they do, so that mitigative measures can be taken.

There has been a large number of surveillance monitoring programs performed around offshore exploratory and production oil/gas rigs to assess short- and long-term, near-field and area-wide impacts of permitted discharges on the marine environment. These have been sponsored by state and Federal agencies or the petroleum industry in voluntary or mandated programs. Field studies of impacts of drilling fluid and cuttings discharges usually have taken place around exploratory rigs, while studies of produced water discharges, by necessity, are restricted to the vicinity of production platforms. In the latter case, it should be kept in mind, that areas around production platforms probably have in the past and may still be receiving drilling effluents. Thus, apparent impacts of produced water discharges may include impacts of earlier drilling fluid and cuttings discharges.

Exploratory Drilling

Because laboratory studies have shown that most drilling fluids have a relatively low acute toxicity to marine organisms (National Research Council, 1983), and therefore would be diluted to nontoxic concentrations before they could produce adverse effects in water column organisms and because field investigations have shown that discharged drilling fluids are diluted rapidly in the water column as 90% or more of the drilling fluid solids settle rapidly to the bottom (Ayers *et al.*, 1980a,b; Meek and Ray, 1980; Houghton *et al.*, 1980), field studies of the biological effects of drilling fluids have focused primarily on possible impacts on the benthos.

Benthic communities in the vicinity of a C.O.S.T. well in Lower Cook Inlet, Alaska were studied before, during and after drilling (Dames and Moore, Inc., 1978; Lees and Houghton, 1980; Houghton *et al.*, 1981). The well was located in 62 m of water in a dynamic high energy environment characterized by 4-5 m tides and tidal currents in the range of 42-104 cm/s. Drill cuttings and elevated levels of barium were not detected in sediments near the rig. Some changes in benthic communities were observed near the drilling rig during drilling. However, the investigators had difficulty in relocating and resampling stations established during the predrilling survey. Because of this and because of extreme patchiness and seasonality of the benthic fauna in the area, the investigators were unable to demonstrate a statistically significant impact that could be attributed to drilling discharges. Pink salmon fry, shrimp and hermit crabs were suspended in live boxes at 100, 200, and 1000 m downcurrent from the drilling fluid discharge. After four days, there were no mortalities or sublethal effects that could be attributed to the mud discharge plume.

Crippen *et al.* (1980) studied the effects of exploratory drilling from an artificial gravel island on benthic fauna of the Canadian Beaufort Sea. Dredging to obtain materials for construction of the island and subsequent erosion of the island caused changes in local hydrographic conditions, and increased suspended sediment loads and rates of sedimentation such that it was not possible to distinguish effects of drilling fluid discharges from those resulting from island construction.

Crippen *et al.* (1980) also measured concentrations of metals in drilling fluids, sediments and benthic animals from the drilling site. Several metals, including mercury, lead, zinc, cadmium, and arsenic were present at elevated levels in the drilling fluids due to use of an impure grade of barite. Concentrations of these metals, as well as barium, increased in sediments near the rig during drilling. However, no correlation was detected between the concentrations of these metals in the sediments and their concentrations in tissues of benthic animals from the site.

More recently, Northern Technical Services (1981) investigated the effects of above-ice and below-ice disposal of drilling fluids and cuttings on the near-shore benthos of the U.S. Beaufort Sea off Prudhoe Bay, Alaska. Experimental and reference sites were located in 5-8 meters of water. The maximum amount of material collecting on the bottom immediately after both types of test discharges of drilling fluid and cuttings ranged from 1 to 6 cm. Analyses of grain size and metals concentrations in bottom sediments indicated that the drilling fluids and cuttings were swept out of the area rapidly. The abundance of some species of benthic animals changed in the 3 to 6 months after the experimental discharges. In particular, the numbers of polychaete worms and harpacticoid copepods decreased at a discharge site in comparison to a nearby reference site. However, sediment grain size was different at experimental and reference sites and may have been the main factor responsible for the observed differences in seasonal population fluctuations.

Amphipods and bivalve molluscs were placed in live boxes or trays near the discharge sites before the discharges for up to 89 days after the discharge. The amphipods suffered few mortalities. More molluscs died or were missing in the tray from the discharge site than in trays from a reference site. However, the experimental tray had been disturbed, possibly contributing to the differences.

Concentrations of most metals were higher in animals from the reference sites than in those from the disposal sites. Polychaete tubes and macroalgae *Eunephyta rubriformis* from the disposal sites contained elevated levels of barium; however, these values were obtained by atomic absorption spectrometry and may not be reliable. The macroalgae also had slightly elevated concentrations of chromium. Amphipods maintained for 89 days in live boxes near a disposal site had slightly elevated concentrations of copper in their tissues.

Benthic surveys and associated physical and hydrographic measurements were performed immediately before, immediately after and one year after exploratory drilling in New Jersey 18-3 Block 684 on the mid-Atlantic outer continental shelf at a water depth of 120 m off Atlantic City, New Jersey (EG&G, Environmental Consultants, 1982; Gillmor *et al.*, 1981, 1982; Maurer *et al.*, 1981; Menzie *et al.*, 1980). This study focused on effects of drilling discharges on the benthos in the immediate vicinity of and out to a distance of 3.2 km from the rig site. Shortly before the predrilling survey, another exploratory well was drilled approximately 2.8 km north (upcurrent) of the drill site and could have influenced the results of this investigation.

A zone approximately 150 m in diameter of visible drilling discharge accumulations (primarily natural clays from drill cuttings) was observed immediately after drilling just south of the drill site. The center of the pile was nearly 1 m high. Immediately after drilling, elevated levels of clays were detected up to 800 m southwest of the drill site. One year later, elevated levels of clay were not detected in sediments near the drill site. In both postdrilling surveys, concentrations of barium in the upper 3 cm of sediments were elevated (to greater than 1000 ppm, compared to 148-246 ppm in predrilling sediments) near the rig site and decreased with distance from the rig. Elevated concentrations of barium were detected in sediments up to 1.6 km from the drill site. Concentrations of chromium and several other metals were not elevated in sediments near the drill site after completion of drilling.

The abundance of certain motile predatory species, including red hake *Urophycis chuss*, cancer crabs *Cancer* sp. and starfish *Astropecten americanus* increased between the predrilling and first postdrilling surveys in the immediate vicinity and to the south of the well site. These animals may have been attracted to the region by the increased microrelief afforded by the cuttings accumulations (reef effect), or by clumps of mussels *Mytilus edulis* and other fouling organisms that had fallen off the drilling rig or anchor chains. One year later, the abundance of these species had decreased and some species such as the galatheid crab *Munida* sp. showed a gradient of decreasing abundance as the rig site was approached. These animals may have been sensitive to drilling muds and cuttings solids that collected on the bottom, or more likely were subject to increased predation pressure by the large predators which were attracted to the site by the increased microrelief. Within about 150 m of the rig site, sessile benthic epifauna such as sea pens *Stylatula elegans* were subject to burial by drill cuttings. One year after drilling, sea pens were still completely absent from the main cuttings pile, although some were observed among small patches of cuttings away from the main pile.

There was an apparent nearly four-fold decrease in the abundance of macroinfauna throughout the study area between the predrilling and first postdrilling surveys. This decline in abundance was the same for the four major taxonomic groups (polychaetes, echinoderms, crustaceans, and molluscs). At some stations within about 100 m downcurrent of the rig site, there was a very substantial drop in abundance of most species and particularly the burrowing brittle star *Amphioplus macilentus* and polychaetes. These declines near the rig site persisted for at least a year, and because they were associated with elevated levels of clay in the sediments, probably were due to burial and changes in sediment texture by drill cuttings. The area-wide declines in all major taxa were not correlated to gradients of elevated barium concentrations. They probably were due primarily to natural causes or to sampling/analysis errors. The results of the analysis of benthic infaunal samples from the predrilling survey appear anomalous. Benthic faunal abundance and composition, based on samples collected earlier from a nearby Bureau of Land Management benchmark station, Station A3 (Boesch, 1979), were similar to those reported by EG&G, Environmental Consultants (1982) for the two postdrilling surveys, but were different from those of the predrilling survey.

The authors concluded that physical and biological effects of exploratory drilling discharges on the benthic environment of a low-energy area of the mid-Atlantic outer continental shelf persisted for at least a year after cessation of drilling activities. If the Station A3 samples give a better picture than the predrilling samples of the normal benthic faunal composition and density of the area, then there was substantial recovery during the year following cessation of drilling.

Some samples of mixed species assemblages of brittle stars, molluscs, and polychaetes collected during the first and second postdrilling survey, approximately two weeks and one year, respectively, after drilling-related operations were terminated, had significantly elevated concentrations of barium and chromium in comparison to animals collected in the predrilling survey nearly a year before drilling started (EG&G, Environmental Consultants, 1982). The reported increase in mercury concentration in tissues of animals from the first postdrilling survey (Mariani *et al.*, 1980) was later found to be in error (EG&G, Environmental Consultants, 1982). Recalculation of the range of mercury concentrations in postdrilling mollusc, brittle star and polychaete samples revealed that there was not a statistically significant increase in mercury concentration between pre- and postdrilling biota samples.

During both postdrilling surveys, concentrations of barium in tissues of molluscs from the immediate vicinity of the drill site were within the range observed during the predrilling survey. Barium concentrations in tissues of polychaete worms and brittle stars from the vicinity of the drill site were statistically significantly higher in samples from the first postdrilling survey than in those collected before drilling started. Mean barium concentrations in polychaetes and brittle stars were 23.5 and 15.2 ppm, respectively, before drilling, and 87.8 and 217.8 ppm, respectively, during the first postdrilling survey. One year after completion of drilling, barium concentrations in all but a few polychaete and brittle star samples had returned to the predrilling range.

Concentrations of chromium were elevated, compared to the predrilling range, in tissues of polychaetes during the first postdrilling survey, and in tissues of molluscs, polychaetes and brittle stars during the second postdrilling survey. Concentrations of barium and chromium in tissues of benthic organisms were not correlated with concentration gradients of these metals in bottom sediments.

A similar investigation is being performed on Georges Bank, southeast of the Massachusetts coast (Bothner *et al.*, 1982, 1983; Payne *et al.*, 1982, 1983; Battelle/Woods Hole Oceanographic Institution, 1983, 1984). Exploratory drilling began on Georges Bank in July, 1981 and continued to September, 1982. Eight wells were drilled, all reported to be dry holes. No additional drilling is scheduled in the Lease Sale 42 area in the near future.

The monitoring program was designed to determine the fate and effects on the benthos of exploratory drilling activities. Near-field impacts were monitored near two wells in Blocks 312 and 410 in approximately 80 and 140 m of water, respectively. Area-wide impacts were monitored at 15 regional stations on the southern flank of the bank and south and west (downcurrent) of it.

Approximately 750 metric tons of drilling fluid solids containing 500 tons of barite were discharged from the rig in Block 312, and approximately 600 tons of drilling fluid solids containing 250 tons of barite were discharged from the rig in Block 410 (Bothner *et al.*, 1983; Battelle/Woods Hole Oceanographic Institution, 1984). Approximately 16,200 liters of diesel fuel were used in drilling fluids on the rig in Block 312, and Payne *et al.* (1982) estimated that approximately 525 liters of diesel fuel were discharged in drilling fluids. Several samples of drilling fluid collected at different times during drilling in Block 312 contained 23-1130 mg/l total hydrocarbons (Payne *et al.*, 1982). Approximately 1200 tons of drill cuttings were discharged during drilling of each of these two exploratory wells. Neff (1985) estimated that a total of approximately 9200 metric tons of drill cuttings and 5000 metric tons of drilling fluid solids were discharged from the eight exploratory wells on Georges Bank in 1981-1982.

Small amounts of cuttings were detected in the gravel fraction of sediments within about 200 m of the rigs in Blocks 312 and 410, following drilling (Bothner *et al.*, 1982). No evidence of a cuttings pile was observed in any bottom photographs.

Elevated concentrations of barium, and by inference drilling mud solids, were detected in the upper 2 cm of bulk bottom sediments near the two monitored rigs after drilling (Bothner *et al.*, 1982, 1983). The maximum increases in barium concentration in bulk surficial sediments between predrilling and postdrilling surveys occurred in samples collected within 200 m of the two rigs and were 4.7-fold (from 28 to 131.6 ppm) in Block 312 (Figure 10.3) and 5.9-fold (from 32 to 189 ppm) in Block 410. A 1.2-1.5-fold increase in barium concentrations in surficial sediments between predrilling and immediate postdrilling surveys in Block 312 were detected up to 4-6 km downcurrent from the rig site. Barium

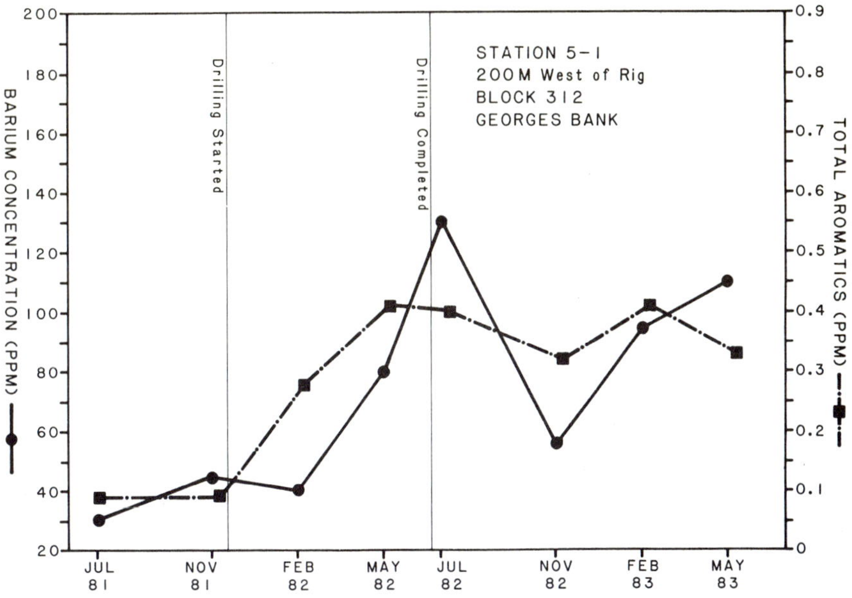

Figure 10.3. Concentration of barium and total aromatic hydrocarbons in sediments approximately 200 m from an exploratory drilling site on Georges Bank, before, during and after exploratory drilling. Aromatic hydrocarbon values for the first four cruises are based on a single analysis of a pooled sample of three replicates. Other data points are the average of three replicate samples (data from Bothner *et al.*, 1983 and Payne *et al.*, 1983 as summarized by Battelle/Woods Hole Oceanographic Institution, 1984).

concentrations in these samples had returned to near background levels ten months after termination of drilling. No increments were detected in the concentration of chromium or other metals in bulk surficial sediments.

In the silt-clay fraction (representing less than 5% of the total mass of most sediment samples), an increment in barium concentration associated with drilling was detected up to 65 km west (downcurrent) of the rig in Block 312. Barium concentration in the clay portion of sediment at this station rose from 240 ppm in July 1981 to 1014 ppm in May 1983. Some of this barium could have been derived from the other exploration rigs on Georges Bank. Bothner *et al.* (1983) estimated that the half-time for the retention of discharged barite within 6 km of the drill site in Block 312 is 4.8 months. Payne *et al.* (1983) reported a small but statistically significant increase in the concentration of aromatic hydrocarbons (by UV fluorescence) in sediments collected within 200 m of the rig between February and July, 1982 (Figure 10.3).

Sediment grab samples for biological analysis have been collected four times per year on a seasonal basis from 46 sampling stations upcurrent, in the vicinity, and downcurrent of the lease blocks (Battelle/Woods Hole Oceanographic Institution, 1983, 1984). The first sampling took place in July, 1981 before drilling began.

Drilling began in Block 410 in July, 1981 and continued until March, 1982. With the methods of analysis used thus far, no biological impacts which could be attributed to drilling activities were detected. Differences between stations were always greater than temporal differences at any one of the three stations.

Drilling began in Block 312 on December 8, 1981 and continued until June, 1982. At the site-specific array of stations in this block, the separation of February, 1982 and May, 1982 samples into discrete clusters may be a result of the decline in total densities at many of the stations in February, followed by a recovery in May. The density declines in February may be related to changes in sediment composition due to accumulation of drill cuttings or to a severe winter storm shortly before the February cruise, or to normal seasonal population cycles. An analysis of the change in densities over time of 24 infaunal species revealed that at stations near the rig site where the greatest increment in barium concentration between July, 1981 and May, 1982 occurred, the densities of many species declined in November, 1981 before drilling began and increased in February, 1982 shortly after the start of drilling (Figure 10.4). At several stations 1 to 2 km from the drill site where smaller amounts of barium accumulated, species densities declined in February, 1982 and increased in May, 1982. Those species which showed the most marked declines in density in November or February, such as the amphipods *Erichthonius rubricornis* and *Unciola inermis*, showed substantial recovery by May or July, 1982.

In an area of fine-grained sediments approximately 200-250 km west southwest (downcurrent) of the drill sites on Georges Bank, thought to be a major depositional area for sediments swept off the Bank (Bothner *et al.*, 1981; Twichell *et al.*, 1981), there was a marked decline in the abundance of individuals and species of benthic infauna between February and May, 1982. However, both parameters returned to near the February, 1982 values by July, 1982. There was

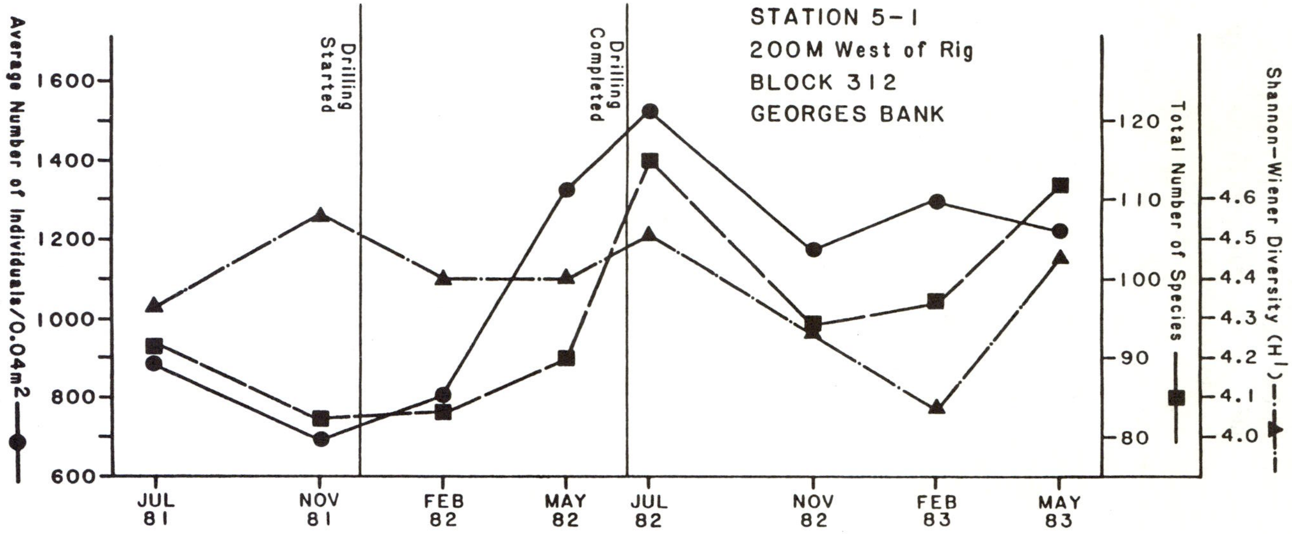

Figure 10.4. Shannon-Wiener diversity, total number of species per 0.04-m^2 (6 replicates combined) and average number of individuals per 0.04-m^2 of benthic infauna approximately 200 m from an exploratory drilling site on Georges Bank, before, during and after exploratory drilling (from Battelle/Woods Hole Oceanographic Institution, 1984).

not a similar decline in the number of individuals and species in this area in May, 1983. Since there was no discernible increase in concentrations of barium (concentration range 240-300 ppm; Bothner *et al.*, 1982, 1983), petroleum hydrocarbons (concentration range 1.0-2.5 ppm total aromatics; Payne *et al.*, 1983) or other tracers of drilling discharges in the sediments of this area during or after drilling, the reason for this decline is unexplained.

Samples of ocean quahogs *Arctica islandica* and four-spot flounder *Paralichthys oblongus* collected from several stations in the lease area on all sampling cruises showed no indication or trend toward increasing tissue body burdens of petroleum hydrocarbons or metals (Al, Ba, Cd, Cr, Fe, Hg, Pb, Ni, V, and Zn) (Payne *et al.*, 1982, 1983).

Benech *et al.* (1980) studied fouling communities on submerged pontoons of a semi-submersible drilling rig off southern California. Pontoons within 10 m downcurrent of the mud/cuttings discharge had different fouling communities than pontoons not exposed to drilling mud, but otherwise located similarly to the impacted pontoons with respect to water depth, currents and exposure to sunlight. Differences were attributed primarily to sedimentation of silt-clay sized particles and not to differences in light or exposure. Species intolerant of fine-grained sediments disappeared. Some sediment-tolerant species increased in abundance. Effects were highly localized.

In order to protect the coral reefs of the Flower Garden Banks off the Texas-Louisiana coast, NPDES permits for exploratory drilling required that all drilling discharges be shunted to within 10 m of the bottom. Shunting resulted in a temporary increase in the amount of mud and cuttings accumulating on the bottom immediately under the discharge pipe (Gettleson, 1978; Gettleson and Laird, 1978). Although some of the discharged drilling fluid and cuttings solids were distributed by currents to distances in excess of 1000 m from the rig, there was no evidence that any discharged solids reached the coral reef zone, which was shallower than the depth of the discharge. No evidence was found of adverse effects of the drilling discharges on corals.

Boland *et al.* (1983) studied the effects of installation of and initial development drilling from a production platform adjacent to the East Flower Garden coral reefs on the reef fish fauna. The platform is located in 120 m of water and approximately 1500 m southeast of the nearest coral reef habitat and 750 m from the nearest live bottom and partially drowned reef habitat areas occurring above the 84-m isobath. The authors concluded that, based upon abundances of individuals and species of reef fishes before and during development drilling, and analysis of distribution patterns of fish around the reef and platform, the discharge of mud and cuttings near the bottom by shunting did not result in any measureable impacts on the spatial density patterns or overall population levels of reef fish. However, the platform structure did provide a substrate for development of a rich fouling community and did attract fish to the area.

The drilling of two exploratory wells very close to one another directly on a coral reef in shallow water off Palawan Island, Philippines caused serious but localized damage to the reef (Hudson *et al.*, 1982). Between 70 and 90% of the foliose, branching and plate-like corals were killed in an area 115 by 85 m around the well heads, probably due to burial by or toxicity of drilling discharges. Communities of small organisms living in crevices and cavities among the coral heads (coelobites) were badly damaged within 40 m of the well heads (Choi, 1982). Lesser changes in coelobite community structure were observed up to 100 m away from the well head. Non-coral animals living on the reef were less seriously affected.

The results of the field studies performed to date support the following conclusions. The severity of impact of drilling fluids and cuttings is directly related to the amount of material accumulating on the substrate, which in turn is related to the amount and physical characteristics of the materials being discharged, and to environmental conditions at the time and site of discharge, such as current speed and water depth. In high-energy environments, little mud and cuttings accumulate and impacts on the benthos are minimal and of short duration. In low-energy environments, more material accumulates, and there may be reductions in abundance of some benthic species due to burial, incompatibility with clay-sized particles, or chemical toxicity of drilling fluid or cuttings components. Impacts of drilling mud and cuttings discharges may be localized and/or patchy in distribution and may be difficult to distinguish from effects of other local changes due to the drilling activity, such as the rain of organic material from the fouling community on the rig and increased predator pressure due to the reef effect (Davis *et al.*, 1982), or sea bed scour around drilling structures placed on the bottom (Carstens, 1976, 1983). Benthic marine animals exposed to drilling fluid solids in the sediments near the rig are unlikely to accumulate sufficient drilling fluid or cuttings-associated metals or hydrocarbons to represent a toxicity hazard to themselves or to prey organisms, including humans.

Development and Production

Mackin (1971) studied the effects of production facilities, and particularly treated produced water effluents, from 13 oil fields along the Texas coast, on estuarine benthic communities of eight Texas Bays. He observed no effects around produced water outfalls in Lavaca and Matagorda Bays. He observed minor highly localized effects in several other bays. In Trinity Bay, part of the Galveston Bay system, he reported a zone of severely depressed abundance and diversity of benthic infauna extending up to 106 m from the submarine outfalls of oil-water separators on the Fishers Reef C-2 platform and the Trinity Bay F-1 platform. From about 150 to several hundred meters downcurrent from the outfalls, there apparently was a zone of enhanced faunal abundance and diversity. The water depth at the platforms is 2-3 m, and the water is highly turbid due to inflow of clay-laden sediments from the Trinity River. The rates of produced water discharge during Mackin's study were 875,000 and 1,810,000 l/day from the F-1 and C-2 platforms, respectively, and produced water salinities were 108 and 63 g/l, respectively. Mackin (1971) did not include any chemical analyses of petroleum hydrocarbons or metals in his study.

Therefore, the study around the C-2 platform in Trinity Bay was repeated in an attempt to verify Mackin's ecological observations and correlate them to the distribution of hydrocarbons in sediments around the produced water outfalls (Armstrong *et al.*, 1977, 1979). During the 20-month time course of this study, produced water with a mean total hydrocarbon concentration of 15 ppm was discharged from the separator platform through an outfall one meter above the bottom at a rate of 650,000 to 1,590,000 l/day. Hydrocarbons were diluted nearly 2500-fold in the water column within 15 m of the outfall (see Chapter 4). Bottom sediments were heavily contaminated with medium molecular weight alkanes (C_{10}-C_{28} *n*-paraffins) and aromatics (C_3 benzenes-trimethylphenanthrenes).

There was a gradient of decreasing naphthalenes concentrations from a mean of about 21 ppm 15 m from the outfall to background 500 to 4800 m from the outfall, depending on direction. There was an inverse gradient of numbers of organisms and numbers of species of benthic infauna with distance from the outfall. Within 15 m of the outfall, the bottom was almost devoid of organisms. Benthic fauna were significantly reduced to approximately 150 m in all directions from the outfall. At stations located 685-1675 m from the outfall, there was an apparent enhancement of the benthic fauna, with greater numbers of individuals and species at these stations than at reference stations 4000-5800 m from the outfall. Thus, the 150 m radius zone of adverse impact with an apparent zone of enhanced faunal abundance and diversity farther out from the C-2 separator platform discharge, first observed by Mackin (1971), was confirmed and correlated to contamination of sediments with petroleum hydrocarbons. Armstrong *et al.* (1979) estimated that a nominal concentration greater than about 2 ppm total naphthalenes was necessary to significantly reduce benthic infaunal populations of Trinity Bay.

Results of these investigations should be extrapolated to offshore situations with extreme caution. The shallow turbid nature of the receiving waters is unlike the situation encountered offshore, with the possible exception of some nearshore areas of the Beaufort Sea. Where water depth is greater and suspended sediment concentrations are lower than those encountered in Trinity Bay, a much smaller fraction of the hydrocarbons in the discharged produced water will be deposited in bottom sediments near the outfall, and adverse effects on the benthos will be much less severe.

There have been three major multiyear multidisciplinary investigations of the impact of intensive long-term offshore oil and gas development and production on the marine environment of the northwestern Gulf of Mexico. These are the Offshore Ecology Investigation, the Central Gulf Platform Study, and the Buccaneer Gas and Oil Field Study. All were ostensibly designed to determine if 10-25 years of intensive offshore oil and gas development and production activities had adversely affected the physical and biological environment of the northwestern Gulf of Mexico. These studies have been discussed in detail in Chapters 9 and 14.

The Offshore Ecology Investigation was performed in eight field sampling efforts conducted on a seasonal basis in 1972-1974. The program was coordinated by the Gulf Universities Research Consortium and consisted of 23 different monitoring tasks and principal investigators (Menzies *et al.*, 1974; Mertens, 1976; Sharp and Appan, 1978; Ward *et al.*, 1979). The program was designed to monitor near-field impacts near platforms in Timbalier Bay, Louisiana and directly offshore, and area-wide impacts in Louisiana coastal and offshore waters.

Elevated concentrations of hydrocarbons were detected in the surface microlayer (air/sea interface) and in bottom sediments, but not in the water column in the study area. The investigators concluded that most of the hydrocarbons were of biogenic origin or from fuel oil or lubricating oil. Concentrations of mercury, zinc, cadmium and lead were higher in water from Timbalier Bay than in offshore water. Only zinc showed a tendency toward a decreasing concentration gradient with increasing distance from platforms. The zinc may have been derived from corrosion of rig structures.

Sediments immediately adjacent to one offshore platform had elevated levels of coarse material identified as drill cuttings. Concentrations of barium, but not other metals, were higher in sediments from Timbalier Bay and from the vicinity (to 6 km) of offshore platforms. In a recent report, Chan and Hanor (1982), reported that concentrations of dissolved barium in water samples collected offshore Louisiana during the Offshore Ecology Investigation were higher than concentrations in water from the open Gulf of Mexico. They suggest that some of the excess barium could be from drilling fluid and produced water discharges or, more likely, from desorption of barium from Mississippi River-borne suspended sediments during estuarine mixing. Some of the excess barium in sediments west of the Mississippi delta, including the study area, could be derived from precipitation of barium from the Mississippi outflow.

Fairly extensive studies were performed of phytoplankton, phytobenthos, zooplankton, benthic meiofauna and macrofauna, demersal fish and crustaceans, and the biofouling community. The conclusions of different principal investigators in their unpublished final reports varied. Waller (1974) reported that stations near platforms had depressed benthic populations compared to nearby reference stations. Other investigators found no differences in benthic communities between production and nonproduction areas (Kritzler, 1974; Farrell, 1974a,b; Fish *et al.*, 1974). The conclusion of the consensus report (Menzies *et al.*, 1974) was that there were no significant ecological changes due to oil industry activities and that the Louisiana coastal environment was in good "ecological health."

Bender *et al.* (1979) and Sanders (1981) have challenged this conclusion for several reasons (see also the critiques of this study in Chapters 9 and 14). The most important relate to the choice of study site. The study area is so subject to wide variations in salinity, temperature, suspended sediment concentrations, and inputs of materials, including anthropogenic contaminants, from the Mississippi River that detection of physical, chemical or biological changes that can be attributed to oil and gas production activities is practically impossible. The designers of the program apparently did not clearly identify and physically and chemically characterize the major perceived impact-producing agents associated with offshore oil and gas development activities. If this had been done at the outset and the results of the physical/chemical analysis compared to a similar analysis of the Mississippi River inflow, a substantially different program might have emerged.

The Central Gulf of Mexico Platform Study was performed in the same general area as the offshore study sites in the Offshore Ecology Investigation and was designed in part to answer many of the questions left from the earlier study (Bedinger, 1981). However, it suffered from many of the problems of the Offshore Ecology Investigation.

Stations between 100 and 2000 m of four primary and 16 secondary platforms were sampled and compared to samples from four control areas during three consecutive seasonal cruises during 1978 and 1979.

Gaseous hydrocarbons (C_1-C_4 alkanes) were detected in water samples near some platforms at concentrations significantly higher than in water from the open Gulf (Nulton *et al.*, 1981). They were ascribed to pipeline leaks, produced water discharges and underwater venting of gas. Volatile liquid hydrocarbons, which are the dominant hydrocarbons in produced water, were not analyzed in water samples. Substantially elevated levels of organic carbon were not detected in any sediment samples. Concentrations of high molecular weight hydrocarbons in sediments were variable. Most of the hydrocarbons were derived from pyrogenic or biogenic sources. Petroleum hydrocarbons (based on presence of an unresolved complex mixture and alkylaromatics) were detected in sediment samples from six locations. In a few cases, gradients of decreasing sediment hydrocarbon concentration with increasing distance from platforms was discerned. Low concentrations (1-20 μg/kg per compound) of aromatic hydrocarbons, probably derived from both petrogenic and pyrogenic sources, were detected in 47% of epifaunal invertebrates and demersal and pelagic fish analyzed. No cases of massive contamination of biota with petroleum hydrocarbons were observed.

Several metals commonly associated with drilling fluids, produced water or corrosion of rig structures (Ba, Cd, Cr, Cu, Fe, Ni, Pb, V, and Zn) were analyzed in sediments, epifaunal invertebrates and demersal and pelagic fish (Tillery *et al.*, 1981). Sediments collected within 100 m, but not elsewhere, of some platforms contained elevated concentrations of Ba, Cr, Cu, Pb and Zn. There was no relation between level of sediment contamination and platform age, number of wells or production volumes. Any evidence of far-field distribution of metals from platform activities was completely masked by metal inputs from the Mississippi River. Concentrations of metals in tissues of marine fauna near the platforms were within the normal range for marine animals and there was no evidence of metal bioaccumulation from drilling or production discharge sources.

Interpretation of the biological data was confounded by three factors: the influence of the Mississippi River outflow on the whole study area; a major tropical storm following the second sampling cruise; and a large area of hypoxic bottom water, "dead bottoms," extending over 54-63% of the sampling stations on the first two cruises.

An increased incidence of histopathological conditions, including parasitism, was observed in several species of fish and, in particular in spadefish *Chaetodipterus faber*, around those platforms in the eastern part of the study area that were more heavily contaminated with hydrocarbons and metals (Sis *et al.*, 1981). Because benthic and demersal invertebrates were absent or severely stressed in areas of hypoxic bottom waters, correlations between incidence of histopathology and distance from production platforms could not be discerned.

Fairly extensive studies were performed of benthic microbiota, meiofauna, macroinfauna, macroepifauna and demersal fishes (Baker *et al.*, 1981; Brown *et al.*, 1981a,b). In a few cases, microbial biomass was higher near platforms than in control areas. Microbes in surficial sediments (top 1-2 cm) throughout the study area had similar oil-degrading abilities. The benthic and demersal meio- and macro-biotal communities, throughout the study area showed evidence of stress from extreme natural environmental conditions. Generally, species diversity of meiofauna, macroinfauna and macroepifauna was higher near the four primary platform sites than at reference sites. An inverse correlation was observed between concentration of unsaturated hydrocarbons (mostly aromatics) in sediments and abundance of a few species of macroinfauna and demersal invertebrates and fish.

Although the designers of this investigation recognized the importance of the Mississippi River in influencing physical, chemical and biological processes in the study area (Bedinger *et al.*, 1981), no attempt was made to quantify the input of materials, including petroleum hydrocarbons and metals, to the study area from the Mississippi. An estimated historical mass balance of inputs of materials from the platforms investigated and from the Mississippi River to the study area at the outset of the investigation would have been useful in designing a more meaningful chemical sampling and analyses program. Since samples were collected once in each of three different seasons, it was difficult to distinguish normal seasonal cycles from effects of catastrophic events such as major storms and hypoxic events. Fewer types of samples and analyses (based on reasonable hypotheses of the sites and modes of impact of offshore development and production activities) taken on a seasonal basis over two-three years would have yielded more meaningful results than the intensive "measure everything" approach taken in this study.

The Buccaneer Gas and Oil Field Study was initiated with a pilot study in 1975 (Harper *et al.*, 1976) and was continued into 1980. The results of the first two years of this program apparently were used to help design the Central Gulf Platform Study. The Buccaneer Field, located 50.5 km south of Galveston, Texas in approximately 20 m of water, had been in production for nearly 15 years (Middleditch, 1981a). Development of the field began late in 1963, and two production platforms with 15 wells each were installed and completed between 1964 and 1966. Although the Buccaneer Field was chosen as a typical oil and gas production field, it actually is primarily a gas field with only a small additional production, primarily of condensate. The rate of treated produced water discharge to the ocean ranged from about 19,000 to 318,000 l/day during the study, less than that from most other fields of similar size in the northwestern Gulf of Mexico. Another reason for choosing this field was that it was relatively isolated from other major offshore development areas in the northwestern Gulf and away from the influence of major fluvial inputs. Although an initial design objective of the program was to focus more narrowly than in the past on an in-depth study of impacts of the field on the marine environment, the final program actually consisted of 23 different research/monitoring tasks with different principal investigators. Many of the results of this program have been summarized in a book (Middleditch, 1981a), a major review on produced water (Middleditch, 1984) and many journal publications. The study also was reviewed by technical representatives of the sponsoring agency (Caillouet *et al.*, 1981) (see also Chapters 9 and 14).

The produced water from the field contained about 20 ppm total petroleum hydrocarbons, a mean of 1.2 ppb benzo(a)pyrene, and 460 ppm elemental sulfur (Middleditch, 1981b). Based on a mean produced water discharge rate of 159,000 l/day, input rates of total petroleum hydrocarbons, benzo(a)pyrene, and sulfur to the ocean would be about 3 kg, 0.19 g and 73 kg per day, respectively.

In the water column near the produced water discharge, gaseous hydrocarbons were at or near background concentrations, and volatile light hydrocarbons (mostly benzenes) were present at concentrations up to 65 ppb (Brooks *et al.*, 1980). These light hydrocarbons were diluted by 104 to 105 within 50 m of the outfall. The concentration of total hydrocarbons (as alkanes) in the water column rarely exceeded 30 ppb (Middleditch, 1981b). Based on analysis of high molecular weight alkanes only, Middleditch (1981b) was able to detect evidence of petroleum contamination in sediments within about 20 m of the platform. The maximum concentration of *n*-alkanes in sediments was about 17 ppm, much of it biogenic. The mean concentration of elemental sulfur in sediments throughout the study area was 3.32 ppm. Middleditch (1981b) also analyzed C_{12}-C_{36} alkanes in tissues of a wide variety of species of zooplankton, fouling organisms from submerged platform structures, benthic macrofauna, shrimp and demersal and reef fish. Several of the organisms or mixed zooplankton assemblages were contaminated with alkanes of apparent petroleum origin. Large and variable fractions of the tissue body burdens of alkanes were of biogenic origin, making it difficult to clearly establish patterns of petroleum hydrocarbon contamination in the biota of the Buccaneer Field, based on alkane concentrations alone. The ecological significance of alkane contamination is questionable, since these compounds are relatively inert biologically and are readily degraded by microbes. A survey of the more toxic and persistent medium molecular weight aromatics and sulfur heterocyclics (naphthalenes, phenanthrenes, dibenzothiophenes) in the biota would have been more useful.

Barium, cadmium and strontium were detected at elevated concentrations in sediments near the platforms (Wheeler *et al.*, 1980). Since some bentonite clay was also identified, the barium and strontium may have been derived from drilling fluids. Produced water is another potential source of these alkaline earth metals. Cadmium probably was derived from corrosion of platform structures. The field is in an area characterized by a complex and dynamic sedimentary environment (Anderson *et al.*, 1981a,b). Waves and wind-driven currents frequently resuspend bottom sediments and carry them out of the area. Behrens (1981) estimated, based on geochemical and radioisotope techniques, that erosion of from one to two meters of sediment had taken place near and up to 1.6 km from the platforms since their installation.

Probably in large part because of the dynamic nature of the near bottom environment, biological impacts of the production field were of a low order of magnitude and difficult to ascribe to particular causal agents. Oil-degrading, sulfur-oxidizing and sulfate-reducing bacteria were more abundant in surficial sediments in the field than in sediments from a reference area 9 km to the north (Sizemore *et al.*, 1981). Otherwise, bacterial populations in the sediments and water column did not show any consistent differences between the study area and a reference area.

Fouling communities on submerged platform structures were extremely dense and dominated by the barnacle *Balanus tintinnabulum* (Fotheringham, 1981). This community was adversely affected only within about 1 m of the produced water outfall. Benthic meiofauna and macrofauna within 100 m of the platforms were characterized by reduced abundance of individuals and species and a slightly different species composition than stations further from the platforms (Harper *et al.*, 1981). This could have been due to the differences in sediment texture and consolidation under the platforms due to scouring, to increased predation from demersal and reef fish and macrocrustaceans attracted to the platform site (Gallaway *et al.*, 1981), or to toxicity of contaminants discharged from the platforms. Although Harper *et al.* (1981) concluded that toxicity of discharged contaminants was the major cause of the altered benthic populations near the platform, they did not attempt to correlate degree or gradients of sediment contamination to magnitude of biological impact.

Several studies were performed on the environmental impact of production platforms Hilda and Hazel located about 3.7 km south of Loon Point in the Santa Barbara Channel, California (Bascom *et al.*, 1976; McDermott-Ehrlich *et al.*, 1978). The platforms are located in about 31 m of water and 24-25 wells were drilled from each. Five to seven years after installation of the platforms, concentrations of hexane-extractable materials, volatile solids, copper and zinc were elevated slightly in some bottom sediment samples collected directly beneath the platforms. Elevated concentrations of highly weathered petroleum also were detected in sediment samples under and adjacent to the platforms. However, tissues of two species of fish, *Sebastes auriculatus* and *S. vexillaris*, crabs *Cancer anthonyi*, and mussels *Mytilus californianus* did not contain concentrations of 11 metals (Ag, Cd, Cr, Cu, Fe, Mo, Ni, Pb, Si, V, and Zn) above normal background levels and did not contain detectable levels of petroleum hydrocarbons.

The cuttings piles under the platforms are about 5 m high and are covered by about 50 cm of mussel shells. More than 200 species of invertebrates were observed on and near the cuttings piles. Many of these were typical of nearby rocky intertidal areas. Large mussels *Mytilus californianus* and sea anemones *Metridium senile* were abundant on the cuttings pile, while crabs *Cancer anthonyi*, bat stars *Patiria miniata*, and sea cucumbers *Parastichopus* sp. inhabited the nearby bottom. A total of 77 species of benthic infaunal and epifaunal species of polychaetes were identified. Fish were extremely abundant under both platforms, with visual estimates ranging from 8000 to 30,000 at different times of year. Totals of 36 and 44 fish species were identified under platforms Hazel and Hilda, respectively, compared to 7 and 21 species at nearby soft bottom and hard bottom reference stations, respectively. Obviously the rigs were acting as artificial reefs, providing habitats for a wider variety of marine animals than occurred on nearby hard and soft bottoms.

Wolfson *et al.* (1979) studied marine communities around platform Eva, located in 18 m of water off Huntington Beach, California. The submerged rig structures have a dense fouling community dominated by mussels, *Mytilus edulis* and *M. californianus*. The authors estimated that approximately one cubic meter (one metric ton) of mussels fall from the platform each day. The fallout of mussels and other fouling organisms supports a very dense population of benthic epifauna, particularly sea stars (six species with an estimated density of 29 individuals/m^2). The sandy substrate adjacent to the shell pile is dominated by the tube-dwelling onuphid polychaete *Diopatra ornata*. Some species of benthic infauna such as the polychaete *Capitata ambiseta* and juvenile brittle stars (family Amphiuridae) are most abundant near the shell pile and abundance decreases with distance from the platform. Others, such as the polychaete *Typosyllis armilaris* and clam *Tellina modesta*, show the opposite trend. Although analyses of contaminants were not made, it is apparent that the major impact of the platform was one of biological enrichment due to a reef effect (Davis *et al.*, 1982).

Several ecological investigations have been performed around production platforms in the North Sea. Biological monitoring of benthic communities in the Ekofisk oil field was performed in August, 1973, 1975 and 1978 (Dicks, 1975; Addy *et al.*, 1978). The field is located in the Norwegian sector of the North Sea in about 70 m of water. The major objective of the study was to assess the impacts on the benthic community of ballast water discharges from the one-million barrel oil storage tank which was emplaced on the sea bottom in the center of the field shortly after the 1973 survey.

Several measures of petroleum hydrocarbon contamination (total organic extractables, total saturated and unsaturated hydrocarbons, unresolved complex mixture, and nC_{18}/nC_{29} ratio) were used to document contamination of sediments. Generally, there was a steep gradient of declining concentrations of petroleum in sediments with distance from the storage tank and platforms in 1977. It should be pointed out that a blowout occurred on April 27, 1977, approximately four months before the August, 1977 survey of Addy *et al.* (1978), on Ekofisk Bravo Platform and in the seven days before it was capped, released an estimated 20,000 tons of crude oil (Grahl-Nielsen, 1978). This may have contributed to the petroleum contamination of sediments reported by Addy *et al.* (1978). It is not known whether oil-based drilling fluids were used to develop the Ekofisk Field. No analyses were performed of metals in sediments. It is interesting to note that the median particle diameters for sediments collected from all stations in 1973 by Dicks (1975) were higher (range 169-199 μm) than those for sediments collected from the same stations in 1977 by Addy *et al.* (1978) (range 130-151 μm).

The composition and abundance of the benthic community was relatively uniform throughout the study area in 1973 (Dicks, 1975). By August, 1975, depressed benthic communities were observed near some platforms. In 1977, affected stations extended out to about 2.5 km from the storage facility. Both numbers of individuals and numbers of species were depressed near the facility. Abundance of several dominant species such as the polychaetes *Myriochele oculata* and *Owenia fusiformis* and the ophiuroid *Amphiura filiformis* was severely depressed near the storage facility and showed a gradient of increasing abundance with distance from the facility. Other species, such as the polychaetes *Chaetozone setosa* and *Pholoe minuta* and juveniles of the clam *Arctica islandica*, showed the opposite distribution trend with highest densities near the central storage facility. There was a good inverse relationship between sediment hydrocarbon concentration and abundance of *Myriochele oculata*. However, as Spies (Chapter 9) has shown, there also was a good inverse correlation between the percent fines in sediments and the abundance of this dominant polychaete. In any event, it is apparent that the various structures in the Ekofisk field and the effluents discharged from them, either intentionally or accidentally, are having a localized effect on the benthos. The severity and areal extent of the impact seems quite small.

Hartley and Ferbrache (1983) recently reported the results of a similar benthic monitoring study in the Forties Field located in the British sector of the North Sea in 100-125 m of water approximately 177 km northeast of Aberdeen, Scotland. The 80 wells drilled to date from four platforms were drilled with water-based muds. Rate of produced water discharge was not stated. It is expected to increase with the age of the field (Read, 1978), and oil-water separators with a nominal capacity of 40 million l/day are being constructed for each platform.

Production from the field began in September, 1975, and three benthic surveys have been performed to date in June, 1975, before production began, and in 1978 and 1981. Concentrations of aliphatic hydrocarbons in sediments measured by gravimetric techniques, have shown a slight rising trend from a mean of 5.7 ppm in 1978 to 8.9 ppm in 1981. Davies *et al.* (1981), using a fluorometric technique, reported higher values for total sediment hydrocarbons in the Forties Field (about 10-60 ppm). Grain size distribution of sediments did not change significantly at any sampling stations during the three surveys. No analyses were performed of metals in sediments.

The relative abundance and composition of benthic fauna in the study area were influenced by water depth and percent silt-clay in the sediments, both of which increased from east to west in the study area. The fauna was rich and diverse throughout the area. However, directly beneath the Platform C and to about 100 m to the west, the benthic macrofauna were severely depressed and there was evidence of hydrocarbon contamination (305-470 ppm aliphatics), possibly from diesel fuel. The abundance of the polychaete *Chaetozone setosa* was much higher near Platforms A and D than at other stations. Abundance of another opportunistic polychaete, the capitellid *Capitomastus minimus*, increased in muddier sediments to the west of the rigs between 1978 and 1981. Rate of uptake and mineralization of ^{14}C-naphthalene by sediment microbiota increased with decreasing distance from platforms in the Forties Field (Saltzmann, 1982). In addition, three species of demersal fish (cod, whiting and haddock) had slightly elevated activity of hepatic aryl hydrocarbon hydroxylase (aromatic hydrocarbon detoxification enzyme system) compared to the same species collected in a reference area about 30 km west of the field (Davies *et al.*, 1981). Overall, it is apparent that impacts of oil production activities from the Forties Oil Field have been localized (primarily within 450 m of the platforms) and of low magnitude.

The Beatrice field lying in Moray Firth in northeast Scotland is only about 19 km from shore (Ferbrache, 1983). A heavy crude oil is produced from two platforms and transported to shore by pipeline. Four benthic surveys were performed in the area: June 1977 (before production), May 1980, May 1981, and February 1982. Oil-based drilling fluids were used after the February 1982 survey.

There were increases in the concentration of barium, chromium, lead and zinc at some stations within 250 m of the platforms in 1981 and particularly 1982. The maximum sediment barium concentration measured was 660 ppm, more than ten-fold above the normal background concentration in the area. Immediately under and out to about 250 m from the platforms, concentrations of total oil were elevated in 1982 (to several hundred ppm). Slightly elevated sediment hydrocarbon concentrations extended out to about 750 m in some directions.

The area supports a very rich and diverse benthic fauna. There were marked gradients of infaunal community structure and abundance over the area which were attributed to gradients of water depth and sediment type. In the 1982 survey, stations were sampled within 250 m of the platforms. Abundance of certain benthic infaunal species (e.g., *Virgularia mirabilis*, *Pholoe minuta*, *Scoloplos armiger*, *Spiophanes bombyx* and *Tellina fabula*) declined near the platforms. The abundance of other species, characteristic of disturbed or polluted habitats (e.g., *Perioculoides longimanus*, *Goniada maculata*, *Chaetozone setosa* and *Caulleriella* sp.), were more abundant near the platforms. The area of depressed and altered benthic infauna extend out for 250-750 m from the platforms. Submerged portions of the platforms themselves support a rich and diverse fouling community (Forteath *et al.*, 1983).

In the Buchan field, where 12 wells were drilled with water-based drilling muds between 1974 and 1981, a floating production facility transfers the oil via a subsea export line to a CALM loading buoy (Ferbrache, 1983). No produced water is discharged. Two benthic surveys have been performed in the area. The closest station to the platform was at 600 m and the closest station to a well site was at 400 m. No biological impacts of the platform or production activities have been detected to date.

Many of the major oil fields in the Norwegian and British sectors of the North Sea were developed with oil-based drilling fluids. The oil-based fluids themselves are not discharged to the ocean; however, drill cuttings are discharged. Although they are washed before discharge, they may still contain significant amounts of adsorbed hydrocarbons. In the Statfjord field in the Norwegian sector of the North Sea, diesel fuel was used as a detergent to wash oil-based drilling fluid from cuttings (Schreiner, 1978). The amount of hydrocarbons remaining on the washed cuttings following this procedure is not known. The amounts of petroleum hydrocarbons discharged with drill cuttings and their subsequent accumulation on the bottom have been documented by Grahl-Nielsen *et al.* (1980), Davies *et al.* (1981) and Law and Blackman (1981), and the environmental impacts of such discharges recently were reviewed by Davies *et al.* (1983).

In 1981-1982, 432 wells were drilled on the British outer continental shelf and about 40% of these wells used oil-based drilling fluids. The estimated amount of diesel fuel hydrocarbons discharged per well with drill cuttings is 91.6 metric tons (89,000 liters) or a total of 14,000-17,000 tons for the two years. The cuttings pile under such platforms may contain 10-15 weight percent oil. Generally, there is a steep gradient of decreasing sediment hydrocarbon concentration with distance from the platform, with background concentrations usually reached within 3000 m, and exceptionally at as much as 18.5 km where fine-grained cuttings are discharged.

Davies *et al.* (1983) identified five zones of chemical and biological impact around platforms discharging diesel-contaminated drill cuttings. Zone I, located immediately under the platform and extending out to 250 m and exceptionally to 500 m from the platform, is characterized by hydrocarbon concentrations in excess of 1000 times background and a severely impoverished and modified benthic community. Zone II is a transition zone, extending roughly from 200-2000 m from the platform. Sediment hydrocarbon concentrations are 10-700 times background, and species diversity and abundance increases with distance from the platform. Opportunistic species such as *Capitella capitata* and *Goniada maculata*, followed at greater distances by *Chaetozone* and *Caulleriella*, reach peak abundance in this area. Zones III and IV have normal benthic communities and decreasing gradients of hydrocarbon contamination. The authors conclude that concentrations of diesel fuel greater than about 100 ppm (or total naphthalenes concentrations greater than about 2 ppm) produce significant adverse effects in the benthos. None of the monitoring studies have continued long enough after cessation of oil-contaminated cuttings discharges for patterns of benthic recovery to be discerned. Davies *et al.* (1983), concluded that the rate of recovery of areas impacted by cuttings would depend on: rates of redistribution and spreading of cuttings; biodegradation or dissolution of the hydrocarbons; burial of the cuttings and recolonization of the surface sediment. Redistribution or burial of cuttings will be most important in low-energy depositional sites.

Although oil-based drilling fluids or cuttings from wells drilled with oil-based drilling fluids are not permitted for disposal in U.S. coastal or continental shelf waters, these studies provide useful insights into the magnitude of "worst case" impacts that might be expected from offshore development and production activities at moderate water depths.

Although the experimental design and overall quality of the several monitoring programs discussed above have varied substantially, some general conclusions do emerge. In offshore oil and gas fields that have been in production for several years, impacts attributable to drilling fluid and cuttings discharges are difficult to identify, except immediately adjacent to platforms where a cuttings pile was formed and has persisted. This is despite the fact that most of the production platforms monitored had drilled multiple wells and had discharged very large volumes of drilling fluids and cuttings. The exception to this generalization is the instances where oil-based drilling fluids were used to develop the field and large amounts of oil-contaminated cuttings were discharged. In these cases, adverse impacts on the benthic fauna may extend out to as much as 500 m from the platforms. In fields that were developed with water-based drilling fluids, impacts are due primarily to reef effects, which may include increased bottom scouring due to platform structures, enhanced production of biomass due to the fouling community and the organisms attracted to it or to the platform structure, and changes in the infaunal community structure due to altered sediment characteristics and influx of predatory epibenthic and demersal organisms. In some cases, adverse impacts on the benthos can be attributed to contamination of sediments near the platform with petroleum hydrocarbons from produced water effluents or accidental spills. No adverse impacts have been described in planktonic or pelagic water column organisms or in marine mammals and birds.

CONCLUSIONS

1. Following discharge to the ocean, drilling fluids are diluted rapidly in the water column to suspended solids concentrations of 1000 ppm within two minutes and below 10 ppm above background within one hour of discharge. In all but very deep or high-energy environments, much of the drilling fluid and cuttings solids settle rapidly to the bottom near the rig site. Concentrations of barium in surficial sediments may be 10 to 20 times above background near the discharge and decrease to background within 2000 m downcurrent of the discharge. Higher concentrations of barium from drilling fluids may be observed in sediments near multiwell development platforms. Produced water is diluted very rapidly following discharge. Significant elevations in salinity or concentrations of hydrocarbons or metals, or decreases in dissolved oxygen usually are not observed at distances greater than 100-200 m from the discharge. In shallow turbid waters, elevated concentrations of hydrocarbons may be detected in surficial sediments up to about 1000 m from the discharge. Very little radium becomes adsorbed to sediments near the discharge.

2. Most of the major ingredients of drilling fluid have a low toxicity to marine organisms. Only chrome and ferrochrome lignosulfonates and sodium hydroxide are slightly toxic. A few specialty chemicals sometimes added to drilling fluids to solve certain problems are toxic. These include diesel fuel, chromate salts, surfactants and paraformaldehyde biocide. Because of rapid mixing with sea water, most physical/chemical features of produced water (low dissolved oxygen and pH, elevated salinity and metals) do not pose a hazard to water column biota. The low molecular weight aromatic hydrocarbons and some metals in produced water are toxic. The toxicity of the soluble organic fraction of produced water is not known.

3. Acute lethal toxicity of more than 70 used water-based drilling fluids has been evaluated in more than 400 bioassays with at least 62 species of marine organisms from the Atlantic and Pacific oceans, the Gulf of Mexico and Beaufort Sea. Nearly 90% of LC_{50} values were above 10,000 ppm drilling mud added, indicating that the drilling muds were practically nontoxic. Only two LC_{50} values

were below 100 ppm. The most toxic drilling muds are those that contain high concentrations of hexavalent chromium, diesel fuel or surfactant. More than 88% of the 54 bioassays performed to date with produced water gave results indicating that the produced water was practically nontoxic. The most toxic produced water samples had been treated with biocides.

4. Chronic and/or sublethal effects of drilling fluids have been performed with at least 40 species of marine animals. In most cases, sublethal responses in marine animals were observed at drilling mud concentrations only slightly lower than those that were acutely lethal. In some species, sublethal responses were observed at drilling fluid concentrations up to two orders of magnitude lower than acutely lethal concentrations. Sensitive species included reef corals, lobster larvae, and scallop embryos and larvae. Recruitment of planktonic larvae to sandy sediments in laboratory microcosms was decreased by high concentrations of drilling mud mixed with or layered on the sediments. Based on laboratory studies of acute and chronic/sublethal toxicity of drilling muds and field observations of rates of dilution of drilling muds in the water column, it is concluded that water column organisms will never be exposed to drilling fluids long enough and at sufficiently high concentrations to elicit any acute or sublethal responses. Where drilling fluid solids settle on the bottom, there could be localized adversed impacts on the benthos, through chemical toxicity, change in sediment texture or burial.

5. Practically no laboratory studies have been performed on the sublethal or chronic effects of produced water in marine organisms.

6. In experimental field studies, accumulation of petroleum hydrocarbons has been demonstrated from produced water but not from drilling fluids. A statistically significant bioaccumulation of barium and chromium and an indication of a slight accumulation of copper, cadmium and lead from drilling mud have been demonstrated in laboratory and field studies. Bioavailability of petroleum hydrocarbons from drilling fluids and of metals from produced water has not been investigated.

7. The field studies performed to date of the impacts of drilling fluids and cuttings discharges from exploration rigs have shown that the severity of impact of drilling fluid and cuttings on the benthos is directly related to the amount of material accumulating on the substrate, which in turn is related to the amount and physical characteristics of the materials being discharged, and to the environmental conditions at the time and site of discharge, such as current speed and water depth. In high-energy environments, little mud and cuttings accumulate and impacts on the benthos are minimal and of short duration. In low-energy and depositional environments, more material accumulates and there may be reductions in abundance of some benthic species.

8. Several field studies have been performed around multiwell development and production platforms to determine long-term biological impacts of all discharges associated with development and production. In shallow water, hydrocarbons from produced water accumulated in bottom sediments and benthic fauna were severely depressed within about 150-200 m of the outfall. Few impacts of offshore production activities in deeper water of the Gulf of Mexico or off southern California have been documented in several large investigations. Rain

out of organic material from the fouling community on submerged parts of the platform structure and the increased bottom microrelief afforded by accumulations of cuttings on the bottom may attract fish and other motile animals to the vicinity and alter the character of the epibenthic and infaunal communities. In the North Sea, accumulation of small amounts of hydrocarbons in sediments and alterations of benthic community structure have been described within about 150 m of some platforms where water-based drilling fluids were used and discharged. In fields where oil-based drilling fluids were used and oil-contaminated cuttings were discharged, benthic communities were severely damaged within 200 m of the platforms and altered out to about 2000 m from the platforms.

RECOMMENDATIONS

As indicated above, a great deal is known about many aspects of the fate and environmental effects of drilling fluids, drill cuttings and produced water discharges to the ocean. Based on the information discussed in this chapter, I have identified several data gaps or information needs and have developed some general strategies for design of long-term monitoring programs.

Long-Term Monitoring Programs

1. Additional long-term investigations of the fate and biological effects of drilling fluids and cuttings discharges from exploratory rigs are not warranted, except when exploratory drilling is proposed for a particularly vulnerable area or there is substantial reason to believe an inherently valuable resource is in jeopardy.

2. A long-term effects program should be performed during intensive development of a new offshore oil field in a frontier area. The area should be on the continental shelf in a low- to medium-energy depositional area, preferably away from major influence of other sources of pollution or riverine inputs. Candidate areas include the Santa Maria Basin, California and the east Texas-west Louisiana shelf. Candidate areas in the Bering Sea off Alaska may be identified in the future.

At an early stage in the design of such a program, a clear definition should be developed of what constitutes a significant environmental impact. Particular attention should be paid to defining impacts that, if they occurred, would be of consequence relative to alternate uses or values of the receiving waters. The monitoring program then should be designed to identify these impacts.

If a major objective of the monitoring program is to document long-term trends in the magnitude or areal extent of environmental change attributable to the development activity, a fairly modest program may be adequate. A carefully-selected set of physical, chemical, and biological samples and measurements could be taken on an annual basis at a relatively small number of stations along a potential "pollution gradient" in the development field. The nearest stations should be within 200 m of the development platform in order to maximize the likelihood of detecting a "signal." Observations would be made and samples taken at the same time each year for at least the duration of development of the field. Selection of parameters to measure should be based on the definitions of impact as described above and might include characterization of rates and long-term trends of change in benthic community structure, recruitment and age-size structure; and measures of reproductive success, fecundity and growth in representative indigenous populations, possibly including sentinal organisms (i.e., the mussel watch concept). Attempts should be made to correlate long-term trends in biological parameters with trends of change in concentrations of contaminants in sediments and biota. Contaminants of concern include metals (Ba, Cr, Pb, Zn), aromatic hydrocarbons, and if produced water is discharged, radium isotopes.

Information Needs

1. Little research has been performed on the fate of organic components of drilling mud following discharge of drilling mud to the ocean. Of particular concern are lignosulfonates and their degradation products and hydrocarbon lubricants, which have been identified as the most toxic components of water-based drilling fluids. Laboratory and field studies should be performed to determine if these additives remain with the light surface plume or with the rapidly-settling fraction of discharged drilling fluid. If a significant fraction of these materials remains with the settleable fraction, what are their persistence in and bioavailability from bottom sediments?

2. Our knowledge of the composition of produced water from different coastal and continental shelf sources is inadequate. More careful analyses of metals in produced water should be performed, with adequate correction for or consideration of interference from the brine matrix, or using methods that are less sensitive to matrix effects. More information is needed on the concentrations of aromatic hydrocarbons with molecular weights greater than that of naphthalene in treated produced water, particularly from oil fields. The composition of the large nonvolatile soluble organic fraction of produced water is largely unknown and should be characterized by sophisticated techniques, probably gas chromatography/mass spectrometry.

3. The environmental fate of produced water ingredients, particularly in shallow coastal areas, should be investigated further. Of particular concern are several metals (arsenic, copper, lead, mercury, and zinc), radionuclides (^{226}Ra, ^{228}Ra), aromatic hydrocarbons (naphthalenes, phenanthrenes, dibenzothiophenes, etc.), and any toxic chemicals identified in the nonvolatile soluble organic fraction. Speciation and ultimate chemical form of produced water metals following discharge to the ocean should be investigated. Radium isotopes may be present in produced water at concentrations more than two orders of magnitude higher than in ambient seawater. ^{226}Ra may be useful as a tracer of the dilution and fate of the soluble fraction of produced water.

4. Long-term bioavailability of metals, radionuclides, aromatic hydrocarbons and soluble organics from produced water should be investigated. Additional, more carefully performed acute and chronic, sublethal bioassays should be performed with produced water from different sources.

5. Since it is generally agreed that any impacts of drilling fluid discharges will be restricted to the benthos where drilling fluid solids accumulate, solid phase bioassay protocols should be developed to more realistically evaluate the toxicity of drilling fluid solids to benthic organisms. Environmentally realistic and interpretable standard bioassay protocols are needed for certification and acceptance of generic muds and new mud additives and formulations as specified in NPDES permits.

LITERATURE CITED

Abel, P.D. 1974. Toxicity of synthetic detergents to fish and aquatic invertebrates. J. Fish. Biol. 6: 179-298.

Addy, J.M., D. Levell and J.P. Hartley. 1978. Biological monitoring of sediments in Ekofisk Oilfield. Pages 515-539 *in* Proc. Conference on Assessment of Ecological Impacts of Oil Spills. American Institute of Biological Sciences, Washington, D.C.

Amiard, J.C., C. Amiard-Triquet, C. Metayer and R. Ferre. 1983. Etude du transfert de quelques oligo-elements metalliques entre le milieu sedimentaire estuarien et les poissons plats "mangeurs de sediments". Mar. Environ. Res. 10: 159-171.

American Petroleum Institute. 1982. Quarterly Review of Drilling Statistics. April 1982. American Petroleum Institute, Washington, D.C.

Anderson, J.B., E.C. Tsivoglou and S.D. Shearer. 1963. Effects of uranium mill wastes on biological fauna of the Animas River (Colorado - New Mexico). Pages 373-383 *in* V. Schultz and A.W. Klement, Jr. (eds.), Radioecology. Reinhold Publ. Corp., New York.

Anderson, J.B., R.B. Wheeler and R.R. Schwarzer. 1981a. Sedimentologic and geochemical results of the Buccaneer oil/gas field study. Pages 721-724 *in* Proc. 1981 Oil Spill Conference. American Petroleum Institute, Washington, D.C.

Anderson, J.R., R.B. Wheeler and R.R. Schwarzer. 1981b. Sedimentology and geochemistry of recent sediments. Pages 59-67 *in* B.S. Middleditch (ed.), Environmental Effects of Offshore Oil Pollution. The Buccaneer Gas and Oil Field Study. Plenum Press, New York.

Anderson, J.W. 1982. The transport of petroleum hydrocarbons from sediments to benthos and the potential effects. Pages 165-179 *in* G.F. Mayer (ed.), Ecological Stress and the New York Bight: Science and Management. Estuarine Research Federation, Columbia, South Carolina.

Anderson, J.W., J.M. Neff, B.A. Cox, H.E. Tatem and G.M. Hightower. 1974. Characteristics of dispersions of water-soluble extracts of crude and refined oils and their toxicity to estuarine crustaceans and fish. Mar. Biol. 27: 75-88.

Andreasen, J.K. and R.W. Spears. 1983. Toxicity of Texan petroleum well brine to the sheepshead minnow (*Cyprinodon variegatus*), a common estuarine fish. Bull. Environ. Contam. Toxicol. 30: 277-283.

Armstrong, H.W., K. Fucik, J.W. Anderson and J.M. Neff. 1977. Effects of Oilfield Brine Effluent on Benthic Organisms in Trinity Bay, Texas. API Publ. No. 4291. American Petroleum Institute, Washington, D.C., 82 p.

Armstrong, H.W., K. Fucik, J.W. Anderson and J.M. Neff. 1979. Effects of oilfield brine effluent on sediments and benthic organisms in Trinity Bay, Texas. Mar. Environ. Res. 2: 55-69.

Atema, J., D.F. Leavitt, D.E. Barshaw and M.C. Cuomo. 1982a. Effects of drilling fluids on behavior of the American lobster, *Homarus americanus*, in water column and substrate exposures. Can. J. Fish. Aquat. Sci. 39: 675-690.

Atema, J., E.B. Karnofsky, S. Olszko-Szuts and B. Bryant. 1982b. Sublethal Effects of Number 2 Fuel Oil on Lobster Behavior and Chemoreception. Report to U.S.

Environmental Protection Agency, Environmental Research Lab., Gulf Breeze, Florida, EPA-600/S3-82-013.

Augenfeld, J.M. 1980. Effects of Prudhoe Bay crude oil contamination on sediment working rates of *Abarenicola pacifica*. Mar. Environ. Res. 3: 307-313.

Augenfeld, J.M., J.W. Anderson, R.G. Riley and B.L. Thomas. 1982. The fate of polyaromatic hydrocarbons in an intertidal sediment exposure system: bioavailability to *Macoma inquinata* (Mollusca: Pelecypoda) and *Abarenicola pacifica* (Annelida: Polychaetea). Mar. Environ. Res. 7: 31-50.

Augenfeld, J.M., J.W. Anderson, D.L. Woodruff and J.L. Webster. 1980. Effects of Prudhoe Bay crude oil-contaminated sediments on *Protothaca staminea* (Mollusca: Pelecypoda): Hydrocarbon content, condition index, free amino acid level. Mar. Environ. Res. 4: 135-143.

Ayers, R.C., Jr., T.C. Sauer, Jr. and P.W. Anderson. 1983. The Generic Mud Concept for Offshore Drilling for NPDES Permitting. IADC/SPE 1983 Drilling Conference. New Orleans, Louisiana, February 1983. Society of Petroleum Engineers, Dallas, Texas 75206. IADC/SPE Paper No. 11399, 8 p.

Ayers, R.C., Jr., T.C. Sauer, Jr., R.P. Meek and G. Bowers. 1980a. An environmental study to assess the impact of drilling discharges in the Mid-Atlantic. I. Quantity and fate of discharges. Pages 382-418 *in* Symposium, Research on Environmental Fate and Effects of Drilling Fluids and Cuttings. Lake Buena Vista, Florida, January 21-24, 1980. American Petroleum Institute, Washington, D.C.

Ayers, R.C., Jr., T.C. Sauer, Jr., D.O. Stuebner and R.P. Meek. 1980b. An environmental study to assess the effect of drilling fluids on water quality parameters during high rate, high volume discharges to the ocean. Pages 351-381 *in* Symposium, Research on Environmental Fate and Effects of Drilling Fluids and Cuttings. Lake Buena Vista, Florida, January 21-24, 1980. American Petroleum Institute Washington, D.C.

Baker, J.H., K.T. Kimball, W.D. Jobe, J. Janousek, C.L. Howard and P.R. Chase. 1981. Part 6. Benthic Biology. Pages 1-391 *in* C.A. Bedinger, Jr. (ed.), Ecological Investigations of Petroleum Production Platforms in the Central Gulf of Mexico. Volume 1. Pollutant Fate and Effects Studies. Southwest Research Institute, San Antonio, Texas.

Baker, J.M. (ed.). 1976. Marine Ecology and Oil Pollution. Halsted Press, New York, 566 p.

Bascom, W. 1983. The non-toxicity of metals in the sea. Mar. Techn. Soc. J. 17: 59-66.

Bascom, W., A.J. Mearns and M.D. Moore. 1976. A biological survey of oil platforms in the Santa Barbara Channel. J. Petrol. Technol., Nov. 1976: 1280-1284.

Battelle/Woods Hole Oceanographic Institution. 1983. Georges Bank Benthic Infauna Monitoring Program. Final Report, Year 1. Prepared for U.S. Dept. of the Interior, Minerals Management Service, New York OCS Office, New York, Contract No. 14-12-001-29192.

Battelle/Woods Hole Oceanographic Institution. 1984. Georges Bank Benthic Infauna Monitoring Program. Final Report, Year 2. Prepared for U.S. Dept. of the Interior, Minerals Management Service, Atlantic OCS Office, Vienna, Virginia, Contract No. 14-12-0001-29192.

Bedinger, C.A., Jr. 1981. Volume III - Executive Summary. Pages 1-29 *in* C.A. Bedinger, Jr. (ed.), Ecological Investigations of Petroleum Production Platforms in the Central Gulf of Mexico. Southwest Research Institute, San Antonio, Texas.

Bedinger, C.A., Jr., R.E. Childers, J.W. Cooper, K.T. Kimball and A. Kwok. 1981. Part 1. Background, Program Organization and Study Plan. Pages 1-53 *in* C.A. Bedinger, Jr. (ed.), Ecological Investigations of Petroleum Production Platforms in the Central Gulf of Mexico. Volume 1. Pollution Fate and Effects Studies. Southwest Research Institute, San Antonio, Texas.

Behrens, W.E. 1981. Total organic carbon and carbon isotopes of sediments. Pages 117-132 *in* B.S. Middleditch (ed.), Environmental Effects of Offshore Oil Production. The Buccaneer Gas and Oil Field Study. Plenum Press, New York.

Beller, W.S. 1983. Streamlining the permit process for ocean discharges. Pages 848-852 *in* Proc. Oceans 83 Conference, San Francisco, California, Volume II. Marine Technology Society, Washington, D.C.

Bender, M.E., J.M. Sharp, D.J. Reish, G.G. Appen and C.H. Ward. 1979. An independent appraisal of the Offshore Ecology Investigation. Proc. Offshore Technol. Conf. (Houston, Texas) 11: 2163-2172.

Benech, S., R. Bowker and B. Pimental. 1980. Chronic effects of drilling fluids exposure to fouling community composition on a semi-submersible exploratory drilling vessel. Pages 611-635 *in* Symposium, Research on Environmental Fate and Effects of Drilling Fluids and Cuttings. Lake Buena Vista, Florida, January 21-24, 1980. American Petroleum Institute, Washington, D.C.

Bianchi, V., L. Celotti, G. Lanfranchi, F. Majone, G. Marin, A. Montaldi, G. Sponza, G. Tamino, P. Venier, A. Zantedeschi and A.G. Levis. 1983. Genetic effects of chromium compounds. Mut. Res. 117: 279-300.

Boesch, D.F. 1979. Benthic ecological studies: Macrobenthos. *In* Middle Atlantic Outer Continental Shelf Environmental Studies. Volume IIB. Chemical and Biological Benchmark Studies. Special Report in Applied Marine Science and Ocean Engineering No. 194. Virginia Institute of Marine Sciences, Gloucester Point, Virginia. U.S. Dept. of the Interior, Bureau of Land Management, New York, OCS Office, New York.

Boesch, D.F. and R. Rosenberg. 1981. Response to stress in marine benthic communities. Pages 179-200 *in* G.W. Barrett and R. Rosenberg (eds.), Stress Effects on Natural Ecosystems. John Wiley and Sons, New York.

Boland, G.S. 1980. Morphological parameters of the barnacle, *Balanus tintinnabulum antillensis*, as indicators of physiological and environmental conditions. M.S. Thesis, Texas A&M University, College Station, Texas, 69 p.

Boland, G.S., B.J. Gallaway, J.S. Baker and G.S. Lewbel. 1983. Ecological Effects of Energy Development on Reef Fish of the Flower Garden Banks. Report to National Marine Fisheries Service, Galveston, Texas on Contract No. NA80-6A-00057 from LGL Ecological Research Associates, Inc., Bryan, Texas, 466 p.

Bookhout, C.G., R. Monroe, R. Forward and J.D. Costlow, Jr. 1984. Effects of soluble fractions of drilling fluids on development of crabs, *Rhithropanopeus harrisii* and *Callinectes sapidus*. Water, Air, Soil Pollut. 21: 183-197.

Boothe, P.N. and B.J. Presley. 1983. Distribution and Behavior of Drilling Fluids and Cuttings around Gulf of Mexico Drill Sites. Draft Final Report to American Petroleum Institute. Dept. of Oceanography, Texas A&M University, College Station, Texas, 65 p.

Bothner, M.H., R.R. Rendigs, E. Campbell, M.W. Doughton, P.J. Aruscavage, A.F. Dorrzapf, Jr., R.G. Johnson, C.M. Parmenter, M.J. Pikering, D.C. Brewster and F.W. Brown. 1982. The Georges Bank Monitoring Program. Analysis of Trace Metals in Bottom Sediments. First Year Final Report to U.S. Dept. of Interior, Minerals Management Service, New York OCS Office, New York. U.S. Dept. of Interior, Geological Survey, Woods Hole, Massachusetts. Interagency Agreement No. AA851-IA2-18.

Bothner, M.H., R.R. Rendigs, E. Campbell, M.W. Doughten, C.M. Parmenter, M.J. Pickering, R.G. Johnson and J.R. Gillison. 1983. The Georges Bank Monitoring Program. Analysis of Trace Metals in Bottom Sediments During the Second Year of Monitoring. Final Report to U.S. Dept. of Interior, Minerals Management Service, Washington, D.C. U.S. Dept. of the Interior, Geological Survey, Woods Hole, Massachusetts.

Bothner, M.H., E.C. Spiker, P.P. Johnson, R.R. Rendigs and P.J. Aruscavage. 1981. Geochemical evidence for modern sediment accumulation on the continental shelf off southern New England. J. Sediment. Petrol. 51: 281-292.

Brandsma, M.G., L.R. Davis, R.E. Ayers, Jr. and T.C. Sauer, Jr. 1980. A computer model to predict the short-term fate of drilling discharges in the marine environment. Pages 588-610 *in* Symposium, Research on Environmental Fate and Effects of Drilling

Fluids and Cuttings. Lake Buena Vista, Florida, January 21-24, 1980. American Petroleum Institute, Washington, D.C.

Brannon, A.C. and K.R. Rao. 1979. Barium, strontium and calcium levels in the exoskeleton, hepatopancreas and abdominal muscle of the grass shrimp *Palaemonetes pugio*: Relation to molting and exposure to barite. Comp. Biochem. Physiol. 63A: 261-274.

Breteler, R.J., P.D. Boehm, J.M. Neff and A.G. Requejo. 1983. Acute Toxicity of Drilling Muds Containing Hydrocarbon Additives and Their Fate and Partitioning between Liquid, Suspended and Solid Phases. Draft Final Report to American Petroleum Institute, Washington, D.C., 93 p.

Breteler, R.J. and J.M. Neff. 1983. Waste disposal in the marine environment. 1. Biological availability of heavy metals to marine invertebrates. Pages 905-910 *in* Proc. Oceans 83 Conference, San Francisco, California, Volume II. Marine Technology Society, Washington, D.C.

Brooks, J.M., E.L. Estes, D.A. Wiesenburg, C.R. Schwab and H.A. Abdel-Reheim. 1980. Investigations of surficial sediments, suspended particulates, and volatile hydrocarbons at Buccaneer gas and oil field. *In* W.P. Jackson and E.P. Wilkens (eds.), Environmental Assessment of the Buccaneer Gas and Oil Field in the Northwestern Gulf of Mexico, 1975-1980, Volume 1. NOAA Tech. Memorandum NMFS-SEFC-47.

Brown, L.R., G.W. Childers, R.W. Landers, Jr. and J.D. Walker. 1981a. Part 5. Microbiology and microbiological processes. Pages 123-223 *in* C.A. Bedinger, Jr. (ed.), Ecological Investigations of Petroleum Production Platforms in the Central Gulf of Mexico. Volume 1. Pollutant Fate and Effects Studies. Southwest Research Institute, San Antonio, Texas.

Brown, L.R., K.L. Jeanes, W.F. Greenleaf and G. Frank. 1981b. Oil-degrading potential of surficial sediments near oil-producing platforms. Pages 725-730 *in* Proc. 1981 Oil Spill Conference. American Petroleum Institute, Washington, D.C.

Bryan, G.W. 1983. The biological availability and effects of heavy metals in marine deposits. *In* Proc. Ocean Dumping Symposium. Wiley Interscience, New York.

Caillouet, C.W., W.B. Jackson, G.R. Gitschlag, E.P. Wilkens and G.M. Faw. 1981. Review of the environmental assessment of the Buccaneer Gas and Oil Field in the northwestern Gulf of Mexico. Proc. Gulf Carrib. Fish. Inst. 33: 101-124.

Cantelmo, F.R., M.E. Tagatz and K.R. Rao. 1979. Effect of barite on meiofauna in a flow-through experimental system. Mar. Environ. Res. 2: 301-309.

Capuzzo, J.M. and J.G.S. Derby. 1982. Drilling Fluid Effects to Developmental Stages of the American Lobster. Report to U.S. Environmental Protection Agency, Environmental Research Lab., Gulf Breeze, Florida, EPA-600/S4-82-039.

Carls, M.G. and S.D. Rice. 1980. Toxicity of Oil Well Drilling Fluids to Alaskan Larval Shrimp and Crabs. Research unit 72. Final Rept. Proj. No. R7120822, Outer Continental Shelf Environmental Assessment Program. U.S. Dept. of Interior, Bureau of Land Management, 29 p.

Carr, R.S., W.L. McCulloch and J.M. Neff. 1982. Bioavailability of chromium from a used chrome-lignosulfonate drilling fluid to five species of marine invertebrates. Mar. Environ. Res. 6: 189-204.

Carr, R.S., L.A. Reitsema and J.M. Neff. 1980. Influence of a used chrome lignosulfonate drilling mud on the survival, respiration, feeding activity and net growth efficiency of the opposum shrimp *Mysidopsis almyra*. Pages 944-963 *in* Symposium, Research on Environmental Fate and Effects of Drilling Fluids and Cuttings. Lake Buena Vista, Florida, January 21-24, 1980. American Petroleum Institute, Washington, D.C.

Carstens, T. 1976. Seabed scour by currents near platforms. Mar. Sci. Commun. 2: 69-93.

Carstens, T. 1983. Scour around an offshore platform. Ocean Sci. Engin. 8: 157-172.

Chaffee, C. and R.B. Spies. 1982. The effects of used ferrochrome lignosulfonate drilling fluids from a Santa Barbara Channel oil well on the development of starfish embryos. Mar. Environ. Res. 7: 265-277.

Chan, L.H. and J.S. Hanor. 1982. Dissolved barium in some Louisiana offshore waters: Problems in establishing baseline values. Contr. Mar. Sci. 25: 149-159.

Chesser, B.G. and W.H. McKenzie. 1975. Use of a bioassay test in evaluating the toxicity of drilling fluid additives on Galveston Bay shrimp. Pages 153-168 *in* Environmental Aspects of Chemical Use in Well-Drilling Operations. U.S. Environmental Protection Agency, EPA-560/1-75-004.

Choi, D.R. 1982. Coelobites (reef cavity-dwellers) as indicators of environmental effects caused by offshore drilling. Bull. Mar. Sci. 32: 880-889.

Collins, A.G. 1969. Chemistry of some Anadarko Basin brines containing high concentrations of iodide. Chem. Geol. 4: 169-187.

Collins, A.G. 1975. Geochemistry of Oilfield Waters. Elsevier Scientific Publishers, New York, 496 p.

Conklin, P.J., D.G. Doughtie and K.R. Rao. 1980. Effects of barite and used drilling fluids on crustaceans, with particular reference to the grass shrimp, *Palaemonetes pugio*. Pages 723-738 *in* Symposium, Research on Environmental Fate and Effects of Drilling Fluids and Cuttings. Lake Buena Vista, Florida, January 21-24, 1980. American Petroleum Institute, Washington, D.C.

Conklin, P.J., D. Drysdale, D.G. Doughtie, K.R. Rao, J.P. Kakareka, T.R. Gilbert and R.F. Shokes. 1983. Comparative toxicity of drilling fluids: Role of chromium and petroleum hydrocarbons. Mar. Environ. Res. 10: 105-125.

Crawford, R.B. 1983. Effects of Drilling Fluids on Embryo Development. Final Report to U.S. EPA Environmental Research Laboratory, Gulf Breeze, Florida, EPA 600/S3-83-021.

Crawford, R.B. and J.D. Gates. 1981. Effects of drilling fluid on the development of a teleost and an echinoderm. Bull. Environ. Contam. Toxicol. 26: 207-212.

Crippen, R.W., S.L. Hodd and G. Greene. 1980. Metal levels in sediment and benthos resulting from a drilling fluid discharge into the Beaufort Sea. Pages 636-669 *in* Symposium, Research on Environmental Fate and Effects of Drilling Fluids and Cuttings. Lake Buena Vista, Florida, January 21-24, 1980. American Petroleum Institute, Washington, D.C.

Dames and Moore, Inc. 1978. Drilling Fluid Dispersion and Biological Effects Study for the Lower Cook Inlet C.O.S.T. Well. Report submitted to Atlantic Richfield Co, by Dames & Moore, Anchorage, Alaska, 309 p.

Davies, J.M., R. Hardy and A.D. McIntyre. 1981. Environmental effects of North Sea oil operations. Mar. Poll. Bull. 12: 412-416.

Davies, J.M., J. Addy, L. Blackman, J. Blanchard, J. Ferbrache, D. Moore, H. Sommerville, A. Whitehead and T. Wilkinson. 1983. Environmental Effects of Oil Based Mud Cuttings. Report of Joint Working Group of UKOOA Clean Seas and Environmental Committee/Dept. of Energy/Dept. of Agriculture and Fisheries for Scotland/ Ministry of Agriculture, Fisheries and Food. Great Britain, 24 p.

Davis, N., G.R. van Blaricom and P.K. Dayton. 1982. Man-made structures on marine sediments: Effects on adjacent benthic communities. Mar. Biol. 70: 295-303.

Derby, C.D. and J. Atema. 1981. Influence of drilling fluids on the primary chemosensory neurons in walking legs of the lobster, *Homarus americanus*. Can. J. Fish. Aquat. Sci. 38: 268-274.

Derby, J.G.S. and J.M. Capuzzo. 1984. Effects of drilling fluids on larvae of the American lobster, *Homarus americanus*. Can. J. Fish. Aquat. Sci. (in press).

Dicks, B.M. 1975. Offshore biological monitoring. Pages 325-440 *in* J.M. Baker (ed.), Marine Ecology and Oil Pollution. Applied Science Publ., Barking, Essex, England.

Dodge, R.E. 1982. Effects of drilling fluids on the reef-building coral *Montastrea annularis*. Mar. Biol. 71: 141-147.

Ecomar, Inc. 1978. Tanner Bank Fluids and Cuttings Study. Conducted for Shell Oil Company, January through March, 1977. Ecomar, Inc., Goleta, California, 495 p.

Ecomar, Inc. 1983. Mud Dispersion Study. Norton Sound Cost Well No. 2. Report for ARCO Alaska, Inc. from Ecomar, Goleta, California, 91 p.

EG&G, Environmental Consultants. 1982. A Study of Environmental Effects of Exploratory Drilling on the Mid-Atlantic Outer Continental Shelf - Final Report of the Block 684 Monitoring Program. EG&G, Environmental Consultants, Waltham, Massachusetts. Available from Offshore Operators Committee, Environmental Subcommittee, P.O. Box 50751, New Orleans, Louisiana 70150.

EPA. 1975. Methods for Acute Toxicity Tests with Fish, Macroinvertebrates and Amphibians. U.S. Environmental Protection Agency, Ecological Research Series, April 1975, EPA-600/3-75-009.

EPA. 1978. Bioassay Procedures for the Ocean Disposal Permit Program. U.S. Environmental Protection Agency, EPA-600/9-78-010.

EPA/COE. 1977. Ecological Evaluation of Proposed Discharge of Dredged Material into Ocean Waters. EPA/COE Technical Committee on Criteria for Dredged and Fill Material. U.S. Army Corps of Engineers, Waterways Experiment Station, Vicksburg, Mississippi.

Espey, Huston & Associates, Inc. 1981. Bioassay and Depuration Studies on Two Types of Barite. Document No. 81123. Report to Magcobar Group, Dresser Industries, Inc., Houston, Texas, by Espey, Huston & Assoc., Inc., Houston, Texas, 25 p.

Essertier, E.P. 1984. Federal Offshore Statistics. OCS Report MMS84-0071. U.S. Department of Interior, Minerals Management Service, Washington, D.C., 123 p.

Farrell, D. 1974a. Benthic Communities in the Vicinity of Producing Oil Wells on the Shallow Louisiana Continental Shelf. Final Report to Gulf Universities Research Consortium, Offshore Ecology Investigation. GURC, Galveston, Texas.

Farrell, D. 1974b. Benthic Communities in the Vicinity of Producing Oil Wells in Timbalier Bay, Louisiana. Final Report to Gulf Universities Research Consortium, Offshore Ecology Investigation. GURC, Galveston, Texas.

Ferbrache, J. 1983. A review of the available literature on the marine benthic effects resulting from the discharge of water base mud and cuttings. Appendix II *in* J. Davies, J. Addy, R. Blackman, J. Blanchard, J. Ferbrache, D. Moore, H. Sommerville, A. Whitehead and T. Wilhenson, Environmental Effects of Oil Based Mud Cuttings. UKOOA Clean Seas and Environmental Committee.

Fish, A.G., L.L. Massey, J.R. Inabinet and P.L. Lewis. 1974. A Study of the Effects of Environmental Factors upon the Distribution of the Foraminifera, Nematodes and Polychaetous Annelids fauna of Timbalier Bay, Louisiana. Final Report to Gulf Universities Research Consortium, Offshore Ecology Investigation. GURC, Galveston, Texas.

Flint, R.W., T.W. Duke and R.D. Kalke. 1982. Benthos investigations: Sediment boxes or natural bottom? Bull. Environ. Contam. Toxicol. 28: 257-265.

Folmar, L.C. 1977. Acrolein, Dalapon, Dichlobenil, Diquat, and Endothal: Bibliography of Toxicity to Aquatic Organisms. U.S. Fish and Wildlife Service, Tech. Paper 88. USFWS, Washington, D.C., 16 p.

Forteath, G.N.R., G.B. Picken and R. Ralph. 1983. Interaction and competition for space between fouling organisms on the Beatrice oil platforms in the Moray Firth, North Sea. Intern. Biodeterior. Bull. 19: 45-52.

Fotheringham, N. 1981. Observations on the effects of oil field structures on their biotic environment: Platform fouling community. Pages 179-208 *in* B.S. Middleditch (ed.), Environmental Effects of Offshore Oil Production. The Buccaneer Gas and Oil Field Study. Plenum Press, New York.

Fucik, K.W., H.W. Armstrong and J.M. Neff. 1977. The uptake of naphthalenes by the clam, *Rangia cuneata*, in the vicinity of an oil platform in Trinity Bay, Texas. Pages 637-640 *in* Proc. 5th Conference on the Prevention, Behavior, Control and Clean-Up of Oil Pollution. American Petroleum Institute, Washington, D.C.

Gallaway, B.J., L.R. Martin, R.L. Howard, G.S. Boland and G.S. Dennis. 1981. Effects on artificial reef and demersal fish and macrocrustacean communities. Pages 237-300 *in* B.S. Middleditch (ed.), Environmental Effects of Offshore Oil Production. The Buccaneer Gas and Oil Field Study. Plenum Press, New York.

Gambrell, R.P., R.A. Khalid and W.H. Patrick, Jr. 1980. Chemical availability of mercury, lead and zinc in Mobile Bay sediment and suspensions as affected by pH and oxidation-reduction conditions. Am. Chem. Soc. 14: 431-436.

Gehrs, C.W. 1978. Environmental implications of coal conversion technologies: Organic contaminants. Pages 157-175 *in* J.H. Thorp and J.W. Gibbons (eds.), Energy and Environmental Stress in Aquatic Systems. Technical Information Center, U.S. Dept. of Energy, CONF-771114. National Technical Information Service, Springfield, Virginia.

Gerber, R.P., E.S. Gilfillan, J.R. Hotham, L.J. Galletto and S.A. Hanson. 1981. Further Studies on the Short and Long-Term Effect of Used Drilling Fluids on Marine Organisms. Unpublished. Final Report, Year II to the American Petroleum Institute, Washington, D.C., 30 p.

Gerber, R.P., E.S. Gilfillan, B.T. Page, D.S. Page and J.B. Hotham. 1980. Short- and long-term effects of used drilling fluids on marine organisms. Pages 882-911 *in* Symposium, Research on Environmental Fate and Effects of Drilling Fluids and Cuttings. Lake Buena Vista, Florida, January 21-24, 1980. American Petroleum Institute, Washington, D.C.

Gettleson, D.A. 1978. Ecological impact of exploratory drilling: A case study. *In* Energy/ Environment '78. Soc. of Petroleum Industry Biologists Symposium, 22-24 August, 1978, Los Angeles, California, 23 p.

Gettleson, D.A. and C.B. Laird. 1980. Benthic barium in the vicinity of six drill sites in the Gulf of Mexico. Pages 739-788 *in* Symposium, Research on Environmental Fate and Effects of Drilling Fluids and Cuttings. Lake Buena Vista, Florida, January 21-24, 1980. American Petroleum Institute, Washington, D.C.

Gilbert, T.R. 1981. A Study of the Impact of Discharged Drilling Fluids on the Georges Bank Environment. New England Aquarium, H.E. Edgerton Research Laboratory. Progress Rept. No. 2 to U.S. Environmental Protection Agency, Gulf Breeze, Florida, 98 p.

Gilbert, T.R. 1982. A Survey of the Toxicities and Chemical Compositions of Used Drilling Muds. Annual Report to U.S. Environmental Research Laboratory, Gulf Breeze, Florida from Edgerton Research Lab., New England Aquarium, Boston, Massachusetts, 31 p.

Gillmor, R.B., C.A. Menzie and J. Ryther, Jr. 1981. Side-scan sonar and TV observations of the benthic environment and megabenthos in the vicinity of an OCS exploratory well in the Middle Atlantic Bight. Pages 727-731 *in* Conference Record. Volume 2. Oceans '81 Symposium. Boston, Massachusetts IEEE Publ. No. 81CH1685-7 IEEE Service Center, Piscataway, New Jersey.

Gillmor, R.B., C.A. Menzie, G.M. Mariani, D.R. Levin, R.C. Ayers, Jr. and T.C. Sauer, Jr. 1982. Effects of exploratory drilling discharges on the benthic environment in the middle Atlantic OCS: Biological results of a one-year postdrilling survey. *In* Proc. Ocean Dumping Symposium. Wiley Interscience, New York.

Grahl-Nielsen, O. 1978. The Ekofisk bravo blowout: Petroleum hydrocarbons in the sea. Pages 477-487 *in* Proc. Conference on Assessment of Ecological Impacts of Oil Spills. American Institute of Biological Sciences, Washington, D.C.

Grahl-Nielsen, O., S. Sundby, K. Westrheim and S. Wilhelmsen. 1980. Petroleum hydrocarbons in sediment resulting from drilling discharges from a production platform in the North Sea. Pages 541-561 *in* Symposium, Research on Environmental Fate and Effects of Drilling Fluids and Cuttings. Lake Buena Vista, Florida, January 21-24, 1980. American Petroleum Institute, Washington, D.C.

Grassle, J.F. 1977. Slow recolonization of deep-sea sediment. Nature 265: 618-619.

Grassle, J.F., R.E. Imgren and J.P. Grassle. 1980. Response of benthic communities in MERL experimental ecosystems to low level, chronic additions of No. 2 fuel oil. Mar. Environ. Res. 4: 279-297.

Harper, D.E., Jr., D.L. Potts, R.R. Salzer, R.J. Case, R.L. Jaschek and C.M. Walker. 1981. Distribution and abundance of macrobenthic and meiobenthic organisms. Pages 133-178 *in* B.S. Middleditch (ed.), Environmental Effects of Offshore Oil Production. The Buccaneer Gas and Oil Field Study. Plenum Press, New York.

Harper, D.E., Jr., R.J. Scrudato and C.S. Giam. 1976. Pilot Study of the Buccaneer Oil Field (Benthos and Sediments). A Preliminary Environmental Assessment of the Buccaneer Oil and Gas Field. Unpubl. Report, Texas A&M University, 63 p.

Hartley, J.P. and J. Ferbrache. 1983. Biological Monitoring of the Forties Oilfield (North Sea). Pages 407-414 *in* Proc. 1983 Oil Spill Conference. American Petroleum Institute, Washington, D.C.

Holtzman, R.B. 1969. Concentrations of the naturally occurring radionuclides ^{226}Ra, ^{210}Po in aquatic fauna. Pages 535-546 *in* Proc. 2nd Nat. Symp. Radioecology, U.S. Atomic Energy Commission, Conf. 670503.

Houghton, J.P., D.L. Beyer and E.D. Thielk. 1980. Effects of oil well drilling fluids on several important Alaskan marine organisms. Pages 1017-1043 *in* Symposium on Research on Environmental Fate and Effects of Drilling Fluids and Cuttings. Lake Buena Vista, Florida, January 21-24, 1980. American Petroleum Institute, Washington, D.C.

Houghton, J.P., K.R. Critchlow, D.C. Lees, R.D. Czlapinski, R.C. Miller, R.P. Britch and J.A. Mills. 1981. Fate and Effects of Drilling Fluids and Cuttings Discharges in Lower Cook Inlet, Alaska, and on Georges Bank. Final Report to NOAA and BLM. Dames and Moore, Seattle, Washington.

Howard, R.L., G.S. Boland, B.J. Gallaway and G.D. Dennis. 1980. Effects of gas and oil field structures and effluents on fouling community production and function. *In* Environmental Assessment of the Buccaneer Gas and Oil Field in the Northwestern Gulf of Mexico, 1978-1979. Volume V. NOAA Technical Memo. NMFS-SEFC-39.

Hudson, J.H. and D.M. Robbin. 1980. Effects of drilling fluids on the growth rate of the reef-building coral, *Montastrea annularis*. Pages 1101-1122 *in* Symposium, Research on Environmental Fate and Effects of Drilling Fluids and Cuttings. Lake Buena Vista, Florida, January 21-24, 1980. American Petroleum Institute, Washington, D.C.

Hudson, J.H., E.A. Shinn and D.M. Robbin. 1982. Effects of offshore oil drilling on Philippine reef corals. Bull. Mar. Sci. 32: 890-908.

Hunt, C.D. and D.L. Smith. 1983. Remobilization of metals from polluted marine sediments. Can. J. Fish. Aquat. Sci. 40: 132-142.

Hutchinson, T.C., J.A. Hellebust, D. Tam, D. Mackay, R.A. Mascarenkas and W.Y. Shiu. 1980. The correlation of the toxicity to algae of hydrocarbons and halogenated hydrocarbons with their physical-chemical properties. Pages 577-586 *in* B.K. Afghan and D. Mackay (eds.), Hydrocarbons and Halogenated Hydrocarbons in the Aquatic Environment. Plenum Press, New York.

IMCO/FAO/UNESCO/WMO Joint Group of Experts on the Scientific Aspects of Marine Pollution. 1969. Abstract of first session report. Water Res. 3: 995-1005.

Jenne, E.A. and S.N. Luoma. 1977. Forms of trace elements in soils, sediments, and associated waters: An overview of their determination and biological availability. Pages 110-143 *in* H. Drucker and R.E. Wildung (eds.), Biological Implications of Metals in the Environment. NTIS Conf.-750929, Springfield, Virginia.

Jumars, P.A. 1981. Limits of predicting and detecting benthic community responses to manganese nodule mining. Pages 213-229 *in* J.R. Moore (ed.), Marine Mining. Crane Russak Press, New York.

Kash, D.E. 1983. Domestic options to offshore oil and gas. Oceanus 26: 46-51.

Kendall, J.J., Jr., E.N. Powell, S.J. Connor and T.J. Bright. 1983. The effects of drilling fluids (muds) and turbidity on growth and metabolic state of the coral *Acropora cervicornis*, with comments on methods of normalization for coral data. Bull. Mar. Sci. 33: 336-352.

Kinne, O. 1971. Salinity. Animals. Invertebrates. Pages 821-995 *in* O. Kinne (ed.), Marine Ecology/A Comprehensive, Integrated Treatise on Life in Oceans and Coastal Waters. Volume 1. Environmental Factors. Part 2. Wiley Interscience, New York.

Koons, C.B., C.D. McAuliffe and F.T. Weiss. 1977. Environmental aspects of produced waters from oil and gas extraction operations in offshore and coastal waters. *In* Proc. Offshore Technol. Conf., OTC Paper 2447. Offshore Technology Conference, Dallas, Texas.

Kritzler, H. 1974. Oil Production and Polychaetous Annelids in a Louisiana estuary. Final Report to Gulf Universities Research Consortium, Offshore Ecology Investigation. GURC, Galveston, Texas.

Krone, M.A. and D.C. Biggs. 1980. Sublethal metabolic responses of the hermatypic coral *Madracis decactis* exposed tc drilling fluids, enriched with ferrochrome lignosulfonate. Pages 1079-1100 *in* Symposium, Research on Environmental Fate and Effects of Drilling Fluids and Cuttings. Lake Buena Vista, Florida, January 21-24, 1980. American Petroleum Institute, Washington, D.C.

Law, R.J. and R.A.A. Blackman. 1981. Hydrocarbons in water and sediments from oil-producing areas of the North Sea. ICES. C.M./E:15, 20 p.

Lees, D.C. and J.P. Houghton. 1980. Effects of drilling fluids on benthic communities at the lower Cook Inlet C.O.S.T. well. Pages 209-350 *in* Symposium, Research on Environmental Fate and Effects of Drilling Fluids and Cuttings. Lake Buena Vista, Florida, January 21-24, 1980. American Petroleum Institute, Washington, D.C.

Liss, R.G., F. Knox, D. Wayne and T.R. Gilbert. 1980. Availability of trace elements in drilling fluids to the marine environment. Pages 691-722 *in* Symposium, Research on Environmental Fate and Effects of Drilling Fluids and Cuttings. American Petroleum Institute, Washington, D.C.

Lunz, G.R. 1950. The Effect of Bleedwater and of Water Extracts of Crude Oil on the Pumping Rate of Oysters. Texas A&M Research Foundation, Project 9. Texas A&M University, College Station, Texas, 107 p.

Luoma, S.N. 1983. Bioavailability of trace metals to aquatic organisms - a review. Sci. Tot. Environ. 28: 1-22.

Lyes, M.C. 1979. Bioavailability of hydrocarbon from water and sediment to the marine worm *Arenicola marina*. Mar. Biol. 55: 121-127.

Lysyj, I. 1982. Chemical composition of produced water at some offshore oil platforms. Report to U.S. Environmental Protection Agency, EPA-600/2-82-034. Municipal Environmental Research Lab., Cincinnati, Ohio.

Mackin, J.G. 1971. A Study of the Effect of Oilfield Brine Effluents on Benthic Communities in Texas Estuaries. Texas A&M Research Foundation, Proj. 735. Texas A&M University, College Station, Texas, 72 p.

Malins, D.C. (ed.). 1977. Effects of Petroleum on Arctic and Subarctic Marine Environments and Organisms. Volume II. Biological Effects. Academic Press, New York, 500 p.

Mariani, G.M., L.V. Sick and C.C. Johnson. 1980. An environmental monitoring study to assess the impact of drilling discharges in the mid-Atlantic. III. Chemical and physical alterations in the benthic environment. Pages 438-498 *in* Symposium, Research on Environmental Fate and Effects of Drilling Fluids and Cuttings. Lake Buena Vista, Florida, January 21-24, 1980. American Petroleum Institute, Washington, D.C.

Marine Bioassay Labs. 1982. Drilling Fluids Bioassays. Texaco Habitat Platform Well A-1 Pitas Point Lease Sale Unit: *Acanthomysis sculpta* and *Macoma nasuta*. Report submitted to Texaco, Inc., Los Angeles, California and IMCO Services, Houston, Texas by Marine Bioassay Labs, Watsonville, California.

Maurer, D., W. Leathem and C. Menzie. 1981. The impact of drilling fluid and well cuttings on polychaete feeding guilds from the U.S. northeastern continental shelf. Mar. Pollut. Bull. 12: 342-347.

McCain, B.B., H.O. Hodgins, W.D. Gronlund, J.W. Hawkes, D.W. Brown, M.S. Myers and J.J. Vandermuelen. 1978. Bioavailability of crude oil from experimentally oiled sediments to English sole (*Parophrys vetulus*), and pathological consequences. J. Fish. Res. Bd. Canada 35: 657-664.

McCulloch, W.L., J.M. Neff and R.S. Carr. 1980. Bioavailability of heavy metals from used offshore drilling fluids to the clam *Rangia cuneata* and the oyster *Crassostrea gigas*. Pages 964-983 *in* Symposium, Research on Environmental Fate and Effects of Drilling Fluids and Cuttings. Lake Buena Vista, Florida, January 21-24, 1980. American Petroleum Institute, Washington, D.C.

McDermott-Ehrlich, D., D.R. Young, G.V. Alexander, T.K. Jan and G.P. Hershelman. 1978. Chemical studies of offshore oil platforms in the Santa Barbara Channel. Pages 133-144 *in* Proc. Energy/Environment '78, A Symposium on Energy Development Impacts. Society of Petroleum Industry Biologists, Los Angeles, California.

McFarland, V.A. and R.K. Peddicord. 1980. Lethality of a suspended clay to a diverse selection of marine and estuarine macrofauna. Arch. Environ. Contam. Toxicol. 9, 733-741.

McLeay, D.J. 1976. Marine toxicity studies on drilling fluid wastes. *In* Industry/Government Working Group in Disposal Waste Fluids from Petroleum Exploratory Drilling in the Canadian Arctic, Volume 10. Yellowknife, N.W.T., Canada, 17 p.

Meek, R.P. and J.P. Ray. 1980. Induced sedimentation, accumulation, and transport resulting from exploratory drilling discharges of drilling fluids and cuttings. Pages 259-284 *in* Symposium, Research on Environmental Fate and Effects of Drilling Fluids and Cuttings. Lake Buena Vista, Florida, January 21-24, 1980, American Petroleum Institute, Washington, D.C.

Menzel, R.W. 1950. Report on Oyster Studies in Caillou Island Oil Field, Lake Pelto Oil Field, Dog Lake Oil Field, Lake Felicity, and Bayou Bas Bleu, Terrebonne Parish, Louisiana. Texas A&M Research Foundation, Project 9. Texas A&M University, College Station, Texas.

Menzel, R.W. and S.H. Hopkins. 1951. Report on Experiments to Test the Effect of Oil Well Brine or Bleedwater on Oysters at Lake Barre Oil Field. Texas A&M Research Foundation, Project 9. Texas A&M University, College Station, Texas.

Menzel, R.W. and S.H. Hopkins. 1953. Report on Oyster Experiments at Bay Ste. Elaine Oil Field. Texas A&M Research Foundation, Project 9. Texas A&M University, College Station, Texas.

Menzie, C.A. 1982. The environmental implications of offshore oil and gas activities. Environ. Sci. Technol. 16: 454A-472A.

Menzie, C.A., D. Maurer and W.A. Leathem. 1980. An environmental monitoring study to assess the impact of drilling discharges in the mid-Atlantic. IV. The effects of drilling discharges on the benthic community. Pages 499-540 *in* Symposium, Research on Environmental Fate and Effects of Drilling Fluids and Cuttings. Lake Buena Vista, Florida, January 21-24, 1980. American Petroleum Institute, Washington, D.C.

Menzies, R.J., J.P. Morgan, S.Z. El-Sayed and C.H. Oppenheimer. 1974. The Offshore Ecology Investigation, Final Project Planning Council Consensus Report. Gulf Universities Research Consortium Report No. 138. GURC, Galveston, Texas.

Mertens, E.W. 1976. The impact of oil on marine life: A summary of the field studies. Pages 508-514 *in* Sources, Effects and Sinks of Hydrocarbons in the Aquatic Environment. American Institute of Biological Sciences, Washington, D.C.

Middleditch, B.S. (ed.). 1981a. Environmental Effects of Offshore Oil Production. The Buccaneer Gas and Oil Field Study. Plenum Press, New York, 446 p.

Middleditch, B.S. 1981b. Hydrocarbons and sulfur. Pages 15-54 *in* B.S. Middleditch (ed.), Environmental Effects of Offshore Oil Production. The Buccaneer Gas and Oil Field Study. Plenum Press, New York.

Middleditch, B.S. 1981c. Biocides. Pages 55-58 *in* B.S. Middleditch (ed.), Environmental Effects of Offshore Oil Production. The Buccaneer Gas and Oil Field Study. Plenum Press, New York.

Middleditch, B.S. 1984. Ecological Effects of Produced Water Discharges from Offshore Oil and Gas Production Platforms. Final Report on API Project No. 248. American Petroleum Institute, Washington, D.C., 160 p.

Montalvo, J.G., Jr. and D.V. Brady. 1979. Concentrations of Hg, Pb, Zn, Cd, and As in Timbalier Bay and the Louisiana oil patch. Pages 235-244 *in* C.H. Ward, M.E. Bender and D.J. Reish (ed.), The Offshore Ecology Investigation. Effects of Oil Drilling and Production in a Coastal Environment. Rice University Studies 65: 1-589.

Moore, W.S., S. Krishmaswami and S.G. Bhat. 1973. Radiometric determinations of coral growth rates. Bull. Mar. Sci. 23: 157-176.

Morse, M.P., W.E. Robinson and W.E. Wehling. 1982. Effects of sublethal concentrations of the drilling fluids components attapulgite and Q-Broxin on the structure and function of the gill of the scallop, *Placopecten magellanicus* (Gmelin). Pages 235-259 *in* W.B. Vernberg, A. Calabrese, F.P. Thurberg and F.J. Vernberg (eds.), Physiological Mechanisms of Marine Pollutant Toxicity. Academic Press, New York.

National Research Council. 1983. Drilling Discharges in the Marine Environment. National Academy Press, Washington, D.C., 180 p.

Neff, J.M. 1979. Polycyclic Aromatic Hydrocarbons in the Aquatic Environment. Sources, Fates, and Biological Effects. Applied Science Publ., Barking, Essex, England, 262 p.

Neff, J.M. 1980. Effects of used drilling fluids on benthic marine animals. Publ. No. 4330. American Petroleum Institute, Washington, D.C., 31 p.

Neff, J.M. 1982. Accumulation and release of polycyclic aromatic hydrocarbons from water, food, and sediment by marine animals. Pages 282-320 *in* N.L. Richards and B.L. Jackson (eds.), Symposium: Carcinogenic Polynuclear Aromatic Hydrocarbons in the Marine Environment. U.S. Environmental Protection Agency, Gulf Breeze, Florida, EPA-600/9-82-013.

Neff, J.M. 1985. The potential impacts of drilling fluids and other effluents from exploratory drilling on the living resources. *In* R.H. Backus (ed.), Georges Bank: Text and Atlas. MIT Press, Cambridge, Massachusetts (in press).

Neff, J.M. and J.W. Anderson. 1981. Response of Marine Animals to Petroleum and Specific Petroleum Hydrocarbons. Halsted Press, New York, 177 p.

Neff, J.M. and R.J. Breteler. 1983. Waste disposal in the marine environment. 2. Biological availability of organic contaminants to marine invertebrates. Pages 970-972 *in* Proc. Oceans '83 Conference, San Francisco, California, Volume II. Marine Technology Society, Washington, D.C.

Neff, J.M., B.A. Cox, D. Dixit and J.W. Anderson. 1976. Accumulation and release of petroleum-derived aromatic hydrocarbons by four species of marine animals. Mar. Biol. 38: 279-289.

Neff, J.M., J.W. Anderson, B.A. Cox, R.B. Laughlin, Jr., S.S. Rossi and H.E. Tatem. 1976. Effects of petroleum on survival, respiration and growth of marine animals. Pages 515-539 *in* Symposium on Sources, Effects and Sinks of Hydrocarbons in the Aquatic Environment. American Institute of Biological Sciences, Washington, D.C.

Neff, J.M., R.S. Carr and W.L. McCulloch. 1981. Acute toxicity of a used chrome lignosulfonate drilling fluids to several species of marine invertebrate. Mar. Environ. Res. 4: 251-266.

Neff, J.M., R.S. Foster and J.F. Slowey. 1978. Availability of sediment-adsorbed heavy metals to benthos with particular emphasis on deposit feeding infauna. Technical Report D-78-42 to U.S. Army Engineer Waterways Experiment Station, Dredge Material Program, Vicksburg, Mississippi, 286 p.

Neff, J.M., W.L. McCulloch, R.S. Carr and K.A. Retzer. 1980. Comparative toxicity of four used offshore drilling fluids to several species of marine animals from the Gulf of Mexico. Pages 866-881 *in* Symposium, Research on Fate and Effects of Drilling Fluids and Cuttings. Lake Buena Vista, Florida, January 21-24, 1980. American Petroleum Institute, Washington, D.C.

Norseth, T. 1981. The carcinogenicity of chromium. Environ. Hlth. Persp. 40: 121-130.

Northern Technical Services. 1981. Beaufort Sea Drilling Effluent Disposal Study. Performed for the Reindeer Island stratigraphic test well participants under the direction of Sohio Alaska Petroleum Company. Northern Technical Services, Anchorage, Alaska, 329 p. Available from Sohio Alaska Petroleum Co., Anchorage, Alaska.

Northern Technical Services. 1983. Open-Water Drilling Effluent Disposal Study. Tern Island, Beaufort Sea, Alaska. Report for Shell Oil Co. from Northern Technical Services, Anchorage, Alaska, 87 p.

NOAA, Marine Assessment Division. 1978. Analysis of Brine Disposal in the Gulf of Mexico. Bioassay results. Report to the Dept. of Energy, Strategic Petroleum Reserve Program, Salt Dome Storage. Washington, D.C.

Nulton, C.P., D.E. Johnson, C.F. Bohnstedt and S.J. Martin. 1981. Part 3. Organic Chemical Analyses. Pages 133-224 *in* C.A. Bedinger, Jr. (ed.), Ecological Investigations of Petroleum Production Platforms in the Central Gulf of Mexico. Volume 1. Pollutant Fate and Effects Studies. Southwest Research Institute, San Antonio, Texas.

Offshore. 1983. World offshore records, p. 47; and Shell well to break east coast silence, p. 84. June 1983.

Ogle, J., S.M. Ray and W.J. Wardle. 1977. The effect of depth on survival and growth of oysters in suspension culture from a petroleum platform off the Texas coast. Gulf Res. Repts. 6: 31-37.

Ogle, J., S.M. Ray and W.J. Wardle. 1978. The feasibility of suspension culture of oysters (*Crassostrea virginica*) at a petroleum platform off the Texas coast. Contrib. Mar. Sci. 21: 63-76.

Olla, B.L., W.W. Steiner and J.J. Luczkovich. 1982. Effects of Drilling Fluids on the Behavior of the Juvenile Red Hake, *Urophycis chuss* (Walbaum). II. Effects on Established Behavioral Baselines. Progress Report to U.S. Environmental Protection Agency, Gulf Breeze, Florida. Report No. SHL 82-15 from NOAA/NMFS, Northwest Fisheries Center, Sandy Hook Laboratory, Highlands, New Jersey.

Ostroff, A.G. 1965. Introduction to Oilfield Water Technology. Prentice Hall, Englewood Cliffs, New Jersey, 412 p.

Oviatt, C., J. Frithsen, J. Gearing and P. Gearing. 1982. Low chronic additions of No. 2 fuel oil: Chemical behavior, biological impact and recovery in a simulated estuarine environment. Mar. Ecol. Prog. Ser. 9: 121-136.

Page, D.S., B.T. Page, J.R. Hotham, E.S. Gilfillan and R.P. Gerber. 1980. Bioavailability of toxic constituents of used drilling fluids. Pages 984-996 *in* Symposium, Research on Environmental Fate and Effects of Drilling Fluids and Cuttings. Lake Buena Vista, Florida, January 21-24, 1980. American Petroleum Institute, Washington, D.C.

Palmer, L.L. 1978. Brine shrimp bioassay procedure for determining produced water toxicity. *In* Proc. California Regional Meeting of the Society of Petroleum Engineers of AIME, SPE 7129, 6 p.

Parker, J.H., J.S. Nickels, R.F. Martz, M.J. Gehron, N.L. Richards and D.C. White. 1984. Effect of well-drilling fluids on the physiological status and microbial infection of the reef building coral *Montastrea annularis*. Arch. Environ. Contam. Toxicol. 13: 119-125.

Patrick, W.H., R.P. Gambrell and R.A. Khalid. 1977. Physicochemical factors regulating solubility and bioavailability of toxic heavy metals in contaminated dredged sediment. J. Environ. Sci. Hlth. A12: 475-492.

Payne, J.R., J.L. Lambach, R.E. Jordan, G.D. McNabb, Jr., R.R. Sims, A. Abasumara, J.G. Sutton, D. Generro, S. Gagner and R.F. Shokes. 1982. Georges Bank Monitoring Program. Analysis of Hydrocarbons in Bottom Sediments and Analysis of Hydrocarbons and Trace Metals in Benthic Fauna. First year final report to U.S. Dept. of Interior, Minerals Management Service, New York OCS Office, New York by Science Applications, Inc., LaJolla, California.

Payne, J.R., J.L. Lambach, R.E. Jordan, C.R. Phillips, G.D. McNabb, Jr., M.K. Beckel, G.H. Farmer, R.R. Sims, Jr., J.G. Sutton, C. deOtiveira and A. Abusamura. 1983. Georges Bank Monitoring Program: Year Two. Analysis of Hydrocarbons in Bottom Sediments and Analysis of Hydrocarbons and Trace Metals in Benthic Fauna. Final Report to U.S. Dept. of Interior, Minerals Management Service, Washington, D.C. by Science Applications, Inc., LaJolla, California.

Pearson, R.G. 1981. Recovery and recolonization of coral reefs. Mar. Ecol. Prog. Ser. 4: 105-122.

Petrazzuolo, G. 1983. Draft Final Technical Report Document: Environmental Assessment: Drilling Fluids and Cuttings Released onto the OCS. Submitted to Office of Water Enforcement and Permits, U.S. Environmental Protection Agency, Washington, D.C. by Technical Resources, Inc., Bethesda, Maryland.

Petrilli, F.L. and S. DeFlora. 1982. Interpretations on chromium mutagenicity and carcinogenicity. Pages 453-464 *in* Mutagens in Our Environment. Alan R. Liss, Inc., New York.

Pottle, R.A. and R.W. Elner. 1982. Substrate preference behavior of juvenile American lobsters, *Homarus americanus*, in gravel and silt-clay sediments. Can. J. Fish. Aquat. Sci. 39: 928-932.

Powell, E.N., M. Kasschau, E. Che, M. Loenig and J. Peron. 1982. Changes in the free amino acid pool during environmental stress in the gill tissue of the oyster, *Crassostrea virginica*. Comp. Biochem. Physiol. 71A: 591-598.

Read, A.D. 1978. Treatment of oily water at North Sea oil installations - a progress report. Pages 127-136 *in* C.S. Johnston and R.J. Morris (eds.), Oily Water Discharges. Regulatory, Technical and Scientific Considerations. Applied Science Publishers, Barking, Essex, England.

Rhoads, D.C., P.L. McCall and J.Y. Yingst. 1978. Disturbance and production of the estuarine seafloor. Amer. Scient. 66: 577-586.

Rice, S.D., J.W. Short and J.F. Karinen. 1977. Comparative oil toxicity and comparative animal sensitivity. Pages 78-94 *in* D.A. Wolfe (ed.), Fate and Effects of Petroleum Hydrocarbons in Marine Ecosystems and Organisms. Pergamon Press, New York.

Roesijadi, G. and J.W. Anderson. 1979. Condition index and free amino acid content of *Macoma inquinata* exposed to oil-contaminated marine sediments. Pages 69-83 *in* W.B. Vernberg, F.P. Thurberg, A. Calabrese and F.J. Vernberg (eds.), Marine Pollution: Functional Responses. Academic Press, New York.

Roesijadi, G., J.W. Anderson and J.W. Blaylock. 1978a. Uptake of hydrocarbons from marine sediments contaminated with Prudhoe Bay crude oil: Influence of feeding type of test species and availability of polycyclic aromatic hydrocarbons. J. Fish. Res. Bd. Canada. 35: 608-614.

Roesijadi, G., D.L. Woodruff and J.W. Anderson. 1978b. Bioavailability of naphthalenes from marine sediments artificially contaminated with Prudhoe Bay crude oil. Environ. Pollut. 15: 223-229.

Rose, C.D. and T.J. Ward. 1981. Acute toxicity and aquatic hazard associated with discharge formation water. Pages 301-328 *in* B.S. Middleditch (ed.), Environmental Effects of Offshore Oil Production. The Buccaneer Gas and Oil Field Study. Plenum Press, New York.

Rossi, S.S. 1977. Bioavailability of petroleum hydrocarbons from water, sediments and detritus to the marine annelid *Neanthes arenaceodentata*. Pages 621-626 *in* Proceedings 1977 Oil Spill Conference (Prevention, Behavior, Control, Cleanup). American Petroleum Institute, Washington, D.C.

Rubinstein, N.I., R. Rigby and C.N. D'Asaro. 1980. Acute and sublethal effects of whole used drilling fluids on representative estuarine organisms. Pages 828-846 *in* Symposium, Research on Environmental Fate and Effects of Drilling Fluids and Cuttings. Lake Buena Vista, Florida, January 21-24, 1980. American Petroleum Institute, Washington, D.C.

Saltzmann, H.A. 1982. Biodegradation of aromatic hydrocarbon in marine sediments of three North Sea oil fields. Mar. Biol. 72: 17-26.

Sanders, H.L. 1981. Environmental effects of oil in the marine environment. Pages 117-146 *in* Safety and Offshore Oil: Background Papers of the Committee on Assessment of Safety of OCS Activities. National Research Council, National Academy Press, Washington, D.C.

Sharp, J.M. and S.G. Appan. 1978. Cumulative effects of oil drilling and production on estuarine and nearshore ecosystems. Pages 57-73 *in* M.L. Wiley (ed.), Estuarine Interactions. Academic Press, New York.

Sharp, J.R., R.S. Carr and J.M. Neff. 1984. Influence of used chrome lignosulfonate drilling and fluids on the early life history of the mummichog *Fundulus heteroclitus*. *In* Proc. Ocean Dumping Symposium. John Wiley and Sons, New York, 14 p. (in press).

Sis, R.S., J.M. Neff, V.L. Jacobs, N. McArthur, R. Tarpley, G. Stott, H. Armstrong and C.C. Corkern II. 1981. Part 7. Normal Histology and Histopathology of Benthic

Invertebrates and Demersal and Platform-Associated Pelagic Fishes. Pages 393-527 *in* C.A. Bedinger, Jr. (ed.), Ecological Investigations of Petroleum Production Platforms in the Central Gulf of Mexico. Volume I. Pollutant Fate and Effects Studies. Southwest Research Institute, San Antonio, Texas.

Sizemore, R.K., C.-H. Hsu and K.O. Olsen. 1981. Bacterial community composition and activity. Pages 223-236 *in* B.S. Middleditch (ed.), Environmental Effects of Offshore Oil Production. The Buccaneer Gas and Oil Field Study. Plenum Press, New York.

Smedes, G.W., R.P. Herbst and J. Calman. 1981. Hydrodynamic modeling of discharges. Pages 387-402 *in* B.S. Middleditch (ed.), Environmental Effects of Offshore Oil Production. The Buccaneer Gas and Oil Field Study. Plenum Press, New York.

Smith, G.A., J.S. Nickels, R.J. Bobbie, N.L. Richards and D.C. White. 1982. Effects of oil and gas well-drilling fluids on the biomass and community structure of microbiota that colonize sands in running seawater. Arch. Environ. Contam. Toxicol. 11: 17-24.

Sprague, J.B. 1973. The ABC's of pollutant bioassay using fish. Pages 6-30 *in* Biological Methods for the Assessment of Water Quality, ASTM STP 528. American Society for Testing and Materials, Philadelphia.

Szmant-Froelich, A., V. Johnson, T. Hoen, J. Battey, G.J. Smith, E. Fleishmann, J. Porter and D. Dallmeyer. 1982. The physiological effects of oil-drilling fluids on the Caribbean coral *Montastrea annularis*. *In* Proc. 4th Intern. Coral Reef Symposium, Manila, Volume 1.

Tagatz, M.E. and C.H. Deans. 1983. Comparison of field- and laboratory-developed estuarine benthic communities for toxicant-exposure studies. Water, Air, Soil Poll. 20: 199-209.

Tagatz, M.E., J.M. Ivey, C.E. DalBo and J.L. Oglesby. 1982. Responses of developing macrobenthic communities to drilling fluids. Estuaries 5: 131-137.

Tagatz, M.E., J.M. Ivey, H.K. Lehman and J.L. Oglesby. 1978. Effects of lignosulfonate-type drilling fluids on development of experimental estuarine macrobenthic communities. Northeast Gulf Sci. 2: 25-42.

Tagatz, M.E., J.M. Ivey, H.K. Lehman, M. Tobia and J.L. Oglesby. 1980. Effects of drilling fluids on development of experimental estuarine macrobenthic communities. Pages 847-865 *in* Symposium, Research on Environmental Fate and Effects of Drilling Fluids and Cuttings. Lake Buena Vista, Florida, January 21-24, 1980. American Petroleum Institute, Washington, D.C.

Tagatz, M.E. and M. Tobia. 1978. Effect of barite ($BaSO_4$) on development of estuarine communities. Estuar. Coastl. Mar. Sci. 7: 401-407.

Thompson, J.H. Jr. 1980. Responses of Selected Scleractinian Corals to Drilling Fluid used in the Marine Environment. Ph.D. Dissertation, Texas A&M University, College Station, 130 p.

Thompson, J.H., Jr. and T.J. Bright. 1977. Effects of drilling fluids on sediment clearing rates of certain hermatypic corals. Pages 495-498 *in* Proc. 1977 Oil Spill Conference. American Petroleum Institute, Washington, D.C.

Thompson, J.H., Jr. and T.J. Bright. 1980. Effects of an offshore drilling fluid on selected corals. Pages 1044-1078 *in* Symposium, Research on Environmental Fate and Effects of Drilling Fluids and Cuttings. Lake Buena Vista, Florida, January 21-24, 1980. American Petroleum Institute, Washington, D.C.

Thompson, J.H., Jr., E.A. Shinn and T.J. Bright. 1980. Effects of drilling fluids on seven species of reef-building corals as measured in the field and laboratory. *In* R.A. Geyer (ed.), Marine Environmental Pollution. Volume 1. Hydrocarbons. Elsevier Oceanography Series. Volume 27A. Elsevier/North-Holland, New York.

Thompson, R.R., B.D. Honeycutt and J.C. Parker. 1972. Cooperative environmental projects, High Island Block 246, offshore, Texas. Proc. Offshore Technology Conference, Houston, Texas 4: 543-548.

Tillery, J.B. and R.E. Thomas. 1980. Heavy metal contamination from petroleum production platforms in the Gulf of Mexico. Pages 562-587 *in* Symposium, Research on Environmental Fate and Effects of Drilling Fluids and Cuttings. Lake Buena Vista, Florida, January 21-24, 1980. American Petroleum Institute, Washington, D.C.

Tillery, J.B., R.E. Thomas and H.L. Windom. 1981. Part 4. Trace Metals Studies in Sediments and Fauna. Pages 1-122 *in* C.A. Bedinger, Jr. (ed.), Ecological Investigations of Petroleum Production Platforms in the Central Gulf of Mexico. Volume 1. Pollutant Fate and Effects Studies. Research Institute, San Antonio, Texas.

Tornberg, L.D., E.D. Thielk, R.E. Nakatoni, R.C. Miller and S.O. Hillman. 1980. Toxicity of drilling fluids to marine organisms in the Beaufort Sea, Alaska. Pages 997-1006 *in* Symposium, Research on Environmental Fate and Effects of Drilling Fluids and Cuttings. Lake Buena Vista, Florida, January 21-24, 1980. American Petroleum Institute, Washington, D.C.

Trefry, J.H., R.P. Trocine and K.E. Yhip. 1983. Drilling mud discharges: minimizing environmental mismatches. Pages 1228-1237 *in* Proc. 3rd Symp. Coastal and Ocean Management, ASCE/San Diego, California, June 1-4, 1983.

Trocine, R.P. and J.H. Trefry. 1983. Particulate metal tracers of petroleum drilling mud dispersion in the marine environment. Environ. Sci. Technol. 17: 507-512.

Trocine, R.P., J.H. Trefry and D.B. Meyer. 1981. Inorganic tracers of petroleum drilling fluid dispersion in the northwest Gulf of Mexico. Reprint Extended Abstract. Div. Environ. Chem., ACS Meeting, Atlanta, Georgia, March-April, 1981.

Twichell, D.C., C.E. McClennen and B. Butman. 1981. Morphology and processes associated with the accumulation of the fine-grained sediment deposit on the southern New England shelf. J. Sediment. Petrol. 51: 269-280.

U.S. Dept. of the Interior. 1975. Draft Environmental Statement DES 75-35. U.S. Dept. of Interior, Washington, D.C. June, 1975. III, 22.

van der Borght, O. 1963. Accumulation of radium-226 by the freshwater gastropod *Lymnaea stagnalis* L. Nature 197: 612-613.

Waller, J.W. 1974. Effects of Platforms on Biota (Invertebrates). Final Report to Gulf University Research Consortium, Offshore Ecology Investigation. GURC, Galveston, Texas.

Ward, C.H., M.E. Bender and D.J. Reish (eds.). 1979. The Offshore Ecology Investigation. Effects of Oil Drilling and Production in a Coastal Environment. Rice Univ. Studies 65: 1-589.

Wheeler, R.B., J.B. Anderson, R.R. Schwarzer and C.L. Hokanson. 1980. Sedimentary processes and trace metal contaminants in the Buccaneer oil/gas field, northwestern Gulf of Mexico. Environ. Geol. 3: 163-175.

Wildish, D.J. 1972. Acute toxicity of polyoxyethylene esters and polyoxyethylene ethers to *S. salar* and *G. oceanicus*. Wat. Res. 6: 759-762.

Wolfson, A., G. van Blaricom, N. Davis and G.S. Lewbel. 1979. The marine life of an offshore oil platform. Mar. Ecol. Prog. Ser. 1: 81-89.

Workman, I.K. and C.E. Jones. 1979. Determine the effects of oil field discharges on species composition and abundance of pelagic fishes and demersal fishes and macrocrustaceans in the oil field. Pages 2.3.5-1 to 2.3.5-149 *in* W.B. Jackson (ed.), Environmental Assessment of an Active Oil Field in the Northwestern Gulf of Mexico, 1977-1978. Volume II. Data Management and Biological Investigations National Oceanic and Atmospheric Administration, National Marine Fisheries Service, Galveston, Texas.

Zein-Eldin, Z.P. and P.M. Keney. 1979. Bioassay of Buccaneer oil field effluents with penaeid shrimp. Pages 2.3.4-1 to 2.3.4-25 *in* Environmental Assessment of an Active Oil Field in the Northwestern Gulf of Mexico, 1977-1978. Volume II: Data Management and Biological Investigations. National Oceanic and Atmospheric Administration, National Marine Fisheries Service, Galveston, Texas.

Zingula, R.P. 1975. Effects of drilling operations on the marine environment. Pages 433-448 *in* Environmental Aspects of Chemical Use in Well-Drilling Operations. U.S. Environmental Protection Agency, EPA-560/1-75-004.

CHAPTER 11

OFFSHORE OIL DEVELOPMENT AND SEABIRDS: THE PRESENT STATUS OF KNOWLEDGE AND LONG-TERM RESEARCH NEEDS

George L. Hunt, Jr.

CONTENTS

GENERAL INTRODUCTION

The destruction of seabirds by oil spilled at sea concerned conservationists long before oil exploration began in the cold, biologically rich waters of high-latitudes (Bourne, 1968a, 1979; Cowell, 1976). Early pollution resulted as much from deliberate discharge of oily bilge water and ballast from ships as from accidental spills (Vermeer and Vermeer, 1974). In more recent years, with the development of offshore oil recovery operations in the North Sea and in the shelf waters of northern North America, there has been an increasing focus on the potential for accidents at drilling rigs and at transfer facilities. In thc mid 1970s meetings were held in southern California (Hunt, 1975) and Seattle, Washington (Nisbet, 1979) to delineate the kinds and extent of baseline studies required. Now, nearly eight years later, it is time to take stock of what we have learned since the earlier reviews (Bourne, 1976; Holmes and Cronshaw, 1977; Brown, 1982) and to refocus our research efforts as appropriate.

Much of the early concern over the effect of oil on seabirds was generated by the sight of oil-covered dead or moribund birds on beaches. Reports of massive kills raised the concern of both the public and the scientific community (Bourne, 1968a; Vermeer and Vermeer, 1974). It was feared that oil spilled on the oceans would destroy seabird populations or at least reduce them to a small fraction of their former size. Losses of colonies along the Brittany coast and in southeast England subsequent to the *Torrey Canyon* and *Amoco Cadiz* spills (Monnat, 1969; Bourne, 1971; Cowell, 1976; Hope-Jones *et al.*, 1978) were cited as examples, as were the decreases in Long-tailed Ducks (*Clangula hyemalis*) and scoters (*Melanitta* sp.) in the Baltic Sea (Bourne, 1968b; Clark, 1969; Cowell, 1976). Today, no one will deny the likelihood of high mortality when seabirds encounter oil at sea and that this waste of life should be avoided.

Although these losses are unfortunate, we must ask whether there has been a detrimental effect from oil pollution on the populations of the species involved (Clark, 1969, 1984; Bourne, 1968a; Dunnet, 1982). The answer is maybe yes in the case of the sea ducks of the Baltic (Lemmetyinen, 1966; Vermeer and Vermeer, 1975; Clark, 1984; although Joensen and Hansen [1977] minimized the damage done) and maybe also yes for the Jackass Penguin (*Spheniscus demersus*) in South Africa (Westphal and Rowan, 1970; Vermeer and Vermeer, 1975; Clark 1984). What about other species? Dunnet (1982) has estimated that the natural mortality of seabirds in the North Sea is in the hundreds of thousands, if not over a million birds, annually. In contrast, he estimates the average annual oil related mortality of birds in the North Sea is in the low tens of thousands. The importance of this pollution-related mortality depends on whether it is additional to the natural mortality, or if the natural mortality is reduced proportionately by the number of birds removed by oil pollution. If the substantial winter mortality of seabirds, particularly young seabirds, is density-dependent, then oil-pollution related deaths may only be removing "surplus" birds that would have otherwise died of other causes (Bourne, 1968a). Alternatively, the birds lost may represent or be replaced by a "floating," non-breeding segment of the population that are able to recruit rapidly to colonies after a natural catastrophe, thus reducing the ability of the population to recover rapidly from natural losses.

We can investigate the impact of the oil-related mortality by examining the fate of breeding colonies of seabirds nesting along the North Sea coast, particularly those in northern England and Scotland where the results of long-term studies make comparisons possible. It appears that Northern Fulmars (*Fulmarus glacialis*), Black-legged Kittiwakes (*Rissa tridactyla*) and murres (*Uria* spp.) are all increasing throughout this region (Fisher, 1966; Heubeck, 1981; Stowe, 1982a; Wanless *et al.*, 1982; Coulson, 1983; Clark, 1984; Stowe and Underwood, 1984). Local populations may recover quickly after small spills (Stowe, 1982b), although in other cases, recovery may be delayed (Stowe and Underwood, 1984; Stowe, in litt.). Overall, in the North Sea the effect of oil pollution on seabird populations has not lived up to its disastrous potential, and it is Clark's (1984) view that oil pollution may well have had no significant effect on North Sea bird populations.

At present we are unable to give similar assurances that offshore oil development will have as apparently benign effects in North America. We do not have an adequate data base to make the appropriate comparisons and the ecological setting for marine birds in much of North America is different from that in Europe. In the Bering Sea and along parts of the coast of California and Oregon, seabirds are concentrated in a relatively small number of large colonies of over 100,000 birds each (Brown *et al.*, 1975; Sowls *et al.*, 1978; Varoujean, 1978; Varoujean and Pitman, 1979; Sowls *et al.*, 1980). In contrast, along the coast of the North Sea most species of breeding seabirds are spread out in smaller colonies (Cramp *et al.*, 1974). Localized spills will usually affect only a small portion of any species' population in Europe; a spill near one of the large North American colonies could affect a major segment of a population. In other parts of the United States (Gulf of Alaska, Atlantic seaboard, Gulf coast) colonies are smaller and dispersed, much as in the North Sea.

Arctic North American waters are ice choked for large portions of the year. The ice traps oil, decreases its dispersion by wave action and concentrates birds in the reduced areas of open water (Gaston and Nettleship, 1981). Additionally, cold temperatures may increase the time oil is dangerous to birds by retarding the loss of volatile components (Bourne and Bibby, 1975; Joensen and Hansen, 1977). These conditions increase population-level hazards to birds from oil spills in northern North America relative to the North Sea (Vermeer and Anweiler, 1975; Brown, 1982), or for that matter relative to the warmer waters of North America.

Species of particular concern are those recognized to be vulnerable to oiling (King and Sanger, 1979; Kaiwi and Hunt, 1985). These include among others, the loons, grebes, waterfowl, phalaropes and auks. Loons and grebes are of particular concern because wintering populations are geographically concentrated in coastal areas vulnerable to pollution and because these species have relatively small world populations (Clark, 1984).

The key to the conservation of marine birds is to assure the continued survival and recruitment of new individuals into populations. If we know enough about food demands and resources, including foraging areas, life tables, immigration and sensitivity of both mortality and recruitment to environmental fluctuation, we will be in a position to predict the potential damage of a spill or of long-term chronic impacts resulting from offshore oil development.

Simulation models (Ford *et al.*, 1982; Samuels and Lanfear, 1982; Wiens *et al.*, 1979) are useful for developing these predictions because they allow examination of the importance of many variables, but the models are limited by the lack of data available for generating the required values of input variables (Wiens *et al.*, 1984). These models of seabird population responses to oil pollution can also be used in sensitivity analyses to determine those aspects of seabird ecology that determine population responses to oil pollution. Future oil-related research on seabirds should address the information requirements of these models and use the models for identifying the most critical areas for field research. This coordination between modeling and field research efforts is mandatory if we are to identify areas of critical concern and concentrate our resources on answering the most pressing questions.

We must work to define a biologically significant loss or impact. While some levels of loss probably will not cause noticeable damage to populations, there is likely a point at which further losses will result in a population decline. We need to determine the extent of loss or decline that is biologically important. Since populations may recover after a decline, we need to define an acceptable time period for population recovery.

The goals and potential value of monitoring studies must also be addressed. Monitoring has at least two functions. On the one hand, it provides the means of assessing the impact of large spills or even a series of small events or chronic low-level pollution. We can use this information to test the predictions of our models and to improve them and, when necessary, to affix blame. On the other hand, we can try to separate the damage to bird populations due to oil pollution from other causes of decline. This information may be extremely important as competition between birds and man for fishery resources occurs in areas of offshore oil development.

Competition between man and seabirds for fishery resources has been reviewed by Furness (1983). Seabirds are sensitive to man-caused reductions in food resources and precipitous declines in bird populations in the wake of overfishing are possible (Palmer, 1949; Schaefer, 1970; Bourne, 1976; Frost *et al.*, 1976; Nettleship, 1977). Since major fisheries co-occur with offshore oil development in many areas off the North American coast, we need to separate with acceptable statistical confidence the effects of oil pollution from those of harvesting marine resources. Concurrent studies of reproductive success, foods used and fish abundance is one approach to this problem (Hunt and Butler, 1980; D.W. Anderson, pers. comm.). If we are not successful in this endeavor, we will be unable to manage and conserve the full spectrum of marine resources.

In this chapter I have focused on studies of marine birds breeding in North America and subject to the effects of offshore oil development. The majority of population, reproductive and ecological studies of these birds has been done in Canada, New England, California and Alaska. Additionally, much information on the northern species, especially the particularly vulnerable alcids, is available from European studies. Thus, these species are emphasized in this report. It is also in the waters of northern New England, the West Coast and Alaska that some of our greatest concentrations of seabirds occur, and it is appropriate to focus on these regions.

The bays, estuaries, shores and wetlands of the southeastern Atlantic coast and Gulf of Mexico also support immense numbers of migrating and wintering shorebirds and waterfowl. These species are also vulnerable to oil pollution, but it is beyond the scope of this chapter to deal with their very different reproductive ecology and their potential responses to offshore oil development. Suffice it to say that waterfowl are widely acknowledged to be extremely vulnerable to contamination by floating oil. Likewise, shorebirds, particularly those feeding in the surgeline on open beaches, like Sanderlings (*Calidris alba*), are vulnerable to oiling (A. Amos, unpubl.). Attention must be paid to the direct contamination of these species, their foods and foraging areas in locations where they concentrate.

Offshore oil development can affect marine birds in a variety of ways. Major spills can directly destroy large numbers of adults and indirectly result in the starvation of nestlings deprived of food. Less dramatic long-term, chronic pollution or disturbance may also have detrimental effects on marine birds or their food supplies. Low levels of pollution may increase adult and juvenile mortality through fouling or ingestion, and sub-lethal amounts of ingested oil may lower reproductive success. Finally, disturbance of birds at colonies may reduce reproductive success or cause desertion.

The chapter is organized to focus first on modeling efforts, what models can tell us, and the kinds of data required for these modeling efforts. Subsequent sections deal with the potential chronic effects related to disturbance and the physiological effects of ingested oil on adults and young. A brief discussion of possible monitoring efforts is provided, although it does not cover the subject in depth.

PREDICTIVE MODELS - POPULATION DYNAMICS AND REGULATION

Models

Simulation models are useful for predicting how marine bird populations will respond to large spills, chronic pollution and disturbance of colonies associated with offshore oil development. At present, there are three such models: one for the Pribilof Islands and Kodiak Island (Ford *et al.*, 1982; Wiens *et al.*, 1979); a second by Samuels and Lanfear (1982) for the northern Gulf of Alaska that incorporates some density-dependent features; and a third, modified from the one used by Wiens' group in Alaska, has been developed for the Southern California Bight and is sensitive to density-dependent features (R.G. Ford, pers. comm.). These models primarily address the impact of oil on bird populations during and subsequent to a large spill. The problems of chronic pollution and disturbance associated with shore facilities have not been investigated although the models can be used to examine this problem. Likewise, the models do not address adequately the behavioral flexibility of birds or density-dependent events on the wintering grounds or the effects of recruitment and movements between subpopulations. Density-dependent interactions are likely important in the largest colonies; no information on its importance is available for wintering populations. Studies of population size changes at colonies demonstrate the major importance of movements between subpopulations (Leslie, 1966; Nisbet, 1973, 1978a; Potts *et al.*, 1980; Dunnet, 1982; Harris, 1983; Clark, 1984). The models need to be refined if they are to be of maximum value in deciding priorities for research and conservation efforts.

The accuracy of the model predictions depends on the structure of the models themselves, and on the quality of data available to define the various functions and variables in the model. Wiens *et al.* (1984) have recently evaluated the available data, and concluded that we are lacking a great deal of species-specific information (physiological parameters), as well as population specific data (age structure, life tables, immigration).

Information Available for Use in Modeling

The models predicting seabird population response to oil require estimates of population size, the pelagic distribution of birds around a colony, energy requirements of adult birds for daily existence, the energy requirements of chicks for existence and growth, reproductive success, and mortality rates, density dependence, immigration, etc. The validity and utility of the model predictions depend on the quality of these inputs. These parameters will frequently differ among colonies. There is also variation between individuals at a given colony. There are few colonies in North America where these parameters have been measured and even fewer where long-term studies have provided an estimate of natural variation.

Population Size

Recent censusing efforts have produced catalogs of seabird colonies for most regions in North America (Table 11.1) including Canada, the east, west and gulf coasts of the United States and Alaska. Information on colony size is also available for particular species or particular habitats (Table 11.1). These population estimates are largely based on single, brief visits to colonies and there is little information on the average size of particular colonies or changes in colony size through time.

Census data based on single visits are likely to be unreliable. Regional colony surveys replicated over periods of a decade or more using standardized methods are needed, but are nonexistent in North America today (Erwin *et al.*, 1981). Colonies show annual fluctuations in size (Barrett and Schei, 1977; Ollason and Dunnet, 1983), and there are both diel and seasonal variations in the number of birds present at colonies within a breeding season (Lloyd, 1973; Birkhead, 1978; Dunnet *et al.*, 1979; Slater, 1980; Gaston and Nettleship, 1982). Hence, the estimates of population size available for use in the models may have very broad confidence limits and all but the largest changes will be hard to document with statistical significance given the present quality of data.

TABLE 11.1

Sources of information on the size of seabird populations at North American colonies. The National Audubon Society and the Cornell University Laboratory of Ornithology also compile information on North American colonial birds in the Colonial Bird Registry

Region	*Source*	*Coverage*
Pacific Coast		
Alaska	Sowls *et al.* (1978)	seabird colonies
Canada	Vermeer *et al.* (1983)	marine birds, west coast
	Canadian Wildlife Service (G. Kaiser, K. Vermeer)	British Columbia seabirds
	Provincial Museum, British Columbia (W. Campbell)	Vancouver Is. and Queen Charlotte Is.
Washington	Manuwal *et al.* (1978)	Strait of Juan de Fuca, Puget Sound
	Peters *et al.* (1978)	sea and wading birds
Oregon	Varoujean and Pitman (1979)	seabird colonies
California	Sowls *et al.* (1980)	seabird colonies
Baja California	Anderson (1983)	seabirds
Atlantic Coast		
Canada	Brown *et al.* (1975)	seabirds
	Chapdelaine and Brousseau (1984)	
	Cannell and Maddox (1983)	New Brunswick seabirds
	Freedman and Svoboda (1982)	Alexandra
	Nettleship and Lack (1973)	Fjord, Ellsmere Is., Northwest Territories, breeding birds
United States	Buckley (1978)	sea and wading birds, New Jersey
	Buckley and Buckley (1980)	water birds, Long Island
	Custer *et al.* (1980)	Atlantic coast herons
	DiCostanzo (1980)	Common Tern
	Drury (1973)	New England seabirds
	Drury and Kadlec (1974)	Herring Gulls
	Duffy (1977)	terns and skimmers, Long Island Sound
	Erwin (1979), Erwin and Korschgen (1979)	coastal waterbirds, Maine-Virginia
	Erwin *et al.* (1981)	seabird colony habitats
	Korschgen (1979)	coastal waterbirds, Maine
	Nisbet (1973)	terns, Massachusetts
	Osborn and Custer (1978)	herons
	Parnell and Soots (1980)	waterbirds, North Carolina
	Portnoy (1980)	gulls and terns, North Carolina-Florida
Gulf Coast	Caesar Kleberg Wildlife Research Institute (1981)	waterbirds, Texas
	Nesbitt *et al.* (1982)	herons, Florida
	Portnoy (1977)	sea and wading birds, Louisiana-Alabama
	Sloan (1982)	White Pelicans

Foraging Distribution around Colonies

The foraging distribution of breeding birds around colonies influences the likelihood of birds encountering oil. Birds may be highly aggregated in one or a few favored foraging areas or they may be widely dispersed. Bird distribution has been studied at several colonies in Alaska: Kodiak Island (Lensink *et al.*, 1978), St. Matthew, St. Lawrence and King's Islands (Hunt, unpubl.) the Pribilof Islands (Hunt *et al.*, 1981b), and at colonies in California (Southern California Bight, Hunt *et al.*, 1979; Farallon Islands, Point Reyes Bird Observatory (PRBO), unpubl.). The distance birds venture from their colonies depends on the foraging behaviors of the particular species, stage in the reproductive cycle, and a number of site specific factors, such as colony size, hydrography (Kinder *et al.*, 1983) and food availability (Gaston and Nettleship, 1982). The likelihood of oiling of breeding birds foraging near a colony thus depends not only on the probability of oil entering a given area, but also on the frequency of use of that area by the birds.

A second aspect of the probability of oiling depends upon the behavior of the birds once oil enters a preferred area. A first concern is how birds already present will react to the oil. If they detect the oil and move away, mortality will be light compared to a situation in which they remain, especially if they make repeated passages through the surface film. Recent studies have made a start in describing the behavior of birds in the presence of oil slicks (Nero and Associates, Inc., 1983), but additional work is required to demonstrate whether birds reduce their foraging activity and move away from the area of a spill.

The damage to the population will differ greatly if birds quickly cease arriving at a polluted site and shift foraging activity to an alternate feeding site. The time course of a shift to an alternate site will be affected by whether departing foragers normally retrace the flight paths of foragers returning to a colony or if they go to traditional or favored areas regardless of what other birds are doing. At present we have little information on which behavior pattern is most typical for colonial seabirds (Wittenberger and Hunt, 1985).

Energy Requirements

The estimates of the energetic requirements of seabirds used in the models are based on Lasiewski and Dawson's (1967) equation for nonpasserine birds which predicts basal metabolic rate from body size. Lasiewski and Dawson's data included few measurements from North American seabirds. Basal metabolic rates (BMR) have not been reported for most of the species in large North American seabird colonies. Actual measurements of basal metabolic rate in seabirds have shown substantial deviations from the predicted value (Table 11.2). Daily existence costs are estimated as multiples of the basal metabolic rate, either 2.6 (Weathers and Nagy, 1980) or 2.7 (Kooyman *et al.*, 1982)) times the BMR, as confirmed by measurements of free existence costs using doubly-labeled water. However, Walsberg (1983) reports that daily existence costs scale differently than BMR, and this difference will affect estimates of energy needs.

The cost of growth among North American seabird species has been studied in Double-Crested Cormorants (Dunn, 1975a,b, 1976a), Herring Gulls (Brisbin,

TABLE 11.2
Measured basal metabolism of North American seabirds compared to values predicted by Lasiewski and Dawson's (1967) equation for nonpasserine birds

Order *Species*	*Sample size*	*Mass (g)*	*BMR* (kcal/day)*	*% Predicted value*	*Source*
Charadriiformes					
Glaucous Gull	2	1210	183.0	203	Scholander *et al.*, 1950
Herring Gull	2	912	136.8	170	Baudinette and Schmidt-Nielson, 1974
	6	1000	102.0	130	Lustick *et al.*, 1978
Western Gull	7	761	70.3	110	Obst, unpubl.
Common Murre	5	956	140.4	190	Johnson and West, 1975
Thick-billed Murre	5	989	140.4	192	Johnson and West, 1975
Xantus' Murrelet	3	167	39.8	185	Eppley, 1984
Pelecaniformes					
Brown Pelican	1	3510	264.0	137	Benedict and Fox, 1927
Procellariiformes					
Fork-tailed Storm-Petrel	1	49	13.2	149	Iversen and Krog, 1972

*Basal Metabolic Rate.

1965; Dunn, 1975b, 1976b) and Pigeon Guillemots (Koelink, 1972). While seabird growth has been studied by a number of researchers (see Ricklefs, 1973), there are few estimates of the costs of growth *per se*. The cost of growth will depend on the thermal independence of the young, the ambient temperature and the amount of brooding the young receive. Young seabirds show several different post-hatching developmental patterns (precocial, semiprecocial, intermediate, semialtricial) which are associated with different patterns of energy allocation to growth and temperature regulation. Estimates of the costs of growth for some of these developmental patterns have not been published.

Reproductive Success and Its Variability

There are a modest number of studies of reproductive success at North American colonies which examine reproductive success over a period of three years or more. However, long-term studies are still needed to determine the normal variability in reproductive success for many species in a number of areas. Published long-term studies of reproductive success are listed in Table 11.3. In addition, there are several long-term and as yet unpublished studies: for Western Gulls at Santa Barbara Island, California (Hunt *et al.*, unpubl.), for various species at the Farallon Islands, California (PRBO, unpubl.), for a number of cliff-nesting seabirds (mostly murres, Black-legged Kittiwakes and puffins) at colonies in Alaska (Semidi Islands, Barren Islands, Ugaiushak Island, Hinchinbrook Island, Chiniak Bay (Baird and Gould, in press), for Herring and Great Black-backed Gulls (*L. marinus*) in Maine (Hunt, 1972; Hunt, unpubl.; Drury, unpubl.) and for a number of colonial seabirds at Kent Island (Huntington, unpubl.). These studies demonstrate considerable yearly variation in reproductive success.

Survivorship and Recruitment Rates

We are beginning to accumulate information demonstrating the density-dependence of reproductive success in seabirds (Coulson *et al.*, 1982; Gaston *et al.*, 1983; Hunt *et al.*, 1986). Since limiting resources (nesting space, food availability) very likely differ between colonies, results from one colony cannot with assurance be extended to colonies of different size or to those in different regions or habitats. Thus, future studies will have to focus on representative colonies of different size in each of the regions of critical concern.

Survivorship estimates for adult Northern Hemisphere seabirds are largely based on band recoveries from live birds returning to or dead birds recovered at small European colonies (Table 11.4). These results may be biased by band loss (Kadlec and Drury, 1969; Kadlec, 1975; Couslon, 1976; Hatch and Nisbet, 1983), which is no longer a great problem due to the use of longer-lasting bands, or movement between colonies (Brooke, 1978; Chabrzyk and Coulson, 1976; Harris, 1983). Additionally, it may not be appropriate to apply European survivorship figures to some North American seabird populations since the North American high latitude seabird colonies are much larger than the European colonies where survivorship has been studied, sea temperatures are much lower, and weather patterns and food webs differ. There may also be differences in the

TABLE 11.3
North American studies of reproductive success in seabirds of three or more years duration

Location	*Species*	*References*
Prince Leopold Island, Northwest Territories	Thick-billed Murres	Gaston and Nettleship, 1982
Funk Island, Newfoundland	Common Murres	Tuck, 1960
Massachusetts	terns	Nisbet and Drury, 1972 Nisbet, 1981 Nisbet and Walton, 1984
Great Gull Island	terns	Cooper *et al.*, 1970 Hays, 1970
Chesapeake Bay	Black Skimmer	Burger, 1982
South Carolina, Florida	Brown Pelican	Blus *et al.*, 1979b
Florida	Brown Pelican	Schreiber, 1979
Louisiana	Brown Pelican	Blus *et al.*, 1979a
Cape Thompson, Alaska	murres Black-legged Kittiwake	Swartz, 1966 Springer and Roseneau, 1978 Springer *et al.*, 1979
Bluff, Alaska	murres Black-legged Kittiwake	Drury *et al.*, 1981
Pribilof Island, Alaska	murres kittiwakes Red-faced Cormorant	Hunt *et al.*, 1981
Triangle Island, British Columbia	Rhinocerous Auklet Tufted Puffin	Vermeer, 1979
Mandarte Island, British Columbia	Glaucous-winged Gull	Vermeer, 1963 Ward, 1972 Hunt and Hunt, 1976
Santa Barbara Island, California	Xantus' Murrelet	Murray *et al.*, 1938
Southern California	Brown Pelican	Anderson *et al.*, 1975 Anderson and Gress, 1983

risks to seabirds due to oil or fishing hazards. Additionally, European populations of several species have been increasing more rapidly while North American populations appear more stable. Small variations in the annual rate of survival can have large impacts on the recovery rates of populations subsequent to decline (Ford *et al.*, 1982).

Survivorship has been estimated at some small (Fulmars; Hatch, unpublished) North American seabird colonies: Glaucous-winged Gulls in British Columbia (Butler *et al.*, 1980), Herring Gulls in New England (Kadlec and Drury, 1968),

TABLE 11.4
Sources for survivorship estimates for North American seabird species

Species	*Average adult* survival rate*	*Location*	*References*
Manx Shearwater	80–95%	European	Perrins *et al.*, 1973
Leach's Storm-Petrel	78.9%	North American	Morse and Buchheister, 1977
Northern Fulmar	97%	European	Dunnet and Ollason, 1978a,b Ollason and Dunnet, 1978
Shag	83%	European	Potts *et al.*, 1980
Arctic Tern	82%	European	Coulson and Wooler, 1976 Cullen, 1957
Black-legged Kittiwake	86%F, 81%M 80% (yr 1–6)	Euorpean	Coulson and Wooler, 1976 Flegg and Cox, 1976
Black-headed Gull	64% (yr 7–10) 80% (yr 1–6)	European	Flegg and Cox, 1976 Flegg and Morgan, 1976
Common Tern	92%	North American	DiCostanzo, 1980
Glaucous-winged Gull	42.1% (yr 1)	North American	Butler *et al.*, 1980
Herring Gull	91–96% (2 yrs) 75–80%	North American	Kadlec and Drury, 1968 Kadlec, 1976
Common Puffin	95.0% 96.0%	European	Ashcroft, 1979 Mead, 1974
Common Murre	87.9% 91.5% 87.0% 87.1%	European	Birkhead, 1974 Birkhead and Hudson, 1977 Southern *et al.*, 1965 Mead, 1974
Razorbill	89.0% 80.7–92.1%	European	Lloyd, 1974 Lloyd and Perrins, 1977

*When not adult, age class in parentheses.

Leach's Storm-Petrels in Maine (Morse and Buchheister, 1977) and Common Terns at Great Gull Island (DiCostanzo, 1980) (Table 11.4). Studies are underway on the Farallon Islands in California which should provide valuable information for a number of species in that region. There is only one long-term study of survival for any species in Alaska. Although breeding numbers of murres may be quite stable, the work of Coulson and his students (pers. comm.) on Shags

(*Phalacrocorax aristotelis*) off the Northumberland coast demonstrates that there can be extremely large fluctuations in the number of birds breeding in a given year without the change reflecting mortality (Coulson, 1961). Likewise, Ollason and Dunnet (1983) have recorded wide, non-mortality related fluctuations in the number of adult fulmars attempting to breed.

At present there is little information on the determinants of juvenile survival, recruitment rates or the proportion of breeding individuals in a population. D.W. Anderson and R.W. Schreider are developing life tables for the Brown Pelican in California and Florida. Estimates are available for the survival of young Herring Gulls (Kadlec and Drury, 1968; Chabrzyk and Coulson, 1976), juvenile Common Murres (Birkhead and Hudson, 1977), Atlantic Puffin (Harris, 1983; Kress in Harris, 1983) and Razorbills (Lloyd and Perrins, 1977). The majority of these studies are the result of long-term banding efforts in Great Britain; studies in North America are few and are concentrated in Canada and New England.

Tests of the models

The models have not been empirically tested as there have been no major spills in the waters for which the models provide predictions. While this lack of spills is fortunate, it leaves us without knowing how accurately the models predict population impact and recovery. The potential for a test of the models existed in the *Esso Bernicia* spill at Sullum Voe, Shetlands. A sizable portion of the wintering local population of Black Guillemots was killed (Richardson *et al.*, 1981). Stowe and Underwood (1984) and McKay *et al.* (1981) suggest that recovery of this population has not been complete. The rate of recovery could be matched against population growth estimates based on local reproductive effort and survival. However early censuses of this population may have underestimated its size (M.P. Harris, in litt.) and there is no information on immigration rates. R.G. Ford (pers. comm.) suggests that models may be profitably tested in the Southern California Bight where populations are countable, there exists a data base (Hunt and Ingram, 1982) and spills can be monitored effectively.

Identification of required data

Wiens *et al.* (1984) have indicated the areas in which additional data are needed for successful implementation of their models (Table 11.5). The data required are a mixture of demographic, distributional and physiological parameters, some of which are site specific and some of which are species specific. Observations in the presence of spilled oil are also required (Table 11.5, High Priority 3, 6 and 8). The list of required data can be shortened somewhat if those aspects concerned primarily with energetic considerations (Table 11.5, Intermediate Priority 6 and 7) are dropped. Ford *et al.* (1982) have concluded that their model is least sensitive to variation in these parameters, and we need to focus on the parameters of major sensitivity.

Other investigators directly involved with long-term studies of individual

TABLE 11.5
Priorities for research needed to provide input to the models of Ford *et al.* (1982) and Wiens *et al.* (1978) [from Wiens *et al.* (1984)]

High priority	*Intermediate priority*
1. Size of breeding population	1. Distributional pattern of oil spills in specific areas
2. At-sea distribution of birds from each colony over whole annual cycles	2. The age structure of the populations
3. Probability that an adult encountering oil will die	3. The proportion of the adult "pool" of birds that is not breeding and is not associated with a given colony
4. Age-specific survivorship of populations	4. The reproductive success of individuals in specific colonies
5. Distribution of trip times for a foraging adult	5. Foraging activity patterns and pathways of individuals
6. The time lag of a population in shifting its foraging area in case of a spill	6. Metabolic costs associated with foraging activity
7. The relationship between foraging rate and behavior of prey	7. How changes in food delivery rates to chicks affect the growth and survival of chicks
8. How spills affect prey resource levels	

populations of seabirds (J. Coulson, J. Croxall, G. Dunnet, R. Schreiber, S. Sealy, pers. comm.) stress the need for demographic information that can be derived only from long-term studies, including:

1. What are the natural sex- and age-specific fecundity and mortality rates of seabirds?
2. What determines how many, which individuals and from where, recruits enter the breeding population each year?
3. To what extent are reproductive success and survival density-dependent, and what is its form?
4. What proportion of the population is breeding each year and where are the non-breeding birds spending their time?

The answers to most of these questions will require individually marked birds that are identifiable over a period of twenty years or more, and preferably that can be identified without the need to recapture them. Obtaining information on birds away from colonies, other than general distribution data, will be difficult, expensive, and of uncertain success. Nevertheless, answers to questions about age-specific mortality, determinants of recruitment and the importance of density-dependent interactions are major gaps in our basic knowledge of seabird biology and are critical for understanding the impact of offshore oil development.

Research Needs

Colony Work

There is a strong requirement for long-term studies of the population dynamics and demography of seabirds in the United States. These studies should be done on selected species and in representative colonies in regions subject to oil pollution. A spectrum of colonies of different sizes and in different habitats will have to be examined so that the results can be generalized to colonies of similar size in the same area.

The studies will require massive banding efforts and should include the use of individual markers visible from a considerable distance. Such highly visible markers will be particularly important in studies at large colonies, if we are to obtain estimates of survival of living birds (Lakhani and Newton, 1983) and to detect movement between colonies. Most important, there should be a clear commitment to continue the studies for the length of the generation time (10-20 years minimum) of the species involved. Without such a commitment, there is little value in investing in the banding effort necessary to provide useful demographic information. The lack of a significant banding effort in past baseline studies of seabirds in Alaska is directly attributable to the fact that, without a commitment of more than a one-year contact (or a 3-5 year study), one cannot hope for sufficient band returns to justify a significant investment in banding (Hunt, 1976). There must be sufficient continuity in the studies to ensure that results are comparable throughout the lifespan of the work. Careful examination of sampling design and variance should be conducted at the outset to establish the appropriate level of effort. If variance is too high to allow detection of modest levels of change (20-40%), then this work should be abandoned.

The studies should include, at a minimum, yearly information on:

1. size of population/colony
2. proportion of population that is breeding
3. banding of both young and adult birds (over time this will allow construction of life-tables, estimates of movements, foraging and wintering areas) (see below)
4. reproductive success
5. foods used

Information on foods used and foraging areas, when combined with other demographic data, are necessary for understanding the relative impacts of human fisheries and petroleum development on seabird populations.

Distribution at Sea

"Sensitivity" analyses (R.G. Ford, pers. comm.; Ford *et al.*, 1982) of the output of the Wiens *et al.* (1978) model have shown that the model's predictions are extremely sensitive to the foraging distribution of birds around colonies. Large spills such as the *Torrey Canyon* or *Amoco Cadiz* have resulted in large kills, particularly of alcids, near colonies (Monnat, 1969; Hope-Jones *et al.*, 1978), and even small spills encountering bird concentrations near colonies may destroy many birds (Hope-Jones *et al.*, 1970; Bourne and Johnston, 1971; Stowe,

1982b). The number of birds killed and the effect on the population are much more sensitive to the location of a spill relative to bird concentrations than they are to the size of a spill.

Field workers and modelers have identified at-sea distribution as a gap in our knowledge of the ecology of marine birds (G. Dunnet, J. Coulson, J. Croxall, pers. comm.; Wiens *et al.*, 1984). The concerns are not limited to determining overall densities of birds in various large regions (Hunt *et al.*, 1981a, 1982; Briggs *et al.*, 1981, Clapp *et al.*, 1982; Powers *et al.* 1980; Fritts *et al.*, 1983) which may be used to estimate the number of birds that might be killed by a particular spill (Brown, 1973). There is also the need for information on where adult birds from specific colonies seek food for their young and where they go when they are not attached to the colony. We also lack information on where young go between fledging and their arrival at a colony preparatory to nesting several years later. This information is needed to estimate the impact of oil spills on colonies distant from a spill (Baillie and Mead, 1982). Understanding these specific movements will require banding and other forms of marking and considerable attention to statistical detail (North, 1980). In particular, radio tracking using modern technology capable of interfacing with satellites (G. Dunnet, pers. comm.) may be the most effective way of answering critical questions in remote regions where the potential for band recoveries is low and bird movements are great. The determination of areas of large concentrations of birds, especially near colonies, is a high research priority.

The behavioral response of adult birds when encountering oil will determine their survival chances. Although adults of some seabird species may leave or avoid oiled waters (Casement, 1966; Buck and Harrison, 1967; Bourne, 1968b; Nero and Associates, Inc., 1983) other species fail to avoid or deliberately enter oil polluted areas (Curry-Lindahl, 1960; Bourne, 1968b; Vermeer and Vermeer, 1975; Custer and Albers, 1980). As noted above, the reaction of birds to oil and their ability to move from contaminated foraging areas greatly influences the predicted mortality.

There is a considerable need to learn more about the behavior of birds when confronted with spilled oil. Will they enter the oil, avoid it but remain in the general area or shift to new foraging areas? Also what controls the recruitment of birds to a foraging area and how quickly will recruits shift to a new area or at least desert the contaminated one? Answers to these questions greatly affect the model derived estimates of adult mortality.

Use of Models to Identify Sensitive Parameters

Although the models are useful for predicting the potential impact of a spill, an alternative and perhaps more important immediate use of modeling efforts is to identify aspects of seabird population biology that are particularly sensitive to offshore oil development. These parameters are identifiable through sensitivity analyses of the models. This use of the models provides a mechanism for identifying the times, places and situations in which oil development has the greatest potential to harm seabird populations. Preliminary results (Ford *et al.*, 1982; R.G. Ford, pers. comm.) suggest that adult mortality in a spill swamps all other variables and may account for 55-85% of the variance in population change. A second area of concern is the long term (15-20 years) depression of reproductive success due to chronic effects. These results suggest that studies of where, when and why birds form dense aggregations on the water would be of major importance, as would studies of the control of reproductive success and the way chronic oil pollution or development activity would depress reproduction.

An initial approach to the investigation of long-term effects would be to study colonies already subject to chronic effects and determine which, if any, effects are resulting in reduced reproductive success. On the basis of this preliminary work, priorities could be set for detailed long-term research. This approach would focus research onto what the models predict to be the most important areas. A danger is that, since the models are untested, we might forgo gathering data that would subsequently be found critically important. Depending upon funding commitments, there is no easy solution to this problem.

DISTURBANCE

Introduction

Disturbance of nesting birds may come from several sources such as the close approach to colonies of aircraft or boats, entry of people into a colony or nearby discharge of firearms. These chronic effects may have a severe impact on populations, and modeling efforts suggest that chronic effects are more detrimental to long-term population stability than are spills (D. Heinemann, in litt.). In theory, most forms of disturbance can be eliminated by prohibiting disruptive activities near colonies. However, safety requirements or exploratory work may require aircraft flights close to colonies. The development of shore facilities in close proximity to colonies will almost inevitably result in increased intrusion into colonies. Additionally, virtually all models developed to predict the effect of oil spills on birds depend upon data on reproductive success obtained by entry into colonies. The value of these data for setting the parameters in the models depends upon the accuracy with which they reflect natural conditions, but colony entry by research personnel gathering data may affect the processes being measured.

Aircraft

Background

There is conflicting evidence concerning the effect of the close approach of aircraft to breeding cliff-nesting birds. At the Pribilof Islands, Hunt (1976) reported two instances when large, multi-engine aircraft flying near colonies caused considerable egg and chick loss for murres, and in 1975 D. Heinemann (in litt.) observed kittiwakes, murres and puffins departing from the cliffs of Nunivak Island when a helicopter approached. Likewise D. Nettleship (in litt.) has observed large panic flights of "thousands, if not tens of thousands, of Thick-billed Murres leaving the cliffs at Coburg Island and Cape Hay (Bylot Island) following the presence of a twin engine Otter aircraft (Coburg and Cape Hay) or helicopter (type not specified) (Coburg)." He reports an instance in 1978 when the noise from an Otter, banking about 1.5 km from cliffs resonated along the complete length of a breeding area (ca. 6 km) causing a continuous wave of birds to leave the cliffs as the sound travelled across the area. He observed eggs and chicks falling to the sea but was unable to estimate the extent of the loss in a meaningful way. K. Briggs (in litt.) also found that at the Farallon Islands, the least accessible colony site in central and northern California, it was necessary to remain above 300 m and 100 m laterally from the shores where Common Murres (*Uria aalge*) nested in order to avoid flushing birds.

In contrast, Briggs (in litt.) has not observed significant flushing of Common Murres along the northern California coast during aerial surveys flown at about 100 m. He reports that "in fact, the largest murre colony in the state, Castle Rock in Del Norte County, lies one km off the end of the Crescent City Airport (in 1980 a plane crashed right into the murre colony). On one occasion we watched with horror from an altitude of about 500 ft. as Navy jets flew directly along the shore at 200 ft., right past major murre colonies near Trinidad. No murres or cormorants flushed! We conclude that nesting birds definitely acclimatize (habituate) to airplane traffic..." Likewise E.C. Murphy (in litt.) comments that by late incubation murres on eggs at the Bluff colonies (Norton Sound, Alaska) do not flush in response to aircraft, including helicopters, flying very close to the cliffs. Aircraft disturbance is frequent in this area ("a few flights/day"), and he believes that habituation at Bluff has been greater than at Cape Thompson, where aircraft pass close to the cliffs much less frequently. Dunnet (1977) observed seabird cliffs before and after the passage of aircraft at Longhaven/Buchan, about 40 km north of Aberdeen, Scotland. He found no evidence that aircraft flying above 100 m over the cliff top affected the attendance of incubating or brooding birds. However, groups of kittiwakes resting on the cliffs did take flight. Schreiber and Schreiber (1980) and J. Jehl (in litt.) have found that at colonies of seabirds frequently visited by planes (including military target areas), the birds come to ignore the aircraft and breed successfully. Wanless (1983) also concluded that there is little evidence of damage to seabird populations by low flying aircraft based on an extensive review of British seabird colony overflights, but cautioned that experiments were needed to provide unequivocable answers.

Hunt *et al.* (1978) attempted a series of experiments in which a helicopter (Bell Jet Ranger) was flown directly at a small section of nesting cliff, above cliff top level, at the Pribilof Islands in order to determine permissible approach distance. Murres left the cliff when the helicopter was 180-250 m from the cliff face, depending on helicopter speed. Downdrafts near the cliffs endangered the aircraft and brought it near flying birds, so further trials were abandoned after four approaches. Several passes were also made parallel to the cliffs. At 400 and 350 m no murres left the cliffs, at 200-250 m moderate numbers left, at 180-200 m murres streamed from the cliffs in huge numbers.

Of the seabirds in North America, murres (*Uria lomvia* and *U. aalge*) are the most vulnerable to disturbance by aircraft. These birds lay their eggs directly on cliffs, without benefit of a nest, and they incubate eggs and brood chicks by holding them on their feet. If startled, adult birds will jump from the cliffs, tumbling their eggs and chicks from the ledges. Loss of eggs or chicks by other species is not as great a problem because the use of a nest or a burrow effectively prevents egg or chick loss. However, in colonies where gulls or corvids can steal eggs or chicks, there may be some loss of reproductive output due to predation. Beyond immediate egg loss, there is the possibility that persistent disturbance will lead to colony abandonment or reduction of a colony to a few individuals insensitive to disturbance (Nettleship, in litt.).

The different reactions to aircraft shown by murres in different colonies may be the result of habituation or other factors such as colony size. If colonies can be protected by setting and enforcing airspace closures, then there is little need for further research; the potential problem is solved administratively. If access to air space near colonies is needed, or if there is conflict about the area around each colony that requires closure, then we need research on the habituation of birds, especially at large colonies.

Research Needs

There has been no controlled study of the effect of aircraft on murre reproduction, colony desertion, or habituation to disturbance. While it would be desirable to conduct such experiments or observations on "pest" species such as Herring Gulls, their mode of nesting and very different behavioral responses are unlikely to tell us anything useful about the cliff-nesting species of greatest concern. If we want information about the long-term effects of disturbance, we will have to conduct a series of experiments at colonies of moderate size that are in the geographic regions of concern, even though these studies may result in localized population damage. Since aircraft disturbance can be minimized by regulation and since birds appear to habituate to aircraft, the need for these studies seems to be of moderately low priority.

Entry into colonies

Background

A wealth of studies are now available that show lowering of reproductive success by the intrusion of investigators and others into gull (Paynter, 1949; Vermeer, 1963; Harris, 1964; Brown, 1967; Kadlec and Drury, 1968; Vermeer, 1970a; Hunt, 1972; Gillett *et al.*, 1975; Robert and Ralph, 1975; Davis and Dunn, 1976; Nisbet, 1978b; Anderson and Keith, 1980; Hand, 1980; Burger, 1981; Fetterolf, 1983) and cormorant colonies (Mendall, 1936; Drent *et al.*, 1964; Vermeer, 1970b; Lock and Ross, 1973; Kury and Gochfeld, 1975; Ellison and Cleary, 1978). One study has documented reproductive failures in Fulmars (nesting on the ground) due to the effects of investigators (Ollason and Dunnet, 1980), and the work of Hunt *et al.* (unpubl.) suggests that some cliff-nesting species (murres; see also Tuck, 1960) may be more susceptible to investigator disturbance than others (kittiwakes).

Colony desertion may also be a problem. At the Pribilof Islands, seabirds no longer nest on the cliffs near the village of St. Paul, probably due to intrusion by people (especially children with air guns) (M. Thompson, pers. comm.; G. Hunt, pers. obs.). Erwin (1980) has shown that beaches frequented by people support fewer nesting seabirds than those that are less disturbed. These studies all indicate that individuals should stay out of colonies unless there is a specific need to be there. Again, in the case of onshore oil production or support facilities, it is essential that personnel be instructed to stay away from colonies, whether the individual be on foot or in a small boat close to shore, as the latter is functionally the same as direct intrusion.

These findings also raise a paradox (or "uncertainty principle," Lenington, 1979), because "if one disturbs birds while measuring their nest success and if such disturbance lowers success, then the more 'accurate' (=frequent or thorough) the measurement, the less real the productivity being measured" (Duffy, 1979). This could be a significant problem when using data from reproductive studies in modeling seabird population response to oil spills. If we are significantly underestimating the reproductive output of seabirds, then we are also overestimating the impact that oil-caused mortality will have on populations. Recovery times may be considerably shorter if reproductive output is higher than is currently believed.

Research Needs

Since the absolute values of reproductive success are of considerable importance in modeling the recovery of seabird populations subsequent to oil spills, it would be useful to set up a carefully controlled study for the comparison of reproductive performance of cliff-nesting birds at study sites that are disturbed and those that are left undisturbed. A first step would be the careful reanalysis of previously completed studies, but all future studies should include analysis of the impact of investigator-caused disturbance. Duffy (1979) suggests alternative methods to reduce impact while maintaining accuracy. Probably the best of these is to use observations made at a distance from undisturbed nesting sites to determine reproductive success, while at other nest sites chicks may be weighed and banded to obtain other types of data. It is clear from a number of studies that useful data on reproductive ecology can be obtained with minimal damage to populations if care is taken in planning and execution of work (Fetterolf, 1983).

PHYSIOLOGICAL ASPECTS OF SEABIRD CONTAMINATION BY PETROLEUM

Introduction

Since the early 1960s there has been a growing interest in the study of effects of petroleum oil contamination on a variety of physiological processes. Many of these studies have been conducted with domestic species under controlled laboratory conditions. Given the known variation in susceptibility between species, it may be very difficult to extrapolate from these studies to natural situations. Recently, investigators have begun to employ native species of seabirds and have worked under field conditions. There are trade-offs in using either method, but, if we are to use physiological data in predicting the impact of oil pollution on birds, we will have to be able to apply research results to native species living under natural conditions.

Another problem is that different oils have different levels of toxicity (e.g., Crocker *et al.*, 1975; Cowell, 1976). Toxicity may vary between fields, and even within a well, depending upon the strata being tapped (R.H. Jenkins, British Petroleum, pers. comm.). The properties of crude oil change as the oil weathers, with the lighter, aromatic fractions evaporating or dissolving in the water column. The variability in toxicity among oils creates an extremely complex problem for interpretation of research results on the physiological effects of oil contamination.

An additional area of concern is the trade-off between conducting short-term studies of acute contamination or long term studies of sublethal, low-level contamination. The early emphasis in research was on acute studies, while more recently studies of chronic, sublethal contamination have gained in prominence. Ecological studies can detect major, short-term changes in the numbers of birds or gross changes in reproductive success. Thus, the greatest contributions of physiological studies to understanding the effect of oil pollution may be the prediction of subtle changes in the probability of adult or fledgling survival and changes in future reproductive success. However, Clark (1984) and M.P. Harris (in litt.) feel that studies of sublethal effects of oil are of little value because there is scant evidence that chronic sublethal effects play a substantial role in nature.

Recent reviews (Dieter, 1977; Holmes and Cronshaw, 1977; Szaro, 1977; Ohlendorf *et al.*, 1978; Eastin and Hoffman, 1979; Stickel and Dieter, 1979; Nisbet, 1980; Holmes *et al.*, 1981) have examined the previously mentioned problems and summarized available research results. Physiological studies have concentrated on several separate areas of concern, such as direct effect on adults, reproductive physiology, egg hatchability, and chick survival. These areas will be discussed separately below.

Effects of Oil on Adult Birds

Background

Birds encountering oil lose their buoyancy (Hawkes, 1961; McEwan and Koelink, 1973) and have increased metabolic demands due to the reduced insulation provided by oiled plumage (Giles and Livingston, 1960; Hawkes, 1961; Erickson, 1963; Hunt, 1961; Hartung and Hunt, 1966; Hartung, 1967; McEwan and Koelink, 1973; Erasmus *et al.*, 1981; Lambert *et al.*, 1982). Birds also ingest oil directly from the water or by attempting to clean their plumage (Phillips and Lincoln, 1930; Hawkes, 1961; Hunt, 1961). Some species may be able to tolerate ingestion of certain oils, apparently without adverse effects on the systems studied (Holmes *et al.*, 1979, 1981; McEwan, 1978; McEwan and Whitehead, 1978, 1980; Patton and Dieter, 1980; Rattner, 1981), but in other cases ingestion of oil appears to have adverse effects on one or more physiological systems.

Ingested oils have been shown to cause changes in intestinal absorption (Peakall *et al.*, 1979), hepatic enzyme function (Gorsline *et al.*, 1981; Gorsline, 1982; Gorsline and Holmes, 1981, 1982a) and osmoregulatory abilities (Holmes *et al.*, 1978b; Miller *et al.*, 1976; Peakall *et al.*, 1979). A variety of endocrine effects include changes in gonadal function and steroid levels (Holmes, 1981, 1982; Holmes *et al.*, 1978a, 1981; Cavanaugh, 1982; Cavanaugh and Holmes, 1982) and changes in adrenocortical function and corticosterone levels (Peakall *et al.*, 1979, 1981; Holmes and Gorsline, 1980; Holmes, 1981; Rattner and Eastin, 1981; Gorsline and Holmes, 1981, 1982a,b; Gorsline, 1982). Two studies (Fry and Lowenstine, 1982; Leighton *et al.*, 1983) have found reduced packed cell volumes and Leighton *et al.* (1983) have been able to demonstrate a Heinz-body hemolytic anemia associated with ingestion of oil by Herring Gull and Atlantic Puffin (*Fratercula arctica*) nestlings.

A number of the above studies fail to relate the effects of ingested oil to changes in life expectancy or mortality rates. However, Holmes *et al.* (1978b, 1979) showed that ingestion of oil rendered stressed birds more likely to die. Likewise Rattner and Eastin (1981) found some mortality in ducks ingesting oil due to reduced tolerance of low temperatures as a result of altered corticosterone concentrations. However, there has been little attempt to produce physiological studies of adult systems that can be directly related to changes in survival of free-ranging native species.

Research Needs

Given the less than clear demonstration that most effects of ingested oil being investigated in birds can be related to changes in survivorship in wild species, we need to determine the long term effects of ingested oil on adult survivorship in native species under natural conditions. We need to know the change in probability of survival of a bird subjected to sub-lethal doses of oil when that bird is next stressed. Changes in survivorship of this sort are hard to detect in nature and would not easily be related to exposure to oil. However, if, as some of the previously mentioned studies suggest, birds that have ingested oil are more vulnerable to stress, then we need to know what this increased vulnerability is. To be of use, studies will have to be validated or conducted with native species, using oils and dosages likely to be encountered in the local environment.

Effect on Reproduction of Oil Ingestion by Adults

Background

Grau *et al.* (1977) first described the alteration of egg yolks due to the ingestion of oil (Bunker C) four days prior to egg laying in Japanese Quail (*Coturnix coturnix*) and Canada Geese (*Branta canadensis*). This study was followed by findings of alteration of yolk structure or egg hatchability when females were fed oil at the time of egg production (Wooton *et al.*, 1979; Ainley *et al.*, 1981). Other studies have shown reduced oviduct weight (Coon and Dieter, 1981) and reduced rates of oviposition (Hartung, 1965; Grau *et al.*, 1977; Holmes *et al.*, 1978a; Eastin and Hoffman, 1979; Vangilder and Peterle, 1980; Coon and Dieter, 1981; Harvey *et al.*, 1982a,b; Cavanaugh and Holmes, 1982), reduction in eggshell thickness (Holmes *et al.*, 1978a; Vangilder and Peterle, 1980; Harvey *et al.*, 1982a) and subsequent changes in chick physiology (Gorsline and Holmes, 1982c) when females were dosed with oil prior to egg production. The effects depended upon the kind of oil or fraction of oil administered. Changes in reproductive endocrine function follow ingestion of oil (Harvey *et al.*, 1981, 1982a,b), but the mechanisms linking changed endocrine function and depressed egg production have not been established. Virtually all of these studies, with the exception of that of Ainley *et al.* (1981) on Cassin's Auklets (*Ptychoramphus aleuticus*), have been done on domesticated birds or captive stock, often using oils other than those most likely to be encountered by major concentrations of North American seabirds. At present it is impossible to extrapolate from these laboratory studies to what might be expected in various wild species or sites of concern.

Research Needs

Study of the effect of ingested oil on reproduction provides a possible means for estimating potential decreases in egg production and hatchability given low levels of oil ingestion by adult seabirds. These changes would be hard to identify or assign to oil contamination in ecological field studies. We thus need to know the level of depression of reproductive output in selected native bird species for the oils most likely to be spilled in a given region. This research will require a shift to more field oriented studies.

In concert with these studies there should be a means of detecting changes in the level of oil contamination of tissues (Burns and Teal, 1971; Ohlendorf *et al.*, 1978; Lawler *et al.*, 1978, 1979; Gay *et al.*, 1980; Boersma, 1981). These studies could supplement direct examination of egg yolks for signs of contamination (Grau *et al.*, 1977, 1978; Wooton *et al.*, 1979). Additionally, it would be valuable to compare individuals of a species from an area that is heavily polluted with those from a relatively clean area to see if reproduction is depressed. The following research would be recommended:

1. Studies of the effects of "local" oils on native seabirds to establish the relationship between the amount of contamination and potential decrease in reproductive output.
2. A comparison of contamination levels in avian tissues and reproductive success between areas with and without high levels of pollution.

Native species and oils that are likely to contaminate birds in their normal habitats should be used to validate laboratory studies. These studies will require collaboration between those who study reproduction in the field and those who measure contamination levels in tissues and laboratory workers. Work on domestic species would be de-emphasized until they can be validated as useful models and until effects significant at a population level can be detected in native species. Studies of mechanisms that cannot be directly linked to predicting changes in survival or reproductive output would also be de-emphasized. Such studies are recognized to be important, but it is not clear how they can be used to help predict the impact of oil pollution on marine birds.

Effect of oil on hatchability of eggs

Background

As early as 1950, Gross (1950) was aware that oil spread on eggs would block development and he coated eggs with an emulsion of oil to control Herring Gull populations on islands off the coast of Maine. Soon thereafter Rittinghaus (1956) and Anonymous (1982) observed the contamination of eggs by oil carried on the plumage of adult birds. There have now been a number of studies demonstrating that the application of oil to the shells of intact eggs will significantly lower hatching success (e.g., Albers, 1977; Szaro and Albers, 1977; Hoffman, 1978; Albers, 1978; Coon *et al.*, 1979; Hoffman, 1979a,b,c; King and Lefever, 1979; McGill and Richmond, 1979; Albers, 1979; White *et al.*, 1979; Macko and King, 1980, Szaro *et al.*, 1980; Hoffman and Gay, 1981; Lewis, 1982; Patten and Patten, 1983) as will the application of some dispersant/oil mixtures (Albers, 1979; Albers and Gay, 1982). One laboratory study demonstrated reduced hatching of eggs coated with oil transferred by birds from contaminated water (Albers, 1980). The sensitivity of embryos to treatment by oil depends not only on the type and condition of the oil (aromatic hydrocarbons are embryotoxic while aliphatic hydrocarbons have virtually no effect, Hoffman 1979a), but also on the stage of embryogenesis (younger embryos are more sensitive, Albers, 1978; Hoffman, 1978). However, the findings of Rittinghaus (1956) and Birkhead *et al.* (1973) and Anonymous (1982) notwithstanding, there is apparently little evidence of depression of hatchability in nature due to transfer of oil from polluted water (Bourne, 1979; Nisbet, 1980; Clark, 1984). Birkhead *et al.* (1973) observed birds oiled during the breeding season, cleaning themselves of oil. In two cases clutches failed to hatch, which may have been due to the transfer of oil.

Research Needs

It is low priority to continue research in this area unless significant evidence of depressed hatching success due to transfer of oil can be found in nature. While oiling of eggs will certainly cause mortality when it occurs, it appears to be too rare in comparison to mortality due to direct oiling of adults (Bourne, 1979) to deserve much attention.

Effect of Oil on Chick Survival

Background

As early as 1974 there was interest in the possible effect on development of oil ingestion by chicks (Crocker *et al.*, 1974). This and several later publications document changes in intestinal absorption, osmoregulatory problems, and other pathological effects in captive ducklings (Crocker *et al.*, 1974, 1975; Szaro *et al.*, 1978; Miller *et al.*, 1979; Szaro *et al.*, 1981; Eastin and Murray, 1981; Eastin and Rattner, 1982; Gorsline and Holmes, 1982c) and Herring Gulls (Miller *et al.*, 1978, 1979; Peakall *et al.*, 1982). In contrast, Gorman and Simms (1978) found no effects on Herring Gull chick growth when they administered 0.1 and 0.5 ml of Forties Field crude oil per day, possibly because they used an oil of different toxicity (Miller *et al.*, 1979).

Three experiments have been conducted by dosing young Herring Gulls, Black Guillemots (*Cepphus grylle*) and Leach's Storm-Petrels (*Oceanodroma leucorhoa*) in the wild (Butler and Lukasiewicz, 1979; Peakall *et al.*, 1980; Trivelpiece *et al.*, 1984). Weight gain was reduced in all three species, and chick survival was reduced in storm-petrels in which the oil was given to an adult instead of directly to the chick. Boersma (1981) has shown that adult storm-petrels ingest oil at sea. Thus the experiments on storm-petrels are of particular significance in developing models of the effects of low-level pollution on population dynamics in these birds.

Research on the effects of oil on chicks is complicated by some of the same factors that bedevil other physiological studies such as choice of species and variability in the toxicity of oils. Additionally, there are questions concerning at what age oils should be administered to chicks, as sensitivity varies with age (Gorsline and Holmes 1982b), and whether oil should be given directly to the chick or to an adult that is feeding chicks. We also need to know the extent to which the oil is digested and the speed with which it is removed from the adult birds.

With the exception of the work on storm-petrels by Boersma (1981), there is little or no information on whether adult birds ingest oil and pass it to their chicks. If, as in the case of indirect oiling of eggs, oil contamination of chick foods is rare or nonexistent in nature, then study of chicks dosed with oil contributes little to our ability to predict the impact of oil pollution on the population dynamics of seabirds. In contrast, when ingestion of oil by adults can be documented and when this oil is transferred to chicks, the study of the effects of oil on chick survival is useful.

Research needs

We need to know if adult birds of native species ingest oil and pass it to their chicks under circumstances of chronic or acute pollution. Secondly, if the research is to be of value in modeling changes in population dynamics it should be focused on changes that can be directly related to pre- or post-fledging survival. Since there may be differences in the sensitivity of different species, the use of native species improves the likelihood that results can be used to model natural populations. Again, careful selection of oils will enhance the usefulness of results. Testing for the uptake by adults, transfer to chicks and pre-fledging effects of oil will be relatively easy; ascertaining post-fledging effects on survival of chicks fledging at lower weights or with physiological abnormalities will be exceedingly difficult. Because there are a large number of combinations of oil types and ages and bird species, a few critical species with high probability of ingesting oil should be selected. A broad scale, "shotgun" approach will be wasteful of resources.

MONITORING

Introduction

Present monitoring of seabird populations with respect to the potential effect of oil pollution includes at least three facets: 1) observations of colonies to detect changes in population size or reproductive success; 2) examination of beaches for carcasses to assess changes in mortality at sea and its causes; and 3) examination of tissues for changes in the levels of petroleum hydrocarbons present. A fourth area of study, that of continual reassessment of the pelagic distribution of concentrations of birds could be considered either under monitoring or under "long-term basic research." Its function, unlike the first three types of studies mentioned, would be more for preventing damage and providing advice in case of a spill, rather than documenting changes caused by a spill or chronic pollution.

An important feature of any monitoring study will be to determine the goals of the study and then develop an appropriate experimental design to provide statistically reliable results. The difficulty in making precise observations and the large size of natural variances are such that the effects of moderate damage to populations may be hard to document. We need to decide the extent of change and statistical precision for detecting the changes that are necessary so that sampling effort is designed appropriately at the outset of a monitoring program (see Carney, Chapter 14). As the normal variances in the system become known, sampling procedures will have to be adjusted to provide the ability to detect the appropriate degree of change at the desired probability level. Additionally, proper examination of variance and sampling design will reveal the practicality or lack of practicality of various research efforts.

Colonies - Populations

Background

As discussed earlier, censuses of the number of birds at colonies, particularly those of surface or cliff nesting species have been conducted, off and on, for a number of years in various parts of North America. Unfortunately, for the most part the counts presented represent a one-time-only visit and often just an estimate of colony size based on an aerial survey or a cursory inspection of the colony site. There are very few confidence limits available for any of the estimates and very few sites or species for which multiple within-year or between-year counts are available. Ingram *et al.* (1983) have provided an outline of a possible seabird monitoring program to be implemented in the Channel Islands National Park by the National Park Service.

The statistical problems addressed by Richardson *et al.* (1981), Wanless *et al.* (1982), Harris *et al.* (1983) and Newman (MS) (see also Kish, 1965) are of importance for accurate and statistically meaningful monitoring of cliff nesting (or surface-nesting) species. The difficulty in obtaining accurate counts and factors influencing variation in the number of birds present are discussed. Dunnet (1977) and Harris (1976) have addressed some of the problems of obtaining accurate counts. A number of other papers have addressed factors causing daily and seasonal fluctuations in colony attendance, a matter of importance to monitoring design (Corkhill, 1971; Coulson and Horobin, 1972; Lloyd, 1975; Birkhead, 1978; Slater, 1980; MacDonald, 1980), and methodologies have been developed by the Canadian Wildlife Service for use in their arctic and eastern colonies (Nettleship, 1976; Birkhead and Nettleship, 1980). According to Wanless *et al.* (1982), with careful attention to censusing, one should be able to detect changes of between 10 and 30%. Burrow nesting species present a different and exceedingly difficult challenge to monitoring. Counts of the burrows are difficult because of problems in locating them and these problems are accentuated by the challenge of determining if the burrows are in use without damaging them or their contents (Harris and Murray, 1981; Hunter *et al.*, 1982; Ingram *et al.*, 1983; Savard and Smith, 1982).

Although adequate monitoring programs can be designed with sufficient planning and effort to detect changes in populations in colonies, I am not convinced that these studies are worthwhile without data on reproductive biology and diet. Coulson (1961), Dunnet *et al.* (1979), Ainley (pers. comm.) and Hunt and Sayce (unpubl.), among others, have evidence of very great annual changes in colony attendance due to a variety of natural and perhaps man-caused events. Thus, records of numbers alone are inadequate to demonstrate the causative role of any one environmental factor. Additionally, data on numbers alone cannot show rates of turnover in colonies and the loss of experienced individuals and their replacement by new, young recruits. If we wish to relate changes in colony size to oil pollution we need to be able to separate out other factors. Data on marked individuals, reproductive success and diet will provide much of the required information. Since these data are expensive to obtain and work will have to continue for a long period, a small number of critical colonies of selected species should be chosen for study.

Research Needs

Information on population changes should be sought at colonies selected for more in-depth studies of reproductive biology and diets. Rather than reliance on overall counts of entire colonies, a series of carefully mapped representative study-plots should be used that are selected in a statistically valid fashion (e.g. Harris *et al.*, 1983). This selection process may require prior experience on the colonies involved. Careful attention to experimental design will be critical to the usefulness of the data gathered, and will allow reliable detection of smaller changes.

Beached Bird Surveys

Background

Surveys of beaches for dead and moribund birds have been used as a means of assessing the impact of oil pollution on pelagic bird populations (Moffitt and Orr, 1938; Hawkes, 1961; Bourne, 1968a; Tanis and Morzer Bruyns, 1969; Clark, 1969; Joensen and Hansen, 1977; Powers and Rumage, 1978; Hope-Jones *et al.*, 1978; Hope-Jones, 1980; Mead, 1981; Baillie and Mead, 1982). *Ad hoc* surveys have documented the oiling of large numbers of birds subsequent to certain notorious spills (e.g., *Torrey Canyon*, *Amoco Cadiz*, *Esso Berniia*) and systematic long-term surveys have recorded the large numbers of birds apparently oiled by smaller spills, often of unknown origin, and by chronic pollution resulting from ballast and bilge pumping (Heubeck, 1979; Stowe, 1982b). It should be noted that the vast majority of the spills contributing to these counts of oiled beached birds have come from shipping accidents, accidents at terminals and the pumping of bilges and tanks at sea. Although there may be a long term decrease in the beaching of oiled birds (Mead, 1977; Joensen and Hansen, 1977; Bourne, 1979), more recent observations suggest an increase in the 1980s (Baillie and Mead, 1982). Beached bird surveys have additionally documented natural die-offs of birds in which oil pollution was not a factor (Bailey and Davenport, 1972; Mead and Cawthorne, 1983). Thus, if the goal of beach surveys is to monitor annual changes in seabird mortality and the relative frequency of oiled birds (Stowe, 1982c), then they have done a fairly successful job, but only when surveys are conducted in a systematic, quantifiable way (NERC, 1977; Stowe, 1982c; Page *et al.*, 1982).

A second goal of beached bird surveys, to assess the numbers and distribution of birds affected by oil pollution (Stowe, 1982c), seems to be less well met. Many authors recognize the difficulty in relating the numbers of oiled birds found on beaches to the number actually killed (Clark, 1969; Powers and Rumage, 1978; RSPB, 1979; Page *et al.*, 1982; Stowe, 1982c). Several experiments have been conducted to determine the proportion of dead birds that come to shore (Coulson *et al.*, 1968; Hope-Jones *et al.*, 1970; Bibby and Lloyd, 1977; Hope-Jones *et al.*, 1978; Bibby, 1981; Stowe, 1982c), the most extensive of which is the work of Page *et al.* (1982). These studies show that the proportion of birds beached depends upon where they die, local, short-term wind and current patterns, the size, exposure, and type of beaches available to receive carcasses and the species of bird involved. Additionally, the time elapsed between surveys will greatly affect the numbers of birds found. As a result, it is almost impossible to estimate from beached birds the population consequences of mortality due to oil pollution, although there have been attempts to do so based on recoveries of banded birds (Baillie and Mead, 1982).

Research Needs

Since recoveries of birds are dependent on site and local weather conditions (Page *et al.*, 1982), research on the percentage of carcasses arriving on shore is of virtually no value for estimating total birds killed except at the actual site of a spill. However, beached bird surveys, if rigorously designed and conducted, can provide data that will allow examination of trends in bird mortality and oiling within species over time for a given location (Page *et al.*, 1982; Stowe, 1982c). Monitoring of this sort can be productive in the vicinity of oil fields, shipping lanes and transfer facilities (Richardson *et al.*, 1981). However, due to the inherent limitations, I believe that beached bird surveys should receive low priority, even though they have a high public relations value.

Detection of Oil in Avian Tissues

Background

Modern methods of gas and liquid chromatography, mass spectrometry and various other techniques have made possible the "finger-printing" of petroleum hydrocarbon samples on feathers (Levy, 1980) and the measurement of petroleum hydrocarbon residues in the tissues of birds (Grau *et al.*, 1978; Lawler *et al.*, 1978, 1979; Miller and Connell, 1980; McEwan and Whitehead, 1980; Gay *et al.*, 1980). More recently, Boersma (1981) and associates have developed methods for detecting the presence of petroleum hydrocarbons in the stomach oils of storm-petrels. These methods promise a complementary approach to those of beached bird surveys for assessing the exposure of marine birds to oil pollution. Perhaps most importantly, examination of the tissues of living, apparently healthy birds provides information on chronic, sublethal contamination by petroleum hydrocarbons. If these sublethal levels of contamination can in turn be related to either reduced survival (Holmes *et al.*, 1978b, 1979; Rattner and Eastin, 1981) or reduced reproductive output (Grau *et al.*, 1977; Holmes *et al.*, 1978a; Harvey *et al.*, 1982a,b; Cavanaugh and Holmes, 1982), then we may be able to estimate the significance of low-level chronic pollution on population processes. Before work in this area proceeds, careful examination of the sample sizes needed and the likelihood of success should be made to determine the practicality of obtaining useful results.

Research Needs

There is a need to continue and expand the surveying of living, apparently healthy birds for evidence of chronic, sublethal levels of contamination by petroleum hydrocarbons. To be of value, these studies must be coupled with studies of the effects of comparable levels of contamination on survival and reproductive success in native species of seabirds. As stressed in earlier sections, this research must be performed on native species under natural conditions using oils likely to be encountered, and rates of digestion, excretion and retention of the oils must be obtained. Studies of this sort will have relatively little value if restricted to domestic species under laboratory conditions.

SUMMARY, CONCLUSIONS AND RECOMMENDATIONS

The critical element in the conservation of seabirds is the protection of populations. Early on, the focus was on the large numbers of dead and dying birds coming ashore coated with oil. While this loss of life is not to be condoned and we should do all possible to eliminate the unnecessary mortality, we must also ask what are the population consequences of these losses.

The evidence from the northern British Isles and the North Sea is that, for many species, oil related mortality has not depleted populations. However, we have no way of knowing if it has slowed population growth. Calculations of oil-related mortality in relation to natural mortality of seabirds suggest that the oil-related deaths represent a small fraction of the natural mortality for most species. Likewise, breeding populations of gannets, fulmars, kittiwakes and murres continue to increase in the northern British Isles despite losses to oil pollution. What we do not know is whether the oil related losses are additional to natural losses, or if compensatory (density-dependent) reductions in natural mortality occur. If the oil caused losses are additional, then we may be losing vital "floater" populations, the importance of which is not seen until some natural disaster occurs to which they normally respond with rapid recruitment.

There are other populations or regions in Europe where oil pollution may have had effects on populations. Sea ducks, particularly scoters and Long-tailed Ducks in the Baltic, have suffered tremendous mortality from oil and their populations have declined drastically. Likewise the decreases of colonies in the southeast of England and on the Brittany coast have almost certainly been affected by oil pollution. However, the precise role of oil in these decreases will remain obscure, as these colonies are at the periphery of the ranges of the species involved and fluctuations due to natural causes are to be expected.

Although the experience in the North Sea is perhaps our best source of information on the effect of oil pollution on seabirds, it is difficult to extrapolate from events there to the situation in North America. In Europe, seabirds are spread out in many relatively small breeding colonies, with colonies of more than 100,000 birds being rare. In contrast, the majority of seabirds on the Pacific coast of the continental United States and particularly in Alaska are concentrated in a few large colonies. Under these circumstances, populations are potentially far more vulnerable to a single spill than when populations are sub-divided in many small, widely dispersed colonies, as they are on the Atlantic and Gulf coasts.

If we are to identify the causes of the greatest vulnerability of seabirds to oil development and our most pressing research needs, we will have to depend upon simulation models of interactions between oil and birds and models of the responses of populations to this damage. Present models provide a useful first-step, but they are still of limited utility. They do not adequately examine recruitment between sub-populations, density-dependent interactions at colonies or during the non-breeding season. Little emphasis has been given to how chronic, low levels of pollution affecting both adult survival and reproductive output may influence population processes. However, preliminary results suggest chronic effects might be as important or more important than occasional spills for long-term population stability.

Present modeling efforts have already identified several areas in which the incompleteness of our data make modeling efforts difficult. For instance, we lack adequate information on a) population sizes (precision and accuracy) that would allow us to detect changes in size, b) the at-sea foraging distribution of birds in the vicinity of most major colonies, c) the probability of death or contamination, d) the sex- and age-specific survivorship rates of various species, e) the extent of recruitment between sub-populations, and f) the extent of density-dependent interactions at various seasons. We also know virtually nothing of the winter distribution of birds breeding at specific colonies, important information for interpreting changes in colony size. Although models have used data for survival based on European studies, it is not clear that European results can be transferred to North America given the very great differences in population distribution and environments between North America and Europe. Data on age specific survivorship, recruitment between subpopulations and pelagic distribution, particularly outside the breeding season, will require long-term population studies using individually identifiable (marked) birds of known age. These studies will require twenty years or more (the generation time of some of the species involved).

Physiological research, based primarily on laboratory studies of domestic species, has identified a number of ways in which ingestion of small, sublethal amounts of oil may affect the ultimate survival or reproductive output of birds. This research now needs to focus on whether native species of marine birds frequently acquire sublethal loads of petroleum hydrocarbons from their environment. If these species in areas of known chronic pollution are contaminated with sublethal levels of petroleum hydrocarbons, then experiments with these species should be conducted in their natural environment to determine the effects of chronic pollution on survivorship and reproductive output. If birds from areas with heavy chronic pollution cannot be shown to have physiological or reproductive disfunctions, then it would seem that physiological research is of little value. From work already done, there are major questions whether the sort of sublethal effects studied in the laboratory occur with frequency in nature and whether results can be extrapolated from one species to another.

Increasingly, evidence is pointing to the potential, if not actual, competition between humans and marine birds (and mammals) for fisheries resources. Much of this fisheries activity is taking place in regions where outer continental shelf oil development is most active. If we are to sort out the relative impacts of fisheries management practices and oil development activities, it is essential that oil related seabird studies be closely coordinated with studies of seabird diet, foraging ecology and food supplies. The population biology of seabirds is, not surprisingly, sensitive to changes in food resources, and predictions relating to the effects of oil on seabirds and their ability to recover after a disaster requires that the condition of food resources be taken into account.

In summary, it appears that the single most important type of information useful to management for minimizing seabird mortality is knowledge of where adult birds aggregate on the water. In order to predict how marine birds will respond to both the acute and long-term effects of oil development, we need a great deal of basic information on the population biology of the species involved. Information on their response to disturbance and sublethal contamination by oil may also be useful. While modeling efforts suggest that chronic, sublethal effects could have serious consequences for population stability, evidence to date is that chronic, sublethal physiological responses to oil contamination are rare and of little population consequence in nature.

Finally, this report should be seen as a starting point for dialogue and planning of future research effort and directions. If recommendations herein are to be implemented, it will be important that they are scrutinized by a wide segment of the seabird research community and that consensus is reached on the most profitable directions and mix of research to pursue. To this end, I recommend that, once additional results are available from the modelers, a multidisciplinary workshop be convened to assess progress and future directions.

Recommendations for Long-Term Research

— Revise and update models of population response to oil pollution to include a) density-dependent responses near and away from colonies, b) recruitment between sub-populations, c) the effects of chronic, low levels of pollution on adult survival and reproductive output, and d) the flexibility in bird behavior.

— Perform sensitivity analyses to identify where the greatest impact of oil will be and the most critical research needs.

— Convene an interdisciplinary workshop to review the information available, assess the results of the modeling studies and recommend a program, including the goals and proportion of different types of research needed, to achieve the required results most expeditiously.

— Commence long-term population studies using individually marked birds of known age. These studies should be conducted in both large and small colonies in each of the separate U.S. offshore regions using as wide a variety of species as possible. At the outset, commitment should be made for a minimum length of study of 20 years. These studies will include population monitoring, reproductive biology, food habits and distribution of local foraging concentrations.

— Survey the level of contamination of seabird tissues to determine whether native species in areas of known chronic pollution are carrying significant amounts of petroleum hydrocarbons. If significant levels are found, then studies should be conducted with these species in their natural environments to determine the effect of the contaminants on survival and reproduction.

— Coordinate studies of the effect of oil development on seabirds with studies of the impact of fisheries on seabird food resources and population biology.

ACKNOWLEDGMENTS

I thank the following people for useful discussions of various aspects of this paper and for directing me to areas of the literature with which I was less familiar: L.J. Blus, D. Boersma, K.T. Briggs, R.B. Clark, J.C. Coulson, E. Cowell, J. Cronshaw, J.P. Croxall, G.M. Dunnet, G.L. Edwards, G. Ford, A.J. Gaston, C.R. Grau, D. Heinemann, J. Jehl, R.H. Jenkins, G. Larminie, E.C. Murphy, D.N. Nettleship, D.B. Peakall, T.J. Peterle, K.D. Powers, B.A. Rattner, R.W. Schreiber, S.G. Sealy, R.C. Szaro, and W. Trivelpiece. I thank the following for critical comments on an earlier draft: L.J. Blus, W.R.P. Bourne, C. Conel, E. Cowell, Z. Eppley, G. Ford, D. Heinemann, L. Jarvela, J. Jehl, I. Nisbet, G. Reetz, T. Stowe (who also gave permission to cite unpublished reports) and K. Vermeer. B.M. Braun and Z. Eppley provided invaluable bibliographic aid. Some of my unpublished work and much of my initial bibliographic work in this field was supported in part by the National Oceanic and Atmospheric Administration (NOAA), contract 03-5-022-72 [through interagency agreement with the Bureau of Land Management under which a multi-year program responding to the needs of petroleum development of the Alaskan continental shelf is managed by the Outer Continental Shelf Environmental Assessment Program (OCSEAP)].

LITERATURE CITED

Ainley, D.G., C.R. Grau, T.E. Roudybush, S.H. Morrell and J.M. Uttis. 1981. Petroleum ingestion reduces reproduction in Cassin's Auklets. Mar. Pollut. Bull. 12: 314-317.

Albers, P.H. 1977. Effects of external applications of fuel oil on hatchability of mallard eggs. Pages 158-163 *in* D.A. Wolfe (ed.), Fate and Effects of Petroleum Hydrocarbons in Marine Ecosystems and Organisms. Pergamon Press, New York.

Albers, P.H. 1978. The effects of petroleum of [sic] different stages of incubation in bird eggs. Bull. Environm. Contam. Toxicol. 19: 624-630.

Albers, P.H. 1979. Effects of Corexit 9527 on the hatchability of mallard eggs. Bull. Environm. Contam. Toxicol. 23: 661-668.

Albers, P.H. 1980. Transfer of crude oil from contaminated water to bird eggs. Environm. Res. 22: 307-314.

Albers, P.H. and M.L. Gay. 1982. Effects of a chemical dispersant and crude oil on breeding ducks. Bull. Environm. Contam. Toxicol. 29: 404-411.

Anderson, D.W. 1983. The seabirds. Pages 246-264 *in* T. Case and M.L. Cody (eds.), Biogeography in the Sea of Cortez. University of California Press, Berkeley.

Anderson, D.W. and F. Gress. 1983. Status of the northern population of California Brown Pelicans. Condor 85: 79-88.

Anderson, D.W. and J.O. Keith. 1980. The human influence on seabird nesting success: Conservation implications. Biol. Conserv. 18: 65-80.

Anderson, D.W., J.R. Jehl, Jr., R.W. Risebrough, L.A. Woods, Jr., L.R. DeWeese and W.G. Edgecomb. 1975. Brown Pelicans: Improved reproduction off the southern California coast. Science 190: 806-808.

Anonymous. 1982. Update: North Carolina mystery spill. Oil Spill Intelligence Report. 5: 1 (21 May 1982).

Ashcroft, R.E. 1979. Survival rates and breeding biology of puffins on Skomer Island, Wales. Ornis Scand. 10: 100-110.

Bailey, E.P. and G.H. Davenport. 1972. Die-off of common murres on the Alaska Peninsula and Unimak Island. Condor 74: 215-219.

Baillie, S.R. and C.J. Mead. 1982. The effect of severe oil pollution during the winter of 1980-81 on British and Irish auks. Ringing and Migration 4: 33-44.

Baird, P.A., and P.J. Gould (eds.). In press. The Breeding Biology and Feeding Ecology of Marine Birds in the Gulf of Alaska. Environmental Assessment of the Alaskan Continental Shelf. Final Reports of Principal Investigators, National Oceanic and Atmospheric Administration, Office of Oceanographic and Marine Assessment, Anchorage, Alaska.

Barrett, R.T. and P.J. Schei. 1977. Changes in the breeding distribution and numbers of cliff-breeding seabirds in Sor-Varanger, North Norway. Astarte 10: 29-35.

Baudinette, R.V. and K. Schmidt-Nielsen. 1974. Energy cost of gliding flight in Herring Gulls. Nature (Lond.) 248: 83-84.

Benedict, F.G. and E.L. Fox. 1927. The gaseous metabolism of large wild birds under aviary conditions. Proc. Amer. Philos. Soc. 66: 511-534.

Bibby, C.J. 1981. An experiment on the recovery of dead birds from the North Sea. Ornis Scand. 12: 261-265.

Bibby, C.J. and C.S. Lloyd. 1977. Experiments to determine the fate of dead birds at sea. Biol. Conserv. 12: 295-309.

Birkhead, T.R. 1974. Movements and mortality rates of British guillemots. Bird Study 21: 241-254.

Birkhead, T.R. 1978. Attendance patterns of guillemots *Uria aalge* at breeding colonies on Skomer Island. Ibis 120: 219-229.

Birkhead, T.R. and P.J. Hudson. 1977. Population parameters for the Common Guillemot *Uria aalge*. Ornis Scand. 8: 145-154.

Birkhead, T.R. and D.N. Nettleship. 1980. Census methods for murres, *Uria* species: A unified approach. Occas. Paper 43., Canadian Wildlife Service, Ottawa, Canada.

Birkhead, T.R., C. Lloyd and P. Corkhill. 1973. Oiled seabirds successfully cleaning their plumage. Brit. Birds 66: 535-537.

Blus, L.J., E. Cromartie, L. McNease and T. Joanen. 1979a. Brown Pelican: Population status, reproductive success, and organochlorine residues in Louisiana, 1971-1976. Bull. Environ. Contam. Toxicol. 22: 128-135.

Blus, L.J., T.G. Lamont and B.S. Neely Jr. 1979b. Organochlorine residues, eggshell thickness, reproduction and population status of Brown Pelicans in South Carolina and Florida, 1969-76. Pestic. Monit. J. 12: 172-184.

Boersma, P.D. 1981. Storm-Petrels as Indicators of Environmental Conditions. Environmental Assessment of the Alaskan Continental Shelf. Annual reports of principal investigators. National Oceanic and Atmospheric Administration, Environmental Research Laboratory, Boulder, Colorado 1: 39-70.

Bourne, W.R.P. 1968a. Oil pollution and bird populations. Pages 99-121 *in* J.D. Carthy and D.R. Arther (eds.), The Biological Effects of Oil Pollution on Littoral Communities. Suppl. Vol. 2, Field Studies Council, London, England.

Bourne, W.R.P. 1968b. Observation of an encounter between birds and floating oil. Nature (Lond.) 219: 632.

Bourne, W.R.P. 1971. Atlantic Puffins decline. Smithsonian Institution Center for Short-lived Phenomena Annual Report 83.71. Smithsonian Institution, Washington, D.C.

Bourne, W.R.P. 1976. Seabirds and pollution. Pages 403-502 *in* R. Johnston (ed.), Marine Pollution. Academic Press, London, England.

Bourne, W.R.P. 1979. Seabird pollution research has gone astray. Mar. Pollut. Bull. 10: 149-150.

Bourne, W.R.P. and C.J. Bibby. 1975. Temperature and the seasonal and geographical occurrence of oiled birds on west European beaches. Mar. Pollut. Bull. 6: 77-80.

Bourne, W.R.P. and L. Johnston. 1971. The threat of oil pollution to North Scottish seabird colonies. Mar. Pollut. Bull. 2: 117-120.

Briggs, K.T., E.W. Chu, D.B. Lewis, W.B. Tyler, R.L. Pitman and G.L. Hunt, Jr. 1981. Distribution, numbers and seasonal status of seabirds of the Southern California Bight area, 1975-1978. Summary of marine mammal and seabird surveys of the Southern California Bight area, 1975-1978, Vol. III. Investigators Reports, Part III - Seabirds, Book I, Chapter I. BLM/YM/SR-81/03. University of California Center for Coastal Marine Studies, Santa Cruz, California.

Brisbin, I.L., Jr. 1965. A Quantitative Analysis of Ecological Growth Efficiency in the Herring Gull. M.S. Thesis, University of Georgia, Athens, Georgia.

Brooke, M. de L. 1978. The dispersal of female Manx Shearwaters *Puffinus puffinus*. Ibis 120: 545-551.

Brown, R.G.B. 1967. Breeding success and population growth in a colony of Herring and Lesser Black-backed Gulls *Larus argentatus* and *L. fuscus*. Ibis 109: 502-515.

Brown, R.G.B. 1973. Sea-birds and oil pollution: The investigation of an off shore oil slick. Progress Notes No. 31. Canadian Wildlife Service, 2 p.

Brown, R.G.B. 1982. Birds, Oil and the Canadian Environment. Pages 105-112 *in* J.B. Sprague, J.H. Vandermeulen and P.G. Wells (eds.), Oil and Dispersants in Canadian Seas - Research Appraisal and Recommendations. Environmental Protection Service, Environment Canada, Ottawa.

Brown, R.G.B., D.N. Nettleship, P. Germain, C.E. Tull and T. Davis. 1975. Atlas of eastern Canadian seabirds. Canadian Wildlife Service, Ottawa, Canada.

Buck, W.F.A. and J.G. Harrison. 1967. Some prolonged effects of oil pollution on the Medway estuary. Rep. Yb. Wildfowl Assoc. Gr. Br. Ire. 1966-67: 32-33.

Buckley, F.G. 1978. Use of Dredged Material Islands by Colonial Seabirds and Wading Birds in New Jersey. Technical Report D-78-1. Army Corps of Engineers, Vicksburg, Mississippi.

Buckley, P.A. and F.G. Buckley. 1980. Population and colony-site trends of Long Island waterbirds for five years in the mid 1970s. Trans. Linn. Soc. New York 9: 23-56.

Burger, J. 1981. Behavioural responses of Herring Gulls *Larus argentatus* to aircraft noise. Environ. Poll. (Series A) 24: 177-184.

Burger, J. 1982. The role of reproductive stress in colony-site selection and abondonment in Black Skimmers (*Rynchops niger*). Auk 99: 109-115.

Burns, K.A. and J.M. Teal. 1971. Hydrocarbon Incorporation in the Salt Marsh Ecosystem from the West Falmouth Oil Spill. NOAA-73032608, National Technical Information Service, Washington, D.C.

Butler, R.G. and P. Lukasiewicz. 1979. A field study of the effect of crude oil on Herring Gull (*Larus argentatus*) chick growth. Auk 96: 809-812.

Butler, R.W., N.A.M. Verbeek and R.G. Foottit. 1980. Mortality and dispersal of the Glaucous-winged Gulls of southern British Columbia. Can. Field-Nat. 94: 315-320.

Cannell, P.F. and G.D. Maddox. 1983. Population change in three species of seabirds at Kent Island, New Brunswick. J. Field Ornithol. 54: 29-35.

Caeser Kleberg Wildlife Research Institute. 1981. An Atlas and Census of Texas Waterbird Colonies, 1973-1980. Caesar Kleberg Wildlife Research Institute, Texas A&I University, Kingsville, Texas.

Casement, M. 1966. Seabirds avoiding oil patches. Sea Swallow 18: 79.

Cavanaugh, K.P. 1982. The effects of South Louisiana and Kuwait crude oils on reproduction. Pages 371-377 *in* C.G. Scanes, M.A. Ottinger, A.D. Kenney, J. Balthazart, J. Cronshaw and I.C. Jones (eds.), Aspects of Avian Endocrinology: Practical and Theoretical Implications. Grad. Studies Texas Tech. Univ. 26, Lubbock, Texas.

Cavanaugh, K.P. and W.N. Holmes. 1982. Effects of ingested petroleum on plasma levels of ovarian steroid hormones in photostimulated mallard ducks. Arch. Environm. Contam. Toxicol. 11: 503-508.

Chabrzyk, G. and J.C. Coulson. 1976. Survival and recruitment in the Herring Gull *Larus argentatus*. J. Anim. Ecol. 45: 187-203.

Chapadelaine, G. and P. Brousseau. 1984. Dou zieme inventaire des populations d' oiseaux marins dans les refuges de la Cote-Nord du Golfe du Saint Laurent. Canadian Field Naturalist 98: 178-183.

Clapp, R.B., R.C. Baukes, D. Morgan-Jacobs and W.A. Hoffman. 1982. Marine Birds of the Southeastern United States and Gulf of Mexico. Part I. Gaviformes through Pelecaniformes. FWS/OBS - 82/101. U.S. Fish and Wildlife Service, Washington, D.C.

Clark, R.B. 1969. Oil pollution and the conservation of seabirds. Pages 76-112 *in* International Conference on Oil Pollution of the Sea. 7-9 October 1968 in Rome. Report of proceedings. Advisory Committee on Oil Pollution of the Sea, London.

Clark, R.B. 1984. Impact of oil pollution on seabirds. Environ. Poll. 33: 1-22.

Coon, N.C. and M.P. Dieter. 1981. Responses of adult Mallard ducks to ingested south Louisiana crude oil. Environ. Res. 24: 309-314.

Coon, N.C., P.H. Albers and R.C. Szaro. 1979. No. 2 fuel oil decreases embryonic survival of Great Black-backed Gulls. Bull. Environm. Contam. Toxicol. 21: 152-156.

Cooper, D., H. Hays and C. Pessino. 1970. Breeding of the Common and Roseate terns on Great Gull Island. Proc. Linn. Soc. New York 71: 83-104.

Corkhill, P. 1971. Factors affecting auk attendance in the pre-egg stage. Nature in Wales 12: 258-262.

Coulson, J.C. 1961. Movements and seasonal variation in mortality of Shags and Cormorants ringed on the Farne Islands, Northumberland. Brit. Birds 54: 225-235.

Coulson, J.C. 1976. An evaluation of the reliability of rings used on Herring and Lesser Black-backed Gulls. Bird Study 23: 21-26.

Coulson, J.C. 1983. The changing status of the Kittiwake *Rissa tridactyla* in the British Isles, 1969-1979. Bird Study 30: 9-16.

Coulson, J.C. and J.M. Horobin 1972. The annual reoccupation of breeding sites by the Fulmar. Ibis 114: 30-42.

Coulson, J.C. and R.D. Wooler. 1976. Differential survival rates among breeding kittiwake gulls *Rissa tridactyla* L. J. Anim. Ecol. 45: 205-213.

Coulson, J.C., M. Duncan and C. Thomas. 1982. Changes in the breeding biology of the Herring Gull (*Larus argentatus*) induced by reduction in the size and density of the colony. J. Anim. Ecol. 51: 739-756.

Coulson, J.C., G.R. Potts, I.R. Deans and S.M. Fraser. 1968. Exceptional mortality of Shags and other seabirds caused by paralytic shellfish poisoning. Brit. Birds 61: 381-404.Cowell, E.B. 1976. Oil pollution of the sea. Pages 353-401 *in* R. Johnston (ed.), Marine Pollution. Academic Press, London.

Cramp, S., W.R.P. Bourne and D. Saunders. 1974. The Seabirds of Britain and Ireland. Collins, London, 287 p.

Crocker, A.D., J. Cronshaw and W.N. Holmes. 1974. The effect of a crude oil on intestinal absorption in ducklings (*Anas platyrhynchos*). Environ. Pollut. 7: 165-177.
Crocker, A.D., J. Cronshaw and W.N. Holmes. 1975. The effect of several crude oils and some petroleum distillation fractions on intestinal absorption in ducklings (*Anas platyrhynchos*). Environ. Physiol. Biochem. 5: 92-106.
Cullen, J.M. 1957. Plumage, age and mortality in the Arctic Tern. Bird Study 4: 197-207.
Curry-Lindahl, K. 1960. Serious situation with regard to Swedish populations of the Long-tailed Duck (*Clangula hyemalis*). Int. Waterfowl Res. Bureau Newsl. 10: 15-18.
Custer, T.W. and P.H. Albers. 1980. Response of captive, breeding Mallards to oiled water. J. Wildl. Manag. 44: 915-918.
Custer, T.W., R.G. Osborn and W.F. Stout. 1980. Distribution, species abundance, and nesting-site use of Atlantic coast colonies of herons and their allies. Auk 97: 591-600.
Davis, J.W.F. and E.K. Dunn. 1976. Intraspecific predation and colonial breeding in Lesser Black-backed Gulls *Larus fuscus*. Ibis 118: 65-77.
DiCostanzo, J. 1980. Population dynamics of a Common Tern colony. J. Field Ornithol. 51: 229-243.
Dieter, M.P. 1977. Acute and chronic studies with waterfowl exposed to petroleum hydrocarbons. Pages 35-42 *in* C. Hall and W. Preston (eds.), Environmental Effects of Energy Related Activities on Marine/Estuarine Ecosystems. EPA-600/7-77-LLL. Interagency Energy - Environment Research and Development Program Report, Washington, D.C.
Drent, R., G.F. van Tets, F. Tompa and K. Vermeer. 1964. The breeding birds of Mandarte Island, British Columbia. Can. Field-Nat. 78: 208-263.
Drury, W.H. 1973. Population changes in New England seabirds. Bird-Banding 44: 267-313.
Drury, W.H. and J.A. Kadlec. 1974. The current status of the Herring Gull population in the northeastern United States. Bird-Banding 45: 297-306.
Drury, W.H., C. Ramsdell and J.B. French, Jr. 1981. Ecological Studies in the Bering Strait Region. Environmental Assessment of the Alaskan Continental Shelf. Final reports of principal investigators. National Oceanic and Atmospheric Administration, Office of Marine Pollution Assessment, Anchorage, Alaska. Biological Studies 11: 175-488.
Duffy, D. 1977. Breeding populations of terns and skimmers on Long Island Sound. Proc. Linn. Soc. New York 73: 1-48.
Duffy, D.C. 1979. Human disturbance and breeding birds. Auk 96: 815-816.
Dunn, E.H. 1975a. Caloric intake of nestling Double-crested Cormorants. Auk 92: 553-565.
Dunn, E.H. 1975b. Growth, body components and energy content of nestling Double-crested Cormorants. Condor 77: 431-438.
Dunn, E.H. 1976a. Development of endothermy and existence energy expenditure of nestling Double-crested Cormorants. Condor 78: 350-356.
Dunn, E.H. 1976b. The development of endothermy and existence energy expenditure in Herring Gull chicks. Condor 78: 493-498.
Dunnet, G.M. 1977. Observations on the effects of low-flying aircraft at seabird colonies on the coast of Aberdeenshire, Scotland. Biol. Conserv. 12: 55-63.
Dunnet, G.M. 1982. Oil pollution and seabird populations. Phil. Trans. R. Soc. Lond. B 297: 413-427.
Dunnet, G.M. and J.C. Ollason. 1978a. The estimation of survival rate in the Fulmar, *Fulmarus glacialis*. J. Anim. Ecol. 47: 507-520.
Dunnet, G.M. and J.C. Ollason. 1978b. Survival and longevity in the Fulmar. Ibis 120: 124-125.
Dunnet, G.M., J.C. Ollason and A. Anderson. 1979. A 28-year study of breeding Fulmars *Fulmarus glacialis* in Orkney. Ibis 121: 293-300.
Eastin, W.C. and D.J. Hoffman. 1979. Biological effects of petroleum on aquatic birds. Pages 561-582 *in* C.C. Bates (ed.), Proceedings of the Conference on Assessment of

Ecological Impacts of Oil Spills. American Institute of Biological Sciences, Arlington, Virginia.

Eastin, W.C., Jr. and H.C. Murray. 1981. Effects of crude oil ingestion on avian intestinal function. Can. J. Physiol. Pharmacol. 59: 1063-1068.

Eastin, W.C. and B.A. Rattner. 1982. Effects of dispersant and crude oil ingestion on Mallard ducklings (*Anas platyrhynchos*). Bull. Environm. Contam. Toxicol. 29: 273-278.

Ellison, N. and L. Cleary. 1978. Effects of human disturbance on breeding of Double-crested Cormorants. Auk 95: 510-517.

Eppley, Z.A. 1984. Development of thermoregulatory abilities in Xantus' Murrelet chicks *Synthliboramphus hypoleucus*. Physiol. Zool. 57: 307-317.

Erasmus, T., R.M. Randall and B.M. Randall. 1981. Oil pollution, insulation and body temperatures in the Jackass Penguin *Spheniscus demersus*. Comp. Biochem. Physiol. 69A: 169-171.

Erickson, R.C. 1963. Oil pollution and migratory birds. Atlantic Naturalist 18: 5-14.

Erwin, R.M. 1979. Coastal Waterbird Colonies: Cape Elizabeth, Maine to Virginia. U.S. Fish and Wildlife Service Publ. No. FWS/OBS 79-10. U.S. Fish and Wildlife Service, Washington, D.C.

Erwin, R.M. 1980. Breeding habitat use by colonially nesting waterbirds in two mid-Atlantic U.S. regions under different regimes of human disturbance. Biol. Conserv. 18: 39-51.

Erwin, R.M. and C.E. Korschgen. 1979. Coastal Waterbird Colonies: Maine to Virginia, 1977. *In* Atlas Showing Colony Locations and Species Composition. U.S. Fish and Wildlife Service Publ. No. FWS/OBS 79-08. U.S. Fish and Wildlife Service, Washington, D.C.

Erwin, R.M., J. Galli and J. Burger. 1981. Colony site dynamics and habitat use in Atlantic coast seabirds. Auk 98: 550-561.

Fetterolf, P.M. 1983. Effects of investigator activity on Ring-billed Gull behavior and reproductive performance. Wilson Bull. 95: 23-41.

Fisher, J. 1966. The Fulmar population of Britain and Ireland, 1959. Bird Study 13: 5-76.

Flegg, J.J.M. and C.J. Cox. 1976. Mortality in the Black-headed Gull. Brit. Birds 68: 437-449.

Flegg, J.J.M. and R.A. Morgan. 1976. Mortality in British gulls. Ringing and Migration 1: 65-74.

Ford, R.G., J.A. Wiens, D. Heinemann and G.L. Hunt. 1982. Modelling the sensitivity of colonially breeding marine birds to oil spills: Guillemot and kittiwake populations on the Pribilof Islands, Bering Sea. J. Appl. Ecol. 19: 1-31.

Freedman, B. and J. Svoboda. 1982. Populations of breeding birds at Alexandra Fjord, Ellsmere Island, Northwest Territories, compared with other arctic localities. Can. Field-Nat. 96: 56-60.

Fritts, T.H., A.B. Irvine, R.D. Jennings, L.A. Collum, W. Hoffman and M.A. McGehee. 1983. Turtles, Birds and Mammals in the Northern Gulf of Mexico and Nearby Atlantic waters. U.S. Fish and Wildlife Service Publ. No. FWS/OBS 82-65. U.S. Fish and Wildlife Service, Washington, D.C.

Frost, P.G.H., W.R. Siegfried and J. Cooper. 1976. Conservation of the Jackass Penguin (*Spheniscus demersus* (L.)). Biol. Conserv. 9: 77-99.

Fry, D.M. and L.J. Lowenstine. 1982. Insults to alcids: Injuries caused by food, by burrowing and by oil contamination. Pac. Seabird Group Bull. 9: 78-79.

Furness, R. 1983. Competition between fisheries and seabird communities. Adv. Mar. Biol. 20: 225-307.

Gaston, A.J. and D.M. Nettleship. 1981. The Thick-Billed Murres of Prince Leopold Island. Monograph Series No. 6. Canadian Wildlife Service, Ottawa, Canada.

Gaston, A.J. and D.N. Nettleship. 1982. Factors determining seasonal changes in attendance at colonies of the Thick-billed Murre *Uria lomvia*. Auk 99: 468-473.

Gaston, A.J., G. Chapdelaine and D.G. Noble. 1983. The growth of Thick-billed Murre chicks at colonies in Hudson Strait: Inter- and intra-colony variation. Can. J. Zool. 61: 2465-2475.

Gay, M.L., A.A. Belisle and J.F. Patton. 1980. Quantification of petroleum-type hydrocarbons in avian tissue. J. Chromatography 187: 153-160.

Giles, L.A., Jr. and L. Livingston. 1960. Oil pollution of the seas. Trans. N. Am. Wild. Nat. Resources Conf. 25: 297-302.

Gillett, W.H., J.L. Hayward, Jr. and J.F. Stout. 1975. Effects of human activity on egg and chick mortality in a Glaucous-winged Gull colony. Condor 77: 492-495.

Gorman, M.L. and C.E. Simms. 1978. Lack of effect of ingested Forties Field crude oil on avian growth. Mar. Pollut. Bull. 9: 273-276.

Gorsline, J. 1982. The effects of south Louisiana crude oil on adrenocortical function. Pages 359-364 *in* C.G. Scanes, M.A. Ottinger, A.D. Kenney, J. Balthazart, J. Cronshaw and I.C. Jones (eds.), Aspects of Avian Endocrinology: Practical and Theoretical Implications. Grad. Studies Texas Tech. Univ. 26, Lubbock, Texas.

Gorsline, J. and W.N. Holmes. 1981. Effects of petroleum on adrenocortical activity and on hepatic naphthalene-metabolizing activity in Mallard ducks. Arch. Environm. Contam. Toxicol. 10: 765-777.

Gorsline, J. and W.N. Holmes. 1982a. Adrenocortical function and hepatic naphthalene metabolism in Mallard ducks (*Anas platyrhynchos*) consuming petroleum distillates. Environ. Res. 28: 139-146.

Gorsline, J. and W.N. Holmes. 1982b. Variations with age in the adrenocortical responses of Mallard ducks (*Anas platyrhynchos*) consuming petroleum-contaminated food. Bull. Environm. Contam. Toxicol. 29: 146-152.

Gorsline, J. and W. N. Holmes. 1982c. Ingestion of petroleum by breeding Mallard ducks: Some effects on neonatal progeny. Arch. Environm. Contam. Toxicol. 11: 147-153.

Gorsline, J., W.N. Holmes and J. Cronshaw. 1981. The effects of ingested petroleum on the naphthalene-metabolizing properties of liver tissue in seawater-adapted Mallard ducks (*Anas platyrhynchos*). Environ. Res. 24: 377-390.

Grau, C.R., T. Roudybush, J. Dobbs and J. Wathen. 1977. Altered yolk structure and reduced hatchability of eggs from birds fed single doses of petroleum oils. Science 195: 779-781.

Grau, G.R., T.A. Wootton, T.E. Roudybush, W.N. Holmes, J. Cronshaw and D.G. Ainley. 1978. Detection of eggs from oil-fed birds by ultraviolet fluorescence of yolk extracts. Pages 297-299 *in* J. Lindstedt-Siva (ed.), Proceedings Energy/Environment '78: A Symposium on Energy Development Impacts. Society of Petroleum Industry Biologists, Los Angeles, California.

Gross, A.O. 1950. The Herring Gull-Cormorant control project. Proc. Int. Ornithol. Congr. 10: 532-536.

Hand, J.L. 1980. Human disturbance in Western Gull *Larus occidentalis livens* colonies and possible amplification by intraspecific predation. Biol. Conserv. 18: 59-63.

Harris, M.P. 1964. Aspects of the breeding biology of the gulls *Larus argentatus*, *L. fuscus* and *L. marinus*. Ibis 106: 432-456.

Harris, M.P. 1976. The seabirds of Shetland in 1974. Scott. Birds 9: 37-68.

Harris, M.P. 1983. Biology and survival of the immature Puffin *Fratercula arctica*. Ibis 125: 56-73.

Harris, M.P. and S. Murray. 1981. Monitoring of puffin numbers at Scottish colonies. Bird Study 28: 15-20.

Harris, M.P., S. Wanless and P. Rothery. 1983. Assessing changes in the numbers of Guillemots *Uria aalge* at breeding colonies. Bird Study 30: 57-66.

Hartung, R. 1965. Some effects of oiling on reproduction of ducks. J. Wildl. Manag. 29: 872-874.

Hartung, R. 1967. Energy metabolism in oil-covered ducks. J. Wildl. Manag. 31: 798-804.

Hartung, R. and G.S. Hunt. 1966. Toxicity of some oils to waterfowl. J. Wildl. Manag. 30: 564-569.

Harvey, S., H. Klandorf and J.G. Phillips. 1981. Reproductive performance and endocrine responses to ingested petroleum in domestic ducks *(Anas platyrhynchos)*. Gen. Comp. Endocrinol. 45: 372-380.

Harvey, S., J.G. Phillips and P.J. Sharp. 1982a. Reproductive performance and endocrine responses to ingested North Sea oil. Pages 379-395 *in* C.G. Scanes *et al.* (eds.), Aspects of Avian Endocrinology: Practical and Theoretical Implications. Grad. Studies Texas Tech. Univ. 26, Lubbock, Texas.

Harvey, S., P.J. Sharp and J.G. Phillips. 1982b. Influence of ingested petroleum on the reproductive performance and pituitary-gonadal axis of domestic ducks (*Anas platyrhynchos*). Comp. Biochem. Physiol. 72C: 83-89.

Hatch, J.J. and I.C.T. Nisbet. 1983. Band wear and band loss in Common Terns. J. Field Ornithol. 54: 1-16.

Hawkes, A.L. 1961. A review of the nature and extent of damage caused by oil pollution at sea. Trans. North Amer. Wildl. Nat. Resour. Conf. 26: 343-355.

Hays, H. 1970. Great Gull Island report on nesting species, 1967-1968. Proc. Linn. Soc. New York 72: 63-76.

Heubeck, M. 1979. A Report to the Shetland Oil Terminal Environmental Advisory Group on the Beached Bird Survey Scheme in Shetland, March 1979 to February 1980. PB82-181231. National Technical Information Service, Springfield, Virginia, 25 p.

Heubeck, M. 1981. A Report to SOTEAG on the 1980 Monitoring Counts of Cliff-Nesting Seabirds in Shetland, with a Review of the Results 1976-1980. Shetland Oil Terminal Environmental Advisory Group, Brae, Shetland.

Hoffman, D.J. 1978. Embryotoxic effects of crude oil in Mallard ducks and chicks. Toxicol. Applied Pharmacol. 46: 183-190.

Hoffman, D.J. 1979a. Embryotoxic and teratogenic effects of petroleum hydrocarbons in Mallards (*Anas platyrhynchos*). J. Toxicol. Environ. Health 5: 835-844.

Hoffman, D.J. 1979b. Embryotoxic and teratogenic effects of crude oil on Mallard embryos on day one of development. Bull. Environm. Contam. Toxicol. 22: 632-637.

Hoffman, D.J. 1979c. Embryotoxic effects of crude oil containing Nickel and Vanadium in Mallards. Bull. Environm. Contam. Toxicol. 23: 203-206.

Hoffman, D.J. and M.L. Gay. 1981. Embryotoxic effects of benzo(a)pyrene, chrysene, and 7,12-dimethylbenz(a)anthracene in petroleum hydrocarbon mixtures in Mallard ducks. J. Toxicol. Environm. Health 7: 775-787.

Holmes, W.N. 1981. Sub-Lethal Effects of Ingested Petroleum on Laboratory-Maintained Mallard ducks (*Anas platyrhynchos*): Evidence for the Suppression of Adrenocortical and Ovarian Function. Workshop V, St. George Synthesis Meeting, Potential Effects of OCS Development on Birds. National Oceanic and Atmospheric Administration, Office of Marine Pollution Assessment, Anchorage, Alaska.

Holmes, W.N. 1982. Some common pollutants and their effects on steroid hormone-regulated mechanisms. Pages 365-370 *in* C.G. Scanes, M.A. Ottinger, A.D. Kenney, J. Balthazart, J. Cronshaw and I.C. Jones (eds.), Aspects of Avian Endocrinology: Practical and Theoretical Implications. Grad. Studies Texas Tech. Univ. 26, Lubbock, Texas.

Holmes, W.N. and J. Cronshaw. 1977. Biological effects of petroleum on marine birds. Pages 359-398 *in* D.C. Malins (ed.), Effects of Petroleum on Arctic and Subarctic Marine Environments. Vol. II. Biological Effects. Academic Press, New York.

Holmes, W.N. and J. Gorsline. 1980. Effects of some environmental pollutants on the adrenal cortex. Pages 311-314 *in* I.A. Commings, J.W. Funder and F.A.O. Mendelsohn (eds.), Symposium on Adrenal Steroid Biosynthesis. Proc. Sixth Int. Congr. Endocrinol. Australian Academy of Science, Canberra.

Holmes, W.N., K.P. Cavanaugh and J. Cronshaw. 1978a. The effects of ingested petroleum on oviposition and some aspects of reproduction in experimental colonies of Mallard ducks (*Anas platyrhynchos*). J. Reprod. Fert. 54: 335-347.

Holmes, W.N., J. Cronshaw and J. Gorsline. 1978b. Some effects of ingested petroleum on seawater-adapted ducks (*Anas platyrhynchos*). Environ. Res. 17: 177-190.

Holmes, W.N., J. Gorsline and K. P. Cavanaugh. 1981. Some effects of environmental

pollutants on endocrine regulatory mechanisms. Pages 1-11 *in* G. Pethes, P. Peczely and P. Rudas (eds.), Recent Advances of Avian Endocrinology. Pergamon Press, London.
Holmes, W.N., J. Gorsline and J. Cronshaw. 1979. Effects of mild cold stress on the survival of seawater-adapted Mallard ducks (*Anas platyrhynchos*) maintained on food contaminated with petroleum. Environ. Res. 20: 425-444.
Hope-Jones, P. 1980. Beached birds at selected Orkney Beaches 1976-1978. Scot. Birds 11: 1-12.
Hope-Jones, P., G. Gowells, E.I.S. Rees and J. Wilson. 1970. Effect of *Hamilton Trader* oil on birds in the Irish Sea in May 1969. Brit. Birds 63: 97-110.
Hope-Jones, P., J.Y. Monnat, C.J. Cadbury and T.J. Stowe. 1978. Birds oiled during the *AMOCO CADIZ* incident-an interim report. Mar. Pollut. Bull. 9: 307-310.
Hunt, G.L. 1972. Influence of food distribution and human disturbance on the reproductive success of Herring Gulls. Ecology 53: 1051-1061.
Hunt, G.L., Jr. 1975. Opening statements for SCAS/BLM workshop on design of baseline studies for the outer continental shelf, southern California marine vertebrates: Birds. Pages 113-117 *in* R.J. Lavenberg and S.A. Earle (eds.), Proceedings: Recommendations for Baseline Research in Southern California Relative to Offshore Oil Development. Southern California Academy of Sciences, Special Publication.
Hunt, G.L. 1976. The Reproductive Ecology, Foods and Foraging Areas of Seabirds Nesting on St. Paul Island, Pribilof Islands. Environmental Assessment of the Alaskan Continental Shelf. Annual reports of principal investigators. National Oceanic and Atmospheric Administration, Environmental Research Laboratory, Boulder, Colorado 2: 155-270.
Hunt, G.L. Jr. and J.L. Butler. 1980. Reproductive ecology of Western Gulls and Xantus' Murrelets with respect to food resources in the Southern California Bight. CalCOFI Report 21: 62-66.
Hunt, G.L. Jr. and M.W. Hunt. 1976. Gull chick survival: The significance of growth rates, timing of breeding and territory size. Ecology 57: 62-75.
Hunt, G.L. Jr. and T. Ingram. 1982. Summary of the Historical Data Base for Selected Species of Marine Birds in the Channel Islands National Park. Unpubl. Rept. to the National Park Service, Channel Islands National Park, Ventura, California.
Hunt, G.L., B. Mayer, W. Rodstrom and R. Squibb. 1978. Reproductive Ecology, Foods and Foraging Areas of Seabirds Nesting on the Pribilof Islands. Environmental Assessment of the Alaskan Continental Shelf. Annual reports of principal investigators. National Oceanic and Atmospheric Administration, Environmental Research Laboratory, Boulder, Colorado 1: 570-575.
Hunt, G.L. Jr., R.L. Pitman, M. Naughton, K. Winnett, A. Newman, P. Kelly and K. Briggs. 1979. Distribution, status, reproductive ecology and foraging habits of breeding seabirds. Summary of marine mammals and seabird surveys of the Southern California Bight area 1975-1978, Vol. III. Investigators Reports, Part III - Seabirds of the Southern California Bight, Book II. Reports to the Bureau of Land Management, Regents of the University of California, Irvine, 399 p.
Hunt, G.L. Jr., P.J. Gould, D.J. Forsell and H. Peterson Jr. 1981a. Pelagic distribution of marine birds in the eastern Bering Sea. Pages 649-718 *in* D.W. Hood and J.A. Calder (eds.), The Eastern Bering Sea Shelf: Oceanography and Resources. U.S. Department of Commerce, National Oceanic and Atmospheric Administration, Office of Marine Pollution Assessment, Rockville, Maryland.
Hunt, G.L. Jr., Z. Eppley, B. Burgeson and R. Squibb. 1981b. Reproductive Ecology, Foods and Foraging Areas of Seabirds Nesting on the Pribilof Islands, 1975-1979. Environmental Assessment of the Alaskan Continental Shelf. Final reports of principal investigators. National Oceanic and Atmospheric Administration, Office of Marine Pollution Assessment, Juneau, Alaska. Biological Studies 12: 1-258.
Hunt, G.L., Jr., J. Kaiwi and D. Schneider. 1982. Pelagic Distribution of Marine Birds and Analysis of Encounter Probability for the Southeastern Bering Sea. Environmental Assessment of the Alaskan Continental Shelf. Final reports of principal investigators.

National Oceanic and Atmospheric Administration, Office of Marine Pollution Assessment, Juneau, Alaska. Biological Studies 16: 1-160.

Hunt, G.L. Jr., Z. Eppley and D. Schneider. 1986. Reproductive performance of seabirds: The importance of population and colony size. Auk 103: 306-317.

Hunt, G.S. 1961. Waterfowl losses on the lower Detroit River due to oil pollution. Proc. Conf. Great Lakes Research 4: 10-26. Univ. Mich. Inst. Sci. Tech., Great Lakes Res. Div. Publ. 7.

Hunter, I., J.P. Croxall and P.A. Prince. 1982. The distribution and abundance of burrowing seabirds (Procellariiformes) at Bird Island, South Georgia: I. Introduction and Methods. Br. Antarct. Surv. Bull. 56: 49-67.

Ingram, T., F. Gress, G.L. Hunt, Jr. and D.W. Anderson. 1983. Handbook for Monitoring Selected Seabird Species in the Channel Islands National Park. Unpubl. Rept. to the National Park Service, Channel Islands National Park, Ventura, California.

Iversen, J.A. and J. Krog. 1972. Body temperatures and resting metabolic rates in small petrels. Norw. J. Zool. 20: 141-144.

Joensen, A.H. and E.B. Hansen. 1977. Oil pollution and seabirds in Denmark 1971-1976. Dan. Rev. Game Biol. 10: 1-31.

Johnson, S.R. and G.C. West. 1975. Growth and development of heat regulation in nestlings, and metabolism of adult Common Murres and Thick-billed Murres. Ornis Scand. 6: 109-115.

Kadlec, J.A. 1975. Recovery rates and loss of aluminium, titanium and incoloy bands on Herring Gulls. Bird-Banding 46: 230-235.

Kadlec, J.A. 1976. A re-evaluation of mortality rates in adult Herring Gulls. Bird-Banding 47: 8-12.

Kadlec, J.A. and W.H. Drury. 1968. Structure of the New England Herring Gull population. Ecology 49: 644-676.

Kadlec, J.A. and W.H. Drury. 1969. Loss of bands from adult Herring Gulls. Bird-Banding 40: 216-221.

Kaiwi, J. and G.L. Hunt, Jr. 1985. Assessment of Oil Spill Risk to Birds. Environmental Assessment of the Alaska Continental Shelf. Pages 527-597 *in* Final Reports of Principal Investigators. Office of Oceanography and Marine Services. National Oceanic and Atmospheric Administration, Juneau, Alaska.

Kinder, T.H., G.L. Hunt, Jr., D. Schneider and J.D. Schumacher. 1983. Correlations between seabirds and oceanic fronts around the Pribilof Islands, Alaska. Estuarine, Coastal and Shelf Science 16: 309-319.

King, J.G. and G.A. Sanger. 1979. Oil vulnerability index for marine oriented birds. Pages 227-240 *in* J.C. Bartonek and D.N. Nettleship (eds.), Conservation of Marine Birds of Northern North America. U.S. Fish and Wildlife Service, Washington, D.C.

King, K. and C. Lefever. 1979. Effects of oil transferred from incubating gulls to their eggs. Mar. Pollut. Bull. 10: 319-321.

Kish, L. 1965. Survey Sampling. John Wiley and Sons, New York.

Koelink, A.F. 1972. Bioenergetics of Growth in the Pigeon Guillemot, *Cepphus columba*. M.S. Thesis, University of British Columbia, Vancouver, B.C., Canada.

Kooyman, G.L., R.W. Davis, J.P. Croxall and D.P. Costa. 1982. Diving depths and energy requirements of King Penguins. Science 217: 726-727.

Korschgen, C.E. 1979. Coastal Waterbird Colonies: Maine. U.S. Fish and Wildlife Service Publ. No. FWS/OBS 79-09. U.S. Fish and Wildlife Service, Washington, D.C.

Kury, C.R. and M. Gochfeld. 1975. Human interference and gull predation in cormorant colonies. Biol. Conserv. 8: 23-34.

Lakhani, K.H. and I. Newton. 1983. Estimating age-specific bird survival rates from ring recoveries - can it be done? J. Anim. Ecol. 52: 83-91.

Lambert, G., D.B. Peakall, B.J.R. Philogene and F.R. Engelhardt. 1982. Effect of oil dispersant mixtures on the basal metabolic rate of ducks. Bull. Environ. Contam. Toxicol. 29: 520-524.

Lasiewski, R.C. and W.R. Dawson. 1967. A re-examination of the relation between standard metabolic rate and body weight in birds. Condor 69: 13-23.

Lawler, G., J.P. Holmes, B.J. Fiorito and J.L. Laseter. 1979. Quantification of petroleum hydrocarbons in selected tissues of male Mallard ducklings chronically exposed to South Louisiana crude oil. Pages 583-612 *in* C.C. Bates (ed.), Proceedings of the Conference Assessment of Ecological Impacts of Oil Spills. American Institute of Biological Sciences, Arlington, Virginia.

Lawler, G., W. Loong and J.L. Laseter. 1978. Accumulation of aromatic hydrocarbons in tissues of petroleum-exposed Mallard ducks (*Anas platyrhynchos*). Environ. Sci. Technol. 12: 51-54.

Leighton, F.A., D.B. Peakall and R.G. Butler. 1983. Heinz-body hemolytic anemia from the ingestion of crude oil: A primary toxic effect in marine birds. Science 220: 871-873.

Lemmetyinen, R. 1966. Damage to water fowl caused by waste oil in the Baltic area. Translation Bureau, Dept. of State, Canada, translation of Suomen Riista 19: 63-71.

Lenington, S. 1979. Predators and blackbirds: The "uncertainty principle" in field biology. Auk 96: 190-192.

Lensink, C.J., P.J. Gould, C.S. Harrison and D. Forsell. 1978. Distribution and Abundance of Marine Birds - South and East Kodiak Island Waters. Environmental Assessment of the Alaskan Continental Shelf. Annual reports of principal investigators. National Oceanic and Atmospheric Administration, Environmental Research Laboratory, Boulder, Colorado 2: 614-710.

Leslie, P.M. 1966. The intrinsic rate of increase and the overlap of successive generations in a population of Guillemots (*Uria aalge* Pont.) J. Anim. Ecol. 35: 291-301.

Levy, E.M. 1980. Oil pollution and seabirds: Atlantic Canada 1976-77 and some implications for northern environments. Mar. Pollut. Bull. 11: 51-56.

Lewis, S.J. 1982. Effects of Oil on Avian Productivity and Population Dynamics. Ph.D. Thesis, Cornell Univ., Ithaca, New York, 191 p.

Lloyd, C. 1973. Attendance at auk colonies during the breeding season. Skokholm Bird Obs. Rept. for 1972: 15-23.

Lloyd, C. 1974. Movement and survival of British Razorbills. Bird Study 21: 102-116.

Lloyd, C. 1975. Timing and frequency of census counts of cliff-nesting auks. Brit. Birds 68: 507-513.

Lloyd, C. and C.M. Perrins. 1977. Survival and age of first breeding in Razorbill (*Alca torda*). Bird-Banding 48: 239-252.

Lock, A.R. and R.K. Ross. 1973. The nesting of the Great Cormorant (*Phalacrocorax carbo*) and the Double-crested Cormorant (*Phalacrocorax auritus*) in Nova Scotia in 1971. Can. Field-Nat. 87: 43-49.

Lustick, S., B. Battersby and M. Kelty. 1978. Behavioral thermoregulation: Orientation toward the sun in Herring Gulls. Science 200: 81-83.

MacDonald, M.A. 1980. The winter attendance of Fulmars at land in NE Scotland. Ornis Scand. 11: 23-29.

Macko, S.A. and S.M. King. 1980. Weathered Oil: Effect on hatchability of heron and gull eggs. Bull. Environm. Contam. Toxicol. 25: 316-320.

Manuwal, D.A., T.R. Wahl and S.M. Speich. 1979. The Seasonal Distribution and Abundance of Marine Bird Populations in the Strait of Juan de Fuca and Northern Puget Sound in 1978. NOAA Technical Memorandum ERL MESA-44. National Oceanic and Atmospheric Administration, Environmental Research Laboratory, Boulder, Colorado.

McEwan, E.H. 1978. The effect of crude oils on salt gland sodium secretion of orally imposed salt loads in Glaucous-winged Gulls, *Larus glaucescens*. Can. J. Zool. 56: 1212-1213.

McEwan, E.H. and A.F.C. Koelink. 1973. The heat production of oiled Mallards and scaup. Can J. Zool. 51: 27-31.

McEwan, E.H. and P.M. Whitehead. 1978. Influence of weathered crude oil on liver enzyme metabolism of testosterone in gulls. Can. J. Zool. 56: 1922-1924.

McEwan, E.H. and P.M. Whitehead. 1980. Uptake and clearance of petroleum hydrocarbons by the Glaucous-winged Gull (*Laras* [sic] *glaucescens*) and the Mallard duck (*Anas platyrhynchos*). Can. J. Zool. 58: 723-726.
McGill, P.A. and M.E. Richmond. 1979. Hatching success of Great Black-backed Gull eggs treated with oil. Bird-Banding 50: 108-113.
McKay, C., C. Prentice, and K. Shepard. 1981. Survey of Breeding Seabirds in Yell Sound, Shetland, Summer 1981. Report to Shetland Oil Terminal Environmental Advisory Group, Brae, Shetland.
Mead, C. 1974. The results of ringing auks in Britain and Ireland. Bird Study 21: 45-86.
Mead, C. 1977. Ten years after the *Torrey Canyon*. B.T.O. News 87: 1-2.
Mead, C. 1981. The Black death. B.T.O. News 112: 1.
Mead, C. and A. Cawthorne. 1983. Massive auk wreck. B.T.O. News 125: 1.
Mendall, H.L. 1936. The home-life and economic status of the Double-crested Cormorant (*Phalacrocorax auritus auritus*) (Lesson). Maine Bull. 39: 1-159.
Miller, D.S., C.J. Bunker, E.B. Burnham, M.C. Rattner, D.B. Peakall and W.B. Kinter. 1976. Effects of crude oil ingestion on plasma osmoregulation in salt-stressed white Pekin Ducks (*Anas platyrhychos*). Bull. Mt. Desert Isl. Biol. Lab. 16: 79-80.
Miller, D.S., D.B. Peakall, and W.B. Kinter. 1978. Ingestion of crude oil: Sublethal effects in Herring Gull chicks. Science 199: 315-317.
Miller, D.S., W.B. Kinter and D.B. Peakall. 1979. Effects of crude oil ingestion on immature Pekin Ducks (*Anas platyrhychos*) and Herring Gulls (*Larus argentatus*). Pages 27-40 *in* Animals as Monitors of Environmental Pollutants. Symposium on Pathobiology of Environment Pollutants. National Academy of Sciences, Washington, D.C.
Miller, G.J. and D.W. Connell. 1980. Occurrence of petroleum hydrocarbons in some Australian seabirds. Aust. Wildl. Res. 7: 281-293.
Moffitt, J. and R.T. Orr. 1938. Recent disastrous effects of oil pollution on birds in the San Francisco Bay region. Calif. Fish and Game 24: 244-259.
Monnat, J.-Y. 1969. Status actual des oiseaux marins nicheurs en Bretagne. Ar. Vron 2: 1-24.
Morse, D.H. and C.W. Buchheister. 1977. Age and survival of breeding Leach's Storm-Petrels in Maine. Bird-Banding 48: 341-349.
Murray, K.G., K. Winnett-Murray, Z.A. Eppley, G.L. Hunt, Jr. and D.B. Schwartz. 1983. Breeding biology of Xantus' Murrelet. Condor 85: 12-21.
National Environmental Research Council. 1977. The Report of a Working Group on Ecological Research on Seabirds. N.E.R.C. Pub. Ser. C. No. 18, National Environmental Research Council, London, 48 p.
Nero and Associates, Inc. 1983. Seabird-Oil Spill Behavior Study. Volume II. Technical Report. U.S. Minerals Management Service Contract #SBO-408a-80-C-SSO/AA851-CTO-70, Pacific Outer Continental Shelf Office, Los Angeles, California, 64 p.
Nesbitt, S.A., J.C. Ogden, H.W. Kale II, B.W. Patty and L.A. Rowse. 1982. Florida Atlas of Breeding Sites for Herons and Their Allies: 1976-1978. U.S. Fish and Wildlife Service Publ. No. FWS/OBS 81-49. U.S. Fish and Wildlife Service, Washington, D.C.
Nettleship, D.N. 1976. Census Technique for Seabirds of Arctic and Eastern Canada. Occas. Paper 25. Canadian Wildlife Service, Ottawa, Canada.
Nettleship, D.N. 1977. Seabird resources of eastern Canada: Status, problems and prospects. Pages 96-108 *in* T. Mosquin and C. Suchal (eds.), Canada's Threatened Species and Habitats. Can. Nat. Fed., Ottawa, Canada.
Nettleship, D.N. and A.R. Lack. 1973. Tenth census of seabirds in the sanctuaries of the north shore of the Gulf of St. Lawrence. Canadian Field Naturalist 87: 395-402.
Newman, A. MS. Monitoring Manual - Reproductive Success Study - General Methods. Unpubl. report. U.S. Fish and Wildlife Service, Honolulu, Hawaii.
Nisbet, I.C.T. 1973. Terns in Massachusetts: Present numbers and historical changes. Bird-Banding 44: 27-55.

Nisbet, I.C.T. 1978a. Recent changes in gull populations in the western North Atlantic. Ibis 120: 129-130.
Nisbet, I.C.T. 1978b. Direct human influences: Hunting and the use by birds of man's waste deposits. Ibis 120: 134.
Nisbet, I.C.T. 1979. Conservation of marine birds of northern North America. Pages 305-318 *in* J.C. Bartonek and D.M. Nettleship (eds.), Conservation of Marine Birds of Northern North America. Wildlife Research Report II. U.S. Fish and Wildlife Service, Washington, DC.
Nisbet, I.C.T. 1980. Effects of toxic pollutants on productivity in colonial waterbirds. Trans. Linn. Soc. New York 9: 103-114.
Nisbet, I.C.T. 1981. Biological Characteristics of the Roseate Tern *Sterna dougallii*. Unpubl. Rept. to U.S. Fish and Wildlife Service from Massachusetts Audubon Society, 112 p.
Nisbet, I.C.T. and W.H. Drury. 1972. Measuring breeding success in Common and Roseate Terns. Bird-Banding 43: 97-106.
Nisbet, I.C.T. and M.J. Walton. 1984. Seasonal variations in breeding success of Common Terns: Consequences of predation. Condor 86: 53-60.
North, P.M. 1980. An analysis of Razorbill movements away from the breeding colony. Bird Study 27: 11-20.
Ohlendorf, H.M., R.W. Risebrough and K. Vermeer. 1978. Exposure of Marine Birds to Environmental Pollutants. U.S. Fish and Wildlife Service, Wildlife Res. Rept. 9: 1-40.
Ollason, J.C. and G.M. Dunnet. 1978. Age, experience and other factors affecting the breeding success of the Fulmar, *Fulmarus glacialis*, in Orkney. J. Anim. Ecol. 47: 961-976.
Ollason, J.C. and G.M. Dunnet. 1980. Nest failures in the Fulmar: The effect of observers. J. Field Ornithol. 51: 39-54.
Ollason, J.C. and G.M. Dunnet. 1983. Modelling annual changes in numbers of breeding Fulmars, *Fulmarus glacialis*, at a colony in Orkney. J. Anim. Ecol. 52: 185-198.
Osborn, R.G. and T.W. Custer. 1978. Herons and Their Allies: Atlas of Atlantic coast Colonies, 1975 and 1976. U.S. Fish and Wildlife Service Publ. No. FWS/OBS 77-08. U.S. Fish and Wildlife Service, Washington, D.C.
Page, G.W., L.E. Stenzel and D.G. Ainley. 1982. Beached Bird Carcasses as a Means of Evaluating Natural and Human Caused Seabird Mortality. Final Report for U.S. Department of Energy Contract DE-ACO3-79EV10254. National Technical Information Service, Springfield, Virginia.
Palmer, R.S. 1949. Maine birds. Bull. Mus. Comp. Zool. Harvard 102: 1-656.
Parnell, J. and R. Soots, Jr. 1980. Atlas of Colonial Waterbirds of North Carolina. Sea Grant Publication UNC-SG8006. University of North Carolina, Chapel Hill, North Carolina.
Patten, S.M., Jr. and L.R. Patten. 1983. Evolution, Pathobiology and Breeding Ecology of Large Gulls (*Larus*) in the Northeast Gulf of Alaska and Effects of Petroleum Exposure on the Breeding Ecology of Gulls and Kittiwakes. Environmental Assessment of the Alaskan Continental Shelf. Final reports of principal investigators. National Oceanic and Atmospheric Administration, Office of Oceanography and Marine Services, Juneau, Alaska. Biological Studies 18: 1-352.
Patton, J.F. and M.P. Dieter. 1980. Effects of petroleum hydrocarbons on hepatic function in the duck. Comp. Biochem. Physiol. 65C: 33-36.
Paynter, R. 1949. Clutch-size and egg and chick mortality of Kent Island Herring Gulls. Ecology 30: 146-166.
Peakall, D.B., D.J. Hallett, J.R. Bend and G.L. Foureman. 1982. Toxicity of Prudhoe Bay crude oil and its aromatic fractions to nestling Herring Gulls. Environ. Res. 27: 206-215.
Peakall, D.B., D. Hallett, D.S. Miller, R.G. Butler and W.B. Kinter. 1980. Effects of ingested crude oil on Black Guillemots: A combined field and laboratory study. Ambio 9: 28-30.

Peakall, D.B., D.S. Miller and W.B. Kinter. 1979. Physiological techniques for assessing the impact of oil on seabirds. Pages 52-60 *in* E.E. Kenaga (ed.), Avian and Mammalian Wildlife Toxicology. American Society for Testing and Materials, Philadelphia, Pennsylvania.

Peakall, D.B., J. Tremblay, W.B. Kinter and D.S. Miller. 1981. Endocrine dysfunction in seabirds caused by ingested oil. Environ. Res. 24: 6-14.

Perrins, C.M., M.P. Harris and C.K. Britton. 1973. Survival of Manx Shearwaters *Puffinus puffinus*. Ibis 115: 535-548.

Peters, C.P., K.O. Rishter, D.A. Manuwal and S.G. Herman. 1978. Colonial Nesting Sea and Wading Bird Use of Estuarine Islands in the Pacific Northwest. Technical Report D-78-17. Army Corps of Engineers, Vicksburg, Mississippi.

Phillips, J.C. and F.C. Lincoln. 1930. American Waterfowl. Houghton Miffin Co., New York, 312 p.

Portnoy, J.W. 1977. Nesting Colonies of Seabirds and Wading Birds: Coastal Louisiana, Mississippi and Alabama. U.S. Fish and Wildlife Service Publ. No. FWS/OBS 77-07. U.S. Fish and Wildlife Service, Washington, D.C.

Portnoy, J.W. 1980. Atlas of Gull and Tern Colonies: North Carolina to Key West, Florida. U.S. Fish and Wildlife Service Publ. No. FWS/OBS 80-05. U.S. Fish and Wildlife Service, Washington, D.C.

Potts, G.R., D.C. Coulson and I.R. Deans. 1980. Population dynamics and breeding success of the shag, *Phalacrocorax aristotelis*, on the Farne Islands, Northumberland. J. Anim. Ecol. 49: 465-484.

Powers, K.D. and W.T. Rumage. 1978. Effect of the *Argo Merchan*t oil spill on bird populations off the New England coast, 15 December 1976 - January 1977. Pages 142-148 *in* In the Wake of the *Argo Merchant*. Proc. Symp. January 11-13, 1978. Center for Ocean Management Studies, Univ. Rhode Island, Kingston, Rhode Island.

Powers, K.D., G.L. Pittman and S.J. Fitch. 1980. Distribution of Marine Birds on the Mid- and North-Atlantic U.S. Outer Continental Shelf. Technical Progress Report, Department of Energy contract No. DE-AC02-78EVO4706. U.S. Department of Energy, Washington, D.C.

Rattner, B.A. 1981. Tolerance of adult Mallards to subacute ingestion of crude petroleum oil. Toxicology Letters 8: 337-342.

Rattner, B.A. and W.C. Eastin, Jr. 1981. Plasma corticosterone and thyroxine concentrations during chronic ingestion of crude oil in Mallard ducks (*Anas platyrhynchos*). Comp. Biochem. Physiol. 68C: 103-107.

Richardson, M.G., G.M. Dunnet and P.K. Kimear. 1981. Monitoring seabirds in Shetland. Proc. Royal Soc. Edinburgh 80B: 157-179.

Ricklefs, R.E. 1973. Patterns of growth in birds. II. Growth rate and mode of development. Ibis 115: 177-201.

Rittinghaus, H. 1956. On the indirect distribution of the oil pest in a seabird sanctuary. Canadian Wildlife Service Translation (TR-GER-98) of Ornithol. Mitteil. 8: 43-46.

Robert, H.C. and C.J. Ralph. 1975. Effects of human disturbance in the breeding success of gulls. Condor 77: 495-499.

Royal Society for the Protection of Birds. 1979. Marine Oil Pollution and Birds. RSPB, Sandy, Bedfordshire, England, 126 p.

Samuels, W.B. and K.J. Lanfear. 1982. Simulations of seabird damage and recovery from oil spills in the northern Gulf of Alaska. J. Environm. Mgmt. 15: 169-182.

Savard, J.-P.L. and G.E.J. Smith. 1982. Comparison of survey techniques for burrow-nesting seabirds. Pacific and Australian Seabird Groups Meeting, Honolulu, Hawaii, 9: 65-66 (Abstract).

Schaefer, M.B. 1970. Men, birds and anchovies in the Peru Current—Dynamic interactions. Trans. Am. Fish. Soc. 99: 461-467.

Scholander, P.F., R. Hock, V. Walters and L. Irving. 1950. Adaptation to cold in arctic and tropical mammals and birds in relation to body temperature, insulation, and basal metabolism. Biol. Bull.: 259-271.

Schreiber, R. 1979. Reproductive Performance of the Eastern Brown Pelican. Nat. Hist. Mus. Los Angeles County, Contr. Sci. 317, 43 p.
Schreiber, E.A. and R.W. Schreiber. 1980. Effects of impulse noise on seabirds of the Channel Islands. Pages 138-162 *in* J.R. Jehl and C.F. Cooper (eds.), Potential Effects of Space Shuttle Sonic Booms on the Biota and Geology of the California Channel Islands: Research Reports. Center for Marine Studies, San Diego State University Tech. Report 80-1, San Diego, California.
Slater, P.J.B. 1980. Factors affecting the numbers of guillemots *Uria aalge* present on cliffs. Ornis Scand. 11: 155-163.
Sloan, N.F. 1982. Status of breeding colonies of White Pelicans in the United States through 1979. Amer. Birds 36: 250-254.
Southern, H.N., R. Carrick and W.G. Potter. 1965. The natural history of a population of guillemots (*Uria aalge* Pont.). J. Anim. Ecol. 34: 649-665.
Sowls, A.L., A.R. DeGange, J.W. Nelson and G.S. Lester. 1980. Catalog of California Seabird Colonies. U.S. Fish and Wildlife Service Publ. No. FWS/OBS 80-37. U.S. Fish and Wildlife Service, Washington, D.C.
Sowls, A.L., S.A. Hatch and C.J. Lensink. 1978. Catalog of Alaskan Seabird Colonies. U.S. Fish and Wildlife Service Publ. No. FWS/OBS 78-78. U.S. Fish and Wildlife Service, Washington, D.C.
Springer, A. and D. Roseneau. 1978. Ecological Studies of Colonial Seabirds at Cape Thompson and Cape Lisburne, Alaska. Environmental Assessment of the Alaskan Continental Shelf. Annual reports of principal investigators. National Oceanic and Atmospheric Administration, Environmental Research Laboratory, Boulder, Colorado 2: 839-960.
Springer, A., D. Roseneau and M. Johnson. 1979. Ecological Studies of Colonial Seabirds at Cape Thompson and Cape Lisburne, Alaska. Environmental Assessment of the Alaskan Continental Shelf. Annual reports of principal investigators. National Oceanic and Atmospheric Administration, Environmental Research Laboratory, Boulder, Colorado 2: 516-574.
Stickel, L. and M.P. Dieter. 1979. Ecological and Physiological/Toxicological Effects of Petroleum on Aquatic Birds. U.S. Fish and Wildlife Service Publ. No. FWS/OBS-79/23. U.S. Fish and Wildlife Service, Washington, D.C.
Stowe, T.J. 1982a. Recent population trends in cliff-breeding seabirds in Britain and Ireland. Ibis 124: 502-510.
Stowe, T.J. 1982b. An oil spillage at a guillemot colony. Mar. Pollut. Bull. 13: 237-239.
Stowe, T.J. 1982c. Beached Bird Surveys. Royal Society for the Protection of Birds, Sandy, Bedfordshire, England, 138 p.
Stowe, T.J. and L. A. Underwood. 1984. Oil spillages affecting seabirds in the United Kingdom, 1966-1983. Mar. Pollut. Bull. 15: 147-152.
Swartz, L.G. 1966. Sea-cliff birds. Pages 611-678 *in* N.J. Wilimovsky and J.N. Wolfe (eds.), Environment of the Cape Thompson Region, Alaska. U.S. Atomic Energy Commission, Oak Ridge, Tennessee.
Szaro, R.C. 1977. Effects of petroleum on birds. Trans. No. Amer. Wildlife and Nat. Res. Conf. 42: 374-381.
Szaro, R.C. and P.H. Albers. 1977. Effects of external applications of No. 2 fuel oil on Common Eider eggs. Pages 164-167 *in* D.A. Wolfe (ed.), Fate and Effects of Petroleum Hydrocarbons in Marine Ecosystems and Organisms. Pergamon Press, New York.
Szaro, R.C., M.P. Dieter, G.H. Heinz and J.F. Ferrell. 1978. Effects of chronic ingestion of south Louisiana crude oil on Mallard ducklings. Environm. Res. 17: 426-436.
Szaro, R.C., N.C. Coon and W. Stout. 1980. Weathered petroleum: Effects on mallard egg hatchability. J. Wildl. Manag. 44: 709-713.
Szaro, R.C., G. Hensler and G.H. Heinz. 1981. Effects of chronic ingestion of No. 2 fuel oil on mallard ducklings. J. Toxicol. Environ. Health. 7: 789-799.
Tanis, J.J.C. and M.F. Morzer Bruyns. 1969. The impact of oil-pollution on sea birds in

Europe. Pages 67-113 *in* International Conference on Oil Pollution of the Sea. 7-9 October 1968 in Rome. Report of Proceedings. Advisory Committee on Oil Pollution of the Sea, London.

Trivelpiece, W.Z., R.G. Butler, D.S. Miller and D.B. Peakall. 1984. Reduced survival of chicks of oil-dosed adult Leach's Storm-petrels. Condor 86: 81-82.

Tuck, L.M. 1960. The murres. Canadian Wildlife Service, Ottawa, Canada.

Vangilder, L.D. and T.J. Peterle. 1980. South Louisiana crude oil and DDE in the diet of Mallard hens: Effects on reproduction and duckling survival. Bull. Environ. Contam. Toxicol. 25: 23-28.

Varoujean, D. 1978. Seabird Colony Catalog: Washington, Oregon and California. U.S. Dept. Interior, Fish and Wildlife Service. U.S. Fish and Wildlife Service, Washington, D.C.

Varoujean, D. and R.L. Pitman. 1979. Oregon Seabird Colony Survey, 1979. Unpubl. Rept. U.S. Fish and Wildlife Service, Washington D.C., 150 p.

Vermeer, K. 1963. The breeding ecology of the Glaucous-winged Gull (*Larus glaucescens*) on Mandarte Island, British Columbia. Occ. Pap. B.C. Prov. Mus. 13: 1-104.

Vermeer, K. 1970a. The Breeding Biology of California and Ring-billed Gulls. Can. Wildlife Rept. Serv. 12, Canadian Wildlife Service, Ottawa, Canada, 52 p.

Vermeer, K. 1970b. Some aspects of the nesting of Double-crested Cormorants at Cypress lake, Saskatchewan, in 1969, a plea for protection. Blue Jay 28: 11-13.

Vermeer, K. 1979. Nesting requirements, food and breeding distribution of Rhinoceros Auklets, *Cerorhinca monocerata*, and Tufted Puffins, *Lunda cirrhata*. Ardea 67: 101-110.

Vermeer, K. and G.G. Anweiler. 1975. Oil threat to aquatic birds along the Yukon coast. Wilson Bull. 87: 467-480.

Vermeer, R. and K. Vermeer. 1974. Oil Pollution of Birds: An Abstracted Bibliography. Canadian Wildlife Service Pesticide Section Manuscript Reports 29, Canadian Wildlife Service, Ottawa, Canada, 68 p.

Vermeer, K. and R. Vermeer. 1975. Oil threat to birds on the Canadian west coast. Can. Field Nat. 89: 278-298.

Vermeer, K., I. Robertson, R.W. Campbell, G. Kaiser and M. Lemon 1983. Distribution and Densities of Marine Birds on the Canadian West Coast. Canadian Wildlife Service Report. Canadian Wildlife Service, Vancouver, British Columbia, Canada.

Walsberg, G.E. 1983. Avian ecological energetics. Pages 161-220 *in* D.S. Farner, J.R. King and K.C. Parkes (eds.), Avian Biology. Volume 7. Academic Press, New York.

Wanless, S. 1983. The Effects of Low Flying Aircraft on Seabird Reserves with Particular Reference to the Isle of May NNR. Unpubl. rept. submitted to the Nature Conservancy Council, November 1983.

Wanless, S., D.D. French, M.P. Harris and D.R. Longslow. 1982. Detection of annual changes in the numbers of cliff-nesting seabirds in Orkney 1976-80. J. Anim. Ecol. 51: 785-795.Ward, J.G. 1972. Studies on the Feeding Ecology and Breeding Success of the Glaucous-Winged Gull (*Larus glaucescens*) on Mandarte and Cleland Islands. Ph.D. Thesis, University of British Columbia, Vancouver, British Columbia, Canada.

Weathers, W.W. and K.A. Nagy. 1980. Simultaneous doubly labeled water ($^3HH^{18}0$) and time-budget estimates of daily energy expenditure in *Phainopepla nitens*. Auk 97: 861-867.

Westphal, A. and M.K. Rowan. 1970. Some observations on the effects of oil pollution on the Jackass Penguin. Ostrich (Suppl.) 8: 521-526.

White, D.H., K.A. King and N.C. Coon. 1979. Effects of No. 2 fuel oil on hatchability of marine and estuarine bird eggs. Bull. Environm. Contam. Toxicol. 21: 7-10.

Wiens, J.A., R.G. Ford and D. Heinemann. 1984. Information needs and priorities for assessing the sensitivity of marine birds to oil spills. Biol. Conserv. 28: 21-49.

Wiens, J.A., R.G. Ford, D. Heinemann and C. Fieber. 1979. Simulation Modeling of Marine Bird Population Energetics, Food Consumption and Sensitivity to Perturbation.

Environmental Assessment of the Alaskan Continental Shelf. Annual reports of principal investigators. National Oceanic and Atmospheric Administration, Environmental Research Laboratory, Boulder, Colorado 2: 1-83.

Wittenberger, J.F. and G.L. Hunt, Jr. 1985. The adaptive significance of coloniality in birds. Pages 1-78 *in* D.S. Farner, J.R. King and K.C. Parkes (eds.), Avian Biology. Volume 8. Academic Press, New York.

Wootton, T.A., C.R. Grau, T.E. Roudybush, M.E. Hah and K.V. Hirsch. 1979. Reproductive responses of quail to Bunker C oil fractions. Arch. Environm. Contam. Toxicol. 8: 457-463.

CHAPTER 12

EFFECTS OF OFFSHORE OIL AND GAS DEVELOPMENT ON MARINE MAMMALS AND TURTLES

Joseph R. Geraci and David J. St. Aubin

CONTENTS

INTRODUCTION

Before the early 1970s, our understanding of how oil might affect marine mammals came from conflicting field reports and popular news accounts. The prevailing notion was that oil would foul the fur of seals, sea lions and otters, plug nasal passages, and intoxicate an animal breathing or ingesting petroleum hydrocarbons. These notions were bathed in an emotionally charged atmosphere in which feelings for dolphins and whales were running at a high pitch. At the same time, there was little concern over marine turtles. Consequently, experimental studies on the effects of oil on marine mammals have been given impetus, while turtles, in keeping with their popular stature, remain unheralded as research subjects.

Our aim is to evaluate the impacts of oil and oil-production activity on marine mammals and turtles, by blending historical accounts and experimental studies with the patchy information on the life history of the various groups. This will lay to rest some fanciful views, elucidate mechanisms by which offshore oil and gas development poses a threat to certain species, and identify areas for fruitful research.

HISTORICAL PERSPECTIVES

Over the past 15 years, reports by the news media and some scientific authorities have implicated oil-fouling as the cause of death of seals, sea otters, small and large whales and, more recently, turtles. Many of the accounts involving marine mammals were evaluated in our previous review (Geraci and St. Aubin, 1980); these and more recent events are summarized in Table 12.1. It is clear that most of the reports are of oil fouling the pelage of seals and otters—the kind of impact easily noticed by even the casual observer. No comparable documentation exists for free-ranging cetaceans. There are only two speculative reports associating oil with the death of a whale and a dolphin (Anon., 1971b; Duguy, 1978). These observations, coupled with accounts of cetaceans swimming and feeding in oil slicks (Shane and Schmidly, 1978; Goodale *et al.*, 1979; Gruber, 1981) represent the extent of our information for this group. The association of oil with manatee (*Trichechus manatus*) mortality is more tenuous, amounting to no more than the incidental recovery of tar in the digestive tract of three animals.

When evaluating these reports, it seems reasonable to assume that some of the deaths, especially of sea otters, could have been due to oil. As for the others, causes of death for most marine mammals found on the beach can seldom be established with certainty. The presence of oil on a carcass usually complicates rather than simplifies the diagnostic process, and for this reason, the association between oil and death, however obvious it may appear, is difficult to validate.

There are few reports of marine turtles encountering oil. Like many cetaceans, they lead a more pelagic existence, and most mortalities would go unobserved. First Diaz-Piferrer (1962), then Rutzler and Sterrer (1970) noted that oil was involved in the death of turtles. Later, Witham (1978) proposed that oceanic residues of oil might pose a continuing hazard to young marine turtles, during the "lost years" when they are rarely seen after leaving the nest. Since Witham's report, 30 green turtles (*Chelonia mydas*), three loggerheads (*Caretta caretta*), two hawksbills (*Eretmochelys imbricata*) and one Kemp's ridley turtle (*Lepi-*

TABLE 12.1
Reports of marine mammals associated with oil

Date	*Location*	*Oil type & quantity*	*Species*	*Impact*	*Reference*
1949	Ramsey Island, Wales Source unknown	Fuel oil Quantity unknown	*Halichoerus grypus*	Pups unaffected by thick coating of fuel oil	Davies, 1949
1953	Antarctic Ship discharge	Fuel oil Quantity unknown	seals	"damage"	Lillie, 1954 in Nelson-Smith, 1970
May 1967	English Channel (*Torrey Canyon*)	Crude oil 100,000$^+$ tons	*Halichoerus grypus*	Seals observed surfacing through oil slicks; 3 confirmed deaths	Gill *et al.*, 1967 Spooner, 1967
	Kurtile Island, U.S.S.R.	Gasoline and diesel	*Enhydra lutris*	Over 100 animals killed	Barabash-Nikiforov *et al.*, 1968
Feb. 1969	Santa Barbara, California	Crude oil 11,000 to 113,000 tons	*Mirounga angustirostris* *Zalophus californianus* *Eschrictius robustus* *Phocoena phocoena* *Tursiops truncatus* *Callorhinus ursinus*	Twenty to 25 dead animals found with surface coatings of oil. No causal relationship established.	McCaull, 1969 Simpson and Gilmartin, 1970 Brownell, 1971 Brownell and LeBoeuf, 1971 LeBoeuf, 1971 Straughan, 1971, 1972 Nicholson, 1972
Feb. 1969	Gulf of St. Lawrence (storage tank)	Bunker C	*Phoca groenlandica*	500–2000 harp seals observed coated with oil. Some mortality—could not be determined whether oil was causal agent	Warner, 1969

(continued)

TABLE 12.1—*contd.*

Date	*Location*	*Oil type & quantity*	*Species*	*Impact*	*Reference*
Nov. 1969	North Dyfed Coast, Wales Source unknown	Unknown type	seals	14 oiled and dead pups. No causal relationship established	Anon., 1970
Feb. 1970	Chedabucto Bay and Sable Island, Nova Scotia (*Arrow*)	Bunker C 3.8 million gal	*Halichoerus grypus* *Phoca vitulina*	500 gray & harbor seals covered with oil, 24 found dead; 2 with oil in mouth; 1 in stomach, 1 in trachea, 2 found heavily oiled in mouth	Mansfield, 1970 Anon., 1971a
Feb. 1970 Mar.	Kodiak Island, Alaska Deliberate discharge	Slop oil and/or oily ballast	"hair seals" "sea lions" *Enhydra lutris*	Est. 500 marine mammals contacted oil—no mortality	Anon., 1971a
Apr. 1970	Alaska Peninsula Unknown source	Light diesel fuel	*Orcinus orca* "hair seals"	400 hair seals exhibited unusual behavior; 1 sick, 1 dead killer whale observed	Anon., 1971b
Aug. 1974	French Coast Unknown source	Fuel oil	*Phoca vitulina* *Halichoerus grypus*	1 harbor seal pup with oil in intestines; 3 gray seals coated with oil; 1 with ingested oil	Duguy and Babin, 1975
1973	Dutch Coast Unknown source	Unknown type	*Phoca vitulina*	Found with patches of oil on fur and inflamed skin. Some had skin lesions associated with oil patches. Larger numbers had lesions without signs of oil fouling	Van Haaften, 1973
Sep. 1974	Pembrokeshire, Wales Unknown source	Unknown type	*Halichoerus grypus*	2 pups, heavily coated and unable to swim drowned when washed off beach	Davis and Anderson, 1976

Oct. 1976	Aransas Pass, Texas Pipeline leak	Crude oil 16,000 gal	*Tursiops truncatus*	Dolphins swam through oil without apparent effect	Shane and Schmidly, 1978
Aug. 1977	Potomac, Greenland Vessel struck ice	Bunker C 380 tons	*Phoca hispida* and other seals	540 seals observed—none oiled or dead. Oil taken from pelts 5 mo. later had different hydrocarbons due to biodegradation	Grose *et al.*, 1979
Mar. 1978	French Coast (*Amoco Cadiz*)	Light crude 200,000 tons	*Halichoerus grypus* *Lagenorhynchus acutus* *Delphinus delphis* *Globicephala melaena*	4 seals and 6 cetaceans dead and variably coated with oil. No causal relationship established.	Prieur and Hussenot, 1978
June 1979	Bay of Campeche, Gulf of Mexico (IXTOC-I oil well)	20,000–30,000 barrels crude per day from June 3–Aug. 13		Dolphins sighted bowriding through oiled water. No mortalities or effects of oil were noted.	Anon., 1979a Bergey, 1979
June 1979	Cape Cod, Massachusetts (*Regal Sword*) 2 tankers collided sinking 1	Bunker C 300,000 L; Fuel oil #2 24,000 L	*Megaptera novaeangliae* *Balaenoptera physalus* *Lagenorhynchus acutus* *Eubalaena glacialis* *Balaenoptera acutorostrata*	2 fin whales observed swimming under thick oil; 5–10 humpbacks observed feeding in thin sheen surfacing through it with open mouths. Cetaceans spotted for 2 days in oiled and unoiled areas.	Goodale *et al.*, 1979
Mar. 1979	Cabot Strait, Cape Breton, Nova Scotia (*Kurdistan*)	Bunker C 7500 tons	*Halichoerus grypus* *Phoca vitulina*	At least 4 gray and 6 harbor seals found dead coated with oil. Could not establish causal relationship.	Marston, 1979 (per comm.) Parsons *et al.*, 1980
Sep. 1978	Matagorda Bay, Texas Towboat ran aground	Fuel oil #6 11,000 L	*Tursiops truncatus*	Dolphins, including one calf observed feeding and playing in areas of heaviest concentrations, apparently unaffected.	Gruber, 1981

(*continued*)

TABLE 12.1—*contd.*

Date	*Location*	*Oil type & quantity*	*Species*	*Impact*	*Reference*
Nov. 1980	Pt. Piedras, California Unknown source	Unknown type	*Enhydra lutris*	Body found moderately decomposed—oiled and with shark wounds	Anon., 1980a
Jan. 1981	San Carlos Creek, Jacksonville, Florida Unknown source	Tar-like petroleum substance	*Trichechus manatus*	Male found freshly dead; broken ribs, tar-like substance in caecum and large intestine	Anon., 1981a
Feb. 1981	Pacific Valley, California Unknown source	Tar	*Enhydra lutris*	1 live abandoned pup with small amount of tar on fur	Anon., 1981b
Feb. 1981	Jacksonville, Florida Unknown source	Tar-like petroleum substance	*Trichechus manatus*	Body badly decomposed with tar-like substance in caecum and upper large intestine	Anon., 1981c
Mar. 1981	Jacksonville, Florida Unknown source	Tar-like petroleum substance	*Trichechus manatus*	Found badly decomposed, plastic in stomach, tar-like substance in stomach	Anon., 1981d
Mar. 1982	Rodanthe, North Carolina Unknown source	Dry tar	*Globicephala* sp.	Animal found stranded. Advanced stage of decomposition. 10 × 20 cm patch dry tar on skin	Anon., 1982a
Mar. 1982	Del Monte Beach, California Unknown source	Tar	*Enhydra lutris*	Found stranded. Advanced stage of decomposition. 10 cm diameter mat of tar on fur.	Anon., 1982b

dochelys kempi) have been reported with evidence of oil contact (Anon., 1980b, 1981e-k, 1982c; Rabalais and Rabalais, 1980; Hall *et al.*, 1983). Many were recovered from the Atlantic coast of Florida as young turtles having ingested tar which sealed their mouths and interfered with normal feeding. Associated with the massive IXTOC-I oil spill in 1979-1980, 11 green and one Kemp's ridley turtle were found ashore fouled with oil (Rabalais and Rabalais, 1980; Hall *et al.*, 1983).

RESPONSES TO OIL

Detection and Avoidance

Marine mammals may or may not be able to detect oil. Yet such capacity is crucial to their ability to avoid it. Historical accounts, by no means conclusive, show that in some cases pinnipeds and sea otters do not avoid oil. Moreover, certain features of their life history may force them into repeated exposure to shoreline accumulations of oil, even if they are able to detect it. For example, Japanese poachers used petroleum products to repel sea otters (*Enhydra lutris*) from shore rocks—presumably oil was repugnant (Barabash-Nikiforov *et al.*, 1947). In the stressful confines of a laboratory study, otters demonstrated their aversion to oil, yet did not avoid it enough to prevent fouling of their fur (Williams, 1978; Siniff *et al.*, 1982). European otters (*Lutra lutra*) apparently made no effort to avoid a spill of Bunker C oil (Baker *et al.*, 1981).

Cetaceans have not been found coated with oil, perhaps because their smooth skin does not allow oil to adhere, or alternatively they may be able to detect and avoid it. We tested the latter possibility using the bottlenose dolphin (*Tursiops truncatus*) as a representative odontocete. Dolphins were presented with up to 12 different petroleum substances in as many as 31 configurations (Geraci *et al.*, 1983). By varying the thickness of each oil slick, or using combinations of light and dark oils, we determined the threshold of their detection ability. We reduced the visual properties of each test substance to a common measurable parameter, optical density. This gave us a basis on which to compare the animals' responses to various oils.

We determined that under optimum conditions of light and water clarity, dolphins could visually detect oil with an optical density greater than 0.2 to 0.34, and with experience, could reliably detect substances with an optical density of 0.05 or less, corresponding to a 1-mm film of dark crude oil. Using echolocation alone, a dolphin was able to detect thick (12-mm or greater) patches of heavy oil, particularly if the substance contained air bubbles—the type of properties one might expect of oil that has been churned by wind and waves.

Following the detection studies, we determined whether the animals would avoid a controlled slick of non-toxic, colored mineral oil that we knew they could detect (Smith *et al.*, 1983). Three dolphins were placed individually into a seawater pen subdivided into three areas, one of which contained the oil. The dolphins initially avoided the oiled area for up to 53 min, then came in contact with it only a few times within the first two hours, and thereafter, completely avoided it. Each time a dolphin contacted the mineral oil, it responded overtly by abruptly diving and quickly returning to an oil-free area. Under these conditions, the avoidance behavior was clear and consistent.

We then sought to establish the threshold for detecting and avoiding oil by presenting the dolphins first with colorless mineral oil, and then with a thin sheen of refined motor oil, both during the day and again at night under a canopy which excluded 92% of moon- and starlight (St. Aubin *et al.*, 1985). The dolphins avoided thicker oil slicks regardless of color or ambient light. During the day, they avoided the sheen, but at night, their response was inconsistent, suggesting that such conditions represent the limits of their ability to detect and avoid oil.

The strong response that the dolphins displayed after contacting the thicker oil slicks highlights our impression that tactile stimulation plays an important role in their avoidance of oils which are difficult to detect visually. Other senses, such as chemoreception and echolocation were probably of little value in this setting. Memory had a strong influence on the dolphins' behavior, causing them to temporarily avoid areas where oil had been during previous sessions. Such behavior might be displayed by coastal species capable of recognizing the boundaries of a previously oil-fouled area, yet their dependence on the region for food and social interaction might override any reluctance to reoccupy it. Perhaps the few observations of cetaceans feeding and "playing" in oil (Shane and Schmidly, 1978; Goodale *et al.*, 1979; Gruber, 1981) may be explained by these influences.

Mysticetes are as much a subject of concern as are the odontocetes. Unfortunately, studies from one group cannot be generalized to the other because of differences in their sensory capabilities. Since mysticetes could not be placed in the same experimental setting as the dolphins, an alternative was to observe the reaction of migrating California gray whales (*Eschrichtius robustus*) to naturally occurring oil seeps (Kent *et al.*, 1983). Typically, the whales would swim through oil, sometimes modifying their speed, but without a consistent pattern. Aerial observers occasionally noted a radical change in the whales' directions when approaching oil, but this was not accompanied by any alteration in respiratory pattern or swimming speed, and in fact may not have been a response to oil. There were some differences, however, in the respiratory behavior of gray whales when in oil-contaminated areas. There they spent less time at the surface and breathed at a faster rate. If this reaction is interpreted as an avoidance response, it suggests that gray whales can detect oil. Whales showing no response either could not detect the amount or type of oil present or were indifferent to it. It should be noted that comparisons are tenuous, as it was not possible to follow specific whales into and out of oiled areas. Such are limitations of field studies.

Ultimately, the ability of a marine mammal to avoid oil rests on its dependence on the area and avenues of escape. On one hand, pelagic dolphins have unlimited mobility, whereas seals and otters seasonally depend on inshore waters where oil tends to accumulate. Manatees confined in rivers and whales in ice leads would presumably be most vulnerable to noxious and toxic properties of oil.

Green turtles may be unique in their reaction to oil. In this case, the threat presented by oil may differ with the various stages of the animal's life history. Newly hatched green turtles leave the beach to forage in the open ocean. During their first year, they feed opportunistically at or near the surface before making the transition to grazing on underwater grasses. The young are thus exposed to oil residues in the form of floating lumps of tar which are found at a concentration of up to 3.5 mg/m^2 in the Caribbean and Gulf of Mexico, and up to 10 mg/m^2 in the Gulf Stream and over the Atlantic continental shelf (Morris, 1971; Jeffrey, 1973; Sherman *et al.*, 1973). It is unclear whether green and ridley turtles found with tar in their mouths were selectively eating tar lumps or did so accidentally in the process of feeding on organisms and vegetation bound by tar (A. Amos, Univ. of Texas, Port Aransas, Texas; N. Rabalais, Louisiana Univ. Mar. Cons.; pers. comm., 1984).

Under controlled laboratory conditions, turtles are able to detect dissolved organic compounds such as alcohols, aldehydes and esters, at concentrations of 10^{-6}M (Manton *et al.*, 1972). Perhaps they may even be attracted to the source of such substances (Kleerekoper and Bennett, 1976). The impact of spilled oil or ubiquitous tar residues would thus be heightened if attraction rather than avoidance were the turtles' basic response. This question is particularly relevant for young turtles, as older ones are less often confronted with floating tar.

Behavioral Effects

Oil exposure can alter normal behavioral patterns, and thereby have immediate and long-term effects on some marine animals. Davis and Anderson (1976) noted reduced growth rate in oiled gray seal (*Halichoerus grypus*) pups, but could not determine whether this was due to interruption of nursing behavior. Ringed seals (*Phoca hispida*) experimentally exposed to crude oil became irritable and aggressive and assumed atypical postures (Geraci and Smith, 1976). Within four days after the 24-h oil exposure, the seals' behavior had returned to normal.

Some attempt has been made to monitor activity of sea otters released after their pelage was fouled with oil. The animals were significantly more active than unoiled control otters, but there was no difference in dive patterns, movements or interactions with other otters (Siniff *et al.*, 1982). A similar study on northern fur seals (*Callorhinus ursinus*) was inconclusive (Kooyman *et al.*, 1976).

One obvious behavioral response to oil fouling of some fur bearing marine mammals is an increase in grooming activity. Sea otters contacting oil on the surface of a holding tank spent "75% of their time underwater trying to clean their pelage" (Williams, 1978). The increase in non-feeding activity of oiled otters after release was therefore presumed to be due to the additional time spent grooming (Siniff *et al.*, 1982). Polar bears (*Ursus maritimus*) show a similar grooming response (Oritsland *et al.*, 1981).

Oil might have a more indirect effect on the behavior of marine turtles. In the nest, young green turtles can become imprinted on chemical cues which are detected through their permeable eggs (Manton, 1979). Adult turtles returning to nesting beaches may be guided by olfactory stimuli associated with the beach where they were born (Carr, 1972). Assuming olfaction is critical to the process, oil-fouling of a nesting area might disturb imprinting of hatchling turtles, or confuse the turtles on their return migration after a 6-8 year absence. The effect on reproductive success could therefore be significant.

SURFACE CONTACT

Thermal Effects

All evidence indicates that animals which rely on hair or fur for insulation will be adversely affected by surface contact with oil; matted fur cannot trap air needed to form a thermal barrier (Hurst and Oritsland, 1982). Conductance of heat through sea otter and northern fur seal pelts can double after oiling (Kooyman *et al.*, 1977). To compensate for the loss, otters must increase their metabolic rate (Costa and Kooyman, 1982) and consequently their consumption of food. Yet in the field, otters released after minor oil fouling apparently did not increase their feeding activity; it could not be determined whether the animals lost weight while presumably under thermal stress (Siniff *et al.*, 1982). Oil-fouled polar bears rapidly became hypothermic when exposed to wind and low temperature (Oritsland *et al.*, 1981).

By contrast, cetaceans, phocid seals, walruses and some sea lions would be resistant to the thermal effects of oil, since their skin or pelts have little intrinsic insulative value. They rely instead, on blubber and vascular control to retain heat. Pelts from ringed harp (*Phoca groenlandica*), bearded (*Erignathus barbatus*), and Weddell (*Leptonychotes weddelli*) seals and California sea lions (*Zalophus californianus*) show little or no change in heat conductance after oiling (Oritsland, 1975; Kooyman *et al.*, 1977). Body temperatures remained stable within the normal range in ringed seals immersed in sea water covered with light crude oil for 24 h, and in newly weaned harp seal pups coated with crude oil for seven days (Smith and Geraci, 1975). Comparable studies have not been performed on cetaceans, but we presume that contact with oil would have no significant effect on their ability to thermoregulate.

Tissue Damage Due to Oil

All marine mammals would undoubtedly experience irritation and inflammation of eyes and sensitive mucous membranes following contact with oil. The nature of the damage was adequately demonstrated in ringed seals immersed in oil-covered sea water. Within minutes, they began to lacrimate profusely, yet made no attempt to close their eyes to avoid the oil (Geraci and Smith, 1976). Eventually, their conjunctival membranes became inflamed, and by 24 h, the seals had developed severe conjunctivitis, swollen nictitating membranes and corneal erosions. The symptoms subsided within 3 h of the seals' return to clean water and were no longer apparent 20 h later. There was no evidence of damage or inflammation of the skin of these seals, nor have any oil-induced lesions in skin been observed in pinnipeds in the natural environment (Davis and Anderson, 1976; Grose *et al.*, 1979).

Cetacean skin has unique properties which may make it vulnerable to oil. The epidermis is composed of numerous tiers of viable cells, is non-glandular and the external layer is not fully keratinized. The cells are surprisingly rich in enzymes (Geraci and St. Aubin, 1979a) and vitamin C (St. Aubin and Geraci, 1980), suggesting that skin may perform important hydrodynamic (Sokolov *et al.*, 1969) and physiological functions beyond that of a simple barrier against the sea. Any substance which affects the skin may have far-reaching consequences for these animals.

With these properties in mind, we undertook a study in which small cuplike discs containing various petroleum products were placed on discrete areas of the skin of captive bottlenose dolphins (Geraci and St. Aubin, 1982). Exposure to gasoline and crude oil for up to 75 min produced no evidence of damage or loss of integrity. Normal color was restored within two hours. In marked contrast, the skin of human subjects who exposed their arms to the same procedure for up to 35 min became distinctly red for up to ten days, and in one case remained discolored for as long as seven months. Using an infrared monitor, we noted that the dolphins did not have a vascular response to the gasoline, whereas humans reacted by generating heat at the site of contact.

Our histological and ultrastructural studies showed that petroleum hydrocarbons can produce mild and transient damage to cells of the epidermis, primarily in the external and intermediate layers, whereas the germinal layer and dermis were unaffected by exposures of less than 75 min to lead-free gasoline. Within three to seven days, epidermal cells showed signs of recovery. The skin of a live-stranded sperm whale (*Physeter catodon*) exposed to crude oil and gasoline for 17 h showed damage to the mid- and outer layers, but not to the basal layer and underlying dermis. This may be some indication of the resiliency of cetacean skin, although we cannot generalize on the basis of a single test.

The skin of cetaceans is often damaged by parasites (Pike, 1951; Humes, 1964; Perrin, 1969), microorganisms (Migaki *et al.*, 1971; Geraci *et al.*, 1979), and predators (Ridgway and Dailey 1972), as well as aggressive social encounters. In some, the skin is roughened by the presence of callosities. To determine how petroleum hydrocarbons might affect already damaged skin in dolphins, we made a number of shallow cuts in the epidermis, deliberately contaminated some with oil, and studied the progress of healing. We observed no gross or microscopic differences in healing between uncontaminated cuts, and those made in skin which had been previously exposed to gasoline or oil for up to 75 min, or wounds contaminated for up to 60 min with crude oil. After 15 days, all wounds had healed, leaving only a dark black halo. Based on the study, it appears that breaks in the continuity of epidermis do not necessarily magnify the reaction of skin to petroleum products.

We examined biochemical processes in cetacean epidermal cells for evidence of functional damage due to oil (Geraci and St. Aubin, 1982, 1985). We measured synthesis of phospholipids fundamental to cell membrane structure and stability, the concentration of α-tocopherol (vitamin E), which protects lipids from oxidants, the activity of creatine kinase, an enzyme involved in energy transfer, the rate of oxygen consumption, as an index of metabolic activity, the concentration and composition of epidermal intracellular lipid energy stores, and the uptake of tritiated thymidine, to assess the rate of cell division. The only consistent effect of oil was to depress phospholipid synthesis *in vitro*, which perhaps correlated with the ultrastructural defects in cell membranes that we observed following 75 minutes exposure to gasoline. However, no qualitative or quantitative changes in phospholipids could be detected even after skin had been exposed for 16 hours *in vitro*. In all, the biochemical changes in epidermis were minor and reversible.

Oil can adhere to the body surface of marine turtles. As yet there is no evidence that it results in tissue damage (Hall *et al.*, 1983). Periocular tissues and other mucous membranes would presumably be most sensitive to oil contact.

Developmental anomalies and embryonic mortality would be an expected consequence of oil contamination of marine turtle eggs, as it is in birds (see Hunt, Chapter 11). Fritts and McGehee (1982), in a series of field and laboratory studies, demonstrated that the age of the oil and exposure time are two variables which determine the survival of hatchlings from contaminated eggs. Fresh oil was highly toxic, especially during the last quarter of the incubation period, whereas aged oil produced no detectable effects. The researchers concluded that oil contamination of nesting beaches would have its greatest impact on nests that were already constructed; nests made on fouled beaches are less likely to be affected, if at all.

INHALATION

Marine mammals and turtles surfacing in an oil spill will inhale petroleum vapors. Numerous reports detail the effects of such substances on terrestrial mammals, and we use this information as an approach to identify hazards to marine species.

Inhalation of highly concentrated vapors, such as gasoline in excess of 10,000 ppm, is rapidly fatal (Machle, 1941). At lower concentrations (up to 1,000 ppm), humans and laboratory animals can develop inflammation, hemorrhage and congestion of the lungs (Nau *et al.*, 1966; Rector *et al.*, 1966; Valpey *et al.*, 1978). Yet such damage to the respiratory system is not consistently associated with exposure to vapors (Carpenter *et al.*, 1975, 1976, 1978). The central nervous system can also be affected, with signs ranging from hallucinations (Tolan and Lingl, 1964) to convulsions, coma and death (Petrie, 1908; Machle, 1941; Ainsworth, 1960; Wang and Irons, 1961). Repeated exposure to gasoline can produce degenerative changes in the brain (Valpey *et al.*, 1978) and peripheral nerves (Machle, 1941; Knave *et al.*, 1978), although such damage may be due more to the effects of additives, such as tetraethyl lead, than to the petroleum fractions (Robinson, 1978). Other sensitive organs include liver (Nau *et al.*, 1966), adrenals (Case, 1972), and the hematopoietic system (Nau *et al.*, 1966), although the effects on the latter are highly variable.

We attempted to determine the relationship between vapor concentration and duration of exposure which, together, influence the type and severity of damage (Figure 12.1). The effects were graded into four broad categories: (a) death due to destruction of lung and nervous tissue, (b) disorders of the central nervous system, (c) irritation of mucous membranes, and (d) no effect. We assumed that the consequences of a given set of exposure conditions would be the same for marine mammals as they are for other mammals.

We attempted to predict the threat presented by petroleum vapors at sea. In

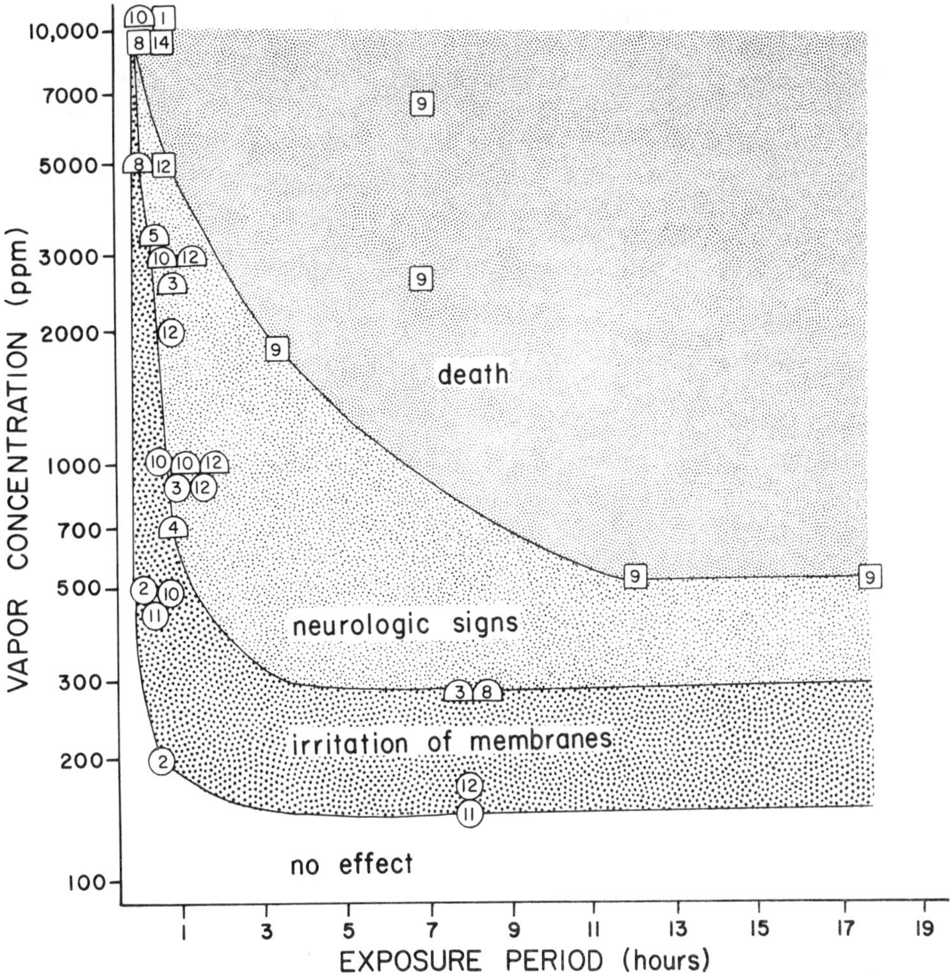

Figure 12.1. Summary of the effects on mammals of exposure to and inhalation of various petroleum vapors, principally those of gasoline. The duration of exposure and concentration of the vapors have been integrated to predict four levels of effects. Data are based on the reports of: 1) Ainsworth, 1960; 2) Davis *et al.*, 1960; 3) Drinker *et al.*, 1943; 4) Gamberale *et al.*, 1975; 5) Haggard, 1920; 6) Lykke and Stewart, 1978; 7) Lykke *et al.*, 1979; 8) Machle, 1941; 9) Nau *et al.*, 1966; 10) Poklis and Burkett, 1977; 11) Runion, 1975; 12) Stewart *et al.*, 1979; 13) U.S. Dept. Health, Education and Welfare, 1981; 14) Wang and Irons, 1961.

the absence of published data, we measured the concentrations of vapors in a laboratory simulation of a l-cm thick slick of West Texas crude oil over seawater within a confined air space l-m high. The system was equilibrated for 15 h at 5° or 20°C, and vapors were analyzed using gas chromatography. Approximately 50% of the vapors represented short chain hydrocarbons (C_4-C_9). The concentration of benzene was in the range of 350 to 500 ppm, which is well above the safety threshold for inhalation, even for short periods. In the natural setting, low molecular weight compounds dissipate within hours, leaving longer chain (C_9-C_{16}) fractions in individual concentrations of 20 ppm or less. When taking into account diffusion, dispersion of the slick and wind, it is unlikely that vapor concentrations would be harmful for more than a few hours.

Nevertheless, marine mammals near the source of the spill, or confined to a contaminated lead or bay, would undoubtedly inhale vapors. Studies on ringed seals showed that during a 24-h immersion in oil-covered water some volatile hydrocarbons were absorbed through the respiratory tract (Geraci and Smith, 1976). That is, petroleum hydrocarbons were found in tissues, with no evidence of ingestion (Engelhardt *et al.*, 1977). Transient kidney and possible liver lesions were observed, but there was no associated lung pathology. The experimental setting within the confined holding pen (Smith and Geraci, 1975) provided a more concentrated exposure to volatile fractions than would normally be encountered in an oceanic spill. Thus, short-term inhalation is not necessarily harmful either in terms of structural damage or gas exchange.

Ultimately the effect of such exposure would probably depend on the health of the animal. Thus, cetaceans that are already stressed by lung and liver parasites and adrenal disorders (Geraci and St. Aubin, 1979b) might be particularly vulnerable to low levels of hydrocarbon vapors. Animals away from the immediate area, or exposed to weathered or residual oils would not likely suffer any consequences from inhalation, regardless of their condition.

Turtles may respond to petroleum vapors in a rather unusual way. Strong odors appear to be obnoxious to turtles, which react by becoming apneic (Maxwell, 1979). In this way, they are able to resist the effects of vaporized anesthetics for long periods. Using this type of avoidance behavior, a turtle may be able to minimize exposure to petroleum vapors. If not, it would presumably be as vulnerable to inhaled vapors as other air-breathing vertebrates.

INGESTION

Oral Obstruction

Young turtles face a peculiar threat from oil spills. Tar becomes lodged in their mouths in such a way as to impair feeding (Witham, 1978). The obstruction persists, and unable to eat, affected turtles eventually languish ashore in poor body condition. In this situation, the effects of petroleum are clear and unequivocal.

It is perceived that mysticetes face a comparable threat—that oil would irreversibly obstruct water flow through baleen, thereby impairing food-gathering efficiency. To test this premise, we monitored water flow through fin (*Balaenoptera physalus*), sei (*B. borealis*), humpback (*Megaptera novaeangliae*) and gray whale baleen before and after contaminating it with various crude oils over a range of temperatures (Geraci and St. Aubin, 1982, 1985). Bunker C at 5°C caused a 140-250% increase in resistance to flow in humpback baleen, whereas sei and fin whale samples showed average increases of 35-40% and 55-75% respectively; gray whale samples were relatively unaffected. Medium weight crude oil had little effect on the properties of fin, sei and gray whale samples, even at low temperatures.

Continuous flushing with sea water rapidly removed the oil coating from preparations fouled *in vitro*. More than 70% of the oil was lost within 30 minutes, and over 95% after 24 hours. The oil residue did not affect resistance to flow as measured in our system, nor do we presume that it would impair function in the living whale. Oil coated fibers however might contaminate ingested plankton, which would then serve as a vehicle for entry of oil. Moreover, oil coating of plates may cause plankton to adhere, further reducing flow and delaying recovery. Such an effect would be more critical during the short seasons when mysticetes feed intensively. As yet, there has been no recorded sighting of a whale with its baleen fouled by oil, and our analysis of baleen fibers from whales taken from Icelandic waters has not revealed the presence of petroleum residues.

Toxicity of Ingested Oil

Consumed crude oil, particularly the lighter fractions, can be toxic to a wide variety of mammals (Coale, 1947; Gerarde, 1959; Narasimhan and Ganla, 1967). Small doses, when aspirated, can cause acute fatal pneumonia. Larger quantities, as much as 140 times the aspirated dose (Gerarde, 1964), can be tolerated if the oil remains in the gastrointestinal tract, but this too can be harmful. Petroleum hydrocarbons irritate or destroy epithelial cells lining the stomach and intestine, thereby affecting motility, digestion and absorption (Anon., 1979b,c, 1980c-f; Narasimhan and Ganla, 1967; Rowe *et al.*, 1973). In the bloodstream, lower molecular weight fractions are carried to various organs, affecting them in the same way as is petroleum absorbed through inhalation.

There is no reason to assume that marine mammals would differ substantially in their response to ingested oil than laboratory mammals, although we can expect the quantity required to intoxicate a larger animal may be less on a weight basis than that established for small laboratory species. Nevertheless, we used the only data available (Anon., 1979b,c, 1980c-f) to construct estimates of the quantity of fuel oil that various marine mammals would have to consume to be at risk (Table 12.2). We presume the oil is not regurgitated and aspirated into the lungs, where even small quantities can prove fatal. This would reasonably apply to cetaceans which are protected from aspiration by their peculiar laryngeal anatomy.

It is unrealistic to assume that most of the animals represented in Table 12.2 would, during the normal course of feeding, ingest the quantities of oil estimated to be toxic. Otters, bears, seals and odontocetes are predatory, and lessons from captivity suggest they would not likely scavenge food that is tainted. Neither do they drink substantial quantities of sea water, whether it contains oil or not. Mysticetes feeding in the area of a spill are more likely to ingest oil-coated or oil-contaminated food, particularly zooplankters which actively consume oil particles. Assuming toxic oils to comprise 10% of the estimated 1600 kg of food consumed in a day by a 40-ton fin whale, the total quantity of ingested oil would be 160 kg. This approaches the critical dose calculated for highly toxic fuel oils. The question is, would fin whales feed around a spill of fresh volatile oil long enough to accumulate such quantities? Whales have been observed feeding amid oil (Goodale *et al.*, 1979), yet there are no reports from stranding records or the whaling industry linking oil ingestion with mortality.

On a smaller scale, Duguy and Babin (1975) suggested that ingestion of oil may have been responsible for the death of harbor seals (*Phoca vitulina*) along the coast of France. Following an oil spill at Sullom Voe Oil Terminal, Shetland, at least 13 otters died, five of which had developed hemorrhagic gastroenteritis associated with oil presumably ingested while grooming (Baker *et al.*, 1981). An experimentally oiled polar bear presumably ingested toxic quantities of oil (Anon., 1981) while licking its fur (Oritsland *et al.*, 1981). Thus, grooming activity essential to restore fouled fur presents another route of exposure to otters and bears.

TABLE 12.2

Estimated quantities of ingested fuel oils calculated to pose a threat to selected marine mammals. Values are based on studies on terrestrial mammals[a]

Species	*Average adult body weight (kg)*	*Volume of fuel oil (1)*	
		Mean	*Range*
Enhydra lutris	35	0.5	0.2–0.9
Phoca vitulina	70	1.0	0.35–1.75
Tursiops truncatus	250	3.0	1.0–5.0
Ursus maritimus	400	6.0	2–10
Balaenoptera acutorostrata	1000	15	5–25
Orcinus orca	3000	45	15–75
Balaenoptera physalus	40,000	600	200–1000

[a]an average LD_{50} of 15 ml/kg determined for 6 types of fuel oil, with a range of 5–25 ml/kg (Anon., 1979b,c, 1980c-f).

Small lumps of tar have occasionally been found in the gut of manatees (Anon., 198la,c,d), with no implication that the material may have been harmful. Yet, these herbivores might be particularly susceptible to disruption of gut flora or interference with secretory activity of gastric glands (Kenchington, 1972). For example, petroleum ingestion in cattle leads to loss of appetite and bloat (Rowe *et al.*, 1973).

There have been three oil ingestion experiments in seals and one in cetaceans. Harp seals given a single dose of up to 75 ml (1-3 ml/kg) of crude oil began to excrete oil in the feces within 1.5 h, suggesting increased gastrointestinal motility (Geraci and Smith, 1976). Some was undoubtedly absorbed into blood and tissues, as was shown in studies of ringed seals given smaller doses (0.2 ml/kg/day for 5 days) of oil (Engelhardt *et al.*, 1977; Engelhardt, 1981). There was no gross, microscopic or biochemical evidence of tissue damage in either species. A bottlenose dolphin given small quantities (2.5-5 ml) of machine oil daily for over three months also showed no clinical signs of organ damage or intoxication (Caldwell and Caldwell, 1982).

Bioaccumulation

Petroleum hydrocarbons accumulate in the food chain, particularly in species which have low capacity for metabolizing and excreting ingested compounds. Molluscs and other benthic invertebrates continue to accumulate residues from contaminated sediments. Fish, on the other hand, metabolize up to 98% of tissue hydrocarbons within two months (McCain *et al.*, 1978). Thus we can expect that bottom feeders such as otters, walruses and bearded seals would ingest more contaminated food than pelagic and surface feeders. However, even the latter are exposed to compounds such as naphthalene and tetramethylbenzene which persist in fish tissues (McCain *et al.*, 1978).

To determine the extent of accumulation, we analyzed over 250 tissue samples from marine mammals stranded along the Atlantic coast of the United States and Canada, and from whales taken as part of the Icelandic fishery. We found detectable levels of naphthalene in most of the tissues, with highest concentrations in blubber (Geraci and St. Aubin, 1982). It is difficult to correlate tissue burdens with type or quantity of hydrocarbons consumed, because cetaceans and pinnipeds have enzyme systems, such as cytochrome P-450 (Geraci and St. Aubin, 1982) and aryl hydrocarbon hydroxylase (Engelhardt, 1981) which in other species help to detoxify and eliminate petroleum hydrocarbons. Whether turtles possess similar mechanisms has yet to be determined, although the high concentrations of hydrocarbons found in their tissues after oil exposure (Hall *et al.*, 1983) may indicate a limited capacity to metabolize such substances, perhaps because of low metabolic rate.

Effects of long-term accumulation of pollutants often become apparent only after decades of surveillance. In an historical analysis of cetacean strandings on the coast of the Netherlands, van Bree (1977) has associated an increase in pollutant levels in the North Sea with reduced population size, as reflected by decreased numbers of stranded animals. A sudden decline took place in 1946, which van Bree suggested was "linked to the dumping of war chemicals at that period or by the increase in the use of oil ....(a) second decrease is clearly related to the increase of pollution of the North Sea."

The conclusion is intriguing, because it draws on 45 years of carefully documented observations on the frequency of strandings. Although pollution may have been involved, it is necessary to recognize that other factors affecting the environment may also have played a role. For example, the decline in strandings noted after 1940, coincides with hydrographic changes which resulted in lower overall productivity of the North Sea (Reid, 1977).

Helle *et al.* (1976) found that about 40% of a sample of Baltic ringed seal females of reproductive age showed pathological changes in the uterus, associated with unusually high levels of DDT and PCB substances. Premature parturition in California sea lions has been correlated with high tissue levels of DDT and PCBs in association with disease agents and trace element imbalances (Gilmartin *et al.*, 1976). We need to consider whether petroleum, especially polycyclic aromatic fractions, might act synergistically with other pollutant residues to induce metabolic disturbances and other pathological effects. Some marine turtles occupy and feed in areas which are known to contain high levels of hydrocarbon pollutants, for example, Kemp's ridley turtle in the northern Gulf of Mexico (Pritchard and Marquez, 1973). We know of no attempt to determine tissue burdens of hydrocarbons in marine turtles.

NOISE—BEHAVIORAL, PHYSIOLOGICAL AND PSYCHOLOGICAL EFFECTS

Noise is associated with all phases of offshore petroleum exploration and production—seismic surveying, drilling, air and ship support, construction, and the operation of onshore and offshore facilities. Gales (1982) showed that marine mammals can hear these sounds at distances to 100 nautical miles (185 km), and farther under favorable conditions, and determined that within 15 m of offshore platforms, the levels of sound in the high sensitivity region would presumably cause discomfort in the harbor seal, bottlenose dolphin, beluga (*Delphinapterus leucas*) and mysticete whales.

Under most circumstances, animals habituate to low level background noise. Some species, such as the humpback (*Megaptera novaeangliae*) and gray whales, harbor and elephant seals (*Mirounga* sp.), bottlenose dolphins, Pacific walruses (*Odobenus rosmarus*), and sea lions, seem to coexist well with human activities. Such habituation, in fact, forms the underlying basis for the success of whale watching cruises.

Nevertheless, Nishiwaki and Sasao (1977) suggested that increased ship traffic in Japanese waters disturbed migration routes of minke whales (*Balaenoptera acutorostrata*) and Baird's beaked whales (*Berardius bairdii*). Migrating gray whales, on hearing recorded sounds of drilling platforms and ships, maintained their heading but reduced their speed (Malme *et al.*, 1983). Bowhead whales (*Balaena mysticetus*) may swim rapidly away from small boats approaching within 2 to 3 km. However, there is no clear evidence that they will avoid seismic ships (Richardson *et al.*, 1983). Undoubtedly, some species will be more sensitive to the effects of noise. Experience will also influence the response. For example, reaction to vessel noise is particularly strong in hunted marine mammals such as belugas (M. Fraker, pers. comm.).

Most marine mammals use sound to communicate, navigate, and locate prey. Background noise may interfere with these sounds, thereby disrupting social interaction and the ability of odontocetes to echolocate effectively. Gales (1982) calculated that sound levels within 800 m of a platform would partially mask lower frequency communication signals. Bowhead whales sometimes cease to call in the presence of air-gun blasts, but usually maintain the same level of vocal activity (Richardson *et al.*, 1983). Without knowing the significance or effectiveness of the calls, it is impossible to judge the impact of noise on social interaction of bowheads.

Odontocetes use high frequency sounds for echolocation. Platforms and ships radiate little noise at these frequencies, which are, in any event, transmitted poorly in sea water (Gales, 1982). At least one odontocete has shown the ability to compensate for increased high frequency noise. A beluga whale, transported from one facility to another where ambient noise was 20-30 dB greater, increased the frequency of its echolocation signals (Au *et al.*, 1983).

SHOCK WAVES

Shock waves are generated during exploration, development and production of offshore oil and gas. Most are associated with seismic surveys carried out using high intensity sound from air-guns and explosives. The latter produce a steep-fronted detonation wave which is transformed into a high-intensity pressure wave (shock wave) and an outward flow of energy in the form of water movement. There is an instantaneous rise in maximum pressure followed by an exponential pressure decrease and drop in energy.

Hill (1978) reviewed the physical effects of underwater shock waves on fish and speculated on their possible effects on marine mammals. Powerful shock waves harm living organisms by destroying tissue at air-fluid interfaces. In mammals, gas-containing organs are affected, principally lungs and hollow viscera.

A formula to calculate safe distance from an explosive charge in water has been developed using fish and land animals as subjects (Yelverton *et al.*, 1973). Its application requires knowledge of target (animal) depth, detonation depth, and charge weight. When applied to a relatively small marine mammal such as a ringed seal at a depth of 25 m, minimum safe distance from a 5 kg charge detonated at a depth of 5 m is calculated to be 359 m (Hill, 1978). The safety range is modified by the nature of the sea floor, ice cover and water depth. If the animal is in shallow water with a rocky bottom, or if the charge is detonated under thick ice, the impact of an explosion is increased.

This formula (Yelverton *et al.*, 1973) was derived using land mammals and fishes as experimental subjects. Hill (1978) concluded that marine mammals would be less vulnerable to underwater shock waves than are land mammals of comparable size. This is due primarily to pressure adaptations and increased protection afforded by their thick body walls. Furthermore, large size is inherently protective.

That is not to say that marine mammals would be entirely resistant to earth-shaking blasts. Some were killed by an underground nuclear detonation on Amchitka Island in the North Pacific (Rausch, 1973). Ten sea otters and four harbor seals were recovered dead from the beach nearest the blast site. They were killed instantly by an estimated overpressure of 200-300 psi (14-21 bar), which was 50-70 times greater than the calculated safe limit. Such pressures greatly exceed those generated during conventional seismic blasting operations.

On a more realistic note, explosives used to clear a navigational waterway killed fish within 800 m of the blast. Killer whales, porpoises, sea lions and fur seals within 5 km were apparently unaffected (Thompson, 1958). Fitch and Young (1948) reported that sea lions were killed by underwater explosions used in seismic explorations; gray whales in the area were apparently undisturbed.

Air-guns are now preferred for marine seismic exploration. Several guns of different sizes are usually fired simultaneously. The shock waves differ from those of explosives in that peak pressures are low, and both the rise time of the shock pulse and the time-constant of the pressure decay are comparatively long. The procedure is harmless to fish (Falk and Lawrence, 1973; Hill, 1978) and would not appear to be physically injurious to marine mammals. One study on cetaceans has shown that air-guns fired intentionally in the presence of bowhead whales produced subtle, inconsistent changes in their surfacing and respiration behavior. There was no conspicuous startle reflex, as is observed in response to low-flying aircraft and rapidly approaching boats (Richardson *et al.*, 1983). Gray whales, on the other hand, alter their swimming speed, milling behavior and respiration rates during and after exposure to air-gun noise (Malme *et al.*, 1983).

Sudden disturbances elicit a startle reflex in some cetaceans, which react by sounding, aggregating or dispersing, with subsequent regrouping of the social structure (Leatherwood, 1977). This is particularly true of the more gregarious odontocetes. These behaviors, although adaptive and obviously designed to protect against a sudden threat, may in some cases be detrimental. Van Bree and Kristensen (1974) reported that a small herd of Cuvier's beaked whales (*Ziphius cavirostris*) stranded in response to an underwater explosion.

The startle reflex in pinnipeds has not been studied *per se*. Field observations show that sudden disturbances cause some animals to disperse from rookeries, by mass movement, or "stampede," into the water (Loughrey, 1959; Salter, 1979). This may disrupt mother-pup pair bonds and injure or kill young animals. Further injury may accompany territorial aggression during recolonization of the rookery. Pinnipeds most vulnerable to these effects might be perinatal females, nursing pups and calves, molting animals and those stressed by parasitism and disease.

INDIRECT EFFECTS OF OIL AND GAS PRODUCTION ACTIVITIES

Indirect effects of oil and oil and gas production activities are difficult to detect, cannot be tested experimentally, but may have the greatest impact on populations of marine mammals and turtles. For example, noise can influence non-auditory physiology (Fletcher, 1971) by driving the stress response toward lowering resistance to disease and promoting hypertension and endocrine imbalance.

Observations on free-ranging marine mammals suggest that stress may be a limiting factor in certain populations. Spinner and spotted dolphins (*Stenella* sp.) and harbor porpoises (*Phocoena phocoena*) succumb to capture and handling stress (Dudok van Heel, 1966; Coe and Stuntz, 1980). Atlantic white-sided dolphins (*Lagenorhynchus acutus*) have a high incidence of adrenal defects which could compromise their adaptability to stressful situations (Geraci and St. Aubin, 1979b). Similarly, electrolyte imbalance in free-ranging ringed seals signals the inability of the adrenal cortex to maintain homeostasis during critical periods in the animal's life history (Geraci *et al.*, 1979).

These conditions along with pre-existing disease may ultimately determine an animal's ability to accommodate to an additional disturbance, such as that presented by noise or oil. This was dramatically illustrated in a study of ringed seals immersed in light crude oil-covered water (Geraci and Smith, 1976). In the more natural Arctic setting, they were relatively unaffected after 24-h exposure. On the other hand, when seals from the same population were tested identically, but under more stressful captive conditions, they died within 71 min. It was concluded that oil may have a selective effect on stressed or otherwise weakened members of a population.

It may be possible to predict some stress-mediated indirect effects of offshore oil production on turtles. These animals are vulnerable to human disturbance, particularly during nesting periods (Philobosia, 1976; Frazier, 1980). Inappropriate lighting can interrupt nest construction and egg laying by adults and cause disorientation in hatchlings. These reactions should alert us to the possible consequences of installing shore-based support operations in the vicinity of nesting beaches and also to the detrimental effects of overzealous oil-spill countermeasures in those areas. The strategy for clean-up operations should vary, depending on the season, recognizing that disturbance to the nest may be more detrimental than the oil (Fritts and McGehee, 1982).

Litter is a hazard to turtles which may mistake plastic bags for jellyfish. The consequences of ingestion of plastic bags are well-documented (Fritts, 1982). The sources of such flotsam are, of course, numerous, and the degree to which offshore oil and gas activities contribute is unknown. Existing regulations concerning the disposal of solid wastes from rigs and platforms should, if observed, minimize this source.

Offshore oil and gas activity would be detrimental if it resulted in a reduction in population size, shift in distribution away from a preferred habitat, or deterioration of the health of a significant number of individuals. In simplest terms, population decline may follow long-term reproductive failure or excessive mortality due directly to oil. Areas of high productivity or prime breeding sites might be abandoned if animals fail to habituate to oil production activities. This and associated stress of accumulated toxic compounds could compromise health, leaving the group more susceptible to pathogens and other short-term insults.

RECOMMENDATIONS

Detection/Avoidance

The question of detection and avoidance has been answered for a representative odontocete. The same approach is not applicable to mysticetes, whose behavior toward oil can best be assessed through observations at the site of a spill. An experimental study of avoidance in pinnipeds might help to clarify their apparently equivocal response to oil. The greatest need is to determine the basic reaction of turtles to oil slicks and tarballs.

Contact/Ingestion/Inhalation

There is no need to test further the hypothesis that fur-bearing marine mammals (otters and polar bears) are adversely affected by contact with oil. Similarly, thermoregulatory studies are not required for pinnipeds or cetaceans. There are still unanswered questions on the effects of oil on biochemical properties of cetacean skin, and nothing is known of its effect on turtles.

There are numerous reports detailing acute effects of oil ingestion and inhalation in terrestrial mammals. This information serves as a guide for predicting effects on marine mammals, and there is little to be gained by continuing research along these lines. However, we need to determine toxicity of ingested or inhaled fractions in turtles. Such studies should focus on respiratory behavior toward vapors and mechanisms leading to oral obstruction.

Effects of oil-fouling of baleen have been examined for representatives of two of the three major groups of mysticetes. Right and bowhead whales have not yet been examined due to limited availability of specimens. Such a study will complete our understanding of this potential impact.

Reproductive Success

Long-term consumption of pollutants can affect reproductive success. There is a critical need to determine how oil fits into the scheme. Oil residues should be included in any program designed to monitor oceanic pollutants in marine mammals, bearing in mind that pollutants are one of many interacting components affecting the health of populations. The data should be correlated with detailed examinations of reproductive organs.

There has been a recent increase in efforts to recover and examine stranded turtles (Rabalais and Rabalais, 1980; Shoop and Ruckdeschel, 1982), which should provide an opportunity to compare petroleum hydrocarbons in animals from polluted and pristine regions. Studies of the effects of oil fouling on hatching success in turtles, should be expanded to include different species and exposure conditions.

Noise and Disturbance

Considerable data have been gathered on types and intensity of sounds associated with offshore activity. Whether these sounds impair communication and hearing in marine mammals has been a matter of speculation. Controlled experiments are needed to determine the animals' needs and capacities to adjust their acoustic behavior in the face of high ambient noise. For mysticetes, these studies would have to be conducted at sea, analyzing vocal behavior in response to controlled sound emissions.

Behavioral response to disturbance can take many forms, some more obvious than others. For example, reproductive success can be influenced by stress-induced hormonal imbalance and interruption of mother-pup pair bonds and nesting behavior. It has been difficult to determine basic reactions to well-controlled disturbances in marine mammals, especially the great whales. We see the need to continue such studies recognizing that even our best efforts might only establish direct short-term impacts.

Long-term consequences on stress and reproductive success are more elusive, and may better be understood by detailed examinations of carcasses, both stranded and those taken as part of commercial fisheries operations. For some marine mammal and turtle species, the critical periods and habitats have been suitably defined. This information should serve as a basis to determine the regions and time when offshore activity would have its greatest impact. This type of analysis should retain high priority.

SUMMARY

During the past five years, studies on marine mammals have brought us closer to an understanding of basic behavioral and physiological responses to oil. For example, experiments have shown that dolphins can detect oil and, under certain circumstances, will avoid it. Oil can cause subtle damage to their skin, the full impact of which is still being assessed. The threat to otters and polar bears is unequivocal. Oiled fur is ineffective as an insulator, and attempts to groom can lead to oil ingestion. Fouling of baleen has short-term effects on water flow and feeding efficiency, although the consequences may not be as great as was predicted. Noise and disturbance associated with offshore production may be within the limits of tolerance for some species.

The full range of effects on turtles is poorly understood. Young turtles can eat tarballs which seal their mouths and interfere with normal feeding. Oil fouling of nests can lead to embryonic abnormalities and hatchling mortality. Turtles are particularly vulnerable to disturbances during the nesting season.

The greatest impact of offshore oil and gas activities may result not from direct mortality, but rather through subtle alterations of habitat, in association with intrinsic stressors within the environment. We provide recommendations which reflect our interpretation of the most significant data gaps and emphasize the need for selective long-term monitoring.

ACKNOWLEDGMENTS

We thank K. Geraci, W. Martin, and C. Robinson (Univ. of Guelph) for assistance with data collection and preparation of the manuscript. R. Witham (Florida Dept. of Natural Resources) and T.H. Fritts (U.S. Fish and Wildlife Service) kindly provided resource material on sea turtles. V. Lounsbury (Guelph) and L. Tinnin (LUMCON) prepared the illustration. We especially thank N.N. Rabalais (LUMCON), for her critical review and for generously providing resource material and D.M. Smith (Guelph) for her indomitable patience through numerous revisions of the manuscript. Our studies on oil effects on cetaceans have been supported by a contract from the U.S. Department of the Interior, Minerals Management Service.

LITERATURE CITED

Ainsworth, R.W. 1960. Petroleum vapour poisoning. Br. Med. J. 1: 1547-1548.

Anon. 1970. Nature in Wales. Mammals 12: 110.

Anon. 1971a. Kodiak Islands oil pollution. No. 15-70. Pages 134-136 *in* Smithsonian Institution Center for Short-lived Phenomena. Annual Rep. 1970.

Anon. 1971b. Alaska Peninsula oil spill. Event No. 36-70. Pages 154-157 *in* Smithsonian Institution Center for Short-lived Phenomena. Annual Report 1970.

Anon. 1979a. Bay of Campeche. Ocean Sci. News 21: 1.

Anon. 1979b. Acute Toxicity Tests of API 78-3 No. 2 Home Heating Oil (10% Cat). Elars Bioresearch Laboratories, Inc. Project No. 1443. American Petroleum Institute, Med. Res. Publ. 2732773, 52 p.

Anon. 1979c. Acute Toxicity Tests of API 78-4 No. 2 Home Heating Oil (50% Cat). Elars Bioresearch Laboratories, Inc. Project No. 1443. American Petroleum Institute, Med. Res. Publ. July 17, 1980, 52 p.

Anon. 1980a. Dead, oiled sea otter. Smithsonian Inst. Scient. Event Alert Network (SEAN) Bull. 5: 34.

Anon. 1980b. Oiled green turtle. Smithsonian Inst. Scient. Event Alert Network (SEAN) Bull. 5: 34.

Anon. 1980c. Acute Toxicity Tests of API 78-2 No. 2 Home Heating Oil (30% Cat). Elars Bioresearch Laboratories, Inc. Project 1443. American Petroleum Institute Med. Res. Publ. 2732771, 48 p.

Anon. 1980d. Acute Toxicity Tests of API 78-7 No. 6 Heavy Fuel Oil (API Gravity 17.1/0.8% S). Elars Bioresearch Laboratories, Inc. Project No. 1443. American Petroleum Institute, Med. Res. Publ. 2732774, 45 p.

Anon. 1980e. Acute Toxicity Tests of API 79-2 No. 6 Heavy Fuel Oil (API Gravity 5.2/1.2% S). Elars Bioresearch Laboratories, Inc. Project No. 1443. American Petroleum Institute, Med. Res. Publ. 2732813, 52 p.

Anon. 1980f. Acute Toxicity Tests of API 78-6 No. 6 Heavy Fuel Oil (API Gravity 11.7/2.7% S). Elars Bioresearch Laboratories Inc. Project No. 1443. American Petroleum Institute, Med. Res. Publ. 2732814, 47 p.

Anon. 1981a. Manatee mortality. Smithsonian Inst. Scient. Event Alert Network (SEAN) Bull. 6: 26.

Anon. 1981b. Abandoned tarred sea otter pup. Smithsonian Inst. Scient. Event Alert Network (SEAN) Bull. 6: 30.

Anon. 1981c. Dead manatee with ingested tar found. Smithsonian Inst. Scient. Event Alert Network (SEAN) Bull. 6: 31.

Anon. 1981d. Dead manatee found with ingested tar and plastic. Smithsonian Inst. Scient. Event Alert Network (SEAN) Bull. 6: 33.

Anon. 1981e. Tarred sea turtle mortalities. Smithsonian Inst. Scient. Event Alert Network (SEAN) Bull 6: 22.

Anon. 1981f. Dead, tarred loggerhead turtles. Smithsonian Inst. Scient. Event Alert Network (SEAN) Bull 6: 16.

Anon. 1981g. Tarred green turtle. Smithsonian Inst. Scient. Event Alert Network (SEAN) Bull. 6: 17.

Anon. 1981h. Tarred hawksbill turtle dies. Smithsonian Inst. Scient. Event Alert Network (SEAN) Bull. 6: 18.

Anon. 1981i. Oiled green turtle. Smithsonian Inst. Scient. Event Alert Network (SEAN) Bull. 6: 22.

Anon. 1981j. Dead, tarred green turtle. Smithsonian Inst. Scient. Event Alert Network (SEAN) Bull. 6: 15.

Anon. 1981k. Tarred green turtle mortality. Smithsonian Inst. Scient. Event Alert Network (SEAN) Bull. 6: 14.

Anon. 1981l. Effects of Crude Oil on Polar Bears. Summary Report No. QS-8283-010-EE-A1. Ministry of Indian and Northern Affairs, Ottawa, Canada, 17 p.

Anon. 1982a. Stranded, tarred pilot whale. Smithsonian Inst. Scient. Event Alert Network (SEAN) Bull. 7: 26.

Anon. 1982b. Dead sea otter with tar patch on fur. Smithsonian Inst. Scient. Event Alert Network (SEAN) Bull. 7: 33.

Anon. 1982c. Green turtle with ingested tar. Smithsonian Inst. Scient. Event Alert Network (SEAN) Bull. 7: 14.

Au, W.W.L., D.A. Carder, R.H. Penner and B.L. Scronce. 1983. Measurements of beluga whale (*Delphinapterus leucas*) echolocation signals in two different ambient noise environments. Page 3 *in* Fifth Biennial Conference on the Biology of Marine Mammals, Boston, Massachusetts. Society for Marine Mammalogy (Abstract).

Baker, J.R., A.M. Jones, T.P. Jones and H.C. Watson. 1981. Otter, *Lutra lutra* L. mortality and marine oil pollution. Biol. Conserv. 20: 311-321.

Barabash-Nikiforov, I.I., V.V. Reshetkin and N.K. Shidlovskaya. 1947. The Sea Otter (Kalan). Translated by Isr. Program Sci. Transl., 1962, Natl. Tech. Inf. Serv., Springfield, Virginia, OTS 61-31057, 227 p.

Barabash-Nikiforov, I.I., S. Marakov and A. Nikolaev. 1968. Kalan (Izd-vo Nauka, Leningrad) p. 115 (cited in Van Blaricom, G.R. and R.J. Jameson. 1982. Lumber spill in central California waters: Implications for oil spills and sea otters. Science 215: 1503-1504).

Bergey, M. 1979. The Ixtoc Oil Spill - Effects on Marine Mammals and Turtles. (An observational report from 6-28 September 1979. Report by Michael Bergey aboard the Research Vessel *Oregon II* on a 22-day Shrimp Tagging Expedition), 49 p.

Brownell, R.L., Jr. 1971. Whales, dolphins and oil pollution. Pages 255-276 *in* D. Straughan (ed.), Biological and Oceanographical Survey of the Santa Barbara Channel Oil Spill 1969-1970. Vol. 1, Biology and Bacteriology. Allan Hancock Found., Univ. of South. Calif., Los Angeles, California.

Brownell, R.L., Jr. and B.J. LeBoeuf. 1971. California sea lion mortality: Natural or artifact? Pages 287-306 *in* D. Straughan (ed.), Biological and Oceanographical Survey of the Santa Barbara Channel Oil Spill 1969-1970. Vol. 1, Biology and Bacteriology, Allan Hancock Found., Univ. of South. Calif., Los Angeles, California.

Caldwell, M.C. and D.K. Caldwell. 1982. A study of the effects of oil ingestion on a bottlenose dolphin, *Tursiops truncatus*. Pages 224-236 *in* J.R. Geraci and D.J. St. Aubin, Study of the Effects of Oil on Cetaceans. Final Report. Prepared for U.S. Dept. of the Interior, Bureau of Land Management, Washington, D.C. Contract #AA551-CT9-29.

Carpenter, C.P., E.R. Kinkead, D.L. Geary, Jr., L.J. Sullivan and J.M. King. 1975. Petroleum hydrocarbon toxicity studies. V. Animal and human responses to vapors of mixed xylenes. Toxicol. Appl. Pharmacol. 33: 543-548.

Carpenter, C.P., D.L. Geary, Jr., R.C. Myers, D.J. Nachreiner, L.J. Sullivan and J.M. King. 1976. Petroleum hydrocarbon toxicity studies. XI. Animal and human response to vapors of deodorized kerosene. Toxicol. App. Pharmacol. 36: 443-456.

Carpenter, C.P., D.L. Geary, Jr., R.C. Myers, D.J. Nachreiner, L.J. Sullivan and J.M. King. 1978. Petroleum hydrocarbon toxicity studies. XVII. Animal response to n-nonane vapor. Toxicol. Appl. Pharmacol. 44: 53-61.

Carr, A. 1972. The case for long-range chemoreceptive piloting in *Chelonia*. Pages 469-483 *in* S.R. Galler, K. Schmidt-Koenig, G.J. Jacobs and R.E. Belleville (eds.), Animal Orientation and Navigation. National Atmospheric and Space Administration, Washington, D.C.

Case, A.A. 1972. Toxicosis of public health interest: Children and animals poisoned by a volatile agent in discarded motor oil. Clin. Toxicol. 5: 267-270.

Coale, A.J. 1947. Drinking of crude oil by cattle. North Amer. Vet. 28: 221.

Coe, J.M. and W.E. Stuntz. 1980. Passive behavior by the spotted dolphin, *Stenella attenuata*, in tuna purse seine nets. Fish. Bull. 78: 535-537.

Costa, D.P. and G.L. Kooyman. 1982. Oxygen consumption, thermoregulation, and the effect of fur oiling and washing on the sea otter, *Enhydra lutris*. Can. J. Zool. 60: 2761-2767.

Davies, J.L. 1949. Observations on the grey seal (*Halichoerus grypus*) at Ramsey Island, Pembrokeshire. Proc. Zool. Soc. Lond. 119: 673-692.

Davis, A., L.J. Schafer and Z.G. Bell. 1960. The effects on human volunteers of exposure to air containing gasoline vapors. Arch. Environ. Health 1: 548-554.

Davis, J.E. and S.S. Anderson. 1976. Effects of oil pollution on breeding grey seals. Mar. Pollut. Bull. 7: 115-118.

Diaz-Piferrer, M. 1962. The effects of an oil spill on the shore of Guanica, Puerto Rico. Pages 12-13 *in* Fourth Meeting, Assoc. of Island Marine Laboratories. Univ. Puerto Rico, Mayaguez, Puerto Rico.

Drinker, P., C.P. Yaglou and M.F. Warren. 1943. The threshold toxicity of gasoline vapor. J. Ind. Hyg. 25: 225-232.

Duguy, R. 1978. Researches on the mortality factors of the cetaceans on the coast of France. Aquatic Mammals 6: 9-12.

Dudok van Heel, W.H. 1966. Sound and Cetacea. Neth. J. Sea Res. 1: 407-507.

Duguy, R. and P. Babin. 1975. Acute Intoxication by Hydrocarbons Observed in a Harbor Seal (*Phoca vitulina*). Conseil International pour L'Exploration de la Mer, Comite de Mammiferes Marins, CM. 1975/N:5.

Engelhardt, F.R. 1981. Hydrocarbon metabolism and cortisol balance in oil-exposed ringed seals, *Phoca hispida*. Comp. Biochem. Physiol. 72C: 133-136.

Engelhardt, F.R., J.R. Geraci and T.G. Smith. 1977. Uptake and clearance of petroleum hydrocarbons in the ringed seal, *Phoca hispida*. J. Fish. Res. Bd. Can. 34: 1143-1147.

Falk, M.R. and M.J. Lawrence. 1973. Seismic Exploration: Its Nature and Effect on Fish. Fish. Mar. Serv., Winnipeg. Tech. Rep. Serv. No. CEN T-73-9, 51 p.

Fitch, J.E. and P.H. Young. 1948. Use and effect of explosives in California coastal waters. Calif. Fish Game 34: 53-70.

Fletcher, J.L. 1971. Effects of Noise on Wildlife and Other Animals. U.S. Environmental Protection Agency, Washington, D.C. EPA-NTID 30.5, 74 p.

Frazier, J.G. 1980. Marine turtles and problems in coastal management. Pages 2395-2411 *in* B.L. Edge (ed.), Coastal Zone '80. Vol. III, Proc. Second Symp. Coastal Ocean Manage.

Fritts, T.H. 1982. Plastic bags in the intestinal tracts of leatherback marine turtles. Herp. Rev. 13: 72-73.

Fritts, T.H. and M.A. McGehee. 1982. Effects of Petroleum on the Development and Survival of Marine Turtle Embryos. U.S. Dept. Interior Publ. No. FWS/0BS-82/37, 41 p.

Gales, R.S. 1982. Effects of Noise on Offshore Oil and Gas Operations on Marine Mammals - An Introductory Assessment. Research Report: 1980-1981. Prepared for the Bureau of Land Management, Department of the Interior. Naval Ocean Systems Center, Technical Report NOSC TR844, Vol. 2. San Diego, California.

Gamberale, F., G. Annwall and M. Hultengien. 1975. Exposure to white spirit. II. Psychological functions. Scand. J. Work Environ. Health 1: 31-39.

Geraci, J.R., B.D. Hicks and D.J. St. Aubin. 1979. Dolphin pox: A skin disease of cetaceans. Can. J. Comp. Med. 43: 399-404.

Geraci, J.R. and T.G. Smith. 1976. Direct and indirect effects of oil on ringed seals (*Phoca hispida*) of the Beaufort Sea. J. Fish. Res. Bd. Can. 33: 1976-1984.

Geraci, J.R. and D.J. St. Aubin. 1979a. Tissue sources and diagnostic value of circulating enzymes in cetaceans. J. Fish. Res. Bd. Can. 36: 158-163.

Geraci, J.R. and D.J. St. Aubin. 1979b. Stress and disease in marine mammals. Pages 223-233 *in* J.R. Geraci and D.J. St. Aubin (eds.), Biology of Marine Mammals: Insights through Strandings. National Technical Information Service Report No. PB-293-890.

Geraci, J.R. and D.J. St. Aubin. 1980. Offshore petroleum resource development and marine mammals: A review and research recommendations. Mar. Fish. Rev. 42: 1-12.

Geraci, J.R. and D.J. St. Aubin. 1982. Study of the Effects of Oil on Cetaceans. Final Report. Prepared for U.S. Department of the Interior, Bureau of Land Management, Washington, D.C., Contract #AA551-CT9-29, 274 p.

Geraci, J.R. and D.J. St. Aubin. 1985. Expanded Studies on the Effects of Oil on Cetaceans. Final Report, Part I. Prepared for U.S. Department of the Interior, Minerals Management Service, Washington, D.C., Contract #14-12-0001-29169, 144 p.

Geraci, J.R., D.J. St. Aubin and R.J. Reisman. 1983. Bottlenose dolphins, *Tursiops truncatus*, can detect oil. Can. J. Fish. Aquat. Sci. 40: 1515-1522.

Geraci, J.R., D.J. St. Aubin and T.G. Smith. 1979. Influence of age, condition, sampling time and method of plasma chemical parameters in free-ranging ringed seals, *Phoca hispida*. J. Fish. Res. Bd. Can. 36: 1278-1282.

Gerarde, H.W. 1959. Toxicological studies on hydrocarbons. V. Kerosene. Toxicol. Appl. Pharmacol. 1: 462-474.

Gerarde, H.W. 1964. Kerosene—experimental and clinical toxicology. Occup. Health Rev. 16: 17-21.

Gill, C., F. Booker and T. Soper. 1967. The Wreck of the *Torrey Canyon*. David and Charles, Newton Abbot, U.K., 128 p.

Gilmartin, W.G., R.L. Delong, A.W. Smith, J.C. Sweeney, B.W. DeLappe, R.W. Risebrough, L.A. Griner, M.D. Dailey and D.B. Peakall. 1976. Premature parturition in the California sea lion. J. Wildl. Dis. 12: 104-115.

Goodale, D.R., M.A. Hyman and H.E. Winn. 1979. Cetacean responses in association with the *Regal Sword* oil spill. Pages XI-1-XI-6 *in* Cetacean and Turtle Assessment Program, University of Rhode Island. Annual Report for 1979. U.S. Dept. of the Interior, Washington, D.C.

Grose, P.L., J.S. Mattson and H. Petersen. 1979. Marine mammals and sea-birds. Pages 9-1 to 9-8 *in* USNS Potomac Oil Spill Melville Bay, Greenland August 5, 1977. National Technical Information Service Publ. No. PB80-173727.

Gruber, J.A. 1981. Ecology of the Atlantic Bottlenosed Dolphin (*Tursiops truncatus*) in the Pass Cavallo Area of Matagorda Bay, Texas. M.S. Thesis, Texas A&M University, College Station, 182 p.

Haggard, H.W. 1920. The anesthetic and convulsant effects of gasoline vapor. J. Pharmac. Exp. Therap. 16: 401-404.

Hall, R.J., A.A. Belisle and L. Sileo. 1983. Residues of petroleum hydrocarbons in tissues of sea turtles exposed to the Ixtoc I oil spill. J. Wildl. Dis. 19: 106-109.

Helle, E., M. Olsson and S. Jensen. 1976. PCB levels correlated with pathological changes in seal uteri. Ambio 5: 261-263.

Hill, S.H. 1978. A Guide to the Effects of Underwater Shock Waves on Arctic Marine Mammals and Fish. Unpublished manuscript, Inst. Ocean Sci., Patricia Bay, Sidney, B.C. Pac. Mar. Sci. Rep. 78-26, 50 p.

Humes, A.G. 1964. *Harpaticus pulex*, a new species of copepod from the skin of a porpoise and a manatee in Florida. Bull. Mar. Sci. Gulf Carib. 14: 517-528.

Hurst, R.J. and N.A. Oritsland. 1982. Polar bear thermoregulation: Effect of oil on the insulative properties of fur. J. Therm. Biol. 7: 201-208.

Jakosky, J.J. and J. Jakosky, Jr. 1956. Characteristics of explosives for marine seismic exploration. Geophysics 21: 969-991.

Jeffrey, L.M. 1973. Preliminary Report on Floating Tar Balls in the Gulf of Mexico and Caribbean Sea. Unpublished Report, Sea Grant Project 53399, Texas A&M University, College Station.

Kenchington, R.A. 1972. Observations on the digestive system of the dugong, *Dugong dugon* (Erxleben). J. Mammal. 53: 884-887.

Kent, D.B., S. Leatherwood and L. Yohe. 1983. Responses of Migrating Grey Whales, *Eschrichtius robustus*, to Oil on the Sea Surface. Hubbs-Sea World Research Institute, San Diego, California, 63 p.

Kleerekoper, H. and J. Bennett. 1976. Some effects of the water soluble fraction of Louisiana crude on the locomotor behavior of juvenile green turtles (*Chelonia mydas*) and sea catfish (*Arius felis*). Preliminary results. *In* Proceedings NOAA Symposium on the Fate and Effects of Petroleum Hydrocarbons in Marine Ecosystems and Organisms. Nov. 10-12, 1976. Seattle, Washington.

Knave, B., B.A. Olson, S.E. Lofsson, J. Gamberale, A. Isaksson, P. Mindus, H.E. Persson, G. Stuve, A. Wennberg and P. Westerholm. 1978. Long term exposure to jet fuel. II. A cross-sectional epidemiologic investigation on occupationally exposed

industrial workers with a special reference to the nervous system. Scand. J. Work. Environ. Health. 4: 19-45.
Kooyman, G.L., R.L. Gentry and W.B. McAllister. 1976. Physiological Impact of Oil on Pinnipeds. Processed Rep., Northwest Fisheries Center, Natl. Marine Fisheries Service, National Oceanic and Atmospheric Administration, Seattle, Washington, 23 p.
Kooyman, G.L., R.W. Davis and M.A. Castellini. 1977. Thermal conductance of immersed pinniped and sea otter pelts before and after oiling with Prudhoe Bay crude. Pages 151-157 *in* D.A. Wolfe (ed.), Fate and Effects of Petroleum Hydrocarbons in Marine Ecosystems and Organisms. Pergamon Press, New York.
Leatherwood, S. 1977. Some preliminary impressions on the numbers and social behavior of free-swimming bottlenosed dolphin calves, *Tursiops truncatus* in the northern Gulf of Mexico. *In* S.H. Ridgway and K. Benirschke (eds.), Breeding Dolphins, Present Status, Suggestions for the Future. National Technical Information Service, PB-273-673, Washington, D.C., 29 p.
LeBoeuf, B.J. 1971. Oil contamination and elephant seal mortality: A "negative" finding. Pages 277-285 *in* D. Straughan (ed.), Biological and Oceanographical Survey of the Santa Barbara Channel Oil Spill 1969-1970. Vol. 1, Biology and Bacteriology. Allan Hancock Found., Univ. Southern California, Los Angeles, California.
Lillie, H. 1954. Comments in discussion. Proc. Int. Conf. Oil Pollut. Sea, London (1953): 31-33.
Loughrey, A.G. 1959. Preliminary Investigation of the Atlantic Walrus, *Odobenus rosmarus rosmarus* L. Dept. Northern Affairs and Natural Resources, Can. Wildl. Serv., Wildl. Manage. Bull. Ser. 1, No. 14, 175 p.
Lykke, A.W.J. and B.W. Stewart. 1978. Fibrosing alveolitis (pulmonary interstitial fibrosis) evoked by experimental inhalation of gasoline vapours. Experientia 34: 498.
Lykke, A.W.J., B.W. Stewart, P.J. O'Connell and S.M. LeMesurier. 1979. Pulmonary responses of atmospheric pollutants. I. An ultrastructural study of fibrosing alveolitis evoked by petroleum vapor. Pathology 11: 71-80.
Machle, W. 1941. Gasoline intoxication. J. Am. Med. Assoc. 117: 1965-1971.
Mansfield, A.W. 1970. Field Report of Seal Investigations in Chedabucto Bay and at Sable Island, Nova Scotia 2 March-7 April, 1970. Fish. Res. Bd. Can., Arctic Biol. Stn., Ste. Anne de Bellevue, P.Q., 3 p.
Malme, C.I., P.R. Miles, C.W. Clark, P. Tyack and J.E. Bird. 1983. Investigations of the Potential Effects of Underwater Noise from Petroleum Industry Activities on Migrating Gray Whale Behavior. Final Report for the Period June 7, 1982 - July 31, 1983. #5366. Prepared for U.S. Dept. of the Interior, Minerals Management Service, Anchorage, Alaska.
Manton, M.L. 1979. Olfaction and behavior. Pages 289-301 *in* M. Harless and H. Morlock (eds.), Turtles: Perspectives and Research. Wiley-Interscience, New York.
Manton, M.L., A. Karr and D.W. Ehrenfeld. 1972. Chemoreception in the migratory sea turtle, *Chelonia mydas*. Biol. Bull. 143: 184-195.
Maxwell, J.H. 1979. Anesthesia and surgery. Pages 127-152 *in* M. Harless and H. Morlock (eds.), Turtles: Perspectives and Research. Wiley-Interscience, New York.
McCain, B.B., H.O. Hodgins, W.D. Gronlund, J.W. Hawkes, D.W. Brown, M.S. Myers and J.H. Vandermeulen. 1978. Bioavailability of crude oil from experimentally oiled sediments to English sole (*Parophrys vetulus*), and pathological consequences. J. Fish. Res. Bd. Can. 35: 657-664.
McCaull, J. 1969. The black tide. Environment 11: 2-16.
Migaki, G., M.G. Valerio, B. Irvine and F.M. Garner. 1971. Lobo's disease in an Atlantic bottlenosed dolphin. J. Amer. Vet. Med. Assoc. 159: 578-582.
Morris, B.F. 1971. Petroleum: Tar quantities floating in the northwestern Atlantic taken with a new quantitative neuston net. Science 173: 430-432.
Narasimhan, M.J., Jr. and V.G. Ganla. 1967. Experimental studies on kerosene poisoning. Acta Pharmacol. (Kobenh) 25: 214-224.

Nau, C.A., J. Neal and M. Thornton. 1966. C9-C12 fractions obtained from petroleum distillates. An evaluation of their potential toxicity. Arch. Environ. Health 12: 382-393.

Nelson-Smith, A. 1970. The problem of oil pollution of the sea. Adv. Mar. Biol. 8: 215-306.

Nicholson, N.L. 1972. The Santa Barbara Oil Spills in Perspective. Calif. Mar. Res. Comm., Cal. Cooperative Fisheries Invest. Rept. 16: 130-149.

Nishiwaki, M. and A. Sasao. 1977. Human activities disturbing natural migration routes of whales. Sci. Rep. Whales Res. Inst. 29: 113-120.

Oritsland, N.A. 1975. Insulation in marine mammals: The effect of crude oil on ringed seal pelts. Pages 48-67 *in* T.G. Smith and J.R. Geraci (eds.), The Effect of Contact and Ingestion of Crude Oil on Ringed Seals of the Beaufort Sea. Beaufort Sea Tech. Rep. 5.

Oritsland, N.A., F.R. Engelhardt, F.A. Juck, R.A. Hurst and P.D. Watts. 1981 (released 1982). Effect of Crude Oil on Polar Bears. Environmental Studies No. 24, Northern Affairs Program, Dept. Indian Affairs and Northern Development, Ottawa, 268 p.

Parsons, J., J. Spry and T. Austin. 1980. Preliminary observations on the effect of Bunker C fuel oil on seals on the Scotian shelf. Pages 193-202 *in* J.H. Vandermeulen (ed.), Scientific Studies During the "Kurdistan" Tanker Incident: Proceedings of a Workshop. Bedford Inst. Oceanography Report Series, BI-R-80-3, Dartmouth, Nova Scotia.

Perrin, W.F. 1969. The barnacle, *Conchoderma auritum*, on a porpoise *Stenella graffmani*. J. Mammal. 50: 149-151.

Petrie, A.S. 1908. Toxic effect of petroleum fumes. British Med. J. 1: 987.

Philobosia, R. 1976. Disorientation of hawksbill turtle hatchlings, *Eretmochelys imbricata*, by stadium lights. Copeia 4: 824.

Pike, G.C. 1951. Lamprey marks on whales. J. Fish. Res. Bd. Can. 8: 275-280.

Poklis, A. and C. Burkett. 1977. Gasoline sniffing: A review. Clin. Toxicol. 11: 35-41.

Prieur, D. and E. Hussenot. 1978. Marine mammals stranded during the *Amoco Cadiz* oil spill. SEPNB, Vallon du Stangalarc'h, 29200 Brest, France, Extrait de Penn ar Bed 11(9): 361-364.

Pritchard, P.C.H. and R. Marquez. 1973. Kemp's Ridley Turtle, or Atlantic Ridley *Lepidochelys kempi*. I.U.C.N. Monogr. 2 Marine Turtle Series. Morges, Switzerland, 30 p.

Rabalais, S.C. and N.N. Rabalais. 1980. The occurrence of sea turtles on the South Texas coast. Contrib. Mar. Sci. 23: 123-129.

Rausch, R.L. 1973. Post Mortem Findings in Some Marine Mammals Following the Cannikin Test on Amchitka Island. Manuscript prepared for U.S. Atomic Energy Commission, Las Vegas, Nevada, 86 p.

Rector, D.E., B.L. Steadman, R.A. Jones and J. Siegel. 1966. Effects on experimental animals of long-term inhalation exposure to mineral spirits. Toxicol. Appl. Pharmacol. 9: 257-268.

Reid, P.C. 1977. Continuous plankton records: Changes in the composition and abundance of the phytoplankton of the north-east Atlantic Ocean and North Sea, 1958-1974. Mar. Biol. 40: 337-339.

Richardson, W.J., R.S. Wells and B. Wursig. 1983. Disturbance responses of bowheads, 1982. Pages 117-215 *in* W.J. Richardson (ed.), Behavior, Disturbance Responses and Distribution of Bowhead Whales, *Balaena mysticetus* in the Eastern Beaufort Sea, 1982. Unpubl. Rept. from LGL Ecol. Res. Assoc., Inc., Bryan, Texas. Prepared for the U.S. Dept. of the Interior, Minerals Management Service, Reston, Virginia.

Ridgway, S.H. and M.D. Dailey. 1972. Cerebral and cerebellar involvement of trematode parasites in dolphins and their possible role in stranding. J. Wildl. Dis. 8: 33-43.

Robinson, R.O. 1978. Tetraethyl lead poisoning from gasoline sniffing. J. Am. Med. Assoc. 240: 1373-1374.

Rowe, L.D., J.W. Dollahite and B.J. Camp. 1973. Toxicity of two crude oils and of kerosene to cattle. J. Am. Vet. Med. Assoc. 162: 61-66.

Runion, H.E. 1975. Benzene in gasoline. Am. Ind. Hyg. Assoc. J. 36: 338-350.
Rutzler, K. and W. Sterrer. 1970. Oil pollution. Damage observed in tropical communities along the Atlantic seaboard of Panama. BioScience 20: 222-224.
St. Aubin, D.J. and J.R. Geraci. 1980. Tissue levels of ascorbic acid in marine mammals. Comp. Biochem. Physiol. 66A: 605-609.
St. Aubin, D.J., J.R. Geraci, T.G. Smith and T.G. Friesen. 1985. How do bottlenose dolphins, *Tursiops truncatus*, react to oil film under different light conditions? Can. J. Fish. Aquat. Sci. 42: 430-436.
St. Aubin, D.J., R.H. Stinson and J.R. Geraci. 1984. Aspects of the structure and composition of baleen, and some effects of exposure to petroleum hydrocarbons. Can. J. Zool. 62: 193-198.
Salter, R.E. 1979. Site utilization, activity budgets, and disturbance responses of Atlantic walruses during terrestrial haul-out. Can. J. Zool. 57: 1169-1180.
Shane, S.H. and D.J. Schmidly. 1978. The Population Biology of the Atlantic Bottlenose Dolphin, *Tursiops truncatus*, in the Aransas Pass Area of Texas. Marine Mammal Commission, Report No. MMC-76/11. Washington, D.C.
Sherman, K., J.B. Colton, R.L. Dryfoos and B.S. Kinnear. 1973. Oil and Plastics Contamination and Fish Larvae in Surface Waters of the Northeast Atlantic. Unpublished manuscript, MARMAP Operational Test Survey Report: July-August 1972, January-March 1973.
Shoop, C.R. and C. Ruckdeschel. 1982. Increasing turtle strandings in the southeast United States: A complicating factor. Biol. Conserv. 23: 213-215.
Simpson, J.G. and W.G. Gilmartin. 1970. An investigation of elephant seal and sea lion mortality on San Miguel Island. BioScience 20: 289.
Siniff, D.B., T.D. Williams, A.M. Johnson and D.L. Garshelis. 1982. Experiments on the response of sea otters *Enhydra lutris* to oil contamination. Biol. Conserv. 23: 261-272.
Smith, T.G. and J.R. Geraci. 1975. The Effect of Contact and Ingestion of Crude Oil on Ringed Seals of the Beaufort Sea. Beaufort Sea Tech. Rep. 5, 67 p.
Smith, T.G., J.R. Geraci and D.J. St. Aubin. 1983. The reaction of bottlenose dolphins, *Tursiops truncatus*, to a controlled oil spill. Can. J. Fish. Aquat. Sci. 40: 1522-1525.
Sokolov, V.E., I. Bulina and V. Rodinov. 1969. Interaction of dolphin epidermis with flow boundary layer. Nature 222: 267-268.
Spooner, M.F. 1967. Biological effects of the "Torrey Canyon" disaster. Pages 12-19 *in* J. Devon Trust Nat. Conser. (1967 supplement).
Stewart, B.W., S.M. LeMesurier and A.W.J. Lykke. 1979. Correlation of biochemical and morphological changes induced by chemical injury to the lung. Chem. Biol. Interact. 26: 321-338.
Straughan, D. 1971. Biological and Oceanographical Survey of the Santa Barbara Channel Oil Spill 1969-1970. Vol. 1. Biology and Bacteriology. Sea Grant Publ. No. 2. Allan Hancock Found., Univ. Southern Calif., Los Angeles, 426 p.
Straughan, D. 1972. Biological effects of oil pollution in the Santa Barbara Channel. Pages 355-359 *in* M. Ruvio (ed.), Marine Pollution and Sea Life. Fishing News, Ltd., London.
Thompson, J.A. 1958. Biological effects of the Ripple rock explosions. Progress Report of the Pacific Coast Station. J. Fish. Res. Bd. Can. 111: 3-8.
Tolan, E.J. and F.A. Lingl. 1964. "Model psychosis" produced by inhalation of gasoline fumes. Am. J. Psychiatr. 120: 757-761.
U.S. Department of Health, Education, and Welfare. 1981. National Institute for Occupational Safety and Health. Registry of Toxic Effects of Chemical Substances. Cincinnati, Ohio.
Valpey, R., S.M. Sumi, M.K. Copass and G.J. Goble. 1978. Acute and chronic progressive encephalopathy due to gasoline sniffing. Neurology 28: 507-510.
van Bree, P.J.H. 1977. On former and recent strandings of cetaceans on the coast of the Netherlands. Z. Saeugetierkd 42: 101-107.

van Bree, P.J.H. and I. Kristensen. 1974. On the intriguing stranding of four Cuvier's beaked whales, *Ziphius cavirostris* G. Curvier, 1823, on the Lesser Antillean Island of Bonaire. Bijdr. Dierkd. 44: 235-238.

Van Haaften, J.L. 1973. Die Bewirtschaftung von Seehunden in den Niederlanden. Beitrage zur Jagd- und Wildforschung 8: 345-349.

Wang, C.C. and G.U. Irons. 1961. Acute gasoline intoxication. Arch. Environ. Health 2: 714-716.

Warner, R.E. 1969. Environmental Effects of Oil Pollution in Canada. An Evaluation of Problems and Research Needs. Manuscript report prepared for the Canadian Wildlife Service, Ottawa, Canada, 30 p.

Williams, T.D. 1978. Chemical Immobilization, Baseline Hematological Parameters and Oil Contamination in the Sea Otter. U.S. Marine Mammal Commission Report No. MMC-77/06. National Technical Information Service, Springfield, Virginia, 27 p.

Witham, R. 1978. Does a problem exist relative to small sea turtles and oil spills? Pages 630-632 *in* The Proceedings of the Conference on Assessment of the Ecological Impacts of Oil Spills. 14-17 June 1978, Keystone, Colorado. American Institute of Biological Sciences, Washington, D.C.

Yelverton, J.T., D.R. Richmond, E.R. Fletcher and R.K. Jones. 1973. Safe Distances from Underwater Explosions for Mammals and Birds. Lovelace Found. Med. Educ. Res., Alburquerque, New Mexico, 63 p.

CHAPTER 13

PHYSICAL ALTERATION OF MARINE AND COASTAL HABITATS RESULTING FROM OFFSHORE OIL AND GAS DEVELOPMENT ACTIVITIES

Donald F. Boesch and Gordon A. Robilliard

CONTENTS

INTRODUCTION

Concerns regarding the environmental effects of offshore oil and gas development typically focus on pollutants, either resulting from an oil spill or routine discharges from drilling or production. Frequently overlooked, but potentially much longer lasting, are effects resulting from the physical destruction or alteration of marine or coastal habitats to accommodate exploration, development, production or transportation. In this context, it is significant to note that in Louisiana, the most heavily developed offshore petroleum region in the world, physical alterations of coastal environments are generally perceived by fishermen, environmentalists and public officials to have greater negative effects than oil spills and discharges.

Various operations may physically alter benthic habitats offshore (for example, by deposition of drill cuttings, emplacement of platforms or islands, and laying of pipelines) or sensitive coastal habitats (for example, by laying of pipelines through marshes and dredging of channels to provide port access). While some of these alterations may be trivial in spatial extent, of short duration, or without significant adverse biological consequences, other physical alterations may have more pervasive effects, may be essentially permanent, and may be deleterious to living resources.

In comparison to the other reviews of long-term environmental effects in this book (Chapters 5, 6, 8, 9, 10, 11, 12, and 14), consideration of the effects of physical alterations is characterized by a paucity of relevant literature to synthesize. This results from the perceived minor significance of resulting effects; the highly site-specific nature of effects, particularly in the coastal zone; difficulties in separating coastal effects related with offshore energy activities with those due to other human activities and natural processes; and the lack, until recently, of organized or official expressions of concern where physical alterations have been most extensive, the northern Gulf of Mexico.

As a result of the paucity of information from the literature, we have had to rely heavily on reports of limited distribution or outside of the formal scientific literature (such as environmental impact statements and personal experience). In addition, in some cases extant data is reanalyzed.

In this chapter, we consider the effects of physical alterations on offshore environments generally and then discuss effects on coastal environments by major geographic regions of the United States. The regional orientation of the discussion is to accommodate the great regional differences in coastal environmental sensitivity, history of development and likely future development activities. We conclude with recommendations for future studies and management strategies.

OFFSHORE ENVIRONMENTS

Effects of Structures

The emplacement of a production platform, smaller well jacket or subsea connection results in a habitat change in a small area as a result of the introduction of hard substrate and vertical structure in a shelf environment which is typically level and blanketed with sediments. The distribution of oil and gas platforms and artificial islands (three in the Beaufort Sea of Alaska and six off California) in both Federal and state offshore waters of the United States as of the end of 1983 is given in Table 13.1. Of a total of 4301 structures, 17 are in Alaska (most in Lower Cook Inlet) and 30 in southern California. All others are in the Gulf of Mexico, and 89% of those are off Louisiana. Offshore oil and gas structures are in place in other parts of the world (e.g., North Sea, West Africa, Persian Gulf, Mexico and Indonesia), but nowhere does the number or density of

TABLE 13.1
Offshore oil and gas platforms and artificial islands in place in U.S. Federal and state waters as of December 31, 1983 (Essertier, 1984)

Region	*Federal waters*	*State waters*	*Total*
Alaska	3	14	17
Atlantic Ocean	0	0	0
California	15	15	30
Gulf of Mexico			
Louisiana			
Southwest	1028	238	1266
South-Central	1338	185	1523
Southeastern	435	377	812
Total	2801	800	3601
Texas			
South Texas	67	98	165
Upper Texas	188	102	290
Total	239	200	455
Mississippi, Alabama, Florida	0	0	0
Total	3074	1029	4103

structures approach that in the Gulf of Mexico. Furthermore, because of deeper water depths, platform costs and the availability of directional drilling technology (in which scores of wells may be drilled from a single platform), the number of offshore structures expected in as yet undeveloped (frontier) regions will be small in comparison to the contemporary Gulf of Mexico.

Because the vast majority of the seabed of continental shelves and slopes is covered by sand to clay sediments, metal or concrete structures provide rare habitat for encrusting epibiota. Furthermore, either as a result of the food provided by this epibiota or because of the refuge from predators offered, structures attract motile animals, particularly fishes. The resulting dense aggregations of animal life create an artificial reef effect well-known around platforms in the Gulf of Mexico (Gunter and Geyer, 1955; Shinn, 1974; Sonnier *et al.*, 1976; Gallaway and Lewbel, 1982) and southern California (Carlisle *et al.*, 1964; Bascom *et al.*, 1976; Simpson, 1977; Wolfson *et al.*, 1979).

Gallaway and Lewbel (1982) estimated that a typical major platform in the Gulf of Mexico provides approximately 8000 m^2 of hard substrate and 4000 m^2 of bottom strewn with discarded or lost equipment and dislodged shelled organisms that colonize the structure. Wolfson *et al.* (1979) also found an area underneath and adjacent to a southern California platform in which the bottom substrate was modified by dislodged shells and biodeposits from filter-feeding epibiota. This affected the distribution of benthos (e.g., densities of seastars were a thousand times greater underneath the platform). Similarly, in some cases currents may cause selective erosion of bottom sediments around the legs of a platform and result in a very localized effect on the benthos (Harper *et al.*, 1981).

Gallaway and Lewbel (1982) further estimated that single-well and quarters platforms in the Gulf of Mexico provide about 550 m^2 of submerged hard substrate and an additional 75 m^2 of debris-covered bottom. On this basis they concluded that some 1602 ha (3957 acres) of artificial reef habitat is provided by oil and gas structures in the Gulf of Mexico. This is obviously a very small fraction of the total benthic habitat of the Louisiana and Texas shelf. Furthermore, the effect of this substrate alteration is generally viewed as beneficial, by increasing the diversity of biota, providing habitat for game and commercial fishes, and possibly increasing productivity of higher consumers.

Offshore oil and gas structures are popular spots for recreational and some commercial fishing in the Gulf of Mexico (Ditton and Auyong, 1984). Although there is little doubt that the platforms make certain fishes more available to fishermen, it is not clear whether this is due to an increase in their populations as a result of the additional food or habitat provided by the structures or merely from concentrations around the structure of otherwise diffuse stocks. In the latter case, it could be argued that the increased fishing susceptibility of fish aggregated around platforms makes their populations vulnerable to overfishing (e.g., Gallaway *et al.*, 1981).

The degree to which fish populations are enhanced or merely concentrated by offshore structures certainly varies from species to species. Gallaway and Lewbel (1982) concluded that despite the presence of a frequently dense encrusting epibiota on the structures, most associated fishes in the Gulf of Mexico feed either on zooplankton (spadefish, lookdowns, moonfish and creole fish), benthic fauna around the platform (red snapper, tomtate and some groupers) or on other fishes (large transient predators, barracuda, cobia and jacks). A more limited group of fishes (sheepshead, blennies and small tropical grazers) feed primarily on encrusting epibiota. Thus it appears that for most species fished the attractiveness of the structure is more a function of the refuge or motile prey it provides than it is of the presence of epibiota as food.

At offshore structures in the Gulf of Mexico, adults of estuarine-dependent species such as seatrouts (*Cynoscion nebulosus* and *C. arenarius*) and croaker (*Micropogonias undulatus*) are the most intensely fished species, followed by widely migrating pelagic species such as king mackerel (*Scomberomorus cavalla*) and cobia (*Rachycentron canadum*) and juvenile and adult red snapper (*Lutjanus campechanus*) (Ditton and Auyong, 1984). It is generally thought that postlarval recruitment or juvenile survival rather than availability of adult habitat limits population size in most fish species. Thus, for species which frequent structures only as large adults, it appears unlikely that the presence of platforms increases their total populations. Only a few species actively fished seem to associate with structures primarily when young. Gallaway *et al.* (1981) suggested that recreational overharvest of pre-adult red snapper at platforms may be resulting in declines in the commercial catch. However, based on additional data and analyses, they later changed their opinion, suggesting that the large numbers of small snapper which are heavily fished around platforms result from an unexploited and perhaps stable stock of adults occurring over sediment bottoms (Gallaway and Lewbel, 1982).

Oil and gas structures produce spatially restricted alterations of the offshore environment which are generally considered to have beneficial rather than deleterious effects on living resources and the utilization of these resources. The only possible long-term deleterious effects may result from overexploitation of certain fishes which aggregate around the structures. In the Gulf of Mexico there is concern about overfishing in several species which are actively fished around structures (speckled seatrout, red snapper and king mackerel), but it is unknown whether such fishing pressure has a significant effect on their populations. Recreational and commercial harvest of structure-associated, migratory fishes should be taken into account in population models which predict optimum sustainable yield of those species. The resulting insight should be a major factor in decisions regarding the deposition of obsolete structures.

Effects of Drill Cuttings

Although the effects of disposal of drilling fluids and cuttings are considered in more detail by Neff (Chapter 10), brief consideration is given here to the deposition of drill cuttings on the bottom in so far as this may result in a long-term physical alteration of benthic habitat. Most studies have focused on cuttings discharges from exploratory wells rather than from production platforms (National Research Council, 1983). Zingula (1975) and Gallaway *et al.* (1981) made some qualitative observations under production platforms in the Gulf of Mexico, and Carlisle *et al.* (1964) and Wolfson *et al.* (1979) monitored colonization of drill cuttings off California.

Substantial accumulation of drill cuttings is generally limited to within 200 m of the well site. The cuttings range from pebble to sand size and thus tend to be coarser than the ambient sediment. Under swift current regimes such as in Lower Cook Inlet and on Georges Bank, there is no visible accumulation of cuttings. Under more quiescent regimes, coarser cuttings deposited on the bottom may resist resuspension or tractive transport by currents and waves and may persist as a mound under the platform for a period of years. This physical alteration of surface sediments is coupled with the aforementioned structure-associated effects of industrial debris and skeletal material from the fouling community. As a result, a benthic community somewhat different from the surrounding bottom may develop and motile fauna such as seastars, crabs and fishes may be attracted to the elevated and coarser-grained habitat (Zingula, 1975; Wolfson *et al.*, 1979; Gallaway and Lewbel, 1982). As with the structure-associated effects, the affected habitat, even in the structure-rich Gulf of Mexico, will be a very small fraction of the available habitat.

Effects of Artificial Islands

In the Arctic, forces of sea ice restrict the use of conventional drilling rigs and platforms. Artificial islands have been used and are projected for future exploratory and development drilling and oil and gas production. In addition to gravel islands which project above the sea surface at least 3 m, the use of caissons placed atop a subsea sand or gravel berm is also planned (Minerals Management Service, 1983a). Gravel islands built for exploratory drilling are approximately 100 m in diameter. A gravel island in 6 m of water requires 192,125 m^3 of gravel, in 9 m of water 384,250 m^3 of gravel, and in 18 m of water 1,900,000 m^3 of gravel (Jackson *et al.*, 1981). Islands built for production must be larger (almost 200 m in diameter) and would require much larger fill volumes.

Potential long-term marine environmental effects associated with artificial islands include: 1) elimination of the marine benthic habitat covered by the island; 2) altered patterns of sediment erosion and deposition in some unknown surrounding area as a result of modification of current flows; 3) disruption of benthos in any offshore area from which the sand and gravel may be mined; and 4) interference with migration routes of whales and anadromous fishes. The full spatial extent of these effects will depend on the intensity of development. Minerals Management Service (1983a) estimated that if eight islands were constructed in 10-m depths and seven in 20-m depths for exploratory drilling as a result of its 1984 Diapir Field (Beaufort Sea) leasing, 370 ha would be disrupted by dredging and 270 ha would be covered by gravel islands. That is, less than 0.01% of the lease area would be directly impacted, although the size of the indirectly-affected area is not known. However, it was argued that, although there may be localized effects, on a regional basis these would be negligible.

Effects of Pipelines

Pipelines have been the conventional means of transporting oil and gas produced offshore to shore-based processing plants, storage facilities or distribution networks. Although tanker transport from offshore collection facilities is possible in some frontier areas depending on distance from shore and the volume of production, seafloor pipelines can be expected in essentially all regions where oil or gas is produced. Over 25,000 km of offshore pipelines have been emplaced in federal waters of the Gulf of Mexico (Figure 13.1); 90% of that off Louisiana (Minerals Management Service, 1983b). A peak of 2100 km of pipeline was installed in 1972, with an average rate of new pipeline construction since then of 935 km/yr in the Gulf. Compare this with an estimated 965 km of pipeline which would be emplaced over seven years in a development scenario in the Beaufort Sea (Minerals Management Service, 1983a) and a maximum 1121 km of pipeline for development of the St. George Basin in the Bering Sea (Minerals Management Service, 1982). Unitization, or combining product streams of several companies in shared pipelines, will probably be required or be economically prudent in many frontier regions. This will have the effect of minimizing the amount of pipeline laid in comparison to the developed regions of the Gulf of Mexico where as much pipeline is laid every year as would be laid in many frontier areas during their full development.

In water depths less than 61 m, it is usually required that pipelines be laid in a trench 3 ft (0.9 m) deep unless it is in an area congested with pipelines, near a platform, or the bottom is rocky and the disturbance due to trenching is expected to be greater than laying the pipe on the surface with some rip rap cover. Trenching is accomplished by hydraulic jetting or cutting a trench under the pipe after it has been laid on the seafloor. Typically the trench is not backfilled by the operators, but will usually fill up with sediment due to wave and sediment action. As water depth increases, natural backfilling may not occur as quickly

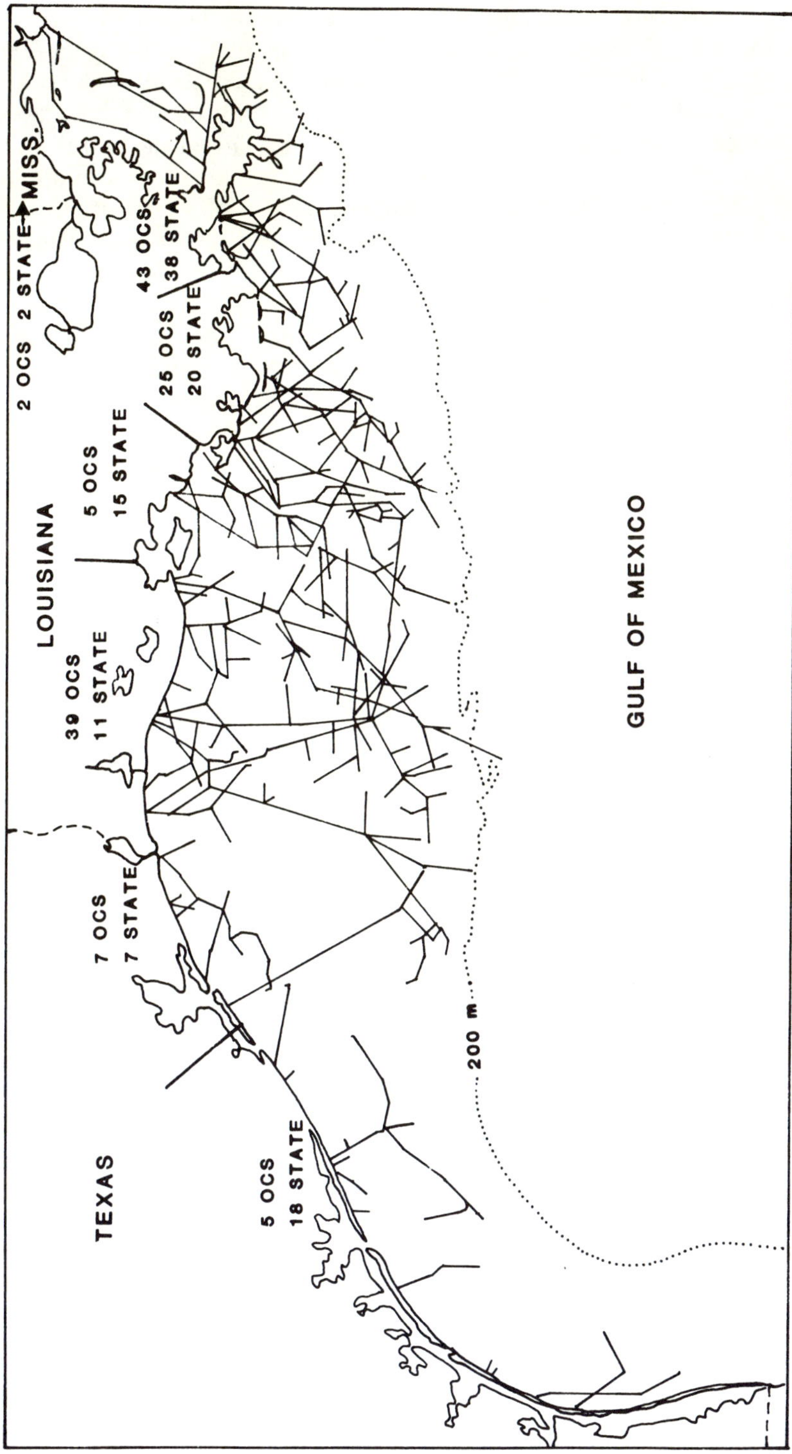

Figure 13.1. Location of major offshore oil and gas pipelines in the Gulf of Mexico. Numbers refer to the number of landfalls in the coastal segments indicated emanating from Federal (OCS) and state waters.

because of the decrease in the influence of surface waves and velocity of bottom currents (Minerals Management Service, 1983b). Pipeline rights-of-way are 200-ft (approximately 60-m) wide in the Gulf of Mexico, but the width of the swath of seabed affected by the pipeline trenching has not been studied. Estimates of the area of benthic habitat affected by pipeline trenching made by Minerals Management Service (1983b) for the Gulf of Mexico assumed that a swath 46-m wide could be affected. On this basis, assuming the 935 km/yr rate of offshore pipeline construction, an area of 43 km^2 would be disturbed each year. Mineral Management Service's (1983a) Beaufort Sea assessment, on the other hand, assumed a path of effects only 15-m wide.

Placement of a pipeline on the seabed obviously has some short-term effect on the local benthos, especially if it involves jetting a trench which could disturb sediments beyond the immediate area of the pipeline. Of interest here, however, is whether a long-term effect on the benthos results. This depends on the temporal duration of the substrate modification and the inherent recovery rates of disturbed communities. Assuming no lasting modification of the sediment substrate, recovery of benthic communities should proceed similarly to that following natural sediment disturbances by storms and large animals or following the disposal of unpolluted dredged material (Rhoads *et al.*, 1978). Recovery of macrobenthos from small scale disturbances ranges from a period of weeks for temperate, shallow water communities, to a year or more in continental shelf environments, and to many years for bathyal (continental slope) communities (Boesch and Rosenberg, 1981). Thus, except in deep-water environments where pipelines are usually not buried, long-term biological effects are not expected, provided no long-lasting substrate alteration occurs.

The lingering effects of trenching on benthic substrates has not been studied. These effects would depend on the amount of bottom sediment transport available to fill the trench and the nature of the excavated substrate. Large lumps of consolidated clay may remain on the bottom and provide surface relief, which would attract motile animals similar to a pile of drill cuttings. Recovery of hard substrate communities or those which depend on biogenous structure, such as corals and seagrass and kelp beds, may require longer periods than sediment-dwelling communities in comparable depths. Generally, such environments are avoided in pipeline routing in the United States. In some cases, such as the West Florida shelf where there are continuous seagrass beds and intermittent, but widespread, reef outcroppings (Chapter 3), this might not be completely possible.

Exposed pipelines may attract encrusting epibiota and motile animals much like platforms do. In the Gulf of Mexico, trawlers sometimes tow their nets parallel with the pipeline, ostensibly enhancing their catch. Overall, however, pipelines and other structures on the seabed are viewed as hazardous for trawling by fishermen both in the Gulf of Mexico and in frontier areas.

COASTAL ENVIRONMENTS

General Considerations

Many activities undertaken to develop oil and gas offshore may affect coastal environments. Most attention is usually directed at oil spills and their potential for reaching shore. Similarly, some wastes generated offshore, including used drilling fluids and produced waters, may be brought ashore for disposal. The effects of oil spills and nearshore discharges are considered in other chapters. In addition to these effects, physical environmental modifications may result from the placement of pipelines which bring the petroleum product ashore, and consequently must cross coastal environments, and the construction of coastal supply facilities and processing plants. Coastal facilities include service bases, which provide platforms, machinery, pipe, drilling fluids, fuel, food and manpower, terminals, and navigation channels. Although these shore-based facilities may have significant effects in terms of modification of terrestrial habitats and in consequences to society, this discussion will be limited strictly to those environmental effects in marine and estuarine habitats up to the high tide line. These effects may result from direct physical modification (such as "reclamation" of a marine environment to provide a support base, laying a pipeline through an intertidal zone, or dredging a channel for navigational access) or indirectly by alteration of natural water flow and animal migration.

The potential for significant long-term effects on coastal ecosystems varies widely among regions potentially producing oil and gas and depends on the nature of the coastal ecosystems and the number and type of pipelines and onshore bases. Some regions of the U.S. are characterized mainly by open coasts and beaches (e.g., California), others by estuaries and wetlands (e.g., Louisiana and Georgia). In some regions, because of the concentration or remote nature of the resource, single or few supply bases, very few pipeline landfalls, and no platform construction facilities are expected (e.g., most potential offshore petroleum provinces in Alaska). On the other hand, in the northwestern Gulf of Mexico there are at least 30 supply bases in Louisiana and Texas, 13 platform fabrication yards, and over 200 pipeline landfalls already in existence (Minerals Management Service, 1983c).

Most of the following discussion emphasizes the northwestern Gulf of Mexico, where most of the concern regarding coastal effects is centered. This region may constitute a worst case for coastal habitat modification because of the extensive offshore and coastal oil and gas development and the great extent of sensitive coastal wetlands, particularly in Louisiana. Nonetheless, understanding habitat modifications there is important not only for guiding future activities in that important oil and gas producing area but also for providing lessons useful in other regions of the world with extensive coastal wetland habitats.

Gulf of Mexico

The coast of the Gulf of Mexico is characterized by a paucity of rocky shores, but otherwise by considerable geomorphologic and environmental heterogeneity, ranging from limestone platforms to one of the world's largest fluvial deltas, from xeric to mesic moisture conditions, and from temperate to tropical marine climatic conditions. To date, offshore oil and gas development has been limited to the north central and northwestern Gulf, from Alabama to South Texas. The vast majority of this development has taken place off Louisiana (Table 13.1,

Figure 13.1), where coastal environments have been formed under the influence of the Mississippi River.

The Mississippi Deltaic Plain in southeastern Louisiana represents the active deltas of the Mississippi River and those previously occupied during the present stillstand of sea level (approximately 7000 years ago). The Chenier Plain of Louisiana and the upper Texas coast represents the coastal environment down-drift of the Mississippi River, having been shaped by alternating depositional and erosional episodes during this same period. Much less offshore development has taken place to the east (off Mississippi and Alabama) and southwest (off South Texas), which are characterized by barrier islands and lagoons. The

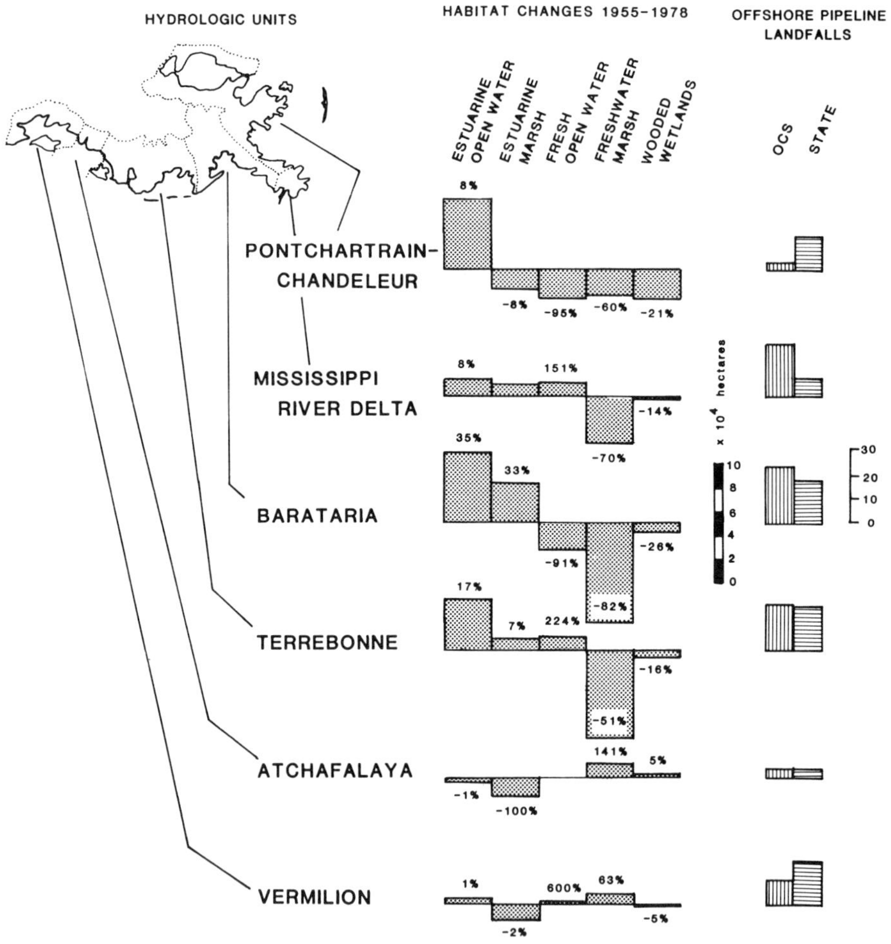

Figure 13.2. Wetland habitat changes within the six hydrologic units of the Mississippi Deltaic Plain of Louisiana between 1955 and 1978 (from Wicker, 1980). Height of bars is proportional to area gained or lost and is additive over habitats or hydrologic units; numbers refer to the percent change in habitat area from 1955 conditions. Numbers of offshore oil and gas pipeline landfalls for each hydrologic unit are also indicated.

Mississippi Deltaic and Chenier Plains are characterized by shallow estuaries and very broad intertidal zones (mostly tidal marshes) which extend inland as much as 60 km from the coast, while the other regions have narrower intertidal zones and less extensive wetlands.

The coastal wetlands of the Mississippi Deltaic Plain, and to a lesser extent, the Chenier Plain have undergone severe deterioration during the last 30 years, resulting in an estimated wetland loss rate in excess of 100 km^2/yr (Gosselink *et al.*, 1979; Wicker, 1980). Related to these wetland changes is a dramatic increase in the salinity of many of the estuaries. The net effect has been an increase in estuarine open water habitat, relatively little change in area of estuarine marsh which has retreated inland, and dramatic reductions in tidal fresh marshes as a result of encroachment by brackish water. Figure 13.2 illustrates this pattern for the six hydrological units (estuarine complexes with separate drainage basins) of the Mississippi Deltaic Plain. Large areas of estuarine marsh have been converted to open water in Pontchartrain-Chandeleur, Barataria and Terrebonne units, which represent older, abandoned Mississippi River lobes and interlobe basins. Furthermore between 51 and 82% of the fresh marsh in the hydrological units of southeastern Louisiana was lost between 1955 and 1978. The two hydrological units of south central Louisiana (Atchafalaya and Vermilion) actually showed an increase in fresh marsh during that time as a result of increased diversion of the flow of the Mississippi River down the Atchafalaya River (control structures now divert 30% of the Mississippi down the Atchafalaya).

The causes of wetland loss and saltwater intrusion are many and complex (reviewed in Gagliano *et al.*, 1981; Boesch, 1982; Boesch *et al.*, 1984): the natural decay due to subsidence and erosion of long-abandoned Mississippi River delta lobes; channelization of river flow, such that sediments and river water are directed offshore rather than allowed to broadly disperse over the wetlands during floods; dredging of navigation channels and oil and gas transportation lines; filling or draining of wetlands for land use; and, possibly, enhanced subsidence resulting from formation fluid withdrawals. Of these, the last four are relevant to offshore oil and gas development, but it is important to appreciate that all of these causes interact, often synergistically. Consequently, it is very difficult to isolate and quantify the effects of a specific pipeline crossing or navigation channel.

Pipeline Crossings

At least 237 oil and gas pipelines emanating from offshore waters under Federal or state jurisdiction have landfalls in Texas, Louisiana or Mississippi (Figure 13.1). Of these 82% (88% of those from Federal waters) strike land first in Louisiana: 146 in the Mississippi Deltaic Plain and 50 in the Chenier Plain. In addition, there are hundreds of other pipelines which serve coastal oil and gas production sites and cross wetlands and estuaries. Pipelines traveling through the coastal environment are operated by over 30 companies, including both major oil and gas producers and pipeline or transmission companies. As a result much "duplication" occurs in the coastal pipeline corridors in order to keep the product streams of various companies separate. In many cases, however, oil or gas from coastal, offshore state and offshore Federal production merges, making it difficult to attribute pipeline-associated impacts to one type of development. Although it is environmentally irrelevant whether a pipeline, navigation channel or supply base serves coastal or offshore development, this may be of considerable importance in terms of regulation.

When pipelines are laid across coastal bays and water bodies, conventional trenching techniques discussed for offshore pipelines are used. There are two methods of constructing pipelines across intertidal landfalls and wetlands: flotation canals and push ditches (Minerals Management Service, 1983b). The flotation canal method has been most commonly used in wetlands. This requires the excavation of a canal into which barges and other floating equipment are maneuvered for construction of the canal but also for installation of the pipeline therein. The material excavated from the canal is placed in canal-side spoil banks, which usually are continuous on either side of the canal but may be intermittent or furnished with breaks. In the push ditch method, a narrower trench is excavated by draglines working from support mats. Flotation devices are attached to the pipe which is then pushed or pulled through the trench by machinery based on higher ground or nearby barges. The ditch is then backfilled.

The area affected by coastal pipeline construction and the duration of impact varies with the construction methods used and the environments affected. Minerals Management Service (1983b, 1984) has used estimates of 12 and 25 acres/mile for wetland directly affected; this equals a swath about 30- or 60-m wide. Johnson and Gosselink (1982), however, found that for a variety of oil and gas flotation canals in coastal Louisiana, the total impact width (canal plus spoil bank) ranged from 70 to 150 m, in part because the actual width of a completed canal (30 m) was generally wider than the permitted width (20 m). The average total impact width was 100 m, i.e., 40 acres/mile. A pipeline canal traversing 20 km (not uncommon) of coastal salt marsh could directly, and probably permanently, destroy 200 ha (494 acres) of marsh. Using a more conservative length of 10 km, the approximately 200 offshore pipelines crossing coastal wetlands could have resulted in loss of 20,000 ha of marsh.

The push (or pull) ditch method may result in less long-term damage than flotation canals which remain without backfilling, although, as with other facets of coastal habitat alterations related to oil and gas development, there has been little investigation of its effectiveness. Sasser *et al.* (1983) have monitored the effects of the 159-km Louisiana Offshore Oil Port (LOOP) pipeline which traverses all major wetland habitats in coastal Louisiana, from salt marshes to bottomland hardwoods. Two years after completion of backfilling, there was little revegetation of the beach at the landfill because of washovers on this rapidly retreating coastal segment. Over the same time period they also found relatively little revegetation of the backfilled ditch (10%) and spoil areas (22%) compared with surrounding salt and brackish marsh, which averaged 50% vegetation cover. Revegetation was more advanced in fresh marshes and swamps.

Although the direct effects of excavation and spoil deposition result in long-term and usually permanent loss of wetland habitat (wetlands in rapidly subsiding coastal Louisiana tend to be unable to repair such scars), potentially far more significant are long-term consequences of the remaining canals and spoil banks on surrounding wetland and estuarine habitats. Such indirect effects result from accelerated shoreline erosion and modification of natural hydrological flow and tidal flooding patterns. The effects of such hydrological modifications include saltwater intrusion into the estuary and die off of adjacent wetlands due to sinking or ponding.

Dredged canals tend to rapidly widen because of the highly erodable nature of the wetland soils placed on spoil banks. Johnson and Gosselink (1982) measured canal widening rates of 1 to 2.5 m/yr, depending on the amount of boat traffic in the canals. In order to limit boat access and reduce the potential for saltwater intrusion, pipeline canals are frequently plugged by a mud or shell berm where they cross major water bodies.

The density of canals, primarily oil and gas well access and pipeline canals, varies considerably over coastal Louisiana as a function of the location of oil and gas resources and transportation routes. Scaife *et al.* (1983) statistically examined the relationship between canal density and wetland loss within $7\frac{1}{2}$-minute quadrangle habitat maps throughout the Mississippi Deltaic Plain. To control the effects of natural processes and larger scale human activities on their analyses, quadrangles within the same delta lobes and a similar distance from the coast were compared. Correlations between canal density and wetland loss were generally highly significant. Furthermore, when the actual canal and spoil bank area was subtracted from the wetland loss rate, there was a residual loss rate which was itself correlated with canal density. Canal density could explain 48 to 97% of the wetland loss, but the actual canal surface area accounted for less than 10%. Craig *et al.* (1979) had also earlier estimated the indirect effects of channelization on wetland loss as four times the area of the canals themselves.

The exact mechanisms of these indirect effects are yet poorly known. Bank erosion due to increased tidal flow is certainly a factor. In addition, numerous shallow ponds open up in marsh adjacent to canals and spoil banks (Turner *et al.*, 1982; Turner, in press). Similar pond development is not seen adjacent to natural channels. These ponds seem to be the result of a disruption of natural marsh hydrology. Spoil banks interfere with overbank flooding of the marsh, disrupting the supply of suspended sediments which subsidize the aggradation necessary to counteract subsidence (Boesch *et al.*, 1983). Further, spoil banks may also impound standing water over the marsh surface and decrease subsurface flows by gravity compression of the marsh deposits underneath the spoil banks (Turner, in press). In either case, marsh grasses succumb to continuous inundation because of the lack of oxygen or presence of high sulfide levels to which the roots are exposed (Mendelssohn *et al.*, 1981).

Navigation Channels

A variety of channels which traverse shallow estuaries and wetlands in the northwestern Gulf of Mexico are used to provide access by supply vessels to inland bases and to transport supplies along the coast to these ports. These craft range from small crew boats which bring personnel to offshore platforms to huge barges which are used to transport large platforms from fabrication yards. The craft generally do not have very deep drafts, but there are few natural harbors along this coast. High land served by roads and rail and suitable for construction and supply bases is available in Louisiana only well inland from the coast. Consequently, vessels supporting offshore oil and gas activities extensively use dredged channels or deepened natural waterways.

Figure 13.3 shows the distribution of major canals used in service of the offshore oil and gas industry along the Louisiana and Texas coasts. These include a major coastwise waterway, the Gulf Intracoastal Waterway (GIWW), which is heavily used by barges transporting equipment and supplies between the coastal service bases, and several channels perpendicular to the coast. Of these, several were built for large commercial shipping and are used incidentally for oil and gas transportation (Mississippi River Gulf Outlet, Calcasieu Ship Channel, Houston Ship Channel, Corpus Christi Ship Channel). These channels would probably exist without offshore oil and gas development in the Gulf, although several primarily serve the petrochemical industry. Oil and gas related traffic may, nonetheless, contribute to bank erosion by wakes where that is a problem (e.g., the Mississippi River Gulf Outlet is now two to three times wider than it was when constructed in the 1960s). Other channels, such as the Houma Navigation Canal, were built for and are primarily used by the offshore oil and gas industry. In the future still others, such as the Atchafalaya River Navigation Channel, will be influenced by their use in support of offshore exploration and production.

Although there may be environmental problems associated with dredged material disposal for some of these channels, their most serious and widespread effects concern the alteration of natural hydrological processes. Most commonly, the effect is saltwater intrusion which results in the elimination of salt intolerant wetlands, including floating fresh marshes (called flotant in southern Louisiana) and wooded swamps. This is most evident in those portions of the Mississippi Deltaic Plain most crossed by navigation channels and pipelines (Figure 13.2), but this is also a major cause of wetland alterations in Chenier Plain estuaries influenced by salinity intrusion through the Calcasieu and Sabine Ship Channels (Gosselink *et al.*, 1979). For the Atchafalaya River Navigation Channel and associated waterways, however, problems emanate from attempts to deal with the rapid sedimentation and flood risks brought by increased river flows. These threaten the huge offshore supply and construction infrastructure which has developed around Morgan City. A closer look at this system and the nearby Houma Navigation Canal will illustrate the complexity and magnitude of the problems.

Figure 13.4 shows the coastal regions below Houma and Morgan City, Louisiana, including the Houma Navigation Canal and the lower Atchafalaya River, the principal navigational access routes to these major supply and construction bases. Superimposed on this map are lines marking the lower limits of fresh marshes in the early 1950s and in 1978, and similar lines marking the upper limits of saline marsh for those same time periods. These marsh types are characterized

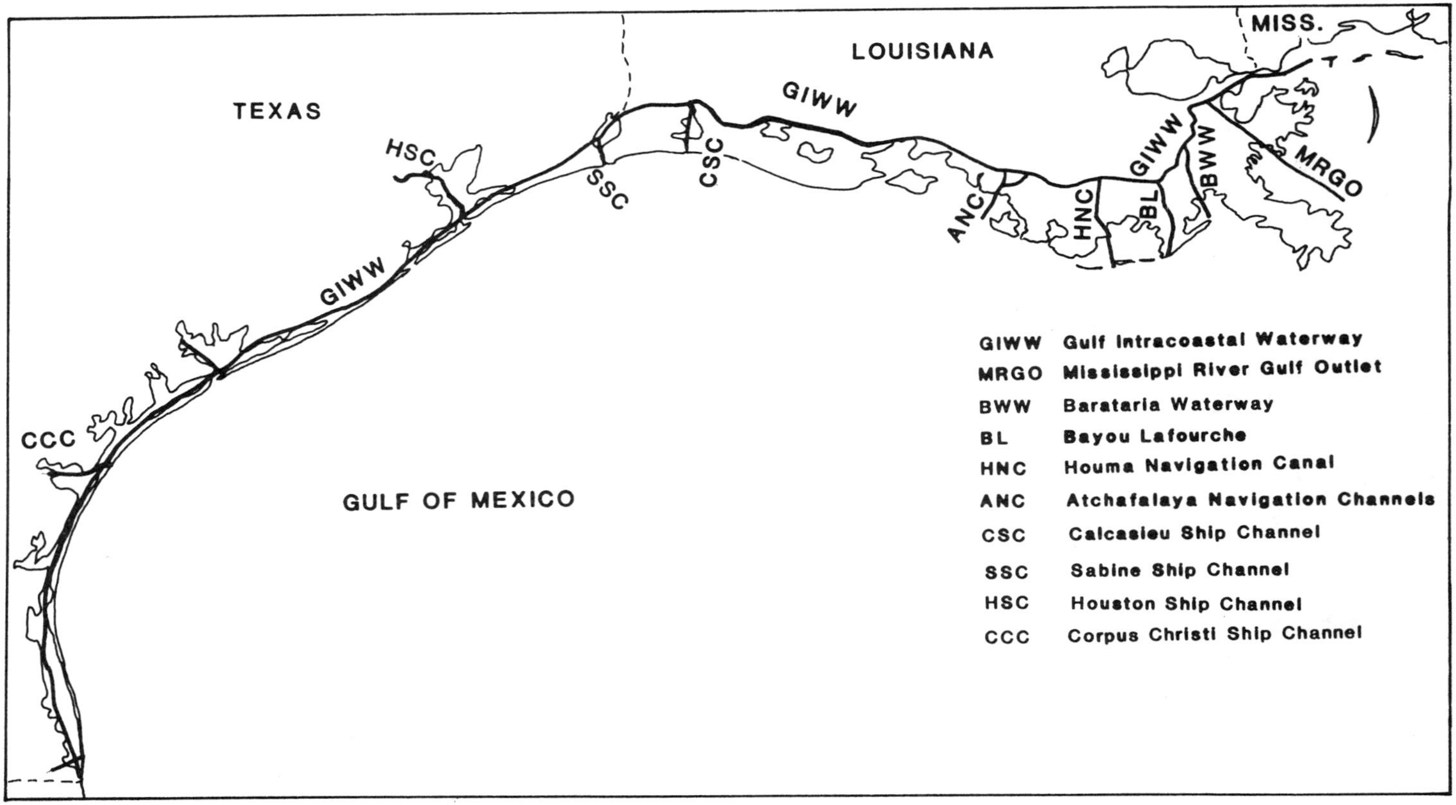

Figure 13.3. Major dredged or deepened navigation channels in the north central and western Gulf of Mexico which serve offshore oil and gas related transportation.

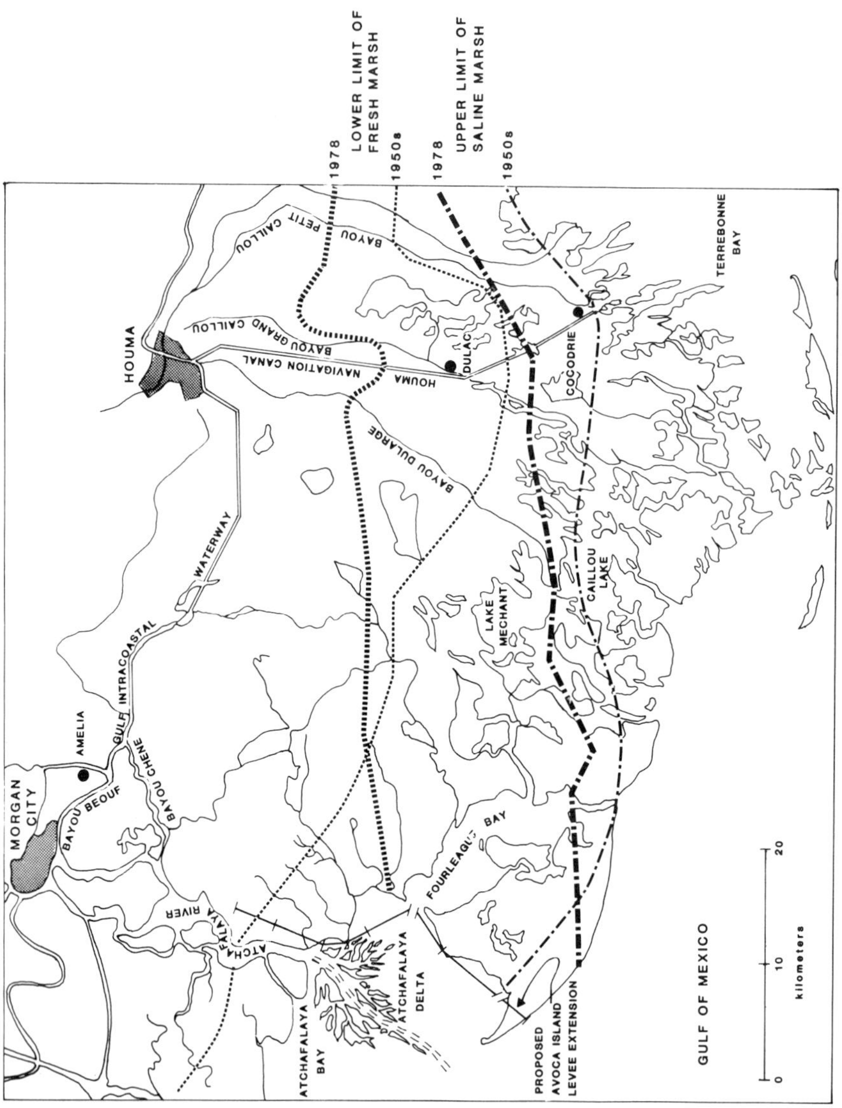

Figure 13.4. Changes in the distribution of coastal wetland vegetation types from the early 1950s to 1978 in relation to the major navigable waterways of the region below Houma and Morgan City, Louisiana.

by particular vegetation (see Wicker, 1980), but generally saline marsh may experience salinity of 18 ppt and more, while fresh marshes tolerate only a few ppt for short periods. The area between the two lines for a given time period supported brackish marsh (brackish and intermediate marsh of Wicker).

As a result of increased flow of the Mississippi River down the Atchafalaya River, stabilized at 30% since 1962, the marshes in the western portion of the area have been freshened. In addition, much of the volume of Atchafalaya Bay has been filled with fluvial sediments and a subaerial delta began to emerge beginning in 1973 (van Heerden and Roberts, 1980). This has caused substantial problems for the offshore construction and support infrastructure in Morgan City and Amelia. Active dredging is required to maintain navigational access through the delta and, as the bay has filled, backwater flooding from the south is an increasing problem during peak river flows. This is caused as the flood waters enter the shallow bay and are unable to drain rapidly. On the other hand, the Atchafalaya River's influence has been a boon to coastal wetlands; the new delta and the surrounding wetlands are the only regions where new marshes are being formed to counteract the tremendous losses elsewhere in Louisiana (Figure 13.2). Baumann and Adams (1982) showed that for the marshes east of Atchafalaya Bay and around Fourleague Bay there was a reversal in the trend from wetland loss between 1955 and 1972 to gains between 1972 and 1978.

To reduce the problem of backwater flooding, the U.S. Army Corps of Engineers has proposed the construction, in incremental sections, of an extension of the existing Avoca Island levee down the east side of the river and Atchafalaya Bay (Figure 13.4). However, this would diminish the flow of fresh water, sediments and nutrients currently nourishing the marsh growth to the east. Fish and Wildlife Service (1981) estimated that a loss of 17,000 acres (6900 ha) of fresh, brackish and saline marsh would result. Even though some flow could be introduced via flood gates to maintain salinity levels, this is unlikely to transmit sufficient sediments to enhance marsh growth at present rates. In addition, in order to provide navigational access during flood periods while maintaining the integrity of the flood barrier, a new navigation channel would be needed, running east of the levee extension. This channel could serve as a conduit for salinity intrusion during low flow.

The Avoca Island levee extension illustrates the complexity of offshore oil and gas development issues in this heavily developed region. On the face of it, this controversy seems a conflict between flood protection and coastal resource interests. Public works of this magnitude could not be justified, however, if the Morgan City area were not such a valuable support base for the offshore industry. But because the levee proposal does involve existing facilities and populations, it has not been considered as an "OCS impact" in the Minerals Management Service's evaluation of environmental consequences of offshore leasing. In a broad historical view, the effects of such coastal engineering, completed and proposed, seem no less a consequence of offshore oil and gas development than an oil spill.

The relationship of the habitat modifications attributable to the Houma Navigation Canal to offshore oil and gas development is more readily apparent. This linear channel was completed in 1962 by local interests to provide the city of Houma with strategic access to the then newly developing offshore industry. It connects the Gulf Intracoastal Waterway with Terrebonne Bay, 50 km to the south, where a dredged channel covers the final 25 km to the open Gulf. Control depths are 4.6 m for a 46-m width, but bank erosion, primarily as a result of vessel wakes, has widened the canal to 200 m or more in its lower reaches. Most of this bank erosion is due to high speed crew and supply boats traveling to or from oil and gas facilities offshore or in Terrebonne Bay (D.F. Boesch, pers. obser.).

Substantial salinity intrusion has occurred since construction of the canal (Figure 13.4). This is particularly noticeable in the vicinity of Dulac, where cypress swamps have been killed by brackish waters. During the fall of 1984, the city of Houma which draws its water supply from local surface waters, was confronted with brackish water (3 ppt) at its intakes and had to switch to alternate sources. Although detailed hydrological studies of the Houma Navigation Canal have not been conducted, it is suspected that the large cross-sectional area of the canal allows greater tidal penetration of salty bay waters and less retention of inland runoff than the natural water bodies. The natural water bodies are much shallower and complexly nonlinear. The more effective flushing of inland areas to the north during falling tides is also thought to contribute to sewage contamination of oyster grounds in the Caillou Lake-Lake Mechant area, which has resulted in occasional closure of these areas to direct harvesting of shellfish. These areas are otherwise well removed from human settlements.

Supply and Service Bases

Because of the scarcity of high, well-drained land along parts of the coast of the Gulf of Mexico, there may be pressures to fill wetlands and shallow coastal waters to provide space for docks, warehouses, pipe storage, processing plants or fabrication yards. There are at least 37 coastal supply bases serving offshore oil and gas development in the region (Minerals Management Service, 1983c). There are 160 gas processing plants in coastal counties or parishes of the Gulf of Mexico region, although most of these are not located in truly coastal situations. Again, it is in Louisiana where coastal uplands are scarce or non-existent where filling of wetlands to provide bases is most common. There have been moderate to extensive physical alterations to coastal marine habitats at such bases as Venice, Grand Isle, Fourchon, Cocodrie, Dulac, Morgan City-Amelia and Intracoastal City.

Subsidence

Withdrawal of fluids from subsurface formations may induce subsidence, or depression of the ground surface as a result of a decline in pore pressure. This phenomenon is well-known in the case of ground water withdrawals from shallow aquifers, but can also take place as the result of withdrawal of oil, gas or formation waters. For example, parts of Long Beach, California subsided as much as 50 cm/yr as a result of oil production until a program of reinjection of water was used to slow the rate of subsidence (Castle *et al.*, 1970). Subsidence from oil withdrawal is also known from coastal Texas (Yerkes and Castle, 1970) and Lake Maricaibo, Venezuela (van der Knaap and van der Vlis, 1967).

In the context of marine environmental effects, withdrawal-induced subsidence may be a concern if coastal or nearshore fluid removal causes rapid subsidence, resulting in inundation and death of wetland vegetation (Boesch *et al.*, 1983). Although some have speculated that subsidence resulting from oil, gas and produced water withdrawals has been a factor in the loss of wetlands in Louisiana (Gagliano *et al.*, 1981), petroleum field subsidence has not been specifically studied there. A key factor in determining the degree of surface subsidence is the depth of the withdrawal. The most dramatic examples of subsidence induced by oil field withdrawals resulted from withdrawals from depths shallower than 1000 m, whereas in coastal and offshore regions of the Gulf of Mexico most producing zones are at 2000 to 5000 m. Some subsidence has been observed in oil fields producing from as deep as 3800 m (Yerkes and Castle, 1970).

Of course, fluid withdrawals from regions well offshore would not be expected to have subsidence effects on coastal environments. Therefore, the effects of subsidence from offshore petroleum production (excluding that in enclosed waters and wetlands themselves) would be limited to withdrawals from nearshore production from relatively shallow reservoirs. Such conditions may exist in the older nearshore fields in the Gulf of Mexico, in southern California and in the Beaufort Sea.

Interaction of Coastal Alterations

For the sake of organization, the effects of pipeline crossings, navigation channels, and dredging and filling for coastal support bases have been separately discussed. Furthermore, the discussion has made only passing reference to the effects on coastal wetlands and estuaries of oil and gas development within these coastal environments themselves. In reality, of course, all of these activities may take place in the same regions. Their effects may be compounded.

An example of multiple impacts related to different oil and gas development activities is presented in Figure 13.5, wherein the changes in coastal wetlands near Cocodrie, Louisiana, between 1955 and 1978 are depicted. Within this relatively small (27.5 km^2) area, there are examples of impacts on wetlands of offshore and coastwise collector pipelines, a navigation channel supporting offshore activity (Houma Navigation Canal), product processing and supply bases, and well location and access canals. The offshore pipeline canal serves two large gas lines and extends 15 km through coastal marshes to the southwest. These lines serve large networks of outer shelf (OCS) wells to water depths in excess of 140 m.

During the 23 years between 1955 and 1978, 441 ha of wetlands (primarily *Spartina* salt marshes) were lost in the area shown (Table 13.2). Of this loss, 321 ha could be accounted for as a direct result of oil and gas development activities: 15% of this due to pipeline crossings (canal and spoil bank), 31% due to dredging, spoil disposal, and bank erosion of the Houma Navigation Canal, 9% due to supply and processing bases, and 20% due to well location and access canals and spoil banks. The remaining losses appeared to result mainly from erosion of

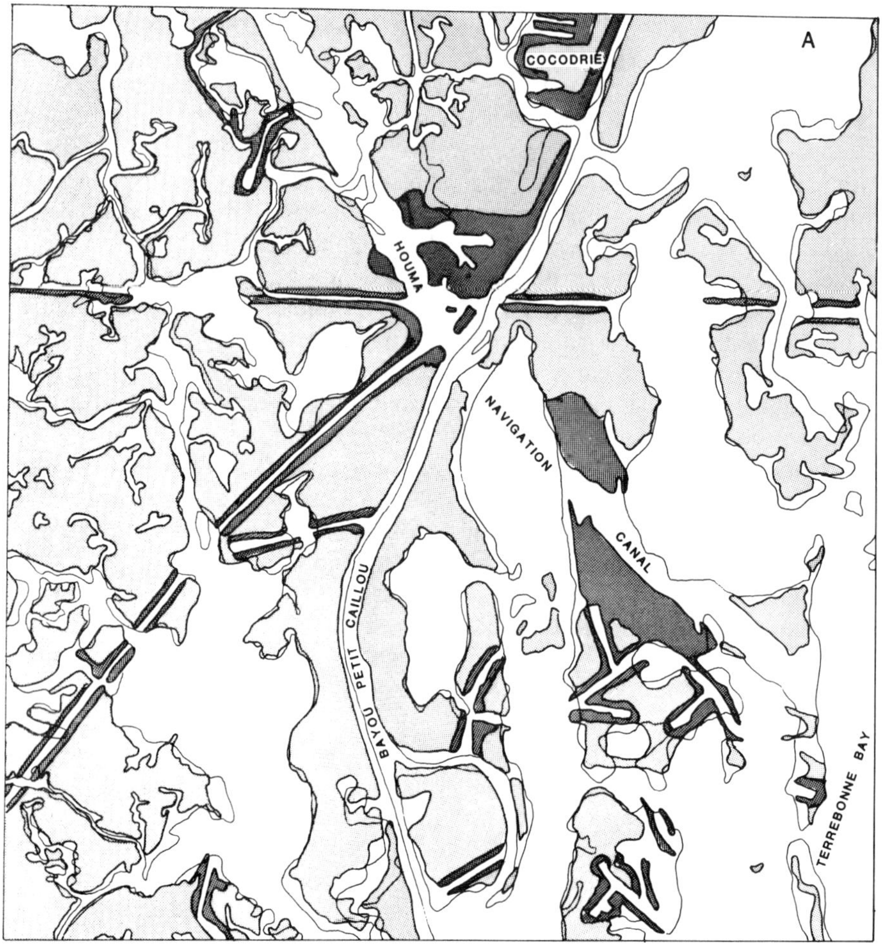

Figure 13.5. The impact of oil and gas related dredging activities on coastal salt marshes in the vicinity of Cocodrie, Louisiana. A, Distribution of salt marsh (light shading) and filled areas (dark shading) in 1978 and salt marsh in 1956 (free-standing lines) (from habitat maps of Wicker, 1980).

shorelines to the southeast which are exposed to the open waters of Terrebonne Bay. An unknown portion of these remaining losses is attributable to the indirect effects of channelization and spoil bank deposits. This results from increased tidal circulation (notice the general widening of natural channels), bank erosion of shorelines, and ponding within the marsh.

Although the Cocodrie region is an admittedly isolated example, it serves to illustrate the complexity of physical habitat alterations which may result from intense oil and gas development activities in coastal regions with extensive intertidal wetlands. It also furnishes an appreciation that, from an environmental perspective, it makes little difference whether a canal serves offshore or coastal development, or whether it was constructed by industry or government. The resulting problems are cumulative and interactive.

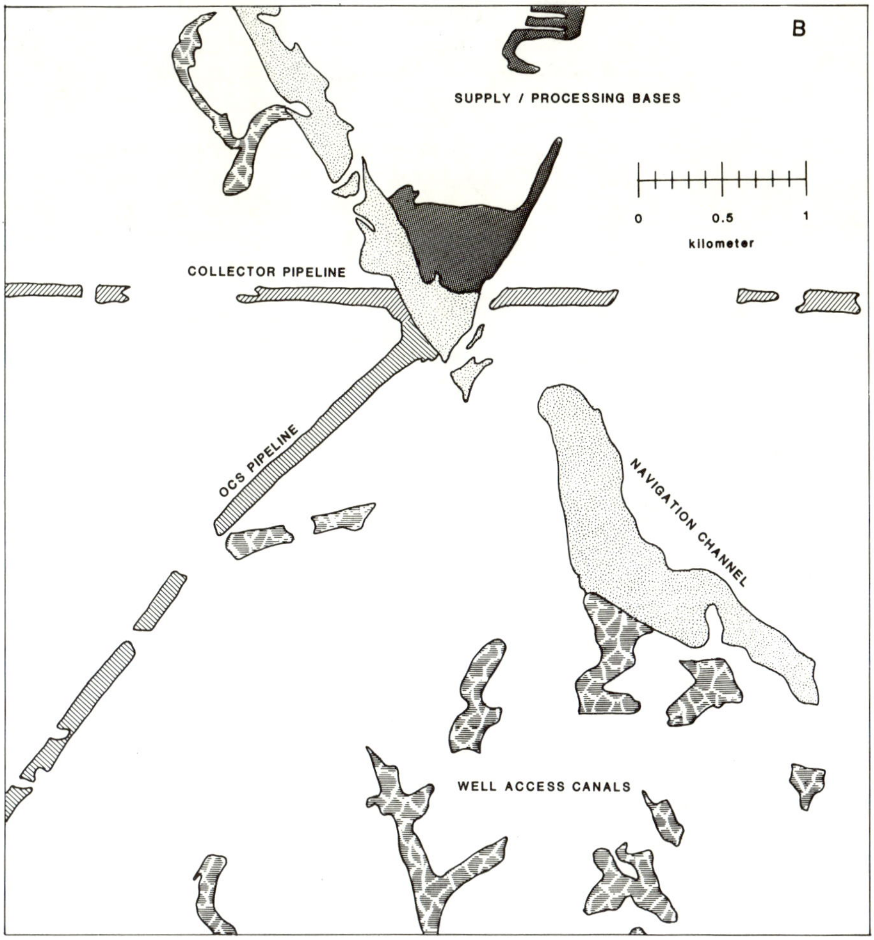

B, Marsh area directly lost due to dredging of canals, bank erosion of canals and filling by major activity type.

Effects on Living Resources

As with contaminant-related effects of offshore oil and gas development, the concerns regarding physical alterations are based ultimately on their consequences to the living resources the affected ecosystems furnish to society. Although it is similarly difficult to relate subtle effects on the ecosystems to significant effects on the resources, effects on coastal ecosystems are of high concern for several reasons. First, the available habitat is generally much more limited in extent in coastal than in offshore ecosystems, thus the potential for affecting a significant part of the resource base is greater. Second, living resources (including fisheries, birds and mammals) are generally more concentrated in coastal habitats. Third, valued species which migrate (e.g., waterfowl and some fishery species) may be dependent on these coastal ecosystems for part of their

TABLE 13.2
Vegetated wetland habitat losses directly attributable to dredging activities in an area depicted in Figure 13.5, south of Cocodrie, Louisiana, for the period between 1955 and 1978

Area considered	*Area (hectares)*	*Percent loss*	*Percent of loss attributable to activity*
Total area mapped	2750		
Vegetated wetland area, 1956	1577		
Vegetated wetland area, 1978	1136	26.3	
Loss directly attributable to:			
OCS pipeline	37		8.8
Collector pipeline	27		6.4
Navigation channel	131		31.5
Well access canals	89		20.1
Supply and processing bases	37		8.8
Total	321		72.8
Losses due to indirect and "natural" causes	120		27.2

lives. This is particularly true in the southeastern United States, where the dominant commercial fisheries are estuarine-dependent. The catch of estuarine-dependent fish and shellfish from the southeast represents over one-half of the total U.S. landings in volume (35% from the Gulf of Mexico alone) and over one-third of the ex-vessel value.

Although the overall relationship of estuarine and wetland environments to the regional production of resources is clear, the consequences of localized modification or destruction of a particular coastal habitat to the resource are difficult to predict. Is there is a direct, linear relationship between the area of wetlands destroyed by a dredge and fill project and the resource or is the relationship more complex and nonlinear? Boesch and Turner (1984) addressed this question and suggested that the interrelationships between wetland and shallow-water habitats were very important in determining the value of the habitats for estuarine-dependent fishery resources. Access to protected shallow waters and marsh edges is important to juvenile fish and shrimp, both in terms of refuge from predators and food supplies. Conceivably, some wetland alterations may actually enhance accessibility and, consequently, support of living resources. The effects on harvestable secondary productivity may be decidedly nonlinear, with some losses enhancing productivity initially, but resulting in dramatic reductions in productivity as deterioration proceeds.

Boesch and Turner (1984) caution, however, that knowledge of the relationship of fishery productivity and wetland conditions is embryonic and it is premature to pursue a strategy of creative wetland modification. When new dredge and fill projects are permitted, however, the options for design and mitigation should be evaluated in terms of the functional values to living resources. Furthermore, there remains the question of mitigation of existing alterations in wetlands modified by oil and gas development activities. As discussed earlier, many physical alterations have a long-term legacy of indirect effects on surrounding coastal environments. These indirect effects may be subject to some control using enlightened mitigative strategies. The functional relationship of living resources and affected coastal habitats is obviously an area where further research is needed for efficacious management and mitigative strategies.

Undeveloped Regions of the Gulf of Mexico

Physical alterations of coastal environments resulting from offshore oil and gas development in yet-undeveloped (frontier) regions are difficult to predict because of uncertainties about the development potential and intensity, location of support bases, and transportation strategies. The northern Gulf of Mexico, with its intensive offshore development over large areas (much of which took place during an era when there were few coastal management regulations) and its extensive coastal wetlands, represents a worst case for coastal impacts, at least in the United States.

Elsewhere in the Gulf of Mexico, offshore oil and gas resources may be developed off South Texas (where some development has already taken place), Alabama and Florida. Exploration and production will be supported from existing ports such as Corpus Christi, Texas; Pascagoula, Mississippi; Mobile, Alabama; and Pensacola, Panama City and Port Manatee, Florida (Havran *et al.*, 1982). Consequently, dredging or deepening of navigation channels should not be required. The availability of existing port facilities and the proximity of upland expansion sites to water should greatly reduce the pressures to fill wetlands to provide coastal supply and construction bases.

With regard to pipeline crossings, concern has been expressed for barrier islands and wetlands, including both salt marshes and mangroves (Havran *et al.*, 1982). The physical instability and migrating tendencies of coastal barriers could pose a hazard to pipelines crossing them. Most of the existing pipeline landfalls in Texas and several in Louisiana are on barrier islands. In addition to the risk of pipeline rupture, the placement of a pipeline across a barrier island may enhance washovers (Sasser *et al.*, 1983) and, possibly, even encourage breaches of narrow islands. In contrast to the Mississippi Deltaic and Chenier Plains, the coastal wetlands around the rest of the Gulf of Mexico are less extensive and discontinuous. Pipelines can be more easily routed to avoid permanent damage to these habitats and should be planned in this manner.

Atlantic Coast

Offshore oil and gas development off the Atlantic coast of the United States is anticipated only at sites well offshore. As a consequence, it is expected that exploration and production would be supported by a limited number of centralized coastal bases and that, should pipelines be laid to shore, there would be a few collective pipeline corridors crossing the coastal zone. Onshore support for exploratory drilling in the Middle Atlantic Bight and on Georges Bank was based at Davisville, Rhode Island (where ample port facilities exist), and Atlantic City, New Jersey (Macpherson and Bookman, 1980; Dorrier, 1981). Similarly, existing port facilities at Jacksonville, Florida; Savannah, Georgia; Charleston, South Carolina; and Wilmington and Moorehead City, North Carolina can be expected to accommodate the modest amount of onshore support facilities required for exploration and development activities in the South Atlantic Bight. Little or no coastal marine impact should result from new channel dredging or filling of wetlands and shallows.

Pipeline crossings of the coastal zone can be routed so as to avoid wetlands (except for some fringing marshes) along most of the Atlantic coast. Only where there is a broad and nearly continuous band of wetlands along the coast might there be the potential for pipeline effects of the type seen in Louisiana. Such conditions exist along the Sea Islands of Georgia and southern South Carolina and the Eastern Shore of Virginia. Should oil and gas production take place off these coastal segments, great care should be taken to avoid trenching pipelines across intertidal marshes.

Pacific Coast

Offshore oil and gas production has taken place off southern California since 1894. The early drilling and production was accomplished from piers extending into the ocean. In some cases wells have been located on artificial offshore islands. Pipelines may be placed on the piers themselves or are buried at the landfall. Because almost all of the coast of southern California consists of high energy beaches and rocky shores, buried pipelines traverse a short intertidal zone and are often covered by rocks to prevent exposure of the pipe. The marine environmental effects of piers and open shore pipeline landfalls are localized.

Offshore oil and gas development in the Southern California Bight and Santa Barbara Channel is serviced from well-established ports, some of which (e.g., Long Beach Harbor) were developed by filling wetlands or shallow water environments. These port developments were mainly driven by commercial interests other than oil and gas production, although some areas were filled to provide oil storage facilities. New offshore development, initially in the Santa Maria Basin area, north of Point Conception, and potentially farther to the north may utilize smaller fishing ports (e.g., Morro Bay and Port San Luis) for supply and operations bases. Although harbor dredging may be required to support these activities, the limited wetlands and strict protection policies suggest that no damage to wetlands would result.

Alaska

With a coastline extending nearly 11,000 km, Alaska has over one-half of the coastline of the United States. Coastal environments range widely in type, including sand beaches and barrier islands and lagoons, fjords, steep cliffs, tidal flats, river deltas and retreating tundra shorelines. In contrast to most of the rest of the U.S. coast, ice plays a major part in the coastal ecology and geomorphology of much of Alaska.

As with other regions yet to experience offshore oil and gas development, it is difficult to discuss coastal effects in other than hypothetical generalities. Only in Lower Cook Inlet has there been any oil and gas production. At this writing, only in the shallow Beaufort Sea have there been any new discoveries. Development in Lower Cook Inlet has resulted in little coastal impact (Jackson and Dorrier, 1980), but there is concern regarding onshore development which might accompany offshore production in more remote regions of the Gulf of Alaska. Collins and Stadnychenko (1981) discussed potential developments at Yakutat. Unlike the "lower 48 states" where oil and gas is generally transported via national pipeline networks, discoveries in the Gulf of Alaska or Bering Sea may require construction of marine terminals and liquid natural gas plants (for example at Yakutat) for transshipment of the oil and gas to markets by tanker.

Development of the St. George Basin, in the Bering Sea north of the Aleutian Island chain, would probably be serviced from St. Paul in the Pribilof Islands, Dutch Harbor in the Aleutian Islands or Cold Bay on the Alaska Peninsula (Minerals Management Service, 1982). Although these ports are relatively undeveloped, construction of support bases would not be expected to affect coastal marine environments. Oil may be transported by pipeline to either St. Paul, Makushin Bay in the Aleutian Islands or near Cold Bay, where it would then be processed and loaded on tankers. Although the pipeline landfalls would probably have minor impact, development of tanker terminals would be required.

The Norton Sound coast has extensive wetland environments associated with the Yukon River delta. Although not particularly rich in commercial fisheries, this area supports large numbers of migratory waterfowl and is important to migrating salmon. A Norton Sound synthesis meeting (Zimmerman, 1982) recommended "that because of geomorphology, substrate stability, and the critical resident wildfowl populations, no support, loading, storage, or transfer facilities should be permitted on the delta or within 60 km of its shore." The support facilities for Norton Sound oil development are expected to be located at the port of Nome, perhaps also relying on large supply barges in the areas of active drilling (Bureau of Land Management, 1982).

The region of Alaska for which coastal effects have been most discussed is the Beaufort Sea in the Arctic. This is because of the use of gravel islands to support drilling and production. If located close to shore, such islands may be connected to the mainland by causeways. Furthermore, causeways may be built to serve as supply docks for vessels operating offshore. As of 1985 there were 22 artificial gravel islands (four of them in Federal waters) located in the Alaskan Beaufort Sea in water depths from 1 to 13 m (Collins and Lynch, 1986). Minerals Management Service (1983a) estimates that four causeways may be constructed in development in the Diapir Field. The causeways alter nearshore water currents, and there is concern that they will interfere with longshore migration of anadromous fish. Studies of the Prudhoe Bay causeway showed that a deflection of the longshore current altered temperature and salinity around the causeway (Mungall, 1978; Bendock, 1979; Robilliard and Colonell, 1983). A model

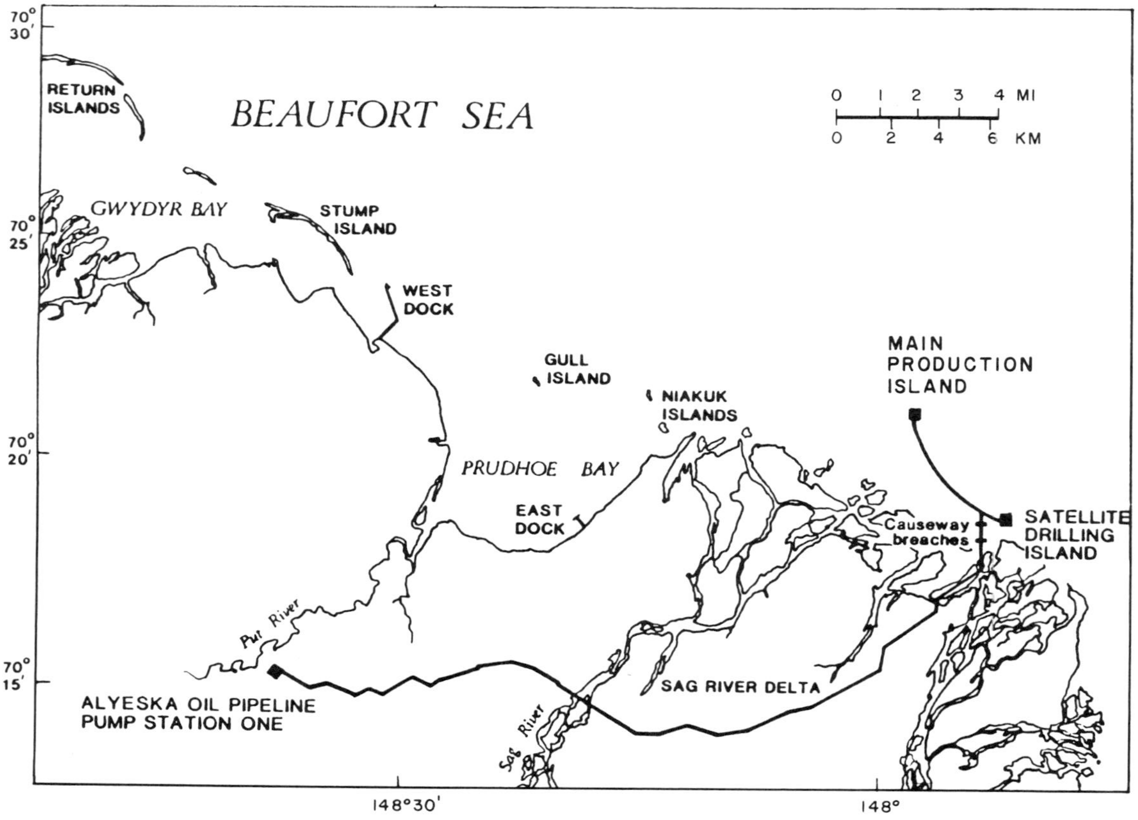

Figure 13.6. Existing causeways (West Dock and East Dock) in the Prudhoe Bay region of Alaskan Beaufort Sea and the planned production islands and connecting causeways of the Endicott field off the Sag River delta (modified from Collins and Lynch, 1986).

developed by Neill *et al.* (1982) suggested that the observed changes in salinity and temperature could result in a reduction in density of the arctic cisco (*Coregonus autumnalis*) based on its observed temperature and salinity preferences. Moulton *et al.* (1985 and pers. comm.) have also shown that the Prudhoe Bay West Dock does affect movements of small fish, primarily Arctic cisco, possibly affecting their availability to local subsistence fisheries. Although it seems that causeways could have significant local effects, it is questionable whether fish populations of the entire region would be affected by the few causeways planned, except perhaps where they may block migration of anadromous fishes to riverine spawning grounds. In this regard, there have been major controversies about the effects of causeways planned for the Endicott field development off the Sag River delta on fish migration and the distribution of salinity and temperature (Collins and Lynch, 1986). As a consequence, the inclusion of two breachways in the causeways (Figure 13.6) and the initiation of a long-term environmental monitoring plan were made as conditions of the causeway construction permit.

Finally, pipeline crossings through the shore zone may pose some environmental risks. First, there is the problem of ice push and ice override on the shoreline rupturing a pipeline and resulting in an oil spill. Second, the warm oil running through the pipe may cause a thaw of the surrounding permafrost, both in subtidal sediments and at the shore face. This could cause pipeline failure because of a weakened foundation and may also result in accelerated shoreline erosion locally. Third, trenching the pipeline across the shore may leave a long-lasting "scar." Although the intertidal zone is not broad (there are no marshes as such), it is heavily ice scoured and in many places rapidly retreating due to natural erosion of the tundra (Reimnitz and Maurer, 1979; Owens and Harper, 1983).

CONCLUSIONS AND RECOMMENDATIONS

Of the many effects on offshore and coastal habitats resulting from oil and gas related physical modifications which have been discussed in this perspective, many are minor in spatial extent or in duration. Others, while potentially serious, can be avoided by wise environmental planning, for example, by avoiding coastal wetlands where possible in pipeline routing, using existing harbors and navigation channels, and seeking alternate approaches to filling marine habitats for support bases. The long-term effects which remain are either worsening legacies of past developmental activities (the effects of navigation channels and pipeline crossings on Louisiana wetlands) or are economically unavoidable in the extraction of energy resources from the environment in question (gravel island and causeway construction, new pipelines across areas with nearly continuous coastal wetlands, and the effects of aggregation around platforms on fish populations).

The most extensive, lasting, and potentially deleterious of these effects are the effects on coastal wetlands in the northern Gulf of Mexico. This area has been and will continue to be (at least over the next 10 years) the site of the majority of the offshore oil and gas development in the United States (Chapter 1). New developments, including pipelines and coastal facilities (Havran *et al.*,

1982; Wiese *et al.*, 1983) continue to affect this area and future projects which facilitate offshore oil and gas development, such as the Avoca Island levee extension, may have considerable coastal environment effects. In addition, there is the question of what can be done to mitigate the compounding effects of past activities. There is little technical consensus on the efficacy of many of the proposed mitigative approaches: backfilling canals, leveling spoil banks, levees, water control structures, locks on navigation canals and freshwater diversions. Research is needed, not simply aimed at apportioning the causes of wetland deterioration among the many natural and human factors involved, but in establishing a technical basis to manage future oil and gas development activities and the environmental legacy of past activities. In this regard, process-oriented research is needed to provide requisite understanding of the hydrology, sedimentology and natural resource utilization in the wetland-estuarine complex. We need to understand the ramifications of coastal habitat modifications on the long-term continuity of the valued living resources. The results of this research would not only be applicable to the northern Gulf of Mexico (where offshore development is heavily concentrated) but also to other wetland ecosystems potentially affected by oil and gas development. These include coastal regions in the South Atlantic Bight and many other regions of the world experiencing similar development.

Of somewhat lower priority for research is that on the effects of gravel islands and causeways in arctic environments and on the effects of offshore structures on fish stocks. Under likely development scenarios, both the effects of gravel islands and causeways and the living resources at risk by these developments appear modest. In addition, this issue has been the subject of much recent, and as yet unpublished, research and ongoing monitoring. An objective appraisal of results and evaluation of predictive models seems in order before embarking on an expanded research agenda.

Finally, although there is little question that offshore structures are beneficial to fishermen (at least to anglers and spear fishermen), there are few data which relate to their benefits to the fish. For those species which frequent platforms but for which population levels are limited by survival in other environments, population models and the data to support them are required to evaluate the influence of rig-associated harvest on the stocks. Such information becomes particularly important in weighing options for the disposition of obsolete platforms (whether they should be completely removed, as is now required, left partially in place, or deposited as fishing reefs).

LITERATURE CITED

Bascom, W., A.J. Mearns and M.D. Moore. 1976. A biological survey of oil platforms in the Santa Barbara Channel. Pages 27-35 *in* Proceedings 8th Annual Offshore Technology Conference, Volume 2. Houston, Texas.

Baumann, R.H. and R. Adams. 1982. The creation and restoration of wetlands by natural processes in the lower Atchafalaya River system: Possible conflicts with navigation and flood control objectives. Pages 8-24 *in* R.H. Stovall (ed.), Proceedings Annual

Conference on Wetlands Restoration and Creation, Volume 8. Hillsborough Community College, Tampa, Florida.

Bendock, T.N. 1979. Beaufort Sea estuarine fish study. Pages 670-729 *in* Environmental Assessment of the Alaskan Continental Shelf. Final Reports of Principal Investigators, Research Unit 233. National Oceanic and Atmospheric Administration, Outer Continental Shelf Environmental Assessment Program, Boulder, Colorado.

Boesch, D.F. (ed.). 1982. Proceedings of the Conference on Coastal Erosion and Wetland Modification in Louisiana: Causes, Consequences and Options. U.S. Fish and Wildlife Service, Biological Services Program, Washington, D.C., FWS/OBS-82/59, 256 p.

Boesch, D.F. and R. Rosenberg. 1981. Response to stress in marine benthic communities. Pages 179-200 *in* G.W. Barrett and R. Rosenberg (eds.), Stress Effects on Natural Ecosystems. John Wiley & Sons, New York.

Boesch, D.F. and R.E. Turner. 1984. Dependence of fisheries on salt marshes: The role of food and refuge. Estuaries 7: 460-468.

Boesch, D.F., J.W. Day, Jr. and R.E. Turner. 1984. Deterioration of coastal environments in the Mississippi Deltaic Plain. Pages 447-466 *in* V.S. Kennedy (ed.), The Estuary as a Filter. Academic Press, Orlando, Florida.

Boesch, D.F., D. Levin, D. Nummedal and K. Bowles. 1983. Subsidence in Coastal Louisiana: Causes, Rates and Effects on Wetlands. U.S. Fish and Wildlife Service, Division of Biological Services Program, Washington, D.C., FWS/OBS-83/26, 30 p.

Bureau of Land Management. 1982. Norton Sound Final Environmental Impact Statement, OCS Proposed Oil & Gas Lease Sale 57. U.S. Department of the Interior, Bureau of Land Management, Anchorage, Alaska, 332 p.

Carlisle, J.G., Jr., C.H. Turner and E. Ebert. 1964. Artificial Habitat in the Marine Environment. California Dept. Fish and Game Bull. 124, 93 p.

Castle, R.O., R.F. Yerkes and F.S. Riley. 1970. A linear relationship between liquid production and oil-field subsidence. Pages 162-173 *in* Land Subsidence. International Association of Scientific Hydrology, Gentbrugge, Belgium.

Collins, J.H. and C.W. Lynch, 1986. Alaska Summary Report, June 1984 - December 1985. OCS Information Report MMS 86-0023. Minerals Management Service, U.S. Department of the Interior, Vienna, Virginia, 114 p.

Collins, K.M. and A. Stadnychenko. 1981. Gulf of Alaska and Lower Cook Inlet Summary Report 2. Open-File Rep. 81-607. U.S. Geological Survey, Reston, Virginia, 49 p.

Craig, N.J., R.E. Turner and J.W. Day. 1979. Land loss in coastal Louisiana (USA). Environ. Manage. 3: 133-144.

Ditton, R.B. and J. Auyong. 1984. Fishing Offshore Platforms Central Gulf of Mexico—An Analysis of Recreational and Commercial Fishing Use at 164 Major Offshore Petroleum Structures. OCS Monograph MMS 84-0006, U.S. Department of the Interior, Minerals Management Service, Metairie, Louisiana, 158 p.

Dorrier, R.T. 1981. North Atlantic Summary Report. Open-File Report 81-601, U.S. Geological Survey, Reston, Virginia, 65 p.

Essertier, E.P. 1984. Federal Offshore Statistics. OCS Report MMS84-0071. U.S. Department of the Interior, Minerals Management Service, Washington, D.C., 123 p.

Fish and Wildlife Service. 1981. Atchafalaya Basin Reports. Avoca Island Levee Extension and Water Management and Land Use Controls. U.S. Fish and Wildlife Service, Lafayette, Louisiana.

Gagliano, S.M., K.J. Meyer-Arendt and K.M. Wicker. 1981. Land loss in the Mississippi River Deltaic Plain. Trans. Gulf Coast Assoc. Geol. Soc. 31: 295-300.

Gallaway, B.J. and G.S. Lewbel. 1982. The Ecology of Petroleum Platforms in the Northwestern Gulf of Mexico: A Community Profile. U.S. Fish and Wildlife Service, Office of Biological Services, Washington, D.C., FWS/OBS-82/27. Bureau of Land Management, Metairie, Louisiana, Open-File Report 82-03, 92 p.

Gallaway, B.J., L.R. Martin, R.L. Howard, G.S. Boland and G.D. Dennis. 1981. Effects on artificial reef and demersal fish and macrocrustacean communities. Pages 237-299 *in* B.S. Middleditch (ed.), Environmental Effects of Offshore Oil Production. The Buccaneer Gas and Oil Field Study. Plenum Press, New York.

Gosselink, J.G., C.L. Cordes and J.W. Parsons. 1979. An Ecological Characterization of the Chenier Plain Coastal Ecosystem of Louisiana and Texas. U.S. Fish and Wildlife Service, Office of Biological Services, Washington, D.C., FWS/OBS-78/9.

Gunter, G. and R.A. Geyer. 1955. Studies of the fouling organisms of the northwestern Gulf of Mexico. Publ. Inst. Mar. Sci. Univ. Texas 4: 37-67.

Harper, D.E., Jr., D.L. Potts, R.R. Salzer, R.J. Case, R.L. Jaschek and C.M. Walker. 1981. Distribution and abundance of macrobenthic and meiobenthic organisms. Pages 133-177 *in* B.S. Middleditch (ed.), Environmental Effects of Offshore Oil Production. The Buccaneer Gas and Oil Field Study. Plenum Press, New York.

Havran, K.J., J.D. Wiese, K.M. Collins and F.N. Kurz. 1982. Gulf of Mexico Summary Report 3. Open-File Report 82-242. U.S. Geological Survey, Reston, Virginia, 99 p.

Jackson, J.B. and R.T. Dorrier. 1980. Outer Continental Shelf Oil and Gas Activities in the Gulf of Alaska (Including Lower Cook Inlet) and Their Onshore Impacts: A Summary Report. Open-File Report 80-1028, U.S. Geological Survey, Reston, Virginia, 78 p.

Jackson, J.B., B.F. Golden, A. Stadnychenko and S. Kolasinski. 1981. Arctic Summary Report. Open-File Report 81-621. U.S. Geological Survey, Reston, Virginia, 137 p.

Johnson, W.B. and J.G. Gosselink. 1982. Wetland loss directly associated with canal dredging in the Louisiana coastal zone. Pages 60-72 *in* D.F. Boesch (ed.), Proceedings of the Conference on Coastal Erosion and Wetland Modification in Louisiana: Causes, Consequences and Options. U.S. Fish and Wildlife Service, Biological Services Program, Washington, D.C., FWS/OBS-82/59.

Macpherson, G.S. and C.A. Bookman. 1980. Outer Continental Shelf Oil and Gas Activities in the Mid-Atlantic and Their Onshore Impacts: A Summary Report, November 1979. Open-File Report 80-17, U.S. Geological Survey, Reston, Virginia, 63 p.

Mendelssohn, I.A., K.L. McKee and W.H. Patrick, Jr. 1981. Oxygen deficiency in *Spartina alterniflora* roots: Metabolic adaptation to anoxia. Science 214: 439-441.

Minerals Management Service. 1982. Final Environmental Impact Statement, Proposed Outer Continental Shelf Oil and Gas Lease Sale, St. George Basin. U.S. Department of the Interior, Minerals Management Service, Anchorage, Alaska.

Minerals Management Service. 1983a. Final Environmental Impact Statement, Proposed Diapir Field Lease Offering. U.S. Department of the Interior, Minerals Management Service, Anchorage, Alaska.

Minerals Management Service. 1983b. Regional Environmental Assessment/Gulf of Mexico Pipeline Activities. U.S. Department of the Interior, Minerals Management Service, Metairie, Louisiana, 195 p.

Minerals Management Service. 1983c. Final Environmental Impact Statement, Proposed OCS Oil and Gas Lease Offerings, Central Gulf of Mexico (April 1984), Western Gulf of Mexico (July, 1984). U.S. Department of the Interior, Minerals Management Service, Metairie, Louisiana, 474 p.

Minerals Management Service. 1984. Final Environmental Impact Statement, Proposed Oil and Gas Lease Sales 94, 98 and 102, Gulf of Mexico Region. U.S. Department of the Interior, Minerals Management Service, Metairie, Louisiana.

Moulton, L. L., B. J. Gallaway, M. H. Fawcett, W. D. Griffiths, K. R. Critchlow, R. G. Fechhelm, D. R. Schmidt and J. S. Baker. 1985. 1984 Central Beaufort Sea Fish Study: Waterflood Monitoring Program. Woodward-Clyde Consultants and Entrix, Inc., Anchorage, Alaska.

Mungall, C. 1978. Oceanographic processes in a Beaufort Sea barrier island-lagoon system: Numerical modeling and current measurements. Pages 732-830 *in* Environmental Assessment of the Alaskan Continental Shelf, Annual Reports of Principal Investigators, Vol. 10. National Oceanic and Atmospheric Administration, Outer Continental Shelf Environmental Assessment Program, Boulder, Colorado.

National Research Council. 1983. Drilling Discharges in the Marine Environment. National Academy Press, Washington, D.C., 180 p.

Neill, W.H., R. G. Fechhelm, B. J. Gallaway, J. D. Bryan and S. W. Anderson. 1982. Modeling movements and distribution of arctic cisco (*Coregonus autumnalis*) relative to temperature/salinity regimes of the Beaufort Sea near the Waterflood Causeway, Prudhoe Bay, Alaska. Pages 39-61 *in* University of Alaska, Biological Papers 21, Fairbanks, Alaska.

Owens, E.H. and J.R. Harper. 1983. Arctic coastal processes: A state of knowledge review. Pages 3-18 *in* Proceedings of Canadian Coastal Conference, May 11, 1983. National Research Council, Ottawa, Ontario.

Reimnitz, E. and D.K. Maurer. 1979. Effects of storm surges on the Beaufort Sea coast, northern Alaska. Arctic 32: 329-344.

Rhoads, D.C., P.L. McCall and J.L. Yingst. 1978. The ecology of seafloor disturbance. Amer. Sci. 66: 577-586.

Robilliard, G.A. and J.M. Colonell. 1983. Ecological impacts of a 4-km causeway at Prudhoe Bay, Alaska: How could government and industry benefit? Pages 895-899 *in* Proceedings Oceans '83 Conference. Marine Technology Society, Washington, D.C.

Sasser, C.E., G.W. Peterson, R.K. Abernathy and J.G. Gosselink. 1983. LOOP Inc. Environmental Monitoring Program, Louisiana Offshore Oil Port Pipeline, 1982 Annual Report. LSU/CEL-83-10. Louisiana State University, Center for Wetland Resources, Baton Rouge, Louisiana, 231 p.

Scaife, W., R.E. Turner and R. Costanza. 1983. Indirect impact of canals on recent coastal land loss rates in Louisiana. Environ. Manage. 7: 433-442.

Shinn, E.A. 1974. Oil structures as artificial reefs. Pages 91-96 *in* L. Colunga and R. Stone (eds.), Proceedings of an International Conference on Artificial Reefs. Center for Marine Resources, Texas A&M University, College Station, Texas.

Simpson, R.A. 1977. The Biology of Two Offshore Oil Platforms. Institute of Marine Resources, University of California, IMR Ref. 76-13.

Sonnier, F., J. Teerling and H.D. Hoese. 1976. Observations on the offshore reef and platform fish fauna of Louisiana. Copeia 1976: 105-111.

Turner, R.E., R. Costanza and W. Scaife. 1982. Canals and wetland erosion rates in coastal Louisiana. Pages 73-84 *in* D.F. Boesch (ed.), Proceedings of the Conference on Coastal Erosion and Wetland Modification in Louisiana: Causes, Consequences and Options. U.S. Fish and Wildlife Service, Biological Services Program, Washington, D.C., FWS/OBS-82/59.

Turner, R.E. In press. Coastal Land Loss, Canals, and Canal Levee Relations in Louisiana. U.S Fish and Wildlife Service, Biological Services Program, Washington D.C.

van der Knapp, W. and A.C. van der Vlis. 1967. On the cause of subsidence in oil-producing areas. Pages 85-105 *in* Rock Mechanics and Oilfield Geology, Drilling and Production. 7th World Petroleum Congress, Mexico City.

van Heerden, I.L. and H.H. Roberts. 1980. The Atchafalaya Delta: Rapid progradation along a traditionally retreating coast (south-central Louisiana). Zeitschrift fur Geomorphologie, Suppl. 34: 188-416.

Wicker, K.M. 1980. Mississippi Deltaic Plain Regional Ecological Characterization: A Habitat Mapping Study. A User's Guide to the Habitat Maps. U.S. Fish and Wildlife Service, Office of Biological Services, Washington, D.C., FWS/OBS-79/07, 84 p.

Wiese, J.D., D.L. Slitor and C.A. McCord. 1983. Gulf of Mexico Summary Report. U.S. Department of the Interior, Minerals Management Service, Reston, Virginia, 106 p.

Wolfson, A., G. Van Blaricom, N. Davis and G.S. Lewbel. 1979. The marine life of an offshore oil platform. Mar. Ecol. Prog. Ser. 1: 81-89.

Yerkes, R.F. and R.O. Castle. 1970. Surface deformation associated with oil and gas field operations in the United States. Pages 55-65 *in* Land Subsidence. International Association of Scientific Hydrology, Gentbrugge, Belgium.

Zimmerman, S.T. (ed.). 1982. Proceedings of a Synthesis Meeting: The Norton Sound Environment and Possible Consequences of Planned Oil and Gas Development. National Oceanic and Atmospheric Administration, Office of Marine Pollution Assessment, Juneau, Alaska, 55 p.

Zingula, R.P. 1975. Effects of drilling operations in the marine environment. Pages 433-488 *in* Environmental Aspects of Chemical Use in Well Drilling Operations. EPA-560/1-75-004. U.S. Environmental Protection Agency, Washington, D.C.

CHAPTER 14

A REVIEW OF STUDY DESIGNS FOR THE DETECTION OF LONG-TERM ENVIRONMENTAL EFFECTS OF OFFSHORE PETROLEUM ACTIVITIES

Robert S. Carney

CONTENTS

INTRODUCTION

In environmental studies the word design is often used in a broad sense encompassing the full range of activities undertaken and schedules followed during the course of a particular project. In effect, design has been taken to mean "course of action." In this review a definition more associated with basic research will be employed. Design is the plan by which a goal will be achieved and includes an unambiguous statement of that goal, a sampling scheme for the collection of appropriate data and an objective method of drawing conclusions from the examination of those data. When reviewing designs, we are most greatly concerned with the connection between original purpose and final conclusion.

In order to remain tractable, this review must have a relatively limited focus. Of all the possible studies that might have been considered, only five in United States territorial waters are included. The benthic fauna is the only component of the continental shelf community which is discussed, and technical procedures are only rarely mentioned. The lack of discussion of sampling techniques should not be taken to mean that this is an unworthy topic. Rather, it reflects the relative urgency of the problems encountered in long-term effects studies. We are now fully capable of going onto the continental shelf, of taking good samples, and of carrying out analyses that range from mundane to remarkable. We are not, however, especially adept at weaving the facts provided by our technology into the fabric of understanding.

It must be stressed from the beginning that the single greatest design problem of long-term effects studies is not simply one of misapplication of statistical analysis. While techniques of design have been poorly utilized, far more serious and harder to deal with is a pervasive lack of understanding of continental shelf ecology. Our present empirical understanding does not allow us to identify an ideal design, and it is not clear if contemporary theoretical ecology provides a practical guide.

When the first long-term effects study in the continental shelf region was planned more than a decade ago, the great difficulty of detecting an impact in a highly variable and poorly understood environment was simply not appreciated. The failure to anticipate natural variation resulted in the adoption of designs that were simple extensions of baseline surveying, the uninformed use of analyses and virtually no dependence upon hypothesis testing. Subsequent studies have shared this common problem to some degree. Due, at least in part, to design weakness, they were only able to detect gross faunal changes within 100 m of offshore platforms. More subtle changes across wider areas, if they did occur, could not be distinguished from natural variation by the designs employed. Left with little substantive evidence, final conclusions tended to invoke mechanisms that had not been studied and rested too heavily upon opinions.

Comments on Points of View

Impact studies cannot be criticized without acknowledgment of a debate as to the best way to do ecology (for a typical exchange, see Roughgarden, 1983; Simberloff, 1983). Central to the argument is the relative merit of imagination allowed to move freely between natural history and theorization versus imagination restrained by formal criteria of proof. When very little is known about the function and structure of an ecosystem, is it ultimately more informative to undertake exploratory studies without prior hypotheses hoping for insightful explanations, or is more gained through a formal approach designed around prior hypotheses?

Unfortunately, the survey-and-explain approach has dominated offshore long-term effects studies. This might be defended on grounds that restrictive formal approaches delay insight, place undue emphasis on analytical detail and are ultimately unproductive. Even in the light of such charges, formal testing has two compensating features that argue strongly for its adoption. First, when a study fails it is possible to isolate the design flaw and attempt improvements. Second, it is possible to estimate the probability of error and to consider that likelihood (large or small) when regulations are issued and enforced.

Throughout this chapter there is liberal reference to Green's (1979) *Sampling Design and Statistical Methods for Environmental Biologists*. Of many titles on ecological data analysis it is distinguished by its lucidness and its advocacy of hypothesis testing and careful design. Many field biologists will find this reference more immediately understandable than the fundamental account by Cochran (1963). It must be remembered, however, that neither Green (1979) nor this chapter are handbooks for long-term impact studies. As noted by Kempton (1981), Green's book (and this chapter) is about the statistical method not about statistical methods.

Organization

The chapter is divided into three major parts: a review, a discussion of design problems and conclusions. Five studies are reviewed individually with emphasis upon five features which reveal major design strengths and limitations: the spatial layout of sampling, the timing and replication of sampling across those sites, the coordination of biological and environmental sampling, major analytical methods employed and conclusions presented. The first three studies were long-term impact studies in the strictest sense. The last two studies, although dealing with short-term impact, were included in the review to illustrate specific points about design.

The discussion sections are directed towards major unresolved design problems. Appropriate operational definitions of impact are considered. The difficulties of finding a significant impact are explored through reference to the general linear model. With attention focused upon dealing with natural variation, the spatial and temporal layout of sampling are discussed. The discussion concludes with consideration of alternate ways of looking at data.

The conclusion section calls for two courses of action. The first recognizes long-term impact as an immediate problem and outlines a course of design based upon best current technologies. The second anticipates that the current best will not be good enough and calls for innovation and research aimed specifically at understanding natural patterns of variation in continental shelf environments.

FIVE EXERCISES

In keeping with contractual requirements, large environmental studies produce numerous reports detailing progress. Many of these "gray" references are hard to find and most are not especially informative. This critique is based on the most comprehensive summary accounts. This approach may seem to omit important information, but it is really the summaries that are most influential and must withstand critical evaluation. For brevity, major references will be introduced in this section and seldom cited again. Spies (Chapter 9) also makes reference to some of these same studies and should be consulted for additional details.

Offshore Ecology Investigation

The Gulf Universities Research Consortium (GURC) Offshore Ecology Investigation (OEI) was the first large investigation of long-term effects in a region of heavy development. The project was specifically concerned with the cumulative ecological effects of prolonged oil and gas activity in a coastal and offshore region. The Louisiana coast was selected because of the history of intense development producing more than 1500 platforms over a 25-year period. The region included all phases of petroleum activity from drilling through production and piping ashore and seemed to pose a worst case situation. Sampling was conducted between 1972 and 1974. A summary report was issued by Menzies *et al.* (1979) (for citations of component reports, see Chapter 9). Due to its pioneer status and some obvious weaknesses, OEI-GURC has already been extensively criticized.

Geography of Sampling

Stations were positioned to provide comparison between "control" or ambient sites (no exposure to petroleum activity) and exposed sites. Locations in Timbalier Bay and adjacent offshore areas were selected for study. The region met the criteria of intense petroleum activity with minor influence of the Mississippi River relative to other potential sites. Additional stations were scattered throughout the region to provide a basis for the assessment of general "ecological health."

Direct comparisons were possible between two sets of platform stations in the bay and nearby, "upstream" controls. Offshore there was another platform and control pair along with assorted sampling sites (Figure 14.1). In addition to the direct comparisons, five transects were established offshore. The most western was considered to be part of the bay study, while the remainder were general survey transects lying along the route to the logistics base.

Replication

The benthic ecology of the OEI-GURC region was investigated by several

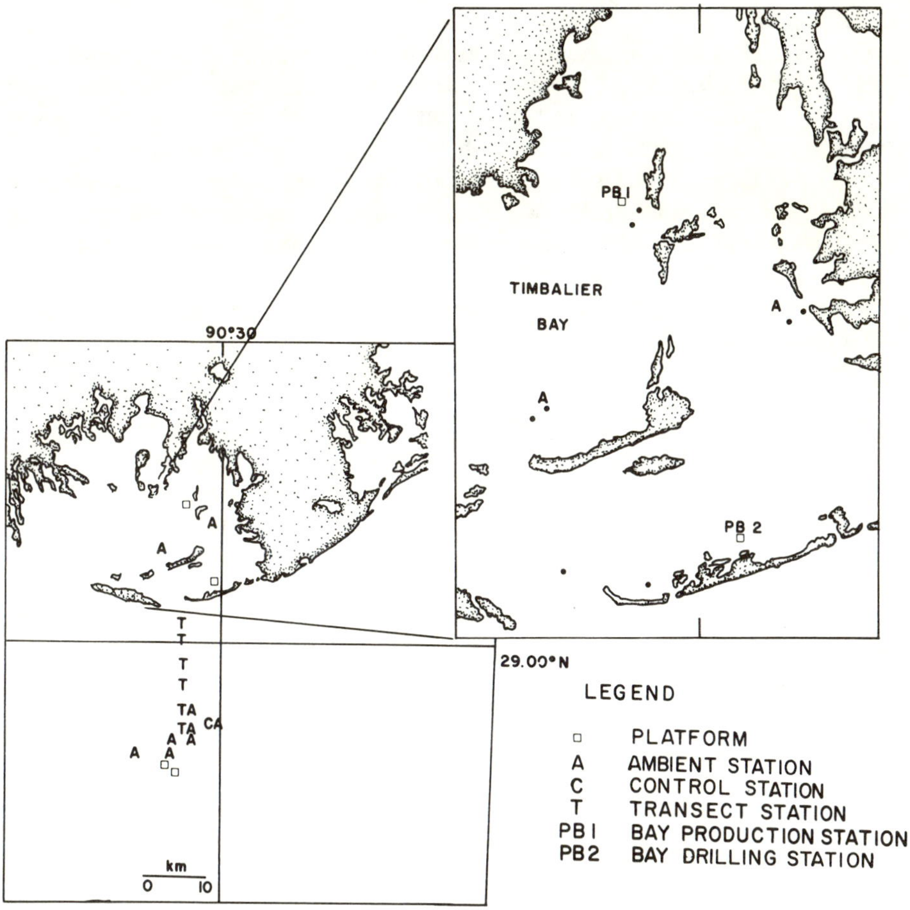

Figure 14.1. Location of stations in the OEI-GURC study in Timbalier Bay and offshore areas of Louisiana. The stations were scattered through the study area in an attempt to characterize the ecology of the region and to compare sites near production activity and in presumedly unaffected, ambient areas.

projects which are not directly comparable due to different methodologies (Farrell, 1974; Fish *et al.*, 1974; Griffin and Ripy, 1974; Humm, 1974; Kritzler, 1974; Ostrom, 1974; Perry, 1974). From these studies, Bender *et al.* (1979) selected the polychaete, crustacean, and molluscan data for synthesis and reexamination. Polychaetes were collected by a corer and air-lift system and removed from the sediment by filtration through a 0.5-mm sieve. In the bay study 16 replicates (total of 1-m^2 area, 10-cm deep) were collected at two stations near platforms and two in control areas during five periods in 1973. A Van Veen grab (2000 cm^2) was used to collect molluscs and crustaceans at six stations, three times during 1973. Apparently no replicates were taken, and the results were simply multiplied by five to estimate the catch for one square meter. The mollusc-crustacean sampling coincided with polychaete sampling on three cruises.

Coordination

OEI-GURC was a poorly coordinated project in execution. Those components gathering biological data and those measuring various environmental factors, especially hydrocarbons, seem to have proceeded with little interaction. As a result, direct comparison of faunal with environmental data was rarely possible. A major criticism of the OEI-GURC project was that it was not possible to determine if the faunal samples were actually from exposed or unexposed areas. In addition, the biological studies were over-subdivided among investigators, making it difficult to assemble a picture of the benthic community. In fact, because no investigator was apparently assigned echinoderms, the presently most obvious infaunal organism in Timbalier Bay, an ophiuroid (D.F. Boesch, pers. comm.), was not mentioned in the reports.

Analyses Undertaken

The projects of OEI-GURC did not employ a single strategy of analysis. In the individual project reports, qualitative assessment played an important role, and the great mass of data collected hindered a coordinated use of data processing. Subsequently, Bender *et al.* (1979) did present a reanalysis; however, even it was rather superficial. General faunal associations were determined with the Bray-Curtis index and group mean cluster analysis. In addition, location groups were structured and then compared with the similarity indices of Morisita and Ono. No attempt to make a more formal comparison of treatments (platform versus control), season, or replication was undertaken. Diversity, as expressed with H', seems to have been the parameter most used to assess ecological "health."

Conclusions

The results of the OEI-GURC have been cited as evidence of two opposing conclusions. First, the final report of the OEI-GURC project clearly made the point that failure to manage high and confounding natural variation made it impossible to make a definitive statement about the existence of an effect of petroleum activity (Bender *et al.*, 1979). However, in spite of the fact that OEI-GURC was an equivocal study, both Oppenheimer *et al.* (1979) and Mertens (1978) concluded that there was no effect, and more recently this claim has been repeated (Sharp and Appan, 1982). This recent reassertion of no impact is based upon a reexamination of data. However, since the design does not allow for the most meaningful comparisons to be made, no amount of multivariate resurrection can produce unequivocal results.

Major Failures

All of the OEI-GURC project's many failings come from four separate origins. First, an elaborate field effort was initiated without sufficient prior information. Second, the central goals were too vague to produce a coordinated effort. Third, in spite of concern for overwhelming natural variation, there was no serious effort to assess and then accommodate that variation in the design. Fourth, opinions that there were no long-term effects were presented as consensus findings while the data were equivocal.

Due to the study's pioneer status and the fact that its equivocal results were sometimes offered as proof of "no effect" (see Oppenheimer *et al.*, 1979; Mertens, 1978), it came under careful scrutiny. Of several critical points made by Sanders (1981), five (here rephrased) are of general interest:

1. If a comparison is to be made between an area that is exposed to petroleum activity and one that is not, then it is necessary to prove the lack of exposure in the "control." To the contrary in the OEI-GURC case, both exposed and control may have been within the affected region. In practice, a totally unaffected location need not be available, if it is possible to find sites for which differences in degree of exposure can be proven (assuming some simple dose-response model).

2. The study failed to give sufficient attention to sediment-bound hydrocarbons. As such it was not possible to carefully examine the relationship between sampled biota and the hydrocarbons to which these organisms were exposed. Failure to collect environmental data critical to interpretation concurrent with biological samples is a repeated problem in other studies.

3. Replicates, time series and spatial series were pooled when parameters such as diversity were estimated. This has the effect of hiding high levels of variance and giving the false impression that comparisons produce meaningful results. This is equally true for simple statistics such as counts, as it is for diversity estimates.

4. There was insufficient sampling. This relates to the problem of variance in point 3. If it is possible to repeatedly sample the same population of animals, then variance could have been reduced through collection of more samples. In practice, additional samples might have led to sampling across an increasing range of habitats with an increase in variance due to the heterogeneity.

5. The Sanders critique also argues that the observed assemblages were not indicative of a "healthy" situation. It can be argued that "ecosystem health" is an ambiguous concept. While it may assume a useful form for some ecologists, the gap between its conceptual bases and an applicable formulation is extremely wide. The OEI-GURC project never sought to produce an operable definition of long-term impact and undertook a study with poorly defined goals as a result.

Central Gulf Platform Study

During 1978-1979 this project investigated a number of sites on the south-eastern Louisiana shelf, including some areas sampled during the previous OEI-GURC study. Although there was a considerable effort expended on faunal surveys, the program stressed the chemical aspects of fate and effects. The biological results were given in the form of a descriptive account with only vague conclusions concerning impact. The principal reference is the report edited by Bedinger *et al.* (1981).

This study combined different design approaches, but was similar to OEI-GURC in that direct comparisons would be made between platform areas and controls, and a general picture would be developed from additional "secondary" sites. Recognizing the problem of heterogeneity, four different environments were compared separately. As with OEI-GURC, no operative definition of impact was presented, the goals were poorly specified and no prior hypotheses were stated.

Geography of Sampling

Of special interest is the use of a circular sampling pattern (bullseye) around each platform (Figure 14.2). In this way, the Central Gulf study made a better effort to assess the geometry of any impact that might be found near a platform. Each control was intended to be similar to a primary platform environment with the exception of less exposure to petroleum activity. A pristine control was considered impossible. Sampling stations were established at fixed distances (500 and 2000 m) and fixed directions (N,E,S,W) from each of four primary platforms. Each companion control site was sampled as a single station. A single N-S transect (stations at 500 and 2000 m) was established for each secondary site. A Smith-McIntyre grab was the primary sampling device for the infauna, and an otter trawl, for the epifauna.

Replication

There was a rather complicated protocol for the handling of grab replicates. A series of ten sequential grabs was taken at each station. Five subcores were taken from the first four grabs to be used for meiofaunal analysis (four analyzed, one archived). The meiofauna was sieved out and then split with a plankton splitter to one-fourth the volume to simplify counting. Six grabs were used for macrofaunal analysis (grabs 5 through 10). A single trawl at one station was taken for megafaunal analysis. Sampling at the primary stations was repeated on three cruises to assess seasonal variation. At total of 560 meiofaunal, 840 macrofaunal, and 40 trawl samples were collected.

In final analysis the replicates were not used to estimate variance within a station, but were taken to assure "representativeness" of samples (Baker *et al.*, 1981). In some instances they were pooled, much as with OEI-GURC, and at other times averages were taken without any mention as to variation.

Coordination

The collecting of faunistic and environmental data was well-coordinated. Of special importance is the fact that samples for sediment texture, sediment chemistry and biota were collected simultaneously. As a result, direct comparison was possible.

Analyses Attempted

The study conducted and presented a bewildering number of analyses following three approaches: (1) inferring process from patterns seen in plots of various summary statistics (diversity, evenness, etc.); (2) using cluster analysis to look for obvious differences in the faunal composition; and (3) using correlations to look for significant relationships between fauna and the physical environment.

The main analytical tool by which controls and primary sites were compared was hierarchical cluster analysis. The project performed a rather straight-forward series of cluster analyses, but produced confusing results due to computational

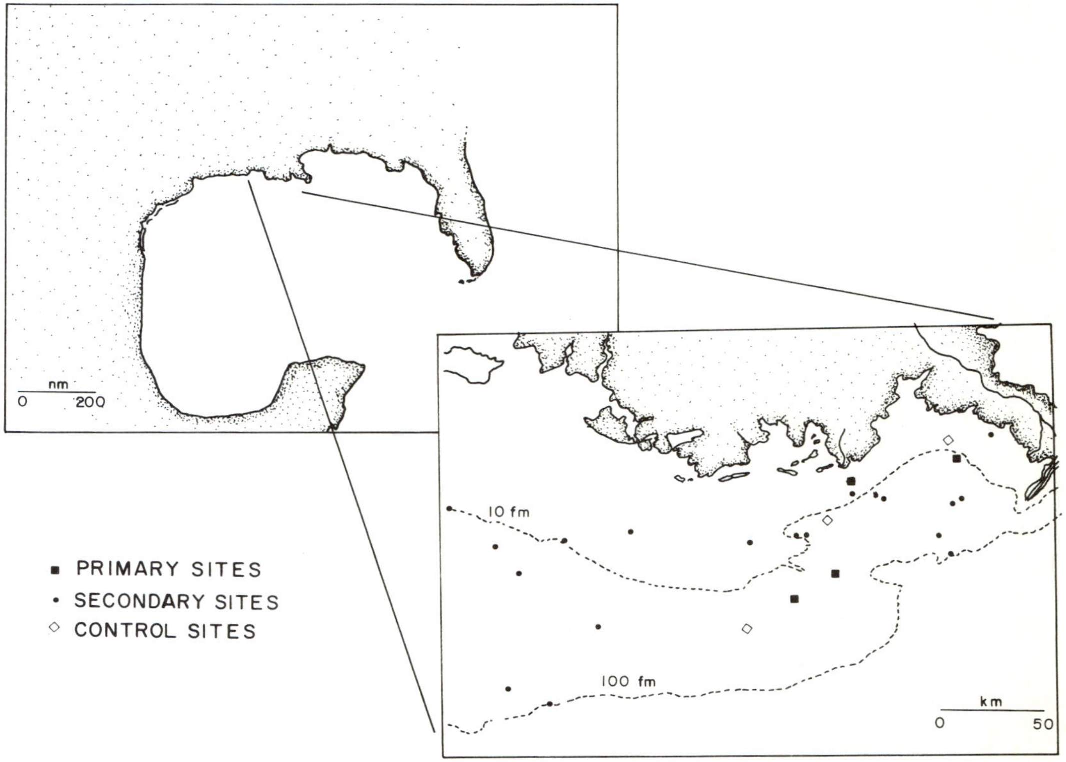

Figure 14.2. Location of stations in the Central Gulf Platform Study off Louisiana. As with the OEI-GURC study, stations were scattered through a wide area with some intended for regions of petroleum activity and some for unaffected regions. Additional secondary sites served to characterize the ecology of the larger area.

overkill. A simpler approach would have been to perform an ordination analysis (Gauch, 1982) followed by a plotting of the components on a map. This would have made it easier to see the relationship between topographic contour and faunal composition. The many comparisons based on diversity were rendered uninformative due to the omission of variances.

A total of 12,000 correlation coefficients were calculated, in an attempt to identify fauna-environment interactions. There are three serious problems with this approach. First, it assumes linearity of response which is seldom the case for organism distributions along a gradient (Gauch, 1982). Second, there will be several hundred spuriously significant correlations. Third, when *n* is large, even very weak linear trends become significant, but are not especially informative.

The basic design, although cluttered by secondary sites, was along the lines of an analysis of variance with several possible factors: impacted area versus control, seasonal factors, position factors, distance from platform and replication. An hypothesis concerning impact could have been stated such as no difference between control and platform or no difference with distance from the platform. However, the only use of analysis of variance (ANOVA) was to justify pooling of replicates on the grounds that within station variation was less than between stations.

When justifying expedient sample processing methods, Baker *et al.* (1981) did employ statistical tests to support claims of no unjustified loss of information. In this manner it was argued that meiofauna samples are best narcotized prior to preservation, that meiofauna samples can be split using a Folsom plankton splitter, and that elutriation techniques are consistently efficient and do not impose a bias associated with sediment type. The analyses used, however, were not convincing.

Most critical is the claim that the elutriation technique is equally efficient over a range of sediment types. If this is untrue, a systematic bias would result which could be confounded with natural sediment effects on abundance, composition and diversity. The analysis presented ignores any within-sediment type variation, and deals only with percentages derived either through pooling or averaging of 3 to 10 samples in each category. The use of Kendall's Coefficient of Concordance W is not explained.

Similarly, the test of the narcotization technique ignores other sources of variation. From the first grab at each station, two cores were taken for meiofaunal analysis. One was treated with a relaxant then both were preserved. Using a signed rank test (without justification), the counts for major taxa were compared between cores. The test did not consider that there might be substantial difference between the two cores due to faunal variation within the grab sample and attributed all observed departures from the mean as due to preservational method. Although it was concluded that relaxation produced a better yield, the results were neither consistent nor convincing. Out of 21 comparisons, only four were significant at the alpha=0.05 level without any allowances for multiple comparisons. The pattern of apparent significance also changed unexplainedly with season.

Conclusions

Although somewhat obscured by lengthy discussion, it was concluded that any faunal variation due to petroleum activity could not be separated from that due to periodic, major environmental fluctuations. Nevertheless, the opinion was expressed that there was an effect due to chronic levels of some chemicals. Such a conclusion cannot be supported by the data collected. Elevated trace metal levels were found within 100 m of platforms, but there was no companion faunal sampling that close.

Major Failures

Relative to its predecessor, the Central Gulf Platform Study was a better organized and focused project. It began with a better informed understanding of the problem and the region. However, it still had four serious failings. First, while the goals were better defined, the lack of prior hypotheses led to largely descriptive and marginally productive data analyses. Second, the faunal components were baseline surveys that made little use of the design. Third, there was no sufficient attempt to accommodate natural variation in the treatment of replicates or in the analyses used. Fourth, there was a tendency to express opinion as fact when the actual findings were equivocal.

Buccaneer Gas and Oil Field Study

The Buccaneer Gas and Oil Field, located 50.5 km south of Galveston, Texas, came under study between 1976 and 1980. Having been in production for about 15 years, it was assumed that impacted communities, if present at all, should be well-established. Unlike the heavily-developed Louisiana coast, it was felt that the relative isolation of the Buccaneer Field would allow unaffected areas to be included in the design. The results of the study are available in the open literature (Middleditch, 1981).

The overall design of the study, as presented by Middleditch and Gallaway (1981), has a strong "systems ecology" flavor, and a modelling effort (Fucik and Show, 1981) was a novel component. Essentially three separate models were developed: hydrodynamic, biological and chemical. The biological model was structured to show carbon flux through a multicompartmentized system and was then used to model the flow of hydrocarbons. In practical field design, the model could not have been especially influential since the modeling effort began in 1978 and the survey effort ended in 1977. However, it may have been the inspiration for two interesting approaches which seem to have been considered in the final year or two of near-field studies. First, faunal groups were classified as to whether they imported or exported carbon. Second, the platform fouling community was singled out for detailed investigation because it appeared to be a net importer of carbon.

There were three separate ecological studies involving the benthos. A general benthic survey of macrobenthos and meiobenthos, similar to those in the other studies, was reported by Harper *et al.* (1981). The fouling community was studied by Fotheringham (1981), and the whole artificial reef community of the platforms was studied by Gallaway *et al.* (1981). In this report discussion will be restricted to the general survey.

Geography of Sampling

The faunal studies of the Buccaneer Field study took the form of a grab survey along transects radiating outward from two production platforms (Harper *et al.*, 1981). As shown in Figure 14.3, the placement of the transects is not symmetrical. Although the full justification for this design was not presented, it was noted that preliminary surveys indicated some faunal change to the northeast and that major bottom currents were in that direction. Stations were taken at increasing intervals with the explanation that impact should be most prevalent close to the platform. A diver operated Ekman grab of 232-cm^2 area and an ideal penetration of 15 cm was the main sampling device. Meiofauna was collected from the Ekman grabs by subsampling with a 2.54-cm core tube.

Replication

The replication scheme was quite simple. Three grabs were taken at each station, and stations were occupied quarterly over one annual cycle. As with

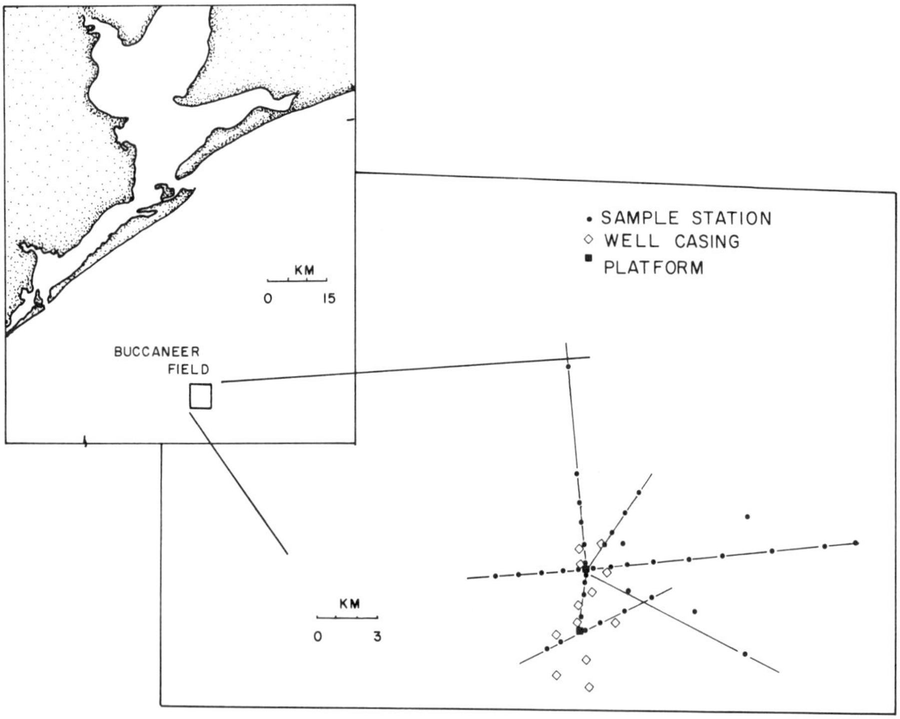

Figure 14.3. Location of stations in the Buccaneer Oil and Gas Field Study off Texas. While this study did employ surveys over wide areas, emphasis was placed upon the region within a few kilometers of platforms. It was assumed that radiating transects would aid in detection of impacts based on gradients in ecological parameters.

the previous projects, unfortunately, replication was not used to estimate the variability of the sampled fauna. In analysis, the samples were either pooled or the mean values used without reference to variation.

Coordination

There are two aspects of project coordination that need to be discussed. The first concerns the use of an ecological systems model as an aid in design. The second is the actual coordination of sampling between the biological and the environmental components.

There can be no doubt that modeling can play a very important role in the design and refinement of long-term effects studies. Used in conjunction with preliminary data, models should make it possible to restrict the choices of sampling design and analytical approaches needed to test hypotheses about effects. However, it is not clear how valuable a model is when it is the final product of a study. In the case of the Buccaneer Field Study, the final model was actually a "fates" model and did not deal with ecological effects. It had the undesired effect of taking attention away from far-field, long-term effects, and directing them to acute near-field impacts.

Coordination of environmental and biological data employed a technique that might be termed "after-the-fact map overlay." Environmental and biological parameters were studied separately, values were determined at a variety of locations, and the resulting map contoured. Sedimentary and geochemical parameters (Anderson *et al.*, 1981), surficial sediments (Brooks *et al.*, 1981), organic carbon (Behrens, 1981), and alkane concentration in surficial sediments (Middleditch, 1981) were determined by a general survey and radiating transects near the platforms. The environmental and faunal transects did not coincide.

Analyses Attempted

Examination of plots along transects and of maps was the principal method of deciding if there was an effect on abundance, diversity, or faunal composition. Within 100 m of both platforms there were consistent decreases in diversity and abundance; however, the omission of information as to the type of variation seen decreases the usefulness of these observations. Only the possibility that the stations had significantly different H′ diversities was tested, but the actual form of the hypothesis and test was not explained.

Cluster analysis and principal component analysis were used, but there was insufficient detail included in the report. Neither the similarity index used, the clustering strategy, initial data manipulation or the exact nature of the principal component analysis were given. Even with these omissions, the results do not seem to support the conclusions. First, the cluster analysis did not reveal four distinct groups, but rather one large cluster and a scatter of poorly related stations within which there may have been a second group. The unlabeled graphical representation of sample position in three principal dimensions seems to suggest the same interpretation. Indeed, there may be a gradual grading from the platform stations (with 100 m) to the more general background condition which can be seen up to 3 km from the platform.

Conclusions

The major conclusions and discussion concerned the near-field effects. There was a consistent decrease in density within 100 m of a platform, but there was no suppression of diversity. This combination was taken to indicate that there was little likelihood of chronic toxicity near the well, and that current scour might be altering the habitat near the well. The actual importance of these mechanisms was not explicitly examined in the design. Far-field effects were left virtually undiscussed, and it is not clear if any could have been detected using the highly descriptive approach taken.

Major Failures

The Buccaneer Field Study continued the failures of the previous efforts along with some new problems. The modeling approach was not successfully extended to the question of wide area, long-term impact. As a result, the faunal surveying was similar to a baseline survey. Faunal and environmental data were collected separately and could not support direct comparisons. Neither the treatment of replicates nor the analyses presented sought to deal effectively with natural variation.

Mid-Atlantic Block 684 Study

This study was a before and after evaluation of the short-term effects of drilling discharges on the benthic community around an exploratory well on the Middle Atlantic continental shelf. Although not a long-term effects study, the detailed development of the circular symmetrical sampling pattern warrants its inclusion in this review. As it turned out, the elegant design was too ecologically uninformed and was abandoned before final analysis.

The study was initiated after and conducted during the other projects in 1978-1979. The final report (EG&G, Environmental Consultants, 1982), especially the appendix on design considerations (Robson *et al.*, 1982), is the principal reference. A 0.1-m^2 Smith-McIntyre or a modified ponar grab was the main benthic sampling gear.

Geometry of Sampling

The selection of sample sites around the well was based upon an explicit model which defined impact as a change in abundance about a point source and sought to produce an efficient estimate of that impact (Robson *et al.*, 1982). The model was essentially the same used in traditional microbial bioassay design in which a drug of unknown potency diffuses across an agar plate from a central point, affecting or impacting the bacterial biota (see Finney, 1964). While the formal development is rather tedious, the important points can be considered here without resorting to analytical geometry.

The study defined impact as a change in faunal abundance occurring after a potentially impacting event, over a circular area about a point source. The change would manifest itself as an increase or decrease when compared with data collected before the possible impact, and when compared with data collected far removed from the possible circle of influence. In order to estimate the impact at different distances, it was necessary to sample radially about the point source. Two basic options were considered. First, for the sake of simplicity, a radially symmetrical design was selected rather than to duplicate the current pattern around the platform. Second, fixed distances and fixed directions were selected rather than randomization within sectors of annuli due to a greater interest in the actual shape of the distance-effect curves. A total of 48 stations were occupied.

Stations were established along six lines through the well site (Figure 14.4). The most distant stations were at 2 nautical miles and then progressive halves (1 mile, $\frac{1}{2}$ mile, $\frac{1}{4}$ mile, etc.) nearer the well. The resulting pattern of stations was seven concentric hexagons. The selection of six radii was largely pragmatic. It is more dense than four radii, but less dense than a full grid system. The logarithmic progression in distance was based upon the assumption that the impact would diminish rapidly away from the point source.

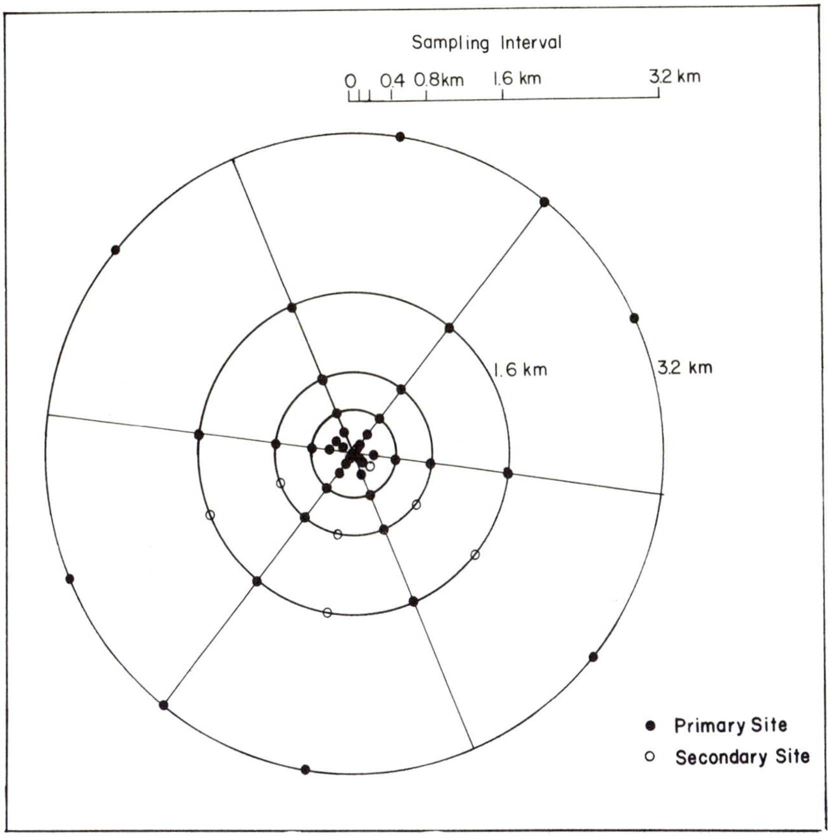

Figure 14.4. Station placement around an exploratory drilling site in the Mid-Atlantic Block 684 Study. While not a long-term effects study in the strictest sense, this project represented a much more carefully designed sampling of an area within a few kilometers. The number of radiating transects was selected to detect the directional pattern of an effect, and the position of stations along a transect was selected in anticipation of an exponentially decreasing effect.

Replication

The benthic fauna was surveyed three separate times, once before drilling, shortly after cessation, and a year later. In the predrilling survey, six replicates were taken from 40 stations in the sampling array. Of these, 22 were analyzed and the rest archived. In the first postdrilling project 48 stations were sampled with 41 being analyzed and seven archived. In the final survey 41 stations were sampled and analyzed. While six replicates were taken at each station, only two were analyzed while the other four were archived. This was due to a deliberate choice between the costs of replicate processing and covering a larger area.

Coordination

The collection of biological and environmental data was closely coordinated because the same six replicate cores were used for macrobenthos census, trace metals, hydrocarbons, and granulometric parameters. Tissue analyses were performed on animals from separate grabs, but taken at some of the same stations as the survey grabs. Other physical measurements were not really applicable to the long-term impact situation, and additional discussion will be omitted.

Analyses Undertaken

Although the circular symmetrical design was used for the positioning of stations, the statistical analyses required by the distance-effect model were not considered in the final report. With little explanation, a simple two way analysis of variance, cluster analysis and principal components analysis were substituted. Apparently, the detailed data were pooled into fewer groups according to distance from the well, sampling period and direction. Then the hypothesis of no differences was tested and not rejected. Unfortunately, details of this pooling and testing were omitted.

Overall faunal density, diversity, species per unit area, and evenness were treated as simple variables and examined over the whole sample set for each of the three surveys. Cluster analysis and a form of principal components analysis were also used to establish the existence of faunal groups and their location. Along with some density data for ophiuroids, these community analyses provided the evidence for near-field effect.

The study made use of changes in clusters to conclude that there was a near-field impact although there was little discussion of the validity of the approach. The station data were examined using a flexible sorting strategy on Bray-Curtis similarities. Principal components were based upon a Gower similarity. Separate analyses were done for each of the three surveys. To give a spatial interpretation, cluster membership of each station was shown on the sample chart. The implicit criteria for determining impact appears to have been a change in the faunal affinity near the platform.

There were some minor irregularities and missed opportunities. Clusters are characterized by their species components, yet there was no discussion of the species groups that contributed to the clusters. It was noted that the faunal clusters were also characterized by similar values of diversity, evenness and abundance. While it was commented that these are related to similarity "in practice" it was missed that they are related rather explicitly in algebraic terms (e.g., it was a redundant comparison). Finally, an ordination approach could have been used to map trends in faunal associations. Had this been done, faunal patterns could have been more easily related to physical attributes through inspection.

Conclusions

Even though only short-term effects of exploratory drilling were investigated, the conclusions were quite similar to those of the long-term studies. First, there seems to be an easily noticed effect within 100 m of the platform which is attributable to gross physical disturbance. Second, there is a suggestion that other spatial patterns of impact may exist, but that the design employed could not identify them.

The report mentions several changes in the fauna and specifically attributes the following as due in whole or part to the drilling efforts. Megabenthos increased near the drilling site in response to increased environmental heterogeneity. Local infaunal abundance and diversity near the site was reduced by burial and reduced larval settlement (larval settlement was not actually examined). Results from a study of an abundant ophiuroid were taken as the strongest evidence for a persistent (over one year) effect.

Major Failures

The serious failures of this study all stem from a design effort that was undertaken without consideration to the type of field data that might be expected. The model employed was simple in structure, elegantly developed, but too unrealistic to be of use. The envisioned distance-effect curves either did not exist, or they were hidden in an unexpectedly variable fauna. Unable to pursue the analyses dictated by the model, the study degenerated into little more than a good faunal survey with an elaborate sampling scheme.

Georges Bank Monitoring Program

At this writing the Georges Bank study is still underway, having been initiated in 1981. Like the Mid-Atlantic Block 684 study, it is concerned with the effects of exploratory drilling discharges (drilling muds and cuttings) and is not a long-term effects study in a strict sense. However, it is an important project from three perspectives. First, it combines circular symmetrical sampling around a platform with wider sampling that will serve as the starting point for any long-term studies if the field develops. Second, it seems to be proceeding without a fully-developed operational definition of impact, but is doing so cautiously. Third, even this early in the project, it is obvious that replication is being used properly to estimate variance rather than for "representativeness." The first annual report (Battelle/Woods Hole Oceanographic Institution, 1983) was the reference consulted; a second has since been produced (Battelle/Woods Hole Oceanographic Institution, 1984).

Geography of Sampling

The positioning of near-platform and far-field stations was based upon the supposition that most pronounced effects would be adjacent to drilling activities with a spatial pattern reflecting average transport. High levels of exposure and effects away from point sources were considered possible in places where topography and transport serve to accumulate particulate material.

Two groups of stations were established to examine near-field effects. Of greatest interest in this review were the stations around an exploratory rig in Block 312. Twenty-nine stations were arrayed circularly within 6 km of the well site (Figure 14.5). These made up two crossing transects with some intermediate samples. The axes of the array were aligned with the major axis of the tidal current ellipse and the mean current vector (which were essentially orthogonal). The first circle of stations was within 200 m of the rig with successive rings at 0.5, 1.0, 2.0, 4.0, and 6.0 km radius.

The regional stations were selected on the basis of two assumptions and an important bit of prior information on faunal distributions. First, it was known that the greatest natural faunal variation occurs with depth, so stations were established to sample three levels of depth (called transects in the report). Then, it was assumed that average transport would be the primary factor controlling exposure to discharges; therefore, three levels of potential exposure were sampled, upstream, through the site, and downstream (the report refers to three transects). Lastly, it was assumed that topographic features would trap and funnel bedload particulates which might bear pollutants, thus ten stations were located in areas where this might occur.

Replication

The regional and near-site stations were sampled seasonally, four times during the first year. Drilling began at the test well after the first two cruises. At the regional stations six replicate large (0.10 m^2) and six small (0.04 m^2) Van Veen grab samples were taken. Epifauna was surveyed photographically (at least 20 m^2 cumulative area) at each station. In the near-site array six replicate small grab samples were taken together with the same 20 frames of epifaunal photographs.

Coordination

The collection of environmental and biological data was closely coordinated. Chemical and granulometric data were either collected from a subsample of the faunal grabs, or obtained from the same series of samples. Trace metals derived from drilling muds (barium, chromium, zinc, cadmium and lead) received special emphasis. Due to their high concentrations in drilling fluids, barium and chromium were taken as the principal indicators of exposure.

Preliminary Analyses Undertaken

The analyses presented in the first report were appropriately descriptive for their preliminary manner. Noteworthy was the fact that they were concisely described with supportive references so that if they prove to be inappropriate in the future the need for reevaluation will be obvious and easily accomplished.

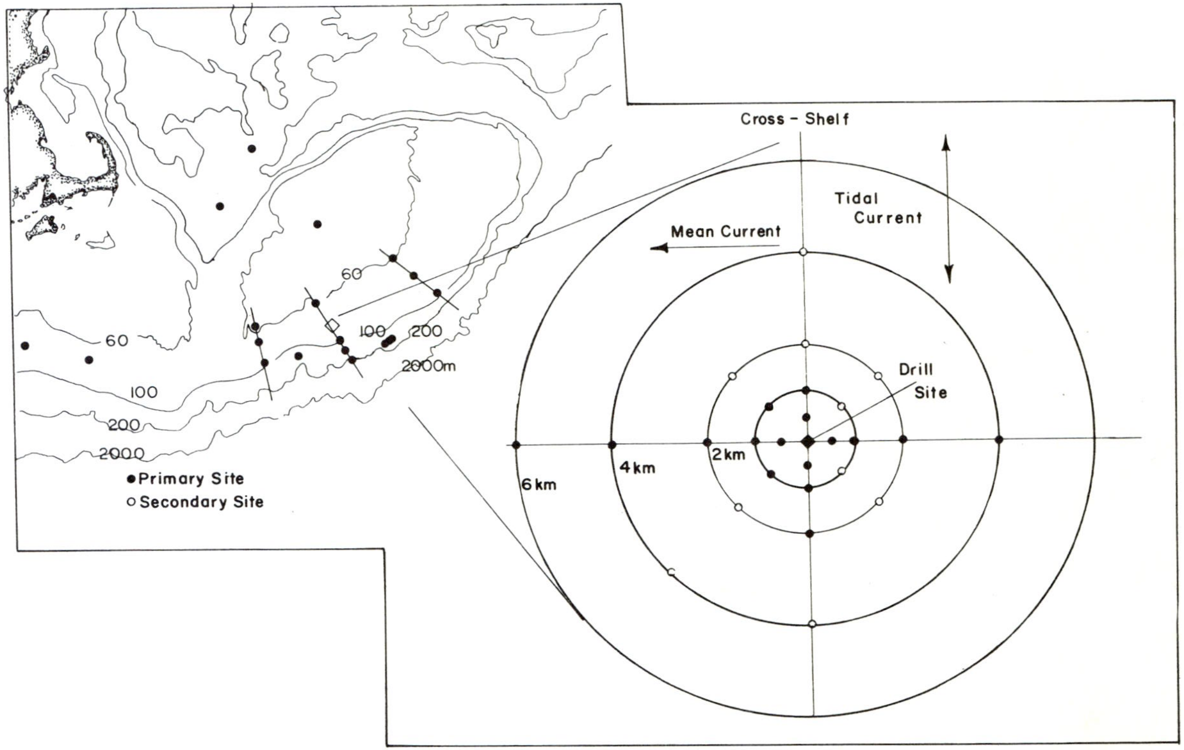

Figure 14.5. Location of stations in the Georges Bank Monitoring Program. As the most recent of the reviewed studies, the Georges Bank Program has the most comprehensive sampling program. Three transects should provide information on natural cross-shelf faunal variation. Near-platform transects are aligned with previously determined transport axes, and sample spacing anticipates a response which decreases with distance. Rather than attempt a comprehensive regional coverage, a limited number of stations have been selected at locations which are of particular oceanographic interest.

Data summaries were presented with standard errors, and unexplained *ad hoc* testing kept to a minimum. Cluster analysis produced results analogous to a factorial analysis of variance. For instance, when it was noted that replicates clustered first according to station, it was analogous to showing that within-station variance was less than between.

Major Failures

Clearly, it is too early to identify failures. The great strength of this study is that it seems to be carefully trying to deal with natural variation, but two problems can be anticipated. First, without a prior definition of impact it is easy to become distracted by an unanticipated natural pattern. Second, the marked faunal variation associated with depth may give the illusion that natural variation on smaller scales can also be easily partitioned in final analysis.

Both the near-field and regional sampling could support analyses of variance, and it can be hoped these will be conducted later in the project. As will be touched on in the discussion, such analyses can lead to definitive statements about the relative importance of natural factors. When used in power analyses, such information can also lead to definitive statements about the levels of impact which can go undetected.

DESIGNS OF LONG-TERM IMPACT STUDIES

From a long list of errors and minor irritations, two serious common design failures emerge from this review. The first is the lack of an operational definition of long-term impact, and its cause is associated with the uncertain state of theoretical ecology. The second is the failure to effectively deal with natural variation, and its cause would seem to be a combination of misunderstanding of statistics with a lack of understanding of the system under study. We cannot, at this time, resolve the problems associated with what we do not understand about nature. We can, however, more carefully consider the statistical approach and use it to identify areas that need special attention.

There are a great many references available on statistical design in theory and specific application. However, the few that deal with ecology emphasize exploration as most appropriate at this time and omit the type of testing required in impact studies. Green (1979) is a useful exception that advocates hypothesis testing and analysis of variance. Green gives ten principles of design (drawn largely from Cochran, 1977) that can serve as a practical guide for long-term impact studies in offshore regions.

Of these principles, the first concerns prior definitions, and the second through ninth all address problems of dealing with variance. In subsequent sections these problems will be examined in detail, and in the conclusion we will return to Green's principles to see what is possible in the continental shelf environments.

1. Be able to state concisely to someone else what question you are asking. Your results will be as coherent and as comprehensible as your initial conception of the problem.

2. Take replicate samples within each combination of time, location and any

other controlled variable. Differences among can only be demonstrated by comparison to differences within.

3. Take an equal number of randomly allocated replicate samples for each combination of controlled variables. Putting samples in "representative" or "typical" places is not random sampling.

4. To test whether a condition has an effect, collect samples both where the condition is present and where the condition is absent but all else is the same. An effect can only be demonstrated by comparison with a control.

5. Carry out some preliminary sampling to provide a basis for evaluation of sampling design and statistical analysis options. Those who skip this step because they do not have enough time usually end up losing time.

6. Verify that your sampling device or method is sampling the population that you think you are sampling and with equal and adequate efficiency over the entire range of sampling conditions to be encountered. Variation in efficiency of sampling from area to area biases among-area comparisons.

7. If the area to be sampled has a large scale environmental pattern, break the area up into relatively homogeneous subareas and allocate samples to each in proportion to the size of the subarea. If it is an estimate of abundance over the total area that is desired, make the allocation proportional to the number of organisms in the subarea.

8. Verify that your sample unit size is appropriate to the size, densities, and spatial distributions of the organisms that you are sampling. Then estimate the number of replicate samples required to obtain the precision you want.

9. Test your data to determine whether the error variation is homogeneous, normally distributed and independent of the mean. If this is not the case for most of the field data, then a) appropriately transform the data, or b) use a distribution free (non-parametric) procedure, or c) use an appropriate sequential sampling design, or d) test against simulated null hypothesis data.

10. Having chosen the best statistical method to test your hypothesis, stick with the result. An unexpected or undesired result is not a valid reason for rejecting the method and hunting for a "better" one.

DEFINITIONS OF IMPACT: WHAT ARE WE LOOKING FOR?

In the Introduction it was stressed that theoretical ecology is not sufficiently developed to provide applied ecology (i.e., impact studies) with foolproof designs for assessment. This problem is most obvious when an operational definition of long-term impact must be stated. In spite of the critical requirement of good design that the type of phenomena under investigation be known, the reviewed long-term effects studies lacked such definitions. It is the uncertainty as to what an impact is that leads to much of the subsequent uncertainty in long-term effects studies.

Implied Definitions

There are actually two distinctly different types of impacts implied in the discussions and analyses of the reviewed studies. Each leaves a mark upon the benthic assemblage that can be detected by faunal survey, but each requires a different design. The first type can be called "pathological," following the analogy of environmental health in the OEI-GURC, Central Gulf Platform, and Buccaneer Field studies. The second type I will call "discordant." Alternately, these might be called "easy" and "difficult" impact.

Pathological Impact

A long-term impact might alter the fauna into an assemblage that is unambiguously abnormal. Whatever the "symptom" of impact, exotic species, local extinctions, etc., it would be so distinct that little analysis and no formal testing of alternate hypotheses would be needed. If long-term impact manifests itself this way, design would not need to accommodate the management of natural variation. Rather, the main concerns would be the timing, spacing, and intensity of sampling needed to find the symptom. Statistical analyses could be used to determine the adequacy of the search, but the presence of the symptom would be self evident.

Discordant Impact

A long-term impact might produce an assemblage not especially unlike those sometimes encountered under natural conditions, but which is out of accord with the prevailing natural factors. Such an impacted assemblage might easily be confused with a normal assemblage if the designs did not consider the type and magnitude of natural factors. Statistical analyses would be required to partition variance and to determine the certainty with which an effect can be attributed to impact.

Impacts producing pathological assemblages were the form implied by the designs and analyses undertaken in the OEI-GURC, Central Gulf Platform, and Buccaneer Field studies. The absence of designs which manage variation, the use of replication to assure representativeness rather than to estimate variation, and the heavy dependence upon pattern recognition analyses lead to this conclusion. Even in the Mid-Atlantic Block 684 and Georges Bank Studies, there is a wishful and understandably persistent attempt to find an obvious "indicator of impact" which would simplify design and circumvent the need to manage variance.

The temperate coasts of the U.S. are subject to natural, catastrophic environmental changes such as storms, water mass shifts, and widespread hypoxia recurring over a few years to tens of years. The catalogue of natural marine mass mortality compiled by Brongersma-Sanders (1957) is an informative illustration of this point which should have been considered during the planning of OEI-GURC; less dramatic variation is discussed by Jones (1982). Therefore, it is unrealistic to expect pathological assemblages that cannot be attributed to natural events occurring over the long time span of long-term effects studies. If there are petroleum related long-term impacts, they may be manifest in the response of communities to a naturally varying environment. Detection of such impacts requires adoption of the discordant impact approach with emphasis on designs which can deal with multicomponent variation in time and space.

Something Other Than Census

All of the studies worked with the assumption that impact (pathological or discordant) results in faunal change, and each sought to explain possible impacts in terms of ecological processes which were not directly studied. Even in the better studies, Mid-Atlantic Block 684 and Georges Bank, there was a sense of dissatisfaction with being restricted to survey data. In both cases there were attempts to explain impacts possibly seen in the assemblage data in terms of an alteration of life history events (such as larval recruitment) that had not been measured in the project. This leads to the suggestion that assemblage-based definitions should not be the primary focus of long-term effect studies. Rather, direct evaluation of community or life history phenomena (i.e., fecundity, mortality from all causes, immigration, emigration and recruitment) should serve as the foci for assessment.

Of all the possible community or life history phenomena that could be studied, recruitment from the pelagic larval to the benthic stages has attracted special attention (see Menzie, 1984 for discussion) and can be used to make an important point. Laboratory studies have produced a list of natural factors that seem to have some influence upon settlement (Chia and Rice, 1978). However, we do not really know the type and magnitude of natural effects in a field setting. Therefore, if we select to focus upon recruitment, we will have a better definition of impact, but we will still have a poorly understood, highly variable system to study. As desirable as a process-oriented approach may be, there is none that can be taken without a very substantial research effort.

Some of the European community's ideas about long-term oil impacts on marine communities were presented in a symposium volume (Clark, 1982a). In the introduction (Clark, 1982b), the conclusion (Clark, 1982c), and especially in the published discussion of each paper, we find the same theme developed above. These studies must begin to bridge the gap between laboratory studies which easily demonstrated effects and the field situation, but the most thoughtful ecologists see no easy solutions.

STATISTICAL MODELS, ERROR AND POWER

In this section four critical points about the problem of recognizing long-term impacts will be made through a brief discussion of the general linear model. Because these points are so important, and many readers may have an understandable aversion to the jargon of statistics, I shall begin by stating my conclusions in terms directly related to the assessment of long-term effects.

1. However a long-term impact manifests itself, it will just be one more source of variation in a biota that varies in number and composition in response to many natural factors in the continental shelf environment.

2. We will never be able to find a significant impact as long as we are unable to explain a large percentage of the total variation seen in the field.

3. We are restricted to testing the hypothesis that there is no impact. This is equivalent to assuming innocence until guilt is proven, and the usually selected significance levels, combined with the high unexplained variation, make it more

likely that real impacts will go undetected rather than nonexistent impacts erroneously detected.

4. The adoption of statistical models prior to sampling allows us to critically evaluate sample allocation, replication and our ability to resolve an impact.

The term "model" has such a wide and casual use in both basic and applied ecology that initial attention to definitions is required. First, it must be strongly asserted that the fundamental task of recognizing long-term impacts in the presence of great natural variation on the continental shelf cannot be accomplished without the use of statistical models. A statistical model is one which predicts the amount of variation in a parameter under certain conditions. Usually, those conditions include the assumption that the null hypothesis is true. Second, it must be pointed out that a statistical model is, outside of any philosophical discussion of general systems, not an ecosystem model.

I do not want to detract from the value of ecosystem models, such as produced in the Buccaneer Field study, that are intended to show the functional interrelations among components. While they will become increasingly important in the design of impact studies, as yet they offer no greater ecological insight and predictive capacity than non-modeling approaches. Indeed, the present funding imbalance between application and research seen in impact studies may be the result of having expected too much too soon from "predictive models" back during the years when the National Environmental Policy Act was in draft.

General Linear Model

Regression analysis and analysis of variance are the two most common forms of statistical analysis used for hypothesis testing in ecology and appear in various forms in the reviewed studies. Although most general statistical texts treat these analyses separately, they are both forms of the same mathematical model commonly called the General Linear Model (GLM) and are identical in their essential parts. It is premature, given how little we know about the ecology of the continental shelf regions, to propose the GLM as the only course by which to seek long-term impacts. Nevertheless, a discussion of the GLM serves to clarify the problem of detecting impact in a highly variable environment where the biota responds to many factors. The GLM has the added advantage of being the most fully developed statistical tool available, yet it has been no more than superficially explored in offshore long-term effects studies.

Full development of the General Linear Model can be found in many of the newer statistical texts which take a more mathematical approach. In actual application some multivariate form might be selected. All that is intented here is to illustrate the basic problem. I found the account by Horton (1978) especially lucid and have followed the notation of that volume. The complete general form of the GLM may be written

$$y_i = \Theta_1 x_{i1} + \Theta_2 x_{i2} + \ldots + \Theta_m x_{im} + e_i$$

$$(i = 1,2,\ldots,N)$$

where y_i is a measured value (e.g., count of a particular species in a box core). The formula is an hypothesis which states that the value measured in the field is determined by a linear combination of m measured independent variables (e.g., environmental factors, other species, etc.), each weighted (multiplied) by one of m unknown constants. There is in addition e_i, the residual error not explained by the independent variables; this term is randomly distributed.

It is the residual error term which is most critical with respect to our ability to detect an impact. Most often, as scientists, we first encounter the GLM and residual error in the experimental situation where major causative factors are known and controlled. As a result, residual error is small and well-behaved (normally distributed with an expected value of zero), making it possible to detect relatively minor effects of experimental variables. In the situation of continental shelf benthos, we do not yet know what factors are the major source of variation. Therefore, our model is not fully specified and the residual error is very large.

The main purpose of analysis varies with the final form of the GLM adopted, but it is usually to determine which Θ_is are nonzero (i.e., which independent variables have an effect) and the values of those Θ_is (i.e., the extent of influence). The significance of an effect is tested by a ratio which compares the variability (mean square) of the residual error term alone to the combined variability due to an effect plus the residual. The value which the ratio must exceed to reject the null hypothesis (no effect) is dependent upon the degrees of freedom, but a value around 2.0 might be typical in an impact study. In other words, the variation in the faunal count due to the factor being tested must be about twice the variation left unexplained by the model.

Typically in field surveys of benthic communities in which sampling does not cross major bathymetric zones, there is a large residual variability which cannot be attributed to linear combinations of various physical parameters. Unless we can more fully specify our models so as to drastically reduce the residual error, it may be impossible to prove the significance of an impact. The designs which we adopt must allow us to remove a large percentage of the variability due to natural causes which may exceed variability due to an environmental impact.

The comment that surveys which do not cross major bathymetric zones have high residual error needs additional explanation. It is actually quite simple to produce a survey which gives the illusion of having a low residual error. You simply sample over such a broad range of physical conditions, typically depth, that the dominant species being sampled completely change at least once. The huge amount of variation due to having sampled different communities will be almost completely explained by the broad environmental factor, and the residual will be proportionately small. Unfortunately, this inflation of a natural effect does not increase the magnitude of impact effect relative to residual. In this way, one is drawn to the familiar conclusion that natural factors have a greater effect than the subject activity.

In the context of any form of the GLM, the purpose of replication is to provide a means of estimating residual error. However, as used in faunal surveying, replication is often undertaken to assure the completeness with which the species composition has been determined. Cuff and Coleman (1979) discuss this approach and demonstrate its relative lack of usefulness in determining optimal levels of replication. Replication in this sense is obvious in OEI-GURC and the Central Gulf Platform Study. Perhaps if some rare species, indicative of pathological impact, were being sought, completeness would be important and pooling of replicates prior to analysis would be acceptable. In the context of testing for discordant assemblages, it suggests a basic misunderstanding of design.

Cuff and Coleman (1979) provide an excellent review of attitudes and facts about replication in benthic surveys. Building upon the observation of Saila *et al.* (1976) that random sampling did not require inordinate random samples, they considered optimal design using data from a stratified random survey conducted in Australia. The main point from that study is that models provide grounds for design optimization. The second point is equally informative. The degree of replication usually advocated for benthic surveys (3, 4, 5, etc.) often expends too much effort in characterizing each station while losing sight of the overall intent of the design. In the particular case examined, a single sample at a greater number of stations would have been optimal.

Power and Errors

Stated simply, power is the ability of an analysis to prove that a null hypothesis is false. In our case it is the ability of the analysis to detect the possible presence of impact. A discussion of the two types of error inherent in statistical hypothesis testing can be found in any general statistics text (see Dixon and Massey, 1969, for a good example); the concept was first developed by Neyman and Pearson (1928). It is informative to repeat the basic explanation in terms of the long-term effects problem (Table 14.1).

There are two types of error. Type I error, wrongly rejecting a true null hypothesis, is considered especially critical in research, and the probability of having made such an error (1-α) is required evidence. In the offshore development case, this type of error could lead to unnecessary retrictions upon petroleum development. Type II error, wrongly accepting a false null hypothesis, has received little attention, and its probability (β) is rarely reported. However,

TABLE 14.1

Types of error and the associated probabilities in parentheses. Traditional research places most emphasis upon minimizing Type I error which can have the effect of increasing the likelihood of Type II error. If this same priority is applied in long-term impact studies, then we would rather not find actual impact than to come to the false conclusion that there were impacts

Truth	*Conclusion*	
	No effect	*Effect*
No effect	No error (1-α)	Type I error (α)
Effect	Type II error (β)	No error (1-β)

again in the offshore development case, type II error would cause impact to go unnoticed and development to proceed without attempts at mitigation.

An examination of the relationship between Type I and Type II errors, shows that the traditional emphasis upon employing small values of alpha (i.e., requiring strong evidence for the rejection of the null hypothesis) combined with the highly variable nature of marine communities has the effect of making it unlikely that any effect could be detected and judged significant.

Power ($1-\beta$) is the probability that an existing impact will be statistically demonstrable. Power can be determined by a variety of methods, all of which depend upon integration of the area under all possible alternate (non-central) probability functions (F distribution in the case of analysis of variance) from the central distribution (that assuming the null hypothesis to be true) to some specified extreme. Fortunately, tables and nomographs have been produced to facilitate the task. Cohen (1977) provides a very convenient compilation, although some of the discussion of testing of behavioral hypotheses may be distracting.

Without going into detail it is sufficient to say that power is determined by the specified level of alpha, the sample size, and the degree of departure from the probability distribution due to the null hypothesis. The greatest increase in power is produced by a reduction in the residual error term. To some degree this can be accomplished through increased sample size. However, when the residual term is large, far greater increase in power can be obtained by identifying additional sources of variation and removing them from the residual by design. This means we must know more about causes of natural variation.

SPATIAL CONSIDERATIONS

How do we go about trying to reduce the residual variation that will otherwise mask any long-term impacts? Again, our lack of basic understanding prevents a simple answer. We can begin with what we know about benthic faunal variation and try to employ that knowledge in our designs and seek new information when its need becomes apparent. We know that the benthic fauna changes from place to place and from time to time, so we will begin with a discussion of space and then include some comments on temporal phenomena.

The Radially Symmetric Design

A symmetrical series of stations, with increasing spacing, radiating out from a site of petroleum activity was a key feature in four of the studies. On initial consideration this design has two appealing features. First, it does not depend upon a simplistic assumption of exactly where upstream or downstream lie relative to the potential source of impact. Second, it seems to allow a large area to be covered with relatively little sampling effort. However, this design implicitly assumes a continuity of response and is not well-suited for the detection of impact in a patchy environment. In pursuing the symmetrical circular design to an extreme, the Mid-Atlantic Block 684 study provides a good illustration of the weaknesses of the approach.

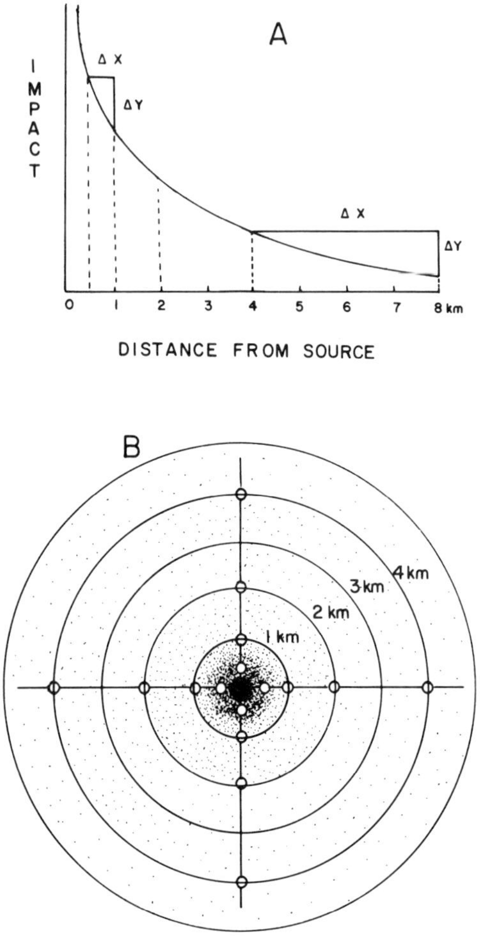

Figure 14.6. Simple impact-distance relationship. A, When a pollutant diffuses from a point source, the concentration decreases exponentially with distance. This could manifest itself as an exponentially decreasing effect on the biota. Sampling which locates stations at increasing distances along a transect are best suited at estimating the slope of such an impact curve. Ideally, samples should be spaced (ΔX) so that the same change in impact (ΔY) is found between each pair. B, If this simple impact-distance relationship is applied to the situation of a platform, then a simple radiating transect pattern is appropriate. Density of impact is indicated by stippling.

The main problems are easily demonstrated by comparing the ideal case with a more probable field situation. The ideal case requires some simple development. First, let us assume that any impact of a point source of pollution dose will manifest itself as a measurable response in the biota. Second, due to some combination of eddy diffusion, advection, and time dependent decay, both dose and resultant response will decrease with distance from the source. Third, the rate at which dose and response decrease with distance, will itself decrease with distance (most of the measureable change will be near the source). This can all be shown as a simple curve (Figure 14.6, A).

If the purpose of the statistical analysis is to obtain good estimates of the parameters of the response/distance curve, then the sampling locations should be at distances which correspond with equal amounts of change in the response (dependent variable). In the case of a response that decreases exponentially with distance from a source, then the samples should be placed at exponentially increasing distances (Figure 14.6, A).

The simplist two dimensional model is a rotation of the one dimensional case above (Figure 14.6, B); response is radially symmetrical about the source. Of course, nobody expects potential pollutants to be radially symmetrical about a point source on the continental shelf, and all of the projects except OEI-GURC established radial transects so that the directional differences in the response curve could be determined. However, the spacing of stations was such that a similar type of response curve in each direction was assumed.

What if the response of the biota was, in fact, patchy on scales larger than the area occupied by a station, but small relative to the circular area included in the design? Such patchiness could have many sources. Whatever the cause, that patchiness will severely limit the precision with which impact can be measured.

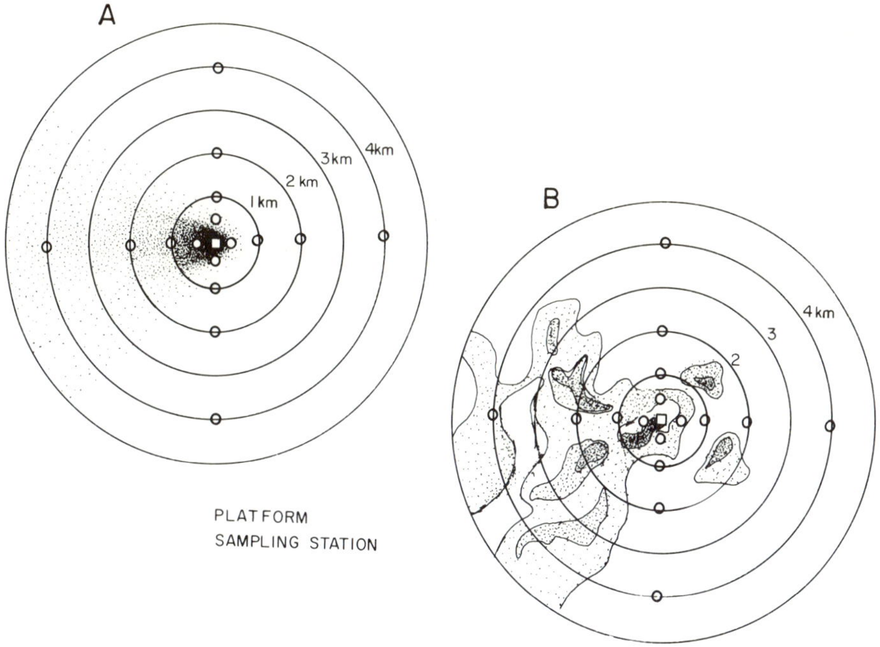

Figure 14.7. Possible departures from a simple impact-distance relationship. The simple diffusion model implicit in the radiating transect approach is unlikely in the coastal environment. A, If there is strong advective transport, then a plume of pollutants and possible impact can be expected. In the case shown, about three quarters of all samples lie outside of the affected (stippled) region, and evidence of impact will come from only one transect. B, A more realistic view of possible impact in the coastal environment must consider that advection is variable, that bottom processes will alter distribution of pollutants, and that impact will be overlain on a mosaic pattern of biota passing through various natural progressions. Severity and pattern of impact may be quite unlike that anticipated by the radial transect design.

This can be shown in an example. Let us assume that physical transport initially spreads the pollutant out so that there is an asymmetric, exponential distribution on bottom (Figure 14.7, A), but that over the long-term this "dusting" of contaminant is reworked into patches. While the density and scale of reworking is independent of distance from the site, patches with high enough concentration to cause an effect decrease in density with distance (Figure 14.7, B).

Keeping things as simple as possible, the probability that a dimensionally small station will fall within a response patch is equal to the proportion of bottom covered by the patches. The radially symmetrical design with increasing sample spacing results in greatest sampling intensity (samples/unit area) in small inner rings which have the greatest likelihood of possessing an effect response. In the much larger far-field area, sampling intensity is greatly reduced, the area occupied by response patches may be less, and the combined effect is to make it highly unlikely that a far-field effect will be noted.

The above model of decreasing degree and decreasing coverage of impacted patches does not have to be very close to nature to make its main point. Patchiness is a reality that is not ignored in field ecology, where it is realized that naturally varying parameters have heterogeneous dispersion patterns (patterns of location in space, Pielou, 1969). There is no reason to assume in the case of environmental impacts that the distribution of pollutants, the susceptibility of the biota, the manifestation of impact and the degree of that manifestation will not also have a complex dispersion pattern.

Faunal Patches in Space

Good design demands that we identify the spatial pattern of faunal variation so that sampling can be stratified within patches, and the "patch effect" removed from the variability of the data. The topics of patchiness and pattern are complex and spread through the literature of many fields which deal with processes in two dimensions. As yet, there have been only a few applications in field ecology, but more can be anticipated.

Due largely to modern geography, the spatial patterns are no longer sought by subjective mapping and "eyeballing," and the old ecological question of under or overdispersed pattern is no longer considered especially informative (Bartlett, 1973). There are now two, not altogether unrelated, approaches being taken to the study of variation over space. The older, spectral analysis, examines the frequency domain of faunal patterns (Platt and Denman, 1975, Dingle, 1979, and Ord, 1979). The newer technique, spatial autocorrelation, is used to test for patterns which are included in a working model (Cliff and Ord, 1981). Although not used in a wide area study of the continental shelf, analysis of spatial autocorrelation has been successfully used in studies of smaller scale patterns (Jumars *et al.*, 1977; Jumars and Eckman, 1983).

Fortunately, the literature on analysis of spatial pattern is not yet burdensome. Cliff and Ord's (1981) second book on the subject is the best single reference. Ord (1979), presents an abbreviated account with reference to ecology. Sokal and Oden (1978a,b) have presented a good general development and discussion of ecological relevance. Jumars and Ekman (1983), although dealing with the deep sea, offer a good brief account. The topic is, however, relatively new and sophisticated, and all of these references will be more comfortably read by someone with an understanding of multivariate statistics or linear algebra.

Before leaving the topic of spatial pattern, it is appropriate to mention the important work of Whittaker (1973 and references therein) on the distribution of plants. This approach, termed Direct Gradient Analysis, is used to examine the distribution of organisms along environmental gradients. The most important aspect of this approach is that it has depended upon formal models of population distribution along gradients (the Gausian model for coenoclines) and then explored the properties of different analyses when applied to the model communities (see Gauch, 1982 and references therein). In the case of long-term impacts on continental shelves, when the potentially impacted area contains a few important environmental gradients, such as depth, then modeling of distributions along that gradient would be of definite help in designing a full field program. When such major gradients are absent, the approach has no particular value.

Faunal Patches in Time

Accommodating temporal changes in the composition of the benthic fauna will be the most difficult design problem encountered in long-term effects studies. The North American continental shelf environment is subject to many physical disturbances that will greatly alter benthic faunal composition. Among these are storms (Boesch *et al.*, 1976), depleted oxygen (Santos and Simon, 1980) and red tides (Dauer and Simon, 1976), all of which were observed in the reviewed studies. Long-term effects studies will be especially subject to disruption by these events. It is highly probable that an area under study will have regions undergoing recovery or that will be disturbed during the course of investigation. If spatial patches are persistent over time, then they can be located with relative ease, characterized and accommodated in design as a source of natural variation. If, however, they change unpredictably, it will be very hard to eliminate the variation due to those changes.

Study of sequential changes in communities, succession, has a long history in terrestrial plant ecology. It is becoming especially popular in those benthic studies that explore MacArthur's (1972) extension of predation and competition concepts to problems of geographic ecology. The general application of the concept of succession is now undergoing reevaluation (see Connell and Slatyer, 1977), and its value in understanding natural variation on continental shelves has yet to be determined.

Views of succession fall between two wide extremes. It is hoped by some that succession is the ecological equivalent of morphogenesis and is a deterministic phenomena. Taking this point of view, succession could be treated as a continuous function that might be parameterized and easily studied. At the other extreme is the view that succession is a purely stochastic process, and it is this view which seems to be most near the truth (see Horn, 1976). As aptly stated in Gallagher *et al.* (1983), succession is a process "...more amenable to the mathematical analysis of the casino than that of the calculus."

A somewhat deterministic model of benthic succession applicable to continental shelves has been proposed by Rhoads and Boyer (1982). It goes beyond an initial descriptive system proposed by McCall (1978) and views biological effects upon sediment chemistry as very important. The model stresses the physical alteration of the environment caused by different organisms. The basic ideas are related to feeding type amensalism (Rhoads and Young, 1970) and the role of large sediment "bulldozers" on the geological time scale (Thayer, 1983). This model has not been presented in an analytical form. It does, however, suggest that successional stages could be identified by chemical gradients, sediment properties and functional groups of organisms.

Under a stochastic model, classification of successional stage is of little predictive value, and the focus must shift from the community as a whole to the individual interactions within the successional process. A useful framework for the study of interactions has been provided by Connell and Slatyer (1977) with the suggestion of three principal mechanisms—facilitation, inhibition or tolerance. Facilitation is the process whereby the early invaders of a disturbed area make colonization by other species easier through physical or chemical modification of the habitat. Inhibition is the process whereby an individual inhibits the colonization of its own or other species. Tolerance is the process by which numerous species may colonize a disturbed area, but there is a hierarchy determined by interspecies competition or predation by "third parties."

Exploiting the framework established by Connell and Slatyer, Usher (1979) produced simple Markov models, and Gallagher *et al.* (1983) applied it to nearshore marine soft bottom communities with the finding that facilitation seemed to predominate under the conditions of the study. In the case of long-term impact in offshore regions, a similar approach could be taken. If dominant mechanisms of interaction between individuals in a succession can be identified, then it will be possible to see if these mechanisms have become altered in areas exposed to chronic petroleum activity.

ALTERNATIVES TO THE INDIVIDUAL SPECIES APPROACH

During the ten-year span of the reviewed studies there have been many attempts to replace burdensome analysis of species census data with some type of more informative parameter which might be easily estimated. Clearly, such efforts must be encouraged because of our great need for useful ideas. However, it is important to consider the rationale behind each approach currently in use. In some cases there are sound ecological principles involved, while in others there may be little more than a desire to avoid complex analyses.

To simplify discussion, two stratagies of data treatment can be recognized. First, there are those analytical schemes which seek to characterize the entire rank abundance distribution with a small set of "community parameters" which can then be used for comparison. Second, there are those approaches which seek to partition the species data into fewer, more meaningful groups. While neither approach carries an automatic assumption about the type of effect one is attempting to detect, in application there seems to be a leaning towards the impact producing an obvious pathology which may be unambiguously detected by the unusual value taken by the parameter or by the presence of a definite impact-indication group.

Community Parameters

Diversification is that combination of ecological and evolutionary processes which lead to the diversity of species within a particular area on an ecological time scale and to the diversity of species within higher taxa on the geological time scale. Although the actual processes remain only sketchily understood, the possibility that they might be detrimentally affected by long-term exposure to polluting activities is a legitimate concern, and one which is examined by looking at species diversity, the result of diversification.

A point made by Pielou (1981) is worth repeating here. We have now arrived at a peculiar situation in which there are two distinct types of diversity studies. Continuing the original purpose, there are still those studies which seek to understand the diversity patterns around us. These are, however, becoming displaced by studies which study the behavior of the various indices which are in use. The current emphasis upon index rather than process is not necessarily harmful if it eventually leads to a more informative investigation of basic questions. Until such time, it is especially important in impact studies that we keep the original purpose in mind and not use an index uncritically.

The use of an index to express diversity was first proposed by Fisher *et al.* (1943) as a variance-like term which could be used to compare rank abundance patterns in multispecies samples. Much of the subsequent work in diversity indices has extended the diversity-variance analog to produce a variety of useful indices. H' or information content (Shannon and Weaver, 1949) has become, in practice, the most common index in use, although there are no convincing arguments for its superiority over other measures.

The finding that species rank abundance from numerous pooled samples often have a truncated log-normal distribution (see Preston, 1948) has gained recent attention in impact work (Gray and Mirza, 1979; Preston, 1980). Gray (1981) proposed that the log-normal distribution was due to the equilibrium between immigration and emigration in an area and that changes in the distribution could reflect pollution-induced changes. This is an interesting idea, but it has been shown by May (1975) that many different feasible processes can give rise to the log-normal distribution. Its existence or absence alone, therefore, is not evidence for Gray's or any other single explanation.

Taking the critical view of diversity indices there are four important points to remember until the processes of diversification are understood:

1. Particular diversity values, no matter which formulation is used, are not unambiguous indicators of stress, stability or impact. There are two origins for the unjustified use of indices as symptoms of impact. Theoretically, it was once thought that high diversity was related to stability and that healthy systems should be stable. Unfortunately, attempts to prove either conjecture have proven equivocal. Empirically, it has been shown that when pollution alters the proportional composition of species, diversity indices will show the change.

However, the direction of change bears no fixed relation to the type and degree of impact (Smith *et al.*, 1979; Chapter 9). (See May, 1973 and Goodman, 1975 for a critique of these attempts.)

2. Diversity indices can be used as a simple method of detecting changes in the relative abundance of species within a studied region. However, as with any other estimated parameter, the significance of an apparent difference can only be established if the natural variation is known. Fortunately, there has been work done to determine the proper methods of estimating the variance of diversity indices (Heltshe and Forrester, 1983; Tong, 1983; Zahl, 1977).

3. Stripped of a proven theoretical and empirical basis for interpretation, diversity indices may not be especially informative. The possibility that impact has caused a change in the relative abundance of some or all species over part of a sampled region can be rigorously tested with some form of the general linear model either one species at a time, or in the multivariate form. In fact a multivariate analysis of proportional data can be designed which is, in effect, an analysis of the variance of diversity.

Indices, "mathematical combination(s) of two or more parameters which have utility at least in the interpretive sense" (Pikul, 1974), are a long standing tradition in ecology. Their popularity is so great, and so many ecologists are so used to them that their abandonment is unlikely. However, when dealing within a hypothesis testing context, their appropriateness must be seriously questioned; they may be based upon sound principles or important empirical evidence, but their form obscures their foundations and makes formal testing difficult. As is briefly discussed in Sokal and Rohlf (1983), ratios are to be avoided in statistical analyses and are often not needed.

Groupings

The possibility that subsets of the total faunal assemblage might respond in a similar manner to natural and anthropogenic environmental factors is especially attractive and deserving of careful investigation in the future. If successful, grouping could lead to generalized tests for impact which are not dependent on local species composition, and would allow for meaningful pooling of low abundance forms. Various criteria for the formation of these subsets have been suggested. Here we will briefly consider four: ecological, taxonomic, analytical and practical groups.

Ecological Groups

The principal evidence that ecological groups could be useful is that some have recognized distribution patterns which can be reasonably well explained in terms of natural environmental factors. Secondarily, contemporary ecological theories can be used to argue that some groups are more important. Here we will consider the usefulness in impact studies of three benthic groups: feeding-mobility guilds, important species and size-biomass.

The observation that different feeding groups interacted to affect community structure (Rhoads and Young, 1970) was an extension of a much older generality that feeding groups had distinctive distributions. The development of a polychaete classification system based on both locomotion and feeding (Fauchald and Jumars, 1979) was an important advance over older classification because the concepts of foraging theory can be used to develop testable hypotheses. If a potentially impacting activity affects the availability of food resources, then it is reasonable to look for impact upon a feeding-locomotion group. A development along these lines can be found in Jumars (1981). This was attempted as an ancillary part of the Mid-Atlantic Block 684 study. Maurer *et al.* (1981) conducted such an analysis and found that the feeding guilds were stable near the drill site, in spite of short-term species changes. Beyond the immediate area around drilling platforms, however, there is no reason to expect an effect upon resource availability. Therefore, there is no obvious reason to adopt this approach.

The established generality that some species have more effect upon community structure than others was most convincingly demonstrated and popularized by the work on "keystone" species by Paine (1966) and others in the marine rocky intertidal. While the exact application of this idea to the subtidal soft bottom has yet to be worked out, in some sense it will probably prove applicable (see Hargrave and Thiel, 1983, for a good, brief discussion). Following ideas developed by Rhoads and Young (1970), it has been suggested that larger organisms which plow through sediments have had such a persistent influence as to be evolutionarily important (Thayer, 1983). The ecological validity of this observational conclusion has yet to be determined, but the idea already makes an important point. If we are to restrict our focus to that group of animals which have the greatest influence upon the community, then we first need to think about possible mechanisms and conduct research to identify those species in continental shelf systems.

The use of the size and biomass distribution of benthic fauna to characterize the state of the community is an untried suggestion (see Hargrave and Thiel, 1983). On practical grounds it is very appealing because there is already a *de facto* size grouping imposed by the sampling methods in common use. Empirically, it is supported by the observations that size biomass proportions do shift with depth (equated with diminished food supply) and oxygen supply. Theoretically, the relationships among size, respiration and food availability have yet to be established. Until the theoretical development becomes convincing, there is no strong reason for this approach to be adopted.

Taxonomic Groups

Indices based on higher taxonomic groupings have a peculiar status, falling somewhere between "quick and dirty" and "theoretically sound." Obviously, if the labor and expertise needed to sort and identify specimens to species could be avoided, then impact studies would be greatly simplified. However, it is not established if this extension of the indicator species approach really works or why it should. The Nematode to Harpacticoid Copepod ratio in a meiobenthic sample is an actively debated use of such taxonomic groupings. Raffaelli and Mason (1981) first proposed this ratio as a general pollution indicator on the basis of a study of organic pollution. It has, subsequently been applied to a variety of impacted environments. Coull *et al.* (1981) challenged the general applicability of the approach and questioned its underlying explanation. In a rebuttal, Raffaelli (1981) agreed that the technique may not be valid, but that it deserves to be studied.

Analytically Determined Groups

Cluster analysis and the category of multivariate analyses called ordination procedures offer two means of partitioning complex environmental data sets into a fewer number of combinations. Typically, as in the reviewed projects, they are part of the final product of a study. However, it is not clear exactly how they might be used to detect a subtle impact. An impact would have to be quite gross in order to form an obvious "impact" group in cluster analysis or an impact factor (eigenvector) in ordination. If impact was of the pathological type, perhaps a cleverly derived similarity coefficient might allow it to be resolved. Discordant impact would still face the problem of natural factors and residual variation.

The real value of analytically derived groups lies in the preliminary efforts to identify the sources of natural variation and the magnitude of the residual. From the mass of data, a limited set of species and physical factors can be identified and made the focus of more intensive sampling to support hypothesis testing. As the following discussion indicates, I am prejudiced towards the use of ordination in preference to cluster analysis. Both can be highly informative about natural variation, but ordination produces a more directly usable product.

Cluster analyses are a category of data analysis which produce classifications (groups) of data based upon an index of similarity. Judging by the results of the reviewed reports, they were probably the single most important analyses used in the search for long-term impacts. Therefore it is important to consider what is their most appropriate use. There are several different types of cluster analysis and numerous references (the bibliography of Blashfield and Aldenderfer, 1978, should keep the curious occupied). For unexplained reasons, hierarchical agglomerative techniques have become especially popular in marine environmental work and were used in all the reviewed reports. In spite of a burgeoning literature, there are dangerously few critical accounts. Boesch (1977) is one of the better accounts and deals with benthic assemblages. Gauch (1982) presents an informative critique, but must be supplemented by papers which include technical details.

Hierarchical cluster analyses are a good tool for reducing data so that patterns of association can be seen, but become confusing when a hundred or more samples are analyzed. They impose clusters upon continuous data which produces an illusion of discrete assemblages (Hill, 1977, treats this in detail) if not cautiously interpreted. As yet, hierarchical cluster analysis does not support hypothesis testing.

Detection of any long-term impact requires partitioning of total variation, which is very hard to do with cluster analysis. Therefore, when the sampling scheme is essentially that of a factorial ANOVA (as in the Georges Bank study), it seems inappropriate to rely on cluster analysis to determine the relative importance of factors. Cluster analyses can, however, be used to examine the relationship of data within an ANOVA if Euclidean distance or a related index is used.

Ordination techniques are a class of multivariate analyses that produce a few uncorrelated linear equations from the original numerous variables, which can make pattern seeking easier. Less popular than cluster analysis because of a more complex algebraic and geometric basis, several techniques are frequently used in faunal surveying (see Gauch, 1982 for critical evaluations and technical references). Since there is virtually no limit to the potential combinations of rescalings, translations, rotations, and projections, new techniques of greater or lesser utility can be expected in the future.

Ordination should assume a prominent role in the early stages of any long-term impact study. Ordination techniques are multivariate extensions of the general linear model, particularly well suited for looking at patterns in space. Variants of principal components analysis will partition a large number of species into fewer linear combinations which vary together. Factor analysis (this term has a rather broad usage) produces linear combinations of biotic and physical parameters which are correlated, and cannonical correlation shows the relationships amongst linear combinations of biotic variables and physical factors. All of these techniques determine the amount of variance (or vector length) explained by the combinations found; which is part of the information needed to design a field sampling program.

Ordination procedures do have limitations; of which, two are especially troublesome. First, they usually are linear models and can only produce linear approximations of major natural relationships. The departures from linearity in these major relationships then appear as spurious patterns in the data (Goodman, 1979, presents a short concise discussion of this and related problems). Second, variance is rarely independent of the mean in faunal counts (see Taylor *et al.*, 1978), and if the data are not appropriately transformed, ordination may produce little more than a very elaborate abundance ranking.

Practical Groups

As desirable as a complete faunal inventory may seem to be, it must be realized that this is an impossible and pointless goal. All sampling equipment is biased in favor of certain sizes. Capture of the rarest of species is fortuitous. The nematodes and smaller organisms pose a distracting taxonomic problem, and there are no grounds to the belief that even more complex data will produce greater insight than has been produced from present massive compilations. Our efforts and resources will be far better spent by trying to make optimal use of that part of the continental shelf benthic biota with which we can easily study.

The more we seek to learn about populations, the more restricted will be the number of species which can be included. For statistical analysis at a population level rare species serve little purpose. For determination of life history parameters, very small organisms may be too time consuming, and when chemical analyses are required, small organisms may not provide sufficient biomass. The suggestion put forward by Hargrave and Thiel (1983) that assessment might be restricted to a single large and abundant species, is perhaps too extreme. Nevertheless, studies could be more conclusive and less overwhelmed by useless data if following preliminary surveys, a limited group of reliably present species which met the requirements of all desired analyses were selected as the focus of intense examination.

CONCLUSIONS

Taken chronologically, the reviewed studies reflect a growing scientific maturity, evidenced by more thoughtful selection of sampling programs, more knowledgable use of data analyses, and diminished use of unsubstantiated explanation. However, all of the completed projects share the common problem that they could not define *a priori* what it was that they were looking for and could not employ optimal design techniques. As a result, none was able to confidently conclude that long-term impact did or did not occur.

Superficially (ignoring all the ecological complexity), detection of long-term impacts seems to call for an analysis of variance approach in which petroleum activities and natural factors are possible sources of variation in the structure and functioning of the local biota. Therefore, it seems odd that not a single project has successfully stated an hypothesis and then carried out a sampling program and a test with an appropriate analysis of variance.

On closer examination, past failure to employ analyses of variance has been due to a pervasive uncertainty as to exactly what a long-term effect is and how the existence of one in the open marine environment can be rigorously assessed. Operating under real and imagined constraints, projects seeking long-term impacts have been forced to proceed cart before horse, looking for evidence of a phenomena before, or while, deciding exactly what it is that is being sought. Rather than adopt designs and analyses which test specific alternatives, exploratory methods have often been adopted and then interpreted as if a rigorous test had been performed.

If we are to be able to detect long-term impacts, then two courses must be followed:

1. For the future, research must be undertaken which examines natural variation in benthic populations. More than any other factor, our inability to explain natural variation places a limit on our ability to resolve anthropogenic changes. In the previous discussion a few important topics have been identified. First, can impact be defined in terms of process changes that can be effectively studied in the field? Second, if a faunal census approach is taken, is something other than a species by species approach more informative and cost effective? Third, can we substantially increase our understanding of the relationship between faunal and environmental spatial and temporal variation? Fourth, what form of the concept of faunal succession is most applicable to the continental shelf benthos, and how can it be used to reduce the apparent variation in fauna data? Fifth, what types of powerful and robust statistical models might be most applicable to long-term effects studies?

2. For the time being we need to continue the faunal census approach. However, far more effort needs to be expended upon making an informed use of good statistical design. This will entail an end to the over-dependence on diversity and similarity indices and greater use of multivariate and univariate

techniques. Whenever possible the adopted designs must result in hypothesis testing.

RECOMMENDATIONS FOR IMPROVED DESIGN WITH OUR PRESENT LEVEL OF UNDERSTANDING

If oil and gas development activities are having a long-term detrimental impact upon continental shelf communities at this moment, then we cannot sit by and await the luxury of a more predictive ecology which increased research might eventually produce. Therefore, we need to proceed with field evaluations making pragmatic decisions and pursuing the consequences in the best available ways. In order to sketch out the design for such a project, we can return to seven of the ten principles of Green (1979), here rephrased, renumbered and rearranged to better suit our purposes.

1. Concise statement of the problem.

As much as it is desirable to deal directly with the phenomena that regulate assemblages, that cannot be done without substantial research. Therefore, we make explicit that which is often implied; if present, an impact will manifest itself as a change in faunal composition that is not attributable to natural factors. These impacts may be of less magnitude than natural changes.

2. Carry out preliminary sampling.

The success of the actual test for impact will be so critically dependent upon choices from the results of preliminary sampling that this initial activity must be carefully planned and well supported. Preliminary sampling must serve four main purposes: A. High density surveying must establish the spatial pattern of faunal and environmental variables in the absence of any theory which allows us to predict or determine such patterns simply. B. Long-term surveying must establish temporal changes in the benthic fauna. C. Preliminary data analysis will identify those groups of species and those environmental variables which can be most productively and efficiently studied in the subsequent testing. D. Power analysis of preliminary data will allow for the informed choice of final design and tests.

3. Verify appropriateness of sampling unit and estimate replication needed to obtain required precision.

This step makes use of the preliminary data to determine if the original definition is still feasible given field realities and to determine the precision to cost ratio.

4. Select and stick to the adopted design and live with the results obtained.

With the definition of impact established and verified and information on natural variation, a final design can be selected. Whatever design is adopted, it must allow for testing of specific hypotheses and not be heavily dependent upon descriptive analyses. The process of selection might be guided by computer modeling and best involve the talents of three types of scientists: applied statisticians, quantitative ecologists, and descriptive field ecologists. The quality assurance mechanisms of the supporting agency must confirm that the design is something more than a schedule of events and that it offers a high likelihood of success in the hands of the selected scientists.

5. In the presence of large scale environmental variation, adopt a stratified approach.

The initial sampling will determine if there are major faunal changes associated with gradients in the study area. In addition, the presence of faunal patches (as distinct from simple changes along a physical gradient) will be recognized and stratification within patches included if necessary.

6. Take randomly allocated replicates within each combination of controlled variables.

Unless the important environmental factors are independent of each other, this will be very hard to accomplish. Hopefully, a location could be found where there were relatively few important environmental factors. It is critical that there be an unambiguous indicator of exposure to petroleum activity such as a unique chemical tracer. In a complex environment, simple proximity to a platform is not sufficient.

7. Use replication to estimate variability.

When dealing with a small set of target species and having preliminary data in hand, the number of replicates needed can be determined by power analysis. It must be remembered that replicates are not just collected to insure representativeness or completeness of inventory.

SUMMARY

1. OEI-GURC, the Central Gulf Platform and the Buccaneer Field studies lacked the ability to detect long-term impact due to the lack of an operational definition of impact, the implicit assumption that any impact would be easily distinguishable from natural variation, and a failure to use the techniques of design afforded by population survey statistics.

2. Even if good designs are adopted and adhered to, at our current level of ecological understanding we can still expect to be faced with high levels of unexplained natural variation. This residual variation will severely limit our ability to detect subtle impacts.

3. Replacement of the old "survey and explain" approach by statistical models and good design is highly desirable, even though we cannot yet fully describe the system under study. A well-designed statistical study has fewer ambiguities, allows for the use of powerful analytical techniques, and provides a good basis for future improvements.

4. Statistical models should be adopted carefully and with considerable thought. The most formally designed study, Mid-Atlantic Block 684, had to abandon its design because the model was ecologically unrealistic.

5. Ultimately, the success of the models employed requires a much greater understanding of natural variation in the benthos. A process-oriented definition of impact which could lead to more fully stated models of community structure is desirable, if not critical to the success of long-term impact studies. However, at this time there seems to be none sufficiently developed to warrant extensive application. As a result, well focused research is needed that will produce alternatives to costly species-by-species surveying.

LITERATURE CITED

Anderson, J.B., R.B. Wheeler and R.R. Schwarzer. 1981. Sedimentology and geochemistry of recent sediments. Pages 59-67 *in* B.S. Middleditch (ed.), Environmental Effects of Offshore Oil Production. The Buccaneer Gas and Oil Field Study. Plenum Press, New York.

Baker, J.H., K.T. Kimball, W.D. Jobe, J. Janousek, C.L. Howard and R.P. Chase. 1981. Part 6. Benthic Biology. Pages 1-317 *in* C.A. Bedinger Jr., (ed.), Ecological Investigations of Petroleum Production Platforms in the Central Gulf of Mexico. Volume 1. Pollutant Fate and Effects Studies. Rept. to Bur. Land Management, Contract No. AA551-CT8-17, Southwest Research Institute, San Antonio, Texas.

Bartlett, M.S. 1973. Statistical Analysis of Spatial Pattern. Chapman and Hall, London, 90 p.

Battelle/Woods Hole Oceanographic Institution. 1983. Georges Bank Benthic Infaunal Monitoring Program. Final Report Year 1. Contract No. 14-12-001-29192, U.S. Dept. of Interior, Minerals Management Service, New York OCS Office, New York, 153 p.

Battelle/Woods Hole Oceanographic Institution. 1984. Georges Bank Benthic Infauna Monitoring Program. Final Report, Year 2. Contract No. 14-12-001-29192, U.S. Dept. of Interior, Minerals Management Service, Atlantic OCS Office, Vienna, Virginia, 173 p.

Bedinger, C.A., Jr., R.E. Childers, J.W. Cooper, K.T. Kimball and Alan Kwok. 1981. Part 1. Background, Program Organization and Study Plan. Pages 1-53 *in* C.A. Bedinger Jr. (ed.), Ecological Investigations of Petroleum Production Platforms in the Central Gulf of Mexico. Volume 1. Pollutant Fate and Effects Studies. Rept. to Bur. Land Management, Contract No. AA551-CT8-17, Southwest Research Institute, San Antonio, Texas.

Behrens, E.W. 1981. Total organic carbon and carbon isotopes of sediments. Pages 117-131 *in* B.S. Middleditch (ed.), Environmental Effects of Offshore Oil Production: The Buccaneer Gas and Oil Field Study. Plenum Press, New York.

Bender, M.E., D.J. Reish and C.H. Ward. 1979. Independent appraisal. Reexamination of the Offshore Ecology Investigation. Pages 35-116 *in* C.H. Ward, M.E. Bender and D.J. Reish (eds.), The Offshore Ecology Investigation, Effects of Oil Drilling and Production in a Coastal Environment. Rice University Studies 65: 1-589.

Blashfield, R.K. and M.S. Aldenderfer. 1978. The literature on cluster analysis. Multivariate Behavioral Research 13: 271-95.

Boesch, D.F. 1977. Applications of Numerical Classification in Ecological Investigations of Water Pollution. E.P.A. Ecological Research Series, EPA-600/3-77-033, 115 p.

Boesch, D.F., R.S. Diaz and R.W. Virnstein. 1976. Effects of tropical storm Agnes on soft-bottom macrobenthic communities of the James and York estuaries and the lower Chesapeake Bay. Chesapeake Sci. 17: 246-259.

Brongersma-Sanders, M. 1957. Mass mortality in the sea. Pages 941-1010 *in* J.W. Hedgpeth (ed.), Treatise on Marine Ecology and Paleoecology. Volume 1, Ecology. Geological Society of America, Memoir 67.

Brooks, J.M., D.A. Wiesenburg, C.R. Schwab, E.L. Estes and R.F. Shokes. 1981. Surficial sediments and suspended particulate matter. Pages 69-115 *in* B.S. Middleditch (ed.), Environmental Effects of Offshore Oil Production. The Buccaneer Gas and Oil Field Study. Plenum Press, New York.

Chia, F. and M.E. Rice (eds.). 1978. Settlement and Metamorphosis of Marine Invertebrate Larvae. Elsevier/North Holland Biomedical Press, New York, 290 p.

Clark, R.B. 1982a. The long-term effect of oil pollution on marine populations, communities and ecosystems: some questions. Phil. Trans. R. Soc. London B 297: 185-192. Reprinted in R.B. Clark (ed.). 1982. The Long-Term Effects of Oil Pollution on Marine Populations, Communities and Ecosystems. The Royal Society, London.

Clark, R.B. 1982b. The impact of oil pollution on marine populations, communities and

ecosystems: a summing up. Phil. Trans. R. Soc. London B 297: 433-443. Reprinted in R.B. Clark (ed.). 1982. The Long-Term Effects of Oil Pollution on Marine Populations, Communities and Ecosystems. The Royal Society, London.
Clark, R.B. (ed.). 1982c. The Long-Term Effects of Oil Pollution on Marine Populations, Communities and Ecosystems. Proceedings of a Royal Society Discussion Meeting Held on 28 and 29 October 1981. The Royal Society, London.
Cliff, A.D. and J.K. Ord. 1981. Spatial Processes, Models and Applications. Methuen, Andover, Hampshire, England, 266 p.
Cochran, W.G. 1977. Sampling Techniques, 3rd edition. John Wiley and Sons, Inc., New York, 428 p.
Cohen, J. 1977. Statistical Power Analysis for the Behavioral Sciences. Academic Press, New York, 474 p.
Connell, J.H. and R.O. Slatyer. 1977. Mechanisms of succession in natural communities and their role in community stability and organization. Amer. Natur. 111: 1119-1144.
Coull, B., G.R.F. Hicks and J.B.J. Wells. 1981. Nematode/copepod ratios for pollution: A rebuttal. Mar. Pollution Bull. 12: 378-380.
Cuff, W. and N. Coleman. 1979. Optimal survey design: Lessons from a stratified random sample of macrobenthos. Jour. Fisheries Res. Bd. Canada 36: 351-361.
Dauer, D.M. and J.L. Simon. 1976. Repopulation of the polychaete fauna of an interstitial habitat following natural defaunation: Species equilibrium. Oecologia 22: 99-117.
Dingle, P.J. 1979. Statistical methods for spatial point patterns in ecology. Pages 95-140 *in* R.M. Cormak and J.K. Ord (eds.), Spatial and Temporal Analysis in Ecology. Statistical Ecology Series 8. International Cooperative Publishing House, Fairland, Maryland.
Dixon, W.J. and F.J. Massey. 1969. Introduction to Statistical Analysis, 3rd edition. McGraw-Hill, New York, 638 p.
EG&G, Environmental Consultants. 1982. A Study of Environmental Effects of Exploratory Drilling on the Mid-Atlantic Outer Continental Shelf - Final Report of the Block 684 Monitoring Program. Prepared for Offshore Operations Committee, Exxon Production Research Company, by EG&G, Environmental Consultants, Waltham, Massachusetts.
Farrell, D.H. 1974. Benthic communities in the vicinity of producing oil wells in Timbalier Bay, Louisiana. Pages 14-95 *in* Gulf Universities Research Consortium, Offshore Ecology Investigation Report No. 138, appended, Galveston, Texas.
Fauchald, K. and P.A. Jumars. 1979. The diet of worms: A study of polychaete feeding guilds. Oceanogr. Mar. Biol. Ann. Rev. 17: 193-284.
Finney, D.J. 1964. Statistical Method in Biological Assay. Charles Griffin and Co., London, 668 p.
Fish, A.G., L.L. Massey, J.R. Inabinet and P.L. Lewis. 1974. A study of the effects of environmental factors upon the distribution of selected sandy beach organisms of Timbalier Bay, Louisiana. Gulf Universities Research Consortium, Offshore Ecology Investigation Report No. 138, appended, Galveston, Texas.
Fisher, R.A., A.G. Corbett and C.B. Williams. 1943. The relation between number of species and the number of individuals in a random sample of an animal population. J. of Anim. Ecol. 12: 42-58.
Fotheringham, N. 1981. Observations on the effects of oil field structures on their biotic environment: platform fouling community. Pages 179-208 *in* B.S. Middleditch (ed.), Environmental Effects of Offshore Oil Production. The Buccaneer Gas and Oil Field Study. Plenum Press, New York.
Fucik, K.W. and I.T. Show. 1981. Environmental synthesis using an ecosystems model. Pages 329-353 *in* B.S. Middleditch (ed.), Environmental Effects of Offshore Oil Production. The Buccaneer Gas and Oil Field Study. Plenum Press, New York.
Gallagher, E.D., P.A. Jumars and D.D. Trueblood. 1983. Facilitation of soft-bottom benthic succession by tube builders. Ecology 64: 1200-1216.
Gallaway, B.J., L.R. Martin, R.L. Howard, G.S. Boland and G.S. Dennis. 1981. Effects of artificial reef and demersal fish and macrocrustacean communities. Pages 237-299

in B.S. Middleditch (ed.), Environmental Effects of Offshore Oil Production. The Buccaneer Gas and Oil Field Study. Plenum Press, New York.

Gauch, H.G. 1982. Multivariate Analysis in Community Ecology. Cambridge Univ. Press, New York, 298 p.

Goodman, D. 1975. The theory of diversity-stability relationships in ecology. Quart. Rev. Biol. 50: 237-266.

Goodman, D. 1979. Applications of eigenvector analysis in the resolution of spectral pattern in spatial and temporal ecological sequences. Pages 139-155 *in* G.P. Patil and M. Rozenzweig (eds.), Contemporary Quantitative Ecology and Related Econometrics. Statistical Ecological Series 12. International Cooperative Publishing House, Fairland, Maryland.

Gray, J.S. 1981. Detecting pollution changes in communities using the log-normal distribution on individuals among species. Mar. Pollution Bull. 12: 173-175.

Gray, J.S. and F.B. Mirza. 1979. A possible method for the detection of pollution-induced disturbance on marine benthic communities. Mar. Pollution Bull. 10: 142-145

Green, R.H. 1979. Sampling Design and Statistical Methods for Environmental Biologists. John Wiley and Sons, Inc., New York, 257 p.

Griffin, G.M. and B.J. Ripy. 1974. Turbidity, suspended sediments, and the origin of the turbid near-bottom layer — Louisiana shelf south of Timbalier Bay — August 1972 - January, 1974 — with comments on a process model for turbid layer transport. Pages 75-79 *in* Abstracts, Summaries and Conclusions from the Offshore Ecology Investigation, 1972-1974, Gulf Universities Research Consortium (GURC), Galveston, Texas.

Hargrave, B.T. and H. Thiel. 1983. Viewpoint: Assessment of pollution-induced changes in benthic community structure. Mar. Pollution Bull. 14: 41-46.

Harper, D.E., Jr., D.L.Potts, R.R. Salazer, R.J. Case, R.L. Jaschek and C.M. Walker. 1981. Distribution and abundance of macrobenthic and meiobenthic organisms. Pages 133-177 *in* B.S. Middleditch (ed.), Environmental Effects of Offshore Oil Production. The Buccaneer Gas and Oil Field Study. Plenum Press, New York.

Heltsche, J.F. and N.E. Forrester. 1983. Estimating species richness using a jack knife procedure. Biometrika 39: 1-11.

Hill, M. 1977. Use of simple discriminant functions to classify quantitative data phytosociological data. Pages 181-199 *in* E. Diday, L. Lebart, J. Pages and R. Tomassone (eds.), First International Symposium on Data Analysis and Informatics, Volume 1. Institut de Recherche d'Informatique et d'Automatique.

Horn, H.S. 1976. The ecology of secondary succession. Ann. Rev. Ecol. System. 5: 25-37.

Horton, R.L. 1978. The General Linear Model: Data Analysis in the Social and Behavioral Sciences. McGraw-Hill, New York, 274 p.

Humm, H.J. 1974. The effect of the offshore oil and gas wells on the benthic marine plants of Louisiana-I, and Timbalier Bay-II. Gulf Universities Research Consortium, Offshore Ecology Investigation Report No. 138, appended, Galveston, Texas.

Jones, R. 1982. Population fluctuations and recruitment in marine populations. Phil. Trans. R. Soc. London B 297: 353-368. Reprinted in R.B. Clark (ed.). 1982. The Long-Term Effects of Oil Pollution on Marine Populations, Communities and Ecosystems. The Royal Society, London.

Jumars, P.A. 1981. Limits in predicting and detecting benthic responses to manganese nodule mining. Marine Mining 3: 213-229.

Jumars, P.A., D. Thistle and M.L. Jones. 1977. Detecting two dimensional spatial structure in biological data. Oecologia 28: 109-123.

Jumars, P.A. and J.E. Eckman. 1983. Spatial structure within deep-sea benthic communities. Pages 399-452 *in* G.T. Rowe (ed.), The Sea, Volume 8. Deep-Sea Biology. Wiley-Interscience, New York.

Kempton, R.A. 1981. Review of Green, R.H., Sampling and Statistical Methods for Environmental Biologists. Biometrics 37: 202-203.

Kritzler, H. 1974. Oil production and polychaetous annelids in a Louisiana estuary. Gulf

Universities Research Consortium, Offshore Ecology Investigation Report No. 138, appended, Galveston, Texas.
MacArthur, R.H. 1972. Geographical Ecology - Patterns in the Distribution of Species. Harper and Row, New York, 269 p.
Maurer, D., W. Leathem and C. Menzie. 1981. The impact of drilling fluid and well cuttings on polychaete feeding guilds from the U.S. northeastern continental shelf. Mar. Pollution Bull. 12: 342-346.
May, R.M. 1973. Stability and Complexity in Model Ecosystems. Princeton University Press, Princeton, New Jersey, 235 p.
May, R.M. 1975. Patterns of species abundance and diversity. Pages 81-120 *in* M.L. Cody and J.M. Diamond (eds.), Ecology and Evolution of Communities. Belknap Press of Harvard University, Cambridge, Massachusetts.
McCall, P.L. 1978. Spatial-temporal distributions of Long Island Sound infauna: The role of bottom disturbance in a nearshore marine habitat. Pages 191-219 *in* M.L. Wiley (ed.), Estuarine Interactions. Academic Press, New York.
Menzie, C.A. 1984. Diminishment of recruitment: A hypothesis concerning impacts on benthic communities. Mar. Pollution Bull. 15: 127-128.
Menzies, R.J., J.P. Morgan, C.H. Oppenheimer, S.Z. El-Sayed and J.M. Sharp. 1979. Design of the Offshore Ecology Investigation. Pages 19-32, *in* C.H. Ward, M.E. Bender and D.J. Reish (eds.), The Offshore Ecology Investigation. Effects of Oil Drilling and Production in a Coastal Environment. Rice University Studies 65: 1-589.
Mertens, E.W. 1978. The impact of oil on marine life: A summary of the field studies. Pages 508-514 *in* Source, Effects and Sinks of Hydrocarbons in the Aquatic Environment. American Institute of Biological Sciences, Arlington, Virginia.
Middleditch, B.S. 1981. Hydrocarbons and sulfur. Pages 15-54 *in* B.S. Middleditch (ed.), Environmental Effects of Offshore Oil Production. The Buccaneer Gas and Oil Field Study. Plenum Press, New York.
Middleditch, B.S. and B.J. Gallaway. 1981. Prologue. Pages 1-14 *in* B.S. Middleditch (ed.), Environmental Effects of Offshore Oil Production. The Buccaneer Gas and Oil Field Study. Plenum Press, New York.
Neyman, J. and E.S. Pearson. 1928. On the use and interpretation of certain test criteria for the purposes of statistical inference. Biometrics 20A: 175-240.
Oppenheimer, C.H., R. Miget and H. Kator. 1979. Ecological relationships between marine organisms and hydrocarbons in the OEI study area, Louisiana. Pages 287-324 *in* C.H. Ward, M.E. Bender and D.J. Reish (eds.), The Offshore Ecology Investigations. Effects of Oil Drilling and Production in a Coastal Environment. Rice University Studies 65: 1-589.
Ord, J.K. 1979. Time series and spatial patterns in ecology. Pages 1-94 *in* R.M. Cormak and J.K. Ord (eds.), Spatial and Temporal Analysis in Ecology. Statistical Ecology Series 8. International Cooperative Publishing House, Fairfield, Maryland.
Ostrom, C. 1974. Relationships between the distribution of selected taxa of recent benthic Foraminifera and measured environmental parameters in Timbalier Bay, La. Gulf Universities Research Consortium, Offshore Ecology Investigation. Report No. 138, appended, Galveston, Texas.
Paine, R. 1966. Food web complexity and species diversity. Amer. Natur. 100: 65-76.
Perry, A. 1974. A bottom/surface trawl and bottom grab study of areas of oil production activity and areas of no activity in estuarine and offshore environments. Effects of platforms on biota (fishes) offshore. Gulf Universities Research Consortium, Offshore Ecology Investigation Report No. 138, appended, Galveston, Texas.
Pielou, E.C. 1969. An Introduction to Mathematical Ecology. Wiley, New York, 286 p.
Pielou, E.C. 1981. A review of Grassle, J.F., G.P. Patil, Ecological Diversity in Theory and Practice S.E.S. vol. 6. 365 pp. Biometrics 36: 742-743.
Pikul, R. 1974. Development of environmental indices. *In* J.W. Pratt (ed.), Statistical and Mathematical Aspects of Pollution Problems. Satellite Symposium on Statistical

Aspects of Pollution Problems. Harvard Business School, Marcel Dekker, New York.
Platt, T. and K.L. Denman. 1975. Spectral analysis in ecology. Ann. Rev. System. and Ecol. 7: 189-210.
Preston, F.W. 1948. The commoness and rarity of species. Ecology 29: 254-283.
Preston, F.W. 1980. Noncanonical distributions of commoness and rarity. Ecology 61: 88-97.
Raffaelli, D.G. 1981. Monitoring with meiofauna—a reply to Coull, Hicks, and Wells (1981) and additional data. Mar. Pollution Bull. 12: 381-382.
Raffaelli, D.G. and C.F. Mason. 1981. Pollution monitoring with meiofauna, using the ratio of nematodes to copepods. Mar. Pollution Bull. 12: 158-162.
Rhoads, D. and L. Boyer. 1982. The effects of marine benthos on physical properties of sediments: A successional perspective. Pages 3-52 *in* P.L. McCall and M.J. Tevesz (eds.), Animal-Sediment Relations. The Biogenic Alteration of Sediments. Plenum Press, New York.
Rhoads, D. and D.K. Young. 1970. The influence of deposit-feeding organisms on sediment stability and community trophic structure. J. Mar. Res. 28: 150-178.
Robson, D.S., C.A. Menzie and Hugh F. Mulligan. 1982. An environmental monitoring study to assess the impact of drilling discharges in the mid-Atlantic. II. An experimental design and statistical methods to evaluate impacts on the benthic environment. *In* EG&G, Environmental Consultants, A Study of Environmental Effects of Exploratory Drilling on the Mid-Atlantic Outer Continental Shelf - Final Report of the Block 684 Monitoring Program. Prepared for Offshore Operators Committee, Exxon Production Research Company, by EG&G, Environmental Consultants, Waltham, Massachusetts.
Roughgarden, J. 1983. Competition and theory in community ecology. Amer. Natural. 122: 583-601.
Saila, S.B., R.A. Pikanowski and D.S. Vaughan. 1976. Optimum allocation strategies for sampling benthos in New York Bight. Estuar. Coastal Mar. Sci. 4: 119-128.
Sanders, H.L. 1981. Environmental effects of oil in the marine environment. Pages 117-146 *in* Safety and Offshore Oil: Background Papers of the Committee on Assessment of Safety of OCS Activities. National Research Council, National Academy Press, Washington, D.C.
Santos, S.L. and J.S. Simon. 1980. Response of soft bottom benthos to annual catastrophic disturbance in a south Florida estuary. Mar. Ecol. Prog. Ser. 3: 347-355.
Shannon, C.E. and W. Weaver. 1949. The Mathematical Theory of Communication. Univ. of Illinois Press, Urbana, Illinois, 117 p.
Sharp, J.M. and S.G. Appan. 1982. The cumulative ecological effects of normal offshore petroleum operations contrasted with those resulting from continental shelf oil spills. Phil. Trans. R. Soc. London B 297: 309-322. Reprinted in R.B. Clark (ed.). 1982. The Long-Term Effects of Oil Pollution on Marine Populations, Communities and Ecosystems. The Royal Society, London.
Simberloff, D. 1983. Competition theory, hypothesis testing, and other community ecological buzzwords. Amer. Natural. 122: 626-635.
Smith, W., V.R. Gibson, L.S. Brown-Leger and J.F. Grassle. 1979. Diversity as an indicator of pollution: Cautionary results from mesocosm experiments. Pages 269-277 *in* Ecological Diversity in Theory and Practice. International Cooperative Publishing House, Fairfield, Maryland.
Sokal, R.R. and N.L. Oden. 1978a. Spatial autocorrelation in biology. 1 Methods. Biol. J. Linn. Soc. 10: 199-228.
Sokal, R.R. and N.L. Oden. 1978b. Spatial autocorrelation in biology. 2 Some biological applications of evolutionary and ecological interest. Biol. J. Linn. Soc. 10: 229-250.Sokal, R.R. and F.J. Rohlf. 1983. Biometry, 2nd edition. Freeman, San Francisco, 776 p.
Taylor, L.R., I.P. Woiwood and J.N. Perry. 1978. The density-dependence of spatial behavior and the rarity of randomness. J. Anim. Ecol. 47: 383-406.

Thayer, C.W. 1983. Sediment-mediated biological disturbance and the evolution of marine benthos. Pages 480-595 *in* M.J.S. Tevesz and P.L. McCall (eds.), Biotic Interactions in Recent and Fossil Benthic Communities. Plenum Press, New York.

Tong, Y.L. 1983. Some distribution properties of the sample species diversity indices and their applications. Biometrika 39: 997-1008.

Usher, M.B. 1979. Markovian approaches to ecological succession. J. Anim. Ecol. 48: 413-426.

Whittaker, R.H. 1973. Direct gradient analysis. Pages 277-336 *in* R.H. Whittaker (ed.), The Ordination of Plant Communities. The Handbook of Vegetation Science, Part 5. Junk, The Hague.

Zahl, S. 1977. Jackknifing an index of diversity. Ecology 58: 907-913.

INDEX